The Thalamus

The Thalamus

EDWARD G. JONES

California College of Medicine
University of California
Irvine, California

Plenum Press • New York and London

Library of Congress Cataloging in Publication Data

Jones, Edward G., 1939–
      The thalamus.

      Bibliography: p.
      Includes index.
      1. Thalamus—Anatomy. I. Title. [DNLM: 1. Thalamus. WL 312 J76t]
QL938.T45J66   1985   599'.0188   85-3437
ISBN 0-306-41856-8

©1985 Plenum Press, New York
A Division of Plenum Publishing Corporation
233 Spring Street, New York, N.Y. 10013

Printed in the United States of America

For Sue, Pippa, and Chris

"Hic rudis et castris . . . ,
Qui tetigit thalamos praeda novella tuos,
Te solam norit, . . ."

—Ovid, *Ars Amatoria* 3, 559–661

Preface

It is now more than fifty years since Sir Wilfrid Le Gros Clark (1932a) published his Arris and Gale lectures on the structure and connections of the thalamus. This authoritative overview came at a time when thalamic studies were passing from a descriptive to an experimental phase and, in his review, Le Gros Clark was able to cover virtually every aspect of the organization and development and much of the comparative anatomy of the thalamus then known.

It is also approaching a half-century since A. Earl Walker (1938a) wrote *The Primate Thalamus*, which was strongly experimental, but with many clinical insights, and which he described as "an attempt to elucidate the role of the thalamus in sensation." The intervening years have seen published a few reports of conferences on aspects of thalamic organization and function but no monographs comparable to those of Le Gros Clark or Walker. Perhaps this is understandable when one considers, not so much the enormity of the new data that have been added, but rather the emphasis upon individual thalamic nuclei as components of separate functional systems, not all of them sensory. It is probably also true to say that studies in the commoner experimental animals such as the rat, cat, and monkey have been so productive in their own right that there was little interest in making an across-species synthesis. Studies of the human thalamus virtually ceased with the introduction of L-dopa and the decline of interest in stereotaxic thalamotomy. Overall, too, looms the lateral geniculate nucleus, from which such an enormous body of fascinating new information continues to come that virtually all other thalamic nuclei fall in its shadow. It is interesting to reflect that Walker virtually ignored the lateral geniculate nucleus, seeing it as presenting "few new problems that need to be solved." In a sense perhaps he was right: the problems are old ones but the solutions keep advancing! Unlike in

ix

Walker's book, therefore, the lateral geniculate nucleus poses the threat of domination of the present work.

In attempting to survey virtually all past and most existing knowledge on the mammalian thalamus (and much of that on the nonmammalian thalamus as well), I have been mindful of Sir Charles Bell's (1811) comment: "I have found some of my friends so mistaken in their conception of the object of the demonstrations which I have delivered in my lectures, that I wish to vindicate myself at all hazards. They would have it that I am in search of the soul; but I wish only to investigate the structure of the brain. . . ." To the last part of his comment, I would add function as well as structure, for I am also cognizant of William Rushton's (1977) dictum: "The great chapters on minute anatomy—those deserts of detail without a living functional watercourse, only a mirage from unverified speculation—are nearly unreadable." Although my book has, justifiably, a strong anatomical content, it surveys the physiology and, where relevant, the clinical pathology of the thalamus as well. I have tried to bring together what I see as principles of mammalian thalamic organization, function, and development, drawing examples from whatever nucleus and species seemed relevant. These principles, or information that seems best suited to lead to new principles, are surveyed in Chapters 3–6. The individual nuclei or constellations of related nuclei are given separate treatment in Chapters 7–16. In these chapters, the basic format is as follows: structure including species variations, followed by terminology, connections, and functional characteristics. The reader should be able to find in these chapters reference to and often a detailed consideration of most mammalian thalami. However, in order to provide a kind of anatomical baseline, brief descriptions and ample photographs of sections through the thalami of six representative mammalian species are provided in Chapter 2. Given the current comparative anatomical climate, I would not dare attempt a synthesis of the nonmammalian thalamus with that of the mammal; therefore, a survey of the thalamus in nonmammals appears separately in Chapter 17. Despite my timidity and some past skepticism, having reviewed the nonmammalian literature, I cannot help but feel that the nonmammalian thalamus may hold some principles in common with that of the mammal after all.

My approach in virtually all the chapters has been a strongly historical one, coupled with a certain degree of didacticism, leading up to what I see as some of the currently exciting issues in thalamic research. This historical emphasis, which starts with a 2000-year perspective in Chapter 1, seemed necessary in all chapters in view of the long period that had elapsed since publication of the works of Le Gros Clark and Walker. But I am also conscious that some of the truly seminal works on thalamic anatomy by European workers of the half-century prior to Le Gros Clark and Walker are virtually lost to view nowadays. The same can probably even be said of the works of several more recent scientists, despite their contributions to fundamental knowledge. I hope that the major contributors are now given the credit they deserve, and the reader who wishes to know who named what and in which species ought to be able to find it in these pages. The didactic element also stems in part from the long period without a comprehensive treatment of the thalamus. However, it also arises from a desire to formulate principles unconstrained by accounts of individual thalamic nuclei or species and to provide a baseline of knowledge for a student or for a neuroscientist entering the area from another discipline. I trust that the reviews of new information will speak for themselves.

This work has occupied me for more time than I care to admit, and I am particularly grateful to my fellows and students, whose forebearance has given me the opportunity to complete it: Stewart Hendry, Chen-Tung Yen, Blair Clark, Michael Conley, Javier DeFelipe, May Kay Floeter, and David Schreyer kept the laboratory going as I became more and more preoccupied and saved me from many potential solecisms in the text. The photography is largely my own, with much assistance from Margaret Bates, but it could not have reached its standard if it had not been for the consistently high quality of the histological material provided by Bertha McClure. The experimental material prepared by Ms. McClure in my laboratory has been generated by the group of colleagues mentioned above as well as by past collaborators and students, who include Randi Leavitt, Maxwell Cowan, Harold Burton, John Krettek, Larry Swanson, Thomas Thach, Nancy Berman, Steven Wise, Joe Dan Coulter, Jean Graham, David Friedman, James Fleshman, Karen Valentino, David Tracey, Robert Porter, Chisato Asanuma, Lorraine Yurkewicz, and Todd Rainey.

Over the last 12 years, my experimental work has been supported by Research Grants NS-10526 and NS-10570, and by various training grants from the National Institutes of Health, United States Public Health Service, and briefly at Washington University by the McDonnell Center for Studies of Higher Brain Function and the George H. and Ethel Ronzoni Bishop bequest. Many colleagues have kindly provided brains of uncommon animals to be sectioned or other materials. For their help I am most grateful, and I have acknowledged their contributions at appropriate places in the text. My greatest debt of thanks is to Margo Gross who, with equanimity, typed, retyped, and typed yet again, while simultaneously dealing with all the secretarial demands of a busy laboratory.

Edward G. Jones

Irvine, California

Contents

The Thalamus

The scientist may admire but cannot accept the paradox that the belief and knowledge of antiquity have been superseded only by the more rational ignorance of today.

F. J. Cole (1949)

I

History

We shall not cease from exploration
And the end of all our exploring
Will be to arrive where we started
And know the place for the first time.
  T. S. Eliot, *Little Gidding*

1

The History of the Thalamus

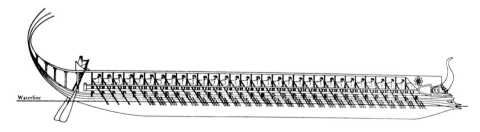

Sketch of a trireme, 4th–5th century B.C. From Casson (1971).

1.1. Galen and the Origin of the Word *Thalamus*

Sir Wilfrid Le Gros Clark relates in his autobiography (1968) that reference to him as one who had "put in some good work on the thalamus," once led to considerable amusement in an Oxford senior common room. With their classical background those dining that evening appreciated that the Greek word *thalamos* not only referred to an inner room, but commonly also to a bridal chamber or bridal couch. Used in reference to marriage, consummation, and the continuity of the tribe, the word appears commonly in Virgil,* and this sense has come down to us in the poetic, *epithalamion*. The connotation of sexual behavior was rendered very explicitly in the works of the Roman poets Ovid, Horace, Propertius, and Petronius.† Rabelais,‡ writing in the 16th century, calls the flagship of the Pantagruelists on their voyage to seek advice from the Oracle of the Holy Bottle regarding the virtue of Panurge's future wife, the "Thalamége." There can be no doubt about his meaning either. The old meaning of *thalamus* survives in the French synonym *couche optique* and was no doubt in Burdach's mind when he named a posterior protrusion of the human thalamus, the *pulvinar* or pillow.

The first extant anatomical usage of *thalamus* has been traced to Galen, who wrote in Greek in the 2nd century A.D. Since he wrote 100 years or more after Petronius, the last of the poets mentioned above, we may assume that he was familiar with its well-established meaning. It seems clear, however, that he used the word in a derivative sense and that he applied it to something other than the large diencephalic mass that we call the thalamus. Though in Book IX of his *Anatomical Procedures*, Galen seems to imply that he has seen the lateral geniculate nucleus by dissecting the optic tract to it, he did not refer to thalamus. In *De Usu Partium*, he makes it clear that he believed that the optic tracts arose from a region where the lateral ventricles come together at the back of the diencephalon and adjacent to the lateral geniculate bodies. This region, he says, is "a thalamus of the ventricles . . . made for the sake of . . . [the op-

---

* Virgil, *Aeneid*, Book 6, lines 519 and 623; Book 7, line 252; Book 10, line 650.
† Ovid, *Ars Amatoria*, Book 2, line 617; Book 3, lines 560 and 592; *Remediorum Amoris*, line 592; Horace, *Odes*, Book 1, lines 13–16; Propertius, *Poems*, Book 2, poem 15, line 14; Petronius, *The Satyricon*, paragraph 26.
‡ Rabelais, *Gargantua and Pantagruel*, Book 4, Chapter 1.

tic] . . . nerves." By means of this communication with the nerves, he felt the pneuma, infused with animal spirits in the rete mirabile, brain, and ventricles could pass via a lumen that he postulated in the optic nerves, to the eyes. According to Dr. Marcus Singer, in a note to May's (1968) translation of Galen's *De Usu Partium,* the communication that Galen observed in the ox brain was probably the choroid fissure of the descending part of the lateral ventricle, torn open where it lies on top of the lateral geniculate body. The appearances that seem to have misled Galen can be obtained by dissecting the brain from behind, as in Fig. 1.1 taken from the work of Polyak (1957).

Galen was obviously referring to a reservoir or to a funnel through which the pneuma could reach the optic nerve. It does not seem unreasonable to refer to this by a term that implies an inner chamber. He may also have seen in the

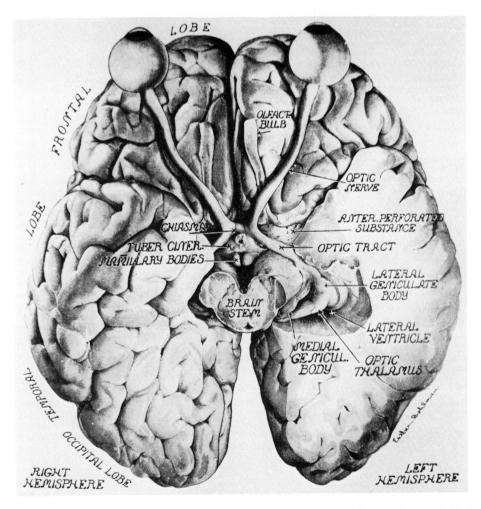

FIGURE 1.1. Human brain with the left temporal lobe dissected away. A preparation of this kind displays the optic tract and inferior horn of the lateral ventricle in the manner in which Galen may have seen them and in a form which may have led him to believe that the optic nerve arose from the ventricle. From Polyak (1957).

dissection of the optic nerve springing from this chamber a fanciful resemblance to an oar of a galley springing from one of the inner chambers or *thalami* of the vessel. Elsewhere, he uses the analogy of the different lengths of oars in a trireme in considering the different lengths of the fingers. We may note here that the rower in what is usually supposed to have been the lowest bench of a trireme was referred to as a *thalamite* and his oarport as *thalamia* (Morrison and Williams, 1968; Sleeswyk, 1982) (Fig. 1.2).

Some have found the term *thalamus* inappropriate to Galen's description. Simon (see Galen reference, 1906) suggested that *thalamos* may have been a misreading of a different word when Galen's Greek text was carried into Arabic by Avicenna. The alternative reading given by Simon—*thalame* or lurking place— would be appropriate if Galen indeed had in mind the deepest compartment of a galley. Walker's (1938a) suggestion of the possibility of derivation of *thalamos* from an Egyptian word for an antechamber also seems plausible and *thalamus* for an antechamber still commonly appears in archeological writings. However, as pointed out above, classical Latin writers had used the word very clearly in the sense that has come down to us. These must have been known to Galen. For one who had referred to the colliculi as nates and testes, the point of attachment of the pineal body as an anus, and the infundibular recess as a pelvis, it would be in character for him to introduce a note of levity in naming a new structure or region. Perhaps he saw the chamber at what he regarded as the commencement of the optic nerves as a center for regeneration, renewal, and continuity. Such a sense is often implied by the contexts in which Virgil uses the word

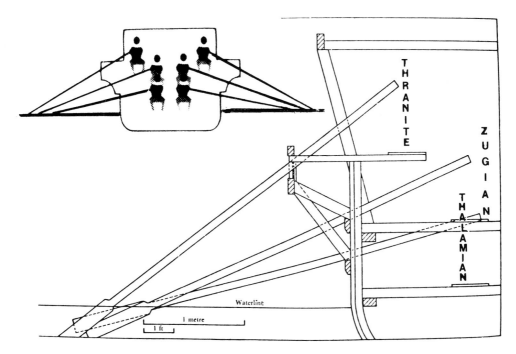

FIGURE 1.2. Modern interpretations of the arrangements of oars and rowers in a Greek trireme. The lowest seated rower was the thalamian or thalamite and his oarport a thal- amia. Large figure from Morrison and Williams (1968); inset from Foley and Soedel (1981).

thalamus and is present in the occasional use of the word in botany to refer to a receptacle from which carpels arise and within which embryonic seeds may develop.

What Galen had in mind when he used the word will probably never be known to us. More than a thousand years later, at the rebirth of anatomical investigation, thalamus indubitably came to mean the large mass of gray matter in the dorsal part of the diencephalon. Though still thought to be intimately associated with the optic tract and thus referred to as *optic thalamus,* it no longer meant a part of the ventricular system.

According to Walker (1938a), the first of the new identifications of the thalamus was by Mondinus in the 14th century (Singer, 1925). Mondinus refers to structures shaped like "anche" between the lateral ventricles and the third ventricle and Singer reads this word as "anchae" and translates it as "buttocks." In the passage referred to by Walker, the distinction between the anche and the superior colliculi, which Galen had called nates or buttocks, seems reasonably clear.*

The thalamus, corpus striatum, and internal capsule are clearly depicted in the horizontal sections of the head seen in plates 7 and 8 of Book VII of Vesalius's *Fabrica* (1543). But the drawing of this part of the brain has a curiously unfinished look and the structures are not named. This probably misled Walker (1938a) into believing that Vesalius had not observed the thalamus.

1.2. Thomas Willis

By the time of Thomas Willis (1664, 1681), *thalamus* was well entrenched. Willis uses it as a synonym for "the chambers of the Optick Nerves," an expression that on at least two occasions he attributes to Galen. Possibly, he obtained the

---

* "Ma avanti che tu pro cedi al ventriculo di mezo considera li mezi fra questo e quel di mezo li quali sono tre. cio e lanche le quali sono come basi over posamento di questo ventriculo anteriore dextro & sinistro: & sono dela sustantia del ceruello ad forma & figura dele anche. & dal lato di ciasche uno degli ventriculi gia decti e una sustantia rossa sanguigna facta a modo di un verme longo overo terreno cio e di quelli che si trovano sotto terra legata con legamenti & nervi da luna & laltra banda: la quale alla dilogatione di se constinge & serra le anche & la via over tra sita da lo anteriore al mezo & dal ventriculo di mezo alo anteriore & quando lhuomo vol cessare dal pensare & considerare di se:eleva le parete & dilata le anche accio chel spirito possi passare ad un ventriculo allaltro: & pero si chiama el verme perche si somegla al vermenella substantia & nela figura & mel moto contractivo & extensiuo."

"Before thou dost proceed to the mid ventricle consider the parts between the fore and mid ventricle. They are three, to wit the *anchae,* which are the base, as it were, of this fore ventricle right and left. They are of the substance of the brain and are shaped like buttocks (*anchae*). At the side of each *ancha,* between the ventricles already mentioned, is a red blood-like substance made like a long or subterranean worm. Ligaments and small veins bind it on both sides. This *worm* can lengthen itself by constriction and block the *anchae* closing the way or passage from the fore to the mid part and contrariwise. When a man doth wish to cease from cogitation and consideration, he doth raise the walls and expand the *anchae* so that the spirit may cross over from one ventricle to the others. It is called *vermis* both for that it doth resemble a subterraneous worm in substance and shape and also by reason of this motion of contraction and extension" [Singer's translation].

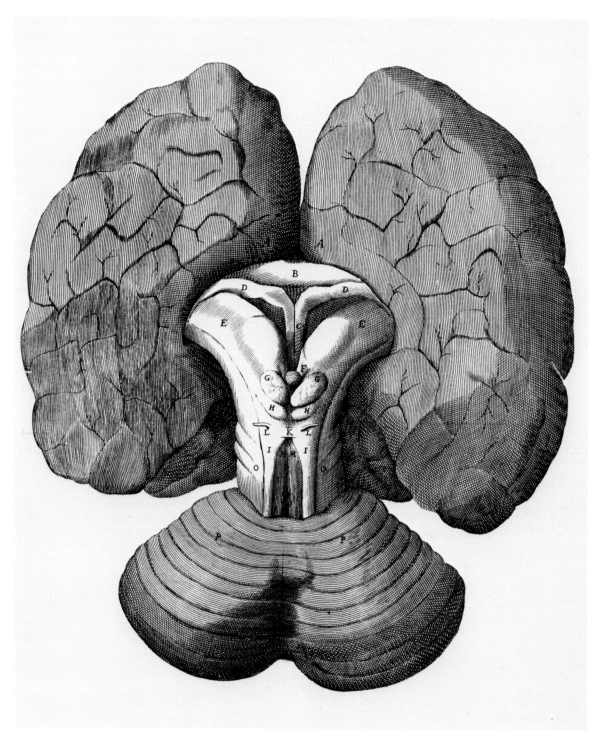

FIGURE 1.3. One of Thomas Willis's drawings of the human brain showing the thalami (E) which he calls "the shanks of the oblong Marrow," out of which (situated further out of sight) are "the streaked Bodies" (corpora striata). From Willis (1664).

name and the attribution indirectly through Riolan who had made a similar comment in 1610. To Willis, the "oblong marrow" (brain stem) was a Y-shaped structure with the arms of the Y ("shanks of the oblong marrow") inserted into the cerebral hemispheres and ending there as the striated (or chamfered) bodies. Below these, "where the streaked bodies end, the chambers or *Thalami,* as they are termed, of the Optick Nerves, possesseth the next part to the oblong marrow; to wit, in this place, its shanks rise into unequal prominences, out of the ridges of which the Optick Nerves arise. . . ." Below the thalami in the stem of the Y come the nates and testes (superior and inferior colliculi). The thalamus is clearly illustrated in four views of human or sheep brains drawn as seen from behind by Christopher Wren and there can be no doubt that it is the thalamus of modern anatomy that is being depicted (Fig. 1.3). In two of the plates what are probably the geniculate eminences are also seen.

Though he remarked upon the lack of an obvious cavity in the chambers of the Optick Nerves, and did not describe the penetration of the optic nerves into these chambers, Willis leaves us in no doubt about his adherence to the Galenic view of a flow of psychic spirit into the nerves. ". . . these Nerves are inserted into the medullar trunk, as branches of a Tree to the stock, that so they may receive by that means the influence of the Spirits. . . ." He noted the disproportionately large size of the diencephalon in relation to the cerebral hemispheres in birds and fish. This led him to remark that " . . .in this place, the animal spirits seem to have their chief Mart or Empory in a most large medullar chamber, or the Sphere of their Expansion. And so, when from hence the animal Spirits are derived from so full and plentiful a Store-house, it is for this reason that Fowls are furnished with so curious an Eye, and with so highly perspicacious and acute a sight." To Willis, the whole oblong marrow was permeated by the animal spirits which flowed from there into the various nerves arising from its different parts. "We have already shewed that the animal Spirits are procreated only in the Brain and Cerebel, from which they continually spring forth, inspire and fill full the medullar Trunk: (like the Chest of a musical Organ, which receives the wind to be blown into all the Pipes) but those Spirits being carried from thence into the Nerves, as into so many Pipes hanging to the same, blow them up and actuate them with a full influence; then what flow over or abound from the Nerves, enter the Fibres dispersed every where in the Membranes, Muscles, and other parts, and so impart to those bodies, in which the nervous Fibres are interwoven, a motive and sensitive or feeling force. And these Spirits of every part are called Implanted, forasmuch as they flow not within the Nerves, as the former, with a perpetual flood; but being something more stable and constant, stay longer in the subject bodies; and only as occasion serves, *viz.* according to the impressions inwardly received from the Nerves, or impressed outwardly by the objects, are ordained into divers stretching or carryings out for the effecting of motion or sense either of this or that manner or kind."

1.3. The Recognition of the Thalamic Nuclei

In the 150 years succeeding publication of Willis's work the thalamus was illustrated many times, the internal capsule was dissected past it, and some ascending tracts were dissected toward it. Several authors noted the anterior tubercle of the thalamus, thrown up by the underlying anterior nuclei, and the pulvinar can often be seen in their drawings (Vieussens, 1684; Soemmering, 1778; Vicq d'Azyr, 1786; Gall and Spurzheim, 1809, 1810) (Fig. 1.4). The geniculate bodies had also been recognized since Willis though often joined together and the term *geniculate* seems to have been first applied by Santorini (1724, 1775) who traced the optic tract to the lateral member of the pair.

The first clear indications of nuclear subdivisions were given by Karl Friedrich Burdach (1822) in his account of the human thalamus. Until recently, the numerous contributions of Burdach to brain anatomy had probably not been adequately recognized. With the publication of Alfred Meyer's *Historical Aspects of Cerebral Anatomy* (1971), Burdach is now reestablished as one of the major figures in the history of neuroanatomy. By examining slices of alcohol-hardened human brains with a hand lens, Burdach identified the internal medullary lamina and recognized that it divided the thalamus into superior (anterior), inner (medial), and external (lateral) nuclei. He also clearly recognized that what he called the external and internal geniculate bodies were parts of the thalamus and seems to have gained at least a superficial impression of lamination in the former. As well as the optic tract, he traced the brachium of the superior colliculus to the lateral geniculate body. Burdach named the *pulvinar* and a *stratum corneum* which seems to correspond to the external medullary lamina and possibly to the reticular nucleus. The reticular nucleus (stratum reticulatum) received its name later from Arnold (1838).

Burdach made very little comment about the possible functions of the thalamus. Full recognition of its important role as a sensory way station en route to the cerebral cortex was to come much later in the century.

After Burdach, perhaps the next significant contributions on the anatomy of the thalamus were made by Luys (1865). In the interim Stein (1834) had published what Walker (1938a) and Fulton (1949) regarded as the first thesis on the thalamus, the title page of which serves as the frontispiece to Walker's book. Luys emphasized the subdivision of the thalamus into four centers each of which he felt was isolated from its neighbors and made up of nerve cells communicating with special groups of afferent fibers. Basing his interpretation on comparative and pathological anatomy, he believed that the centers were independent foci through which different kinds of sensory impressions are relayed to the cerebral hemisphere. From dissections of the hemisphere, he recognized that the different centers are connected to different regions of the hemisphere though these were not very specifically nor, from a modern standpoint, very accurately identified.

The four centers of Luys (Fig. 1.5) were: (1) *centre antérieur*, equivalent to the anterior nuclei. He noted the relatively large size of the anterior thalamic tubercle in macrosmatic animals, felt that it was connected to the basal olfactory areas via the stria terminalis, and therefore considered it to be olfactory in function. (2) *Centre moyen*, from his illustration clearly equivalent to the internal (medial) nucleus of Burdach but regarded by Luys as the terminus of the optic tracts and, therefore, visual in function. (3) *Centre médian*, discovered by him in the human brain and still bearing this name, though, as indicated in his figure, probably including part of the mediodorsal nucleus. He considered it to be the center for "the condensation of sensory impressions," i.e., the terminus of the

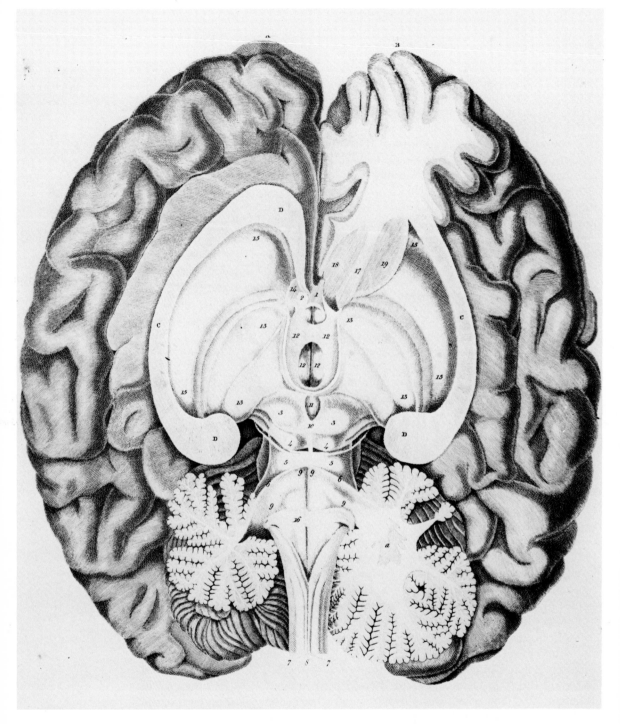

FIGURE 1.4. Plates copied by Knox from Gall and Spurzheim (left) and from Vicq d'Azyr (right) showing the thalamus (13 in left figure) and related blood vessels and other structures (right figure). In the right figure note the striae medullares (15–18) and massa intermedia (13, 14). From Knox (1832).

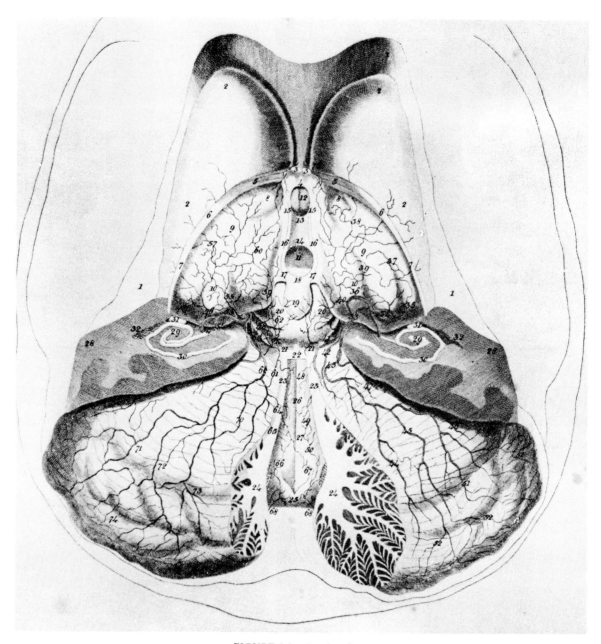

FIGURE 1.4. *(continued)*

somatic sensory pathways. Destructive lesions affecting it, he noted, were associated with contralateral hemianesthesia. (4) *Centre postérieur,* seemingly a part of the pulvinar in his illustration but regarded by him as auditory in function, mainly because it appeared degenerated in the brains of two deaf mutes. Luys also recognized the central gray matter component of the thalamus, considering it a continuation of the gray matter of the spinal cord and probably containing ascending sensory fibers. This was a not uncommon belief, dating at least to Gall and Spurzheim. Luy's illustrations also delineate Burdach's external (lateral) nucleus, but, curiously, he simply labels this "couche optique" and does not mention it further.

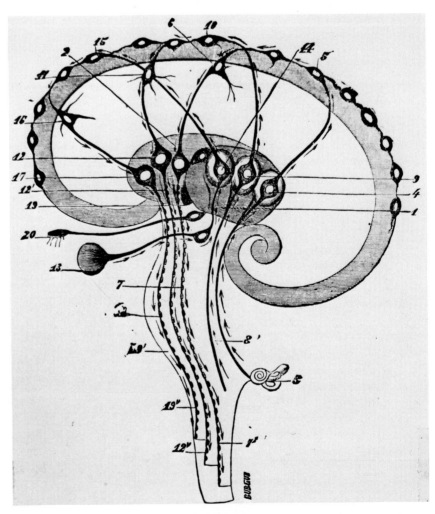

FIGURE 1.5.   Luys's scheme of "the sensori-motor processes of cerebral activity." 1 is the thalamus and 4, 9, and 14 its centers for relaying auditory (4), somatic sensory (9), and visual (14) impressions to the sensorium (5, 10, 15). From the sensorium, the large cells convey messages to the corpus striatum (2) and its corresponding centers (12) for the initiation of muscular activity. From Luys (1881).

With the lack of modesty characteristic of the time, Luys writes in his English book (1881) ". . . I have insisted on the fact, which ten years ago I was the first in France to bring to light, namely, that the optic thalamus, with the isolated grey ganglions of which it is composed, represents a place of passage and re-inforcement for excitations radiated from the sensorial periphery, while the corpus striatum, with its different compartments, and arches one within another, is on the contrary directly related to the passage of voluntary-motor excita-tions. . . ." He goes on to say "From a physiological standpoint, the optic thalami are intermediary regions interposed between the purely reflex phenomena of the spinal cord and the activities of psychical life. . . ." And finally "By their isolated and independent ganglions they serve as points of condensation for each order of sensorial impressions that finds in their network of cells a place of passage and a field for transformation. It is there that these are for the first time condensed, stored up and elaborated by the individual metabolic action of the elements that they disturb in their passage. It is thence, as from a penultimate stage, that, after having passed through ganglion after ganglion, along the cen-tripetal conductors which transport them, they are launched forth into the dif-ferent regions of the cortical periphery in a new form—*intellectualized* in some way, to serve as exciting materials for the activity of the cells of the cortical substance. These are, then, the sole and unique open gates by which all stimuli from without, destined to serve as *pabulum vitae* for these same cortical cells, pass; and the only means of communication by which the regions of psychical activity come into contact with the external world. . . ." There can be no clearer statement of the new emphasis on the thalamus as a sensory relay station.

1.4.   The Thalamus and Sensory Function

The first suggestions that the thalamus was involved in sensory processes had come from clinicopathological reports in man. Possibly one of the earliest of these was by John Hunter (not a relative of the great anatomist-surgeon). In 1825, he reported the case of a patient with progressive loss of sight, hearing, and touch over 3 years. At autopsy the brain of the patient displayed a lesion confined to the two thalami. Reporting a similar case in 1837, Richard Bright remarked that it corresponded "with our preconceived notions of the influence exerted by this portion of the brain," implying that the idea of the thalamus as a sensory center was widespread. Türck (1859a,b), Luys (1865), J. Crichton Browne (1875), and J. Hughlings Jackson (1864, 1866, 1875) all reported uni-lateral thalamic lesions associated with contralateral hemianesthesia. The case reported by Jackson (1866) is characteristically well documented: destruction of the right pulvinar and adjacent parts of the thalamus but without damage of other regions of the brain, was associated with diminished tactile sensation con-tralaterally, weakness of the contralateral leg, a contralateral hemianopsia due to "paralysis of the right sides of both retinae," and possible diminution of smell, taste, and hearing contralaterally.

Although the idea of the thalamus as a sensory center gradually gained prominence, it was not without its detractors. Many authorities of the middle years of the 19th century, among them such influential figures as Magendie (1823, 1841) and Vulpian (1866), were convinced that the thalamus, along with the corpus striatum, was part of the efferent, motor apparatus of the cerebral hemisphere. Magendie based his case primarily upon the experimental work of himself and Flourens (1824) which showed disturbances of motion resulting from thalamic lesions. Careful review of their reports, however, reveals that not only were their lesions extremely large but also that they relied heavily upon observations made on nonmammals, particularly birds and frogs. The organization of the diencephalon in these animals is such that it would even now be difficult to make an extrapolation to mammals.

None of the contemporary experimental work on mammals can be seen today as supporting the viewpoint of Magendie and Vulpian, though along with the results of experiments on frogs and birds, negative findings were often quoted in its favor. Perhaps one should bear in mind, however, the subthalamic syndrome of Foix and colleagues (Chiray et al., 1923; Foix and Bariéty, 1926; Foix et al., 1926) in which thrombosis of the more medial thalamostriate arteries in man, presumably because of interference with the ansa lenticularis and terminations of the brachium conjunctivum, may lead to athetoid movement. In 1873 Fournié injected a sclerosing solution of zinc chloride into the thalami of cats and produced bilateral sensory loss, apparently without much interference with motor behavior. But when his experiments were repeated in dogs by Veyssiére (1874), it was reported that a sensory disturbance could only be induced by spread of the injected material into the posterior limb of the internal capsule. Though Veyssiére produced little in the way of a motor disturbance, his results were frequently quoted as ruling out involvement of the thalamus in sensory activities.

In 1873, Nothnagel injected chromic acid into the thalamus of rabbits or destroyed it with an expanding stylet introduced through a trochar. He reported that a visual defect always ensued and that there could be some circling movements in the immediate postoperative period; but he emphasized the absence of a lasting defect of somatic sensation or of motion. Ferrier (1876) among others was quick to criticize this conclusion, pointing out that the animals showed withdrawal responses only to very strong cutaneous stimulation and allowed their limbs to be placed for long periods in abnormal positions. To Ferrier, these two features were indicative of a significant loss of cutaneous and position sense.

The abnormal postures adopted by Nothnagel's rabbits served to support a somewhat modified view of thalamic function proposed by Meynert (1884). Meynert had adopted the principle of "sensations of innervation" which was then experiencing a brief popularity, notably in the writings of Helmholtz, Wundt, Bain, and others (see Jones, 1972). In this view, all outgoing nerve impulses, especially those from motor neurons to muscle, were accompanied by sensation. To Meynert, the sensations of innervation accompanying motoneuron activity (and leading to an appreciation of the position of the part of a limb moved) were relayed back to the motor cortex after an interruption in the thalamus. In the motor cortex, they served as guides to its control of the body musculature. The more conventional sensations, mediated by impulses entering the spinal cord in the dorsal roots, bypassed the thalamus en route to sensory centers in more posterior parts of the cerebral hemisphere. The circling movements and the abnormal limb postures adopted by Nothnagel's rabbits were, therefore, primarily due to the motor cortex overacting and producing excessive muscle contractions in an attempt to induce a return flow of sensations of innervation that it was no longer receiving. The rest of Meynert's account of his theory of thalamic function is rather confused. At times he seems to imply that movements, particularly of the arms, may be initiated from the thalamus. In places he clearly states that all sensory pathways, except those mediating sensations of innervation, bypass the thalamus. Yet in others he mentions that pain messages ascending to the sensory cortex are interrupted in the thalamus and states also that the thalamus may be set in action by visual impulses. The verdict of history must be that Meynert's contributions to cerebral anatomy were far more substantial than his contributions to the analysis of thalamic function.

Contemporaneously with Nothnagel, the English neurologist David Ferrier (1874) commenced his studies on the effects of electrical stimulation and ablation of different parts of the brain of monkeys and other animals. By the time that he wrote the first edition of *The Functions of the Brain* (1876), Ferrier could only remark that although the corpus striatum seemed confirmed as a motor center, the functions of the thalamus were controversial. To him, there appeared to be neither constancy nor uniformity of symptoms reported in association with thalamic lesions and he felt that the cases reported by Luys in support of his concept of the thalamus as a sensory center were neither particularly suitable nor well documented.

In his own studies in monkeys, Ferrier knew that fibers leaving the thalamus are distributed to "posterio- and temporo-sphenoidal regions of the hemisphere" in which he had localized (in part incorrectly) the principal sensory areas. And he found that direct electrical stimulation of the thalamus elicited no motor responses. On the other hand, when an expandable trochar similar to Nothnagel's was introduced from behind and laterally, through the cerebral hemisphere, so as to cause severe destruction of the thalamus (and, incidentally, of the adjoining internal capsule) (Fig. 1.6), he noted that the animal was blind in the contralateral visual field, held its contralateral limbs motionless except in struggling, and reacted neither to cutaneous nor noxious heat stimulation applied to the contralateral side of the body.* Ipsilateral ptosis and contralateral pupillary dilation were also present suggesting involvement of the midbrain. No mention was made of the animal's auditory capacity. Though Ferrier recognized that the lesion was extensive and had caused considerable damage of the auditory and visual radiations, he felt that the lack of responses to cutaneous stimuli was sufficient to indicate an association of regions "in and around the optic thalamus" in tactile sensation.

On the basis of this single experiment, backed by the work of others that he had earlier criticized, Ferrier proceeded to elaborate a theory of sensory function for the thalamus, a theory that had many parallels with the theory then in vogue regarding the motor function of the basal ganglia. It was the common belief that the motor pathways of the cerebral hemisphere, including the py-

---

* Ferrier, in fact, never specifically mentions loss of sensation to light cutaneous stimulation but implies it in his ensuing text.

ramidal tract, emanated from the corpus striatum and many considered that the corpus striatum was the principal motor center of the brain. J. Burdon Sanderson (1874), who had submitted Ferrier's first communication (1874) to the Royal Society, actually felt obliged to attack Ferrier's views in a companion article. Having shown that movements could be elicited by electrical stimulation of the motor cortex and believing that these were "purposive" movements, Ferrier was obliged to make the corpus striatum subservient to the cortex. He, therefore, proposed that the corpora striata integrated the activities of all cortical motor centers and became the predominant centers controlling learned movements that had became habitual or automatic. Adopting a similar line of reasoning in regard to the thalamus, and believing that the "optic thalamus bears the same relation to the *tegmentum,* or sensory tracts of the crus cerebri, which the corpus striatum has to the foot [pes pedunculi] or motor tracts," Ferrier suggested that although the thalamus should remain subordinate to the sensory centers that he had demonstrated in the cerebral cortex, with learning and habituation a sensory stimulus leading to a motor response could become unconscious and, therefore, relayed directly from thalamus to corpus striatum, bypassing both the sensory and the motor areas of the cortex.

The theories of Ferrier seem to represent early forerunners of theories of cerebral function based upon reflex action. And when he goes on to stress that the basal ganglia and thalamus may play a more predominant role than the

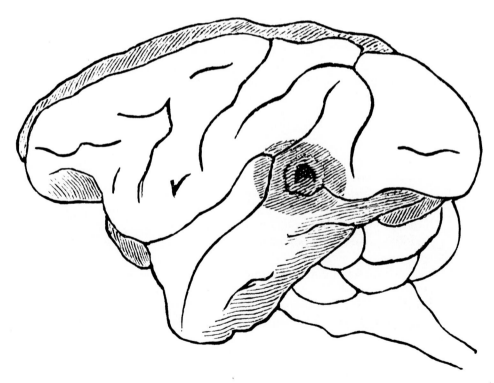

FIGURE 1.6.   Ferrier's drawing of a monkey brain in which he caused contralateral hemianesthesia by driving a trochar through the darkened area into the thalamus. From Ferrier (1876).

cortex in sensory motor behavior in animals than in man, his theory has much in common with later views on progressive corticalization of function.

1.5. Meynert and Forel

In his studies on the brains of man, monkeys, and some other species, Meynert (1872) used a combination of gross dissection, slices, and sections fixed

FIGURE 1.7. August Forel (1848–1931), Professor of Psychiatry, University of Zürich. Portrait made in 1910 by Oskar Kokoschka and currently in the Institut für Hirnforschung, University of Zürich. Kindly provided by Dr. M. Cuénod.

with potassium dichromate and stained with gold chloride or with carmine. He recognized most of the subdivisions of Burdach as well as the reticular nucleus ("a kind of claustrum") and the internal medullary lamina. He also described at some length the fiber lamination in the lateral geniculate nucleus of several species. His major new contribution, however, was the recognition of the alternating gray and white lamellae in the ventral part of the lateral nucleus of Burdach. This enabled him to divide the lateral nucleus into a dorsal, lightly myelinated and a ventral, heavily myelinated field (Chapter 2). He also mentions the extensive system of fibers joining the thalamus, including the geniculate bodies, to the cerebral hemisphere and noted that particular regions of the thalamus were connected by what are still sometimes called thalamic peduncles to different lobes. The optic tract, ansa peduncularis, fornix, and brachium of the inferior colliculus were also noted to terminate in the thalamus. Among Meynert's other contributions was one of the first descriptions and naming of the habenular region, and of the fasciculus retroflexus or habenulopeduncular tract. The latter was almost immediately named after him by his student, August Forel, in his doctoral thesis of 1872.

Forel (Fig. 1.7) himself added little that was new to the descriptive anatomy of the thalamus. However, his papers, particularly that of 1877, not only reviewed the literature in considerable depth, concluding in favor of a sensory function for the thalamus, but also provided some of the best illustrations of the sectioned thalamus published up to that time. He strongly advocated the use of Burdach's subdivisions but transcribed Burdach's terms *external, internal, superior,* and *inferior* into *lateral, medial, anterior,* and *posterior*. Since that time, these have become the more widely used reference words for describing the thalamus. It is interesting to note that this authoritative and influential paper was written at the age of 29 when he had only just become Privatdocent.

FIGURE 1.8. Bernard von Gudden (1824–1886), Professor of Psychiatry, Munich. From K. Kolle, *Grosse Nervenärzte* (1959), with permission.

Forel's reviews of the thalamus came as thalamic studies passed from an era of rather gross description to one that was to be predominately microscopic. During this new era, the first experimental studies were conducted on the thalamus. At the time he wrote his paper, Forel had been von Gudden's assistant at Munich for approximately 4 years. von Gudden's influence in this phase of thalamic history is pervasive. Not only did he perform the first experimental studies on the thalamus, but in addition to Forel, the two other major contributors to thalamic anatomy of the period, Nissl and von Monakow, worked with him.

As early as 1870, von Gudden (Fig. 1.8) had noted that if rabbits were permitted to survive for many months after removal of an eye in infancy, the contralateral optic tracts, lateral geniculate nucleus, adjacent parts of the thalamus and superior colliculus underwent detectable atrophy and few or no cells could be stained with carmine in these structures. By 1881 both he and von Monakow (1882) (Fig. 1.9), who had studied with him on a visit in 1876–1877, had observed thalamic atrophy ensuing from lesions of the occipital cortex in infant rabbits. By this time von Gudden had invented a form of sliding microtome (1875) and was regularly using the celloidin embedding technique, introduced by Duval in 1879. In the course of his studies, von Gudden had mentioned the histological features of certain thalamic nuclei in the rabbit, but it is to his assistant, Nissl (Fig. 1.10) that we owe the first full account of the thalamus of any mammal. While still a medical student in about 1884, Nissl had apparently been encouraged by von Gudden's then assistant, Ganser (himself the describer of certain nuclei in the thalamus of the hedgehog in 1882), to write an essay on the application of aniline dye stains to the cerebral cortex (Spatz, 1961). Presented as a paper at a meeting of the German Society for Natural Science and Medicine

FIGURE 1.9. Constantine von Monakow (1853–1930), Director of the Hirnanatomisches Institut, Zürich. From a newspaper photograph published at the time of his death.

in 1885, this formed the basis for his subsequent development of the "Nissl stain" (Nissl, 1894).

In 1889, 3 years after von Gudden's tragic death, Nissl presented another paper at the Heidelberg meeting of the same society. In this, on the thalamus of the rabbit, he provided not only the first complete histological description of the thalamus of a single species, but also, on the basis of cytoarchitecture, he identified most of the thalamic nuclei that we know today. He divided the anterior nucleus of Burdach into dorsal, ventral, and medial divisions; the medial nucleus he divided into anterior, middle, and posterior divisions; in Burdach's lateral nucleus he recognized ventral and lateral nuclei and redivided the ventral nucleus into the approximate equivalents of the modern ventral posterior lateral and ventral posterior medial nuclei; the lateral nucleus he redivided into its anterior and posterior subdivisions; he recognized the geniculate nuclei, redividing the lateral into its dorsal and ventral nuclei; he described the large cells (grosszellige Kern) of what would now be called the central lateral nucleus, a nucleus of the midline, the reticular nucleus (ventrale GitterKern), and the habenular nuclei. He also described two divisions of a posterior thalamic nucleus which though still sometimes called that, represents mainly the pretectum.

No illustrations accompanied Nissl's one-page abstract and he did not publish his definitive paper on the rabbit thalamus until 1913. In the meantime, however, he had made his preparations available to von Kölliker who used them as the basis for his description of the rabbit thalamus (Fig. 1.11); in the sixth edition of his *Handbuch der Gewebelehre des Menschen* (1896), von Kölliker reprinted in full a short description made by Nissl in 1890 and accompanied it with a modified description of his own, some excellent drawings of fiber-stained preparations, and some of the earliest drawings of Golgi-impregnated thalamic cells (Chapter 3). Apart from converting Nissl's German terms for some of the thalamic nuclei into latinized forms, von Kölliker renamed the medial division of Nissl's ventral nucleus, *nucleus arcuatus,* but added little else. The arcuate or ventral posterior medial nucleus had, in fact, already been identified in the

FIGURE 1.10. Franz Nissl (1860–1919), Professor of Psychiatry, Heidelberg. From K. Kolle, *Grosse Nervenärzte* (1959), with permission.

human brain by Flechsig (1886) who called it *Schalenförmige Nucleus*, though the Vogts (1941) give credit for this name to Tschisch. Dejerine and Dejerine-Klumpke (1895) had also observed it, calling it the *semilunar nucleus*.

By the time Nissl published his definitive account, many descriptions of the thalamus in a variety of species had appeared. They include: studies on rodents and lagomorphs by Haller (1900), Münzer and Wiener (1902), Bianchi (1909), Ramón y Cajal (1911), Winkler and Potter (1911), and D'Hollander (1913); studies on the dog by daFano (1909); a study on the opossum by Röthig (1909); studies on monkeys and lemurs by Mann (1905), Vogt (1909), Sachs (1909a,b), and Friedemann (1911); studies on man by von Monakow (1895), Dejerine (1901), Marburg (1904), Malone (1910), and Jakob (1911). Many of these are based upon Nissl-stained preparations but others, such as those by von Monakow and Vogt, are myeloarchitectonic studies based upon Weigert's stain, introduced in 1882.

Nissl's delay in publishing caused his contribution to be overshadowed by that of von Monakow (Fig. 1.12). By 1895, in cats, dogs, and rabbits he had not only confirmed Nissl's delineation of the thalamic nuclei but also, by means of

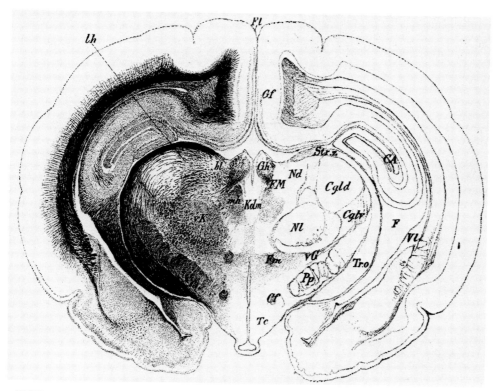

FIGURE 1.11.   An early drawing of a Weigert-stained section through the thalamus of a rabbit by von Kölliker (1896) in which nuclei are named after Nissl (1889). Nuclei shown are the lateral (Nl) or ventral (vK), the dorsal and ventral lateral geniculate (Cgld, Cglv), medial (mh), reticular (vG), lateral posterior (Nd, hl), and habenular (Gh). Kdm are "nuclei of the midline." From von Kölliker (1896).

von Gudden's method of retrograde atrophy, had worked out the broad outlines of thalamocortical organization as we know it today. von Monakow (1885, 1889, 1895) found that destruction of cortex at the frontal pole resulted in degeneration of the medial thalamic nucleus and that damage to cortex on the medial aspect of the hemisphere caused degeneration of the anterior nuclei. Ablation of motor or anterior parietal cortex caused degeneration of anterior and posterior halves respectively of the ventral nucleus. Posterior parietal cortex damage led to degeneration of the lateral and posterior nuclei; occipital or temporal cortex ablations caused atrophy of the lateral or medial geniculate body. Parietotemporal ablations caused degeneration of the reticular nucleus, but both he and Nissl noted that the degeneration in it was less severe than in the other nuclei.

Although von Monakow's delineation of the thalamic nuclei is essentially the same as that of Nissl, he designated some of the finer subdivisions by letters rather than by names. The medial nucleus was divided into nuclei medialis a, b, and c. These appear to represent the mediodorsal, centre médian, and central lateral nuclei, respectively. The ventral and anterior groups were similarly divided into subnuclei a, b, and c. Traces of this nomenclature, which did not survive, can be seen in the once popular but rather inaccurate atlases of the rabbit and cat brain of Winkler and Potter (1911, 1914).

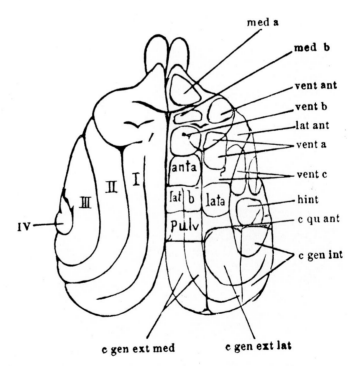

FIGURE 1.12.   von Monakow's scheme of thalamocortical relationships in the cat, based upon his studies of thalamic degeneration ensuing from localized lesions of the cortex. The nuclei related to a particular cortical region are named at the right. From von Monakow (1895).

1.7. The Thalamic Pain Syndrome

The 24 years intervening between Nissl's abstract and his definitive paper had also seen the recognition of the "thalamic pain syndrome," the result of thrombosis of the thalamogeniculate artery and concomitant destruction of the posterolateral part of the ventral thalamic nuclei. The resulting symptom complex is still enigmatic. It is characterized by a transient hemiplegia, sometimes succeeded by motor incoordination, a severe disturbance of both cutaneous and deep sensation, associated with intolerable spontaneous pain and with particularly unpleasant sensations initiated by trivial stimuli. Some of the symptoms can be attributed to involvement of the terminations of the cerebellar dentatothalamic pathway and of the medial lemniscus. Accompanying vasomotor symptoms are, likewise, probably attributable to hypothalamic involvement. But the reason for the hyperpathia still eludes us.

Spontaneous pain arising in association with thalamic lesions may have been first described by Edinger (1891).* But the now classical thalamic syndrome and its precise etiology were undoubtedly first recognized by Dejerine and Egger (1903) and by Dejerine and Roussy (1906), after whom it is commonly named. A further influential account was that of Head and Holmes (1911), though Head's (1920) view of the cause of the hyperpathia has now been discredited. It was his belief that "protopathic sensibility" rose to consciousness in the thalamus whereas "epicritic sensibility" did so in the cerebral cortex. He further believed that corticothalamic fibers, which he thought terminated only in the lateral thalamus, permitted the epicritic centers to exert a suppressive influence over the protopathic. Putative destruction of these fibers by thrombosis of the thalamostriate artery was therefore thought to release the protopathic medial regions of the thalamus from this suppression. This view was strongly attacked by Lhermitte and his associates (1921–1936) and the modern viewpoint is that the hyperpathia is probably due to some kind of imbalance between lemniscal and spinothalamic inputs.

1.8. The Modern Descriptive Period

The first 30 years of the 20th century were dominated by descriptive and comparative accounts of the mammalian thalamus. In that time the human thalamus was described at length by Foix and Nicolesco (1925), though their account largely recapitulates that of Dejerine and Dejerine-Klumpke (1895) (Fig. 1.13). Starting then and continuing up until the present day, the thalamus of at least one representative of virtually every mammalian order has been described at least superficially and sometimes at great length. Apart from those mentioned

---

* The symptomatology had obviously been described earlier. Spillane (1981), for example, interprets John Cooke's (1824) report of "the celebrated Dr. Saussure" (who described his own illness) as a case of thalamic syndrome.

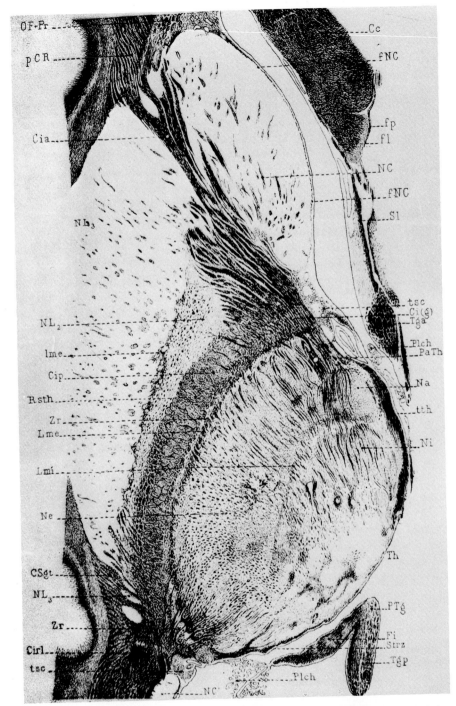

FIGURE 1.13. Drawing of a Weigert-stained, horizontal section through the human thalamus showing the reticular nucleus, internal medullary lamina, and certain dorsal thalamic nuclei. From Dejerine (1901).

elsewhere in this book, a sample listing of references would include: Abe (1952), Andô (1937, 1938), Arai (1939), Atlas and Ingram (1937), Bachmann (1950), Bauchot (1959), Bodian (1939), Brunner and Spiegel (1918), Campbell and Ryzen (1953), Chu (1932), Le Gros Clark (1928, 1929a,b, 1930a, 1931), Feremutsch (1963), Gerebtzoff (1939), Glorieux (1929), Goldby (1941), Graef and Volker (1968), Gröschel (1930), Grünthal (1945), Heiner (1960), Herbert (1963), Hess (1955), Hines (1929), Hirokawa (1941), Holmes (1953), Ibrahim and Shanklin (1941), Ingram *et al.* (1932), Kaelber (1966), Kanagasuntheram and Wong (1968), Kanagasuntheram *et al.* (1968), Kruger (1959), Kurespina (1966), Mayeda (1941), Miura (1933), Mussen (1923), Olszewski (1952), Oswaldo-Cruz and Rocha-Miranda (1967), Papez (1932), Přecechtěl (1925), Sachs (1909,a,b), Simma (1957), Sloane (1951), Solnitzky (1938), Wahren (1957), and Walker (1938c).

Possibly the only significant experimental studies in the period subsequent to Nissl's 1913 paper were those of Minkowski (1913, 1920) on the visual system of cats and monkeys. Making use of the transneuronal degeneration that occurs in the lateral geniculate nucleus following removal of an eye, he was able to show the segregation of left and right eye inputs in separate laminae of the monkey lateral geniculate nucleus. He also demonstrated the topographic organization of the geniculostriate projection by observing the systematic shift in the position of the patch of retrograde cell degeneration seen in the lateral geniculate nucleus of cats, following small lesions in different parts of the visual cortex.

Some of the best descriptive accounts of the thalamus appeared in this period. From the laboratory of Cécile and Oskar Vogt (Fig. 1.14) in Berlin came descriptions of the myeloarchitecture (Vogt, 1909) (Fig. 1.15) and cytoarchitecture (Friedemann, 1911) in cercopithecine monkeys* and lemurs (Pines, 1927). These works are illustrated by photographs as elegant as any being produced today and the subdivisions made are generally the same as those currently recognized. The terminology used was, unfortunately, a rather cumbersome extension of that introduced by von Monakow and has generally not retained its popularity. Some terms introduced by the Vogt school, however, do survive. These include: *nucleus limitans, parafascicular nucleus, parataenial nucleus, dorsomedial nucleus, pregeniculate nucleus, submedial nucleus* (but used for the nucleus reuniens), and *paralamellar nucleus.* Also recognized for the first time were the subsidiary nuclei of the pulvinar and medial geniculate body. The magno- and parvocellular layers of the lateral geniculate body, and the nuclei within the internal medullary lamina were also clearly identified. Dr. Jerzy Rose in 1972 drew attention to the skepticism with which the new architectonics was greeted by Ludwig Edinger, at the time certainly the leading comparative neuroanatomist and one of the most influential scientists in Europe. To Edinger and probably to many of his contemporaries it seemed scarcely credible that something they had come to regard as functioning holistically could consist of 35 or more different subdivisions.

A later member of the Vogt school was Maximilian Rose who, in 1935, published a particularly well-illustrated account of the cytoarchitecture of the

---

* These monkeys, which may have been the same as those used by Brodmann (1909) for his cytoarchitectonic studies of the cerebral cortex, were later identified by von Volkmann (1928) as the "Meerkatz," *Cercopithecus mona,* a guenon.

rabbit thalamus. In this he subdivided into finer and finer divisions many of the nuclei originally delineated by Nissl and he introduced an exceedingly complicated new terminology. It was probably this, as much as its publication in a relatively inaccessible journal, that caused Rose's account to fade from view. Nevertheless, traces of his terminology survive in words such as *ventrobasal complex* popularized by his nephew, J. E. Rose.

Another excellent account of the rabbit thalamus is that of D'Hollander (1913). His contributions and those of his students have been surprisingly neglected and I can find no serious consideration of his paper on the thalamus since the publications of Gurdjian (1927) and Papez (1932). D'Hollander recognized three groups of thalamic nuclei: the internal or medial group included anteromedial, medial (i.e., mediodorsal), parataenial, parafascicular, habenular, lamellar (i.e., paracentral), and paramedial (i.e., medioventral) nuclei, together with the réunissant or midline nuclei numbered 1 to 4; the middle group included anterodorsal, anteroventral, ventral, posterior, and reticular nuclei; the external group included lateral, magnocellular (i.e., central lateral), and the two geniculate nuclei. D'Hollander's account represents the first systematic application of the standard, latinized terms of descriptive anatomy to the thalamic nuclei, though some earlier attempts had been made by von Kölliker (1896) and by Münzer and Wiener (1902) to adapt Nissl's German terms. In the new nomenclature, *vordere dorsale Kern* of Nissl becomes *nucleus anterior dorsalis*.

Though, with modifications, D'Hollander's style of terminology was to be-

FIGURE 1.14. (Left) Cécile (1875–1962) and Oskar Vogt (1870–1959) on the steps of the Kaiser Wilhelm Institut für Hirnforschung at Berlin-Buch in the early 1930s. Photograph in the author's collection. (Right) Maksymilian Rose (1883–1937), Professor of Psychiatry and Neurology, University of Wilno, who worked with the Vogts and established himself as one of the foremost cytoarchitectonists of his age. Photograph taken about 1930. Courtesy of Dr. J. E. Rose.

come the most popular, there were other contemporary accounts that adopted quite different terminologies. The terminology of the Vogts has already been mentioned. One account of the human thalamus that had some popularity was that of Malone (1910). His parcellation of the thalamus, however, depended upon classifying together cell types with similar staining properties so that some nuclei named by Malone bear little resemblance to those of other authors. Malone has been credited (Meyer, 1971; Berman and Jones, 1982) with the introduction of the term *nucleus reuniens* though I have since noted it in an earlier paper of Röthig (1909).

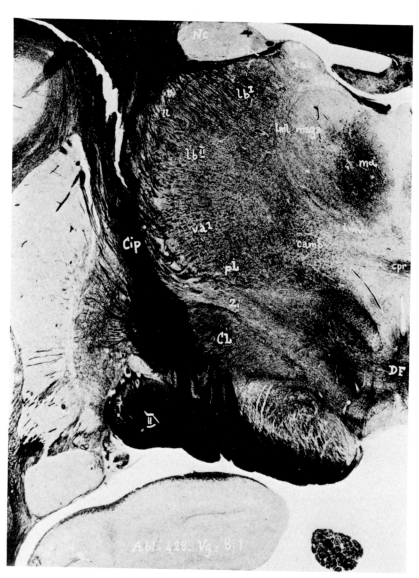

FIGURE 1.15. Copy of one of Cécile Vogt's photographs showing a Weigert-stained section of the monkey thalamus. From Vogt (1909).

Ramón y Cajal (1904, 1911) in Golgi as well as Nissl stains accurately identified many of the nuclei of the rabbit thalamus. But his names tend to describe the morphological appearance rather than the relative disposition of the nuclei and few have come into common use. Ramón y Cajal classified the nuclei into three classes or "ranges." The external range included internal and external geniculate nuclei and a pulvinar. The middle range included a posterior or prebigeminal nucleus (probably the pretectum), a dorsal (i.e., anterior) nucleus, and a sensory nucleus; the sensory nucleus was equivalent to the ventral nucleus of Nissl and had two accompanying satellites: the anterior semilunar nucleus is possibly part of the reticular nucleus but the posterior semilunar or trapezoid nucleus is difficult for a modern reader to identify. Ramón y Cajal's internal range of nuclei included habenular, internal (or medial), intermediate (or centre médian), rhomboid, and commissural nuclei. The commissural nucleus had a winglike expansion, the falciform nucleus, clearly equivalent to Nissl's magnocellular nucleus and to the central lateral nucleus of modern terminology. Ramón y Cajal in his Golgi preparations noted the terminations in particular nuclei of many of the afferent pathways to the thalamus but possibly because of its nonexperimental nature, he has been given surprisingly little credit for this aspect of his work. His contributions on the morphology of individual thalamic cells, especially the distinction he made between relay neurons and interneurons, has been more widely recognized (Chapter 3).

Given the different nomenclatures available for the thalamic nuclei, "the terminological atmosphere," as Le Gros Clark (1932a) put it, was anything but clear. Gradually, however, a terminology similar to that of D'Hollander and formulated mainly by what Walker (1938a) called "the Michigan School of Anatomy" took hold and has become that in commonest use today. The Michigan school consisted mainly of the students of G. C. Huber (Figs. 1.16–1.18) who carried out some of the earliest yet still valuable comparative cytoarchitectonic studies of the thalamus in the opossum (Tsai, 1925; Bodian, 1939), rat (Gurdjian, 1927), cat, and dog (Thuma, 1928; Rioch, 1929a). Concurrently, Huber's other students such as Elizabeth Crosby (Fig. 1.19) and, elsewhere, Herrick (Fig. 1.19)

FIGURE 1.16. E. Stephen Gurdjian (1900–   ), Professor of Neurosurgery, Wayne State School of Medicine, Detroit, Michigan, who while an intern at the University of Michigan published the first major descriptive work on the thalamus of the rat. Photograph courtesy of Doctor Gurdjian.

and Kappers (Fig. 1.19) were engaged in describing the nonmammalian thalamus (see Chapter 17). The terminology of the Michigan school was quickly adopted by investigators working on other mammalian species, first by Le Gros Clark (1929a,b, 1930a, 1932a) who in his earlier studies (1928) had used different terms, and later by Crouch (1934) and Walker (1936, 1938a,b). Among major studies of the period, only that by Grünthal (1934) on the thalami of several species used a different nomenclature. The Michigan terminology is still in widest use in animals and has been adapted to the human thalamus (Sheps, 1945; Toncray and Krieg, 1946; Dekaban, 1953; Kuhlenbeck, 1954), though many descriptions of the human thalamus, coming as they do from the continent, tend to follow the terminology of the Vogts (1941) (Hassler, 1950, 1959; Feremutsch and Simma, 1953, 1954a,b, 1955, 1958, 1959; Simma, 1957; Hopf, 1971; Macchi, 1971).

1.9. The Rebirth of Experimentation

After 1932, investigations of the thalamus again became predominately experimental. The year 1932 is marked by the publication of a lengthy review article by Le Gros Clark (1932a) (Fig. 1.20) in which he surveyed much of the antecedent descriptive literature and mentioned his early experiments on thalamocortical connections in rats. Thereafter, and up until about 1940, numerous studies were carried out with the method of retrograde cell degeneration which enlarged in detail on the relationships of individual thalamic nuclei to the cerebral cortex as originally defined by von Monakow. The period saw numerous studies on the primate thalamus by Le Gros Clark and his collaborators (Le Gros

FIGURE 1.17. David McK. Rioch (1900–   ), Director, Neuro-Psychiatry Division, Walter Reed Army Institute of Research, who while a postdoctoral fellow at the University of Michigan published the first major works in English on the thalamus of the cat and dog. Photograph in the Washington University, Neurology Library.

FIGURE 1.18. David Bodian (1910–   ), Professor of Anatomy, The Johns Hopkins University, who while a postdoctoral fellow at the University of Michigan published a series of important papers on the opossum thalamus and its connections. Photograph courtesy of Doctor Bodian.

FIGURE 1.19. A bevy of comparative neuroanatomists at the University of Michigan in 1937. From left, Cornelius U. Ariëns Kappers (1877–1946), C. Judson Herrick (1866–1960), Olof Larsell (1886–1964), and Elizabeth C. Crosby (1888–1983). Photograph courtesy of Dr. David Bodian.

Clark, 1936a; Le Gros Clark and Boggon, 1933a,b, 1935; Le Gros Clark and Northfield, 1937), by Polyak (1927, 1932, 1933), and by Walker (Fig. 1.21) (Walker, 1935, 1936, 1938,a,c,d, 1940b; Walker and Fulton, 1938). But complementary to these were investigations of thalamocortical relations in the rat, cat, and opossum by D'Hollander and his students (De Haene, 1936; Gerebtzoff, 1937; D'Hollander and Gerebtzoff, 1939), by Waller (Waller, 1934, 1938, 1940a,b; Waller and Barris, 1937), and by Bodian (1942).

Just as Le Gros Clark's paper of 1932 ushered in this experimental period, so it culminated with the publication in 1938 of Walker's *The Primate Thalamus* (1938a) which commences with a historical introduction and thereafter sets forth most of the existing knowledge about afferent and efferent connections of the primate thalamus. A notable omission is the absence of any detailed consideration of the visual system about which Walker felt "few questions remain to be solved." This was no doubt a compliment to Minkowski and Polyak but, in the light of recent developments, a major misapprehension.

From the time that Nissl introduced his stain (1885) and showed its value in connection tracing (1892), virtually all significant experimental work on thalamocortical connections was carried out with the retrograde degeneration method. von Monakow (1889, 1895) had used a combination of carmine staining and the Weigert method to identify the cell and fiber loss of von Gudden's atrophy but

FIGURE 1.20. Sir Wilfrid E. Le Gros Clark (1895–1971), Doctor Lee's Professor of Anatomy, University of Oxford. Photograph taken on a visit to New Zealand in about 1950.

this method was supplanted by that of Nissl. The Marchi method had been introduced in 1885 (Marchi and Algeri, 1886) and was used widely in attempts to localize the distributions of most of the afferent pathways in the thalamus (Fig. 1.22), but few significant studies were carried out on thalamocortical connections with it.

The Marchi technique served to resolve many of the disputes that had arisen regarding the terminations of the afferent pathways in the thalamus. The optic tract, medial and lateral lemnisci, and brachium conjunctivum had, of course, been traced by gross dissection to the vicinity of the thalamus or upper midbrain from the earliest times (Gall and Spurzheim, 1809, 1810; Reil, 1809; Burdach, 1822; Arnold, 1838; Gratiolet, 1857; Stilling, 1857; Meynert, 1867; Henle, 1871; Luys, 1876; Forel, 1877). But many details were lacking. In 1876, for example, Ferrier could still write that the optic tract terminated in both the medial and the lateral geniculate bodies and, like many of his contemporaries, he still thought that the tegmentum of the midbrain merely represented a continuation of the sensory pathways of the spinal cord up to the thalamus. The widespread use of the von Gudden method in infant animals and study of the brains of human patients with long-standing lesions also tended to promote controversy. Under these circumstances, degeneration is commonly not confined to the axons and cells connected to the lesion site but may spread anterogradely and retrogradely

FIGURE 1.21. A. Earl Walker (1907–), Professor of Neurosurgery, The Johns Hopkins University, whose famous book *The Primate Thalamus*, representing postdoctoral work done at the Universities of Chicago and Iowa, Yale University, and the Amsterdam University, was published in 1938. Photograph courtesy of Doctor Walker.

across several synapses into other cells and axons. Hence, in the early work of von Gudden, von Monakow and others, it is sometimes possible to see atrophy of the medial lemniscus accompanying that of the thalamus after neonatal destruction of the cerebral cortex. In his book of 1905, Campbell remarks upon degeneration of neurons in layer V of the cerebral cortex in long-standing cases of tabes dorsalis or of amputation of a limb (Jones, 1983c). Not surprisingly, therefore, such results were sometimes taken to confirm the old belief, based upon gross dissection, that the medial lemniscus, optic tract, and other afferent pathways bypassed the thalamus and terminated directly in the cerebral cortex (e.g., Meynert, 1884; von Monakow, 1885; Flechsig and Hösel, 1890). Mahaim (1893) and Bechterew (1895) pointed out that the degeneration in the medial lemniscus secondary to destruction of the pre- and postcentral gyri in man was significantly less than in the internal capsule and, thus, seem to have been the first to recognize the secondary (transneuronal) character of the effect. By the time of Campbell, it was well recognized.

Mott (1892, 1895) and von Monakow (1895, 1914) by use of the Marchi technique were able to demonstrate in cats and monkeys that the ascending degeneration secondary to lesions of the dorsal column nuclei does not extend beyond the ventral nuclei of the thalamus. This was soon confirmed by Probst (1898, 1900a–c) in cats and dogs and even in 1895, in human pathological material, Dejerine and Dejerine-Klumpke had demonstrated that the medial

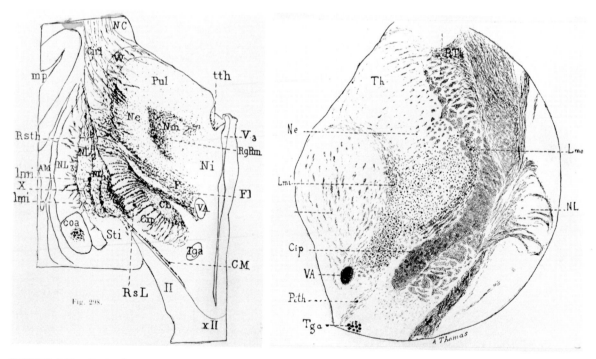

FIGURE 1.22. Early drawings of Marchi-stained preparations showing (left) degeneration in the human thalamus following a large subcortical lesion (from Dejerine, 1901) and (right) degeneration in the thalamus of a dog after hemicerebellectomy (from André-Thomas, 1912).

lemniscus probably terminated in the ventral posterior nucleus. The many other early authors who contributed experimental or clinicopathological studies to this subject included: Bechterew (1895), Bielschowsky (1895), Jacob (1895), Lasurski (1897), von Sölder (1897), Long (1899), Wallenberg (1900), and Vogt (1909).

The trigeminothalamic terminations in the medial part of the ventral posterior nucleus were first reported in rabbits by van Gehuchten (1901) and Wallenberg (1896, 1900) and in monkeys by von Economo (1911).

Mott (1895) in monkeys and Quensel and Kohnstamm (1907) in rabbits showed Marchi-stained degenerating spinothalamic fibers in the posterior part of the ventral thalamic nuclei after spinal cord lesions. Gowers (1886) in Marchi preparations from human cases with spinal cord wounds and Edinger (1890, 1900) in silver-stained preparations from the embryos of cats and humans had earlier traced fibers of spinal origin to the midbrain.

The terminations of the ascending auditory pathways in the medial geniculate nuclei appear to have been first demonstrated with the Marchi technique by Ferrier and Turner (1894) and by von Monakow (1895).

The delineation of cerebellothalamic terminations (Fig. 1.22) received a great deal of attention, Marchi himself (1891) making one of the first investigations with his method. Ferrier and Turner (1894), von Monakow (1895), and many other investigators traced Marchi degeneration to the thalamus after destruction of the contralateral denate nucleus or after section of the contralateral superior cerebellar peduncle, but they did not clearly describe the region of termination. Wallenberg (1900) in the rabbit, Probst (1900a–c, 1901a–c) in several species, and Vogt (1909) in the monkey were probably the first to trace degenerated cerebellar efferents to the anterior part of the ventral nuclei and to the nuclei of the internal medullary lamina. Vogt stated very explicitly that the terminations in the ventral nuclei lie rostral to those of the medial lemniscus.

The remaining major thalamic afferent pathway studied at this time was the ansa lenticularis. This had been traced from the globus pallidus into the thalamus by von Monakow (1895) and by Vogt (1909) who localized its terminals in parts of the ventral nuclei anterior to the terminations of the dentatothalamic fibers. Experimental studies confirming this with the Marchi method in monkeys were made by Wilson (1914) and by Vogt and Vogt (1920). The Vogts' description of the terminal region in the thalamus is probably more accurate than Wilson's. They localized it in the nucleus Vtl, or noyau ventral orale of Vogt (1909) and Friedemann (1911), anterior to the terminations of the brachium conjunctivum.

The 1930s saw a resurgence of interest in studies with the Marchi technique on the afferent pathways to the thalamus. Brouwer (1923), Brouwer and Zeeman (1926), Overbosch (1927), Lashley (1934a), and Bodian (1937) used it to investigate the topography of the retinogeniculate projection. Polyak (1932) studied the thalamic terminations of the auditory pathway. The terminations of the medial lemniscal, trigeminal, spinothalamic, and cerebellothalamic pathways were investigated by Allen (1924), Polyak (1932), Le Gros Clark (1936b), Foerster and Gagel (1932), Ranson and Ingram (1932), Walker (1934, 1936, 1937, 1938a,b), Ferraro and Barrera (1935), Gerebtzoff (1936), and Papez and Rundles (1937).

Vogt and Vogt (1941), Ranson *et al.* (1941), and Papez (1942) reexamined the pallidothalamic projection.

Some attempts were made to study thalamocortical connections with the Marchi method (e.g., Polyak, 1932; Le Gros Clark and Boggon, 1933a,b; Bodian, 1942) but the lesions used were generally so large that the resolution of the method was far less than that of retrograde, cellular degeneration. Though not used particularly effectively in the study of thalamocortical connections, the Marchi method was instrumental in demonstrating the presence of cortico-thalamic connections (Probst, 1900b, 1906; D'Hollander, 1922, 1928) which came to be recognized as probably arising in widespread areas of the cerebral cortex (Biemond, 1930; Mettler, 1935a–d) though confirmation of their ubiquity and reciprocal relationships with all thalamic nuclei had to wait until the 1960s (Berman and Jones, 1982).

1.10. Rose and Woolsey and the Dawn of the Recent Era

After publication of Walker's book in 1938, the next major contribution and probably one of the most significant landmarks in the recent history of thalamic studies was the work of Jerzy E. Rose (Fig. 1.23) (1942a) on the development of the diencephalon of the rabbit. In this he was able to show that during ontogeny, independent cell masses give rise to the precursors of the dorsal thalamus, epithalamus, and ventral thalamus. These terms had been introduced in the literature of comparative neuroanatomy many years previously (Edinger, 1885; von Kölliker, 1896; Kappers, 1908; Herrick, 1910, 1918; DeLange, 1913) and generally served to homologize parts of the mammalian diencephalon with that of submammalian species (Chapter 17). Even here, there was no complete agreement about the use of the terms and European workers generally preferred the term *subthalamus* to the *ventral thalamus* of Herrick (e.g., Le Gros Clark, 1932a). Kappers *et al.* (1936), in their frequently referenced book of comparative neuroanatomy, regarded *ventral thalamus* and *subthalamus* as synonyms but included parts of the midbrain in the region (the red nucleus, subthalamic nucleus, entopeduncular nucleus, substantia nigra), along with the zona incerta, fields of Forel, and the ventral lateral geniculate nucleus; but they excluded the reticular nucleus. This reflects some of the confusion surrounding the use of the two terms.

It was recognized that the epithalamus, comprising the paraventricular and habenular nuclei (and the pineal body), are primarily connected with the hypothalamus and interpeduncular regions, whereas the rest of the thalamus, especially the dorsal thalamus, was more closely related to the cerebral hemisphere. Nissl had concluded his 1913 paper by stating that *thalamus* is identical with *Grosshirnanteil*, i.e., with the region that is related to the cerebral hemisphere. He remarked that the epithalamus, prethalamic regions, prebigeminal (pretectal) nuclei, and ventral lateral geniculate nuclei are not Grosshirnanteile, noting that the ventral lateral geniculate nuclei should be counted with the hypothalamus.

However, he placed the ventral margin of the cerebral hemisphere-related thalamus at the ventral border of the zona incerta (part of his ventrale Gitterkern), and so made no clear distinction between dorsal thalamus and the components of the ventral thalamus other than the ventral lateral geniculate nucleus. Droogleever-Fortuyn (1912) may have been first to emphasize that the reticular nucleus developed in association with the ventral lateral geniculate nucleus from the ventral thalamus in rabbit embryos, though Bianchi (1909) seems also to have recognized this. Connectional differences between ventral and dorsal thalamus began to emerge, principally from studies of normal fiber preparations (Rioch, 1929b), rather than from experimental work. But even up until publication of Walker's book (1938a), most studies involving retrograde degeneration following cortical ablations made no distinction between dorsal and ventral thalamus.

Rose's contribution (Chapter 6) was to clarify that the ventral thalamus, comprising the reticular nucleus, ventral lateral geniculate nucleus, zona incerta, and fields of Forel, had closer developmental affinities with the subthalamic region than with the dorsal thalamus, which developed in close association with

FIGURE 1.23. Jerzy E. Rose (1909–   ), Professor of Neurophysiology, University of Wisconsin. Photograph taken in 1973 by Terrill P. Stewart.

the cerebral cortex. He also showed, probably for the first time, that, developmentally, the pretectal region is a derivative of the epithalamus, not of the dorsal thalamus. The term *pretectum* is Edinger's (1896a,b), but the region had been referred to as the *posterior nucleus of the thalamus* by Nissl in his paper of 1889, though he later (1913) adopted Ramón y Cajal's (1904, 1911) name, the *prebigeminal nucleus*. Though excluded from the Grosshirnanteile by Nissl, until Rose's paper the pretectum was commonly included in the thalamus (Le Gros Clark, 1928, 1932a; Rose, 1935) and still sometimes appears erroneously labeled as a posterior nucleus of the thalamus (Snider and Niemer, 1961; König and Klippel, 1963).

By studying Nissl-stained preparations of normal fetuses of different ages, Rose was able to identify the time of appearance of the individual nuclei of the dorsal thalamus and to show that some which are separated in the fully developed brain, arise from a common pronuclear mass (see Chapter 6). He identified five pronuclei in the precursor of the dorsal thalamus: a medial pronucleus forms the mediodorsal and medioventral (reuniens) nuclei which become separated by a central pronucleus that grows through them; the central pronucleus forms the anterior and ventral groups of nuclei as well as the intralaminar nuclei and their continuation across the midline (Rose's central commissural nuclei); a dorsal pronucleus forms the lateral and posterior nuclear groups; separate pronuclei give rise to the medial and lateral geniculate bodies.

Rose's developmental study was followed by a series of experimental studies with Clinton N. Woolsey (Fig. 1.24) (Rose and Woolsey, 1943, 1948a,b, 1949b, 1958). In these they showed by the retrograde degeneration method that areas of the cerebral cortex that they defined in terms of cytoarchitecture or of the mapping of surface potentials evoked by peripheral stimulation, and sometimes by both methods, were related to different dorsal thalamic nuclei. The work represented a considerable refinement of that previously carried out for the earlier work had relied largely upon less well-defined, regional ablations of the cortex. The main conclusion of Rose and Woolsey's work, that a definable architectonic field in the cortex receives its input from a specific subdivision of a nuclear grouping in the dorsal thalamus, still forms one of the basic tenets of thalamocortical organization. This conclusion of Rose and Woolsey was soon reinforced in many different mammalian species by Le Gros Clark and Powell (1953), by Akert and co-workers (Roberts and Akert, 1963; Scollo-Lavizzari and Akert, 1963), by Rose and Malis (1965), and especially by Diamond and co-workers (Diamond *et al.*, 1958, 1970; Diamond and Utley, 1963; Hall and Diamond, 1968; Kaas *et al.*, 1972a; Diamond, 1979).

Rose and Woolsey followed Nissl in pointing out that from the point of view of the maintenance of its integrity following experimental procedures, the dorsal thalamus is entirely dependent upon the telencephalon. They divided the dorsal thalamic nuclei into *extrinsic* and *intrinsic* nuclei. Extrinsic nuclei comprised the anterior and ventral nuclear groups and the geniculate bodies. These were held to receive their afferents from extrathalamic sources and to be related to primary projection areas of the cerebral cortex, including the sensory, motor, and limbic areas which they believed were phylogenetically the oldest. Intrinsic nuclei were believed to receive afferents primarily from other thalamic nuclei. Some of these, the medial, lateral, and posterior nuclear groups, were related to the supposedly phylogenetically younger "association" areas of the cortex and were, therefore, more prominent in primates. There is no modern evidence to support the idea of extensive intrathalamic connections. Other "intrinsic nuclei," the midline and intralaminar nuclei, were thought by Rose and Woolsey to be related predominately to anterior rhinencephalic structures. Rose and Woolsey, for reasons that are not particularly clear, did not favor the alternative view, put forward by Gerebtzoff (1940b) and by Vogt and Vogt (1941), that the midline and intralaminar nuclei are connected mainly with the caudate nucleus and putamen. This initiated a mild controversy that was to last for approximately 30 years. Followers of Gerebtzoff and the Vogts believed that after cortical ablations, the cellular degeneration observed in the intralaminar nuclei was less severe than in the principal nuclei and that it only became severe if the caudate nucleus and putamen were damaged (Droogleever-Fortuyn and Stefens, 1951; Stefens and Droogleever-Fortuyn, 1953; Powell, 1958; Powell and Cowan, 1954, 1956; Cowan and Powell, 1955). Walker (1938a) believed that even after hemidecortication, the intralaminar nuclei showed no degenerative changes. Followers of Rose and Woolsey, on the other hand, considered that the cellular degeneration in the intralaminar nuclei after cortical ablations was just as severe as in the principal nuclei and believed that lesions of the striatum affected the intralaminar nuclei

FIGURE 1.24. Clinton N. Woolsey (1904–), Professor of Neurophysiology, University of Wisconsin. Photograph given to Doctor James L. O'Leary in the 1950s.

through involvement of the internal capsule (Murray, 1966). Only with the introduction of more sensitive anatomical techniques in the 1970s did this difference of opinion become resolved when it was shown that the intralaminar nuclei send axons to both cortex and striatum, though the projection to the striatum is by far the heavier (Jones and Leavitt, 1974). Modern studies show, however, that the intralaminar nuclei (including the centre médian and parafascicular) are fundamentally different from the principal nuclei (Chapter 12).

In their studies which went on for more than a decade, Rose and Woolsey also considered the possibility that cells in a particular thalamic nucleus could project by branched axons to more than one cortical field. From this emerged their concept of *essential* and *sustaining projections* (Rose and Woolsey, 1948a, 1949a, 1958). Essential and sustaining projections were defined operationally: an area of the cortex was said to receive an *essential projection* from a given thalamic nucleus "if destruction of such an area alone causes marked degenerative changes in the nucleus"; on the other hand, "if two cortical areas are considered, and if destruction of neither of them leads to degenerative changes in a thalamic nucleus or to only slight alterations but if a simultaneous destruction of both causes a profound degeneration of the thalamic element, we shall say that both areas receive *sustaining projections* from this nucleus."

In considering the possible nature of a sustaining projection, Rose and Woolsey (1958) put forward two possible explanations: (1) "The first would be the collateral type in which the axon of the thalamic cell would simply branch and end in two cortical fields"; or (2) "The axon of a thalamic cell would terminate in one cortical field only but its collateral would impinge on another thalamic cell which in turn projects upon a second cortical field."

Some thalamic nuclei have since been confirmed to have the widespread cortical projections postulated by Rose and Woolsey in their first interpretation. But in other nuclei, as we shall see, subsequent work has not borne out the belief that they have branched projections.

The period subsequent to the decade of Rose and Woolsey's studies saw continuing use of the retrograde degeneration method, notably at the hands of Diamond and his collaborators, as mentioned earlier. But the modern period had begun and more and more investigations of thalamic connectivity came to be conducted first with the axonal degeneration methods (e.g., Nauta and Whitlock, 1954; Mehler *et al.*, 1960) and later with the more sensitive techniques based upon anterograde and retrograde axoplasmic transport (e.g., Hendrickson, 1969; Edwards *et al.*, 1974; Jones and Leavitt, 1973, 1974; Kuypers *et al.*, 1974; Nauta *et al.*, 1974). By means of these methods we now have detailed knowledge of the afferent and efferent connections of virtually every thalamic nucleus (Walker, 1966; Jones, 1981a, 1983a,d; Macchi, 1983).

In addition to the elucidation of the connectional relationships of the intralaminar nuclei, the anatomical methods have provided us with the knowledge that probably all thalamic nuclei receive inputs from subcortical sites, as well as corticothalamic fibers returning from the cortex itself. It is now clear that the earlier idea of some thalamic nuclei being intrinsic and connected only to other thalamic nuclei is no longer tenable. We also know that every dorsal thalamic nucleus projects to the cortex, though not all project to the striatum (Jones,

1981a; Macchi, 1983). The most recent work of all (Shipley and Sørenson, 1975; Herkenham, 1979) tells us that "cortex" in this respect includes not only the neocortex but also the paleocortex of the piriform lobule and even the archicortex of the hippocampal formations. In the words of Nissl, therefore, the dorsal thalamus is truly "identisch mit Grosshirnanteile."

The passing of the retrograde degeneration method also saw the rise in popularity of the microelectrode and later of the electron microscope. Many of the early microelectrode studies were concerned with elucidating the manner in which the receptive surfaces of the body, retina, and cochlea are mapped onto the relevant thalamic nuclei. In these, J. E. Rose's name was again to the fore, notably in his collaborations with Mountcastle (Galambos *et al.*, 1952; Rose and Galambos, 1952; Rose and Mountcastle, 1952, 1954; Bishop *et al.*, 1959a,b; Poggio and Mountcastle, 1960, 1963). The microelectrode and the electron microscope have now given us considerable information about the synaptic organization of many thalamic nuclei and about the synaptic events that occur during passage of afferent impulses through the thalamus (e.g., Galambos *et al.*, 1952; Hubel and Wiesel, 1961; Nelson and Erulkar, 1963; Andersen *et al.*, 1964a,b; Szentágothai *et al.*, 1966; Burke and Sefton, 1966; McIlwain and Creutzfeldt, 1967). From these studies has come confirmation of Ramón y Cajal's (1904, 1911) original description of both relay neurons and interneurons in every thalamic nucleus (Chapter 4).

The thalamus is now being explored electrophysiologically in preparations that range from conscious behaving animals (Strick, 1976b; Macpherson *et al.*, 1980) through human patients (Jasper and Bertrand, 1966) to isolated tissue slices (Kelly *et al.*, 1979; Jahnsen and Llinás, 1984a,b). Its functional classes of neurons are being directly visualized by intracellular injections after physiological characterization (Ahlsén *et al.*, 1978; Friedlander *et al.*, 1979, 1981) and the transmitter characteristics of many of these neurons are being determined by immunocytochemistry (Houser *et al.*, 1980; Graybiel and Elde, 1983; Penny *et al.*, 1983) and by receptor binding techniques (Pert *et al.*, 1974, 1976; Hunt and Schmidt, 1978; Rotter *et al.*, 1979; Palacios *et al.*, 1981).

The past history of thalamic studies, reviewed in this chapter, has been characterized by periods during each of which the work was dominated by a particular methodological approach. It would seem that the era now dawning will bring us closer to understanding the synaptic, biochemical, and perhaps even the molecular characteristics of this region that Le Gros Clark (1932a) termed "the anatomical equivalent of the very threshold of consciousness."

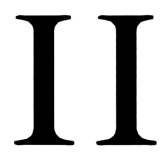

... ce seroit un grand bon-heur pour le genre humain, si cette partie, qui est la plus delicate de toutes, & qui est sujette à des maladies tres-frequentes, & tres-dangereuses, estoit aussi bien connuë, qui beaucoup de Philosophes & d'Anatomistes se l'imaginent.

N. Stensen (1669)

Fundamental Principles

The thalamus is like the Flying Dutchman: many have heard of it, some
believe in it, but few have actually seen it.

Attributed to J. E. Rose

2

Descriptions of the Thalamus in Representative Mammals

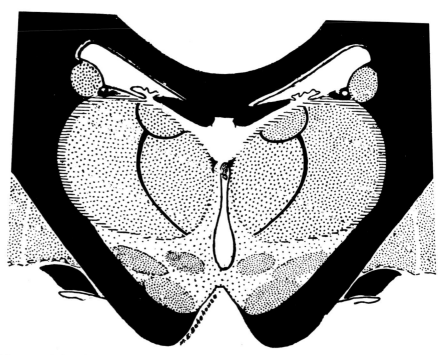

Semidiagrammatic frontal section through the human thalamus and the structures which surround it. Modified from Ranson (1921).

2.1. Cat

Figures 2.1A–L show a series of Nissl-stained frontal sections through the thalamus of a cat to illustrate the general configuration of the thalamic nuclei in a representative mammal. The cat introduces this descriptive account since its nuclear parcellation is clearer than in rodents but less subdivided than in primates. Most of the anterior surface of the thalamus is covered by the reticular nucleus which, more posteriorly, wraps around the lateral and ventral surfaces of the dorsal thalamus. The more medial part of the anterior surface is covered by the anterior paraventricular nucleus of the epithalamus, arching down the front of the thalamus, behind the fornix, to become continuous with the hypothalamus. At anterior levels of the dorsal thalamus (Figs. 2.1B,C), the three anterior nuclei are seen (anterodorsal, anteromedial, anteroventral), together with the parataenial nucleus lying always deep to the stria medullaris and probably best considered a component of the medial group of nuclei. The medial group also includes the mediodorsal nucleus (seen more posteriorly in Fig. 2.1D) and the medioventral or reuniens nucleus (Fig. 2.1C).

The mediodorsal and medioventral nuclei are separated from one another by the central medial nucleus, which lies in the midline capped anteriorly by a small rhomboid nucleus. The rhomboid nucleus expands posteriorly and dorsally, separating the mediodorsal nuclei of the two sides. The central medial nucleus expands laterally within the internal medullary lamina as the paracentral and central lateral nuclei (Figs. 2.1C,D). These three, the rostral group of intralaminar nuclei, at slightly more posterior levels, separate the anterior group from the lateral mass of the dorsal thalamus, which consists of the dorsally placed lateral nuclei and the ventrally placed ventral nuclei. The most anterior of the lateral nuclei, the lateral dorsal nucleus, breaks through the internal medullary lamina to become closely associated with the anterior group (Fig. 2.1C). Posterior to the lateral dorsal nucleus lie the lateral posterior nucleus and the pulvinar nucleus (Figs. 2.1D–G). The former of these consists of several nuclear aggregations (a nuclear complex) (Fig. 2.1F) and some parts of it should be regarded as equivalent to certain of the nuclei of the primate pulvinar.

The ventral nuclei consist of anteriorly situated ventral anterior and ventral lateral nuclei that are not clearly distinguishable in nonprimates (Figs. 2.1B–D), and a ventral posterior nucleus or ventrobasal complex, divisible into a lateral

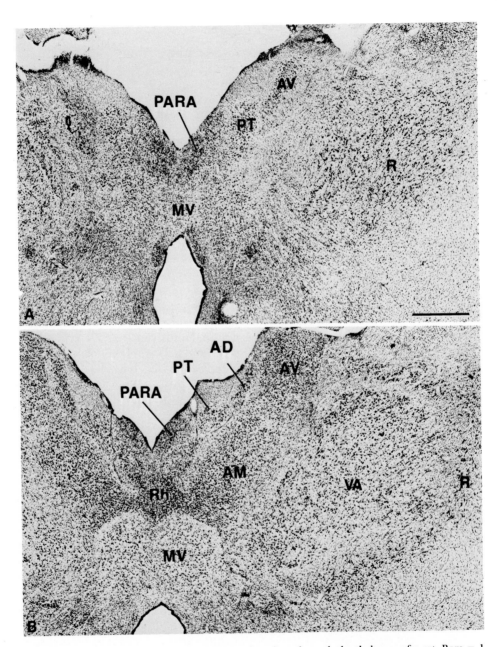

FIGURE 2.1.   A series of thionin-stained frontal sections through the thalamus of a cat. Bars = 1 mm. From material provided by Dr. A. L. Berman.

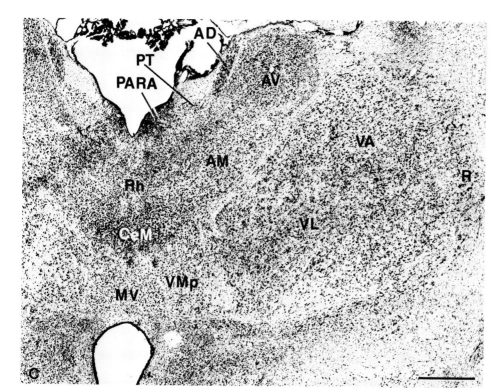

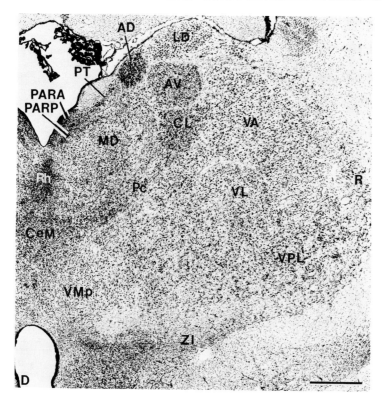

FIGURE 2.2. *(continued)*

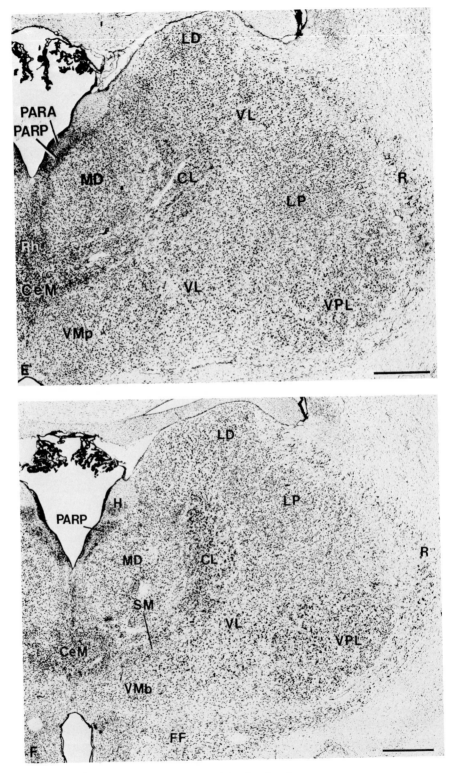

FIGURE 2.1.   (*continued*)

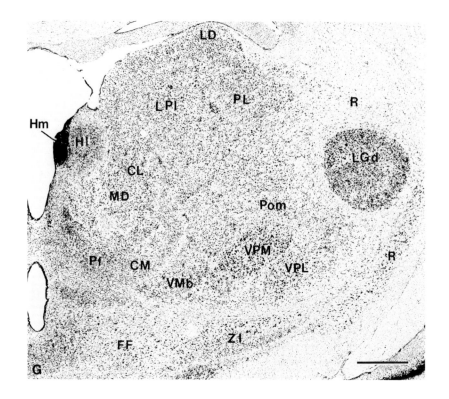

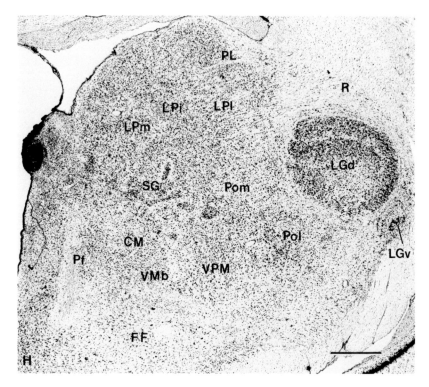

FIGURE 2.1. (*continued*)

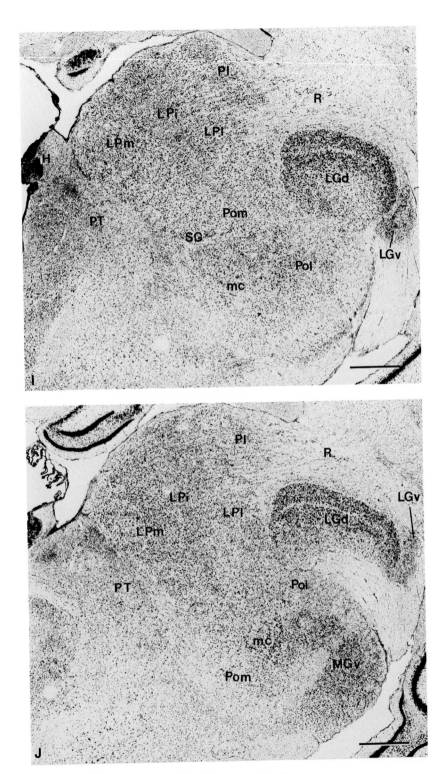

FIGURE 2.1.   (*continued*)

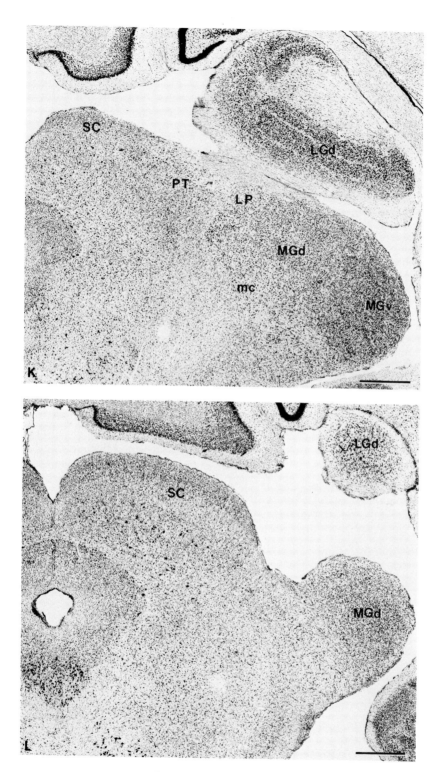

FIGURE 2.1. *(continued)*

(ventral posterolateral or external) subnucleus and a medial (ventral postero-medial or arcuate) subnucleus (Fig. 2.1E).

Medially, at levels posterior to the medioventral nuclei, is the ventral medial complex which consists of a principal ventral medial nucleus, an indistinct sub-medial nucleus, and a basal ventral medial nucleus (Figs. 2.1C–F). Posteriorly, the ventral nuclei give place to the medial geniculate complex of nuclei with an intervening, somewhat ill-defined region referred to as the posterior complex (Figs. 2.1F–H). The borders of the posterior complex with the overlying lateral posterior complex are also difficult to define. The medial geniculate complex consists of a cell-sparse dorsal group of nuclei, a cell-dense ventral nucleus, and a medial magnocellular nucleus lying along the medial lemniscus (Figs. 2.1F–I).

The nuclei of the internal medullary lamina are particularly well defined at posterior levels and form a caudal group consisting of the centre médian and parafascicular nuclei (Figs. 2.1E,F). These, together with several condensations of cells in the posterior complex collectively termed the suprageniculate nucleus (Figs. 2.1F,G), form the posterior surface of the thalamus, separated from the pretectum by a medial medullary lamina. The dorsomedial aspect of the thal-amus is covered by the stria medullaris which eventually gives way posteriorly to the medial and lateral habenular nuclei—components of the epithalamus. The parataenial nucleus does not accompany the stria back this far but a posterior paraventricular nucleus continues along the medial edge of the stria and ends by undershooting the habenular nuclei and penetrating the rostral end of the mesencephalic central gray matter.

The lateral geniculate complex (Figs. 2.1E–J) is an outgrowth of the lateral nuclear mass, only partially separated from it by a thin medullary lamina (the medial ramus of the optic tract). It consists of laminar and medial interlaminar nuclei which are parts of the dorsal thalamus and are collectively termed the *dorsal lateral geniculate complex*. Beneath them is the ventral lateral geniculate nucleus, lying within the optic tract (Figs. 2.1F–I) and more or less continuous with its two fellow components of the ventral thalamus, the reticular nucleus and the zona incerta (Figs. 2.1A–G). All three are separated from the dorsal thalamus by the external medullary lamina. Between the dorsal lateral genicu-late nucleus and the reticular nucleus is a thin strip of cells, the perigeniculate nucleus.

2.2. Rat

Figures 2.2A–N show a sequence of Nissl-stained frontal sections from a rat thalamus at levels comparable to those shown in Figs. 2.1A–L. The chief difference, apart from size, is in the ease with which individual nuclei can be distinguished from one another. Some, such as the three anterior nuclei, the rostral intralaminar nuclei, and the components of the ventral thalamus, have sufficiently distinct cell populations to be easily separable from their neighbors. Others, however, have cells of similar size, intensity of staining, and packing density and they therefore blend with their neighbors without a clear boundary.

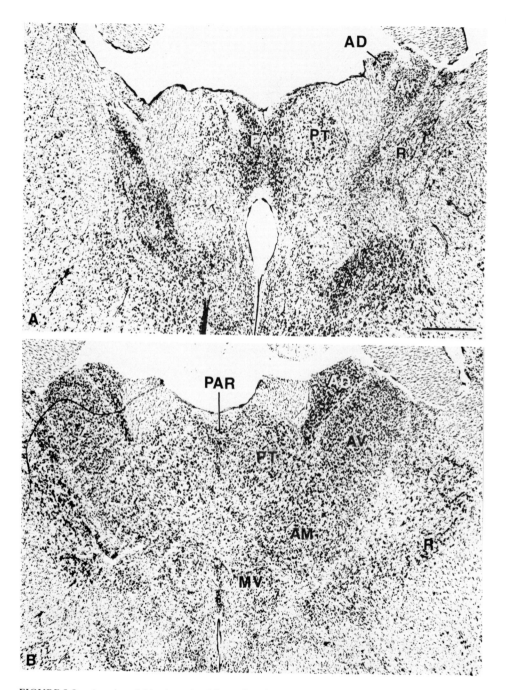

FIGURE 2.2.   A series of thionin-stained frontal sections through the thalamus of a rat. Bars = 500 μm.

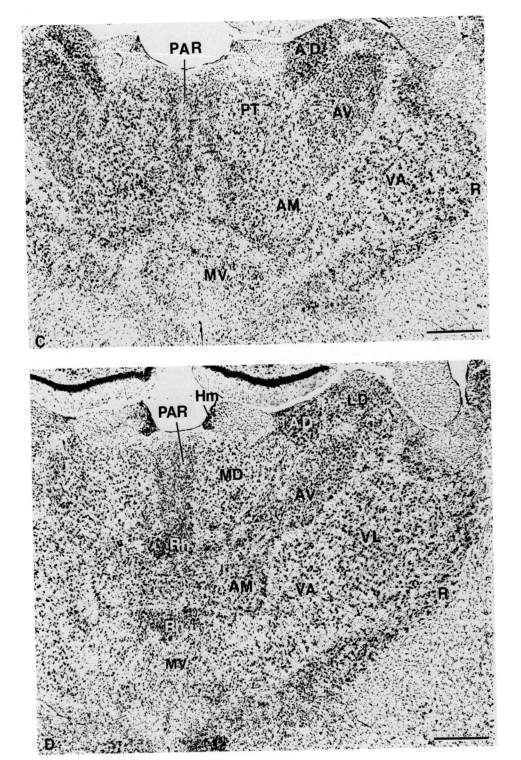

FIGURE 2.2.   (*continued*)

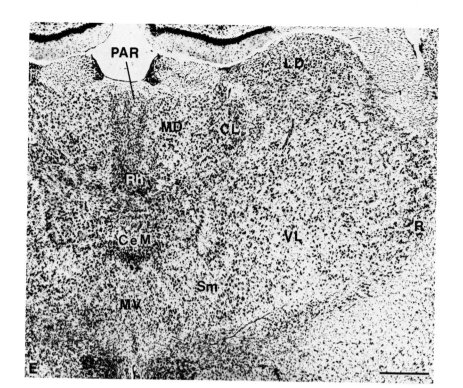

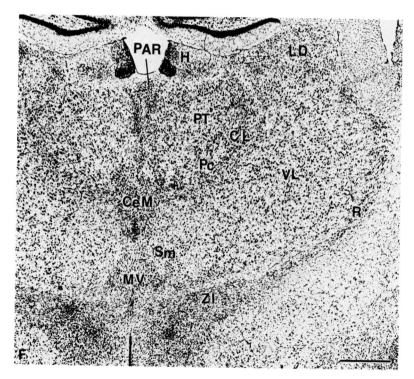

FIGURE 2.2. *(continued)*

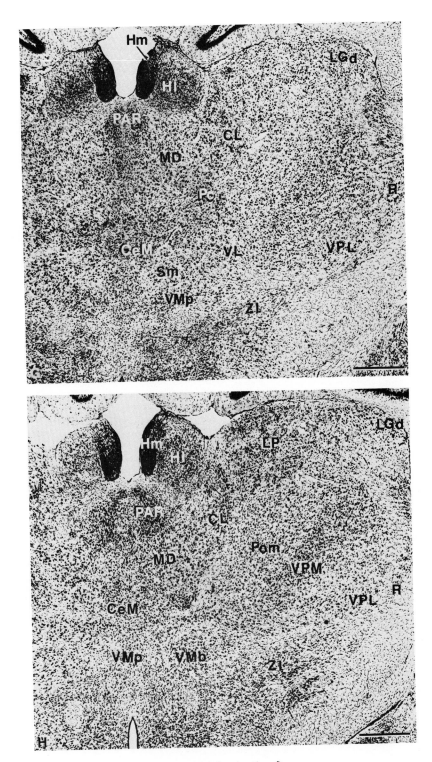

FIGURE 2.2. (*continued*)

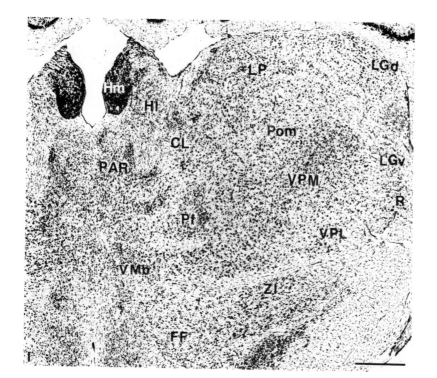

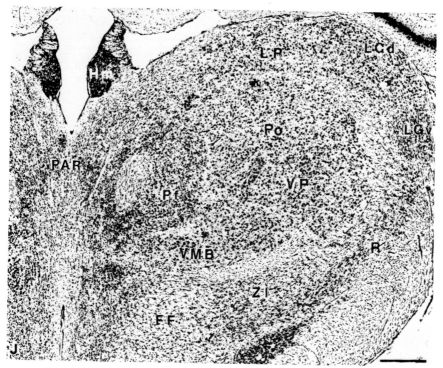

FIGURE 2.2. *(continued)*

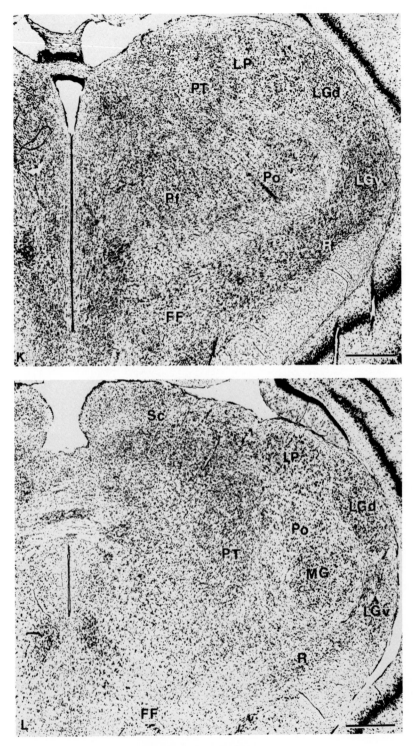

FIGURE 2.2. (continued)

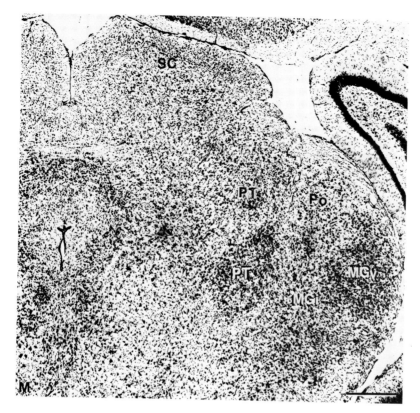

FIGURE 2.2. *(continued)*

The dorsal lateral geniculate nucleus is a case in point, being difficult to separate from the underlying lateral complex. The former is not laminated.

A division of the lateral complex into lateral dorsal and lateral posterior nuclei is seen but the latter is not readily subdivisible in Nissl stains and there is no pulvinar nucleus. Divisions of the ventral and medial geniculate nuclei are not dissimilar from the cat, though the cytoarchitectonic appearances of nuclei of the same name are not identical. A submedial nucleus [called by Krieg (1944) the *nucleus gelatinosus*] is particularly clear, as also is a posterior complex, though it lacks a suprageniculate nucleus.

The paraventricular nuclei form a midline strip, expanded ventrally between the two mediodorsal nuclei.

There is no centre médian nucleus in the intralaminar system, though some workers (e.g., Albe-Fessard *et al.*, 1966) split off a part of the parafascicular nucleus as the centre médian.

The pretectum is particularly large relative to the thalamus and parts of it have sometimes been erroneously labeled as a *nucleus posterior thalami* (e.g., König and Klippel, 1963).

Figures 2.3A–K are a series of frontal sections through a guinea pig thal-

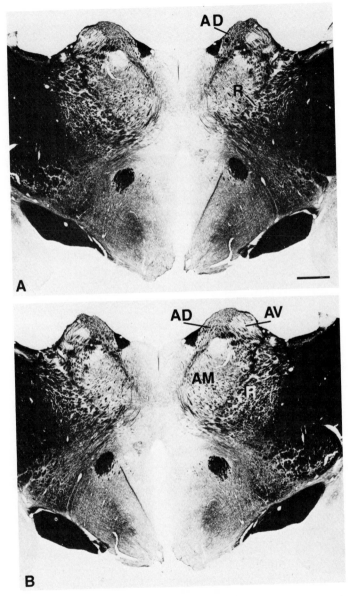

FIGURE 2.3. A series of myelin-stained frontal sections through the thalamus of a guinea pig.
Bars = 1 mm.

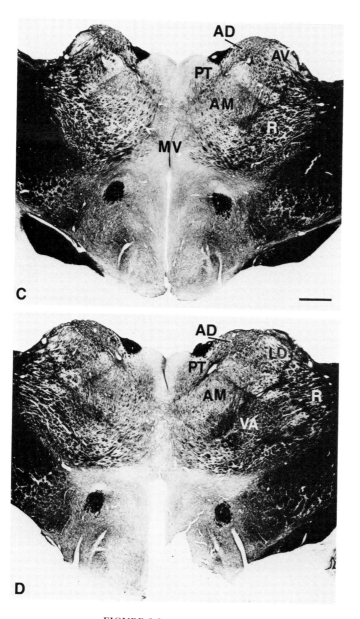

FIGURE 2.3. *(continued)*

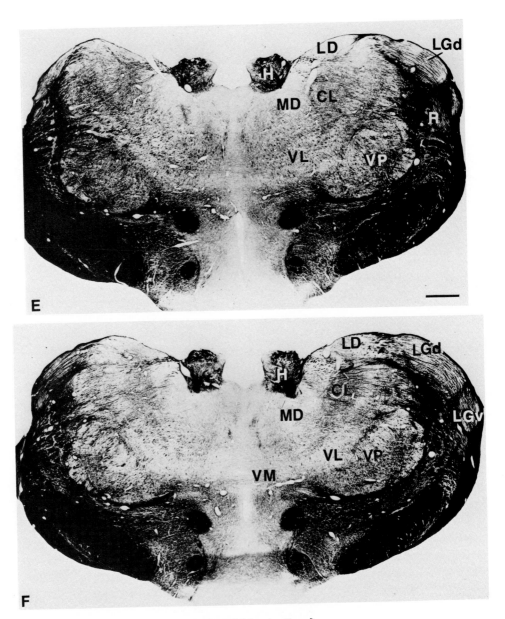

FIGURE 2.3. (*continued*)

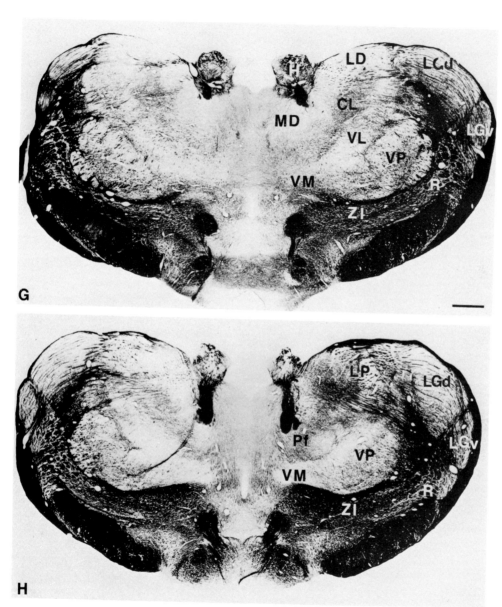

FIGURE 2.3.   (continued)

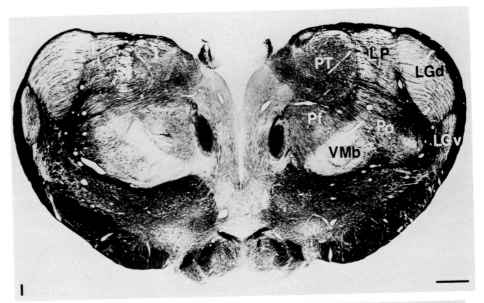

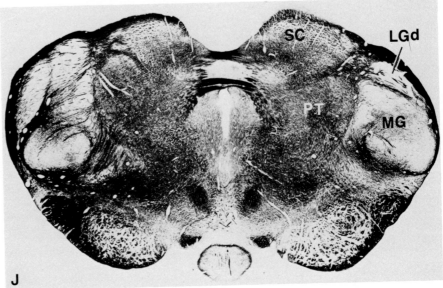

FIGURE 2.3. *(continued)*

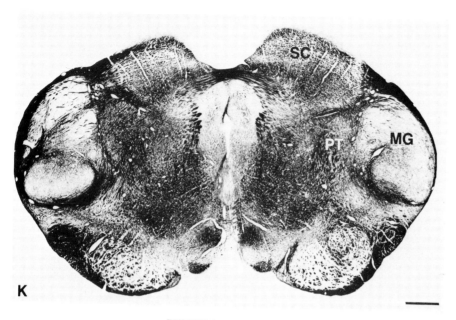

FIGURE 2.3. *(continued)*

amus, at levels comparable to Figs. 2.2A–N, but stained for myelin. In myelin-stained preparations of this type, many of the nuclei, delimited by the Nissl stain, can also be detected owing to differences in myeloarchitecture. The internal and external medullary laminae and several of the fiber tracts entering and leaving the thalamus are identifiable.

2.3. Monkey

Figures 2.4A–L show Nissl-stained sections from the thalamus of a macaque monkey at levels comparable to those shown for the cat and rat. The naming of some nuclei is new to this book but is derived from Olszewski (1952), Burton and Jones (1976), and Asanuma *et al.* (1983a). The chief differences exhibited by the monkey thalamus, in comparison with that of the cat, are the great overdevelopment of the ventral and lateral nuclear complexes. Not only are these greatly enlarged relative to the anterior, intralaminar, and medial nuclei, but they display a greater number of cytoarchitectonically distinct subdivisions than in the cat. The division of the ventral complex into ventral anterior, ventral lateral, and ventral posterior nuclei is particularly obvious, especially in sagittal and horizontal sections (see Chapter 7), and further nuclear subdivisions can be discerned in them. The principal change in the lateral complex is the massive enlargement of the pulvinar in which four major nuclei can be discerned, though at least some of these are more likely to be equivalent to divisions of the lateral posterior complex of the cat, rather than representing subdivisions of the cat's pulvinar nucleus.

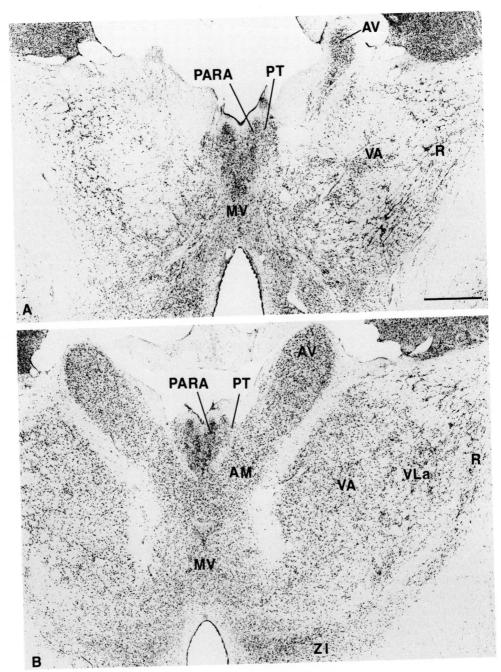

FIGURE 2.4. A series of thionin-stained frontal sections through the thalamus of a cynomolgus monkey (*Macaca fasci-cularis*). Bars = 1 mm.

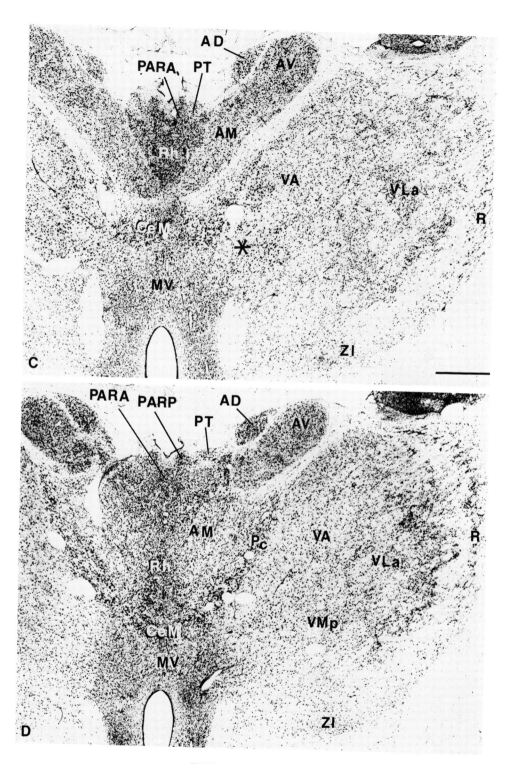

FIGURE 2.4. (*continued*)

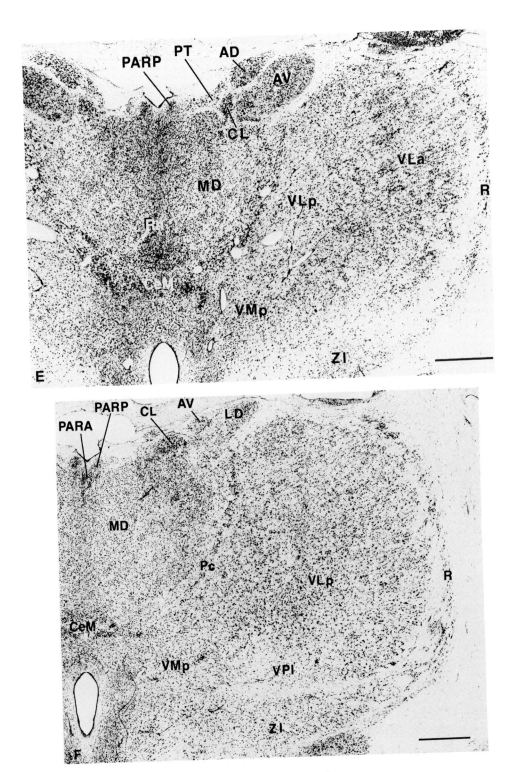

FIGURE 2.4. *(continued)*

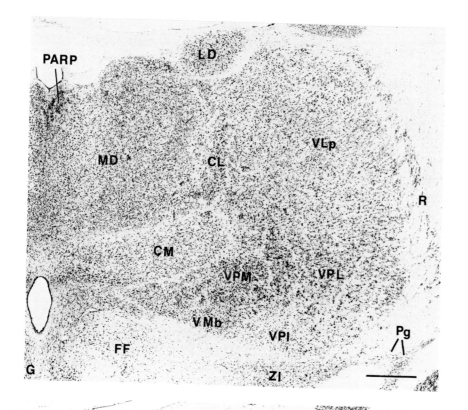

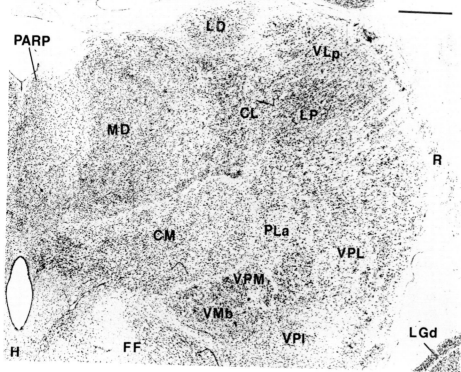

FIGURE 2.4. (*continued*)

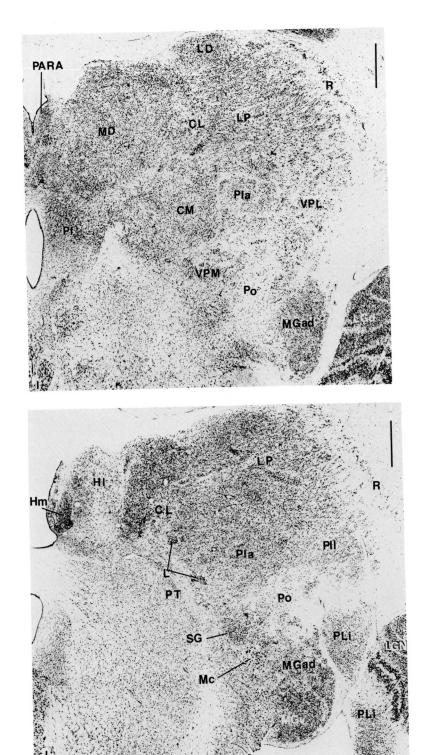

FIGURE 2.4.   (*continued*)

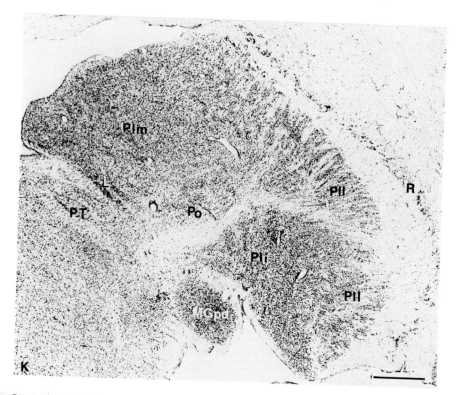

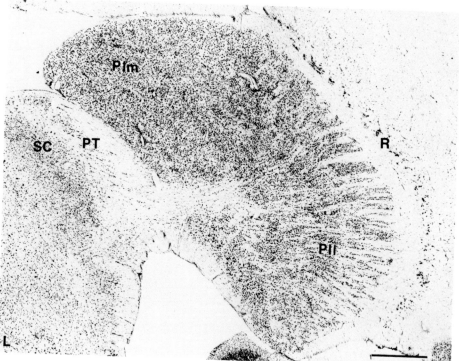

FIGURE 2.4. (*continued*)

The mediodorsal nucleus and the posterior complex display more distinct cytoarchitectonic subdivisions than in the cat and there appears to be subsidiary divisions of some of the subnuclei of the medial geniculate complex. The dorsal lateral geniculate nucleus has, as it were, rotated downwards and laterally and displays a lamination pattern different from that of the cat. The ventral lateral

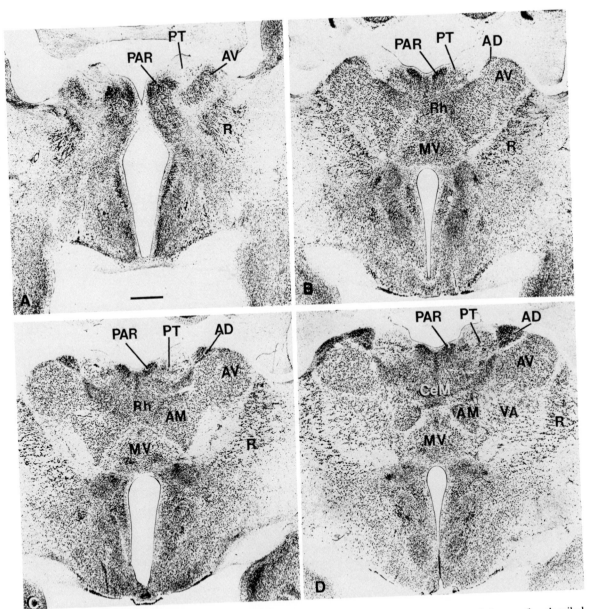

FIGURE 2.5. A series of thionin-stained frontal sections through the thalamus of a marsupial phalanger (brush-tailed possum, *Trichosurus vulpecula*). Bar = 1 mm. Brain provided by Dr. D. J. Tracey.

geniculate nucleus has become the small, pregeniculate nucleus situated anterodorsal to the dorsal lateral geniculate nucleus.

The intralaminar nuclei are still distinct, although those of the rostral group appear thinner in comparison with the other nuclei. The centre médian nucleus, however, is particularly large.

The only nucleus of lower animals that is difficult to identify in the monkey brain is the submedial nucleus of the ventral medial complex. This is probably

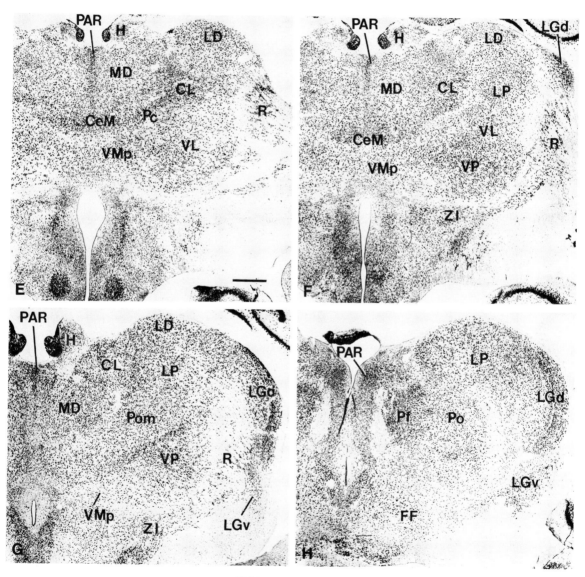

FIGURE 2.5. (continued)

represented by a small portion of the principal ventral medial complex slung beneath the internal medullary lamina (see Chapter 7).

2.4. Other Species

The thalami of most other mammalian species can usually be seen as conforming to one or the other of the basic patterns of organization described in the preceding sections. The thalamus of marsupials (Figs. 2.5A–K) and of many

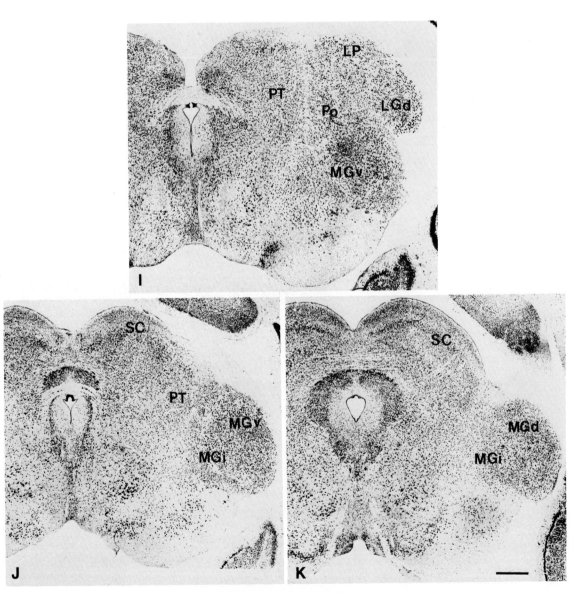

FIGURE 2.5. (*continued*)

insectivores, for example, resembles that of the rat, the thalamus of ungulates resembles that of the cat, and the thalamus of man can be readily described in terms of that of the monkey. The thalami of prosimian primates (Figs. 2.6A–L), for descriptive purposes at least, come somewhere between those of the rat and monkey. Where differences occur within orders, they are usually those of size of the whole thalamus and/or the over- or underdevelopment of a particular nucleus. Seals have a thalamus larger than that of an average monkey but the nuclear parcellation is clearly catlike. Many highly visual marsupials, insectivores, and rodents have thalami resembling that of the opossum, hedgehog, or rat

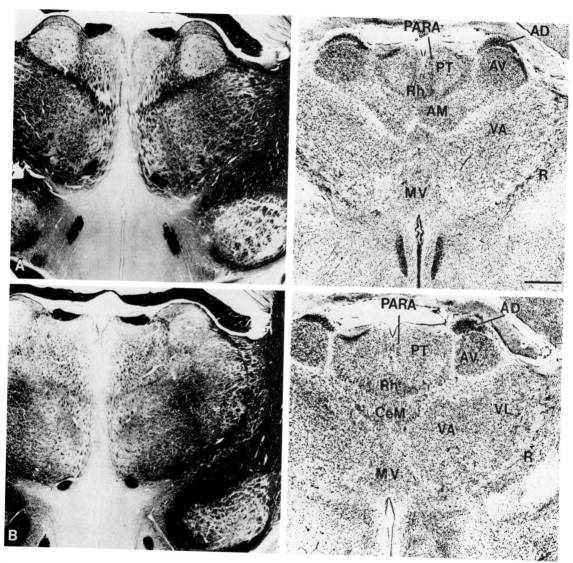

FIGURE 2.6. A series of alternating myelin- and thionin-stained frontal sections through the thalamus of a prosimian, the bush baby (*Galago crassicaudatus*). Bars = 1 mm. Brain provided by Dr. Mary Carlson.

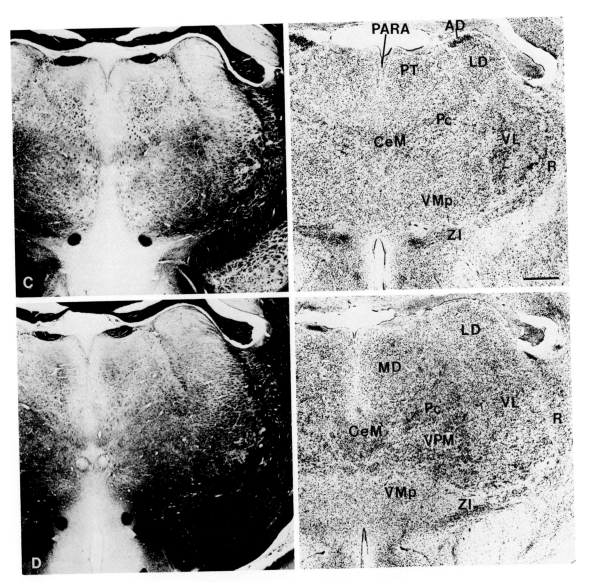

FIGURE 2.6. (*continued*)

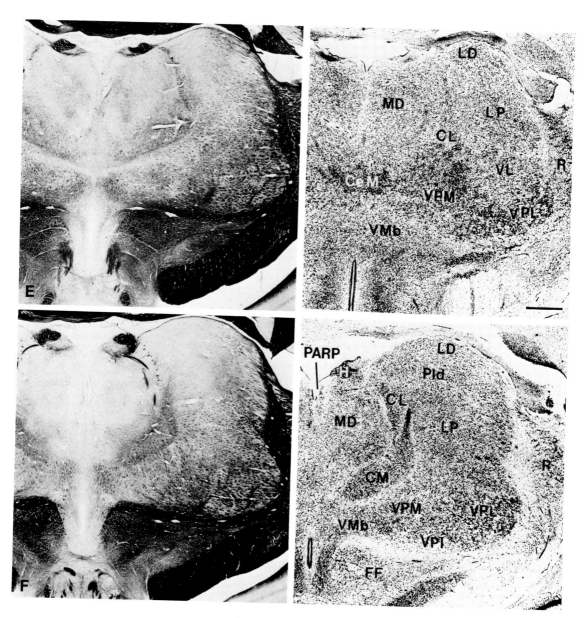

FIGURE 2.6. *(continued)*

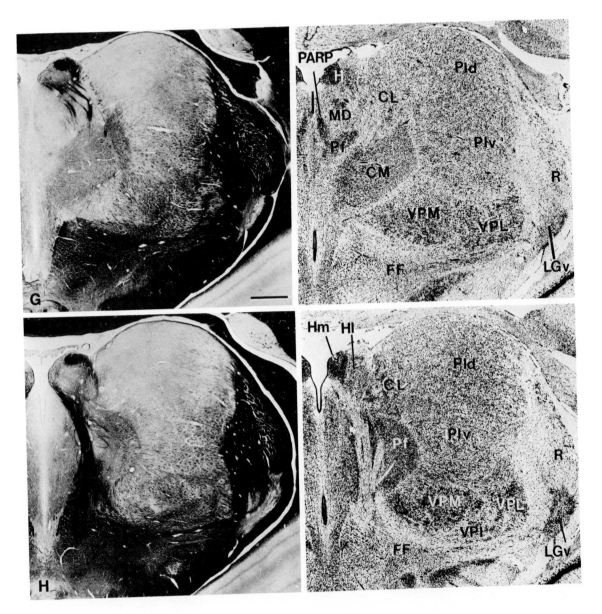

FIGURE 2.6. (*continued*)

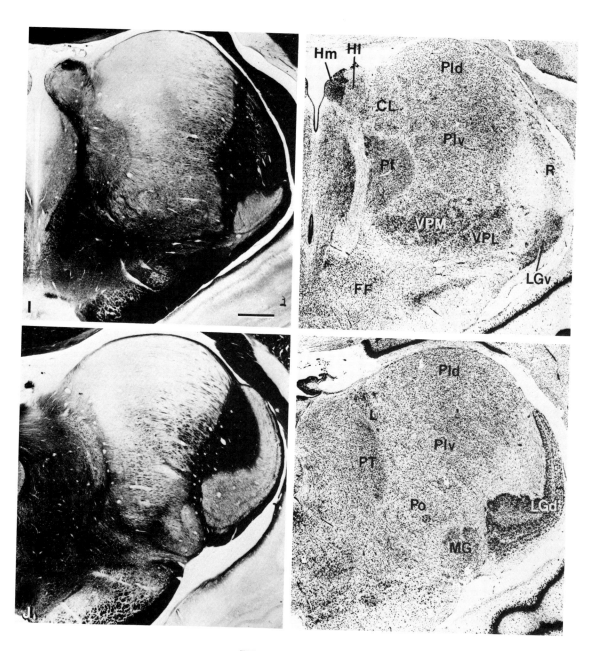

FIGURE 2.6. *(continued)*

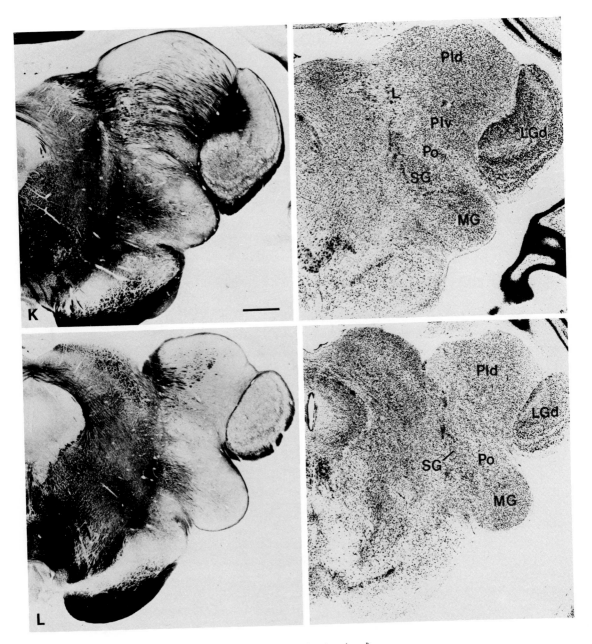

FIGURE 2.6. (continued)

except for a great overdevelopment of the dorsal lateral geniculate nucleus. With a feel for the three arbitrary types of thalamus described here, one can usually examine the thalamus of an unknown mammalian species and fairly readily find one's way around it. The only possible exception where difficulty may be experienced is in the thalamus of the monotremes (Chapter 17) which, though large, shows an unusually homogeneous cytoarchitecture and relatively few overt nuclear distinctions. We shall have occasion to refer to some of the subtle and not so subtle species differences among comparable thalamic nuclei in the chapters of Part IV of this book.

Abbreviations

Unless otherwise indicated, these abbreviations are used throughout the book.

AD	Anterodorsal nucleus
AM	Anteromedial nucleus
AV	Anteroventral nucleus
BIC, B	Brachium of inferior colliculus
CL	Central lateral nucleus
CeM	Central medial nucleus
CM	Centre médian nucleus
CN	Caudate nucleus
CTT	Corticotectal tract
FF	Fields of Forel
GLd, LGd	Dorsal lateral geniculate nucleus
GLv, LGv	Ventral lateral geniculate nucleus
H	Habenular nuclei
Hl	Lateral habenular nucleus
Hm	Medial habenular nucleus
HPT, HT	Habenulopeduncular tract
IC	Inferior colliculus
L, Lim	Limitans nucleus
LD	Lateral dorsal nucleus
LG, LGN	Lateral geniculate complex
A, A1, C	Laminae of cat dorsal lateral geniculate nucleus
M	Medial interlaminar nucleus of cat dorsal lateral geniculate nucleus
LM	Medial lemniscus
LP	Lateral posterior nucleus or complex
LPi	Intermediate nucleus
LPl	Lateral nucleus
LPm	Medial nucleus
MD	Mediodorsal nucleus
MG	Medial geniculate complex
MGd, D	Dorsal nucleus
MGi	Internal nucleus

MG (*continued*)

MGM, M, mc	Magnocellular nucleus
MGV, v	Ventral nucleus
MT, MTT	Mamillothalamic tract
MV	Medioventral (reuniens) nucleus
OC	Optic chiasm
OT	Optic tract
Par	Paraventricular nuclei of thalamus
Para	Anterior paraventricular nucleus
Parp	Posterior paraventricular nucleus
Pc, PC	Paracentral nucleus
Pf	Parafascicular nucleus
Pg	Perigeniculate nucleus (cat)
Pl	Pulvinar nucleus (cat)
Pla	Anterior pulvinar nucleus
Pld	Dorsal pulvinar nucleus (galago)
Pli	Inferior pulvinar nucleus
Pll	Lateral pulvinar nucleus
Plm	Medial pulvinar nucleus
Plv	Ventral pulvinar nucleus (galago)
Po	Posterior complex or nucleus
Poi	Intermediate nucleus
Pom	Medial nucleus
Pol	Lateral nucleus
Pr	Pretectal nuclei
Prg	Pregeniculate nucleus (monkey)
Pt	Parataenial nucleus
PT	Pretectum
R	Reticular nucleus
Rh	Rhomboid nucleus
RN	Red nucleus
SC	Superior colliculus
SG	Suprageniculate nucleus
Sm	Submedial nucleus
SM, ST	Stria medullaris
SPf	Subparafascicular nucleus
VA	Ventral anterior nucleus
VL	Ventral lateral complex
VLa	Anterior nucleus
VLP, VLp	Posterior nucleus
VM, VMp	Principal ventral medial nucleus
VMb	Basal ventral medial nucleus
VPI	Ventral posterior inferior nucleus
VPL	Ventral posterior lateral nucleus
VPM, VPm	Ventral posterior medial nucleus
ZI	Zona incerta

The statement of Hughlings Jackson, in reference to the central nervous system, that "difference of structure of necessity implies difference in function" is no doubt broadly true, but the converse is also true.

Le Gros Clark (1952)

3

Principles of Thalamic Organization

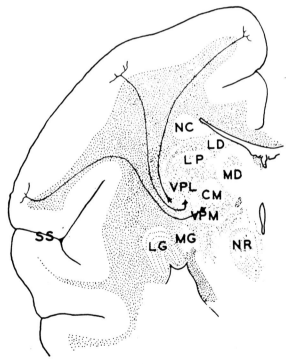

Walker's (1938a) drawing showing the topical projection of the nucleus ventralis posterior on the postcentral gyrus.

3.1. Subdivisions of the Thalamus

From the earliest developmental and comparative studies, it has been customary to divide the thalamus into three parts: epithalamus, dorsal thalamus, and ventral thalamus (Droogleever-Fortuyn, 1912; Herrick, 1918; Le Gros Clark, 1932a) (Fig. 3.1). Though these subdivisions were based in the first instance on the derivatives of three of the cellular masses of the developing diencephalic wall (see Chapter 6), coupled with a certain amount of phylogenetic speculation, as the years have passed, the tripartite division has been given an increasingly strong connectional basis.

The epithalamus comprises the anterior and posterior paraventricular nuclei and the habenular nuclei. From it are derived in ontogeny, the pineal body and the pretectum (Rose, 1942a). The epithalamus proper neither sends fibers to nor receives fibers from the cerebral cortex. It has closer affinities with the hypothalamus. Parts of the pretectum, however, receive corticofugal fibers.

The dorsal thalamus, comprising the majority of the nuclei named in Chapter 2, sends fibers to and receives fibers from the cerebral cortex and striatum (caudate nucleus, putamen, and nucleus accumbens). The projection to and from the cortex includes not only the neocortex but also the paleocortex of the piriform lobule and the archicortex of the hippocampal formation (Shipley and Sørenson, 1975; Herkenham, 1978; Krettek and Price, 1977b; Wyss *et al.*, 1979; Amaral and Cowan, 1980). The only other known projection targets of the dorsal thalamus are, like the cortex and striatum, parts of the telencephalon. They are the lateral amygdaloid nucleus and the olfactory tubercle (Chapters 10 and 13).

A separation of the dorsal thalamic nuclei into principal nuclei and nuclei of the internal medullary lamina (intralaminar nuclei) is still valid, as will emerge below. There seems to be little point, however, in referring among the principal nuclei to relay nuclei and association nuclei. This division was based upon the erroneous assumption that association nuclei, though projecting to the cortex, received their input from other thalamic nuclei (Chapter 1). There is no foundation for this and, in a sense now, every principal nucleus is a relay nucleus, since it receives a subcortical input.*

---

* There are some apparent exceptions to the subcortical input rule among the nuclei of the lateral complex, though these exceptions may rest upon the fact that insufficient study has yet been made of the afferent connections of these nuclei (see Chapter 10).

There is little use either in referring to nuclei of the midline (Chapter 1). Though this term was in vogue in past years, it was never clearly defined and appears to have included components of the epithalamus, of the intralaminar system, and of certain principal nuclei such as the mediodorsal and ventral medial. This is such a diverse group that it seems unduly confusing to link the nuclei together.

The ventral thalamus comprises the reticular nucleus, ventral lateral geniculate nucleus, zona incerta, and, according to some, the nucleus of the fields of Forel. Though receiving fibers from the cerebral cortex and possibly from the striatum, it does not send axons to either of these sites.

3.2. Definition of a Dorsal Thalamic Nucleus

A dorsal thalamic nucleus has been defined anatomically as a circumscribed region of cytoarchitecture receiving a particular set of afferent connections and projecting within the borders of a particular cortical field or fields. This is a traditional definition derived in large part from the early work of von Monakow (1914) and the later work of Rose and Woolsey (1949a). Where it has come under attack in recent years, it has usually been on two fronts: first, it has not always been easy to correlate the terminal distribution of major afferent fiber systems with thalamic cytoarchitecture; second, not all thalamic nuclei project their axons within the confines of one or a few individual cortical fields.

I think that the first criticism has been to a considerable extent overstated.

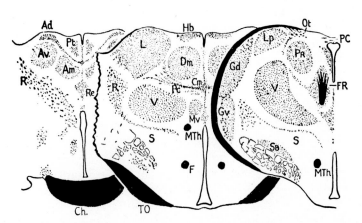

FIGURE 3.1. The primary nuclear groupings in the mammalian thalamus, as drawn by Le Gros Clark (1932a). Ad, Av, Am: the three anterior nuclei; Cm, Pc: intralaminar nuclei; Dm: mediodorsal nucleus; Gd: dorsal lateral geniculate nucleus; Gv: ventral lateral geniculate nucleus; Hb: habenula; L, Lp: lateral nuclei; Mv: ventral medial nucleus, Pt: parataenial nucleus; R: reticular nucleus; Re: reuniens (medioventral) nucleus; S: "subthalamus" (zona incerta and fields of Forel); V: ventral nuclei. Other structures are: Ch: optic chiasm; F: fornix; FR: habenulopeduncular tract; MTh: mamillothalamic tract; Ot, Pr: nuclei of pretectum; PC: posterior commissure; Sb: subthalamic nucleus; TO: optic tract.

Few, if any, thalamic nuclei have cytoarchitectonic borders as clear-cut as the primate or carnivore lateral geniculate nucleus. Some, such as the divisions of the primate ventral complex, are fairly clear but by no means do the borders form a straight line. Other nuclear borders, such as those between the four or five divisions of the cat lateral posterior complex (Chapter 10), are far from clear and, though sometimes hinted at in earlier studies (e.g., Rioch, 1929a), had been generally discounted. Yet when studied in conjunction with connectional patterns (Updyke, 1981b; Asanuma *et al.*, 1983a,b) or with patterns of histochemical staining (Graybiel and Berson, 1980b) these decidedly unclear cytoarchitectonic features assume a new importance, for it becomes clear that the terminations of different sets of afferents overlap little, if at all, at the borders and also that the borders are sharp when defined histochemically (see Chapter 10).

Where thalamic cytoarchitecture is clearly demonstrated, there is usually a clear correlation between the terminal distribution of a major fiber tract and the borders of a thalamic nucleus (Fig. 3.2). If the terminal distribution of a fiber system seemingly transcends the borders of a thalamic nucleus, it may be appropriate to question whether the traditional cytoarchitectonic parcellation is

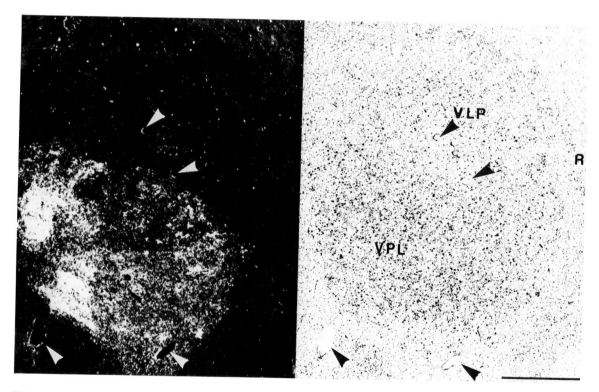

FIGURE 3.2. Correlation of afferent fiber distribution with borders of a thalamic nucleus. Darkfield and brightfield photomicrographs of same sagittal section of a monkey thalamus at same magnification, showing distribution of medial lemniscal fiber ramifications labeled by axoplasmic transport from an injection of tritiated amino acids in the dorsal column nuclei. Fibers ramify within borders of ventral posterior lateral (VPL) nucleus. Arrowheads indicate same blood vessels. Bar = 1 mm. From Tracey *et al.* (1980).

accurate. Where workers have found it hard to correlate connectional data, whether it be derived anatomically or physiologically, with cytoarchitecture (or, even worse, with the pages of an atlas or a set of stereotaxic coordinates!), they have often found it expedient to speak of overlap, transition, or convergent regions. It seems to me, however, that such interpretations often reflect a failure to recognize that architectonic borders are not linear and that cell populations of one nucleus can, over some litle distance, intrude into the territory of another. Often it requires cutting sections in an unusual plane to demonstrate a border. The separation between the ventral lateral and the ventral posterior nuclei in the monkey brain has been argued over repeatedly in frontal sections, yet it is remarkably discrete in horizontal or sagittal sections (Fig. 3.2 and Chapter 7). The line on the drawings of frontally cut sections in an atlas is the bane of this kind of study for many workers strive too diligently to see in their sections what the atlas maker drew on his drawings. They would be better encouraged to practice cutting good sections, doing good Nissl stains, and taking good photographs. As J. E. Rose (1942a) has said: "[Nissl] rendered all his findings by photomicrographs. Thus it is usually easy to find out what [he] meant by his designations . . . ," unlike "Winkler and Potter . . . [who] . . . give only drawings in their work. That makes it almost impossible to see anything from their plates beyond what the authors themselves believed they saw."

One does not wish to argue the case that afferent fiber systems never converge within a thalamic nucleus, nor that a single afferent system may not terminate within more than one nucleus. There are numerous examples to gainsay each of these positions. A case in point is the spinothalamic tract which overlaps the terminal territory of the medial lemniscus in the ventral posterior nucleus and then extends into the ventral lateral nucleus to overlap the terminal territory of the cerebellothalamic fiber system (Asanuma et al., 1983b). But the terminations of the spinothalamic tract do not stray outside the borders of these two nuclei and the cerebellothalamic and medial lemniscal terminations do not overlap at all (Chapter 7).

It would be incorrect to assume that these statements are meant to imply a functional homogeneity in each thalamic nucleus. Nothing could be further from the truth, at least in the sensory relay nuclei. In those that have been thoroughly studied, there is a clear differentiation of relay neurons in terms of their discharge properties, selective afferent inputs, peripheral receptive fields, and cortical projections. It is at this level that thalamic function is expressed physiologically and probably more insights into the functions of the thalamus will be derived from this level of analysis than from arguments over questionable degrees of anatomical overlap between the terminations of long fiber tracts at nuclear borders.

The second criticism mentioned above has been leveled against the old concept of nucleus-to-field specificity in thalamocortical connections. The criticism rests upon the fact that certain dorsal thalamic nuclei (considered in Section 3.3) project rather diffusely upon the cortex, their axons seemingly not constrained by the borders of architectonic fields. But the cells of other dorsal thalamic nuclei do obey the nucleus-to-field principle and the terminal ramifications of *their* axons seem to stop rather abruptly at cytoarchitectonic borders in the cortex (Jones and Burton, 1976b). Rather than causing us to abandon completely the older definition of a dorsal thalamic nucleus, we are now forced into a position of looking for different types of dorsal thalamic nuclei depending on their cortical projections.

3.3. Types of Dorsal Thalamic Nuclei

There is now even less reason than in earlier times to question a fundamental division of the thalamus into epithalamus, dorsal thalamus, and ventral thalamus. However, among the dorsal thalamic nuclei, there are clearly a number of types that have different connectional relationships with the cortex and with certain other telencephalic structures (Jones, 1981a; Macchi, 1983).

3.3.1. Overview

When one thinks of the outflow of the dorsal thalamus, the thalamocortical projection comes immediately to mind. Many would think no further, so much has the interaction between dorsal thalamus and cerebral cortex dominated the field. And to many, "cerebral cortex" probably means "neocortex," for the piriform lobule and hippocampus are rarely thought of as functional dependencies of the thalamus. However, it has to be emphasized that the dorsal thalamus sends axons to several basal regions of the cerebral hemisphere in addition to the cortex. The chief among these is the striatum (caudate nucleus, putamen, and nucleus accumbens) but the olfactory tubercle and parts of the amygdala are also included. Moreover, in the cortex itself, not only the neocortex but also parts of the paleocortex of the piriform lobule and the archicortex of the hippocampal formation receive dorsal thalamic inputs. The dorsal thalamus, thus, gains entrée to virtually the whole cerebral hemisphere. Many of the data upon which these generalizing statements are based result from quite recent work and will be reviewed in the ensuing sections. At present one will merely state that the lengthy arguments of the past, based largely on retrograde degeneration studies (see Chapter 12), and revolving around which nuclei project to cortex or striatum and which do not, seem to have been resolved in favor of a more simplified approach (see Macchi, 1983; Jones, 1983d) which makes the following points: (1) every dorsal thalamic nucleus projects to the cerebral cortex (Table 3.1); (2) virtually the whole cortex (allocortex as well as neocortex) receives a projection from the dorsal thalamus; (3) the nature of the projection upon the neocortex can vary from nucleus to nucleus and even among populations of cells in the same nucleus; (4) some dorsal thalamic nuclei project only to the cerebral cortex; others project to the cerebral cortex and striatum; (5) for completeness, one should perhaps reiterate here that the epithalamus and the ventral thalamus do not project upon the cortex (see Chapters 15 and 16).

TABLE 3.1. Subcortical Inputs and Cortical Outputs of the Dorsal Thalamic Nuclei[a]

Nonprimate	Primate	Subcortical input	Cortical target
		Ventral group	
VA	VAp	?	?Diffuse
	VAmc	?	?Diffuse
VL	VLa	GPi	Area 6
	VLp	Deep cerebellar nuclei Vestibular nuclei Spinothalamic tract (part only)	Area 4
VM	VMp	Substantia nigra	Diffuse, frontal, cingulate
	VMb	Taste, ?vagal	SI and G
	Submed	Spinal, ?others	Medial frontal
VP (VB)	VPL VPM	Medial and trigeminal lemnisci, spinothalamic tract	SI (areas 3a, 3b, 1, 2) and SII
	VPI	?	Dysgranular insular, ?SII
		Lateral group	
LD	LD	Fornix	Cingulate, retrosplenial
LP-Pl	LP	?	Area 5 (post.)
	Pla	?	Area 5 (ant.)
	Pll	?Sup. colliculus and pretectum	Parietotemporal and diffuse to occipital
	Plm	?	Sup. temporal gyrus
	Pli	Sup. colliculus (superficial layers) and pretectum, retina (small)	Prestriate
		Posterior group	
Po	SG-limitans	Sup. colliculus (deep layers)	Granular insular
	Pom	Spinothalamic	Retroinsular
	Pol	Auditory	Postauditory
		Medial geniculate complex	
MGp	MGv	Central nucleus inf. colliculus	AI
	MGd ant.	?Pericentral or external nuclei of inf. colliculus	AII and others
	MGd post.		Temporal auditory field
MGmc	MGmc	Multiple	Diffuse insular-temporal
		Dorsal lateral geniculate nucleus	
LGd	Parvocellular layers	Retina X	Striate area, layers IVA and IVCβ, I
	Magnocellular layers	Retina Y	Striate area, layer IVCα

TABLE 3.1. *(continued)* **93**

PRINCIPLES OF
THALAMIC
ORGANIZATION

Nonprimate	Primate	Subcortical input	Cortical target
C laminae (cat)		Retina W, and superficial layers of sup. colliculus	Diffuse, mainly layers I and III
	Intercalated and S laminae	Sup. colliculus	Striate area, layer III
	Medial group		
MD	MDpl (CL)	?Spinal	Frontal eye field
	MDc	Olfactory cortex	Orbitofrontal
	MDm	and	Lateral frontal
	MDr	amygdala	Basal olfactory structures
MV	MV (reuniens)	Fornix	Hippocampal formation and adjacent regions
Pt	Pt	?	Medial frontal
	Anterior complex		
AV	AV	Subiculum and	Anterior limbic
AM	AM	presubiculum,	Cingulate, subiculum,
AD	AD	mamillary nuclei	retrosplenial, and presubiculum
	Intralaminar complex		
CL post.	CL post.	Spinothalamic	All to striatum and diffuse to: Area 4, SI, parietal
CL ant.	CL ant.	Cerebellar nuclei, reticular formation, substantia nigra, and sup. colliculus	Parietal, frontal
CeM	CeM		
Pc	Pc		
CM	CM	GPi	Frontal, including motor
Pf	Pf	?Periaqueductal gray	Lateral, frontal

[a] Modified from Jones (1981a).

3.3.2.   Types of Thalamocortical Projection

The long-held belief that a dorsal thalamic nucleus always projects its efferent axons within the borders of one or at most a few cortical fields (Chapter 1) seems still to be true for a majority of the nuclei. However, there are clearly some that do not conform to this pattern of organization since they project diffusely over wide areas of cortex. In this, there is one basis for a classification of dorsal thalamic nuclei into types (Figs. 3.3 and 3.4). We shall refer to the diffuse projection as a "nonspecific projection" and the better known nucleus-to-field projection as a "specific projection." For reasons that will become clear in passing, one cannot, however, refer to the parent nuclei as nonspecific and specific, for the two kinds of projection can sometimes arise from the same nucleus.

3.3.2.1. The Nature of the Nonspecific Thalamocortical Projection

Recent anatomical work on the intralaminar nuclei forces us to think in terms of thalamic nuclei that project diffusely upon the cortex in addition to those that project within the boundaries of a cortical field. The intralaminar nuclei (central medial, central lateral, paracentral, centre médian, and probably the parafascicular), unlike the major relay nuclei, have a projection that is neither dense nor confined to particular architectonic fields (Jones and Leavitt, 1974; Jones, 1975a; Macchi *et al.*, 1975, 1977). The axons of intralaminar cells seem to spread rather diffusely over several fields, although there is some disagreement whether their fibers terminate predominately in the outer part of the molecular layer (Jones, 1975a) or in the deepest layers (Herkenham, 1980). The major target of the intralaminar nuclei, however, is not the cortex but the striatum (caudate nucleus and putamen) to which they furnish a particularly dense projection (Fig. 3.5) (Jones, 1975a; Kunze *et al.*, 1979; K. Kalil, 1978; Royce,

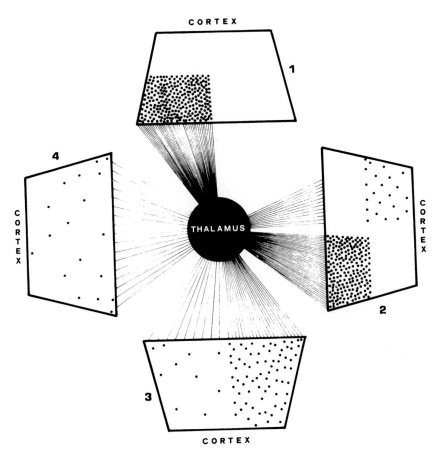

FIGURE 3.3. Macchi's (1983) scheme of patterns of thalamocortical projection in the cat. 1: Nuclei projecting densely on single cortical areas; 2: nuclei projecting densely on one area and diffusely on another; 3: nuclei projecting diffusely on several cortical fields but with a concentration on single fields or regions; 4: nuclei projecting diffusely over widespread areas (see Fig. 3.4).

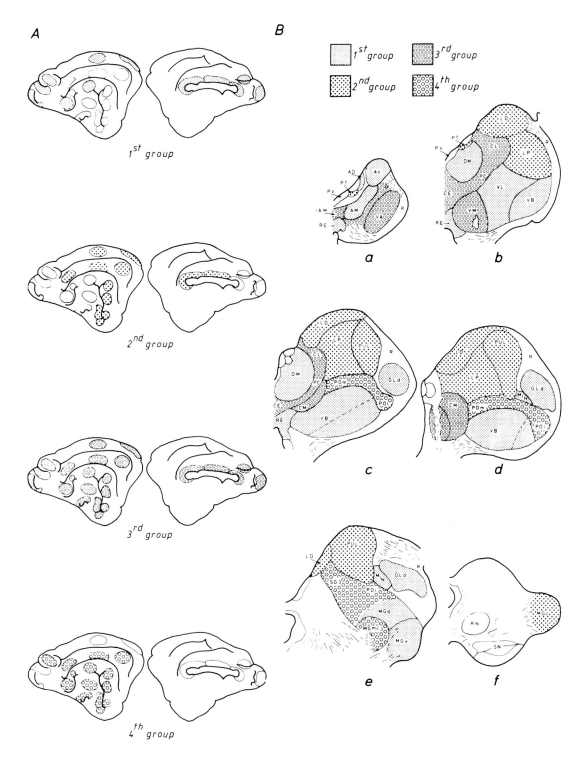

FIGURE 3.4. Macchi's (1983) scheme of the nuclei (right) with the four types of cortical projection (Fig. 3.3) and their cortical targets (left).

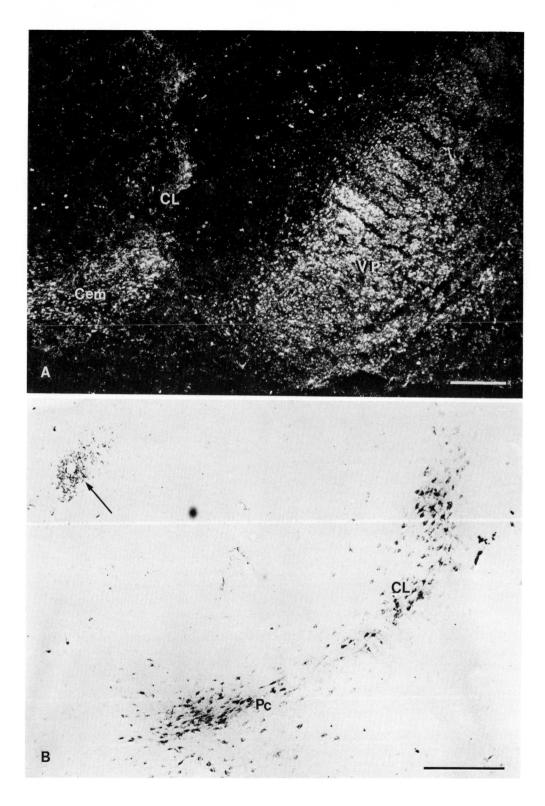

1978a; Bentivoglio *et al.*, 1983). Relatively few of the cortical axons are collaterals of those directed to the striatum (Bentivoglio *et al.*, 1983).

The intralaminar nuclei, although projecting diffusely to the cortex, appear to have regional zones of dominance; as judged by the number of labeled cells following intracortical injections of retrogradely transported tracers, they are particularly related to frontal, motor, and parietal rather than to sensory fields (Table 3.1, Fig. 3.4). Significantly, these regional zones, although transgressing the conventional architectonic field organization, are the regions from which cortical recruiting responses following thalamic or brain stem stimulation are best elicited (Jasper, 1960; Sasaki *et al.*, 1972, 1976; Chapter 12). They also correlate well with regions of maximal barbiturate spindling and with regions of evoked responses following stimulation of the midbrain reticular formation (Thompson, 1967).

The work of Herkenham (1980) in the rat suggests that a number of nuclei additional to the intralaminar nuclei, project in a similarly diffuse manner upon the cerebral cortex. These other thalamic nuclei are not traditionally regarded as part of the intralaminar system, but nevertheless seem to provide a diffuse, mainly layer I (Fig. 3.6A) input to medial, posterior parietal, temporal, and occipital cortex. They include the principal ventral medial nucleus (Herkenham, 1980; Beckstead *et al.*, 1979), anteromedial nucleus (Domesick, 1972; Krettek and Price, 1977b), medioventral nucleus (Herkenham, 1978), the magnocellular nucleus of the medial geniculate complex (Ryugo and Killackey, 1974; Jones and Burton, 1976b), parts of the lateral posterior or pulvinar complexes (Benevento and Rezak, 1975, 1976; Gilbert and Kelly, 1975; Ogren and Hendrickson, 1976, 1977), and even certain components of the lateral geniculate nucleus (see Chapter 9) (Hubel and Wiesel, 1972; LeVay and Gilbert, 1976; Fitzpatrick *et al.*, 1983; Weber *et al.*, 1983).

These layer I inputs overlie cortical areas also receiving the classically recognized, dense, topographically organized projection from a typical relay nucleus, a projection that ends in the middle cortical layers (Fig. 3.6B) within the boundaries of the field. It seems appropriate, therefore, to speak of the diffuse projections as *nonspecific cortical projections,* and the remainder as *specific,* following Lorente de Nó (1949). It should be recognized, however, that the nonspecific fibers arise from several nuclei additional to the intralaminar system. It should also be noted (Chapter 12) that the individual nuclei of the nonspecific system are not totally diffuse in their cortical projection and do have regional zones of projection. Moreover, individual cells in them probably do not have widespread terminations in the cortex (Bentivoglio *et al.*, 1981; Steriade and Glenn, 1982).

All nuclei of the dorsal thalamus providing specific inputs to the cortex as defined in Table 3.1 may be regarded as *relay* nuclei. Some of these, such as the suprageniculate–limitans complex, have sometimes been included in the intra-

---

FIGURE 3.5. (A) Retrograde labeling of cells in a principal nucleus (VP) and in the intralaminar nuclei (CL, CeM) after injection of horseradish peroxidase in the somatic sensory cortex of a rat. (B) Retrograde labeling of cells only in the intralaminar nuclei of a rat after injection of horseradish peroxidase in the striatum. Arrow indicates fiber labeling in stria medullaris due to involvement of prethalamic nuclei in the injection. Bars = 250 μm.

laminar system (Le Gros Clark, 1936a), but this has now been discounted (Burton and Jones, 1976). It may be noted that among the nonspecifically projecting nuclei, there are some that project to the striatum and some that do not (Table 3.1). Hence, striatal connectivity and a diffuse cortical projection are not necessarily correlated.

A set of data that has implied to some a diffuse projection pattern even within the specific thalamocortical projections is that if Kievit and Kuypers (1975) in the monkey. Based upon their findings in retrograde transport studies, these authors think that mediolateral (transverse) bands of cortex receive input from bands of thalamic cells that extend, usually from anterolateral to posteromedial, across several nuclei without regard to nuclear borders. It seems, however, that this is a demonstration of a regional pattern of thalamocortical organization that does not necessarily imply a breakdown in the classic nucleus-to-field pattern of organization. When a mediolateral strip more or less following the precentral

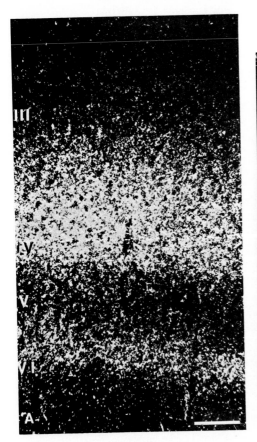
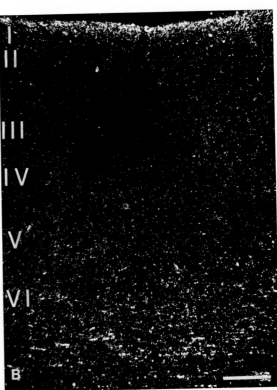

FIGURE 3.6.   (A) Specific thalamocortical projection. Labeling of terminal ramifications of thalamocortical axons in layers IIIB–IV and at the border of layers V and VI in area 3b of a squirrel monkey following injection of tritiated amino acids in the ventral posterior nucleus. From Jones and Burton (1976b). (B) Nonspecific thalamocortical projection. Labeling of thalamocortical fiber ramifications in layers I and VI of the cingulate cortex of a cat following injection of tritiated amino acids in the ventral medial and intralaminar nuclei. (C) Mixed projection. Labeled specific projection

gyrus was injected with a large quantity of horseradish peroxidase in Kievit and Kuypers's experiments, retrogradely labeled cells appeared in a band extending through the anterior nucleus of the ventrolateral complex, the posterior ventral lateral nucleus, the intralaminar nuclei, and the mediodorsal nucleus. But the injected strip included parts of several cortical fields: area 4, a part of area 6, and a part of the frontal operculum. Injections of horseradish peroxidase in the separate fields would be expected to label cells retrogradely only in the separate nuclei. And because there is no overlap in the subcortical inputs to these nuclei

from suprageniculate nucleus stops abruptly at border of granular insular field (curved arrow). Labeled nonspecific layer I projection from adjoining magnocellular medial geniculate nucleus continues into adjoining field (straight arrow). Squirrel monkey; from Jones and Burton (1976b). Bars = 100 μm.

(Chapter 7), the specificity of input to the cortical areas is preserved. Because of the diffuse projection of the intralaminar nuclei, all the injections in the fields of the transverse band should result in labeling in these nuclei, but as pointed out above, this seems to be at another level of organization.

3.3.2.2. The Nature of the Specific Thalamocortical Projection

It is an old adage, traced to the work of Ramón y Cajal (Jones, 1983e), that thalamocortical axons terminate primarily among the cells of the internal granular layer (layer IV) of the cortex. Rather surprisingly, recent studies on the laminar distribution of thalamocortical fibers in the cortical area that forms the target of an individual, specific thalamic nucleus show that the traditionally accepted view is true for only a minority of cortical areas, at least in primates (Jones and Burton, 1976b). Only in the koniocortical areas (area 3b of the somatic sensory cortex, the striate area, and the primary auditory area) does layer IV receive the majority of the thalamic terminations. Even here, substantial numbers terminate in the supervening part of layer III (Fig. 3.6B, Table 3.2). Outside these areas, even in areas lying adjacent to them, such as area 1 in the somatic sensory or area 19 in the visual, thalamic fibers tend to avoid layer IV and terminate almost completely in the deeper half of layer III (Jones, 1975b; Burton and Jones, 1976) (Fig. 3.7). This striking difference is not simply a matter of equivalent layers having different names in different areas because in the parietal–temporal fields, thalamic terminations surround the somata of large pyramidal cells characteristic of layer III, while the deeper-lying small granule cells characteristic of layer IV in all areas are free of terminal ramifications (Fig. 3.7). In all fields in which thalamic afferents terminate in layer III instead of layer IV, the density of thalamic terminations is greatly reduced in comparison with konicortical areas (Jones, 1975b; Jones and Burton, 1976b; Table 3.2). In concert

TABLE 3.2. Laminar Distribution and Relative Density of Thalamic Terminal Plexus[a,b]

Areas	I	II	IIIa	IIIb	IV Sup.	IV Deep	Va	Vb	VIa	VIb
7				+ +						
5				+ + +				+ +		
1–2, Ri			+	+ +	+				+	
3b			+ +	+ + + +	+ + + +	+ +			+ +	
G, SII				+ + +	+ + +					
Id, Ig[c]			+	+ +	+					
RL, Pi				+ + +	+ + +			+		
AI			+ +	+ + + +	+ + + +	+ +			+ +	
Pa			+	+ +	+					
T1, T2				+ + +				+		
T3				+ +				+		

[a] From Jones and Burton (1976b).
[b] +, Low; + +, moderate; + + +, heavy; + + + +, intense.
[c] Layers IIIa and IIIb are not readily separable.

with this, the number of granule cells is reduced. Although their somata lie outside the major zone of thalamic terminations, it is uncertain whether they fail to receive thalamic terminations altogether, for their dendrites extend up into layer III (Jones, 1975d; Hendry and Jones, 1983a).

The nature of the experiments used to define thalamic terminations in cortical fields outside the primary sensory areas seems to rule out the possibility of additional thalamic nuclei providing input to the underlying layer IV. In the visual cortex, however, there is a very clear sublaminar segregation of the terminals of different classes of thalamocortical fibers arising in the lateral geniculate nucleus and elsewhere. Hubel and Wiesel (1969) first showed in the monkey that geniculocortical fibers arising in the parvocellular layers of the lateral geniculate nucleus terminated in area 17 primarily in the deeper half of layer IV (layer IVcβ), with a smaller, more superficial zone of terminations in the upper part of layer IV bordering on layer III (layer IVa). By contrast with this, geniculocortical fibers arising in the magnocellular laminae terminated just above the deeper terminations of fibers from the parvocellular layers, in layer IVcα. These observations have since been repeated many times with autoradiographic

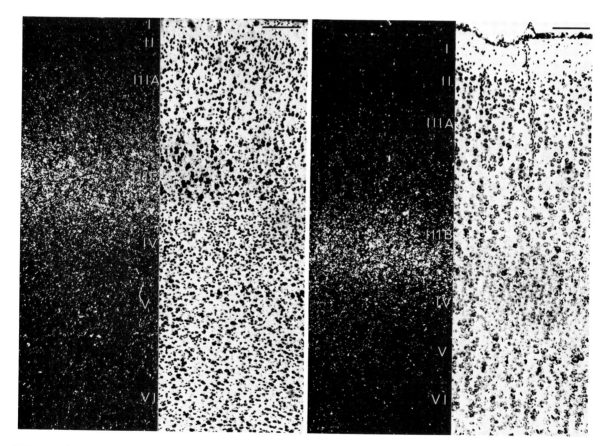

FIGURE 3.7. Darkfield and brightfield photomicrographs from identical slides showing labeled thalamic fiber terminations confined to layer IIIB of areas 5 (left pair) and 7 (right pair) in a squirrel monkey. Bars = 100 μm. From Jones and Burton (1976b).

methods (Fig. 3.8A) (e.g., Glendenning *et al.*, 1976; Kaas *et al.*, 1976; Hendrickson *et al.*, 1978) and have been extended by the demonstration that cells in certain intercalated or subsidiary laminae of the monkey lateral geniculate nucleus have preferential projections to layers I and III (Chapter 9; Fitzpatrick *et al.*, 1983; Weber *et al.*, 1983). In the cat, LeVay and Gilbert (1976) (Fig. 3.8B) have shown a similar segregation of thalamocortical terminations: the deep C laminae of the cat dorsal lateral geniculate nucleus provide the input to layer I and to parts of

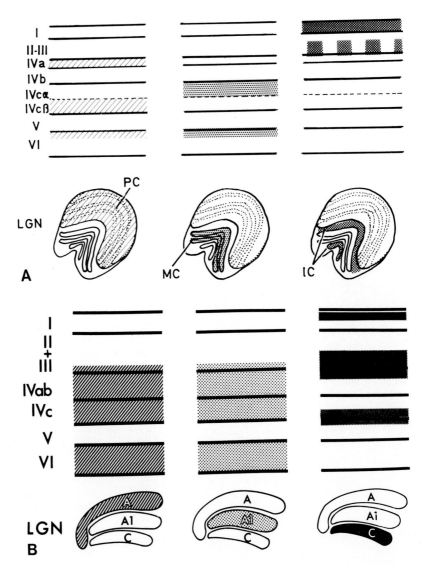

FIGURE 3.8.   (A) Schematic representation of differential laminar distributions in visual cortex of thalamocortical axons from parvocellular (left), magnocellular (middle), and S and intercalated laminae (right) of monkey dorsal lateral geniculate nucleus. (B) Similar scheme showing differential laminar distribution in visual cortex of thalamocortical axons arising in A laminae (left, middle) and C laminae (right) of cat dorsal lateral geniculate nucleus. From LeVay and Gilbert (1976).

layer IV lying superficial and deep to inputs from the A laminae. Unlike the inputs to deeper layers, which were confined to areas 17 and 18 of the cortex, the layer I terminations spread beyond these into surrounding areas. Comparable projections arising from single geniculate laminae may be present in other species (Carey *et al.*, 1979; see Chapter 9) but there is no published evidence for similar laminar segregations of thalamocortical terminations in other cortical areas.

Apart from terminations in layers III and/or IV, thalamocortical fibers consistently provide a deep zone of terminations at the junction of layers V and VI. This observation, first made in the cat visual cortex by Rosenquist *et al.* (1974), has been consistently repeated in most cortical areas and in many species (e.g., Wiesel *et al.*, 1974; Wise, 1975; Hubel, 1975; Robson and Hall, 1975; Jones and Burton, 1976b; Wise and Jones, 1978) (Fig. 3.6B, Table 3.2). Together with the evidence for differential terminations in layers III–IV and I it raises the possibility of differential thalamic inputs to specific classes of cortical neurons. This subject has only been incompletely explored. It is now obvious, however, that contrary to the old established view, thalamic fibers do not terminate solely on a single class of cortical interneurons that then provide a relay to the output (pyramidal) cells. There is substantial morphological and physiological evidence that thalamocortical fibers in a variety of areas terminate directly on pyramidal cells that project both intercortically and subcortically, as well as on several varieties of cortical interneurons (e.g., Peters *et al.*, 1976; White, 1979; Bullier and Henry, 1979; Deschênes *et al.*, 1982; Hendry and Jones, 1983b). For reviews of the potential intracortical circuits whereby thalamic fibers, pyramidal cells, and interneurons may interact, see Jones (1981c, 1983e).

Ferster and LeVay (1978) have made in the cat a strong circumstantial case to warrant a correlation between different anatomical classes of geniculocortical axons and geniculate relay cells types. One class of axon (Fig. 3.9), probably arising from the morphological equivalents (see Chapter 9) of Y-type relay cells, terminated primarily in the upper two-thirds of layer IV and in the supervening part of layer III. Its terminal aborization was widespread (2–5 mm) but was segregated into two or more clumps with intervening clearer patches, both clumps and patches corresponding more or less to the dimensions of an "ocular dominance column" (see Section 3.7 and Chapter 9). A second, thinner class of axon (Fig. 3.9), probably arising from the morphological equivalents of X-type relay cells, terminated primarily in the deeper one-third of layer IV with a much more restricted arborization corresponding to the dimensions of a single ocular dominance column. Both classes of axons, however, had collateral branches to the upper part of layer VI, the terminal arborizations of these branches being mainly in register with those in the supervening layers. To some extent these findings have been corroborated in a more restricted but electrophysiologically controlled study by Gilbert and Wiesel (1979).

As pointed out by Ferster and LeVay, there is a close correspondence between the distributions of the axons they studied and the laminar positions of cortical cells with different categories of receptive fields, once again suggesting the possibility of selective terminations on different classes of cortical cells. It is probable that a corresponding pattern of organization exists in primates, because the parvocellular laminae of the monkey lateral geniculate nucleus, which contain

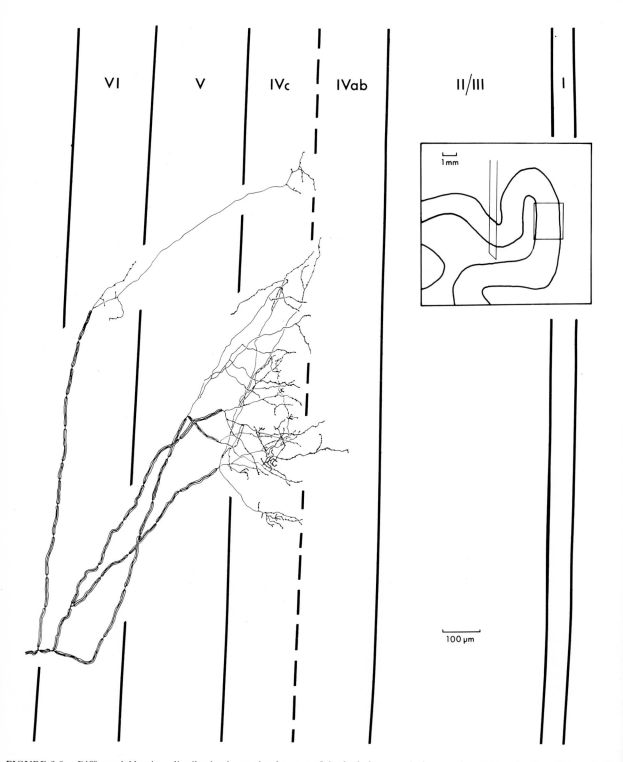

FIGURE 3.9.   Differential laminar distribution in cat visual cortex of single thalamocortical axons thought to arise from X-type (left)

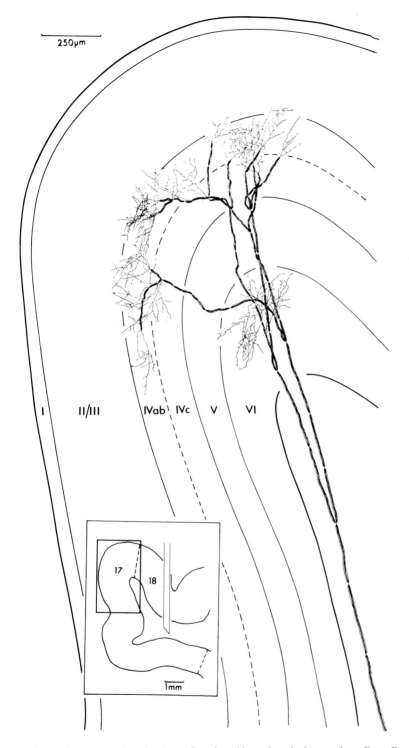

and Y-type (right) relay neurons in A laminae of cat dorsal lateral geniculate nucleus. From Ferster and LeVay (1978).

X cells, project to deeper parts of layer IV in the cortex than the magnocellular laminae, which contain Y cells (see Chapter 9).

Ferster and LeVay described a third, very thin class of intracortical axons, probably arising from the presumed morphological equivalents of W relay cells, in the deepest (C1–C3) laminae of the cat lateral geniculate nucleus (see Chapter 9). These axons ramify rather widely, primarily in layer I, but they have collaterals in layers III and V. This confirms the autoradiographic study of LeVay and Gilbert mentioned above.

Whether these three classes of thalamocortical axons with their individual distribution patterns in the visual cortex are duplicated for other thalamic nuclei is still an open question. As pointed out in other parts of this chapter, data scattered throughout the literature imply projections from different thalamic nuclei to layer I and to layers III or IV of the same cortical area. In rodents, Caviness and Frost (1980; Frost and Caviness, 1980) have suggested that most (although not all) thalamic nuclei have projections to layers III–IV that terminate within the boundaries of the architectonic field(s) related to that nucleus. All

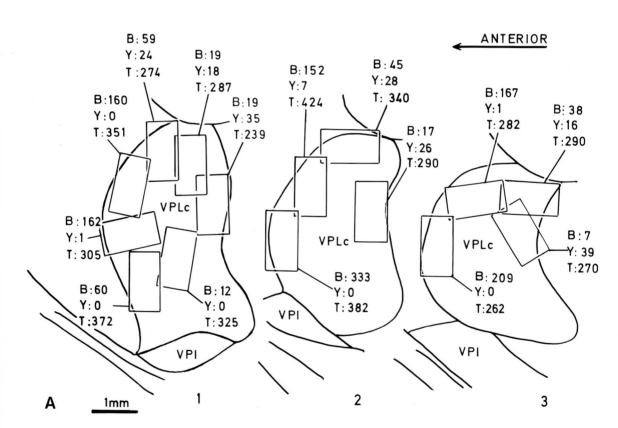

FIGURE 3.10. Three 50-μm-thick sagittal sections at 150-μm intervals from the ventral posterior nucleus of a monkey in which the projection areas of the same muscle nerve in area 3b and in area 2 were injected respectively with dyes fast blue and nuclear yellow. Each box shows the number of cells retrogradely labeled with fast blue (B) or nuclear yellow (Y) and total number of cells (T) in same area as determined by subsequent counterstaining. Note differential

nuclei (without exception) have other projections to layers I and V; some have additional (or predominant) projections to layer VI and with extensive spread over the borders of architectonic fields. Whether this systematization and the finer segregation demonstrable in the cat visual system are applicable to all species and all cortical areas remains to be seen.

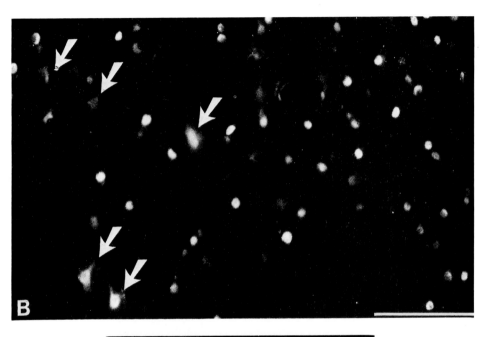

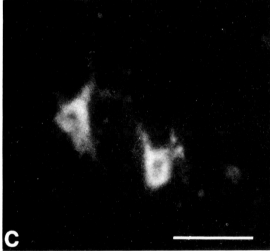

distribution of cells projecting to each area, and lack of double labeled cells. From Jones (1983d). (B) Fluorescence photomicrograph showing fast blue (arrows) and nuclear yellow (stained nuclei) labeled cells in preparation of (A). Bar = 100 μm. (C) Higher-magnification view of the two fast blue-labeled cells at bottom left in (B) showing lack of nuclear labeling. Bar = 50 μm.

3.3.2.3. Collateral Thalamocortical Projections?

Diffusely projecting nuclei, such as the magnocellular medial geniculate or the intralaminar, show a more intense retrograde reaction in experimental animals as larger amounts of cortex are removed (Nissl, 1913; Rose and Woolsey, 1958; Macchi *et al.*, 1975) and more cells in them are retrogradely labeled as cortical injections of tracer become larger. Though doubtless indicating relatively widespread projections from single cells, this type of projection is quite different from that in which a single cell in a relay nucleus provides axon branches that terminate as part of a topographically ordered sequence in each of two specific cortical fields. Suggestive examples of this, based upon retrograde degeneration, were to be found in the putatively branched projection of large lateral geniculate neurons to areas 17 and 18 of the cat (Garey and Powell, 1967) and of ventral posterior neurons to areas 3, 1, and 2 of the monkey (Le Gros Clark and Powell, 1953). Both of these suggestions rested upon the observation that ablation of area 18 with area 17 or of areas 1 and 2 with area 3, led to more severe retrograde degeneration of the relevant thalamic cells than did ablation of area 17 or of area 3 alone. In these cases, retention of a collateral axon to area 18 or to areas 1 and 2 was felt to "sustain" the thalamic cell from the severest reaction. Subsequent work has not always confimed this suggestion. In the cat visual system, area 17 receives both X- and Y-relay cell inputs whereas area 18 receives only Y inputs (see Chapter 9). This might imply that the larger lateral geniculate cells of the A and A1 laminae receiving Y-cell inputs provide branched axons to both areas. Though seemingly confirmed by antidromic techniques (Stone and Dreher, 1973), morphological studies have disagreed over the extent of collateral projections. Geisert (1980) saw a certain number of cells double labeled in the dorsal lateral geniculate nucleus of the cat after injections of two different retrograde tracers in areas 17 and 18 but Norita and Creutzfeldt (1982) saw none.

In the case of the somatic sensory cortex of the monkey, morphological evidence shows that separate populations of thalamic cells project to areas 3a, 3b, 1, and 2. Injection of horseradish peroxidase into areas 1 and 2 some months after ablation of area 3 leads to retrograde labeling of cells that, although within the confines of a ventral posterior nucleus containing many retrogradely degenerated cells, are themselves unshrunken (Jones *et al.*, 1979); and injections of different tracers at physiologically defined equivalent parts of the body representation in any two of these areas result in no double retrogradely labeled cells in the thalamus (Jones, 1983g) (Fig. 3.10). It has also been suggested that some neurons in the thalamic ventral posterior nucleus may project by branched axons to both the first (SI) and the second (SII) somatic sensory areas of the cortex. In a study in the cat, Manson (1969) found that of a population of 113 ventral posterior cells antidromically discharged by electrical stimulation of the cortex, 48% projected to SI alone, 42% to SI and to SII, and 10% to SII alone. This remains the only fairly convincing example of a collateral projection from a thalamic relay nucleus to the cortex. Even this has not been without criticism, for although Andersson *et al.* (1966) thought that cells of the ventral posterior nucleus receiving group I muscle inputs projected by means of a branched axon to SI and SII, this could not be confirmed by Rosén (1969a). Anatomical studies using double retrograde tracing methods (Jones, 1975a; Spreafico *et al.*, 1981) have shown only a relatively small number of ventral posterior cells projecting to SI and SII. The existence of other collateral thalamocortical projections has been suggested, for example, from the gustatory relay nucleus (Benjamin and Burton, 1968) and from the medial geniculate complex (Winer *et al.*, 1977). However, neither has yet been studied at the single-cell level. The position regarding collateral projections in the specific thalamocortical projection system thus remains somewhat unclear. At the moment, the weight of evidence seems to be against widespread collateralization from individual cells.

3.3.2.4. Cortical "Overlap" Zones

"Specific" cortical projections from a thalamic relay nucleus are normally distributed within the borders of an architectonic field, with virtually no overlap into the cortical target fields of other relay nuclei (Jones and Burton, 1976b; Caviness and Frost, 1980). In some instances, however, substantial overlap does seem to occur. In two well-known examples, the overlap of two thalamocortical projections is accompanied by a recognizable physiologic correlate. In the first, in marsupials, the cortical projections of the ventral lateral and ventral posterior nuclei seem to be completely superimposed (Ebner, 1967; Killackey and Ebner, 1973; Ebner *et al.*, 1976; Donoghue and Ebner, 1981a,b), leading to a common sensorimotor representation (Lende, 1963). This occurs despite total segregation of cerebellar and dorsal column–lemniscal terminations, respectively, in the ventral lateral and ventral posterior nuclei (Rockel *et al.*, 1972; Walsh and Ebner, 1973). In the cortex of the opossum and other marsupials, in association with the superimposition of the two inputs, there is complete overlap of a granular layer IV typical of sensory cortex upon a giant pyramidal cell layer V typical of motor cortex (Brodmann, 1906).

A comparable situation is present in the cortex of the rat. There the aggregation of layer IV granule cells that forms the morphological correlate of the hindlimb representation in the sensory cortex (Welker, 1976) is underlain by an extension backwards of giant pyramidal cells of the motor cortex (Jones and Porter, 1980; Fig. 3.11). In this region, unlike other parts of the sensory and motor cortex, there is superimposition of inputs from the ventral lateral and ventral posterior nuclei (Donoghue *et al.*, 1979) and a common motor and sensory representation of the hindlimb (Hall and Lindholm, 1974; Sanderson *et al.*, 1984). There is little evidence for this kind of overlap in sensory or motor representations in larger mammalian brains. Here, the nucleus-to-area plan of specific thalamocortical projections seems well founded.

3.3.2.5. Thalamic Projections to Allocortex

a. Hippocampal Formation and Parahippocampal Regions. Occasional suggestions of a projection from the thalamus to the hippocampal formation or adjacent structures can be found in the earlier literature dealing with the distribution of the cingulum (e.g., Krieg, 1947; Domesick, 1970). In 1974 Segal and Landis, using the then new method of tracing connections by retrograde labeling, reported the presence of retrogradely labeled cells in various nuclei of the anterior and medial thalamus after rather ill-defined, large injections into the hippocampus. In 1975 Shipley and Sørenson presented convincing evidence for a projection from the anterior thalamic nuclei of the guinea pig to the ipsilateral presubiculum. Their study was based upon both axonal degeneration and autoradiography and, though not defining the precise anterior nuclei in which the projection originated, indicated that the fibers terminated in layers I and III of the presubiculum. This laminar pattern of termination in the parahippocampal

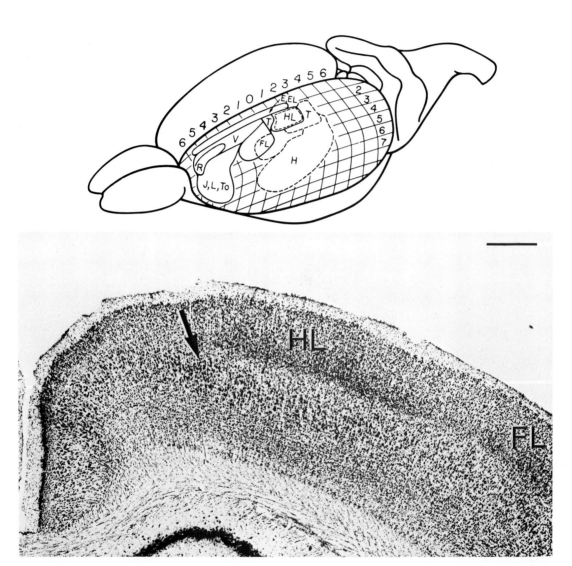

FIGURE 3.11. (Bottom) Photomicrograph of part of a frontal section through brain of a rat. HL and FL are granule cell aggregations of first somatic sensory area in which hindlimb and forelimb are represented. Arrow indicates extension of giant cells of motor cortex beneath hindlimb aggregation. Here, there is an amalgam of both sensory and motor representations (B) and thalamic afferents from ventral posterior and ventral lateral nuclei overlap. Bar = 500 μm. After Jones and Porter (1980). (Top) Hall and Lindholm's (1974) map of the rat cortex showing motor representation (solid lines) based upon electrical stimulation of cortical surface and sensory map based upon recording of multiunit responses to light tactile stimuli (dashed lines). Note overlap in hindlimb area (HL).

region of the cortex is very similar to the pattern of termination of anterior thalamic nuclear fibers in the retrospenial, cingulate, and medial frontal regions (Domesick, 1972; Krettek and Price, 1977b; Robertson and Kaitz, 1981) and, thus, indicates a continuity of thalamocortical projections from neocortex toward allocortex.

Herkenham (1978), in his study of the medioventral (reuniens) nucleus of the rat thalamus (Chapter 13), extended the observations of Shipley and Sørenson (1975) showing that this nucleus, via the cingulum, provided a rather heavy input to the entorhinal and parahippocampal areas of cortex as well as to the subiculum and the CAI field of Ammon's horn (Fig. 3.12).* In the entorhinal and parahippocampal areas the fibers terminate in the bilaminar pattern described by Shipley and Sørenson but in Ammon's horn the fiber terminations are confined to the stratum lacunosum et moleculare. Wyss *et al.* (1979) in the rat and Amaral and Cowan (1980) in the monkey, confirmed Herkenham's results, showing that large injections of horseradish peroxidase anywhere in the hippocampal formation and adjacent areas led to retrograde labeling of cells in the medioventral nucleus. Other labeled thalamocortical cells were localized to

---

* See Berman and Jones (1982) for definition of these terms.

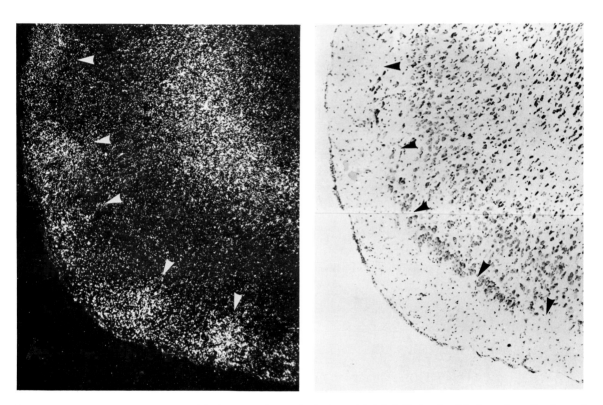

FIGURE 3.12. Pars lateralis of entorhinal cortex of a rat in dark- and brightfield showing labeling of ramifications of thalamic fibers originating in medioventral (reuniens) nucleus. From Herkenham (1978).

the anterior nuclei and to parts of the parataenial and lateral dorsal nuclei as well. The exact projections of the latter two nuclei were not determined so it is uncertain whether they provide inputs to the hippocampal formation proper or to adjacent regions.

b. Olfactory Tubercle, Piriform Cortex, and Adjacent Regions. The projection of the medioventral and anterior nuclei to the hippocampus, parahippocampal regions, and the entorhinal areas can be thought of as forming the thalamic input to posterior allocortex. Anterior allocortical areas certainly include prepiriform and periamygdaloid cortex, and many would also incorporate the olfactory tubercle, surface nuclei of the amygdala, and anterior olfactory nucleus in this definition.* Whether some of these are or are not correctly regarded as cortical structures we leave open. For our purposes, the question is: do any of these areas receive inputs from the dorsal thalamus? Rather surprisingly, this is a question that has not been satisfactorily resolved. From their studies of the retrograde degeneration occurring in the thalamus as the result of lesioning various regions of the rabbit cortex, Rose and Woolsey (1943, 1949a) concluded that "anterior rhinencephalic structures" received inputs from the intralaminar nuclei. In coming to this conclusion, however, they rejected the idea (Gerebtzoff, 1940b; Vogt and Vogt, 1941), since proven correct, that the principal target of the intralaminar nuclei is the striatum.

In the recent literature, I can find few positive statements about whether some parts of the thalamus projects to prepiriform or periamygdaloid cortex. Occasionally, there are reports of a few retrogradely labeled cells in the vicinity of the mediodorsal or adjacent nuclei after injections of tracer in the prepiriform cortex (e.g., Haberly and Price, 1978). The prepiriform and periamygdaloid areas do, however, send fibers *to* the thalamus (Chapter 13). There, they end primarily in the mediodorsal nucleus (Powell *et al.*, 1965; Benjamin *et al.*, 1982; Price and Slotnik, 1983). Could this corticothalamic projection betoken the existence of a thalamocortical one as it does in every other cortical area (Section 3.6)? The only anterior allocortical region that seems confirmed as receiving an input from the dorsal thalamus is the olfactory tubercle (Krettek and Price, 1977b; Chapter 13). It is, however, perhaps debatable whether this is indeed an allocortical area. In any case the terminations are reported to be to the deep multiform layer, rather than to more superficial layers and, as described, the terminations could even be in adjoining parts of the striatum (Heimer, 1972).

Thalamic projections to the anterior allocortex, thus, still remain in doubt. They will need to be demonstrated before one can confidently assert that the whole surface of the cerebral hemisphere is a target of the dorsal thalamus. The evidence for thalamic inputs not only to the whole neocortex but also to the posterior allocortex makes this, however, a possibility worth exploring. The starting point for many studies of the hippocampus or prepiriform cortex has been either their efferent connections or their afferent connections from adjacent cortical areas or the olfactory bulb. With the possible exception of the motor areas, studies of neocortex are more inclined to start with the thalamic input. It is interesting to reflect, therefore, what new insights into allocortical

---

* See Berman and Jones (1982) for definition of these terms.

function might ensue if they, too, were looked at from the perspective of the thalamus.

3.4. The Thalamostriatal Projection

3.4.1. Background

Although evidence of the thalamocortical projection has existed for virtually a century, that confirming the presence of a substantial thalamostriatal projection is of much more recent vintage. One reason for the failure of early anatomists to identify the thalamostriatal projection may lie in the nature of the experimentation carried out. In most early experimental studies, such as those of Nissl (1913) and von Monakow (1914), the basal ganglia were usually damaged in association with the cortex. Under these circumstances, most thalamic nuclei underwent severe retrograde degeneration and, thus, there was apparently no need to postulate a projection elsewhere.

Yet most reports in the 1930s, particularly in primates, reported relatively little or no cellular reaction in certain thalamic nuclei, particularly those in posterior parts of the intralaminar system, following cortical ablations (Le Gros Clark and Boggon, 1933b; Papez, 1938; Le Gros Clark and Russell, 1939; Walker, 1935, 1936, 1938a). Though there was little evidence for it, the idea was clearly in the air (e.g., Walker, 1938a) that these thalamic nuclei, at least, might project to the basal ganglia. Walker thought the globus pallidus was the most likely target though the case examined by Le Gros Clark and Russell (1939) showed that virtually complete degeneration of the centre médian nucleus accompanied a vascular lesion in the putamen.

In 1940 Gerebtzoff (1940b) reported severe retrograde degeneration in the parafascicular and centre médian nuclei of rabbits following destruction of the striatum. This was soon confirmed in a compelling study of human pathological material by Vogt and Vogt (1941). Focusing their attention on cases with long-standing lesions of subcortical white matter, of caudate nucleus and putamen, or of the efferent bundles of the globus pallidus, they noted severe retrograde degeneration in the centre médian nucleus only after lesions of the striatum (caudate nucleus and putamen). They made, therefore, a very strong case that this nucleus at least was connected with the striatum and not with the cortex. Paraphrasing Nissl (1913) who had referred to the thalamus as "Grosshirnanteil," they noted that the centre médian nucleus is "einen *Striatumanteil* des Thalamus." Vogt and Vogt detected no degeneration in the parafascicular nucleus after striatal lesions and made no mention of other intralaminar nuclei. Confirmation of the findings of Vogt and Vogt soon came from similar cases reported by Freeman and Watts (1947), McLardy (1948), Hassler (1949a), and Simma (1950).

Rose and Woolsey (1943, 1949b) in their studies of the rabbit brain, though not specifically rejecting a thalamostriatal projection, were unwilling to accept that such a projection was provided by the intralaminar nuclei. They considered that the principal target of these nuclei was the cerebral cortex. Nevertheless, as the years passed, more and more evidence was marshaled for the existence of a thalamostriatal pathway arising in the intralaminar nuclei. Droogleever-Fortuyn (1950), Droogleever-Fortuyn and Stefens (1951), Powell and Cowan (1954, 1956), and Cowan and Powell (1955), working in various mammals, did not contest that the cells of the intralaminar nuclei showed some pallor and shrinkage after cortical lesions but felt that cellular degeneration was much more severe in these nuclei after lesions of the striatum and, thus, reinforced the idea of a thalamostriatal projection. The last shots in the battle were fired by Murray (1966) and by Powell and Cowan (1967). The first considered that the cellular degeneration ensuing in the thalamus following striatal lesions could be due to undercutting the cortex. Powell and Cowan, on the other hand, reaffirmed their earlier point of view.

There were surprisingly few attempts to demonstrate thalamostriatal fibers directly by anterograde degeneration. Some such studies were in fact carried out (e.g., Ranson *et al.*, 1941; Glees, 1945; Nauta and Whitlock, 1954; Mehler, 1966a) and, together with some suggestive physiology (e.g., Rasminsky *et al.*, 1973), served to convince all but the very skeptical of the existence of thalamostriatal fibers. However, the thalamostriatal projection was really only established irrefutably with the introduction of methods for connection tracing by retrograde axoplasmic transport. Virtually the first three major studies with the new technique showed retrograde labeling of thalamic cells after injections in the striatum of rats, cats, and monkeys (Nauta *et al.*, 1974; Kuypers *et al.*, 1974; Jones and Leavitt, 1974) (Fig. 3.5).

3.4.2. Organization of the Projection

To date, only the intralaminar nuclei seem confirmed as giving rise to the thalamostriatal projection (Table 3.1) (Jones and Leavitt, 1974; Carpenter, 1981). Apart from the intralaminar nuclei, other thalamic nuclei that may project upon the striatum are the ventral anterior nucleus (Royce, 1978b), the magnocellular medial geniculate nucleus (Heath and Jones, 1971a; Ryugo and Killackey, 1974), and the parataenial nucleus which may project to the nucleus accumbens (Swanson and Cowan, 1975; Groenewegen *et al.*, 1980). Some of these reports have not been adequately confirmed, however. Among the intralaminar nuclei, both the rostral group of nuclei (central medial, central lateral, and paracentral) and the caudal group (centre médian and parafascicular) provide thalamostriatal axons. All of these nuclei also project upon the cortex (Jones and Leavitt, 1974; Macchi *et al.*, 1977; Bentivoglio *et al.*, 1981, 1983). Whether many cells have branched axons to both cortex and striatum remains an open question (Bentivoglio *et al.*, 1983). There is some physiological and anatomical support for branching in the cat (Cesaro *et al.*, 1979; Jinnai and Matsuda, 1981) but others have not been able to demonstrate it in an extensive physiological study (Steriade and Glenn, 1982).

The thalamostriatal projection is a systematic, topographically organized one. The rostral group of intralaminar nuclei projects to the head of the caudate nucleus and to anterior parts of the putamen while the centre médian and parafascicular nuclei project to the body of the caudate nucleus and to much of the putamen (see Powell and Cowan, 1956, 1967). It seems probable, though this point has not been extensively investigated, that an intralaminar nucleus projects to the same parts of the striatum that receive corticostriatal fibers from those cortical areas to which that intralaminar nucleus also projects. For example, the part of the monkey putamen receiving fibers from the centre médian nucleus is the same that receives corticostriatal fibers from the motor cortex (Künzle, 1975; Jones *et al.*, 1976; K. Kalil, 1978). The centre médian nucleus projects to motor cortex (Strick, 1975). The nucleus accumbens, also an integral part of the

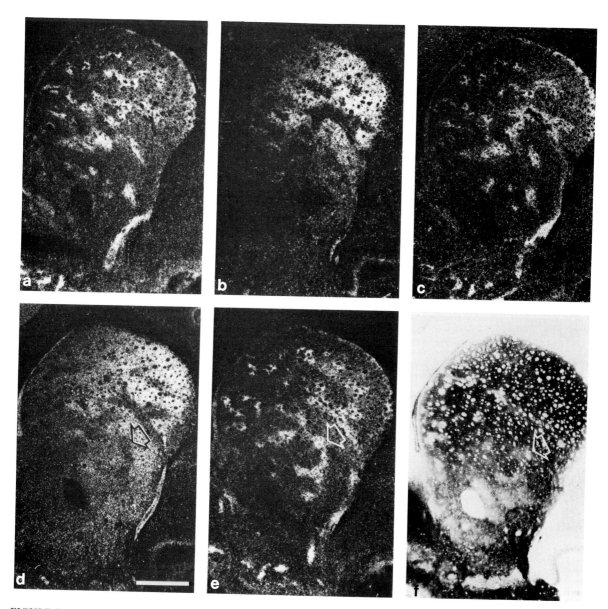

FIGURE 3.13.   Photomicrographs of six sections through the same part of the caudate nucleus of rats, showing patchy distribution of opiate receptor binding (a, c, e), of fibers from the parafascicular nucleus of the thalamus (b, d), and of staining for acetylcholinesterase (f). From Herkenham and Pert (1981).

striatum, may receive its thalamostriatal input from the parataenial nucleus (Swanson and Cowan, 1975; Groenewegen *et al.*, 1980). The parataenial nucleus has not usually been regarded as part of the intralaminar system and the projection may arise from intralaminar cells that extend dorsal and medial to the mediodorsal nucleus (Chapter 13).

The terminal ramifications of thalamostriatal fibers are not diffusely distributed. Instead, bundles of fibers end in small focal aggregations (Fig. 3.13) separated by gaps (K. Kalil, 1978; Royce, 1978a; Herkenham and Pert, 1981). The aggregations show preferential association with foci of diminished acetylcholinesterase staining, and with foci of immunocytochemical staining for enkephalins (Chapter 12) (Graybiel and Ragsdale, 1978; Ragsdale and Graybiel, 1981; Graybiel *et al.*, 1979, 1981a,b). Similar foci of striatal efferent neurons (Graybiel *et al.*, 1979), of corticostriatal terminations (Goldman and Nauta, 1977; Jones *et al.*, 1977), of nigrostriatal, dopamine terminations (Graybiel *et al.*, 1981a), and of opiate receptors (Herkenham and Pert, 1981) have also been described. The interrelationship of the patchlike patterns formed by all these systems together, is, however, not yet totally clear. The physiological interactions between corticostriatal, thalamostriatal, and nigrostriatal inputs have been studied by intracellular recording in rats (Wilson *et al.*, 1983).

3.5. Thalamic Projections to Other Parts of the Basal Telencephalon

Apart from the striatum (Section 3.4) and possibly the olfactory tubercle (Section 3.3), the only other noncortical structure to receive afferents from the dorsal thalamus is the amygdala. The part of the amygdala receiving thalamic fibers is rather small and confined to a dorsolateral portion of the lateral nucleus of the amygdaloid complex (Jones and Burton, 1976a) (Fig. 3.14). To the best of my knowledge this projection has only been demonstrated in monkeys, in which it arises in the medial pulvinar nucleus. There is, however, some older evidence for a comparable projection in the cat (Graybiel, 1970; Heath and Jones, 1971b).

3.6. The Nature of Thalamic Inputs

3.6.1. Specific and Nonspecific Subcortical Inputs

For the purposes of description, the subcortical inputs to the dorsal thalamus can be divided into two general categories. In the first are the classical pathways, such as the optic tract, the medial lemniscus, and the brachium of the inferior colliculus. These have an easily definable, information-bearing function, form part of an obvious functional system, and terminate with a coherent, topographic ordering within the confines of the appropriate thalamic relay nucleus. In the second category is the thalamic input which is part of a diffusely organized pathway that distributes fibers to a number of diencephalic and other basal forebrain centers en route to the cerebral cortex. The obvious representatives of this category are the ascending noradrenergic and serotoninergic systems (see Swanson and Hartman, 1975; Jones and Moore, 1977; and Chapter 4).

The hallmark of the classical system is specificity. Not only do the afferent fibers terminate within the confines of a particular thalamic nucleus but they are distributed systematically as the basis of the topographic ordering and mapping of the particular receptor sheet that they represent. Within this ordered system there is often further specificity, for fibers with different physiological characteristics can end on distinct classes of thalamocortical relay cells (see Chapters 4 and 9).

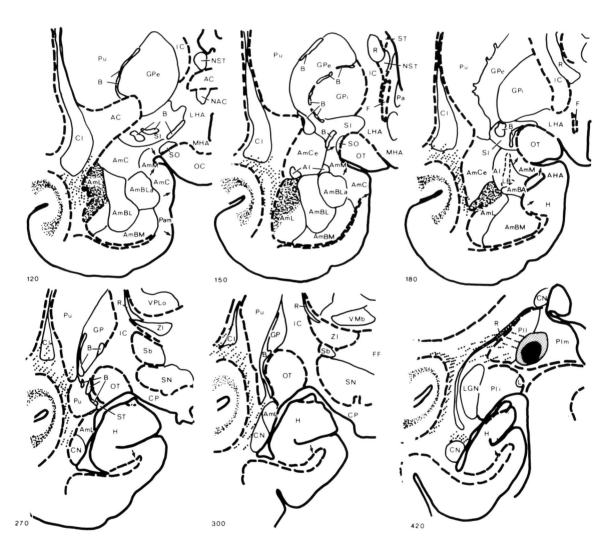

FIGURE 3.14.   Projection drawings of a series of sections through the temporal lobe of a squirrel monkey showing autoradiographically labeled fibers leaving an injection site in the medial pulvinar nucleus and passing to a portion of the lateral amygdaloid nucleus, as well as to the overlying cortex. From Jones and Burton (1976a).

Topographic ordering of inputs within the borders of a defined thalamic relay nucleus, though best described in the three principal sensory relay nuclei (Section 3.6.2), can also be demonstrated in the posterior ventral lateral nucleus which receives fibers from the deep cerebellar nuclei and forms the thalamic relay to motor cortex (Strick, 1976a; Asanuma *et al.*, 1983b; Chapter 7). It is also apparent in the lateral posterior nucleus (Graham, 1977) or inferior pulvinar nucleus (Harting *et al.*, 1980) which receive fibers from the superficial layers of the superior colliculus (Chapter 10). In these cases, a representation of the appropriate half of the body or of the visual field has been correspondingly demonstrated in the nucleus (Allman *et al.*, 1972; Gattass *et al.*, 1978a; Mason, 1978; Strick, 1976b). There are some hints in the literature of topographic ordering in other thalamic inputs, e.g., in the globus pallidus input to the anterior ventral lateral nucleus and in the mamillothalamic tract input to the anterior nuclei (De Vito and Anderson, 1982; Cowan and Powell, 1954).

The nonspecific input systems seem to be organized without this degree of topographic finesse (Chapter 4). Both the ascending noradrenergic system arising principally in the locus coeruleus (Lindvall and Björklund, 1974; Pickel *et al.*, 1974; Swanson and Hartman, 1975; Jones and Moore, 1977), and the ascending serotoninergic system arising principally in the dorsal nucleus of the midbrain raphe (Dahlstrom and Fuxe, 1964; Conrad *et al.*, 1974; Bobillier *et al.*, 1976) show localized concentrations of presumed terminal ramifications in certain thalamic nuclei, but unlike in the specific systems, the terminations are by no means confined to particular nuclei. For example, noradrenergic fibers show a very dense concentration in the anteroventral nucleus but can be demonstrated in many other nuclei as well (Swanson and Hartman, 1975). In these nuclei the noradrenergic fibers would converge on the territory of distribution of some of the classical pathways.

Another widely studied amine, dopamine, does not appear to form a similarly diffuse, nonspecific input to the thalamus, since the pathways containing it largely bypass the thalamus en route to the basal ganglia and cortex (Ungerstedt, 1971; Lindvall *et al.*, 1974). The question of an ascending diffuse cholinergic system from the midbrain is reviewed in Chapters 4 and 15.

3.6.2. Topographic Organization of Specific Inputs and Outputs

The detailed representation of the related receptor surface found in the ventral posterior and medial and lateral geniculate nuclei (Mountcastle and Henneman, 1949; Rose and Mountcastle, 1952; Bishop *et al.*, 1962; Aitkin and Webster, 1972) is dependent in the first instance on the systematic ordering of incoming afferent fibers.

In all three relay nuclei, afferent fibers belonging to what I have called the specific input systems enter in an orderly manner and terminate on comparably ordered groupings of neurons. A striking example of this ordering is found in the ventral nucleus of the medial geniculate complex (Morest, 1964, 1965a; Aitkin and Webster, 1972; see Chapter 8). In the cat and probably in other animals as well, bundles of afferent fibers ascending from the inferior colliculus enter the ventral nucleus of the medial geniculate complex from its ventromedial aspect. From there the fibers run sequentially along the dendritic fields of a series of thalamocortical relay neurons (Fig. 3.15). The neurons are aligned so as to form a number of more or less sagittally oriented sheets extending anteroposteriorly through the length of the nucleus. Along these sheets, and following their contours, run the incoming fibers, many of which are distributed to each sheet.

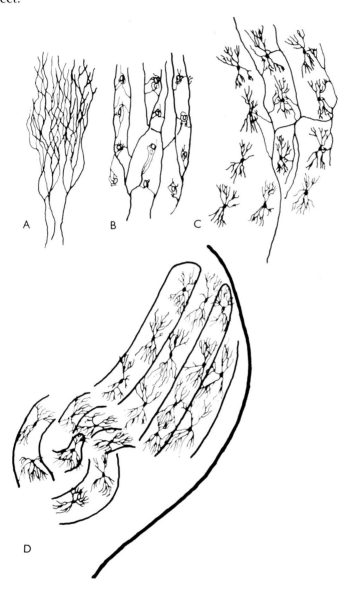

FIGURE 3.15.   Drawings of Golgi preparations from the ventral medial geniculate nucleus of the cat. (A) Bundling of afferent fibers. (B) Overlap of terminal clusters of individual fibers. (C) Distribution of a single fiber to more than one lamellar arrangement of thalamocortical relay cells. (D) Parallel, lamellar configuration of dendritic fields of thalamocortical relay cells, with ventromedial ends of the lamellae coiled in region labeled MGP (OV) in Fig. 3.16. From Morest (1964).

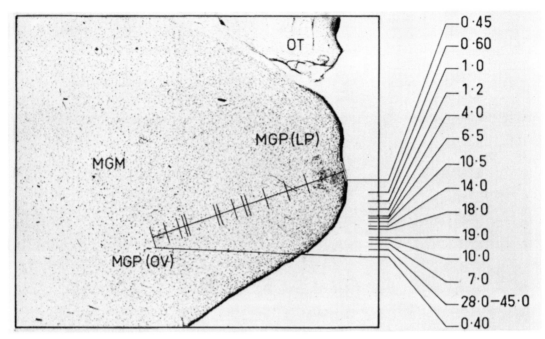

FIGURE 3.16. Systematic, low to high progression of best frequencies recorded from single units in ventral medial geniculate nucleus of a cat, in an electrode penetration oriented across the rows of cells and fibers illustrated in Fig. 3.15. From Aitkin and Webster (1972).

The sheets are not totally independent entities since dendrites of cells in one cross into the adjacent sheets and afferent fibers can branch to run along more than one sheet. However, the systematic relationship between the incoming fibers and chains of thalamocortical relay cells forms the basis of the tonotopic organization that exists in the medial geniculate nucleus. A recording microelectrode driven through the nucleus from dorsal to ventral, thus tending to run along the planes of the fiber and cell sheets, encounters mainly neurons activated (at sound pressure levels close to threshold) by the same or closely similar tones. By contrast, a microelectrode driven from lateral to medial, and thus across the sheets (Fig. 3.16), encounters neurons that are activated by progressively higher pitches of sound (Aitkin and Webster, 1972). From these studies it is clear that a shift in the mediolateral dimension of the medial geniculate nucleus represents a shift along the basilar membrane. The dorsoventral dimension seemingly represents a position along the length of the basilar membrane. There are no data yet to suggest a shift in topographic or other properties with a move in the anteroposterior dimension, i.e., in the plane of the cell and fiber sheets.

FIGURE 3.17. (A) Schematic drawing of horizontal (left) and frontal (right) sections of monkey ventral posterior nucleus, showing division of complex into a large cutaneous and a smaller deep component (from Jones and Friedman, 1982). In cutaneous portion, body parts are represented as a series of lamellae, as determined from microelectrode recordings. (B) Horizontal section showing labeling of terminal ramifications of lemniscal fibers arising from a small group of cells in the dorsal column nuclei, and extending as a rod that follows the contours of a lamella. Horseradish peroxidase, no counterstain. Bar = 500 μm. From Jones et al. (1982). (C) Darkfield photomicrograph from a frontal section showing anterograde labeling of terminal aggregations of lemniscal fibers; each cluster represents a rod, like that in (B), cut in cross-section. Autoradiograph. Bar = 500 μm. From Tracey et al. (1980).

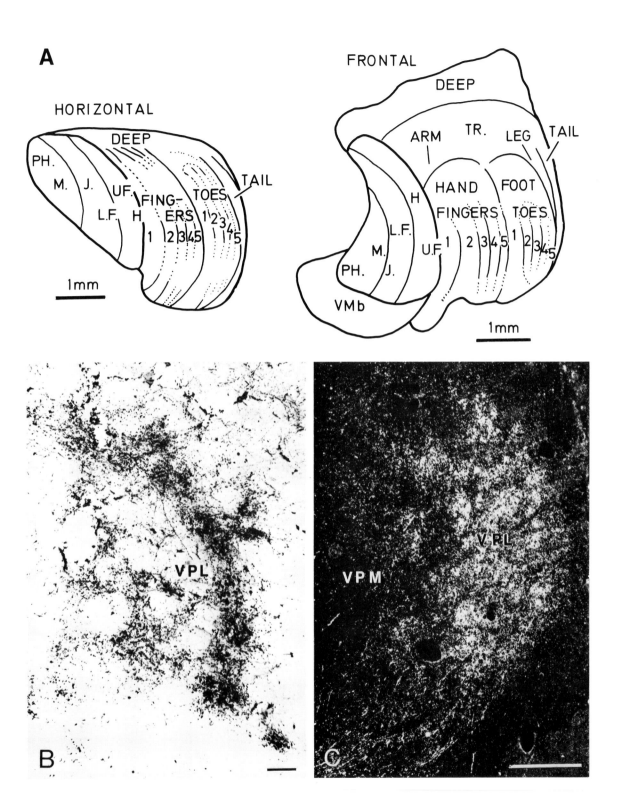

A

HORIZONTAL

DEEP

PH.
M. J. UF. FING- TOES
L.F. H ERS 1 2 3
1 2 3 4 5 1 4 5

TAIL

1mm

FRONTAL

DEEP

ARM TR. LEG TAIL

H HAND FOOT
FINGERS TOES
L.F. U.F. 1 2 3 4 5 1 2 3 4 5
M. J.
PH.

VMb

1mm

VPL

B

VPM VPL

C

The topographic organization of the ventral posterior nucleus appears to be based upon principles similar to those observed in the medial geniculate nucleus. As determined by evoked potential and single-unit studies (Mountcastle and Henneman, 1949, 1952; Poggio and Mountcastle, 1963; Welker, 1974), the body surface is represented as a series of approximately sagittal lamellae passing through the dorsoventral extent of the nucleus, each lamella representing a body part (Fig. 3.17). Those representing the most caudal body parts lie anterolaterally and those representing progressively more rostral body parts and the trigeminal region lie successively more posteromedially (see Chapter 7 for more details). In general, the lamellae extend through the anteroposterior dimensions of the nucleus, thus resembling those of the medial geniculate nucleus. Like those of the medial geniculate nucleus, the lamellae are to some extent conceptual conveniences and are not clearly defined in Nissl-stained properties. However (Ramón y Cajal, 1911; Hand and Van Winkle, 1977; Jones, 1983b), fibers of the medial lemniscus enter the nucleus in serial order from medial to lateral and

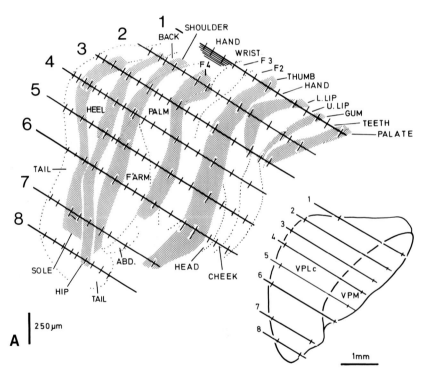

A

FIGURE 3.18. (A) Results of single- and multiunit recording in electrode penetrations made across the monkey ventral posterior nucleus from anterolateral to posteromedial. Clear or stippled strips represent regions at which multiunit activity was recorded in response to cutaneous stimulation of indicated body part. Bars represent positions at which single units were isolated. Note elongated nature of regions of neurons with common receptive field positions. Hatching indicates responses to stimulation of deep tissues. From Jones and Friedman (1982). (B) Receptive fields of selected sequences of units recorded from two electrode penetrations made horizontally from behind into monkey VPL (1) and VPM (2) nuclei. In the two sequences illustrated, units had very similar peripheral receptive fields and responded to the same type of stimulus. Units for which receptive fields are not illustrated were related to other body parts. From Jones *et al.* (1982c). (C) Results of electrode penetrations made vertically from above through monkey VPL nucleus in same parasagittal plane. Same convention as in (A). Elongated anteroposterior arrangement of units with common receptive fields is, again, illustrated. From Jones and Friedman (1982).

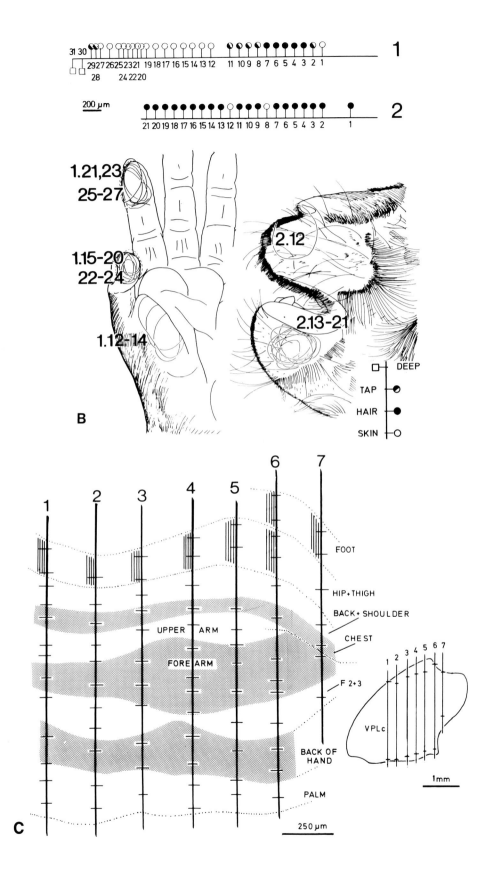

1

2

200 μm

B

1.21,23
25-27

2.12

1.15-20
22-24

2.13-21

1.12-14

DEEP

TAP

HAIR

SKIN

C

1   2   3   4   5   6   7

FOOT

HIP+THIGH

BACK+SHOULDER

UPPER   ARM

CHEST

FORE ARM

F 2+3

BACK OF
HAND

1 2 3 4 5 6 7

PALM

VPLc

250 μm

1mm

run along the lamellae, thus preserving somatotopy and ensuring the representation of a single body part in each functional lamella. Individual lemniscal axons do not terminate throughout the dorsoventral extent of the lamellae of the body representation but at selected levels within a lamella, usually over a relatively long anteroposterior distance. This has its functional correlate in the representation of receptors in deep tissues dorsally and of cutaneous receptors ventrally. Moreover, a microelectrode introduced into the nucleus horizontally from behind encounters units with identical place and modality properties over long anteroposterior distances, whereas one introduced dorsoventrally or horizontally from the side encounters rapid changes (Fig. 3.18) in the receptive field positions and modality properties of the units (Jones *et al.*, 1982c). The terminations of the lemniscal axons encompass narrow, elongated groupings of thalamocortical relay cells whose axons project as a bundle upon the cortex (Jones *et al.*, 1979, 1982c). That is, the bundle projects to a cortical "column." This has important functional considerations that will be taken up in Section 3.6.3 and Chapters 4, 7, and 8.

The lamellar type of organization appears to be basic, but within this there is clearly another smaller unit of organization formed by small, usually elongated groups of cells receiving afferent input of the same type, or from the same part of the peripheral receptive surface, and projecting as a group upon the sensory cortex. It is possible that not all types of afferent input to a sensory nucleus necessarily conform to this pattern. The distribution of individual spinothalamic fibers to the ventral nuclei, for example, has been described as far more diffuse than that of the lemniscal fibers (Scheibel and Scheibel, 1966b).

The dorsal lateral geniculate nucleus also shows a well-defined topological pattern. In many mammalian species, the retinotopic representation is repeated several times in a series of cellular laminae (see Chapter 9). These cellular laminae are not the same as the functional lamellae described in the medial geniculate and ventral posterior nucleus, for each lamina in the lateral geniculate nucleus contains a total representation of the appropriate half visual field. The equivalents of the lamellae would be sequential, concentric crescents or arcs of a lamina, centered on the foveal representation and representing degrees of visual field eccentricity (Fig. 3.19).

The elongated clusters of cells and associated afferent fibers within the topological lamellae of the ventral posterior nucleus have as their counterparts in the lateral geniculate nucleus, the so-called projection columns (Fig. 3.20; Sanderson, 1971). A projection column is a line of cells that is oriented at right angles to all of the sheetlike laminae of the lateral geniculate nucleus; it receives afferent input from ganglion cell axons representing a single point in the visual field and projects to a localized patch of visual cortex. The representation of a single retinal point in one lamina is aligned with other representations of that point in other laminae and, where appropriate, with those of the homonymous point (in the contralateral retina) in intervening laminae. Therefore, a projection column representing a point in the binocular visual field extends as a line across all laminae and projects as a kind of unit to the visual cortex (Fig. 3.20). A projection column is, thus, probably comparable to the elongated cell groupings decribed in the ventral posterior nucleus, and the part of a projection column found in an individual lamina is laid down by the mode of distribution of retinal

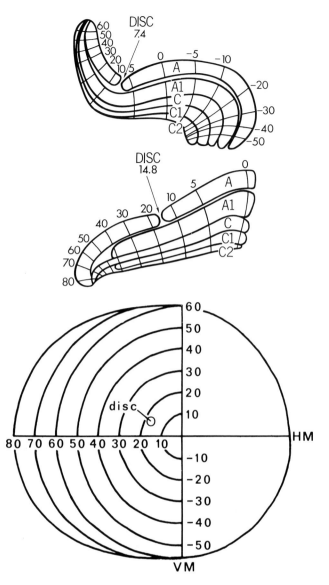

FIGURE 3.19. (Upper and middle) Schematic sagittal (upper) and frontal (middle) sections of the laminar dorsal lateral geniculate nucleus of a cat. Upper figure shows the projection columns representing degrees of visual field eccentricity above and below the horizontal meridian (0). Middle figure shows projection columns representing points of increasing eccentricity from vertical meridian (0). "Disc" indicates zones of absent cells "representing" relative position of the optic disc. From Kaas *et al.* (1972). Lower figure indicates mode of projection of visual half field onto the individual laminae. Each projection column seems to form a part of a lamellar distribution of afferent fibers and the cells upon which they terminate; these in turn represent an arclike portion of the visual field. Note larger representation of parts of field closest to fixation point. Laminae A1 and C1, in receiving fibers only from ipsilateral eye, represent a smaller part of the hemifield and are thus shorter than laminae A, C, and C2 which receive fibers from the contralateral eye. HM: horizontal meridian; VM: vertical meridian.

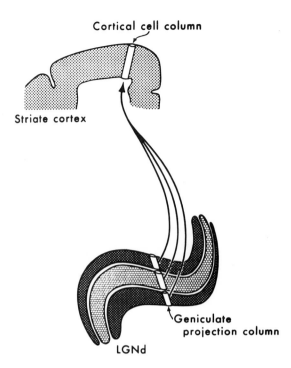

Cortical cell column

Striate cortex

Geniculate
projection column

LGNd

FIGURE 3.20. Schematic representation of how a projection column spanning all layers of the lateral geniculate nucleus of the cat projects to a column in the visual cortex. From Sanderson (1971).

fibers entering that lamina. These fibers give rise to terminal arborizations with their long axes aligned perpendicular to the plane of the lamina (see Chapter 4) and forming synapses on the contained cells of the underlying part of a projection column. There is probably considerable overlap between the terminal arborizations of adjacent optic tract axons but the arrangement forms the basis for the formation of the projection columns, which are, in turn, units of topographic projection upon the cortex.

Comparable topographic ordering of the inputs to other principal thalamic nuclei has also been demonstrated. There is good evidence in the posterior ventral lateral nucleus for a lamellar organization with subsidiary elongated focal clusters comparable to that in the ventral posterior nucleus but based upon cerebellar inputs (Fig. 3.21) (Asanuma *et al.*, 1983c). Pallidal inputs to the anterior ventral lateral nucleus are similarly disposed (De Vito and Anderson, 1982). Similarly, in the lateral posterior nucleus of the cat, afferent fibers from the superior colliculus are distributed in a series of parallel, slablike formations (Graham, 1977) which show increased density of acetylcholinesterase activity (Graybiel and Berson, 1980b). It seems not unreasonable to think that comparable topographic ordering of input–output connections will become evident in other principal thalamic nuclei as well.

FIGURE 3.21. Dysjunctive groupings of fiber terminations (B) in the monkey VLp nucleus (C) labeled autoradiographically following a single injection of tritiated amino acids in the contralateral dentate nucleus (A). From Thach and Jones (1979).

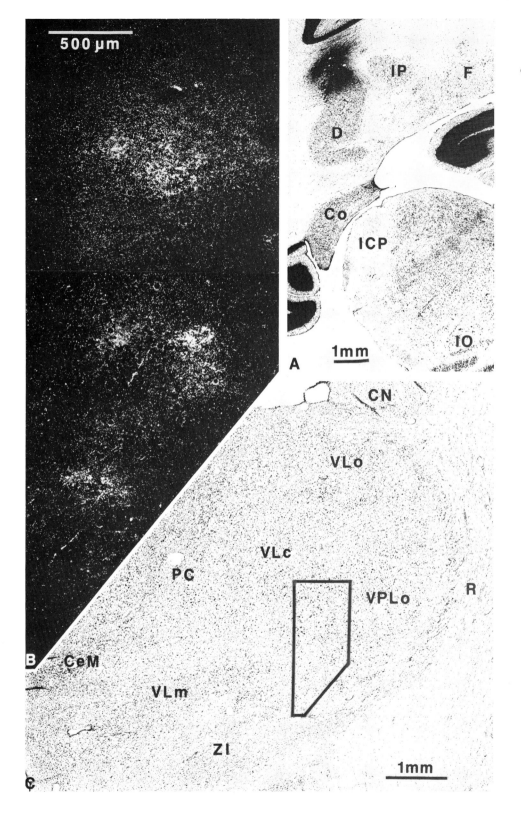

3.6.3. The Rod-to-Column Principle in Thalamocortical Connectivity

The principle of topographic ordering in the specific thalamocortical projection can be readily comprehended. Neighboring portions of a thalamic relay nucleus project upon neighboring regions of their cortical target area, so that the systematic mapping of a peripheral receptive surface seen in the thalamus is projected onto the cortex. The expanded representations of central regions of the retinae in the lateral geniculate nucleus, of the hands, feet, and lips in the ventral posterior nucleus, and of the basal turn of the cochlea in the ventral medial geniculate nucleus, thus, gain comparably enlarged representations in the relevant sensory area(s). In the visual system parallel arcs or crescents representing increasing degrees of visual field eccentricity project upon comparable arcs or crescents of striate cortex centered on the representation of the fovea (Figs. 3.22 and 3.23). In the somatic sensory system parallel lamellae representing sequential body parts project upon parallel bands spanning the postcentral gyrus anteroposteriorly.* In the auditory system, parallel lamellae of the ventral medial geniculate nucleus representing successively higher frequencies project upon parallel isofrequency bands spanning the first auditory cortex. Comparable patterns can be seen in the projection of the posterior ventral lateral nucleus on the motor cortex, in the projection of the inferior pulvinar nucleus on the extrastriate visual cortex, and so on.

The arc-to-arc or lamella-to-band pattern of thalamocortical organization is readily demonstrated by the injection of retrograde tracer to affect a moderate-sized patch of sensory cortex (Fig. 3.24A). Because of the precise reciprocity of corticothalamic connections (Section 3.6.4 and Chapter 4), an injection of anterograde tracer demonstrates essentially the same thing. The labeling in each case extends as an arc through most of the anteroposterior dimension of the dorsal lateral geniculate nucleus or as a lamella through most of the anteroposterior dimension of the ventral posterior or ventral medial geniculate nucleus (Fig. 3.24B).

The arc-to-arc or lamella-to-band pattern really presents only the crudest picture of the order in the thalamocortical projection. Within it, how are subsets of thalamic neurons representing small portions of the receptive periphery and particularly those representing different modalities or similar functional qualities projected on to the cortex? The answer to this question seems to lie in what I have christened "the rod-to-column" principle of thalamocortical organization (Jones *et al.*, 1982c; see also Chapters 7, 8, and 9). The sensory areas of the cortex and probably other areas as well are organized in a series of functional columns (Mountcastle and Edelman, 1977; Jones, 1983e). A microelectrode driven vertically through the thickness of the cortex succesively encounters neurons in each layer that have some commonality, e.g., in receptive field position, modality specificity, eye preference, frequency selectivity, a particular kind of binaural interaction, and so on. Such columns are usually described as about 0.5–1 mm in diameter. If now, an injection of tracer sufficiently small to involve one

---

* Banding is less obvious in the projection on the somatic sensory cortex of lissencephalic animals but a comparable mediolateral sequence is present.

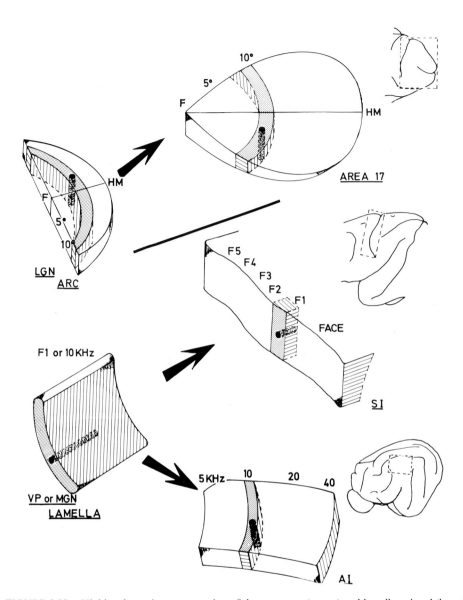

FIGURE 3.22.   Highly schematic representation of the arc-to-arc (upper) and lamella-to-band (lower) pattern of projection of the lateral geniculate (LGN), ventral posterior (VP), and medial geniculate (MGN) nuclei upon the monkey visual cortex (area 17), first somatic sensory area (SI), or cat first auditory area (AI). An arc representing a particular degree of retinal eccentricity in a lamina of the LGN projects to a comparable arclike ocular dominance stripe in area 17. The equivalent arrangement in the other two nuclei is the projection of a lamella representing a particular body part (e.g., a finger, F1) or a narrow range of frequencies (e.g., 10 kHz), to a band in the SI or AI cortex. Within each arc or lamella, a rod of cells, either spanning its thickness (in the LGN, cf. Fig. 3.20) or extending through some part of its length (in the VP and MGN, cf. Figs. 3.17 and 3.18), projects to a column in the cortex and mediates the finer projection of place and modality upon the cortex.

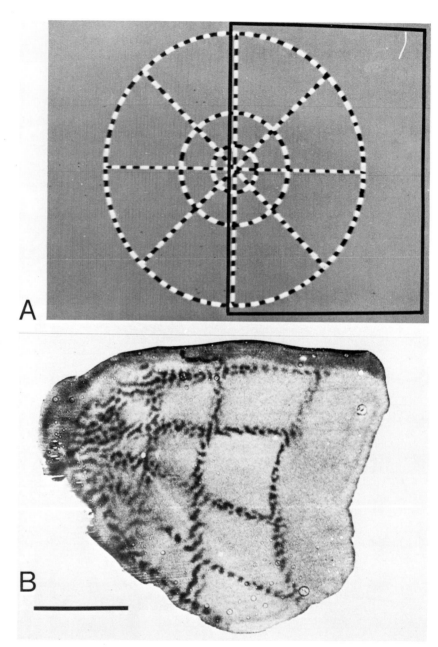

FIGURE 3.23. (A) A computer-generated visual stimulus consisting of eight rays and three concentric rings equally spaced on a logarithmic scale and composed of small blocks that flickered between black and white at 3 Hz. The part enclosed in the rectangle, when presented to one eye of an anesthetized monkey for 30 min, stimulated the region of visual cortex shown in (B) in a tangential histological section passing mainly through layer IV. The striped pattern of cortical activation was revealed by $[^{14}C]$-2-deoxy-D-glucose uptake and autoradiography. The resulting map is a faithful facsimile of the stimulus with distortions being due to the sheetlike nature of the cortex and the magnification of the representation of central vision. Blackened segments of the map correspond to portions of ocular dominance stripes (see Chapter 9) responding to the stimulated eye. Intervening clear segments correspond to portions of ocular dominance stripes related to the unstimulated eye. From Tootell *et al.* (1983).

of these columns is made in the cortex, the retrograde (or anterograde) labeling of cells in the thalamus takes the form, not of an arc or a lamella, but of a narrow rod. In the dorsal lateral geniculate nucleus it is a rod orthogonal to the plane of the laminae and following the lines of the projection columns (Fig. 3.20; Section 3.6.2 and Chapter 9). In the ventral posterior and ventral medial geniculate nuclei it is a rod running anteroposteriorly through most of the nucleus (Fig. 3.25; Chapters 7 and 8). The labeling of multiple thalamic cells after injections more or less confined to a single cortical column implies that the input to a column is made up of a bundle of thalamocortical fibers with similar functional properties, rather than by individual fibers (Jones, 1983b).

In the dorsal lateral geniculate nucleus the thalamocortical rod spans the thickness of the larger arc of visual field eccentricity mentioned above. It clearly specifies a locus in the visual field and in animals with significant binocularity, the alignment of rods in adjacent laminae representing different eyes ensures the projection to a patch of cortex of homonymous points in the visual fields of

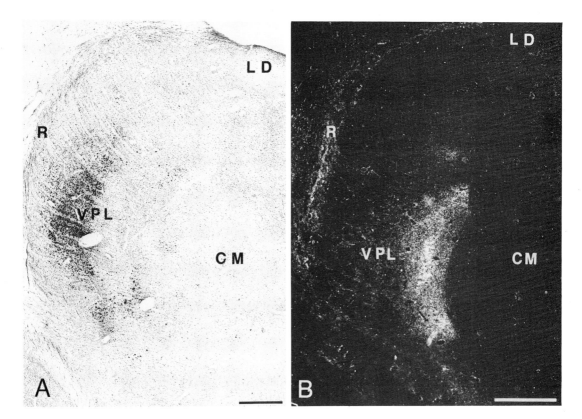

FIGURE 3.24.   Lamellar configurations of thalamocortical relay cells retrogradely labeled with horseradish peroxidase (A) and of corticothalamic fiber terminations anterogradely labeled with tritiated amino acids (B), in the ventral posterior nuclei of monkeys following large injections of tracer in the hand representation of the somatic sensory cortex. Bars = 1 mm.

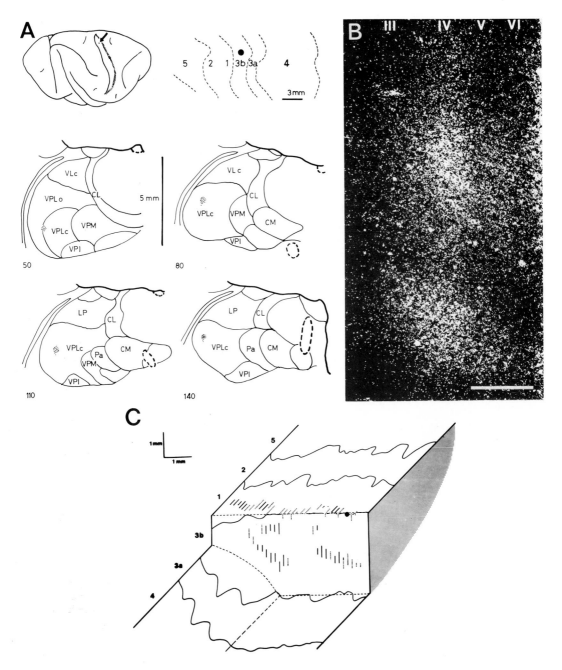

FIGURE 3.25. (A) Punctate injection of retrograde tracer in one field (area 3b) of monkey SI cortex leads to retrograde labeling of a rod of cells in the ventral posterior nucleus of the thalamus. From Jones (1981a). (B) Focal injections of tritiated amino acids in the ventral posterior nucleus lead to anterograde labeling of focal zones of termination in the SI cortex. Bar = 250 μm. From Friedman and Jones (1980). (C) When reconstructed on an unfolded surface map, the foci are seen to form parts of short strips. From Jones *et al.* (1982c).

the two eyes (Chapter 9).* In the ventral posterior and ventral medial geniculate nuclei, the thalamocortical rod runs along a portion of the anteroposterior extent of a lamella of body part representation or of isofrequency. Here it also specifies a locus on the body surface or along the basilar membrane. The rodlike arrangement in these nuclei, however, also seems to provide for the segregation of functional subsets of thalamocortical relay cells.

The evidence is most direct in the ventral posterior nucleus. If a microelectrode is driven through the nucleus horizontally from behind, it successively encounters neurons that respond only to the same type of stimulus and that have receptive fields on the same part of the body (Jones *et al.*, 1982c; Fig. 3.18). Such sequences of neurons of the same type can be demonstrated over long anteroposterior distances, amounting to at least half the length of the ventral posterior nucleus. Dorsoventral and lateromedial electrode penetrations, on the other hand, encounter neurons whose modality selectivity and receptive field positions change rapidly over very short distances, often as little as 50–100 μm. The place and modality specificity of the anteroposterior rods of cells demonstrable electrophysiologically clearly match those of columns in the somatic sensory cortex and, therefore, the conclusion seems justified that the rods of thalamocortical relay cells demonstrable anatomically by injection of tracers in the cortex are the same as those demonstrable by microelectrode recording in the thalamus. That is, place- and modality-specific thalamic rods project to similar place- and modality-specific cortical columns. A lamella in the ventral posterior nucleus, therefore, represents a large body part, such as a finger, but it contains several rods that represent different points and different constellations of receptors on and in that finger. Further details of the organization are discussed in Chapter 7.

The evidence for functional specificity of the thalamocortical rods in the ventral medial geniculate nucleus is indirect. It rests upon the projections of separate rods to cortical columns containing neurons showing different kinds of interactions between inputs from the two ears (Middlebrooks and Zook, 1983). At least in the high-frequency representation, cortical columns of cells receiving excitatory inputs from each ear (EE cells) are connected with rods situated in certain parts of the isofreqency lamellae in the ventral medial geniculate nucleus. Columns containing cells receiving excitatory input from one ear and inhibition from the other (EI cells) are connected with rods situated in different parts of the isofrequency lamellae. Because the EE and EI types of binaural properties appear to be generated in lower brain-stem auditory centers (see Chapter 8), this would imply that there are separate rods of EE and EI cells in the ventral medial geniculate nucleus. No work has been done, however, to demonstrate this (see Chapter 8).

The principal afferent fiber systems entering the thalamic relay nuclei not only follow the arciform or lamellar layout of the representation but also respect the arrangement of thalamocortical rods. Optic tract fibers terminate in narrow

---

* In animals with obvious ocular dominance columns (Chapter 9) in their visual cortex, the projection of parts of a rod in laminae representing each eye are actually to adjacent cortical columns, rather than to the same column. It is more correct, therefore, to say that a rod spanning all layers of the dorsal lateral geniculate nucleus projects to a "hypercolumn" in the visual cortex (Chapter 9).

rodlike zones across the thickness of one or more lateral geniculate laminae (Mason and Robson, 1979; Bowling and Michael, 1980, 1984; Sur and Sherman, 1982b; Michael and Bowling, 1982; Chapter 9) (Fig. 3.26). Medial lemniscal fibers arising from small clusters of place- and modality-specific neurons in the dorsal column nuclei end in narrow anteroposterior rods in the ventral posterior nucleus (Jones *et al.*, 1982; Fig. 3.17). Fibers arising from small groups of neurons in the central nucleus of the inferior colliculus end in similar elongated rods in the ventral medial geniculate nucleus (Andersen *et al.*, 1980b; Fig. 3.27). Comparable rodlike terminations of cerebellothalamic fibers are seen in the posterior ventral lateral nucleus (Asanuma *et al.*, 1983c; Fig. 3.21; Chapter 7).

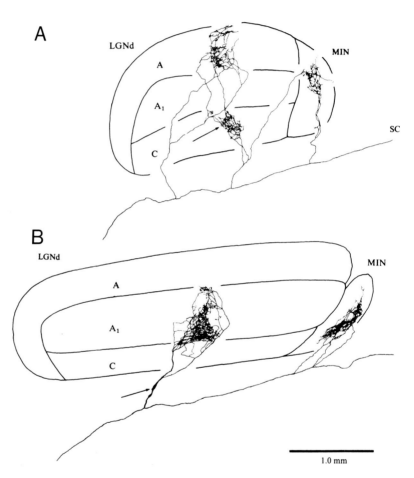

FIGURE 3.26. Camera lucida reconstructions of single optic tract axons, physiologically characterized and injected with horseradish peroxidase, ending in dorsal lateral geniculate nuclei of cats. (A) On-center, Y-type axon from contralateral eye ending across thickness of laminae A and C and with additional branches to medial interlaminar nucleus (MIN) and superior colliculus (SC). (B) Off-center, Y-type axon from ipsilateral eye ending across thickness of A1 lamina and in MIN. (C) Terminals of axon in (B), showing endings spanning thickness of lamina along a projection column. From Bowling and Michael (1980).

The questions, naturally, come up as to whether an individual afferent fiber branches to innervate more than one thalamic rod and whether an individual fiber ends along the full length of a rod, i.e., upon every relay cell in the rod. There is no evidence in the dorsal lateral geniculate nucleus that individual optic tract axons end in more than one rod in the same lamina, though one class in the cat can branch to innervate rods that form part of the same projection column in two laminae related to the same eye (Bowling and Michael, 1980, 1984; Sur and Sherman, 1982b; Chapter 9). In the ventral posterior nucleus of monkeys and cats a proportion of the medial lemniscal axons end in two rods, though the terminations in one rod are invariably denser than in the other (Rainey and Jones, 1983; Jones, 1983b). In the ventral medial geniculate nucleus, the details of the terminations of individual afferent axons are not known. Ramón y Cajal (1911) and Morest (1964) seem to have stained only the preterminal portions of the axons and not the elongated end formations that would be anticipated from the work of Andersen *et al.* (1980b).

Optic tract fibers of the functional class referred to as X fibers (see Chapters 4 and 9) end in smaller foci of terminations than Y fibers in the A laminae of

200 µm

FIGURE 3.26. (*continued*)

the dorsal lateral geniculate of the cat (Sur and Sherman, 1982b) and there is some physiological evidence that X and Y fibers may have their major terminations at different depths in these laminae (Mitzdorf and Singer, 1977). In the similarly constituted dorsal lateral geniculate nucleus of the mink, X or Y axons coming from on or off types of retinal ganglion cells (Chapters 4 and 9) end at different levels (LeVay and McConnell, 1982). This would imply that relay cells at different levels along a thalamocortical rod are differentially innervated and that within the bundle of axons projecting from the rod to its cortical column are axons of different functional classes. Both implications are confirmed by single-unit studies (Chapter 4). Hence, within the place-specific entities of thalamic rods and cortical columns are contained additional functional units.

In the ventral posterior nucleus, individual medial lemniscal fibers clearly

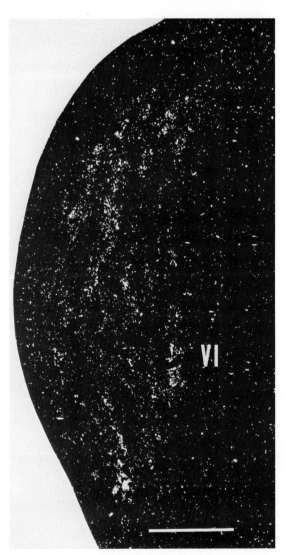

FIGURE 3.27. Anterograde labeling of clusters of afferent fiber terminations in the ventral medial geniculate nucleus following injection of tritiated amino acids in the central nucleus of the inferior colliculus. Note the pattern of lamellae oriented parallel to the surface of the nucleus (to left), with subsidiary clusters which are cross-sections of longer, anteroposterior-oriented, rodlike terminations. Bar = 250 μm. From Andersen *et al.* (1980a).

do not end along the full length of a thalamocortical rod (Fig. 3.28). We know too little about the properties of individual lemniscal axons and ventral posterior cells to be able to say that this implies subtle physiological differences between axons and cells of the same broad place and modality class that innervate the same cortical column. Nor are there any relevant data in the ventral medial geniculate nucleus.

A final question is whether an individual thalamic rod may innervate more than one cortical column and, conversely, whether the same cortical column may receive inputs from more than one thalamic rod. There is no evidence from retrograde tracing, for the projection of single rods to dissociated points in the visual or somatic sensory cortex, nor for the convergence of widely dissociated lateral geniculate or ventral posterior rods on the same patch of cortex. Obviously, with retrograde tracing methods, it is difficult to analyze columns or rods that lie side by side. However, as injections of anterograde tracer grow in size in the thalamus, multiple columns are labeled in the cortex (Jones and Friedman, 1982) and as injections of retrograde tracer grow in size in the cortex, multiple rods are labeled in the thalamus (Jones *et al.*, 1979). In the auditory system, Middlebrooks and Zook (1983) have suggested that at least three thalamic rods may converge upon the bands of auditory cortex that contain columns of EI cells. Columns of EE cells, however, receive inputs from single thalamic rods. This suggests some diversity in the rod-to-column pattern of organization.

3.6.4. Corticothalamic Inputs

No description of thalamic input connectivity would be complete without mention of corticothalamic connections which return faithfully from every cortical area to the dorsal thalamic nucleus or nuclei providing input to that area (Jones, 1981a; Chapter 4). The corticothalamic projection to nuclei providing the topographically ordered, "specific" input to a cortical area is restricted to that nucleus and is itself precisely ordered topographically (Fig. 3.24B), faithfully following the lamellar and rodlike or projection column patterns described in Sections 3.6.2 and 3.6.3. Cortical areas receiving inputs from a diffusely projecting thalamic nucleus will all project back to that nucleus. The principle of thalamocortical and corticothalamic reciprocity (Diamond *et al.*, 1969; Chapter 4) can be extended to include the connections of the allocortex, for the hippocampal formation and parahippocampal areas all return fibers to the dorsal thalamic nuclei from which they receive inputs (e.g., Herkenham, 1978; Price and Slotnik, 1983). The detailed anatomy and possible functions of corticothalamic fibers are considered in Chapter 4.

3.6.4.1. Laterality in Thalamic Organization

The thalamus of one side is primarily related to the contralateral side of the body or of extrapersonal space and accordingly has connections with appropriate parts of the rest of the nervous system. Many major input pathways to the thalamus are unilateral, either crossed or uncrossed, e.g., the medial lemniscus, dentatothalamic, and interpositothalamic paths; pallidothalamic; in-

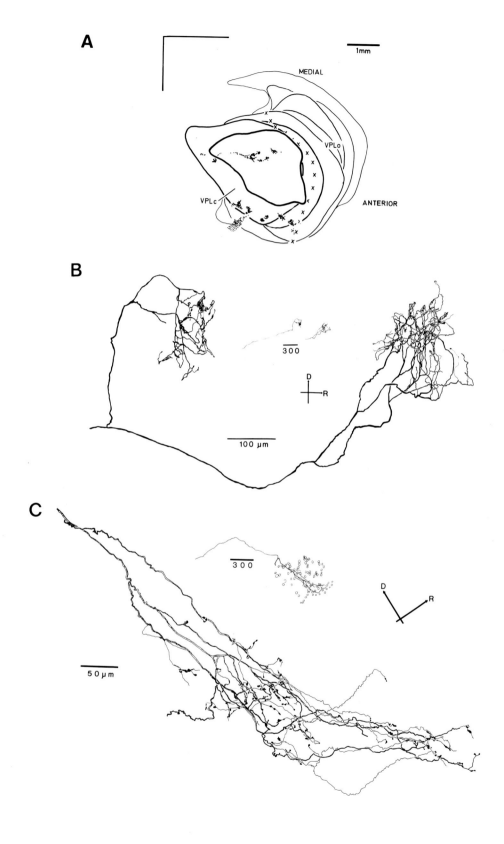

puts from the superior colliculus. Many bilateral input pathways also exist, however. Among the bilateral paths are: the retinal inputs to both lateral geniculate nuclei (Chapter 9); the spinothalamic input to both ventral posterior lateral nuclei; the trigeminal input to medial parts of the ventral posterior medial nuclei in some species (Chapter 7); the input of the central nucleus of the inferior colliculus to both ventral medial geniculate nuclei (Chapter 8); the fastigial nucleus input to both ventral lateral posterior nuclei (Chapter 7); brain-stem inputs to certain parts of the ventral medial and intralaminar nuclei (Chapters 7 and 12); mamillothalamic tract inputs to some of the anterior nuclei (Chapter 14).

Although many of its inputs are bilateral, the output of the thalamus is, to all intents and purposes, strictly unilateral, i.e., the thalamus of one side projects only to cortex and striatum of the same side. Where retrograde labeling of cells has been observed in the contralateral thalamus after an injection of tracer in the cortex of one side, the contralateral cells have invariably been few, close to the midline in nuclei such as the anteromedial, mediodorsal, central medial, medioventral, or ventral medial, and continuous with a zone of many labeled cells in the ipsilateral nucleus (Jones and Leavitt, 1974; De Vito, 1969; Künzle, 1976, 1978; Wyss *et al.*, 1979; Amaral and Cowan, 1980; Berman and Payne, 1982). Though not completely ruling out a bilateral projection from these midline regions of the thalamus, it seems to me that the most parsimonious explanation of such results is simply an irregularity in the position of the midline.

There is much stronger evidence for bilateral corticothalamic connections. Some of the reports of these, again relate to projections that simply seem to "overflow" slightly from one nucleus at the midline into its neighbor especially in the intralaminar system (De Vito, 1969). The fibers contributing to these enter the ipsilateral thalamus. Other reports, however, are of far more substantial connections that pass from one cortex to symmetrical and widely separated parts of two nuclei in each thalamus. The fibers contributing to these may reach the contralateral thalamus by way of the corpus callosum. Among these reports of truly bilateral projections are clear indications of bilateral projections to central parts of the mediodorsal and submedial nuclei from medial and lateral frontal regions in the cat and rat (Rinvik, 1968b; Beckstead, 1979; Reep and Winans, 1982), to lateral parts of both mediodorsal nuclei from lateral frontal regions in the monkey (Künzle, 1978; Goldman, 1979), to lateral parts of the anteromedial nuclei from dorsal retrosplenial and presubicular cortex in the rat (Kaitz and Robertson, 1981), to the principal ventral medial nuclei from lateral areas of cortex in the rat (Molinari *et al.*, 1983). None of these bilateral corticothalamic projections appear to be matched by bilateral thalamocortical projections. This is particularly well demonstrated in

---

FIGURE 3.28. (A) Reconstruction made by stacking tracings of serial sagittal sections through a monkey's thalamus on top of one another and projecting autoradiographic labeling of medial lemniscal terminations (dots) onto the surface. The elongated configurations of labeled terminations result from a punctate injection of tritiated amino acids in a part of the gracile nucleus in which units possessed receptive fields on the sole of the contralateral foot. Crosses represent border between VLp (VPLo) and VPL (VPLc) nuclei. (B, C) Single medial lemniscal axons and their terminal ramifications in VPL anterogradely labeled by injection of horseradish peroxidase in the medial lemniscus of a cat. Inset in (C) shows terminal ramifications in relation to counterstained thalamic cells. D: dorsal; R: rostral. From Rainey and Jones (1983).

the work of Molinari *et al.* (1983). These workers used the same tracer (lectin-conjugated horseradish peroxidase) as both an anterograde and a retrograde marker and found spread of only the anterogradely labeled corticothalamic axons across the midline, not of retrogradely labeled cells (Fig. 3.29). In the same report, bilateral corticothalamic projections to the ventral posterior nucleus are convincingly ruled out and I know of no reliable report of bilateral corticothalamic connections with other principal nuclei such as the ventral lateral or medial and lateral geniculate nucleus.

3.6.5.   Inter- and Intrathalamic Connections?

3.6.5.1.   Lack of Interthalamic Connections

The only anatomically established interthalamic connections is that which joins the ventral lateral geniculate nuclei of the two sides to one another (Edwards

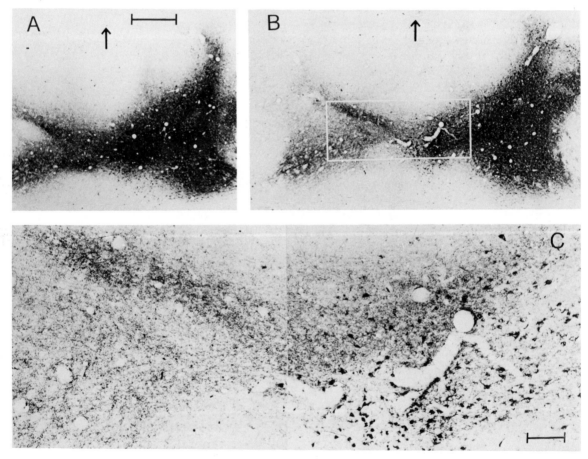

FIGURE 3.29.   (A, B) Anterograde and retrograde labeling in the thalamus of a rat after injection of lectin-conjugated horseradish peroxidase in the frontal cortex of one side (to the right of the figure). The arrows point to the third ventricle. (C) Higher-power view of the labeling at the thalamic midline [box in (B)]. Note extension of anterograde labeling across midline into contralateral thalamus (to left). Retrograde labeling stops at midline. Bars = 500 μm (A and B), 100 μm (C). From Molinari *et al.* (1985).

et al., 1974; Swanson *et al.*, 1974). Reports appear at intervals indicating, from electrophysiological studies, that the dorsal thalami of the two sides may be functionally interconnected. However, there are, to my mind, no reliable anatomical data to support the idea of a monosynaptic connection between the two sides. One assumes, therefore, that physiological responses recorded in one side after stimulating the other must be mediated by indirect routes. Potential pathways would include the cortex, either via the corpus callosum or more directly by bilaterally projecting corticothalamic fibers. Bilateral corticothalamic fibers (Section 3.6.4.1), however, are not ubiquitously distributed and do not appear to pass to many of the nuclei, such as the ventral posterior, in which interthalamic connections have been reported.

3.6.5.2. Intrathalamic Connections

a. Absence in Dorsal Thalamus. It is an old viewpoint (see Chapter 1) that nuclei of the dorsal thalamus projecting to "association cortex" receive their inputs from thalamic nuclei projecting to the primary sensory areas. There has been no evidence to support this point of view for many years. Occasional reports of retrogradely labeled cells in one nucleus subsequent to injection of tracer in another can usually be accounted for on the basis of interrruption of efferent axons traversing the nucleus injected. Probably just about every nucleus of the dorsal thalamus has been attacked anatomically in recent years and the weight of evidence from these studies is against connections joining together dorsal thalamic nuclei. A possible exception is the intralaminar nuclei. Anterograde labeling of efferent axons followng injection of tracer or lesioning of these nuclei at more caudal levels usually spreads forward in the internal medullary lamina for a considerable distance (Nauta and Whitlock, 1954; Jones, personal observations) before the fibers swing off anterolaterally toward the striatum. It is possible that the fibers are merely traversing the other intralaminar nuclei en rou te to their destination. If they have en passant terminations, however, this would represent a form of intrathalamic connection.

b. Reticular Nucleus and Perigeniculate Nucleus. The only well-documented intrathalamic connection is that joining the reticular nucleus of the ventral thalamus (Fig. 3.30) to the nuclei of the dorsal thalamus (Chapter 4). In this, I include the connection between the perigeniculate nucleus and dorsal lateral geniculate nucleus of the cat. The reticular nucleus appears to receive collaterals of all thalamocortical and corticothalamic fibers passing through it (Scheibel and Scheibel, 1966a; Jones, 1975c; Montero *et al.*, 1977; Yen and Jones, 1983). The same is probably true of thalamostriatal and possibly of pallidothalamic fibers (Jones, 1975c). The tendency for fibers passing from and to a particular dorsal thalamic nucleus to traverse the same part of the reticular nucleus leads to certain parts of the reticular nucleus being consistently related to a particular dorsal thalamic nucleus (Jones, 1975c; Montero *et al.*, 1977). There is a loose topography (Montero *et al.*, 1977), and some parts may respond to stimulation of a particular sensory pathway with the cells showing small, nonconvergent receptive fields (Sugitani, 1979; Sumitomo *et al.*, 1966; Hale *et al.*, 1982; Shosaku and Sumitomo, 1983; Pollin and Rokyta, 1982). But there is substantial overlap in the reticular nucleus of terminations of fibers passing to and from different dorsal thalamic nuclei (Jones, 1975c), and since the dendrites of reticular nucleus cells are extremely long (Ramón y Cajal, 1911; Scheibel and Scheibel, 1966a; Yen and Jones, 1983), it is unlikely that a given dorsal thalamic nucleus can be said to dominate totally any part of the reticular complex.

The perigeniculate nucleus of the cat, like the reticular nucleus, receives input from collaterals of geniculocortical and corticogeniculate axons (Ferster and LeVay, 1978; Ahlsén *et al.*, 1978, 1982b; Friedlander *et al.*, 1979; Updyke, 1977).

The major output of the reticular nucleus is back to the dorsal thalamus (Ramón y Cajal, 1911; Scheibel and Scheibel, 1966a; Jones, 1975c; Montero *et al.*, 1977). This input is GABAergic and presumably inhibitory (Chapter 4) (Houser *et al.*, 1980; Ohara *et al.*, 1983). It is organized so that nuclei whose thalamocortical and corticothalamic fibers traverse a particular part of the re-

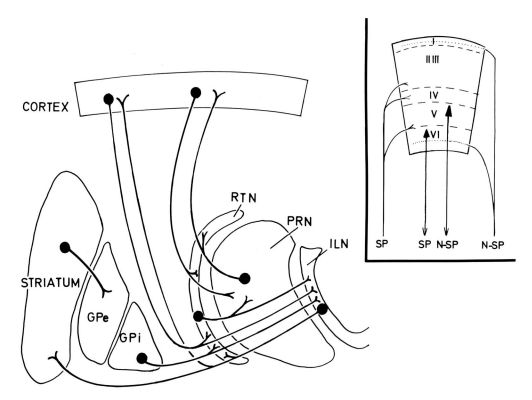

FIGURE 3.30.   Highly schematic outline of the connectivity of the dorsal thalamus and of the associated reticular nucleus (RTN) of the ventral thalamus. Afferent fibers from the cerebral cortex deposit collaterals in the reticular nucleus and continue on to the principal (PRN) or intralaminar (ILN) nuclei. Afferent fibers from the internal segment of the globus pallidus (GPi) also give off collaterals to the reticular nucleus and continue on to the intralaminar nuclei. Efferent fibers from the principal nuclei and from nuclei with nonspecific projections (e.g., the intralaminar nuclei) also give collaterals to the reticular nucleus. The intralaminar nuclei provide the only thalamic input to the striatum, again probably with collaterals to the reticular nucleus. Inset shows laminar pattern of termination of specific and nonspecific afferents in cerebral cortex. Intralaminar nuclei are one of several nuclear groups with nonspecific projections. Corticothalamic fibers to principal thalamic nuclei arise from cells with somata in layer VI. Those reciprocating a nonspecific projection arise from cells with somata in layer V.

ticular nucleus receive fibers also from that part of the reticular nucleus (Chapter 15). But there is, again, substantial overlap (Jones, 1975c). The perigeniculate nucleus projects back to the dorsal lateral geniculate nucleus (Ide, 1982).

In the absence of any evidence for any intrathalamic connections it is assumed that the short-latency responses of neurons in lateral thalamic nuclei to stimulation of rather ill-defined medial and midline thalamic regions (see Chapter 12) could be due to the loop passing through the reticular nucleus. The implications of this and the possible functions of the reticular and perigeniculate nuclei are taken up in Chapters 4, 9, and 15.

3.7. Convergence and Divergence in Thalamic Connectivity

The degree of convergence of afferent fibers upon neurons in a thalamic nucleus has occupied the attention of a number of workers over the years. Because of the topographic organization of the sensory relay nuclei, it can be assumed that the number of cells that potentially receive the terminals of individual afferent fibers is finite and limted. It would be surprising if extensive overlap were the rule and anatomical tracing experiments showing the distribution of single afferent fibers or of small groups of fibers argue against it (Figs. 3.26 and 3.27). Clearly, in attempting to assess the degree of convergence, several types of information are needed. Data are required on the number of cells in a particular thalamic nucleus and the number of afferent fibers terminating in it, so that a ratio of fibers to cells can be established. Differences in cell packing density throughout the nucleus need to be taken into account; the degree of branching and the extent of the terminations of individual afferent axons must be known. Even if all these features can be accurately quantified, any innervation ratios derived from them must still remain tentative in the absence of knowledge about the exact sites of termination of afferent fibers on the soma-dendritic surfaces of the thalamic neurons because a soma, though lying within the terminal ramifications of one fiber, may in fact receive the terminals of other fibers on its dendrites over a much wider area.

If one looks for innervation ratios between thalamic fibers and cortical cells, obviously the same kinds of data are needed. Yet there is really very little information on the subject as a whole and that which is available has often been criticized for potentially being inaccurate. Nevertheless, it would seem useful to have some idea of thalamic and cortical innervation ratios as a basis at least for stochastic evaluation of the transmission process.

3.7.1. Convergence in the Dorsal Lateral Geniculate Nucleus

Probably many more anatomical and physiological data of a quantitative kind are available for the retinogeniculocortical system than for any other neural pathway. For this reason, many workers have addressed the issue of the degree of convergence or divergence in the transmission of visual field information through the dorsal lateral geniculate nucleus. Le Gros Clark (1941a) counted the number of neurons in Nissl-stained sections of the rhesus monkey dorsal lateral geniculate nucleus and concluded that the four parvocellular layers (see Chapter 9) contained 1,590,000 nerve cells and the magnocellular layers 206,000. That is, the nucleus contained a total of 1,796,000 neurons. The number of fibers in an adult rhesus monkey optic nerve, as counted in fiber or myelin preparations, was found by Bruesch and Arey (1942) to be approximately 1.2 million. This has been confirmed electron microscopically by Rakic and Riley (1983). The number of these axons projecting to sites other than the dorsal lateral geniculate nucleus is not known for sure but if it is about 25%, then the ratio of afferents to geniculate cells is of the order of 1 : 2. It seems scarcely credible that one fiber would innervate only two geniculate cells, and, obviously, one needs to take into account the degree of terminal branching of the individual optic nerve fibers. From reduced silver peparations, Glees and Le Gros Clark (1941) concluded that each optic tract axon branched only five or six times in the monkey geniculate and that each branch ended axosomatically on only one cell. Each cell was said to receive only one bouton so that convergence of afferents was thought to be nonexistent. As pointed out by Walls (1953) these data can not be made to fit with the existence of approximately 1,800,000 geniculate cells and 1,200,000 optic axons. More importantly, however, the extent of the terminal distribution and density of terminal boutons on individually stained optic tract axons in monkeys or other species (Mason and Robson, 1979; Bowling and Michael, 1980, 1984; Sur and Sherman, 1982b) (Fig. 3.26) speaks for a far greater number of cells being contacted by individual fibers and for considerable convergence to afferents onto single cells. Sur and Sherman (1982b), for example, show two optic tract axons ending in lamina A of the cat (see Chapter 9) that have terminal distributions of approximately 0.0203 and 0.0825 mm^3 and that possess 793 and 1405 stained boutons terminaux. In the A and A1 laminae of cats, I have counted the number of neurons found in cubes of the same volumes as those subtended by the afferent fibers injected by Sur and Sherman and find that they contain, on average, 250 and 420 neurons. Electron microscopic studies, as well as showing that optic tract axons do not end axosomatically, also show multiple terminals on the dendrites of individual cells (e.g., Mason and Robson, 1979). And physiological studies (e.g., Hubel and Wiesel, 1961) indicate that multiple optic tract fibers can induce synaptic potentials in a single neuron. There is, therefore, no basis for a strict segregation of the type assumed by Glees and Le Gros Clark (1941).

3.7.2. Convergence in the Ventral Posterior Nucleus

The number of lateral geniculate cell somata subtended by the terminal ramifications of single optic tract axons having the dimensions mentioned above is not necessarily an accurate guide to the number of cells innervated by a single geniculate afferent. It ignores potential terminations on dendrites entering the terminal region from somata situated outside it. Rainey and Jones (1983) attempted to assess this in the ventral posterior nucleus of the cat. They made the following analysis of single, horseradish peroxidase-filled, medial lemniscal axons ending in relation to thionin-stained neuronal somata in the nucleus (Fig. 3.28). The terminal ramifications of the lemniscal axons had a dense central region of boutons measuring, on average, 150×75 μm. The assumption was made from correlative electron microscopy that most lemniscal boutons ended on dendrites within 75 μm of the dendrites' origin from their parent somata. Hence, a sphere, with radii 75 μm greater than those of the dense central zone of boutons, should contain all the cells that could potentially receive terminals of the fiber forming the ramification. For nine ramifications with core dimensions of approximately 150×75 μm, the number of stained somata counted in and within 75 μm of the core region ranged from 50 to 120. The numbers seemed higher in dorsal than in ventral parts of the nucleus. These figures are somewhat lower than in the dorsal lateral geniculate nucleus (Section 3.7.1). However, they take no account of lemniscal axons that formed more than one teminal ramification.

The types of analysis just mentioned can not give any indication of the degree of convergence of multiple lemniscal axons on single cells. Tömböl (1967) had attempted to assess the degree of convergence of afferent axons on single cells in an earlier Golgi study of the cat ventral posterior nucleus. She found that the dendritic fields of most neurons measured, on average, 2×106 μm^3. She then estimated that the total terminal spread of single Golgi-impregnated medial lemniscal axons measured 150×200 μm. This led her to conclude that a single thalamic cell could potentially receive terminals from 10 medial lemniscal fibers, even if the terminals of the fibers did not overlap (which they do). Her analysis, however, does not take into account the fact that the majority of lemniscal terminals make synaptic contact preferentially with proximal dendritic segments close to the cell soma (Ralston, 1969; Jones and Powell, 1969c). Hence, the exact degree of convergence still remains uncertain.

A high degree of convergence of afferent terminals on single thalamic cells may not be necessary for effective activation of individual cells. Andersen *et al.* (1966) found that the EPSPs recorded in thalamic cells innervated by the spinocervical tract were not only large at stimulus intensities close to threshold but at threshold were all-or-none and of constant amplitude, suggesting that they were induced by single afferent fibers. Because summation of only two such unitary EPSPs was often sufficient to discharge a cell, they concluded that input from no more than two fibers was adequate to ensure thalamocortical transmission. From this, we would have to assume that a high degree of afferent convergence is not a prerequisite for effective thalamic function.

3.7.3. Convergence in the Thalamocortical Projection

Le Gros Clark (1941a) had also tried to quantify the geniculocortical projection morphologically. From his counts of the total number of cells in the dorsal lateral geniculate nucleus of the monkey, coupled with his estimate of the area of the striate cortex, he concluded that each square millimeter of cortex

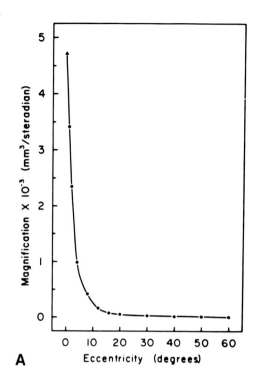

A

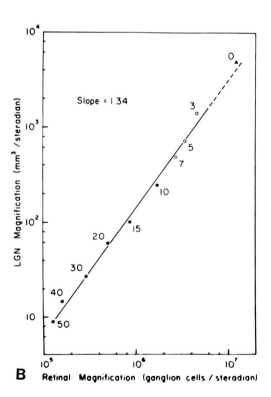

B

probably received the axons of 1350 geniculate cells. Moreover, from counting the number of geniculate cells that showed retrograde atrophy after lesions of the striate cortex, he concluded that the ratio of area of lesion to number of atrophied cells was the same as the ratio of total area of cortex to total number of geniculate neurons, implying a uniform projection of geniculate neurons to the cortex. Le Gros Clark was severely taken to task by Walls (1953) who suggested that his estimates of cortical area could have been inaccurate. And Walls raised the question of differential density of the geniculocortical projection by alluding to differences in the magnification of the cortical representation of different parts of the retina. The idea of a "magnification factor" in cortical representation was introduced by Talbot and Marshall (1941) from their physiological studies on the visual cortex of monkeys, when they attempted to determine the number of degrees of visual angle projected onto one linear milli-

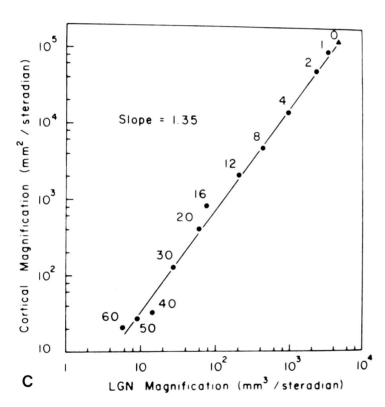

FIGURE 3.31. (A) Magnification of the visual field representation in the monkey lateral geniculate nucleus, as a function of visual field eccentricity from the center of gaze. The magnification factor on the ordinate is expressed in cubic millimeters of nucleus per degree of solid angle. (B) Magnification of the visual field representation in the monkey lateral geniculate nucleus (volume per degree) as a function closely correlated with magnification in the retina (cells per degree). The symbols refer to eccentricities from the center of gaze. (C) Magnification of the visual field representation in the monkey visual cortex as a function closely correlated with magnification in the lateral geniculate nucleus. The symbols refer to eccentricities. All from Malpeli and Baker (1975).

meter of area 17 cortex.* As is well known, this varies considerably across the cortex, ranging from as little as 1–2 min/mm at the foveal representation to many degrees per millimeter toward the peripheral representation (Daniel and Whitteridge, 1961; Hubel and Wiesel, 1974). Using figures derived from the study of Daniel and Whitteridge and correlating them with their own figures for magnification factors in the dorsal lateral geniculate nucleus, Malpeli and Baker (1975) were able to show that the ratio of cortical magnification to lateral geniculate magnification varies with visual field eccentricity (Fig. 3.31). The ratio is ten times greater for central than for peripheral vision. This would imply that the density of projection of geniculocortical cells to the cortex is least for central vision. As Malpeli and Baker point out, a uniformity in the projection could only be achieved if the density of projecting neurons were 10 times greater in the posterior part of the geniculate (representing central vision) than in the anterior part. The only data relating to this are those of Le Gros Clark (1941a) who detected only a twofold increase in cell density in posterior parts of the geniculate.

There has been no direct measurement of the number of geniculocortical axons terminating in a unit area of visual cortex. Sholl's (1956) estimates on the basis of the number of axons crossing the gray–white matter border in the cat are likely to be inaccurate due to the difficulty of assessing the proportion of efferent fibers to be excluded. Looking at the many photomicrographs of autoradiographically labeled, ocular dominance columns that have now been published (Chapter 9), one senses that the density of axonal terminations is likely to be the same at both central and peripheral representations. Hence, the apparently reduced numbers of geniculate cells projecting to a unit area of cortex representing central vision may compensate by branching more profusely than their fellows innervating the peripheral representation. To date there is no evidence for or against such a presumption.

Lashley (1941) in the rat and Chow *et al.* (1950) in the monkey attempted to relate the number of neurons in the dorsal lateral geniculate nucleus to the number in the striate cortex, in an effort to estimate the ratio of cortical cells to afferent fibers. Lashley counted 34,000 cells in the rat lateral geniculate nucleus and approximately 19 times that number in the visual cortex, suggesting an overall thalamus-to-cortex innervation ratio of 1 : 19. For layer IV, which he regarded as the principal termination layer of geniculate afferents, he found a ratio of only 1 : 37. In two "immature" rhesus monkeys, Chow *et al.* (1950) found a mean of 1,084,138 cells in the dorsal lateral geniculate nucleus, and a mean of 145,285,313 in the striate cortex, giving an overall innervation ratio 1 : 134. (They estimated a ratio of 1 : 45 for layer IV.) In the cortical representation of the macula, with 39,227,035 cells, they calculated a geniculocortical ratio of 1 : 72 and in the representation of the periphery, with 106,058,379 cells, they estimated a ratio of 1 : 196. These results seem to be at variance with those of Malpeli and Baker (1975) who based their calculations on a comparison of magnification factors (i.e., volume of geniculate or cortex per unit of visual field solid angle;

---

* As expressed here, this is the reciprocal of the magnification factor, defined by Daniel and Whitteridge (1961) as the linear extent of visual cortex concerned with each degree of visual field.

see above). However, the nature of the analyses are so different that a direct comparison hardly seems valid. Furthermore, given the evidence for specificity in the connections of particular classes of thalamic cells with comparable classes of cortical cells (Chapters 4 and 9), one wonders whether analyses of innervation ratios are of much value.

3.7.4. Number of Cells Projecting to a Given Cortical Area

It is often stated that the degree of granularity of a cortical field (as judged in cytoarchitectonic studies) is a reflection of the density of its thalamocortical innervation (e.g., Rose, 1949). It is extremely difficult to discover who first made this statement or to ascertain his or her reasons for making it. Certain qualitative observations speak in favor of it. For example, the density of thalamocortical terminations in areas 1 and 2 of the monkey somatic sensory cortex, as judged autoradiographically (Table 3.2) or by axonal degeneration, is substantially less than in area 3b (Jones and Powell, 1970a; Jones, 1975b; Jones and Burton, 1976b). The density of terminations is similarly low in many other areas of the parietal cortex in comparison with the primary sensory areas (Jones and Burton, 1976b). Axons of cells projecting to area 2 at least are much thinner than those projecting to areas 3a and 3b (Chapter 7).

The relative number of thalamic cells that can be labeled retrogradely from injections of tracers in the cortex also seems to vary from area to area, though it is hard to control such experiments because of the difficulty of ensuring comparable sizes of injection. In the monkey ventral posterior lateral nucleus, however, many more cells can be retrogradely labeled after injections of areas 3a and 3b than after seemingly comparable injections of areas 1 and 2 (Jones *et al.*, 1979; Jones, 1983g; Fig. 3.10).

In the cat ventral posterior nuclei, the number of cells that can be retrogradely labeled from the vicinity of the second somatic sensory area appears to be rather less than the number that can be labeled from the first somatic sensory area (Spreafico *et al.*, 1981). In this case, however, the density of thalamocortical terminations in the two areas qualitatively appears the same (Jones and Powell, 1969a). Hence, in some areas there may be a compensatory, enhanced branching of thalamocortical fibers that does not occur in others. On the other hand, because the SII area is so much smaller than the SI area, a smaller number of projecting cells may be able to produce a comparably dense terminal pattern. This point may need to be borne in mind when assessing relative densities of thalamic input to other cortical areas as well.

... in biology, the findings of analysis achieve scientific meaning only when
they are synthesized into principles of functional operation.

J. C. Eccles (1977)

4

Synaptic Organization in the Thalamus

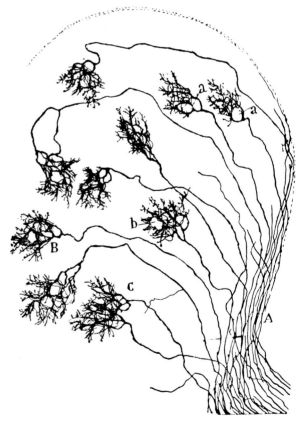

Ramón y Cajal's drawing of the terminal arborizations of medial lemniscal axons
in the ventral posterior nucleus of a mouse.

4.1. Relay Neurons and Interneurons

Nissl (1889), in his initial abstract on the rabbit thalamus, remarked that the nerve cells of one thalamic nucleus often differed greatly in size from those in other nuclei but he made no comment about different classes of neurons within individual nuclei. By 1896, however, von Kölliker had identified three forms of Golgi-impregnated cells in the thalami of cats and rabbits (Fig. 4.1). One cell form, which he termed *Buschzell,* was found throughout the dorsal thalamus. It was of relatively large size with many short radiating dendrites, sometimes adopting a bitufted form, later recognized by many authors. A second large type or *Strahlenzell* he observed only in the ventral lateral geniculate nucleus and in what may have been the reticular nucleus. It was more starlike with relatively few long dendrites, studded with protrusions. From Strahlenzellen, unlike from Buschzellen, he could trace axons. von Kölliker gives priority to Marchi for first describing these cells. The third cell type of von Kölliker was described in the thalami of newborn mice. It was small with dendrites of variable length, often beaded and covered in protrusions. Its axon could be traced for only a short distance but it was seen to give off collateral branches.

Buschzellen and the small cell type of von Kölliker were later recognized by Tello (1904) in the lateral geniculate nucleus of the cat and by Ramón y Cajal (1911) in all the dorsal thalamic nuclei of several species (Fig. 4.2). Though giving credit to Held for first clearly identifying the axons of the bushy cells, Ramón y Cajal emphasized the fundamental differences between these axons and the axons of the small cell type. The axon of the larger bushy cell projected outside the parent nucleus to the cerebral cortex and within the nucleus of origin possessed few or no collaterals.* The axons of these together constituted "la *grande voie sensitive supérieure centrale* ou *thalamo-corticale.*" The small form of cell was a short-axoned cell with multiple axon branches all confined to the parent nucleus. One of Ramón y Cajal's characteristic drawings is reproduced in Fig. 4.2. He considered that the short-axoned form was more numerous in the dog and cat than in other species such as the guinea pig and the mouse. He also

---

* Ramón y Cajal (1911), though drawing many relay cells in the dorsal thalamus, shows intranuclear axon collaterals on only three (his Figs. 182, 251, 253).

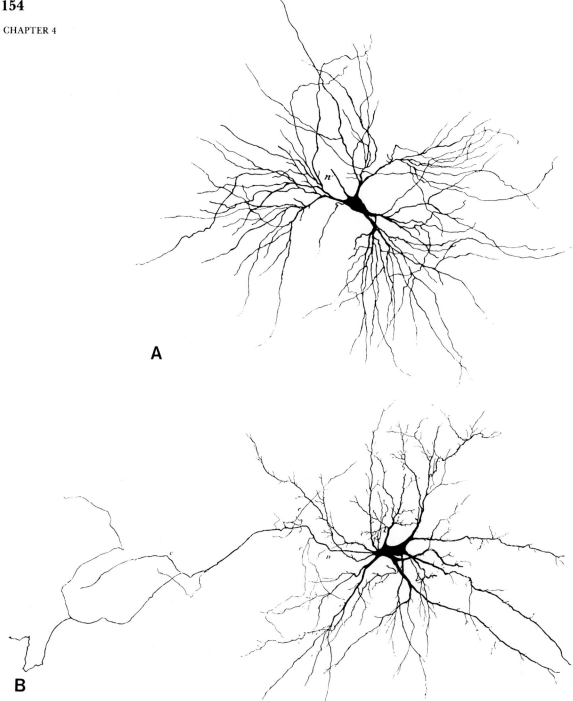

FIGURE 4.1. Drawings of three Golgi-impregnated cells in the thalamus, published by von Kölliker in 1896. (A) Buschzell (relay cell) from the human thalamus; (B) Strahlenzell from the "optic tract of a cat, lateral to the lateral geniculate nucleus" and, therefore, probably in the reticular nucleus.

noted the presence of dendritic appendages on both forms of cell, mentioning their extreme length, branching, and bulbous dilations on the short-axoned form and the reduction in their numbers on the long-axoned form in the rabbit as compared with the mouse.

Very little was added to Ramón y Cajal's description of thalamic cells for more than half a century. The work of O'Leary (1940) and Polyak (1957) on the lateral geniculate nucleus is rather sketchy and adds little to that already noted by Ramón y Cajal. But with the revival of interest in the synaptic organization of the thalamus in the 1960s, attention became focused on the unique dendritic architecture of the cells. Szentágothai and colleagues (Szentágothai, 1963; Szentágothai *et al.*, 1966; Tömböl, 1967, 1969; Tömböl *et al.*, 1969) redescribed the morphology of the short-axoned cells in Golgi preparations of the cat lateral geniculate nucleus. As described by them, the cell possesses a thin, loosely ramifying axon and numerous long, thin, beaded, and often highly branched dendritic protrusions which closely resemble axons and form, with the true axon, a plexus enveloping the parent cells and the dendrites of adjoining principal cells. Gradually, as more and more electron microscopy came to be done on the thalamus, it was recognized that the beaded dendritic appendages of the short-axoned cell were a peculiar form of presynaptic dendrite (Ralston, 1969, 1971;

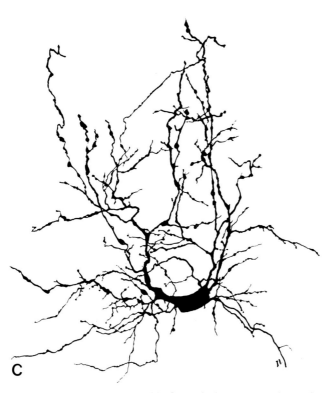

(C) Small cell (possibly an interneuron), probably from the lateral posterior nucleus of a newborn mouse. Cells are at different magnifications.

Jones and Powell, 1969c; Jones and Rockel, 1971; Morest, 1975), having all the ultrastructural characteristics of true dendrites but possessing localized aggregations of synaptic vesicles and making characteristic symmetric synaptic junctions (Fig. 4.3). The short-axoned cell has now been reidentified in other thalamic nuclei and in species such as primates, not described by Ramón y Cajal (Guillery, 1966; Tömböl, 1967; Majorossy and Réthélyi, 1968; Famiglietti and Peters, 1972; Morest, 1975; Ogren and Hendrickson, 1979). Forms with longer and shorter axons have been described in some nuclei (Tömböl, 1969; Ogren and Hendrickson, 1979). In the cat lateral geniculate nucleus, the longer-axoned form is said to connect adjacent laminae whereas the shorter-axoned form is confined to an individual lamina (Tömböl, 1969). The short-axoned cell with presynaptic dendrites is now generally regarded as *the* thalamic interneuron (Szentágothai, 1973; Madarász *et al.*, 1981) (Fig. 4.3). Madarász *et al.* (1981) estimate that some 25% of neurons in any relay nucleus belong to this class.

The principal or long-axoned form of thalamic cell received little attention until Guillery's (1966) now-classic work on Golgi-impregnated neurons in the lateral geniculate nucleus of the cat (Fig. 4.4). In the A laminae, he identified two forms of large, bushy cells mainly on the basis of differences in their pop-

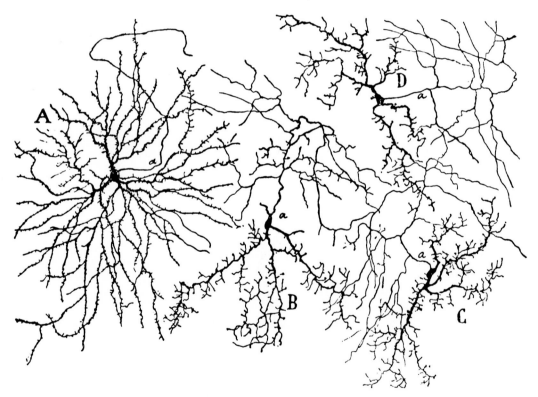

FIGURE 4.2.   Drawings of Golgi-impregnated relay neurons (A) and interneurons (B–D) in the ventral medial geniculate nucleus of a neonatal cat. From Ramón y Cajal (1911).

ulations of dendritic protrusions. The larger form, the class 1 cell, has a soma 25–40 μm in diameter, many more or less equally radiating dendrites, often distinctly tufted, and with large numbers of slender appendages resembling spines. Class 2 cells are medium-sized (15–30 μm somal diameter) with fewer, usually shorter dendrites. They are characterized by the presence of clusters of large, grapelike appendages situated close to the branch points of the larger dendrites. These appendages only become fully developed in mature animals, perhaps accounting for the failure of earlier investigators, who used infant animals, to describe them.

A third type of principal cell was identified by Guillery in the C laminae of the cat lateral geniculate. This, class 4 cell, is characterized by the possession of only a few, long but smooth dendrites and a small soma. Though also recognizing what would now be called the interneuron, and referring to it as a class 3 cell, Guillery did not appear to impregnate its axon or appendages as fully as Szentágothai (1963) and Szentágothai *et al.* (1966) (Fig. 4.3).

The significance of Guillery's identification of three forms of principal neuron lies in their having been correlated with different, electrophysiologically distinguishable classes of thalamocortical relay cells. LeVay and Ferster (1977) first equated Guillery's cells with geniculate cells that selectively receive inputs from the X and Y categories of retinal ganglion cells mentioned in Chapter 3. They found that cells possessing multilamellar bodies in their cytoplasm projected to area 17 of the visual cortex and their numbers declined laterally in the A and A1 laminae, i.e., in the representation of the visual field periphery. Larger cells without the multilamellar organelle projected to areas 17 and 18 and their numbers increased laterally in the A laminae and in lamina C. The differential cortical projections and the distribution in relation to visual field eccentricity, formed the basis for correlating the lamellar body-containing cell type with X-type relay cells and that lacking the lamellar body with Y-type relay cells. Ferster and LeVay (1978) later extended their analysis to correlate Guillery's class 4 cell, found only in the C layers, with a relay cell type receiving input from the third, W type of retinal ganglion cell. A more complete Golgi characterization of the C-laminae cells has recently been published by Hitchcock and Hickey (1983).

There have been relatively few Golgi studies of other thalamic nuclei as thorough as those on the lateral geniculate nucleus. Cells obviously comparable to the typical, bushy relay cells and the putative interneurons of the lateral geniculate are readily identifiable (Fig. 4.5). The interneuron has been well described in the ventral posterior and medial geniculate nuclei and pulvinar (Tömböl, 1969; Morest, 1975; Ogren and Hendrickson, 1979) but there have been no descriptions of subtypes of relay cells as convincing as those of Guillery. One reason for this may lie in the fact that dendritic protrusions seem to be relatively less common on relay cells in nuclei other than the lateral geniculate and thus cannot be used as a systematic variable for classification purposes, though Pearson and Haines (1980) have made an attempt at a classification comparable to Guillery's in the ventral posterior nucleus of a prosimian. One suspects that morphological differences between relay cells may exist in the ventral posterior nucleus if only because some of its cells have different patterns of cortical connections. Some cells, for example, are thought to project only to the first or second somatic sensory area while others project to both (Manson, 1969; Jones, 1975a; Spreafico *et al.*, 1981); large cells are thought to project only to layer IV and small cells only to layer I (Penny *et al.*, 1982). In addition, there is considerable evidence for cells in the ventral posterior complex with different types of peripheral receptive field and projecting to different fields of the first

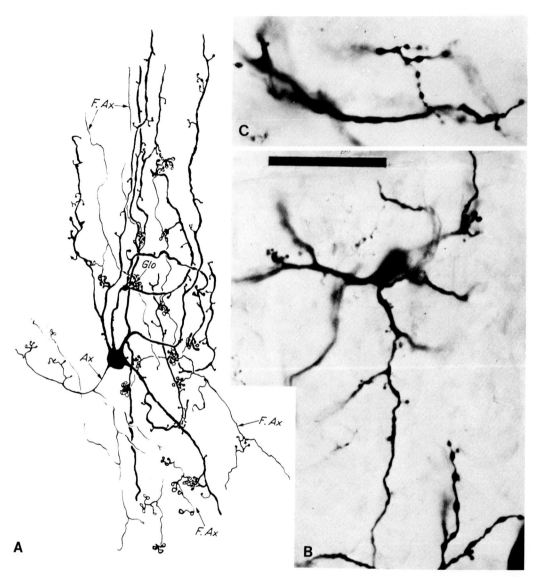

FIGURE 4.3.   (A) Szentágothai's (1973) drawing of a Golgi-impregnated interneuron from the dorsal lateral geniculate nucleus of a cat. The cell is drawn with an axon (F.Ax) and presynaptic appendages (Glo). (B, C) Photomicrographs of a Golgi-impreganted interneuron and some of its dendritic appendages from the ventral posterior nucleus of a cat. Bar = 50 μm (B), 20 μm (C). From Rainey and Jones (1983). (D) Electron micrograph of presynaptic dendrites (T2),

somatic sensory area (Friedman and Jones, 1981; Jones and Friedman, 1982).

Recent studies, making use of intracellular injection of physiologically characterized cells in the lateral geniculate nucleus, have attempted a more direct correlation of structure and function (Ahlsén *et al.*, 1978; Friedlander *et al.*, 1979, 1981; Stanford *et al.*, 1981, 1983) (Fig. 4.6). The results can be seen as generally supporting the equation of Guillery's type 2 cell with X cells, his type

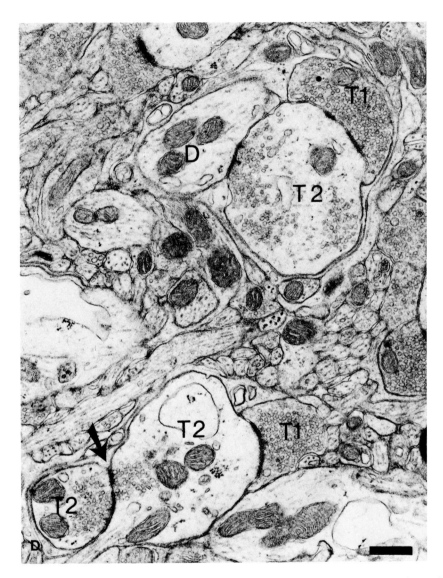

probably corresponding to the dendritic appendages of interneurons in the ventral posterior nucleus of a cat. The presynaptic dendrites are postsynaptic to other conventional axon terminals (T1). Note reciprocal synapse at lower left (arrow). D: dendrite. Bar = 1 μm.

1 cell with Y cells, and his type 4 cell of the C laminae with W cells. However, there are some discrepancies. Among these are the fact that some type 2 cells seem to have either X or Y properties and some type 3 cells not only can have X- or W-type properties but, moreover, can project to the cerebral cortex. Similar conclusions were arrived at by Meyer and Albus (1981) who studied the morphology of the cat lateral geniculate neurons retrogradely labeled by injections of tracer in the visual cortex. In their case, type 3 cells projecting to area 17 were found in laminae A and A1. Friedlander *et al.* (1981), from their extensive intracellular study of the cat dorsal lateral geniculate nucleus, have drawn up a series of morphological criteria for X and Y cells, which extend Guillery's classification (Fig. 4.6). The dendrites of Y cells are thicker than those of X cells and,

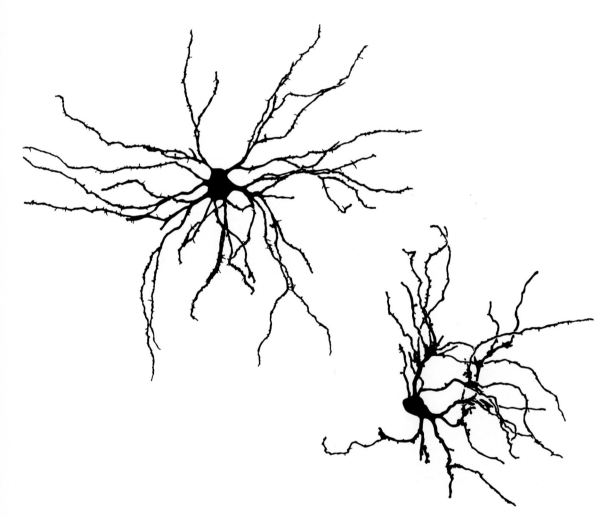

FIGURE 4.4. Guillery's (1966) drawings of Golgi preparations from the lateral geniculate nucleus of a cat showing the two types of thalamocortical relay neurons. Larger type is typical of such neurons in most principal thalamic nuclei. Dendritic protrusions and appendages are major sites of synaptic contact with ascending fibers. × approximately 300. From Guillery (1966).

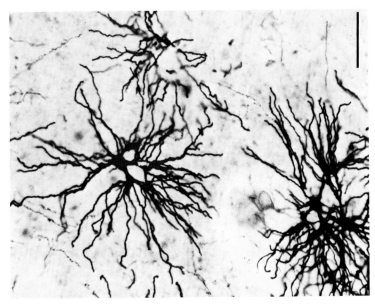

FIGURE 4.5.   Photomicrograph of a Golgi preparation showing typical bushy relay neurons in the ventral posterior nucleus of a rat. Bar = 100 μm.

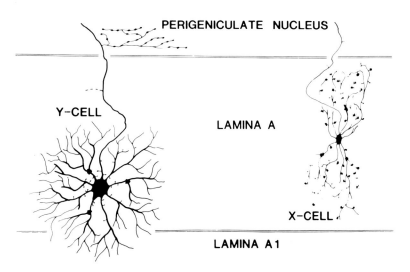

FIGURE 4.6.   Schematic summary of the principal morphological features of an X-type and a Y-type relay neuron in the dorsal lateral geniculate nucleus of cats, as determined by intracellular staining of physiologically identified cells. The Y-type cell, corresponding more or less to Guillery's (Fig. 4.4) large type, has a radially symmetrical dendritic tree with few dendritic appendages and its dendrites can cross into a neighboring lamina. The axon is thick and is more likely to have collateral branches in the perigeniculate nucleus. The X-type cell, corresponding more or less to Guillery's smaller type, has a smaller soma, and thinner dendrites oriented across the lamina in which the cell lies, with many large dendritic appendages. The dendrites do not cross to other laminae; the axon is thin and less likely to have collaterals in the perigeniculate nucleus. From Friedlander *et al.* (1981).

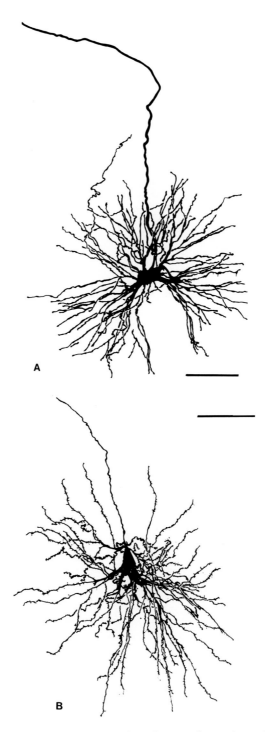

FIGURE 4.7.   Intracellularly injected neurons from the ventral posterior nucleus of cats. Two types of principal cell are found: a large type (A, D) has thick, smooth dendrites branching in bushy tufts and a thick axon; a smaller type (B) has thinner, dichotomously branching dendrites with hairlike

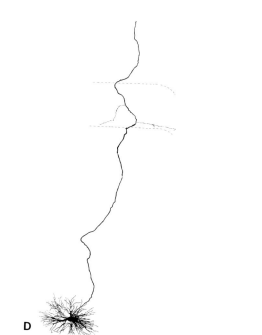

protrusions and a thin axon. The axons of both had collaterals in the reticular nucleus but not in the dorsal thalamus. Cell A responds in a sustained manner to a peripheral stimulus and cell B in a transient manner. The third cell (C) is an interneuron. Bars = 100 μm. From Yen *et al.* (1985).

unlike X cells, cross laminar borders. Y-cell dendrites form a more radially symmetrical field whereas that of X cells is elongated. Y cells have fewer dendritic appendages and a thicker axon than X cells. Preliminary findings with intracellular injections in the cat ventral posterior nucleus (Yen and Jones, 1983; Yen *et al.*, 1985) reveal some morphological variation among cells characterized in terms of their receptive fields but not the same diversity as in the lateral geniculate nucleus (Figs. 4.7–4.8).

A somewhat disconcerting result from the studies of intracellularly injected lateral geniculate cells in the cat is the apparent similarity between some cells, shown by antidromic stimulation or retrograde labeling to project to the cortex, and the interneuron of Szentágothai (the class 3 cell of Guillery). This finding has fueled a controversy that has simmered for 50 or more years and which came to the fore again with the first use of retrogradely transported tracers in the examination of thalamocortical connectivity. Many workers have described cells in the principal thalamic nuclei that survive ablation of the cerebral cortex in man and animals (Minkowski, 1913, 1920; Waller and Barris, 1937; Walker, 1938a,c; Le Gros Clark and Russell, 1940; Sheps, 1945; Coombs, 1949, 1951; Powell, 1952; McLardy, 1963; Chow and Dewson, 1966). Generally speaking, more neurons were thought to survive in the lateral and ventral nuclei than in the geniculate bodies or mediodorsal nucleus. Powell (1952), for example, described a survival of some 40–50% of the cells in the human ventral and lateral nuclei 24 days after hemidecortication in man. In the dorsal lateral geniculate nucleus of man and animals, the survival rate has generally been reported as 5–10% or even lower, with most of the surviving neurons being in the magnocellular layers (Brouwer, 1939; Polyak, 1932; Walker, 1935, 1938a; Glees and Le Gros Clark, 1941; Hassler, 1964; Mihailovic *et al.*, 1971; Pasik *et al.*, 1973). Though the concept of the interneuron was rarely brought up in the reports of these studies, it was occasionally suggested that even neurons not projecting to an area of damaged cortex would degenerate as the result of transneuronal degeneration, either through connections with degenerated cells or through destruction of corticothalamic inputs (Cook *et al.*, 1951; Rose, 1952; Goldby, 1957; Powell and Cowan, 1967).

The modern tendency has been to regard neurons that survive cortical ablations in nuclei other than the intralaminar system, as interneurons, although Madarász *et al.* (1983) have recently suggested that a proportion of the relay neurons in the cat lateral geniculate nucleus may also survive cortical ablation. During the heyday of Golgi and electron microscopic studies in the 1960s any controversy over the relative numbers of interneurons in the principal nuclei tended to remain in the background. However, when horseradish peroxidase first started to be injected into the cerebral cortex in the middle 1970s, it was widely reported that this led to retrograde labeling of every cell in the appropriate part of the related thalamic nucleus (e.g., Jones and Leavitt, 1973; Saporta and Kruger, 1977; Norden and Kaas, 1978; Lin *et al.*, 1978; Norden, 1979).

FIGURE 4.8. (A, B) Photomicrographs of dendrites of cells A and B in Fig. 4.7, at same magnification. C is axon of cell A, E is axon of cell B; D shows collateral of axon of cell A in reticular nucleus. Bars = 50 μm (A, B, C, E), 25 μm (D). From Yen and Jones (1983).

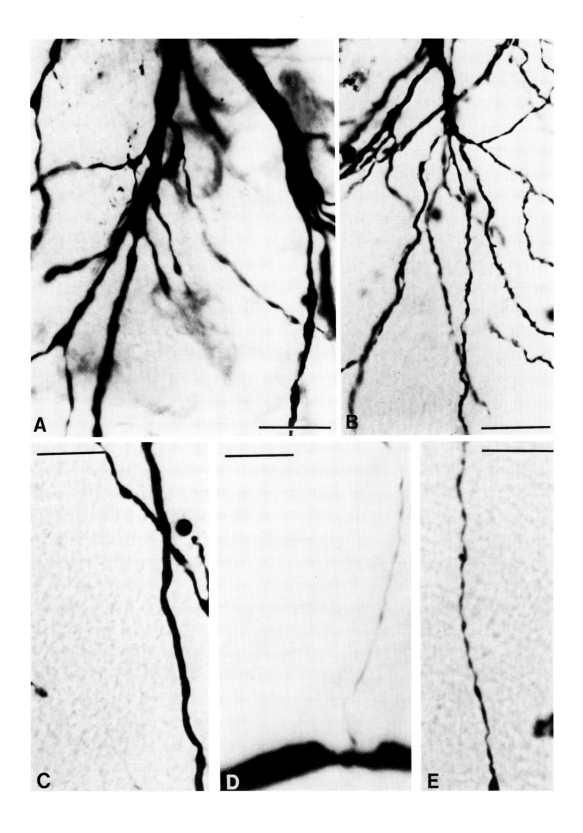

Norden and Kaas estimated that as many as 98.6% of all neurons were labeled in a segment of the magno- and parvocellular layers of the dorsal lateral geniculate nucleus of the owl monkey. Seemingly, in answer to the question "where are the interneurons?", Winfield *et al.* (1975), LeVay and Ferster (1977, 1979), and Weber and Kalil (1983) suggested that thalamic interneurons escape detection in anatomical experiments of this type because, though unlabeled, they are sufficiently small to be mistaken for neuroglial cells or to be obscured by overlying labeled relay cells. LeVay and Ferster (1979) considered that as many as 25% of the neurons in the cat lateral geniculate nucleus may be interneurons. Weber and Kalil (1983) reported 22% in the same nucleus. These figures correlate well with the proportion of small GABAergic neurons that concentrate [^{3}H]-GABA or that stain immunocytochemically for GAD in the dorsal lateral geniculate nuclei of the cat or rat (Sterling and Davis, 1981; Ohara *et al.*, 1983) (Chapter 5). Such cells are unlikely to project to the cerebral cortex (see Hendry and Jones, 1981). A figure of approximately 20% of interneurons also correlates quite well with Tömböl's (1969) Golgi observations of 25% thalamic cells with the form of interneurons.

Against the evidence of LeVay and Ferster, however, other workers placed the fact that their studies were carried out with a substrate for the enzymatic tracer now known to be relatively insensitive. When coupled with the direct

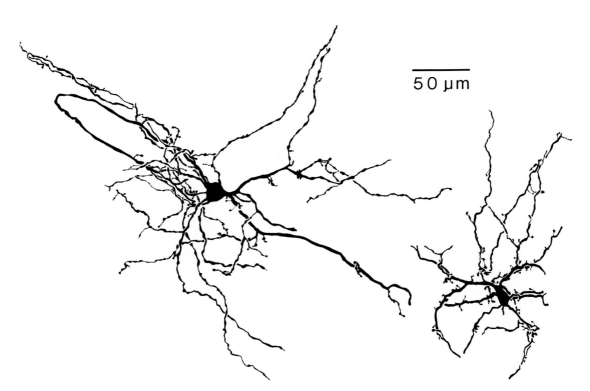

50 μm

FIGURE 4.9. The so-called type V cell (left) described by Updyke in the dorsal lateral geniculate nucleus as possibly representing a form transitional between relay neurons and an interneuron (right). Drawn from a Golgi preparation of the cat ventral posterior nucleus. From Rainey and Jones (1983).

evidence of a few intracellularly labeled, cortically projecting cells resembling the traditional interneuron, these data still leave some room for doubt. A recent study carried out with similar methods on the rat ventral posterior nucleus found less than 5% of neurons unlabeled after cortical injections (Ralston, 1983). It may, however, be invalid to compare results on two different thalamic nuclei, for Ohara *et al.* (1983) were able to stain immunocytochemically far fewer GABAergic neurons in the ventral posterior nucleus than in the dorsal lateral geniculate nucleus of the rat. In the cat ventral posterior nucleus, Penny *et al.* (1983) and Spreafico *et al.* (1983) have described as many as 25% of the neurons as GABAergic.

One possible solution to the controversy over the thalamic interneurons may lie in the existence of a population of cells with a form transitional between the classical relay cell and interneuron. Such a cell has, in fact, been described in a Golgi study of the cat lateral geniculate nucleus by Updyke (1979). He refers to it as a type V cell (Fig. 4.9). In an intracellular injection study, Friedlander *et al.* (1979) reported a few similar cells with X properties and Meyer and Albus (1981) identified a few such cells projecting from the pulvinar of the cat to area 18 of the cortex. The population of transitional cells is still not clear. Another potential resolution of our problem would be the restriction of the interneurons to a part of a thalamic nucleus different from that containing the relay neurons or even to an adjacent nucleus. There is no convincing evidence for the former and, as pointed out in Chapter 3, only the reticular nucleus appears to give rise to intrathalamic connections. It is difficult to envisage this as the sole pool of thalamic interneurons, especially in view of the GABA-uptake studies of Sterling and Davis (1981), and the immunocytochemical studies of Ohara *et al.* (1983) and Penny *et al.* (1983).

The neurons of the intralaminar nuclei, despite some reports to the contrary (Leontovich and Zhukova, 1963; Scheibel and Scheibel, 1966b; Ramón-Moliner, 1975), also appear to belong to the two major classes described in the other nuclei (Hazlett *et al.*, 1976). The dendritic field of the principal cell appears to be less radial or tufted and more elongated, perhaps on account of the compression of the cells within the internal medullary lamina. But despite this appearance, the dendrites bear long appendages and otherwise resemble relay neurons in other nuclei. The other cells are small and have locally ramifying axons and presynaptic dendrites typical of the putative interneurons in other nuclei.

4.2. The Axons

Two principal axon types have been described innervating virtually every dorsal thalamic nucleus. One of these arises in the cerebral cortex and the other forms the subcortical input. Axons of the reticular nucleus, though obviously also present, have not been described in an equivalent amount of detail.

Ramón y Cajal (1911) recognized subcortical afferent fibers in the three principal thalamic sensory nuclei, describing the individual fibers as terminating in single bushy tufts of multiple terminal branches without obvious boutons and enclosing many thalamic cell bodies. In his drawings these tufts appear shortest in the ventral posterior nucleus and longest in the medial geniculate nucleus. In the lateral geniculate nucleus of the cat he showed them confined to single laminae with a form likened by Tello (1904) to a cypress tree (Fig. 4.10). Ramón y Cajal noted less dense tufts of optic axon terminations in the C laminae of the cat lateral geniculate and drew them as though derived from axons different from those ending in the A laminae. This has recently been confirmed (Sur and Sherman, 1982a,b). In the ventral posterior nucleus, Ramón y Cajal identified only lemniscal and not spinothalamic axons.

Ramón y Cajal described corticothalamic fibers in several nuclei as thin, highly branched, and widely ramifying but he stressed that they were usually derived from "enormous" parent axons. In most of his drawings these parent axons often appear thicker than the subcortical afferents. Unlike the subcortical afferents he gives few details of the terminal structure of the corticofugal axons.

Virtually none of the early investigators of axon terminations in the thalamus, from von Kölliker (1896) to O'Leary (1940), described boutonal enlargements on the axons that they described, even though they recognized the dense terminal branching patterns of lemniscal or optic tract axons. In 1941, Glees and Le Gros Clark identified neurofibrillar rings on some optic tract axons, many of which they felt terminated axosomatically in the lateral geniculate nucleus, but they did not visualize the full axonal arborizations. In 1957, a drawing

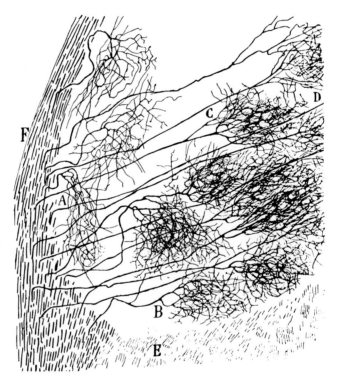

FIGURE 4.10. Cypress tree-like terminal ramifications of optic tract fibers in the dorsal lateral geniculate nucleus of the cat. The full terminal sprays were probably not completely impregnated on each axon. From Ramón y Cajal (1911).

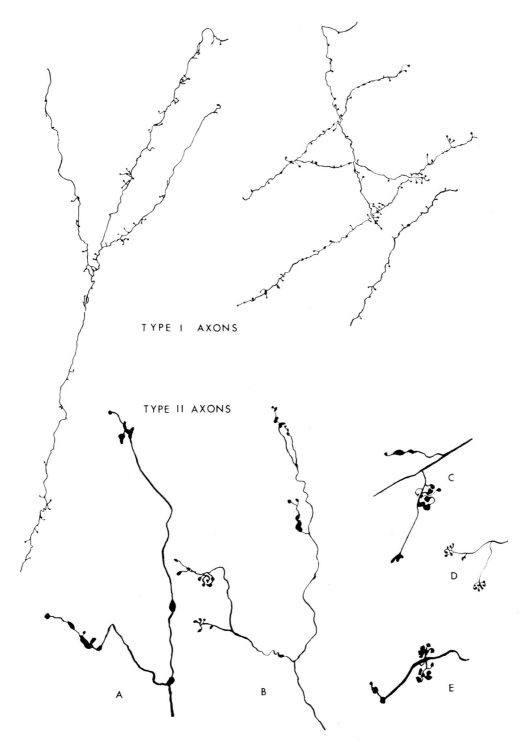

TYPE I AXONS

TYPE II AXONS

A

B

C

D

E

FIGURE 4.11.   Guillery's (1966) original drawings of Golgi preparations showing terminal portions of type I or corticothalamic and of type II or retinal afferent fibers, from the lateral geniculate nucleus of the cat. These are typical of comparable fibers in other principal thalamic nuclei. From Guillery (1966).

by Polyak appeared in his posthumous book, *The Vertebrate Visual System,* in which he showed part of the terminal bush of a putative optic tract axons studded with boutons in the lateral geniculate nucleus of a monkey. Once again, however, realization of the finer details of the axonal arborizations in the thalamus had to await the renewed Golgi studies of Guillery and of Szentágothai and his school, on the cat brain. The recognition of newer and finer details may be attributable in part to modifications of the Golgi technique such as initial aldehyde fixation and perfusion of reagents. But perhaps the major reason is the greater maturity of the animals studied in modern times (Szentágothai, 1963; Guillery, 1966).

Szentágothai (1963), in his initial study on the cat lateral geniculate nucleus, identified two classes of axons, each with extensive terminal enlargements or boutons: a thinner axonal type forming a dense tangled plexus he considered to arise from other parts of the brain, possibly in the cerebral cortex; a thick type, oriented across the laminae of the lateral geniculate and with short side branches studded with boutons, he regarded as arising in the retina. This was later confirmed by Szentágothai *et al.* (1966) in one of the earliest electron microscopic studies of Wallerian degeneration in the central nervous system. The two types of axons were also detected and similarly identified by Morest (1964, 1965a) in the cat medial geniculate complex.

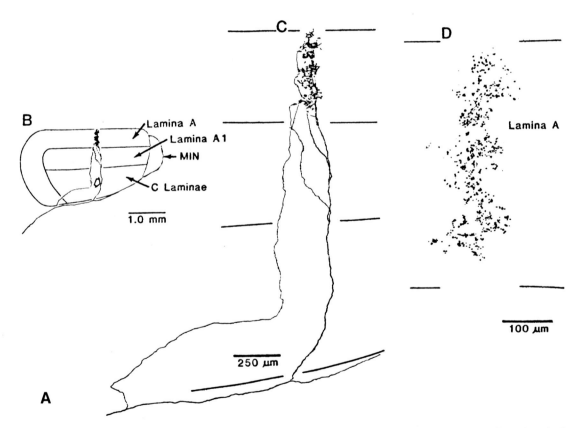

FIGURE 4.12. Examples of intracellularly injected X (A) and Y (B) type axons entering the dorsal lateral geniculate nucleus of the cat. Y axon has broader terminal field than X axon. From Sur and Sherman (1982b).

Guillery (1966) described the terminal portions of the two axonal types somewhat more fully in the cat lateral geniculate nucleus (Fig. 4.11). The thinner axon type runs a relatively straight course, crossing from lamina to lamina and branching occasionally but is characterized by the presence of numerous short side branches each ending in a small spherical bouton about 1 μm in diameter. Similar boutons are seen at intervals along the length of the parent fiber too. The larger axon type has its terminals confined to a single lamina and varies in thickness but gives off multiple branches that bear clusters. The branches form terminal clusters of varying degrees of complexity, sometimes nestlike, sometimes flowerlike, sometimes open, and sometimes closed. In a few instances the large boutons appeared to make contact with dendritic appendages of principal neurons.

Since the early studies of Szentágothai, Morest, and Guillery, there have been numerous Golgi studies of axons in the thalamus. Most of the principal nuclei and the intralaminar nuclei have been sampled. All of the studies serve to affirm the presence of the two types of afferent fibers in all the nuclei (Morest, 1965a; Tömböl, 1969). Gradually, as more and more correlative electron microscopy came to be done, it emerged that the fine, relatively straight axons with the short side branches ending in small boutons, arise in the cerebral cortex

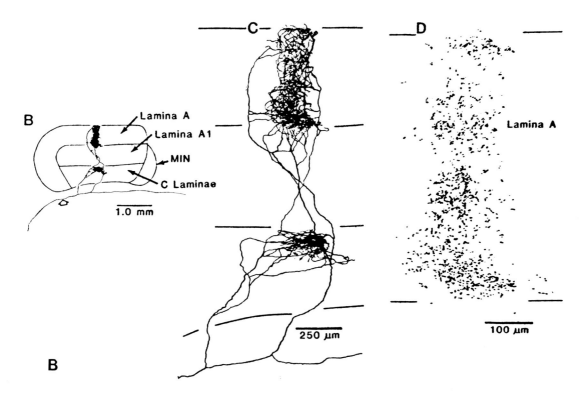

FIGURE 4.12. (*continued*)

(Jones and Powell, 1969b) while the usually thicker type, with complicated terminal branching and grapelike clusters of large boutons, arise in the principal source of subcortical afferents, e.g., the retina, inferior colliculus, dorsal column nuclei (Tömböl, 1967; Morest, 1965a; Ralston, 1969; Guillery, 1966). Even so, the extremely complicated nature of the terminations of the subcortical afferents was not fully appreciated from the Golgi studies. Anterograde labeling of individual afferent fibers by injections of horseradish peroxidase into the optic tract or medial lemniscus has now revealed their total extent (Mason and Robson, 1979; Bowling and Michael, 1980, 1984; Sur and Sherman, 1982b; Jones, 1983b; Rainey and Jones, 1983; Yen and Jones, 1983) (Figs. 4.12 and 4.13). Different

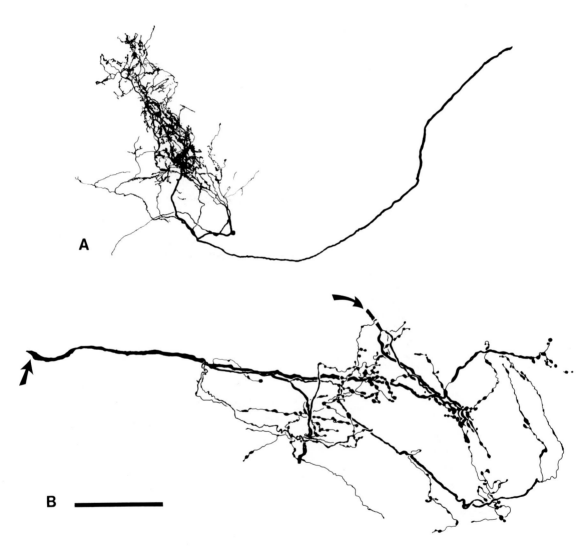

FIGURE 4.13. (A) Terminal portion of an intracellulary injected medial lemniscal axon with a cutaneous receptive field in the ventral posterior nucleus of the cat. From C.-T. Yen and E.G. Jones, unpublished. (B) Terminal portions (arrows) of a single anterogradely labeled medial lemniscal axon in the ventral posterior nucleus of the monkey. Bars = 250 μm (A), 100 μm (B). From Jones (1983b).

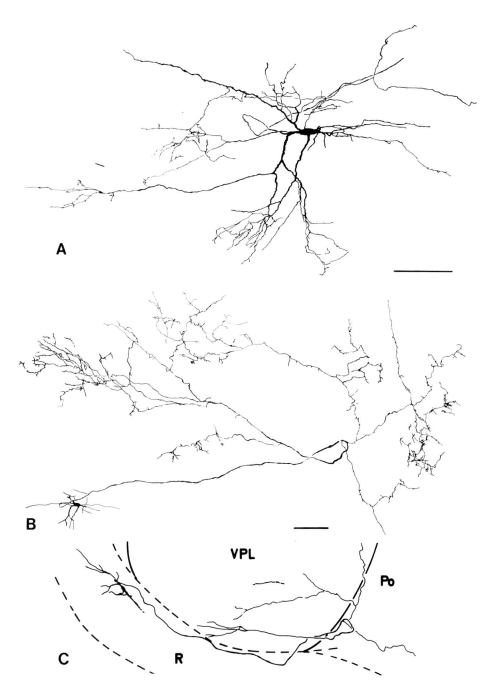

FIGURE 4.14.   Intracellularly injected cell (A) with a somatic sensory receptive field in the reticular nucleus of the cat. Note extensive arborization of its axon (B, C) in ventral posterior and other adjacent thalamic nuclei. Bars = 100 μm (A), 200 μm (B). From Yen and Jones (1983).

subtypes, initially suggested by Polyak (1957) in the lateral geniculate nucleus, are starting to be detected (Chapters 3 and 9).

The axons entering the dorsal thalamus from the reticular nucleus have not been described in as much detail as the subcortical and cortical afferents. Ramón y Cajal did not give any details of their terminations. Scheibel and Scheibel (1966a) showed fine axons of reticular nucleus cells branching extensively throughout the thalamus and this pattern has recently been confirmed by intracellular injections (Fig. 4.14).

4.3. Synaptic Organization

4.3.1. Fine Structure

The 2 or 3 years prior to and after 1970 were a period of intensive electron microscopic studies on the thalamus. Virtually every thalamic nucleus was examined, some in more detail than others but all sufficiently thoroughly to indicate that a basic pattern of synaptic organization is common to all the nuclei, relay as well as intralaminar. Only the reticular nucleus and other elements of the ventral thalamus seem to have escaped the ultramicrotome. In the following account, rather than indicating the contributions of individual workers, I have attempted a synthesis that indicates the fundamental pattern. In it, some will recognize certain rather sweeping generalizations but, hopefully, no errors of fact. (The details can be found in the works of: Szentágothai, 1963; Szentágothai *et al.*, 1966; Peters and Palay, 1966; Guillery, 1969a,b, 1971b; Jones and Powell, 1969b,c; Guillery and Colonnier, 1970; Guillery and Scott, 1971; Wong-Riley, 1972; Famiglietti and Peters, 1972; Jones and Rockel, 1971; Majorossy *et al.*, 1965; Pecci Saavedra *et al.*, 1968; Ralston, 1969, 1971, 1983; Ralston and Herman, 1969; Mathers, 1972a,b; Campos-Ortega *et al.*, 1968; Hámori *et al.*, 1974, 1978; Hajdu *et al.*, 1974; Lieberman and Webster, 1972; Rafols and Valverde, 1973; Pasik *et al.*, 1976; Hámori *et al.*, 1974; Somogyi *et al.*, 1978; Mason and Robson, 1979; Robson and Mason, 1979; Tömböl, 1967; Majorossy and Réthélyi, 1968; Harding, 1973a,b; Špaček and Lieberman, 1974; Morest, 1975; Dekker and Kuypers, 1976; Harding and Powell, 1977; Ogren and Hendrickson, 1979; Hazlett *et al.*, 1976; Madarász *et al.*, 1981; Rainey and Jones, 1983.) It may be noticed that in the list of titles, the lateral geniculate nucleus, again, predominates.

The distinguishing feature of all the dorsal thalamic nuclei, as might be anticipated from the descriptions of afferent axons, given above, is the presence of large aggregations of synaptic terminals (Fig. 4.15). Between them is a somewhat differently organized neuropil. The synaptic aggregations are often partially encapsulated by thin sheets of astrocytic cytoplasm and are sometimes referred to as glomeruli or synaptic islands, though they are not as localized as these names suggest and, rather, tend to form more extensive, sheetlike configurations. The islandlike appearance is also more a feature of cats and other small mammals than of primates in which the individual elements, though all present, tend to be more spread out.

The essential elements of the synaptic aggregations are three: a postsynaptic dendritic component and two presynaptic components. The one or two dendrites present are usually the primary branches of one or more relay cells close to their parent cell somata and, if present, one or more of their dendritic appendages (Fig. 4.16). Around these are gathered in considerable numbers the two types of presynaptic elements. The first type is usually in the minority. It is a conventional axon terminal derived from the principal ascending afferent fiber system to the nucleus. It is usually large, up to 3 μm in diameter and 5–7 μm in length with a rather dense cytoplasm and is packed with spherical synaptic vesicles (Fig. 4.17). Guillery (1969a,b) refers to these in the lateral geniculate nucleus as RLP

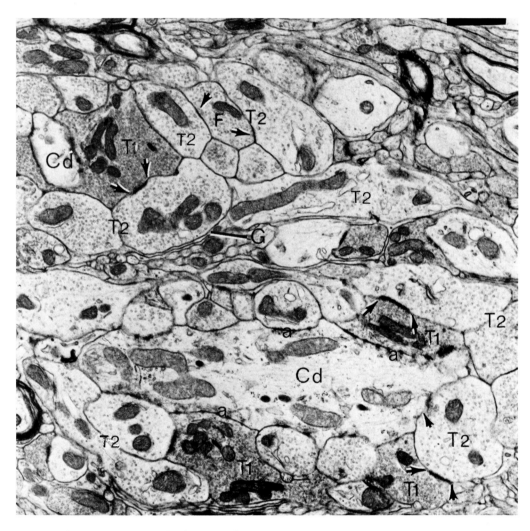

FIGURE 4.15.   Electron micrograph showing synaptic aggregations in the ventral posterior thalamic nucleus of the cat, with the typical central dendrite (Cd), principal afferent terminals (T1), profiles interpreted as presynaptic dendritic terminals (T2) and axon terminals (F) of interneurons, adhesive (a) and synaptic (arrows) junctions, and encircling astrocytic lamellae (G). Bar = 1 μm. After Jones and Powell (1969a).

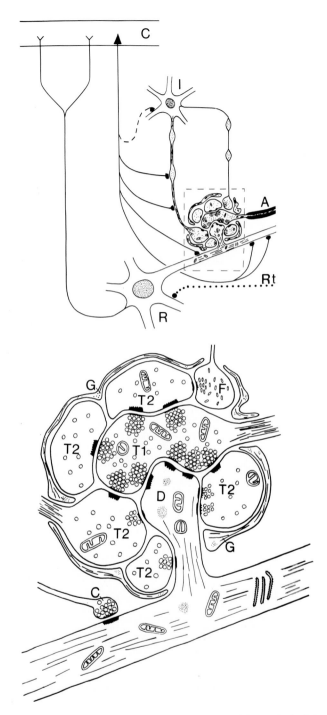

FIGURE 4.16. Schematic drawing indicating synaptic relationships typical of the majority of thalamic nuclei. Dendritic protrusions (D) of thalamocortical relay cells (R) receive terminals (T1) of
ascending afferent fibers (A) and presynaptic dendrites (T2) of interneurons (I). Presynaptic dendrites and probably conventional dendrites of interneurons are also postsynaptic to the afferent
fiber terminal (and sometimes to one another; see Fig. 4.3C). Axons of interneurons also terminate

terminals since they contain round vesicles, are large, and are said to contain pale mitochondria. One or more of these terminals are usually present in a single thin section of a synaptic aggregation. They make multiple, asymmetrical, synaptic contacts with the dendritic components of the aggregation and with the second type of presynaptic component. This second component is a presynaptic dendrite belonging to the thalamic interneurons. It resembles an axon terminal in that it contains synaptic vesicles and makes conventional synaptic contacts. The vesicles tend to flatten or become pleomorphic in aldehyde-fixed material and the synaptic contacts are symmetrical. But also present among the vesicles are free ribosomes, rough-surfaced endoplasmic reticulum, and often microtubules. These features betray a dendritic (and occasionally even a somatic) origin. The presynaptic dendrites usually outnumber the subcortical afferent terminals in a synaptic aggregation by three or four times. They make symmetrical synaptic contacts on the dendritic elements of the aggregation and sometimes on one another, but are themselves postsynaptic, at asymmetrical synapses, to the subcortical afferent terminals (Figs. 4.3 and 4.15). Commonly at the perimeter of a synaptic aggregation, a further type of presynaptic terminal may be seen. This is a conventional axon terminal containing flattened or pleomorphic synaptic vesicles and terminating in a symmetric membrane contact on one or more presynaptic dendrites (Fig. 4.18). It is generally thought to arise from the thin, conventional type of axon possessed by the interneurons. Such terminals can resemble the presynaptic dendritic terminals very closely so that if ribosomes are not particularly obvious in some of the latter, the proportions of conventional flattened vesicle-containing terminals and presynaptic dendritic terminals in a synaptic aggregation are difficult to estimate.

A final feature of the whole synaptic aggregation is the presence of numerous specialized, but apparently nonsynaptic types of contacts between the various elements of the aggregation. They have been variously called adhesive contacts, filamentous contacts, and even desmosomes. To some extent they do resemble the zonulae adhaerentes of epithelia. They consist of localized patches of symmetrically increased electron density on short portions of two membranes apposed to one another across a slightly widened extracellular cleft (Fig. 4.15). From the deep surface of each thickening, wisps and strands of filamentous material extend deeply into the underlying cytoplasm on each side but with no associated accumulation of synaptic vesicles. The filamentous contacts are seen joining dendrites to other dendrites or to dendritic protrusions and joining the various presynaptic elements to the dendrites or to one another. Profiles of dendritic appendages may appear filled with whorls of filamentous material derived from multiple adhesive contacts on them. Nowhere else in the nervous system are contacts of this kind found in such large numbers. It seems doubtful that they serve any synaptic function, and their large numbers and restriction

---

(F) mainly on the presynaptic dendrites. The complex synaptic aggregation tends to be ensheathed in astrocytic processes (G). Outside this, corticothalamic terminals (C) end on relay cell dendrites and on presynaptic dendrites of interneurons. In the case of the relay cell, most cortical terminals are distally situated. Terminals of reticular nucleus axons (Rt) also terminate on or close to somata of relay neurons.

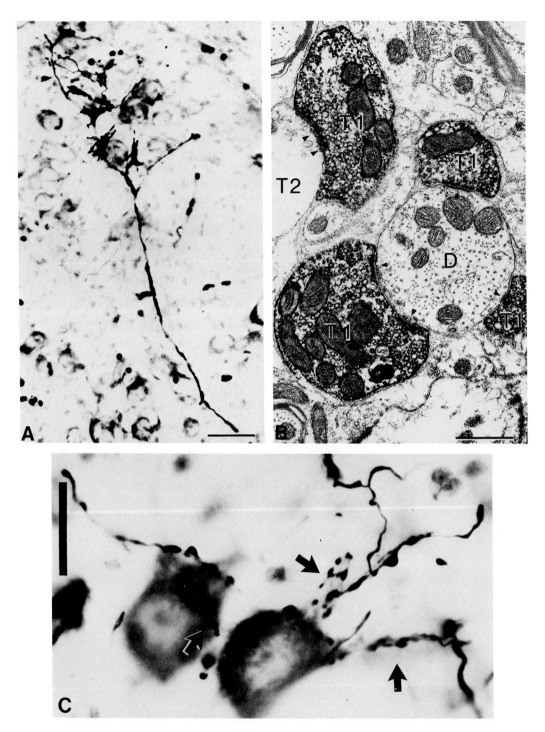

FIGURE 4.17.   (A) Photomicrograph of one of the terminal branches of a medial lemniscal axon filled with horseradish peroxidase (HRP) in the cat ventral posterior nucleus. (B) Electron micrograph of labeled terminals (T1) from a similar axon. Arrowheads indicate points of synaptic contact on a dendrite (D) and on a dendritic terminal (T2). (C) Terminal portions of HRP-filled medial lemniscal axons ending (arrows) on dendrites of relay neurons close to the soma. Bars = 10 μm (A), 0.5 μm (B), 25 μm (C). From Rainey and Jones (1983).

to the synaptic aggregations make it unlikely that they represent some kind of embryological remnant reflecting the past epithelial history of the neural elements of the thalamus. Possibly, they are indeed adhesive in nature and are necessary for binding together all the elements of such a topographically rigidly organized system as the thalamus.

There are some variations in the synaptic aggregations among different nuclei. In the lateral geniculate nucleus an optic tract axon terminal is more commonly the central element, with dendrites, dendritic appendages, and presynaptic dendrites surrounding it. In other nuclei, dendrites or dendritic protrusions are more commonly the central elements with subcortical afferent terminals and presynaptic dendritic terminals arranged around them. In either case, a single subcortical afferent terminal can contact several dendrites, potentially of different cells, and can make multiple synaptic contacts on all of them. In all nuclei the large terminal is nearly always subcortical in origin, though in the centre médian nucleus the fibers derived from axons of the globus pallidus are more typical of corticothalamic axon terminals (Harding, 1973a,b). In the inferior pulvinar of the squirrel monkey, Mathers (1972a) has indicated that although most of the large terminals are on axons arising in the superior colliculus, some may come from corticothalamic fibers. In the anterior nuclei, the mamillothalamic tract axons give rise to the large terminals and the fornix to those typical of corticothalamic terminals (Dekker and Kuypers, 1976). There are more large terminals in the relay nuclei of the cat than in nuclei of the lateral complex (Madarász *et al.*, 1981).

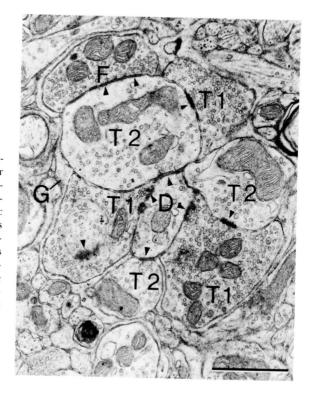

FIGURE 4.18. Electron micrograph from the cat ventral posterior nucleus showing some of the characteristic features of the synaptic aggregations found in most thalamic nuclei. Dendritic protrusion (D) is postsynaptic (arrowheads) to terminals (T1) of ascending afferent fibers and to presumed presynaptic dendrites (T2). Presynaptic dendrites are themselves postsynaptic to the ascending afferent terminals and F-type terminals appear to be derived from interneurons of the type illustrated in Figs. 4.2, 4.3, and 4.7. G indicates ensheathing astroglial processes. Bar = 1 μm.

The synaptic aggregations, though descriptively convenient, are not rigidly defined entities. As localized, glial unsheathed islands, they are far more obvious in the thalami of cats, rats, and other small mammals than in primates. In primates, the same pre- and postsynaptic elements appear to be present but are more dispersed so that focal aggregations are less prominent. Even in cats the synaptic aggregations are less obvious in some regions such as the C laminae of the lateral geniculate nucleus, than in others. Even in nuclei where they are common they are not as well defined as some descriptions imply, for one or two elements can often be traced for some distance beyond an encapsulated region. Moreover, all the terminals of a subcortical afferent need not end in aggregations. Frequently, optic tract or lemniscal axons will bear a number of isolated terminals that end on isolated dendritic profiles in the neuropil outside the synaptic aggregations.

Outside the synaptic aggregations, the neuropil of the thalamic nuclei is dominated by very large numbers of small axon terminals, 1–1.5 μm in diameter (Fig. 4.19). Many of these are the terminals of corticothalamic axons (Jones and Powell, 1969a,b). These terminals tend to be rather electron dense, contain large

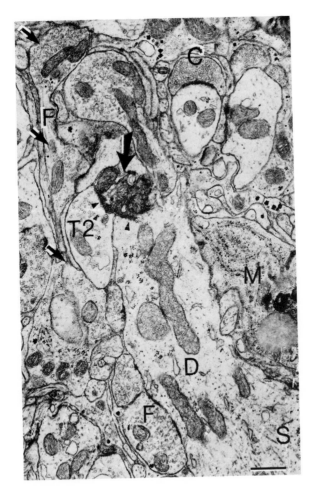

FIGURE 4.19. Electron micrograph from the cat ventral posterior nucleus showing a medial lemniscal terminal (large arrow) degenerating 4 days after destruction of the contralateral dorsal column nuclei. This terminal is making synaptic contact (arrowheads) with a presynaptic dendrite containing ribosomes (T2) and with a proximal dendrite (D) close to its point of origin from its parent cell soma (S). Presumed axon terminals (F) of interneurons arise as dilations (small arrows) of a single axon. Small terminal (C) is a presumed corticothalamic fiber terminal. M indicates a microglial cell. Bar = 1 μm.

numbers of spherical synaptic vesicles, and make asymmetric synaptic contacts. In the lateral geniculate nucleus, Guillery refers to them as RSD terminals (round vesicles, small terminals, and dark mitochondria). They appear to be more concentrated in the relay nuclei than in nuclei of the lateral complex (Madarász *et al.*, 1981). The small dense terminals arise from thin, myelinated axons 1–2 μm in diameter and terminate, outside the synaptic aggregations, both on conventional dendrites and on the parent dendrites of the presynaptic dendritic terminals before these enter the aggregations (Fig. 4.16); occasionally one of the small dense terminals can be seen ending on a presynaptic dendrite at the periphery of an aggregation. More commonly, however, the small dense terminals end in rows, often of considerable length, on thin dendritic profiles. In most studies, these thin dendrites have been shown to be thinner, more peripheral branches beyond the primary dendrites and, therefore, distal to the principal sites of termination of the subcortical afferents on relay neurons. In the cat medial geniculate nucleus, however, Morest (1975) sites the axodendritic contacts made by the small dense axon terminals proximal to those made by the lateral lemniscal axon terminals. Apart from their terminations on the presynaptic dendrites, the extent of terminations of the small dense terminals on conventional dendrites of the putative interneurons has not been determined.

Another question that remains to be answered is the proportion of small dense terminals that belong to corticothalamic axons. It has virtually become dogma that all such terminals are those of corticothalamic axons. But the few studies (Jones and Powell, 1969b; Guillery, 1971b; Morest, 1975) that have demonstrated degeneration of these terminals following ablation of the cerebral cortex, have reported that although many degenerate, others remain intact. Conceivably, some of the latter were simply captured at a stage before overt degenerative changes had set in or their parent axons had escaped the surgical lesion. But it remains possible that those remaining intact are derived from sources other than corticothalamic axons. Intranuclear collaterals of thalamocortical axons once seemed a possible source. Thalamocortical axon terminals in the cortex, though usually larger (Jones and Powell, 1970b), are in all other respects similar to the small dense terminals of the thalamus and one would expect the terminals of collaterals to resemble those of the primary axon. Guillery (1971b) has also suggested that some of the degenerating small dense terminals observed after cortical lesions could be those of thalamocortical axon collaterals degenerating retrogradely. Some recent studies have revealed a few intranuclear axon collaterals (Stanford *et al.*, 1983); others have revealed none (Yen and Jones, 1983). Sources such as the reticular nucleus and the perigeniculate nucleus also seem ruled out on account of the axons of their cells ending in symmetrical synapses (Ohara *et al.*, 1980; Montero and Scott, 1981). The only other potential sources of the small dense axon terminals other than the cortex are some of the nonspecific afferent systems (Chapter 3) though the morphology of the terminals is also against this.

The final synaptic elements in the dorsal thalamus have only recently been positively identified. They are the terminals of axons entering from the reticular nucleus (Fig. 4.16). These are relatively small, contain vesicles that flatten, and end in symmetrical membrane complexes. Some such terminals in the cat lateral geniculate nucleus appear to belong to axons arising in the perigeniculate nucleus rather than in the reticular nucleus. In the cat lateral geniculate nucleus, Montero and Scott (1981) and Ohara *et al.* (1980) describe the reticulogeniculate terminals as ending only on dendrites or somata outside the synaptic aggregations. None have yet been described on the presynaptic dendrites. The number of reticular nucleus axon terminals has not been assessed. Light microscopic immunocytochemistry shows the dorsal thalamus to be filled with large numbers of terminal boutons containing GAD (Houser *et al.*, 1980; Ohara *et al.*, 1983) and the number of high-affinity receptor sites for GABA is correspondingly high (Palacios *et al.*, 1981). However, until the relative proportions of GABAergic terminals derived from reticular nucleus cells and from intrinsic dorsal thalamic neurons (Sterling and Davis, 1981; Ohara *et al.*, 1983) are determined, the density of reticular nucleus synapses remains also unclear. One thing that is obvious is that the number of axosomatic synapses in the thalamus is rather low (Jones and Powell, 1969c). However, it is doubtful that this can yet be taken as indicative of the proportion of reticular nucleus axon terminals.

4.3.2. How Do the Cells with Presynaptic Dendrites Work?

As pointed out earlier, it is likely that the neurons with presynaptic dendrites in the thalamic relay nuclei are a form of GABAergic interneuron, though the question of their collateral projection to the cortex remains unresolved. The principal effect attributed to the operations of these neurons is the prolonged inhibition of relay cells that succeeds their activation by an afferent volley. Over the same time period, the interneurons are discharging repetitively (Fig. 4.20). The rebound phase of enhanced excitability [the "synaptic exaltation" of Andersen *et al.* (1964)] that follows the inhibition of the relay cells is thought to promote the rhythmicity of discharge, characteristic of the relay neurons (Fig. 4.20). There is still, however, an element of conjecture in all this and we are obliged to confess that we know little about the means whereby the neurons with presynaptic dendrites exert their effects.

One of the major sources of the afferent drive to these neurons is presumably the extrinsic afferent fibers of the medial lemniscus, optic tract, etc. The terminals of the afferent fibers end on the presynaptic dendrites close to the points of synaptic contact made by the afferent terminals and the presynaptic dendrites on dendrites of relay neurons. The morphological and physiological evidence already alluded to leaves little doubt that the afferent synapse is an excitatory one. What is not known, however, is whether depolarization of individual presynaptic dendritic terminals is sufficient to cause these to release their transmitter directly and independently on terminals arising from other dendrites of the same cell. An alternative condition for release might be that synaptic potentials should first invade the soma and lead to an axon potential in the cell. This, on invading the dendrites, might only then cause transmitter release but in a synchronous fashion from all presynaptic dendrites. The first of the alternatives offers the possibility of a more graded release of transmitter, especially if synaptic potentials were insufficiently large to invade the soma. The thinness of some of the stalks of the presynaptic dendrites might well predispose to this (Ralston, 1971).

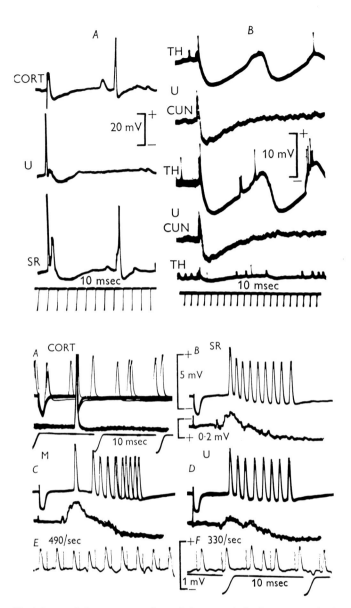

FIGURE 4.20.   (Top) Intracellular responses from thalamocortical relay neurons in the cat ventral posterior nucleus. (A) Electrical stimulation of the contralateral ulnar (U) and superficial radial (SR) nerves sets up large spike potentials and later IPSPs. Cortical stimulation (CORT) produces a small initial depolarization and a later IPSP. (B) Electrical stimulation of the contralateral ulnar nerve sets up EPSPs and later rhythmic IPSPs in thalamocortical relay cells (TH) with signs of burst discharges on the depolarization between successive IPSPs. CUN indicates simultaneous recorded trace over cuneate nucleus, and in lowest TH record, electrode has moved virtually to an extracellular position. (Bottom) Extracellular records of responses attributed to an interneuron in the ventral posterior nucleus. Cortical stimulation (A) fails to excite the cell or to inhibit spontaneous discharge. Electrical stimulation of the contralateral median (M), superficial radial (SR), and ulnar (U) nerves results in characteristic high-frequency discharges. E and F show following of high-frequency stimuli at frequencies indicated. Both from Andersen et al. (1964b).

The two alternatives are not mutually exclusive, however, and could serve to reinforce one another. The repetitive firing of interneurons in response to an afferent volley suggests that synchronous release of transmitter might eventually predominate. This might explain the relatively slow rise time but prolonged course of the IPSPs that occur in the relay cells that are postsynaptic to the interneurons.

The role of the conventional axons of the neurons with presynaptic dendrites has not yet been mentioned. Their terminals are usually described as terminating in symmetrical synapses on the presynaptic dendrites. Conceivably, these synapses could serve to effect self-inhibition of the interneurons which would then release the relay neurons from the inhibition imposed on them by the interneurons and this could lead to the characteristic postinhibitory rebound excitation.

All that has been written here is largely conjectural and much work remains to be done before it can be validated or disproved.

4.3.3. Analysis of Synaptic Events

Many early studies of the thalamic relay nuclei were concerned with the mapping of receptive fields of individual neurons, especially in an effort to determine the nature of the representation pattern of a peripheral receptive surface such as the body, the retina, or the basilar membrane. These studies are referred to in Part IV devoted to the individual thalamic nuclei. Some of the earliest studies, however, were devoted to the analysis of synaptic events during the passage of afferent impulses through a relay nucleus (e.g., Bishop *et al.,* 1959a, 1962; Erulkar and Fillenz, 1960; Nelson and Erulkar, 1963; Fuster *et al.,* 1965a,b; Burke and Sefton, 1966; Aitkin and Dunlop, 1969).

Perhaps one of the most influential of the earlier studies was that carried out in Sir John Eccles's laboratory (Fig. 4.21) by Andersen *et al.* (1964a,b; Andersen and Sears, 1964) on the ventral posterior nucleus of the cat. It still remains the classic work to which all other studies have tended to defer. In a combined intra- and extracellular study, Andersen *et al.* first identified thalamocortical relay cells by antidromic invasion following a stimulus applied to the sensory cortex or subcortical white matter. After electrical stimulation of peripheral nerves, relay cells were usually powerfully excited, as would be expected, but the excitation was followed within 1 msec by a profound IPSP (Fig. 4.20) lasting as long as 100 msec and commonly succeeded by a heightened excitability ("postanodal exaltation") that led to reverberatory burst discharges. These features of relay cell activation have been confirmed by every subsequent study. It was reported by Andersen *et al.* that relay cells were subject to both pre- and postsynaptic inhibitory effects. Presynaptic inhibition was not measured directly but it was found that volleys in peripheral nerves often led to prolonged increases in the excitability of presynaptic lemniscal axon terminals. This, in turn, was thought to indicate a level of presynaptic depolarization that could be responsible for presynaptic inhibition. The increase in lemniscal terminal excitability was measured by observing the sizes of antidromic spikes recorded in the cuneate nucleus in response to a thalamic stimulus of a fixed size. Comparable excitability changes could not be detected following cortical stimulation.

A set of presynaptic inhibitory interneurons was then tentatively identified on the basis of their displaying prolonged, high-frequency discharges in response to volleys in peripheral nerves (Fig. 4.22). The cells continue to be identified by this characteristic. They have high resting rates of discharge that are not influenced by cortical stimulation; they follow high frequencies of peripheral nerve stimulation at latencies comparable to those of relay cells but, unlike the relay cells, the EPSPs induced in them are not followed by the characteristic prolonged IPSPs. The response of these putative interneurons to a volley in one peripheral nerve is depressed by succeeding volleys in other nerves and this was suggested to indicate that lemniscal axons terminate on them as well as on relay cells. In the scheme envisaged by Andersen *et al.*, lemniscal terminals were thought to terminate on both relay neurons and presynaptic inhibitory interneurons and the axons of the interneurons were thought to terminate in axoaxonic contacts on lemniscal terminals. Putative presynaptic inhibition of this kind has also been described in the lateral and geniculate nucleus (e.g., Burke and Sefton, 1966; Fuster *et al.*, 1965a; Singer and Creutzfeldt, 1970). There is, unfortunately, no morphological basis for it, as has been outlined in an earlier section, and several

FIGURE 4.21. Sir John C. Eccles (1903–   ), formerly Professor of Physiology, The John Curtin School of Medical Research, Canberra. Photographed by the author in 1983.

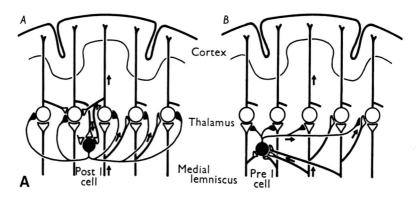

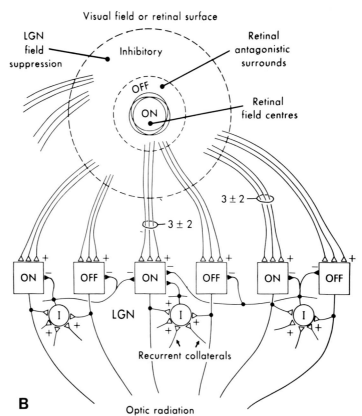

FIGURE 4.22. (A) Diagrams by Andersen *et al.* (1964b) showing their postulated pathways for what they interpreted as pre- and postsynaptic inhibition in the ventral posterior nucleus. Scheme A for postsynaptic inhibition postulates effects mediated by collaterals of thalamocortical relay cells which are now thought to occur in the reticular nucleus rather than in the ventral posterior nucleus. Part B, for putative presynaptic inhibition, is still acceptable where it shows excitation of inhibitory interneurons by medial lemniscal terminals but the idea of their axons ending presynaptically on lemniscal terminals has been ruled out (cf Fig. 4.16). (B) Schematic figure showing typical concentric, antagonistic receptive field of dorsal lateral geniculate neurons and the connections that probably underlie it. An essential feature was thought to be the recurrent collaterals of thalamocortical fibers activating inhibitory interneurons. The interneurons shown are now thought to lie in the reticular or perigeniculate nucleus and others are activated directly by retinal axons. From Levick *et al.* (1972).

methodological criticisms have been leveled at the interpretation of the results in terms of presynaptic inhibition (Burke and Cole, 1978; Singer, 1977). In the eyes of these reviewers, the depolarization of optic nerve terminals that accompanies passage of an afferent volley through the lateral geniculate nucleus can be accounted for entirely in terms of a buildup of potassium ions in the extracellular space, primarily as the result of the active relay cell discharges (Singer and Lux, 1973).

A second category of interneuron with a postsynaptic inhibitory effect was reported by Andersen *et al.* (Fig. 4.22). These neurons were thought to be activated by intranuclear collaterals of the cortically projecting axons of the relay cells, for they could be induced to fire repetitively by an afferent volley or, antidromically, by stimulation of the white matter beneath the cortex (the cortex was usually removed some days earlier in an effort to deactivate corticothalamic cells). The putative postsynaptic inhibitory interneurons were thought to be responsible for the prolonged IPSPs occurring in the relay neurons within 2 msec of their initial excitation by an afferent volley or by subcortical stimulation. The prolonged rise time of the IPSP (seen in Fig. 4.20), it was felt, could be attributed to the repetitive firing of the interneurons. No convincing reasons were given for ruling out direct lemniscal activation of the interneurons, other than the latency of onset of the IPSPs in the relay cells. Apart from this, the pattern of organization postulated demanded the existence of an extensive system of thalamocortical axon collaterals which has also been difficult to define morphologically (see below). Nevertheless, the prolonged, slowly rising IPSPs continue to be reported in relay cells of all sensory thalamic nuclei following an afferent volley or cortical stimulation and require explanation (Burke and Sefton, 1966; Aitkin and Dunlop, 1969; Singer and Creutzfeldt, 1970; Singer and Bedworth, 1973; Lindström, 1982).

Subsequent to the work of Andersen *et al.*, several reports of work carried out in both the ventral posterior and the lateral geniculate nucleus of the cat have indicated that the latency difference between the onset of the EPSP and the subsequent IPSP in a relay cell in response to an afferent volley is sufficiently short (ca. 0.8 msec) to be consistent with direct lemniscal or optic tract activation of both relay neurons and interneurons (Sumitomo *et al.*, 1969; Coenen *et al.*, 1972; Singer *et al.*, 1972; Iwamura and Inubushi, 1974a,b; Tsumoto and Nakamura, 1974; Baldissera and Margnelli, 1979; Eysel, 1976; Stevens and Gerstein, 1976; Dubin and Cleland, 1977; Mooney *et al.*, 1979).

Eysel (1976) reported a bimodal distribution of IPSP latencies in relay cells of the cat lateral geniculate nucleus following electrical stimulation of the optic chiasm. He pointed out that the longer latencies (3–4.8 msec) would be consistent with activation of inhibitory interneurons by relay cell collaterals but that the shorter latencies (2–2.6 msec) could indicate either monosynaptic activation of interneurons by slower-conducting (X-type) optic tract axons or their activation by collaterals of relay cells receiving faster-conducting (Y-type) axons. Singer and Bedworth (1973) and Eysel (1976) found that stimulation of the optic chiasm at intensities below the threshold for X axons but above the threshold for Y axons led to EPSPs with latencies of 1–1.5 msec in Y-type relay cells and later IPSPs with latencies of 1.6–2.6 msec in X-type relay cells. Lindström (1982) saw a similar latency difference to stimulation of the optic tract in a population of

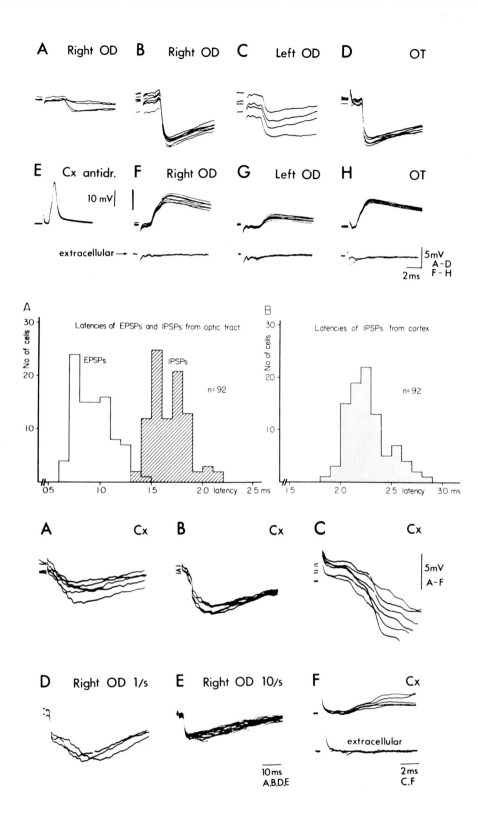

mainly Y cells. He noted that the EPSPs were always engendered by stimulation of one eye but IPSPs could be engendered in the same cell by stimulation of either eye. As the IPSPs engendered by each eye do not facilitate or occlude one another, they must be mediated by separate populations of interneurons. Moreover, the IPSP engendered by stimulation of the excitatory eye was probably mediated by retinal ganglion cells other than those providing the excitatory drive, because the IPSPs usually consist of several components independent of the more nearly unitary EPSPs. Dubin and Cleland (1977) have also adduced evidence for interneurons selectively activated by X or Y inputs, with those receiving Y inputs activated by branches of Y axons innervating Y relay cells. All of these facts point to the presence of a variety of patterns of inhibitory interneuronal mechanisms in the lateral geniculate nucleus. Moreover, as Dubin and Cleland (1977), Mooney *et al.* (1979), and Lindström (1982) have pointed out, the latency differences between the onset of relay cell excitation and inhibition in the cells studied are too short to involve more than one intervening synaptic step and, therefore, too short to be mediated via thalamocortical axon collaterals. This evidence, therefore, points to monosynaptic afferent inputs to the inhibitory interneurons.

Lindström (1982) has also noted that Y relay cells in the cat lateral geniculate nucleus can be antidromically activated by stimulation of the cerebral cortex at latencies of 0.5 to 1 msec, yet IPSPs in the same cell occur only at latencies of approximately 2.6 msec. This he suggested is appropriate for it to be mediated by intranuclear collaterals of the antidromically activated axons and an inhibitory interneuron. The effect is too short to be mediated by the slowly conducting corticothalamic axons (Gilbert, 1977). Because IPSPs induced by stimulation of the cortex rise slowly and stepwise whereas those induced by optic tract stimulation are fast-rising and smooth, Lindström considers that two inhibitory pathways are in operation on relay neurons following passage of afferent impulses; both are mediated by inhibitory interneurons but one is a feedforward effect, via direct afferent connections on interneurons, while the other is a feedback effect, via recurrent collaterals of thalamocortical relay axons ending on interneurons (Fig. 4.23). In the latter case, at least, he feels that the inhibitory in-

---

FIGURE 4.23. (Upper) IPSPs induced in a relay cell of the cat dorsal lateral geniculate nucleus following threshold (A) and maximal (B) electrical stimulation of the right optic disc and of the right optic tract (D). The right eye was the dominant eye, as indicated by the small EPSPs in B and D. Stimulation of the left optic disc (C) leads to an IPSP only. Records A–D were taken with the cell depolarized by injection of current through the intracellular electrode. Responses F–H correspond to A–D but without the depolarizing current and with the IPSPs reversed (in F and H) due to leakage of chloride ions from the KCl electrode. Lowest traces are extracellular field potentials and E shows antidromic activation from cortex. (Middle) Latencies of EPSPs and IPSPs recorded in geniculate relay cells by optic tract stimulation (A) and of IPSPs evoked antidromically by stimulation of the visual cortex (B). Stimulus strength at two times threshold. Both sets of IPSPs indicate a disynaptic inhibitory effect on the relay cells. (Lower) IPSPs evoked in the cell shown in upper part of figure by stimulation of the visual cortex. Stimulation strengths were subthreshold for antidromic activation of the cell and cell was depolarized by current injection. IPSP is reversed in F. D and E show the IPSPs induced in the same cell by stimulation of the optic disc at 1 and 10 stimuli per second. From Lindström (1982).

terneurons are situated in the perigeniculate nucleus, for the perigeniculate nucleus is known to receive collaterals of relay cell axons from the geniculate (Dubin and Cleland, 1977; Ahlsén *et al.,* 1978, 1982b; Ferster and LeVay, 1978; Friedlander *et al.,* 1981; Ahlsén and Lindström, 1982a; see Ide, 1982, for fine structural details). It is also known to send its axons into the geniculate (Ide, 1982), and there is a close temporal correspondence between discharges in perigeniculate cells and the onset and shape of IPSPs in geniculate relay cells following antidromic stimulation of the latter's axons (Ahlsén and Lindström, 1982a). The anatomical identification of relay cell axon collaterals in the perigeniculate nucleus (Fig. 4.6) but not in the geniculate itself, also points to the perigeniculate nucleus as having a crucial role in intranuclear regulatory mechanisms. In the rabbit, Lo (1981, 1983) feels that feedback inhibition from a part of the reticular nucleus comparable to the perigeniculate nucleus may be the only source of afferent inhibition in the dorsal lateral geniculate nucleus.

All of this recent work has led us away from the necessity for presynaptic inhibition and, through the perigeniculate nucleus, from the necessity for an extensive intranuclear system of thalamocortical axon collaterals in explaining synaptic phenomena in the thalamus. It seems likely that the observations made in the lateral geniculate nucleus will turn out to have parallels in the other relay nuclei as the synaptology of the nuclei is virtually identical. Though accepting, however, that intranuclear axon collaterals are particularly uncommon in the other relay nuclei as in the lateral geniculate (Fig. 4.7), we are obliged to confess that we know of no nucleus other than the reticular nucleus comparable to the perigeniculate nucleus as a source of interneurons for these nuclei. It is possible that the perigeniculate nucleus is simply part of the reticular nucleus (see Chapter 15). Otherwise, we must seek either in the other relay nuclei themselves or in some additional nucleus for the pool of inhibitory interneurons comparable to the perigeniculate nucleus.

Tsumoto and Nakamura (1974), from intracellular records and poststimulus time histograms, felt that they could identify interneurons in the cat ventral posterior nucleus on the grounds of the synaptic delay following arrival of a lemniscal volley in the nucleus, and by their orthodromic but not antidromic activation from stimulation of the sensory motor cortex. On these grounds, they postulated the existence of two types of interneurons: one innervated by fast lemniscal fibers with a mean conduction velocity of 30 m/sec and with a synaptic delay of 0.91 msec, suggesting a single synapse; a second innervated by slower lemniscal fibers with a mean conduction velocity of 16.7 m/sec and with a synaptic delay of 1.93 msec, suggesting the presence of two intervening synapses. They suggested that joint units in the VPL were subjected to feedforward inhibition because stimulation of the medial lemniscus but not of the cortex caused depression of their activity. This would presumably be a function of the monosynaptically innervated interneurons. Hair units in the VPL were thought to be subjected to "backward postsynaptic" (i.e., recurrent) inhibition because the degree of depression of these cells was the same following stimulation of either the medial lemniscus or the sensory motor cortex. This would presumably be a function of the disynaptically activated interneurons. In the absence of intranuclear collaterals of thalamocortical relay cells, one must suppose that these interneurons are activated via the reticular nucleus, lie in the reticular nucleus itself or are situated in some hitherto unsuspected pool of neurons comparable to the perigeniculate nucleus.

191

SYNAPTIC
ORGANIZATION IN
THE THALAMUS

4.3.4. Integrative Action in the Thalamus

The thalamus cannot be regarded as a mere relay station that transfers information to the cerebral cortex in an untransformed state. It was evident to some of the early anatomists of the modern period that a considerable degree of convergence and divergence had to occur in the innervation of a thalamic nucleus by its afferent fibers (Chapter 3). And in the last two decades, physiological studies have given clear indications of some of the transformations that are brought about by interactions between afferent fibers, by the activities of thalamic interneurons, and by the influence of corticothalamic and reticulothalamic pathways. At the very least, these integrative activities of the thalamus are manifested by modifications in the receptive fields of relay cells, in comparison with the fibers innervating them. At most, they result in an actual gating of flow through the thalamus, a gating that appears to be dependent on the internal "state" of the brain.

The properties of thalamic neurons that might be regarded as "integrative" can probably be brought out in a review of the three principal sensory relay nuclei: the lateral and medial geniculate nuclei and the ventral posterior nucleus. To the extent that other nuclei have been studied, the evidence supports the notion, derived from examination of the three sensory nuclei, that thalamic neuronal behavior is characterized by the relatively secure transfer of some afferent properties to the cortex, coupled with interactive influences of a rather subtle nature that are often state dependent and which seem to depend upon the activities of local interneurons.

4.3.4.1. Lateral Geniculate Nucleus

The classic study of the receptive fields of single geniculate cells is that carried out by Hubel and Wiesel (1961) in the cat. They showed that the receptive fields of most geniculate neurons resemble those of retinal ganglion cells (Kuffler, 1953; Wiesel, 1960) more closely than they do those of the striate cortex (Hubel and Wiesel, 1959). Except for the C laminae, where cells had receptive field centers several times larger, the receptive fields of A and A1 laminae cells were of approximately the same size as those of retinal ganglion cells and exhibited the concentric organization typical of ganglion cells. That is, they had "on" or "off" centers and an antagonistic surround (Fig. 4.24). Like retinal ganglion cells, cells with receptive fields near the area centralis had smaller receptive field centers and stronger suppression from the field periphery. Hubel and Wiesel were unwilling to attribute off (i.e., inhibitory) responses to any particular mechanism, simply stating that they could be due to inhibitory optic tract fibers or to intrageniculate mechanisms.

Already by that time, it was known that the receptive fields of cells in the striate cortex were quite varied and that the center–surround type was relatively uncommon. Lest it be thought, however, that geniculate fibers relayed ganglion cell activity in a virtually untransformed state to the cortex, Hubel and Wiesel pointed out that several of their findings pointed to a certain amount of transformation occurring in the nucleus. First, a single impulse in an optic tract fiber customarily led to a burst of discharges in a geniculate cell. Second, in simultaneous recordings from an optic tract fiber and a geniculate cell that it innervated, it could be shown that the suppression from the receptive field surround was far greater for the cell than for the fiber, despite the virtual superimposition and identical size of the receptive field centers of axon and cell. Third, some cells showed prepotentials that were indicative of inputs from more than one optic tract fiber. Fourth, the discharges of the geniculate cells could be modified by the overall state of the animal (Hubel, 1960).

All these features, particularly the second which implies an increased selectivity for a visual stimulus on the part of geniculate neurons, point to integrative mechanisms operating in the thalamus and argue against any concept of the lateral geniculate nucleus being a simple relay.

At about the same time, evidence for binocular interactions of an inhibitory kind, acting at the level of the lateral geniculate nucleus, was forthcoming (Bishop *et al.*, 1959b, 1962). This showed that cells in adjacent laminae and connected to different eyes but processing information from the same portion of visual space exert inhibitory effects upon one another. Because the mutual antagonism survives decortication, it must depend upon intrageniculate mechanisms (San-

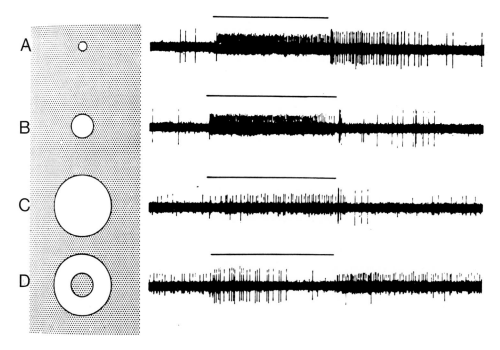

FIGURE 4.24.   Responses of a typcial "on-center" neuron recorded in layer A of the cat dorsal lateral geniculate nucleus. A, B, and C show responses to centered light spots 1, 2, and 14° in diameter. D shows inhibition by an annular stimulus with inner diameter 2° and outer diameter 14°. (Horizontal line indicates when light is on.) From Hubel and Wiesel (1961).

derson *et al.*, 1971; Singer, 1970). Later came evidence for lateral inhibition, as shown by reciprocal inhibition of neighboring cells of the same type, i.e., between on-center cells or between off-center cells (Singer and Creutzfeldt, 1970; Singer *et al.*, 1972). There was also some evidence for inhibition between cells of different types, i.e., between on-center and off-center cells (Singer and Creutzfeldt, 1970), though this is now contested (see below). In more recent times, it has been reported that the X and Y systems relaying through the lateral geniculate nucleus of the cat (see Chapters 3 and 9) also interact so that Y inputs can under certain circumstances suppress the responses of X cells in the geniculate, presumably by acting through an inhibitory interneuron (Hoffmann *et al.*, 1972; Singer, 1973b; Singer and Bedworth, 1973; Eysel, 1976). A similar interaction between fast and slow retinal ganglion cell inputs could not be demonstrated in the rabbit dorsal lateral geniculate nucleus (Lo, 1983).

Against the positive evidence for intralaminar interactions must be balanced that which shows that the channels from the retina that mediate on-center or off-center receptive fields probably do not converge in the lateral geniculate nucleus (Schiller, 1982; and Chapter 9). There is also the evidence that the inputs of the X, Y, and W ganglion cell classes are relayed to the cortex with little or no modification in the geniculate (see Chapter 9). X and Y cells with on-center and off-center receptive fields which are segregated from one another in the dorsal lateral geniculate nucleus of the mink (LeVay and McConnell, 1982) may even project to separate cortical columns (McConnell and LeVay, 1983). Hence, in the lateral geniculate nucleus at least, the thalamus appears to perform some integrative operations but it also appears to let a considerable amount of information gain access to the cortex in an untransformed state.

In other thalamic relay nuclei a combination of modulation and relatively direct transmission of information also appears to be the case.

4.3.4.2. Ventral Posterior Nucleus

Inhibitory interneuronal mechanisms, a dependence of neuronal responses on anesthetic state, and afferent convergence have all been reported in the ventral posterior nucleus. Some properties of ventral posterior neurons such as receptive field size and position and modality specificity, are particularly secure, and reflect those of the incoming medial lemniscal fibers which they relay with little or no transformation to the cerebral cortex. These properties have been termed static properties by Poggio and Mountcastle (1963). Other properties, such as the length of the recovery cycle after activation, the capacity for frequency following and the different levels of excitation observed in a ventral posterior neuron according to the position of a stimulus in its receptive field, are dependent on anesthetic state. These were termed dynamic properties by Poggio and Mountcastle and were thought to be dependent on inhibitory mechanisms.

Interneuronal mechanisms have usually been invoked in the ventral posterior nucleus in order to explain the prolonged depression that follows activation of relay cells by natural or electrical stimulation (Gaze and Gordon, 1954; Rose and Mountcastle, 1954; Poggio and Mountcastle, 1963; Andersen *et al.*, 1964a; Tsumoto and Nakamura, 1974; see Section 4.3.3). Though more than one type of inhibitory interneuron has sometimes been postulated (Andersen *et al.*, 1964a,b; Tsumoto and Nakamura, 1974; see Section 4.3.3), there is little evidence for differential effects of a kind that could be correlated with the interneuronal mechanisms outlined in the preceding section on the lateral geniculate nucleus. Indeed, in the earlier studies, in both anesthetized and unanesthetized animals, lateral inhibition was difficult to detect in the ventral posterior nucleus (Poggio and Mountcastle, 1963; Golovchinsky *et al.*, 1981).

By using "quasi-intracellular" (or "imminent membrane contact") recording, in the cat ventral posterior medial nucleus, Gottschaldt *et al.* (1983) claimed that the long-recognized prepotentials that precede the discharge of a thalamic cell under the influence of afferent driving (Rose and Mountcastle, 1954; Bishop, 1953; Maekawa and Purpura, 1967a,b) are dendritic potentials (possibly even dendritic spikes), rather than, as long assumed, afferent fiber discharges. Large unitary EPSPs possibly corresponding to the dendritic potentials have been recorded in ventral posterior and in medial and lateral geniculate neurons under the influence of afferent driving for some time (Bishop *et al.*, 1962; Andersen *et al.*, 1964a,b; Aitkin and Dunlop, 1969). Gottschaldt *et al.* found that the synaptic transmission of afferent fiber spikes to dendritic prepotentials is very secure but that the conversion of the dendritic potentials to relay cell discharges can be enhanced or diminished depending on the sites and characteristics of mechanical stimuli applied in the excitatory receptive field of the neuron. Using DC recording and feeding the slow-wave forms of the prepotentials through a high-pass filter, they could transform the prepotentials into single, fast signals and thus develop comparable poststimulus time histograms for these input signals and for the somal spikes in the recipient cell.

They were able to show that for some cells a tonic set of input signals is followed by a tonic set of output signals. For others, however, a tonic set of input signals leads only to a transient (phasic) output signal (Fig. 4.25A). Interestingly, the nature of the conversion often varied with the nature of the stimulus. For example, bending a hair in one direction might cause a sustained-to-sustained conversion but bending in the opposite direction might lead to a sustained-to-transient conversion (Fig. 4.25B). Vibratory stimuli superimposed on a bending stimulus (Fig. 4.25B) or stimulation of a different sinus hair could also affect the type of conversion.

For many cells Gottschaldt *et al.* showed that a sustained-to-sustained conversion could be changed into a sustained-to-transient conversion by the iontophoresis of GABA or a sustained-to-transient conversion changed to a sustained-to-sustained conversion by the iontophoresis of the GABA antagonist, bicuculline (Fig. 4.25C). They speculate, therefore, that the conversion of the prepotentials to somal spikes is controlled by GABAergic interneurons whose terminals are focused on the impulse-initiating site of the thalamocortical relay cell. The electron microscopic findings reviewed in Section 4.3 show that the probable GABAergic dendritic terminals of the intrinsic thalamic interneurons and the more conventional GABAergic axon terminals of reticular nucleus cells do not terminate on the axon hillocks or initial segments of the relay neurons. They are, however, clustered in considerable numbers around the principal afferent fiber terminals and so could conceivably be exerting a controlling influence at that site. Whatever the mechanism, the effect is that the output of a set of thalamocortical neurons with the same receptive fields would vary depending on the nature and exact position of a stimulus in the receptive field. Gottschaldt *et al.*, therefore, point out that "the input to the somatosensory cortex, provided by the population of thalamo-cortical fibres, will no longer be a faithful replication of the activity in primary afferent or trigemino-thalamic fibres. Instead, there should be a dynamically changing activity pattern in each individual thalamo-cortical fibre, and of course in the whole fibre population, varying with the site and kind of peripheral stimulation."

It is difficult yet to see where similar mechanisms could be operating at the thalamic level on the relay of other somatic sensory receptors. In the somatic sensory cortex, neurons responding to low-frequency vibrations (below 80 Hz) have discharge patterns phase-locked to the stimulus and very comparable to those of the primary afferents (Mountcastle *et al.*, 1969), suggesting that relay through the thalamus has had but little effect on the signals. Most cortical neurons responding to high-frequency vibrations (above 80 Hz), on the other hand, are reported to follow the stimulus in less than a one-to-one fashion and thus to be different from the primary afferents (from Pacinian corpuscles) which can be phase-locked to the stimulus up to about 1000 Hz (Mountcastle *et al.*, 1969; Ferrington and Rowe, 1980). It is not known, however, whether this transformation is effected at thalamic levels or in the cortex itself. Mountcastle *et al.* (1969) give reasons for believing that it is a cortical effect. Information bearing on this has been reported by Hellweg *et al.* (1977). These authors found that thalamocortical axons entering the whisker representation in the cat's SI cortex could be driven by stimulation of the contralateral infraorbital nerve at much higher frequencies (up to 200 Hz) than the cortical cells among which they terminated (Fig. 4.26). Cortical cells respond with a typical EPSP–IPSP sequence but will follow peripheral stimulation up to only 20–30 Hz. Tonic response components to natural stimuli were also more common in thalamocortical fibers than in cortical cells, i.e., most of the fibers recorded responded to whisker displacement with an onset transient and a sustained discharge; in the cells the sustained component was largely eliminated. The dynamic responses of the fibers to changes in stimulus amplitude and velocity were also greatly reduced in the cortical cell. About half the fibers responded to bending of a single whisker whereas 80% of the cortical cells would usually respond to bending of more than one but in a graded manner, i.e., one hair was more or less effective than its neighbors in discharging the cell. Cortical cells also showed inhibition from many more whiskers than those whose stimulation led to excitation and also from the general muzzle fur. Most of these features, according to the authors, can probably be explained by the intervention of intracortical inhibitory mechanisms in the thalamocortical transformation process. None of their recordings, however, specifically identified cortical interneurons that might be responsible.

The transfer characteristics of neurons in the ventral posterior nucleus that respond to movements of joints have been examined in two studies. Mountcastle *et al.* (1963) and Werner and Mountcastle (1963) concluded from their studies of the ventral posterior nucleus of unanesthetized monkeys that neurons responding to movements of joints could signal joint position to the cortex by the frequency of their discharges. They found that many joint-related neurons discharged with movement in one or other direction, always increasing their average frequency of firing, monotonically, toward the extreme of movement in the

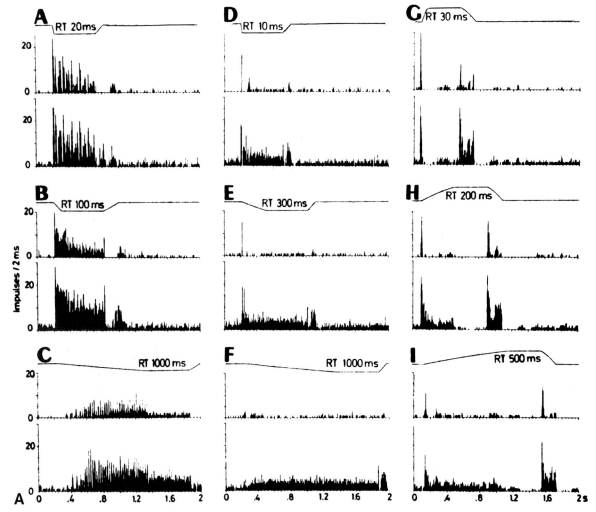

FIGURE 4.25. (A) Responses of three VPM relay neurons to trapezoidal movement of a single sinus hair. Upper trace indicates the movement, lower poststimulus time histogram indicates input response of the neuron, and upper histogram the output signals of the neuron. A–C, D–F, and G–I are from the three different neurons. The input–output responses are different for each neuron. In A–C there is an input to the cell in response to the return movement of the stimulus but this is reduced in the output response. Velocity and amplitude responses, however, are clear in both the input and the output. In D–F the sustained component of the input is suppressed in the output. In G–I the dynamic input response is cut down to a phasic output response. The suppression is less marked in relation to the return movement. From Gottschaldt *et al.* (1983a). (B) Responses of a thalamic neuron in the cat to simple vibratory movements of a single sinus hair on the face (left column) and to combined bending and vibration of the same hair (right column). A in both columns shows original recordings of spikes and prepotentials. In B, from same cell, middle trace indicates stimulus, upper poststimulus time histogram shows soma spikes that represent output of the cell, and lower poststimulus time histogram shows the prepotentials that represent the inputs to the cell. C shows similar data from another cell. In B the prepotentials are transformed into a better soma spike response when vibration is superimposed on a bending stimulus. In C the prepotentials and soma spikes are combined for each of four different types of stimulation. Bending alone elicited only an off response, a vibratory stimulus alone elicited a sustained response in which prepotentials were converted better into soma spikes depending on direction of the added bending stimulus. Off response also disappears. From Gottschaldt *et al.* (1983b). (C) Prepotentials and spikes recorded from the same thalamic cell in response to vibration of a single sinus hair in the cat. Under normal conditions (A) the prepotential to soma spike conversion is a sustained-to-transient one. In the presence of increasing amounts of iontophoresed bicuculline (B, C) the conversion becomes a sustained-to-sustained one. Iontophoresis of GABA (D) to antagonize the bicuculline effect leads to a return to the sustained-to-transient conversion. From Gottschaldt *et al.* (1983b).

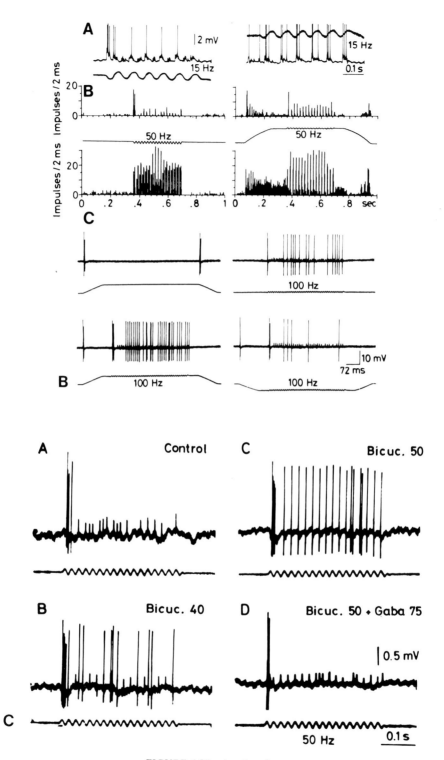

FIGURE 4.25. (*continued*)

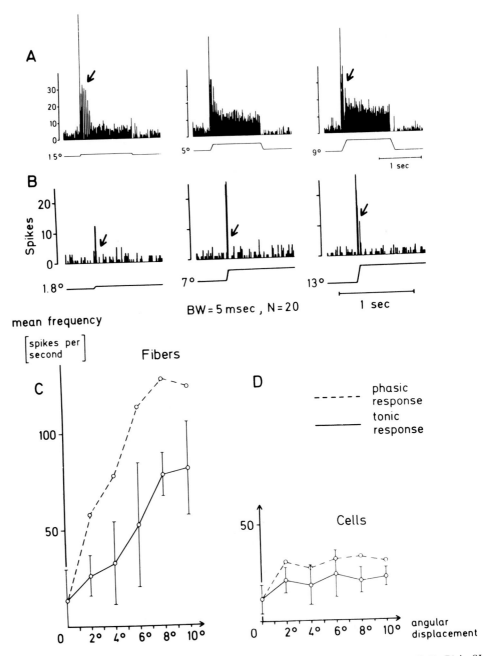

FIGURE 4.26. Dynamic range responses of a thalamocortical fiber (A, C) and a cell (B, D) in SI whisker representation of the cat, to various amplitudes of bending of whiskers. For fibers the amplitude of the tonic response increases with increasing stimulus amplitude; the peak of frequency of the initial phasic response does not change but the integrated discharge rate during the first 30 msec increases with stimulus amplitude. In the cell there is no change in maintained discharge rate. The initial peak frequency shows an increase between 1.8 and 7° of bending but not beyond. A and B show responses of a single fiber and cell; C and D include data from nine fibers and nine cells. From Hellweg *et al.* (1977).

relevant direction. Hence, the position of a joint should be signaled by a particular frequency of discharge. Joint angles over which discharges occurred averaged 73°. At the time, the available evidence from studies of joint afferent nerves in cats suggested that some primary joint afferents were single-ended, as in the thalamus, and increased their discharges in one direction of movement. They were said, however, often to have much narrower excitatory joint angles, on average 20° (Andrew and Dodt, 1953; Boyd and Roberts, 1953; Skoglund, 1956). This suggested to Mountcastle and his associates that convergence and integration were occurring between the periphery and the thalamus so that the spatial definition of joint angle by individual groups of joint afferents became converted to a definition in terms of frequency of discharge. Whether this conversion was a function of the thalamus or of lower centers was not established. As a further indication of integrative action, Mountcastle *et al.* (1963) noted that although joint primary afferent discharges were remarkably steady and uniform over long periods for a fixed joint position, those of thalamic joint neurons tended to occur at irregular intervals though maintaining an overall constant average frequency. The significance of this change in spatial pattern for the cortex was considered by them at some length.

The early work on joint nerves indicated that other primary afferents could have double-ended receptive angles (i.e., they would increase their discharge in two directions of movement). Subsequent work on primary afferents in joint nerves or in the dorsal columns of the spinal cord has shown that most afferents are double-ended and that few possess the narrow excitatory angles formerly attributed to many of them; most instead have large excitatory angles and many increase their discharge toward both extremes of movement of the joint rather than at intermediate angles. Moreover, their rate of discharge can be related to the frequency of movement of the joint (Burgess and Clark, 1969; McCall *et al.*, 1974).

Similar cells have now been reported in the ventral posterior nucleus of the cat thalamus by Yin and Williams (1976). These workers found that most knee joint-related neurons had rapidly adapting (i.e., phasic) discharge patterns, rather than the slowly adapting or tonic patterns demonstrated by the joint-related neurons studied by Mountcastle *et al.* (1963) in the monkey. Yin and Williams found that the phasic cell could be made to fire over a wide range of joint angles if velocity or frequency of movement is high, though most have a receptive angle of about 25° where sensitivity is greatest. As frequency of movement increases, so the discharges become phase-locked, i.e., impulses occur preferentially in a narrow range of joint (or phase) angles and at high frequency a neuron may respond at two different phase angles, 180° apart (Fig. 4.27). This is, therefore, a bidirectional response akin to that seen at the periphery. There is, however, a gradual lag in the peak of the phase-locked response that appears to be additional to the conduction delay from periphery to thalamus. This Yin and Williams see as representing a low-pass filter effect introduced by the thalamic neurons themselves which could serve to filter out the noise components of a signal. Possibly such an effect could be introduced as the result of mechanisms of the kind suggested by the work of Gottschaldt *et al.* (1983a,b) alluded to earlier.

The role of joint afferents in signaling joint position has been called into

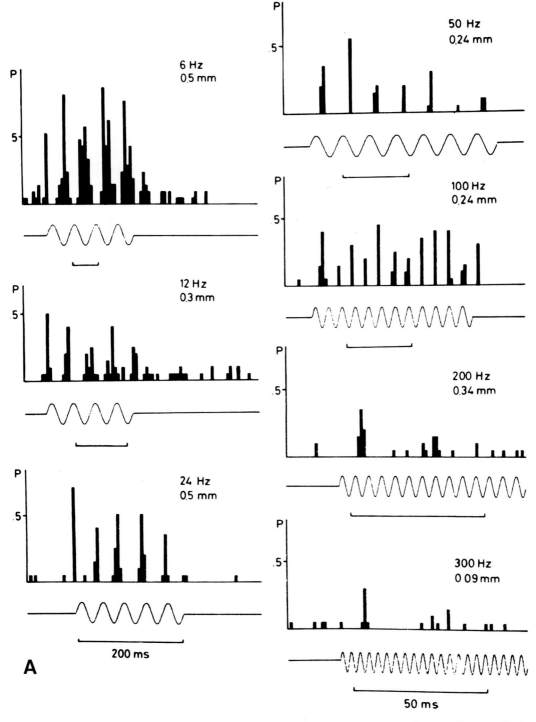

FIGURE 4.27. (A) Poststimulus time histograms showing probabilities of discharge (ordinates) of a cell in the monkey ventral posterior nucleus responding to stimulation of group I muscle afferents by sinusoidal displacement of the muscle at different frequencies (abscissae) and at amplitudes just above threshold (mm). Partial driving of the neuron is seen with frequencies up to 100 Hz. At 200–300 Hz, only on-responses are seen. From Maendly *et al.* (1981). (B) Cycle

question by Burgess and Clark (1969) who consider that muscle spindle afferents innervating muscles working on a joint are much more likely candidates. As we shall see, however (Chapter 7), the neurons studied by Mountcastle *et al.* lay in part of the ventral posterior nucleus in which the central projection pathway of spindle primary afferents does not relay, whereas those studied by Yin and Williams were in a group I relay zone where joint afferents may not relay (Andersson *et al.*, 1966). Any influence of spindle afferents over these thalamic joint neurons would therefore have to occur by convergence at lower levels of the neuraxis. The possible influence of muscle afferents over thalamic neurons responding to joint movement and the different types of movement-related neurons studied by the two groups make it difficult at this time to discern the exact nature of any transformation occurring at thalamic levels between position and movement detectors of the periphery and the cerebral cortex.

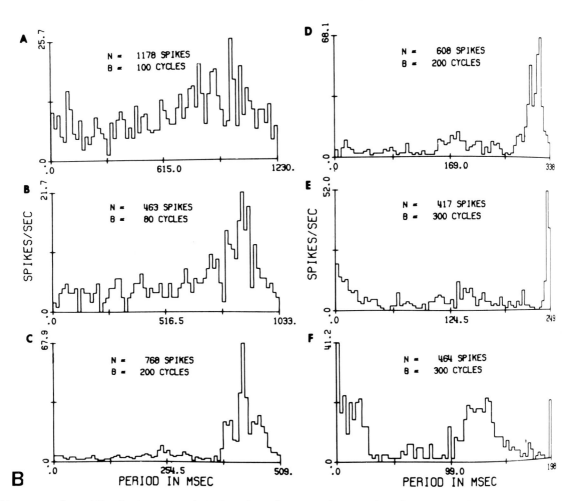

histograms of a rapidly adapting neuron in VPL nucleus of an anesthetized cat responding to different frequencies of sinusoidal movements of the knee joint with a bias angle of 80° and an excursion angle of 5°. N: total number of impulses counted; B: number of complete cycles of input. First half-cycle is extension of limb, second half-cycle flexion. Abscissae show period of input sine wave. From Yin and Williams (1976).

The transfer characteristics of thalamocortical relay neurons responding to group I muscle afferent input appear to be quite faithful reflections of muscle spindle primary afferent discharges. The properties of VPL neurons responding to stimulation of low-threshold muscle afferents and projecting to area 3a have been well studied in cats by Andersson *et al.* (1966) and in monkeys by Maendly *et al.* (1981). Those neurons situated in the forelimb representation were shown to have antidromic latencies in response to stimulation of area 3a, of 1.4 msec or less in monkeys and 1 msec in cats, implying large axons. All responded to group I volleys in the contralateral deep radial nerve and in the study of Maendly *et al.* showed high sensitivity to small controlled stretches of single forearm muscles (Fig. 4.27). As the frequency of sinusoidal stretching increased (up to 300 Hz), so the smallest effective stretch declined to as low as 25 μm. This and certain other properties suggest driving by muscle spindle primary endings. Most neurons displayed dynamic (or "on") discharges that, at least up to 100 Hz, were in synchrony with sinusoidal displacement. A few, however, discharged tonically, in response to maintained stretch. Overall, their properties were rather similar to those of neurons in area 3a, suggesting a very secure relay to the cortex.

In the thalamus single group I relay neurons could be shown to receive inputs from more than one muscle on the back of the forearm, though never from muscles on the back and front of the forearm. Approximately 45% of the group I activated cells in monkeys and cats also received convergent cutaneous inputs, identified by natural stimulation or by electrical stimulation of a cutaneous nerve. In this again, the thalamic cells resemble those of area 3a (Heath *et al.,* 1976). Convergent inputs from the elbow joint nerve could not be detected in the cat. Rather curiously, although afferent volleys (as monitored in the common radial nerve) in the deep and superficial radial nerves of the monkey entered the nervous system at approximately the same time, in the thalamus the latency of the superficial radial nerve response was 1.4 msec longer than the deep radial nerve response in the same cell. A difference of 0.5 msec was observed in the arrival of the comparable afferent volleys in the cat thalamus. The nature and/ or site of the delay are currently unclear.

The convergence of other afferent pathways onto single ventral posterior thalamic neurons has been explored to some extent. In the cat, electrical stimulation of the spinocervicothalamic pathway results in field potentials and single-unit discharges localized to an anterolateral portion of the ventral posterior lateral nucleus. In this region two-thirds of the units responding to spinocervicolemniscal inputs could also be shown to receive convergent inputs at slightly longer latency from the dorsal column lemniscal pathway and/or from the radial nerve (Landgren *et al.,* 1965). These findings were later confirmed by intracellular recording in which it was demonstrated that many of the cells receiving such convergent input project to the first and/or second somatic sensory areas of the cerebral cortex (Andersen *et al.,* 1966). Among the cells receiving convergent input, postexcitatory inhibition following spinocervicothalamic input was much weaker than that following inputs from the medial lemniscus or from peripheral nerves. It is difficult even with the present state of knowledge to suggest intrathalamic pathways by which these interactive effects might be mediated. Nevertheless, the effects are reminiscent of those exerted by Y cells over X cells in the lateral geniculate nucleus (Section 4.3.4.1) and raise the possibility of a similar mechanism operating in the ventral posterior nucleus.

Three aspects of thalamic integration were reviewed under the lateral geniculate and ventral posterior nucleus: afferent convergence, inhibitory interneurons, and state-dependent activity. Afferent convergence has not been explored to any extent in the medial geniculate nucleus and it is difficult to find evidence to show the nature of any transformations that might occur between auditory thalamus and auditory cortex. So far as they have been studied, the responses of neurons in the principal (ventral) nucleus of the medial geniculate complex are remarkably similar both to those of the central nucleus of the inferior colliculus from which the ventral nucleus receives its input, and to those of the auditory cortex to which it projects.

In these three sites most neurons respond to click, pure tone, or other auditory stimuli with a transient, short-latency discharge. The discharges become less transient and more sustained in unanesthetized animals (Aitkin and Prain, 1974; Allon et al., 1981). This is a property reminiscent of firing patterns of cells in the other two nuclei so far considered. Cells in the ventral medial geniculate nucleus and in the other auditory centers typically have "tuning curves." That is, at sound pressure levels close to threshold, most units respond to a range of pure tone stimuli—on average, 1–5 kHz and with a particularly low-threshold "best frequency" (Fig. 4.28) (Galambos et al., 1952; Hind et al., 1963; Adrian et al., 1966; Rose et al., 1966; Brugge et al., 1969, 1970; Aitkin and Webster, 1972; Aitkin et al., 1975; Semple and Aitkin, 1979; Calford and Webster, 1981). The range of frequencies to which a neuron will respond near threshold is greatest for neurons with low best frequencies and smallest for neurons with high best frequencies. That is, the tuning curves are wide and narrow, respectively. These characteristics resemble the "static" properties of neurons in the ventral posterior nucleus (Section 4.3.4.2). It has been shown that the static discharge rates of medial geniculate neurons recorded over relatively long periods in the unanesthetized cat do vary even in the absence of changes in the state of arousal of the animal. Nevertheless, the overall pattern of discharges remains relatively constant despite variations in rate (Imig and Weinberger, 1973), a finding reminiscent of that reported for joint-related neurons in the ventral posterior nucleus (Mountcastle et al., 1963).

Indications of inhibitory mechanisms at work in the medial geniculate nucleus include prolonged postexcitatory suppression of relay cell excitability (Nelson and Erulkar, 1963; Maruyama et al., 1966; Dunlop et al., 1969; Aitkin and Dunlop, 1969), an effect that is often followed by a rebound and reverberatory excitation, resembling the postanodal exaltation observed in the ventral posterior nucleus (Section 4.3.3) and usually interpreted as being due to similar intrinsic inhibitory mechanisms (Aitkin and Dunlop, 1969). If a cell is stimulated at frequencies well above or below its best frequency, it may show no on responses, only an inhibition of its spontaneous discharge (Dunlop et al., 1969; Aitkin and Webster, 1972). Here one can possibly detect a phenomenon similar to that in the dorsal lateral geniculate nucleus in which inhibitory responses can be induced from larger areas of retina than excitatory responses and presumably also reflects the activities of inhibitory intrageniculate neurons. These characteristics appear to be dependent at least in part on anesthetic state and, thus, can be regarded as "dynamic" properties.

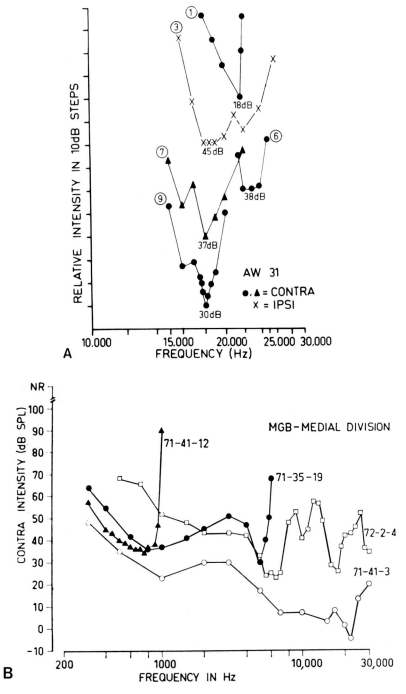

FIGURE 4.28. (A) Narrow tuning curves of five units in the ventral medial geniculate nucleus of the cat. From Aitkin and Webster (1972). (B) Very broad tuning curves of four units in the magnocellular medial geniculate nucleus of the cat. From Aitkin (1973).

One feature in which the ventral medial geniculate nucleus differs from the other thalamic relay nuclei so far considered is that most of its neurons are bilaterally (i.e., binaurally) influenced (Aitkin and Webster, 1972; Calford and Webster, 1981). The majority of units in the ventral nucleus of an anesthetized animal respond to a tonal stimulus delivered to the contralateral ear with a burst of discharges. If, however, a second stimulus is simultaneously applied to the ipsilateral ear, the effect of the second stimulus is often to enhance the response to the contralateral ear, often by as much as 100%, even if stimulation of the ipsilateral ear alone hardly affects the cell (Fig. 4.29). Such a responding unit is customarily referred to as an EE cell (Aitkin and Webster, 1972). Rather interestingly, the best frequency for an effect from the ipsilateral ear can be different from that for the contralateral ear. And some EE cells with low best frequencies have their peak firing rates when a stimulus applied to one ear is delayed with respect to the first. The delay can be characteristic for the cell (100 μsec or longer). Other cells in the ventral medial geniculate nucleus may respond to dichotic stimulation with an ipsilateral induced reduction of the contralateral induced excitation (EI cells). EI cells with high best frequencies will often show peak firing when there is an intensity difference between the tones applied to the two ears. EE and EI properties appear to be relayed rather faithfully to the cerebral cortex for identical binaural responses can be recorded there (Imig and Brugge, 1978; Imig and Adrian, 1977; Middlebrooks and Zook, 1983). Similar responses are typical of all brain-stem auditory centers above the cochlear nuclei

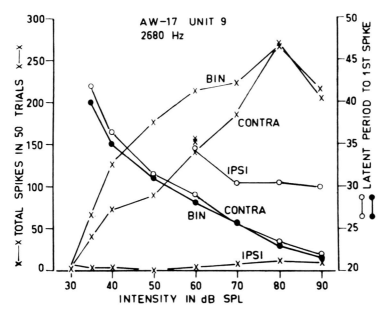

FIGURE 4.29.   Latency and spike count data for a typical EE neuron in the ventral medial geniculate nucleus of the cat. The traces show responses to monaural contralateral (CONTRA), monaural ipsilateral (IPSI), and simultaneous binaural (BIN) stimulation. Note facilitation of discharges for binaural in comparison with contralateral stimulation up to 80 dB, and relative lack of effects with ipsilateral stimulation alone. Latent period to first spike is same with contralateral and binaural stimulation, indicating predominance of contralateral affects. From Aitkin and Webster (1972).

(Rose *et al.*, 1966; Goldberg and Brown, 1968, 1969; Brugge *et al.*, 1969, 1970) so it is not yet clear to what extent intrageniculate mechanisms affect the property of binaural interaction. The efferent axons of many of the lower auditory centers converge on the central nucleus of the inferior colliculus (Roth *et al.*, 1978), whereas this latter structure appears to be the only subcortical center providing input to the principal (ventral) nucleus of the medial geniculate complex (Kudo and Niimi, 1978; Andersen *et al.*, 1980b). Possibly, therefore, the convergence of different afferent systems such as occurs in the ventral posterior nucleus (Section 4.3.4.2) is paralleled by convergence in the central nucleus of the inferior colliculus, rather than in the ventral medial geniculate nucleus.

The magnocellular nucleus of the medial geniculate complex seems to be the only thalamic nucleus which at present has been investigated for the phenomenon of long-term synaptic potentiation. This long-lasting facilitation of synaptic transmission that can follow brief, high-frequency electrical stimulation, has sometimes been implicated in learning and memory (Bliss and Gardner-Medwin, 1973; Bliss and Lømø, 1973; Douglas and Goddard, 1975). The magnocellular medial geniculate nucleus receives a diverse set of subcortical inputs and projects diffusely to widespread areas of the cortex (Chapters 3 and 8). In it, neuronal activity elicited by a conditioning stimulus to one of its inputs (the brachium of the inferior colliculus) systematically increases during the acquisition of a conditioned behavioral response (Ryugo and Weinberger, 1978). Long-term potentiation, manifested by increased amplitude and decreased latency of spike responses to afferent stimulation lasting at least an hour, could be shown to occur following brief high-frequency stimulation of the brachium of the inferior colliculus (Gerren and Weinberger, 1983). These results, if they can be generalized to all thalamic nuclei, may be an important clue to the role of the thalamus as an integrative center.

4.4. The Functional Role of Corticothalamic Fibers

4.4.1. Corticothalamic Reciprocity

Despite the early years of controversy (see Chapter 1 and Berman and Jones, 1982), it is now well established that every cortical area returns fibers to the dorsal thalamus. En route, such fibers give collaterals to the reticular nucleus (Jones, 1975c). In the ipsilateral dorsal thalamus, the fibers terminate in all the nuclei from which the cortical area receives thalamocortical fibers but in no others. So precise is this "principle of reciprocity" (Diamond *et al.*, 1969) that patterns of corticothalamic connectivity can often be used to map thalamocortical relationships (e.g., Jones *et al.*, 1979). Where nuclei have been reported to receive corticofugal fibers that seemingly do not follow this rule, the verdict of time and subsequent work has usually been that the aberrant nuclei actually provide a nonspecific thalamocortical input to the cortical area in question. A case in point is the projection from area 17 to the lateral nuclear complex (cf. Garey *et al.*, 1968; Updyke, 1977) and from several areas to the intralaminar nuclei (Chapters

3 and 12). A further "rule" may in fact be that every cortical area projects corticothalamic fibers to its relay nucleus and to a "nonspecific" nucleus.

The topography of corticothalamic connectivity is as precise as that of thalamocortical connectivity (Fig. 3.24). In the lateral geniculate, medial geniculate, and ventral nuclei, any part of the sensory representation in the associated cortical field or fields projects back only to the region of the nucleus representing the same part. Hence, in the lateral geniculate nucleus, corticothalamic fibers arising from a punctate zone of the visual cortex terminate across all geniculate laminae in a columnlike array comparable to a projection column. In the ventral posterior nucleus, they terminate in an anteroposteriorly elongated "rod" (see Chapter 3). So precise is the topographic reciprocity that simultaneous injection of small amounts of mixed anterograde and retrograde tracers at the same point in the cortex leads to superimposed foci of anterograde axonal and retrograde cellular labeling in the thalamus (Fig. 4.30). In the case of the cat laminar lateral geniculate nucleus, therefore, injection of areas 17 and 18 will give labeled cells and terminals in laminae A, A1, and in the C laminae whereas injection of area 19 will give labeled cells and terminals only in the C laminae (Chapter 9).

4.4.2. Cellular Origins of Corticothalamic Fibers

Corticothalamic axons arise from pyramidal cells in layers V and VI of the cortex (Fig. 4.31). Those arising in layer VI form the projection to the principal relay nucleus (Jacobson and Trojanowski, 1975; Gilbert and Kelly, 1975; Lund *et al.,* 1975; Robson and Hall, 1975; Jones and Wise, 1977; Wise and Jones,

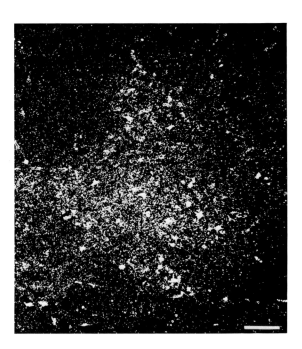

FIGURE 4.30. Darkfield photomiograph from the ventral posterior nucleus of a rat in which tritiated amino acids and horseradish peroxidase were injected at the same site in the SI cortex. Grains representing anterograde labeling of terminal ramifications of corticothalamic fibers have the same distribution as retrogradely labeled thalamocortical relay cells projecting to the same cortical site. Bar = 200 μm.

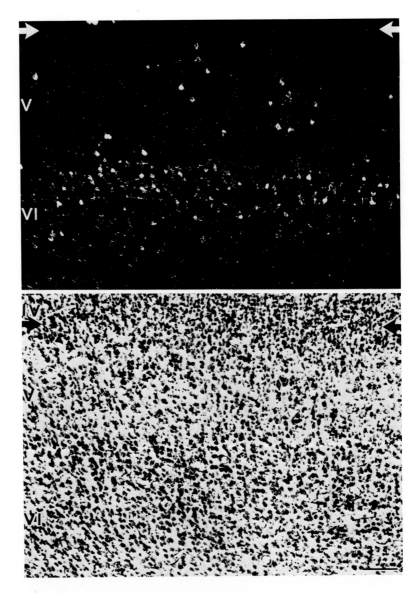

FIGURE 4.31. Darkfield (upper) and brightfield (lower) photomicrographs at same magnification and from same field showing retrograde labeling of many corticothalamic cells in layer VI with a smaller number of layer V. SI cortex of a squirrel monkey. Thionin counterstain; bar = 100 μm. From Jones and Wise (1977).

1977a,b). The parent cells (Fig. 4.32) are the small modified pyramids of that layer (Jones, 1975d, 1983f). The axons are fine and furnished with numerous intracortical collaterals that ascend to layer IV of the cortex (Gilbert and Wiesel, 1979). Though the number of corticothalamic cells residing in layer VI seems particularly high, it has been determined, from double retrograde labeling studies, that these are independent of other layer VI cells projecting to the claustrum (LeVay and Sherk, 1981), to other ipsilateral cortical areas (Jones, unpublished), or through the corpus callosum (Catsman-Berrevoets and Kuypers, 1978; Catsman-Berrevoets *et al.*, 1981). Apart from the reticular nucleus [and, in the case of the visual system, the perigeniculate nucleus (see Chapter 9)], the axons of layer VI corticothalamic neurons do not seem to have collaterals to other subcortical sites.

Within layer VI of the monkey visual cortex there is a segregation of the cells of origin of corticothalamic fibers projecting to different laminae of the dorsal lateral geniculate nucleus. A deeper stratum of slightly larger cells adjacent to the white matter projects to the magnocellular layers and a more superficial stratum adjacent to layer V projects to the parvocellular layers (Lund *et al.*, 1975). There is no evidence for a comparable dissociation of cells projecting to other nuclei.

The corticothalamic axons arising from cells in layer V project to the "non-specific" thalamic nuclei associated with the cortical area or areas in which they lie. Layer V corticothalamic cells in frontal areas thus project to the intralaminar nuclei (Catsman-Berrevoets and Kuypers, 1978), those in visual areas to a part of the lateral nuclear complex (Gilbert and Kelly, 1975; Lund *et al.*, 1975), and so on. The parent cells are more typical pyramids than those of layer VI and may give collaterals to subcortical sites other than the thalamus (Swadlow and Weyland, 1981) though this is not firmly established.

4.4.3. Mode of Activation of Corticothalamic Cells

The basal dendrites of layer VI pyramidal cells branch profusely within a small zone of thalamic axon terminations situated at the junction of layers V and VI (Jones, 1975b; Hendry and Jones, 1983a,b). The apical dendrites of many layer VI cells also branch profusely within the major zone of thalamic terminations in the middle cortical layers (III and IV) (Gilbert and Wiesel, 1979; Hendry and Jones, 1983a) (Fig. 4.32). In either place, they receive the terminations of thalamic axons though in the mouse somatic sensory cortex, at least, the number of thalamic terminations is greatest in layer IV (White and Hersch, 1982).

Some corticothalamic cells identified physiologically in the visual cortex of the cat are activated by stimulation of the optic tract or chiasm at latencies commensurate with their receiving monosynaptic activation from thalamocortical fibers (Bullier and Henry, 1979; Gilbert, 1977; Harvey, 1978). All the layer VI cells, however, are binocularly activated (Schmielau and Singer, 1977; Harvey, 1978; Gilbert, 1977), have particularly large receptive fields—up to 16° of visual angle (Gilbert, 1977), and lack the concentric receptive field organization typical of cells receiving their major afferent drive from the thalamus [the proportions of simple and complex cells is contested (Gilbert, 1977; Harvey, 1978; Tsumoto and Suda, 1980)]. All of these features suggest that the corticothalamic cells are also subjected to significant intracortical mechanisms. Gilbert and Wiesel (1979) consider that the large receptive fields, in particular, could be built up by the extremely long, horizontal-spreading collaterals given off by the axons of layer V cells as these descend through layer VI (Fig. 4.32). Such collaterals should contact large numbers of layer VI cells and, thus, serve to expand the sizes of receptive fields in comparison with those of cells higher in the cortex.

The axons of corticothalamic cells have been found in electron microscopic studies to be myelinated but extremely thin (Jones and Powell, 1969b), though the latter conflicts with the earlier descriptions of Ramón y Cajal (1911). Physiologically, in the cat visual system, their conduction velocities vary, the antidromic latency to stimulation of the lateral geniculate nucleus ranging from less than 1 msec to more than 6 msec (Gilbert, 1977; Harvey, 1978; Tsumoto and Suda, 1980; Ahlsén *et al.*, 1982a).

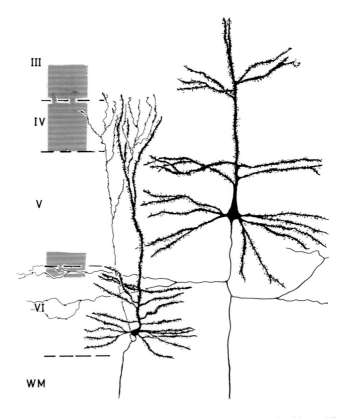

FIGURE 4.32. Semischematic figure showing modified pyramidal cell of layer VI which projects to thalamus and its recurrent collateral axons to layer IV. Layer V cell, projecting to other subcortical targets, has extensive collateral axonal ramifications in layer VI, probably accounting for large receptive fields of layer VI cells in visual cortex. Layer VI cells receive monosynaptic thalamic inputs in layer IV and VI (stipple).

4.4.4. Mode of Termination of Corticothalamic Axons

Corticothalamic axons appear to give off collateral branches to the reticular nucleus of the thalamus as they traverse it (Chapters 3 and 15). In the case of the dorsal lateral geniculate nucleus of the cat, collaterals seem to terminate in the perigeniculate nucleus (Ahlsén and Lindström, 1982a). In both sites it is probable that they end as small, spherical vesicle-containing boutons that make asymmetrical axodendritic contacts (Ohara and Lieberman, 1981; Ide, 1982). In the dorsal thalamus, corticothalamic axons end as large numbers of similar small boutons in the neuropil outside the synaptic islands (Jones and Powell, 1969b) (Section 4.3.1). Guillery (1969a,b) has called these endings RSD terminals because they contain round vesicles, are small, and contain electron-dense mitochondria. From their morphology and the hints that they release aspartate or glutamate (see Chapter 5), it is assumed that these terminals are excitatory in nature. They terminate in large numbers (Figs. 4.17 and 4.19) on secondary and tertiary dendrites of relay neurons and on presynaptic dendrites of interneurons (Jones and Powell, 1969b). In either case, the corticothalamic effect seems to be exerted on dendrites distal to the point of termination of lemniscal, inferior collicular, or optic tract terminations and thus more removed from the somata of relay neurons (Jones and Powell, 1969b; Ogren and Hendrickson, 1979). I feel that the same is true of the interneurons, though in the medial geniculate nucleus, Morest (1975) considers that corticothalamic fibers terminate closer to the somata of interneurons than afferent fibers from the inferior colliculus. Irrespective of where they terminate, the strength of corticothalamic synapses appears much less than that of ascending afferent synapses, for the effects of corticothalamic synapses, in comparison to the ascending synapses, have never been clearly defined. Corticothalamic axons, unlike the ascending afferents, appear to contribute little to the receptive fields of thalamic neurons.

4.4.5. Functions of Corticothalamic Fibers

Attempts to elucidate the functions of corticothalamic axons have a long history but have never been particularly satisfying. One can probably sum up most of the early work by saying that it showed that a weak conditioning stimulation of the cortex could either facilitate or inhibit transmission of a subsequent afferent volley through the thalamic relay nuclei. Some reported the effect as predominately inhibitory (Widén and Ajmone-Marsan, 1960; Iwama *et al.*, 1965; Shimazu *et al.*, 1965; Watanabe *et al.*, 1966), others as predominately facilitatory (Andersen *et al.*, 1967; Steriade *et al.*, 1972). Some reported the effect as powerful and long-lasting (Widén and Ajmone-Marsan, 1960; Shimazu *et al.*, 1965; Watanabe *et al.*, 1966), others as relatively weak and best demonstrated when transmission was impaired by fatigue (Andersen *et al.*, 1967). After reversible cooling of the cortex, effects have, again, been variously reported as inhibitory, facilitatory, mixed, or absent (Hull, 1968; Kalil and Chase, 1970; Singer, 1970; Burchfiel and Duffy, 1974; Richard *et al.*, 1975; Ryugo and Weinberger, 1976; Baker and Malpeli, 1977; Molotchnikoff *et al.*, 1980).

Geisert *et al.* (1981) demonstrated that the responses to photic stimulation

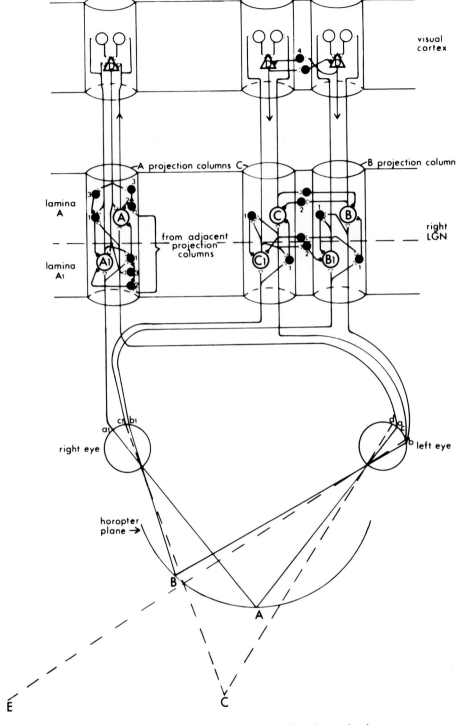

FIGURE 4.33. Singer's conception of how corticothalamic fibers in cat visual system may control binocular inputs to visual cortex. Point A (at bottom) is the fixation point and a and a₁ are ganglion cells that lie in corresponding retinal areas of the two eyes (zero disparity). These are connected to

of most X and Y cells with both on- or off-centers in the A laminae of the cat lateral geniculate nucleus could be affected by cooling the visual cortex. The effects, however, varied and cells could become more or less responsive to stimulation of their receptive field centers or surrounds during cooling. These results, therefore, seem to indicate involvement of corticothalamic fibers with both inhibitory and excitatory mechanisms in the lateral geniculate nucleus.

Schmielau and Singer (1977) and Ahlsén *et al.* (1982a) found that both the central on-response and the surround off-(or inhibitory) response of lateral geniculate relay cells with typical concentrically organized receptive fields could be enhanced by cortical stimulation, suggesting direct activation of both relay cells and inhibitory interneurons by corticothalamic fibers. The enhancement of surround inhibition at thalamic levels would presumably be of importance in discriminating visually between objects and between objects and background. But it could also be envisaged as of comparable significance in somatic sensory and auditory discriminations.

Perhaps the most elaborate theory of the functions of corticothalamic fibers is that proposed for the visual system by Singer (Schmielau and Singer, 1977; Singer, 1977) (Fig. 4.33). It seems to have its origins in brief comments originally made by Pettigrew (1972). Singer's view of the actions of corticogeniculate fibers depends upon four observations: (1) the tight retinotopic coupling of lateral geniculate nucleus and visual cortex already alluded to in Chapter 3; (2) the enhancement of geniculate relay cell responses by simultaneous photic stimulation of homonymous points in the two eyes, provided corticogeniculate mechanisms were not deactivated by cooling the cortex (Schmielau and Singer, 1977); (3) the weak excitation of corticogeniculate neurons by monocular stimulation, by contrast with the powerful effect on them of binocular stimulation (Schmielau and Singer, 1977); and (4) the inhibitory effect exerted by cells in a lamina connected to a stimulated eye over those in a lamina connected to the opposite, nonstimulated eye (see Section 4.3.4).

When corresponding points in the two retinae are stimulated (Fig. 4.33), the activity of cells in adjacent geniculate laminae but situated along the projection column (see Chapter 3) receiving inputs from the corresponding points would be enhanced for two reasons: (1) interlaminar inhibitory mechanisms would be canceled by the simultaneous excitation of cells in ipsi- and contralaterally innervated layers; (2) corticothalamic fibers whose parent cells would also be maximally excited by congruent binocular stimulation, run vertically down the projection column, crossing all laminae; they would make excitatory con-

---

relay cells A and A1 in a lateral geniculate projection column. The relay cells are subjected to interlaminar inhibition by type 1 interneurons, but cortical disparity detector, D, receiving congruent inputs from cells A and A1, leads to corticofugal excitation of relay cells and of type 3 interneurons which serve to inhibit type 1 interneurons as well as additional type 2 interneurons, which mediate intralaminar inhibition. Point C (at bottom) is located behind the horopter plane and activates noncorresponding retinal areas d and c_1 (disparity). Because noncongruent geniculate cells are activated, the corticofugal loop is not, and interlaminar (binocular) inhibition remains unaffected. Similarly, there is intralaminar inhibition: from adjacent facilitated projection column B and B1 and in the cortex (interneuron 4) from adjacent columns activated by B and B1. According to Singer, because point B is on fixation plane and point C is not, the result is improved separation of figures from the background. From Singer (1977).

nections on relay neurons and on inhibitory neurons that would serve to inhibit interlaminar inhibitory interneurons. In normal vision, simultaneous stimulation of homonymous receptive fields and of geniculate cells along the same projection column, would occur only with two objects situated on the horopter plane through the fixation point (Fig. 4.33). Hence, the responses from both eyes would be facilitated and relayed symmetrically to the cortex. Objects situated in front of or behind the fixation plane would not stimulate homonymous receptive fields, nor geniculate cells along the same projection column. Hence, direct interlaminar inhibition of the dominant over the nondominant eye in adjacent laminae of two projection columns would be activated, along with local, intralaminar inhibitory interneurons. Moreover, in not receiving congruent information, corticothalamic cells would be only weakly excited (and possibly even subjected to lateral inhibition in the cortex). This, too, would lead to increased interlaminar inhibition in the geniculate. Hence, when stimuli are too disparately placed to be fused into a single binocular image, relay cells are subjected to maximal binocular inhibition in the lateral geniculate nucleus and, in short, double vision would be prevented.

Corticogeniculate cells are, thus, in a sense, disparity detectors controlling the input to the cortex and occluding background so as to improve the separation of figures on the fixation plane from objects in the background.

It may be noted that Singer's theory does not take into account any potential differential effect exerted by corticogeniculate fibers arising in areas 17, 18, or 19 of the cat and which have different patterns of termination in the laminar dorsal lateral geniculate nucleus (Updyke, 1977; see Chapter 9). In monkeys one would envisage separate corticogeniculate channels acting on X- and Y-type cells, by virtue of the different laminar origins of cortical cells sending axons to the parvo- and magnocellular layers of the geniculate.

Although the fundamental circuitry used by Singer in constructing his hypothesis is present in all thalamic relay nuclei, it is not easy to identify appropriate stimulus parameters that might enable comparable effects of corticothalamic fibers to be demonstrated in other sensory systems. Perhaps, however, by selectively activating certain inhibitory interneurons, corticothalamic cells can under certain circumstances help to set up lateral inhibition in relay nuclei such as the ventral posterior and medial geniculate.

In the ventral posterior nucleus we have an example of corticothalamic axons converging on a single nucleus from more than one cortical area: in the monkey from areas 3a, 3b, 1, and 2 of the first somatic sensory area (SI) and from the second somatic sensory area (SII) as well (Chapter 7). By analogy with the dissociation of corticothalamic inputs to the layers of the lateral geniculate nucleus that contain different classes of relay cells, we might anticipate a similar dissociation on different cell classes in the ventral posterior nucleus.

4.5. The Role of the Brain-Stem Reticular Formation in the Control of Thalamic Transmission

There have been numerous reports of modulation of thalamic sensory transmission or of spontaneous neuronal discharge by high-frequency electrical stimulation of the brain-stem reticular formation. Usually the effect is manifested by an enhancement of both evoked and spontaneous discharges. In the dorsal lateral geniculate nucleus, effects are seen as early as 10 msec after stimulation and last for several hundred milliseconds thereafter (Suzuki and Taira, 1961; Satinsky, 1968; Fukuda and Iwama, 1971; Singer and Bedworth, 1974). Cells identified as relay cells in the dorsal lateral geniculate nucleus are reported to have their activity enhanced by stimulation of the reticular formation even in the phase of synaptic depression ensuing after previous photic stimulation or electrical stimulation of the optic tract (Fig. 4.34) (Singer and Bedworth, 1974; Singer and Phillips, 1974). Neurons identified as interneurons are reported to be inhibited (Singer and Dräger, 1972; Singer, 1973a; Fukuda and Iwama, 1971) with a reduction of the inhibitory effect produced by stimulation of a nondominant eye (Singer and Schmielau, 1976; Fukuda and Stone, 1976). There appears to be some enlargement of receptive fields of geniculate relay neurons (Meulders and Godfraind, 1969), perhaps also due to the reduction in intranuclear inhibition (Fukuda and Stone, 1976).

The most effective site for eliciting effects on the lateral geniculate nucleus by stimulation of the reticular formation appears to be in the center of the rostral part of the mesencephalon, the so-called nucleus cuneiformis (Wilson et al., 1973). Afferent fibers from this region do not, however, project directly to the dorsal lateral geniculate or to other nuclei of the dorsal thalamus except for the intralaminar group (Edwards and De Olmos, 1976). Hence, it is necessary to invoke direct pathways arising in other sites and traversing the nucleus cuneiformis or less direct routes to the relay nuclei. Electrical stimulation of the mesencephalic reticular formation leads to the production of field potentials in the dorsal lateral geniculate nucleus very similar to the PGO waves of desynchronized (REM) sleep (see Chapter 12) and the effect of stimulation is enhanced in sleeping animals (Cohen et al., 1962). Hence, it is conceivable that the same pathways held responsible for mediating the PGO effect are involved; i.e., those from lower in the brain stem and arising in the locus coeruleus, or parabrachial region and others from the midbrain raphe (see Chapter 3). An enhancement of the concentric antagonism in lateral geniculate cells' receptive fields is reported in cats aroused from slow-wave sleep (Livingstone and Hubel, 1981). This, too, may be indicative of a projection from the brain-stem sleep control centers to the dorsal lateral geniculate nucleus.

PGO waves have been suggested to be due to a presynaptic depolarization of optic tract axon terminals (Bizzi, 1965; Sakakura and Iwama, 1965; Angel et al., 1965); though there is a concomitant increase in the responsiveness of thalamic relay neurons to photic or electrical stimulation (see Singer, 1977; Burke and Cole, 1978), presumably because of the reduction in intranuclear inhibition alluded to earlier. It may be noted (as shown in Section 4.3.1) that no afferent axon terminals in the thalamus receive synapses from other axon terminals. Hence, there is no obvious morphological basis for the presynaptic inhibitory effect suggested in early studies of the effects of stimulation of the brain-stem reticular formation. It has been suggested from work with ion-selective electrodes that the depolarization of optic tract terminals in the dorsal lateral geniculate nucleus after stimulation of the reticular formation is the result of extracellular accumulation of potassium, presumably released by depolarization of the terminals of reticulothalamic nerve fibers and/or by the discharging of geniculate cells (Singer and Lux, 1973). The PGO waves may, therefore, reflect synaptic

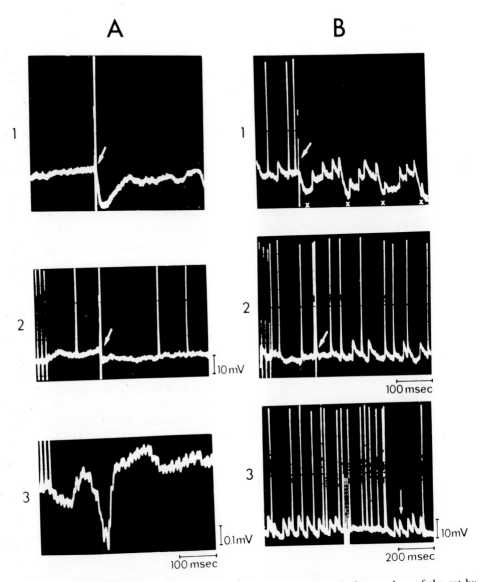

A

B

1

2

10 mV

100 msec

1

2

100 msec

3

0.1 mV

100 msec

3

10 mV

200 msec

FIGURE 4.34.   Facilitation of transmission in the dorsal lateral geniculate nucleus of the cat by stimulation of the midbrain reticular formation. A1 and B1 respectively show IPSPs succeeding relay cell discharge due to antidromic stimulation of the subcortical white matter (A1) or stimulation of the optic chiasm (B1). In A2 and B2, the IPSPs disappear when the stimulation is preceded by conditioning stimulation of the midbrain reticular formation. A3 is the negative field potential evoked in the visual cortex by stimulation of the reticular formation. B3 shows that for about 200 msec following stimulation of the reticular formation all spontaneous EPSPs in a geniculate cell reach firing level, followed (arrow) by a period of inhibition. From Singer (1979).

mechanisms operating at the level of the thalamic relay cells themselves, either by direct excitation or by suppression of inhibitory interneurons. Though there are no direct morphological data on the terminations of reticular formation axons in the dorsal lateral geniculate nucleus, it is the latter which is favored by Singer (1973a, 1977) and others (Burke and Cole, 1978), because as described above, IPSPs induced in geniculate relay cells by stimulation of the optic tract are consistently reduced in amplitude by conditioning stimulation of the mesencephalic reticular formation and over a time period similar to that of the corresponding PGO waves.

In thalamic nuclei other than the lateral geniculate, e.g., the ventral lateral (Purpura *et al.*, 1966), ventral anterior (Sasaki *et al.*, 1972, 1976), ventral posterior (Steriade, 1970), lateral posterior (Steriade *et al.*, 1977a,b) and medial geniculate (Symmes and Anderson, 1967), sensory transmission can also be facilitated at relatively short latency by stimulation of the midbrain reticular formation. Like the lateral geniculate nucleus, however, there is no obvious anatomical pathway leading directly from the reticular formation to these nuclei. Small numbers of neurons can be retrogradely labeled in the locus coeruleus, parabrachial nucleus, and periaqueductal gray matter and dorsal nucleus of the midbrain raphe following injection of tracers in various thalamic nuclei (e.g., Gilbert and Kelly, 1975; Leger *et al.*, 1975; Tracey *et al.*, 1980; Mackay-Sim *et al.*, 1983). Many of these regions, like the nucleus cuneiformis, are thought to be critical for the mediation of reticulothalamic effects (see Chapter 12). Yet anterograde labeling of axons emanating from these or adjacent brain-stem regions does not demonstrate diffuse projections to all thalamic nuclei. The lateral geniculate complex, and intralaminar complex are traversed by fibers from all three sites (see Chapters 3 and 5). Coeruleothalamic fibers have additional concentrations in the anterior nuclei and parabrachial fibers in the basal ventromedial nucleus. Other nuclei receive few or no such fibers (Conrad *et al.*, 1974; Pickel *et al.*, 1974; Swanson and Hartman, 1975; Graybiel, 1977; Jones and Moore, 1977; Saper and Loewy, 1980; Card and Moore, 1982; McKellar and Loewy, 1982; Robertson and Feiner, 1982).

The pathway for the fairly obvious effect on thalamic transmission in, say, the ventral nuclei, therefore, remains something of a mystery. If the nucleus cuneiformis is the origin of the effect, then it could conceivably affect the relay nuclei indirectly via its known projections to the intralaminar nuclei or to the reticular nucleus of the ventral thalamus. Stimulation of "medial" and intralaminar thalamic nuclei have often been reported to affect transmission in the relay nuclei (Chapter 12). However, there is no evidence for an anatomical connection between these and the relay nuclei (Chapter 3). Hence, they too must operate indirectly, possibly via their projections to the reticular nucleus (Chapters 3 and 15) or through their diffuse projection to the cortex which may be involved in controlling levels of cortical excitability (Chapter 12).

The possibility of brain-stem reticular effects on dorsal thalamic nuclei being mediated by direct inputs to the reticular nucleus has often been raised. There is no obvious direct brain-stem input to the reticular nucleus other than that from the rostral mesencephalon [the "nucleus cuneiformis" (Edwards and De Olmos, 1976; Mackay-Sim *et al.*, 1983)] or deep pretectum (Berman, 1977). Hence, for reticular effects from the lower brain stem, intermediate relays in this part of the midbrain have to be involved. It may be noted that the vicinity of the nucleus cuneiformis was found by Wilson *et al.* (1973) to be the most effective site from which electrical stimulation could influence synaptic transmission through the lateral geniculate nucleus of the squirrel monkey. Moreover, Dingledine and Kelly (1977) found that high-frequency stimulation of this region of the cat would inhibit the spontaneous activity of neurons in the reticular nucleus while increasing, at longer latency, the firing of neurons in the adjacent ventral posterior nucleus. The same neurons in the reticular nucleus showed an inhibition of spontaneous activity in response to iontophoretic application of acetylcholine (see Chapter 5). Hence, a cholinergic pathway acting via the reticular nucleus and serving to disinhibit relay neurons in the dorsal thalamus was proposed. Disinhibition of relay neurons would facilitate sensory transmission through the thalamus and presumably lead to desynchronization of the electrocorticogram, with accompanying arousal. Over the years, it is the association of reticular nucleus function with the electrocorticogram that has received a great deal of attention and this will be reviewed in the next section.

4.6. The Role of the Reticular Nucleus

4.6.1. Organization

As a part of the ventral thalamus, the reticular nucleus does not project upon the cortex, yet it appears to be inextricably involved in thalamocortical and corticothalamic activities. The nucleus appears to be a large pool of GABAergic neurons that receive collaterals of all thalamocortical and corticothalamic fibers passing through it (Chapters 3 and 15). The same is probably true of thalamostriatal and pallidothalamic fibers as well (Jones, 1975c). The only other confirmed source of input appears to be the nucleus cuneiformis of the mesencephalic reticular formation (Edwards and De Olmos, 1976; Mackay-Sim *et al.*, 1983). The origin of this input was interpreted by Berman (1977) as certain deeper-seated nuclei of the pretectum. Apart from this connection, however, the reticular nucleus has no particular affinity with the brain-stem reticular formation. The outputs of the nucleus are back into the underlying dorsal thalamus (Ramón y Cajal, 1911; Scheibel and Scheibel, 1966a; Minderhoud, 1971; Jones, 1975c) and, as reviewed in Chapters 3 and 15, each dorsal thalamic nucleus is interrelated with a particular part of the reticular nucleus, so that some parts of the latter can be said to be somatosensory, visual, auditory, and so on. These inputs to the dorsal thalamus are GABAergic and presumably inhibitory (Section 3).

Other evidence indicative of a close interaction between the reticular nucleus and the relay nuclei is now starting to appear in works reporting the results of receptive field mapping in the reticular nucleus. Anatomical studies, it will be recalled (Chapter 3), had shown that the connectivity of different parts of the reticular nucleus was such that these parts had to be related to the visual system, somatic sensory system, auditory system, and so on (Jones, 1975c; Montero *et al.*, 1977). Single-unit studies of the reticular nucleus now have shown neurons in these parts to have visual, somatic sensory, or auditory receptive fields (Hale *et al.,* 1982; Pollin and Rokyta, 1982; Sugitani, 1979; Shosaku and Sumitomo, 1983; Sumitomo *et al.,* 1976; Yen and Jones, 1983). Though the anatomical results had predicted some degree of sensory parcellation, it was felt that the extremely long dendritic fields of reticular nucleus cells (Fig. 4.14) should cause many of them to receive convergent inputs from several sensory systems. The single-unit studies, however, have revealed a great deal of sensory specificity and only where auditory, somatic sensory, or visual regions of the reticular nucleus abut is there significant convergence (Shosaku and Sumitomo, 1983; Sugitani, 1979). Montero *et al.* (1977) were able to show in rabbits that corticothalamic fibers and transneuronally labeled fibers from the dorsal lateral geniculate nucleus were distributed in an orderly fashion in the reticular nucleus, implying a retinotopic organization. This has subsequently been confirmed physiologically in the comparable part of the rat reticular nucleus (Hale *et al.,* 1982). Ahlsén *et al.* (1982b) believe that this part of the rat reticular nucleus is equivalent to the perigeniculate nucleus of the cat (see Chapter 15). Evidence against widespread convergence within the visual part of the rat reticular nucleus was provided by Hale *et al.* (1982) who showed that receptive fields of visual reticular nucleus cells were not much larger than those in the dorsal lateral geniculate nucleus. There appears to be a crude somatotopy in the part of the reticular nucleus adjacent to the ventral posterior nucleus in cats and monkeys (Pollin and Rokyta, 1982), although there, receptive fields are reported to be larger than those in the ventral posterior nucleus and some are even bilateral.

Units in visual, auditory, and somatic sensory parts of the reticular nucleus can usually be driven by appropriate natural stimuli or by electrical stimulation of afferent pathways. Latencies of response are usually sufficiently longer than responses in the associated relay nucleus to imply activation of the reticular cells by collaterals of relay cell axons (Sugitani, 1979; Hale *et al.,* 1982; Pollin and Rokyta, 1982; Shosaku and Sumitomo, 1983; Yen *et al.,* 1985). Hale *et al.* (1982) provided convincing evidence for disynaptic innervation of reticular nucleus cells by showing that an electrical shock applied to the optic tract and coming within the time of postexcitatory depression of dorsal lateral geniculate cells excited by a previous shock, failed to excite reticular nucleus cells.

4.6.2. Function

There is good reason to believe that the reticular nucleus may play a role in gating the relay of information through the dorsal thalamus en route to the cerebral cortex. The evidence for this concept rests upon the anatomical data alluded to, coupled with the interpretations of some rather diverse physiological experiments. Neurons in the reticular nucleus of the barbiturate-anesthetized cat have a high level of spontaneous activity but it is characteristic of them that this spontaneous activity is commonly interrupted by sustained bursts of high-frequency discharges followed by silent periods. Such bursts may last for up to 60 msec during which the neurons may fire at rates of 200–400 per second or higher (Fig. 4.35) (Negishi *et al.,* 1962; Mukhametov *et al.,* 1970b; Schlag and Waszak, 1971; Lamarre *et al.,* 1971; Steriade and Wyzinski, 1972; Waszak, 1974).

These spontaneous burst discharges also appear during behavioral slow-wave sleep (Mukhametov *et al.*, 1970b; Steriade and Wyzinski, 1972; Horvath and Buser, 1976) and can be induced by conditions which lead to EEG synchronization and the development of thalamocortical recruiting responses, e.g., by low-frequency electrical stimulation in medial regions of the thalamus (Purpura and Cohen, 1962; Negishi *et al.*, 1962; Lamarre *et al.*, 1971; Schlag and Waszak, 1971). Bursting can also be set up in neurons of the reticular nucleus by stimulation of the cerebral cortex or by stimulation of the adjacent ventral lateral nucleus (Schlag and Waszak, 1971). During waking and under other conditions in which desynchronization of the EEG is induced, the neurons of the reticular

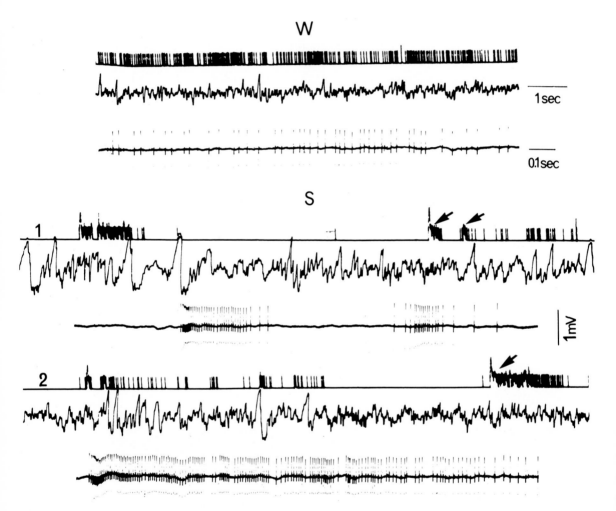

FIGURE 4.35. Spontaneous discharges of a neuron recorded from the anterior part of the reticular nucleus of a cat during simultaneously recorded electroencephalographic patterns of waking (W) and slow-wave sleep (S). In the three sets of traces (one under W, two under S) the upper trace shows unit discharges recorded on an ink writing machine, middle trace the EEG patterns recorded over the premotor cortex, and the lower trace the original neuronal spikes. Note continual firing of the reticular cell in W and long and irregular bursts in S. Arrows indicate bursts reproduced in the lower traces of sets 1 and 2 of the S phases. From Steriade *et al.* (1972).

nucleus are said to be continually active and to show less burst activity (Steriade and Wyzinski, 1972; Waszak, 1974).

Burst discharges also occur in neurons of the thalamic relay nuclei at frequencies of around 8–12 per second during slow-wave sleep (Hubel, 1960; Baker, 1971; Hayward, 1975; Steriade, 1981, 1983). Moreover, some workers have reported that the activity of neurons in relay nuclei of the thalamus during EEG synchronization or desynchronization is the opposite of that in the reticular nucleus, e.g., the relay cells are silent when the reticular nucleus is bursting. Low-frequency stimulation of the medial thalamic region leads to rhythmic EPSPs and intervening IPSPs in neurons of the ventral lateral complex (Purpura and Schofer, 1963; Purpura et al., 1966) and it has been suggested that the temporal characteristics of the IPSPs are the same as the bursts occuring under these circumstances in the reticular nucleus (Frigyesi and Schwartz, 1972). High-frequency electrical stimulation of medial regions of the thalamus or of the mesencephalic reticular formation leads to EEG desynchronization (Moruzzi and Magoun, 1949) and the development of sustained excitatory driving upon cells in the ventral lateral and ventral anterior nuclei (Purpura and Schofer, 1963; Purpura et al., 1966; Sasaki et al., 1976). Transmission is also facilitated in the dorsal lateral geniculate, medial geniculate, and ventral posterior nuclei (Singer, 1973a; Symmes and Anderson, 1967; Steriade, 1970). Intracellular recording indicates that under these circumstances there is a disinhibition of the relay neurons (Purpura and Schofer, 1963). It has been suggested that the disinhibition is due to a cessation of bursting in the reticular nucleus cells (Schlag and Waszak, 1971; Yingling and Skinner, 1976). Not all workers, however, accept that the temporal patterns of reticular nucleus cell discharges can necessarily be correlated with reciprocal rhythms in relay cells set up by stimulation of the cortex or other sites (Mukhametov et al., 1970a,b; Steriade and Wyzinski, 1972; Steriade et al., 1972; Steriade, 1981) (Fig. 4.36). These workers find (Fig. 4.36) that cortical stimulation or the change from slow-wave sleep to desynchronized sleep or wakefulness, leads to exactly the same type of increased excitability in reticular nucleus cells as in the relay cells of the dorsal thalamus and that there is no correlation between reticular nucleus activity and postexcitatory inhibition or rebound. Other workers consider that the rhythmic activities of thalamic relay cells are more likely to depend on intrinsic interneurons (Andersen et al., 1964b; Andersen and Andersson, 1968). The reticular nucleus, however, may exert a controlling influence over these interneurons (Steriade et al., 1983; Deschênes et al., 1983; Domich et al., 1983) because rhythmic bursts of discharges and the underlying intermittent hyperpolarizations that promote them cease if the dorsal thalamic nucleus is disconnected from the reticular nucleus.

Acetylcholine may play a role in mediating the disinhibition of relay neurons in the dorsal thalamus during stimulation of the mesencephalic reticular formation. There is some pharmacological evidence for such a pathway (Chapters 3 and 5) and certainly iontophoresis of acetylcholine extracellularly onto reticular nucleus neurons leads in small amounts and at short latency to suppression of spontaneous discharges (Ben-Ari et al., 1976; Dingledine and Kelly, 1977; Sillito et al., 1983). However, bursting of the neurons is unaffected or even enhanced, so relay cell excitability may be less affected by the bursts of reticular cell discharges than by their overall levels of activity.

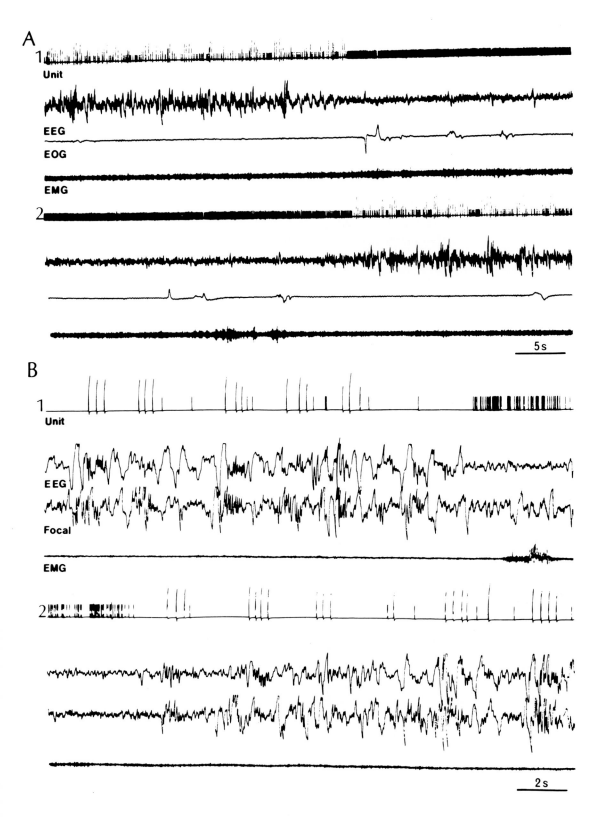

Fortunately, there is now direct physiological evidence for control of the dorsal thalamus by the reticular nucleus. Sumitomo *et al.* (1976) first showed that the recovery from postexcitatory depression of cells in the rat dorsal lateral geniculate nucleus is greatly facilitated after destruction of the visual part of the reticular nucleus. In keeping with this is the observation of Kelly *et al.* (1979) that IPSPs cannot be recorded in slices of the rat dorsal lateral geniculate nucleus from which the reticular nucleus is excluded. Single shock stimulation of the appropriate part of the reticular nucleus can suppress both spontaneous and evoked discharges of neurons in the lateral or medial geniculate nuclei (Sumitomo *et al.*, 1976; Shosaku and Sumitomo, 1983). With this, there is a concomitant reduction in the cortical sensory evoked potentials (Yingling and Skinner, 1976). This may well be one of the strongest points in favor of a gating role for the reticular nucleus in the control of transmission through the dorsal thalamus. How focused the control exerted by the reticular nucleus might be is debated. Labeling of cells *en masse* by means of retrograde tracers, suggests that a functional part of the reticular nucleus projects back only to that dorsal thalamic nucleus from which it receives inputs (Hale *et al.*, 1982). On the other hand, Golgi staining (Scheibel and Scheibel, 1966a) or intracellular dye injections of single cells (Yen and Jones, 1983) reveal that the axon of a single reticular nucleus neuron can ramify through several dorsal thalamic nuclei (Fig. 4.14). These findings would suggest that, although generally focused on the relevant relay nucleus, the output of a functional component of the reticular nucleus may affect adjacent nuclei as well.

FIGURE 4.36. Discharge patterns of thalamocortical neurons projecting to parietal (A) or motor (B) cortex in the cat during transitions from sleep to wakefulness and the reverse. Units monitored as in Fig. 4.35, and electroencephalogram (EEG), electrooculogram (EOG), and electromyogram from neck muscles (EMG) simultaneously recorded. Note the increase in neuronal firing rate as electrographic signs indicate entry into waking state and decrease, with clustered firing, just prior to beginning of sleep. From Glenn and Steriade (1982).

. . . this field lies a little away from the highways on which they travel with their racing cars. Yet what an admirable conquest! What beauty in this varied utilization by living organisms of minute amounts of a few simple molecules!

Z. M. Bacq (1975)

5

Transmitters, Receptors, and Related Compounds in the Thalamus

$$\begin{array}{l} CH_2-CH_2-CH_2-NH_2 \\ | \\ COOH \end{array}$$

GABA, a powerful inhibitory transmitter and the only transmitter whose widespread presence in the thalamus can be confidently asserted at present.

5.1. Introduction

With the possible exception of γ-aminobutyric acid (GABA), the neurochemical transmitter agents released during passage of an afferent volley through the thalamus are unknown. But there is a long history of iontophoretic studies of thalamic neurons and there are many accounts of the localization of transmitter-related compounds and of various kinds of receptor agents in the thalamus. It is to these that we turn our attention in this chapter.

5.2. Acetylcholinesterase

The mapping of the cholinesterase-containing cell and fiber systems of the brain by Shute and Lewis (1967) and Lewis and Shute (1967) represents one of the earliest forays into the field of "biochemical neuroanatomy." In analyzing what they took to be the cholinergic reticular system of the rat forebrain, Shute and Lewis demonstrated a dorsal tegmental system of cholinergic fibers arising in the midbrain reticular formation and extending up through the tectum to the thalamus and possibly beyond. A ventral system essentially bypassed the thalamus en route to the basal forebrain.

In the thalamus, the densest fiber plexuses of the dorsal tegmental system were localized in the intralaminar nuclei, ventral lateral geniculate nucleus, and anteroventral nucleus; somewhat less dense staining was observed in the reticular nucleus, dorsal lateral geniculate nucleus, medial geniculate nucleus, and in parts of the lateral and ventral complexes. The staining in all these nuclei disappeared after destruction of the so-called nucleus cuneiformis, essentially the dorsal part of the midbrain reticular formation. Hoover and Baisden (1980) have since suggested that the cholinesterase-containing pathway to the anteroventral nucleus may actually arise in the adjacent dorsal tegmental nucleus because the only cholinesterase-positive cells that could be retrogradely labeled after injections of horseradish peroxidase in the anteroventral nucleus were in this site.

Shute and Lewis described no cholinesterase-positive cell bodies in the thalamus and this finding was later repeated by Jacobowitz and Palkovits (1974) who also reported the same localization of cholinesterase among the thalamic nuclei of the rat. Parent and Butcher (1976), however, not only reported the presence of cholinesterase-positive cell bodies in several thalamic nuclei of the rat but also demonstrated the highest level of neuropil staining in the anterodorsal rather than the anteroventral nucleus. Heavier staining in the anterodorsal nucleus was also reported in the monkey by Olivier *et al.* (1970). It can be seen in the rat in our Fig. 5.1. Other nuclei showing dense cholinesterase staining in our figures are the anteroventral, intralaminar, reticular, ventral lateral geniculate, rhomboid, paraventricular, habenular, and medioventral nuclei and the zona incerta. Lesser densities are seen in anterior parts of the lateral posterior nucleus and in the dorsal lateral geniculate nucleus. With minor variations, involving mainly the mediodorsal and lateral nuclei, the distribution in the rat is very similar to that demonstrated in the cat by Graybiel and Berson (1980b) and comparable patterns are seen in primates (Fig. 5.3).

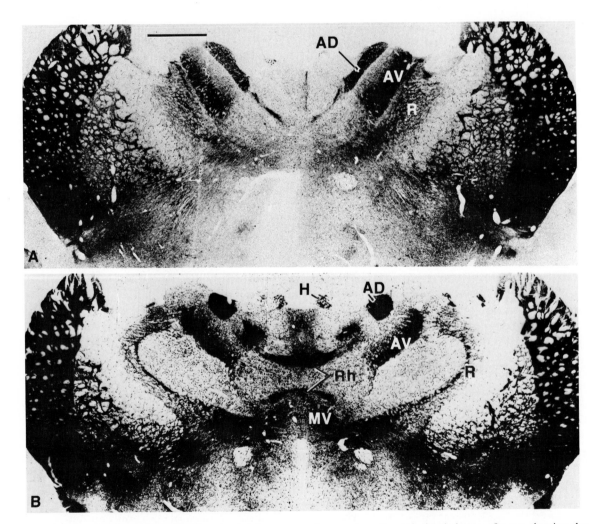

FIGURE 5.1.   A series of frontal sections in anterior to posterior sequence through the thalamus of a rat, showing the distribution of acetylcholinesterase activity. Bars = 1 mm.

The significance of these local concentrations of acetylcholinesterase is not at all clear. Probably the realization that the presence of acetylcholinesterase at a site does not necessarily imply the existence of cholinergic transmission at that site, caused a decline of interest in these patterns of staining. Recently, however, there has been a modest revival of interest because of the correlation of some of the staining patterns with the terminal distributions of certain afferent fiber systems. In the lateral posterior nucleus of the cat, Graybiel and Berson (1980b) have demonstrated that zones of termination of fibers arising from cells in the superficial layers of the superior colliculus and pretectum are cholinesterase positive while zones of termination of fibers arising in the visual cortex are cholinesterase negative (Chapter 10). Fitzpatrick and Diamond (1980) have related acetylcholinesterase staining to the terminations of some but not all corticothalamic fibers in the dorsal lateral geniculate nucleus of the bush baby. In

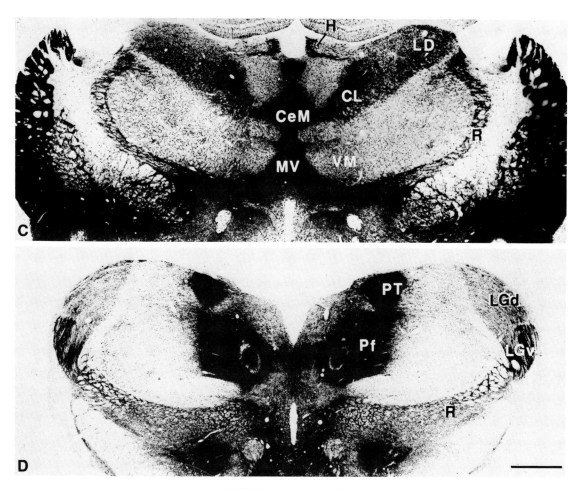

FIGURE 5.1. *(continued)*

this animal, particularly high concentrations of the enzyme staining are found in two of the seven cellular layers of the nucleus. Comparably high levels are found in the parvocellular but not the magnocellular layers of the owl monkey lateral geniculate nucleus. The high intensity of cholinesterase staining in the two layers of the busy baby geniculate was unchanged following destruction of geniculate neurons by injections of kainic acid or following severe transneuronal atrophy induced by long-standing eye enucleation. However, ablation of part of the visual cortex led to loss of cholinesterase staining in the visuotopically related parts of the two layers (Fig. 5.2). These findings not only indicate an association of the enzyme with corticothalamic fibers rather than with geniculate cells or

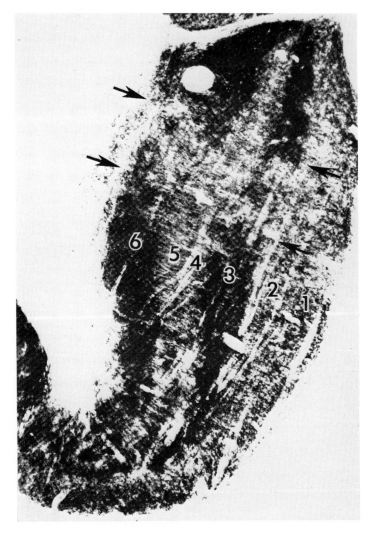

FIGURE 5.2.   Disappearance of acetylcholinesterase staining from topographically related parts of two laminae (3 and 6) of the lateral geniculate nucleus of a bush baby following destruction of a small part of the visual cortex. From Fitzpatrick and Diamond (1980).

optic tract axons but also, since all layers of the geniculate receive corticothalamic axons, they imply an association with only one type of corticothalamic fiber.

In the macaque monkey lateral geniculate nucleus, there is a differential distribution of acetylthiocholinesterase and the pseudocholinesterase, butyrylcholinesterase (Graybiel and Ragsdale, 1982). The former is concentrated in all layers but is particularly high in the magnocellular layers whereas the latter is virtually confined to the parvocellular layers (Fig. 5.3). Moreover, only butyrylcholinesterase is sensitive to the effects of eye removal. Acetylcholinesterase, on the other hand, is the enzyme more closely correlated with the terminal layers of thalamocortical fibers in the visual cortex. Similarly, the layer of dense cholinesterase staining originally reported in the cingulate cortex by Shute and Lewis (1967) is probably also associated with the terminations of thalamocortical fibers, since it disappears after destruction of the anterior nuclei (R. T. Robertson, personal communication, 1983; Fig. 5.12).

5.3. Cholinergic Receptors

Cholinergic receptors of both the muscarinic and the nicotinic type have been identified in the thalamus and their collective distribution corresponds quite well with that of acetylcholinesterase. But the concentrations of receptors appear to be relatively low (Yamamura *et al.,* 1974; Hunt and Schmidt, 1978; Rotter *et al.,* 1979). *In vitro* studies in man, rat, and some other species indicate that the concentration of muscarinic sites in the thalamus is approximately one-quarter that in the caudate nucleus, which has the highest concentration (Hiley and Burgen, 1974; Davies and Verth, 1978; Kobayashi *et al.,* 1978).

Autoradiographic studies of the distribution of muscarinic ligand-binding sites in the rat thalamus (Rotter *et al.,* 1979; Wamsley *et al.,* 1980) are not in complete agreement. Rotter *et al.* (1979), who localized the binding of [^{3}H]propylbenzilylcholine mustard, found the highest concentration of muscarinic sites in the anterior nuclei (Fig. 5.4) with lower concentrations in other nuclei, the lowest being the geniculate nuclei. Wamsley *et al.* (1980), who localized the binding of [^{3}H]-*N*-methylscopolamine, reported high concentrations of high-affinity binding in the lateral and ventral nuclei and lateral geniculate nucleus.

The cause of this discrepancy is not clear though it could be related to the use of lightly fixed tissue by Rotter *et al.* and of fresh tissue by Wamsley *et al.* Another potential factor could be differences in the capacity of the various ligands to recognize high- and low-affinity binding sites.

The highest concentrations of nicotinic cholinergic sites in the thalamus, at least as determined by [^{125}I]-α-bungarotoxin binding, were initially reported by Hunt and Schmidt (1978) and Segal *et al.* (1978) in the ventral lateral geniculate nucleus of the rat. High concentrations can also be demonstrated in the reticular nucleus of the monkey (Jones, 1983h; Fig. 5.5). Moderate concentrations also appear in the rhomboid, parataenial, and parts of the mediodorsal nucleus (Hunt and Schmidt, 1978).

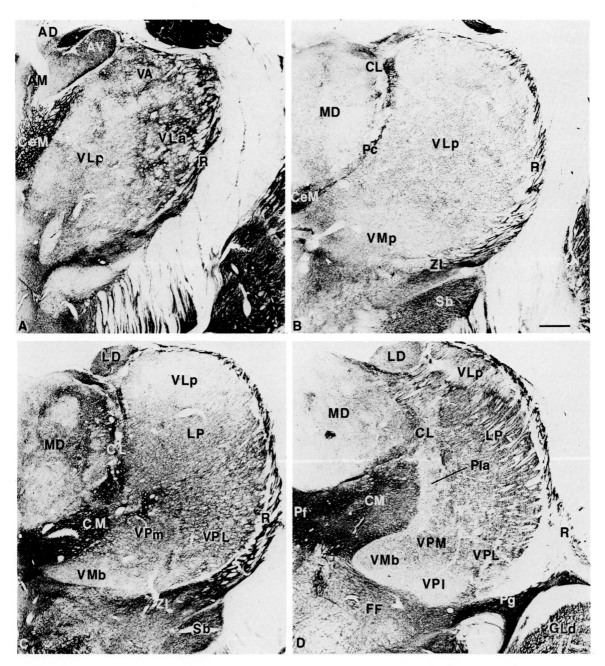

FIGURE 5.3.   A series of frontal sections through the thalamus of a cynomolgus monkey showing the distribution of acetylcholinesterase activity. Bars = 1 mm.

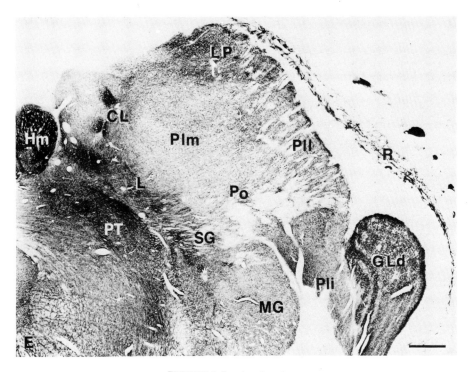

FIGURE 5.3. (*continued*)

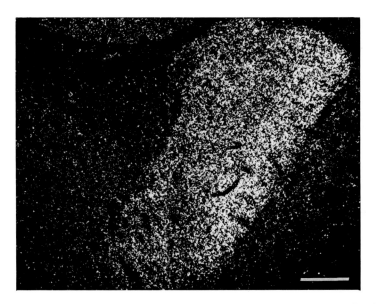

FIGURE 5.4. Autoradiographic demonstration of muscarinic cholinergic receptor sites in the anteroventral nucleus of a rat thalamus by the binding of [³H]propylbenzilylcholine mustard. Bar = 200
μm. From Rotter *et al.* (1979).

5.4. Acetylcholine

Despite the information summarized above, it is still not clear that acetylcholine is a major transmitter in the dorsal thalamus. The only concentration of cholinergic cells demonstrable immunocytochemically in the thalamus is in the medial habenular nucleus of the epithalamus (Houser *et al.*, 1983). These project into the habenulopeduncular tract (Chapter 16). The modest levels of choline acetyltransferase found in the dorsal thalamus normally drop by as much as 60% if lesions are made in the nucleus cuneiformis (Hoover and Jacobowitz, 1979), suggesting that the cholinesterase pathway of Shute and Lewis may indeed be cholinergic.

Until recently, reports of the effects of iontophoretically applied acetylcholine on thalamic neurons have been somewhat contradictory. A weak inhibitory effect was reported on relay cells in some nuclei (Tebēcis, 1972; Godfraind, 1975) but prolonged inhibition was reportedly more common in others (Andersen and Curtis, 1964; Phillis *et al.*, 1967; McCance *et al.*, 1968; Krnjević, 1974). Through indirect reasoning the excitatory effect had been attributed to a presynaptic action (Marshall and McLennan, 1972; Duggan and Hall, 1975). In the reticular nucleus or perigeniculate nucleus a much more pronounced rapid inhibition of spontaneous and glutamate-induced activity has been re-

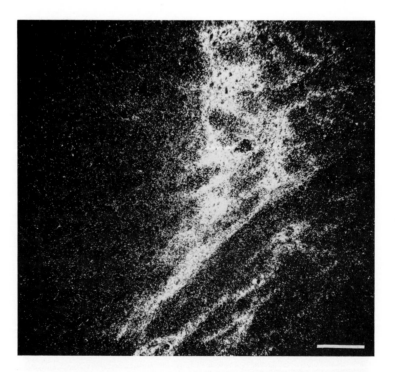

FIGURE 5.5.   Autoradiographic demonstration of the presence of nicotinic cholinergic receptor sites in the reticular nucleus of a monkey thalamus by the binding of $[^{125}I]$-α-bungarotoxin. Bar = 100 μm.

ported even with low doses of acetylcholine (Ben-Ari *et al.*, 1976; Godfraind, 1978).

In a recent study, Sillito *et al.* (1983) have reported that iontophoretically applied acetylcholine strongly facilitates the excitatory responses of X- and Y-type relay neurons in the dorsal lateral geniculate nucleus of cats to appropriate visual stimuli. Stimulus-specific inhibitory effects, such as inhibition due to stimulation of the antagonistic surround of an on-center cell, were also enhanced and commonly background discharge was totally suppressed. The authors attribute these effects to direct excitatory action on the relay cells and on the presynaptic dendrites of the presumed GABAergic interneurons (Chapter 4), the latter leading to inhibition of relay cells. The visually evoked discharges of cells in the perigeniculate nucleus (Chapters 3, 4, and 15) were suppressed by acetylcholine, which might be expected to lead to release of relay cells from perigeniculate inhibition.

The effects of acetylcholine on relay cells and interneurons in the dorsal thalamus are said to be atropine sensitive (Sillito *et al.*, 1983) whereas the effect on the reticular or perigeniculate nucleus is said to be sensitive to high doses of both dihydro-β-erythroidine and atropine, suggesting a mixture of both nicotinic and muscarinic effects (Phillis, 1970; Tebēcis, 1972; Krnjević, 1974).

The nature of any cholinergic pathways converging on the thalamus is still not clear (Chapters 3, 4, and 12) and no anatomical study has reported a pathway other than the principal afferent pathway to a nucleus with the terminal synaptic arrangements implied by the interpretation of Sillito *et al.* (1983). It thus remains to be seen whether the effects of iontophoretically applied acetylcholine are mimicking the effects of stimulation of an afferent pathway.

The reticular nucleus of the thalamus has been suggested as the principal site of action of the ascending cholinergic system (Ben-Ari *et al.*, 1976) and here there is a sounder anatomical basis for such a pathway (Chapters 3, 4, and 15).

5.5. Excitatory Amino Acids and the Corticothalamic Transmitter

Most of the commoner excitatory amino acids have been iontophoretically applied to individual thalamic neurons (Curtis and Johnston, 1974). Of these, glutamate seems to be the most likely candidate for a transmitter agent since its excitatory effect on relay neurons of the cat ventral posterior nucleus is very similar to that elicited by threshold stimulation of peripheral nerves. Both effects are enhanced by the application of substances such as L-glutamic acid dimethyl ester that compete for the uptake of glutamate and other acidic amino acids (Haldeman and McLennan, 1973).

No case has yet been made for glutamate or the other acidic amino acids as the transmitters in the afferent pathways ascending to the thalamus. The absence of a specific antagonist has precluded this. In the retinogeniculate system, taurine has been suggested as a transmitter (Pasantes-Morales *et al.*, 1975) though the evidence is not strong. When iontophoretically applied to thalamic neurons, the effect of taurine is usually a weakly inhibitory one (Krnjević and Puil, 1975).

Glutamate or aspartate are strong candidates for the transmitter used at the corticothalamic synapse. The possible association of these amino acids with cortical efferent fibers, such as the corticospinal and corticostriatal, has been suggested, mainly on biochemical grounds, for some time (Curtis and Johnston, 1974; Divac *et al.*, 1977a). More recently, Lund-Karlsen and Fonnum (1978) showed that the high-affinity uptake of L-glutamate and D-aspartate in the lateral geniculate nucleus of the rat declined substantially within 3 days of destroying the visual cortex. The reduction particularly affected the synaptosomal fraction derived from the nucleus and was associated with a reduction in intrinsic levels of L-glutamate but not of markers for other transmitters, such as choline acetyltransferase or glutamic acid decarboxylase (GAD). These findings suggest that glutamic acid may serve as the corticothalamic neurotransmitter, though aspartic acid also remains a candidate.

Aspartate is concentrated in nerve terminals by the same high-affinity uptake system as glutamate and the use of the D isomer of aspartate, which is not further metabolized, serves as a useful marker for neurons and axon terminals that may synthesize one or other transmitter. Several workers have made use of this phenomenon to show that [³H]-D-aspartate injected into the somatic sensory or

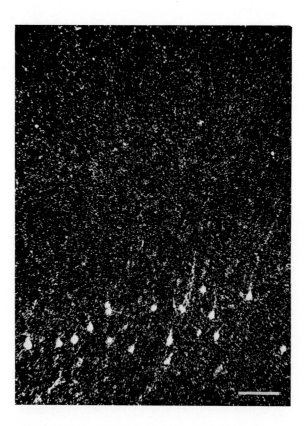

FIGURE 5.6. Retrograde labeling of corticothalamic cell somata in layer VI of monkey somatic sensory cortex following injection of [³H]-D-aspartate in the ventral posterior nucleus. Bar = 100 μm.

visual cortex is selectively concentrated in pyramidal cell somata of layer VI (Søreide and Fonnum, 1980; Baughman and Gilbert, 1981). And when injected into the thalamus the material is retrogradely transported to the somata of corticothalamic neurons that are situated in the same layer (Baughman and Gilbert, 1981; Fig. 5.6). The labeling of corticothalamic cells by retrograde transport appears to be highly specific, for other cell systems cannot be labeled in this fashion. For example, when injected into the cerebral cortex, [³H]-D-aspartate does not retrogradely label thalamocortical relay cells (Baughman and Gilbert, 1981).

5.6. GABA

GABA is, perhaps, the only satisfactorily documented neurotransmitter in the thalamus. It has been known for several years that iontophoretically applied GABA depresses the discharges of relay neurons in the ventral posterior and lateral geniculate nuclei (Curtis et al., 1970; Curtis and Tebēcis, 1972). The depressant effect of GABA is suppressed by the GABA antagonist, bicuculline, which also suppresses the inhibition that succeeds the initial EPSPs induced in thalamic relay neurons by stimulation of peripheral nerves. Such observations argued strongly for a GABA-mediated inhibition in the thalamus and it was usually assumed that GABA was normally released from inhibitory interneurons that lay within the relay nuclei.

Extracellular recording in the dorsal lateral geniculate nucleus of the cat has indicated that inhibitory GABAergic mechanisms play an important role in maintaining the center–surround antagonism in the receptive fields of both X-type and Y-type relay cells (Sillito and Kemp, 1983). Background discharges of relay cells were elevated by iontophoretic application of an excitatory amino acid, such as L-glutamate or DL-homocysteate. Then, the center or surround of a cell's receptive field was stimulated by an appropriate visual stimulus. As would be expected, this led to a suppression of the background discharge when the stimulus was placed in the off component (center or surround) of the receptive field or when a stimulus applied to the on component was turned off. However, these suppressions could be blocked by the iontophoretic application of bicuculline, indicating that the center–surround antagonism of the retinal ganglion cells is maintained by postsynaptic GABA mechanisms in the lateral geniculate nucleus. These postsynaptic inhibitory effects could be mediated by both interneurons in the relay nucleus itself or by neurons entering from the reticular nucleus. Neither can be ruled out by the studies done so far.

In recent times, doubts were raised about the intrathalamic localization of GABAergic interneurons. In the first immunocytochemical study of the localization of GAD, in the rat thalamus Houser et al. (1980) found that the only GAD-positive neuronal cell bodies lay in the reticular nucleus (Fig. 5.7). The dorsal thalamic nuclei were filled with GAD-positive axon terminals, presumably the terminals of GABAergic axons, but even in colchicine-treated animals, where, because of blockage of axoplasmic transport, somal staining should have been enhanced, no stained somata could be visualized. This finding would have implied that GABA-mediated inhibition in the dorsal thalamus is a function of the reticular nucleus, an interpretation that fits nicely with the electrophysiological properties of reticular nucleus cells referred to in Chapters 4 and 15. Against this, however, were the observations of Sterling and Davis (1980) who showed, by electron microscopic autoradiography, that [^{3}H]-GABA and its analogs or

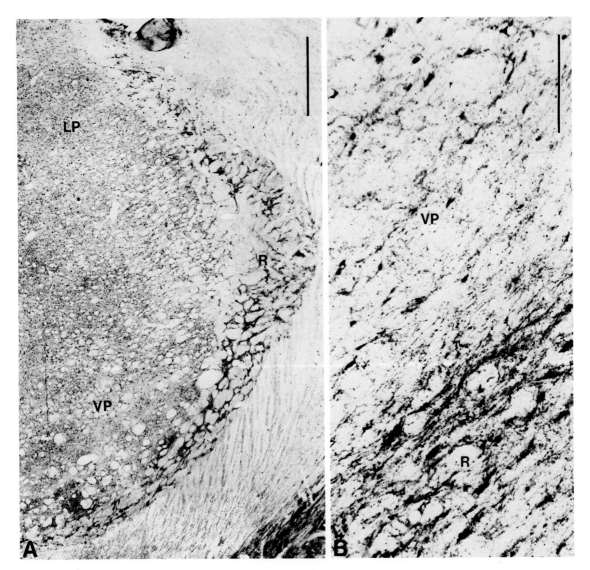

FIGURE 5.7. (A) Immunocytochemical staining of GA-BAergic neurons in the reticular nucleus (R) of a cat, by using antiserum to GAD. Fine dots in dorsal thalamus are GAD-positive somata of interneurons. (B) Higher-powered view of GAD-positive cells in reticular nucleus (R) and ventral posterior (VP) nucleus of a cat. Bars = 1 mm (A), 250 μm (B). From unpublished work of S. H. C. Hendry, C. Brandon, and the author.

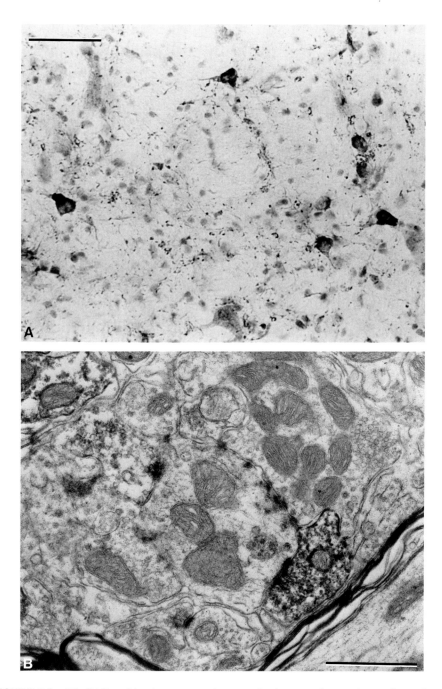

FIGURE 5.8. (A) GAD-positive interneuronal somata in the ventral posterior nucleus of a cat. Unlabeled relay neurons are much larger. From Spreafico *et al.* (1983). (B) Electron micrograph of a GAD-positive presynaptic dendrite in the thalamus of a cat. Bars = 100 μm (A), 0.5 μm (B). Courtesy of Dr. A. Rustioni.

agonists, injected into the dorsal lateral geniculate nucleus of cats, are selectively accumulated by small cell somata and by flattened vesicle-containing profiles that they regard as belonging to intrinsic GABAergic neurons. Following on from this are newer immunocytochemical findings of Ohara *et al.* (1983) who identified a large population of GAD-positive small neurons in the dorsal and ventral lateral geniculate nuclei of the rat. Surprisingly, however, the number of such neurons in other nuclei, such as the ventral posterior, was rather small. By contrast, in the cat, Penny *et al.* (1983) report some 25–30% of the neuronal population in the ventral posterior nucleus is GAD positive, and this has been our own experience in virtually all the dorsal thalamic nuclei of the cat (Figs. 5.8, 5.15B, and 15.7).

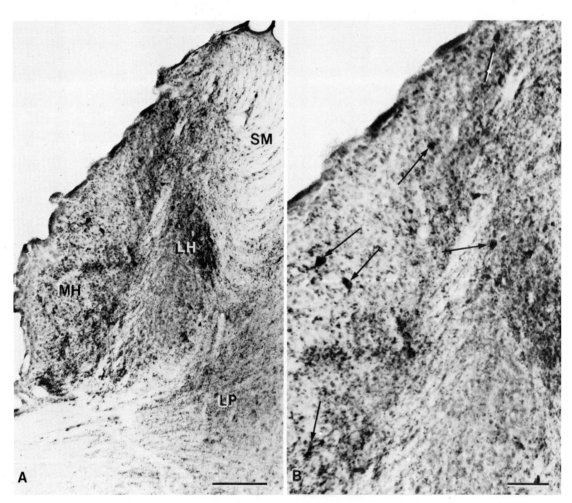

FIGURE 5.9. Lower-power (A) and high-power (B) photomicrographs of GAD-immunoreactive fibers and cells (arrows, B) in parts of the lateral (LH) and medial (MH) habenular nuclei of a cat. SM: stria medullaris. Bars = 250 μm (A), 100 μm (B).

GABA has also been implicated as a transmitter at the synapses of nigro-thalamic fibers in the ventral medial thalamic nucleus, mainly because GABA and GAD levels decline in the thalamus after lesions involving the substantia nigra (DiChiara *et al.*, 1979). There has been no confirmation of this observation with immunocytochemical methods. All dorsal thalamic nuclei display a certain proportion of high-affinity GABA receptors, as detected autoradiographically by binding of agonists, but high densities are found in the ventral anterior and medial geniculate nuclei rather than in the ventral medial nucleus (Palacios *et al.*, 1981).

GABA has also been suggested as the transmitter in the pallidal and basal forebrain inputs to the lateral habenular nucleus (Fig. 5.9) (Gottesfeld *et al.*, 1977, 1981; Nagy *et al.*, 1978; Chapter 16).

5.7. The Amines and "Nonspecific" Afferents

In an earlier chapter, we spoke of specific and nonspecific subcortical afferent systems providing inputs to the dorsal thalamus. The specific systems have an obvious information-bearing content, arise in particular subcortical nuclei, and end in a topographically ordered fashion within the confines of particular thalamic nuclei. The nonspecific systems are arranged without the same degree of finesse and seem to form parts of diffusely organized pathways that distribute to a number of diencephalic and other basal forebrain nuclei en route to the cerebral cortex. The two most obvious representatives of the nonspecific system are the ascending noradrenergic-containing pathway arising mainly in the locus coeruleus of the brain stem (Ungerstedt, 1971; Lindvall and Björklund, 1974; Pickel *et al.*, 1974; Swanson and Hartman, 1975; Jones and Moore, 1977) and the ascending serotonin-containing pathway arising mainly in the dorsal nucleus of the midbrain raphe (Dahlstrom and Fuxe, 1964; Conrad *et al.*, 1974; Bobillier *et al.*, 1976).

The noradrenergic fibers terminating in the thalamus form part of the large dorsal catecholamine bundle in the dorsomedial part of the midbrain tegmentum. In the diencephalon, it ascends beneath the thalamus in close association with the fibers of the superior cerebellar peduncle (Fig. 5.10). In the rat, where they have been localized immunocytochemically with antiserum to dopamine β-hydroxylase (Swanson and Hartman, 1975), the fibers enter the dorsal thalamus through the zona incerta and reticular nucleus. The heaviest concentrations of fibers are found in the anteroventral nucleus, with lesser concentrations in the other nuclei of the anterior complex, in the ventral and dorsal lateral geniculate nuclei, and in the intralaminar system. A few fibers appear in other nuclei of the dorsal thalamus. In the epithalamus, concentrations of fibers, entering from behind, appear in the paraventricular and medial habenular nuclei. There are suggestions that the paraventricular nuclei may receive their noradrenergic innervation from lower brain-stem catecholamine cell groups other than the locus coeruleus, the axons reaching them via a paraventricular system of fibers (Lindvall *et al.*, 1974). The parent bundle of the major thalamic innervation continues into the septum, medial cortical areas, and hippocampal formation.

The terminals of noradrenergic fibers have not been examined electron microscopically in the thalamus and their role in thalamic function has not been revealed. When applied iontophoretically onto thalamic neurons, the effects of noradrenaline are said to be weakly inhibitory (Phillis *et al.*, 1967; Tebēcis and DiMaria, 1972), but a small number of neurons are also reported to show weak facilitatory effects (Nakai and Takaori, 1974; Rogawski and Aghajanian, 1982).

The ascending, serotonin-containing pathways have been traced by fluorescence histochemistry (Dahlstrom and Fuxe, 1964; Ungerstedt, 1971), by conventional anatomical tracing techniques (Gilbert and Kelly, 1975; Leger *et al.*, 1975; Bobillier *et al.*, 1976; Kromer and Moore, 1980), and by immunocytochemistry for the localization of the transmitter itself (Pickel *et al.*, 1974). As defined originally, a few serotonin-containing fibers enter the thalamus from the midbrain reticular formation and are distributed mainly to the ventral lateral geniculate nucleus. A few reach the habenular nuclei also (Kuhar *et al.*, 1972).

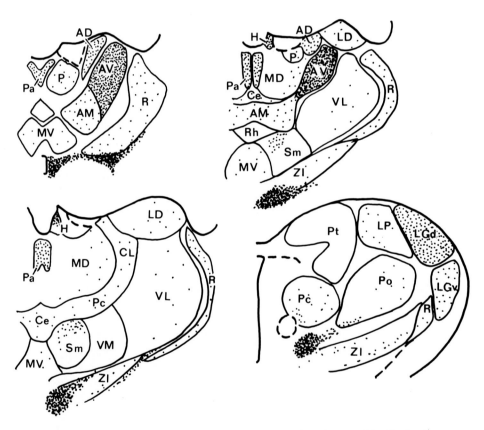

FIGURE 5.10.   Schematic map of the ascending noradrenergic pathway and its distribution among certain thalamic nuclei of the rat, as determined by dopamine β-hydroxylase immunocytochemistry. Redrawn from Swanson and Hartman (1975).

Most emphasis has been placed on the lateral geniculate complex as a terminal site because electrical stimulation in the rostral midbrain leads to a facilitation of synaptic transmission through the lateral geniculate that is associated with field potentials resembling the "PGO" waves of paradoxical sleep (Singer, 1977; Burke and Cole, 1978). While it is true that injections of horseradish peroxidase in the vicinity of the lateral geniculate complex lead to retrograde labeling of cells in the dorsal raphe nucleus (Leger *et al.*, 1975), involvement of fibers passing around the geniculate was not ruled out. Raphe cells can be labeled by injections in other nuclei as well as the lateral geniculate (Gilbert and Kelly, 1975; Tracey *et al.*, 1980; Fig. 5.11). Therefore, it is not clear whether the lateral geniculate complex is the preferred thalamic site of termination for serotonin fibers.

Of the other amines, adrenaline has been suggested as a transmitter in a few thalamic fibers on the basis of the immunocytochemical localization of its associated enzyme, phenylethanolamine *N*-methyltransferase (PNMT). The fibers arise from cells in the lower medulla and ascend in association with the ventral noradrenergic bundle to distribute a moderately dense plexus to the paraventricular nuclei and a much sparser plexus to the small parataenial nucleus (Hökfelt *et al.*, 1974). No further investigations seem to have been made on this system.

Dopamine also does not appear to be a significant transmitter in the thalamus. The large ascending dopaminergic system, arising in the pars compacta

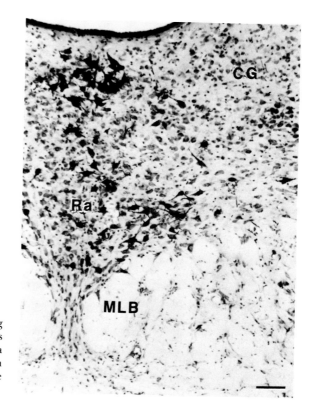

FIGURE 5.11. Retrograde labeling of presumed serotoninergic neurons in the dorsal raphe nucleus (Ra) of a rat following injection of horseradish peroxidase in the lateral part of the thalamus. Bar = 100 μm.

of the substantia nigra and adjacent ventral tegmental area, bypasses the thalamus as it ascends beneath it toward the basal ganglia and cerebral hemisphere. A moderate concentration of dopamine-containing fibers has been demonstrated by fluorescence histochemistry in the paraventricular and medial habenular nuclei of the epithalamus. These probably come from the ventral tegmental area (Chapter 16). However, in the dorsal thalamus there are only sparse plexuses in the rhomboid and parafascicular nuclei. These fibers probably arise from dopaminergic cell bodies in the local periventricular gray matter (Lindvall and Björklund, 1974; Lindvall *et al.*, 1974).

5.8. Thalamocortical and Thalamostriatal Transmitters

The transmitter or transmitters synthesized by thalamocortical and thalamostriatal relay cells are unknown. Injections of [^{3}H]-D-aspartate into the striatum, retrogradely label the cells of origin of the thalamostriatal system in the intralaminar nuclei and also those of the corticostriatal system in layer V of the cerebral cortex (Streit, 1980). Taken together with the observation that glutamate levels fall in the striatum after destruction of the cortex (Divac *et al.*, 1977b; McGeer *et al.*, 1977; Biziere and Coyle, 1978; Reubi and Cuénod, 1979), this suggests that glutamate or aspartate may be the corticostriatal transmitter and, by extension, that of the thalamostriatal system as well.

There are few clues to the nature of the thalamocortical transmitter. Though glutamate, aspartate, and certain other excitatory amino acids have an effect upon all cortical neurons, this appears to be rather nonspecific (Curtis and Johnston, 1974). Neither [^{3}H]glutamate nor [^{3}H]-D-aspartate appears to be selectively transported to thalamic neurons when injected into the cerebral cortex (Streit, 1980; Baughman and Gilbert, 1981).

Acetylcholine has been widely studied as a potential thalamocortical transmitter but there has been no confirmation of its release from thalamocortical fibers. Iontophoretic studies indicate that acetylcholine induces EPSPs particularly in the cells of layer V, including identified pyramidal tract neurons (Krnjević and Phillis, 1963; Crawford, 1970). Neurons in other layers appear to be unaffected or inhibited (Phillis, 1970).

Acetylcholinesterase is concentrated in the middle layers of the visual and certain other areas of the cortex in a laminar pattern that may be coextensive with that of the terminations of thalamocortical fibers (Kristt, 1979; Fitzpatrick and Diamond, 1980; Robertson, 1983a). The enzymatic activity, however, does not disappear in the visual cortex after thalamic lesions, perhaps implying that it is not associated with the thalamocortical axon teminals (Fitzpatrick and Diamond, 1980). By contrast, lesions of the anterior thalamic nuclei cause a profound change in the staining pattern in limbic cortex (Robertson, 1983a) (Fig. 5.12).

Muscarinic cholinergic receptor sites, as localized autoradiographically in the rat cortex by agonist binding, are also distributed in a pattern that resembles

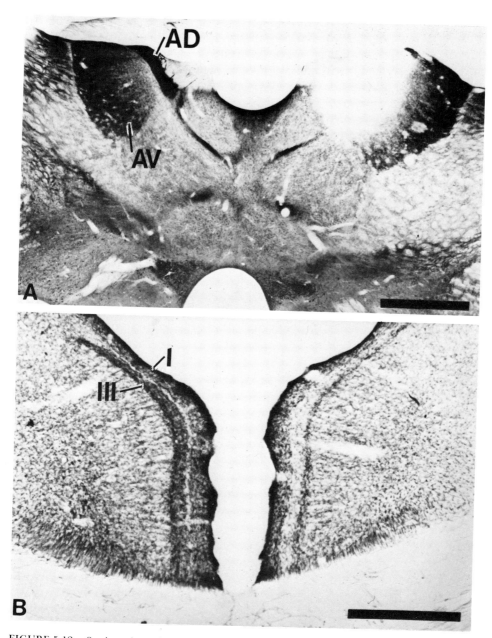

FIGURE 5.12.   Sections through the thalamus (A) and retrosplenial cortex (B) of a rat stained for acetylcholinesterase activity and showing (right) loss of activity primarily in layer III of the retrosplenial area following destruction of the anterodorsal nucleus. Bars = 1 mm (A), 0.5 mm (B). Kindly provided by Dr. R. T. Robertson.

that of the terminations of thalamocortical fibers. Wamsley *et al.* (1980) showed that [^{3}H]-*N*-methylscopolamine bound to sites in layers III and IV and that binding to presumed high-affinity sites in layer IV disappeared after prior treatment with carbachol. Rotter *et al.* (1979) described binding of [^{3}H]propylbenzilylcholine mustard in layers II and III rather than in layer IV but their photomicrographs suggest localization in layer IV as well.

Nicotinic cholinergic receptor sites appear to be present in smaller numbers in the cortex and their distribution does not coincide with the terminations of thalamocortical fibers. In the rat (Hunt and Schmidt, 1978) and monkey (Hendry, Schecter, and Jones, unpublished), α-bungarotoxin binding occurs with moderate density in layers I, V, and VI.

Though some of the features normally associated with cholinergic transmission are demonstrable in the cerebral cortex and may even coincide with thalamocortical fiber terminations, there is still no compelling reason for assuming that thalamocortical innervation is cholinergic. All of the features associated with cholinergic mechanisms may be related, not to the thalamocortical innervation, but to that from the cholinergic cells of the basal nucleus of Meynert in the substantia innominata (Divac, 1975; Jones *et al.*, 1976; Mesulam and Van Hoesen, 1976; Winfield *et al.*, 1975; Johnston *et al.*, 1981; Whitehouse *et al.*, 1982; Pearson *et al.*, 1983). Only in the cingulate region, by virtue of its input from the anterior thalamic nuclei which contain a number of acetylcholinesterase-positive cells (Section 5.1), does cholinergic thalamocortical transmission seem a possibility (Fig. 5.12).

5.9. Opioids and Other Peptides

To date, more attention has been paid to the localization of enkephalins than of other peptides in the thalamus. The principal concentrations of enkephalin-immunoreactive fibers and of opiate receptors are in the ventral lateral geniculate, paraventricular, reticular, and intralaminar nuclei (Elde *et al.*, 1976; Wamsley *et al.*, 1982; Mantyh and Kemp, 1983; Khachaturian *et al.*, 1983) (Fig. 5.13). The paraventricular nuclei appear to be the only nuclei in the vicinity of the thalamus with β-endorphin-containing fibers as well (Pert *et al.*, 1974; Elde *et al.*, 1976; Simantov *et al.*, 1976; Atweh and Kuhar, 1977; Watson *et al.*, 1977; Bloom *et al.*, 1978; LaMotte *et al.*, 1978; Sar *et al.*, 1978). High concentrations of enkephalins and of opiate receptors reported in the "nucleus posterior thalami" of König and Klippel's atlas (1963) are in the pretectum. Reports vary regarding dorsal thalamic nuclei other than the intralaminar. Some workers emphasize the parataenial or mediodorsal nucleus, others parts of the ventral or anterior nuclei. In the epithalamus there is a distinct band of opiate receptors at the border of the medial and lateral habenular nuclei (Pert *et al.*, 1976; Atweh and Kuhar, 1977). The only enkephalin-containing cell bodies that have been reported in the thalamus are a few in the paraventricular nuclei of the epithalamus (Khachaturian *et al.*, 1983) and the cells of origin of the enkephalin-containing fibers in the dorsal thalamus have not yet been identified.

The subcellular distribution of enkephalin and of opiate receptors are also unknown in the thalamus and it is still uncertain whether the enkephalins act as conventional transmitters. When applied iontophoretically, to thalamic neurons, Met-enkephalin depresses both spontaneous discharges and discharges in response to noxious stimuli (Hill *et al.,* 1976). The effect is reversed by the opiate antagonist naloxone.

Elsewhere in the central nervous system, enkephalins are seemingly closely associated with pain pathways, but it is hard to relate their thalamic distribution to any nuclei known to be implicated in nociception. Though high concentrations of enkephalin fibers and opiate receptors occur in the intralaminar nuclei, their distribution is far more extensive than that of the zone of termination of the rather modest number of fibers arising in the spinal cord or periaqueductal gray matter. Far more spinothalamic tract fibers terminate in the ventral nuclei and there the recipient neurons have receptive fields far more typical of spinothalamic neurons than do intralaminar neurons (Applebaum *et al.,* 1979). Paradoxically, however, the ventral nuclei contain only low concentrations of enkephalins and of opiate receptors.

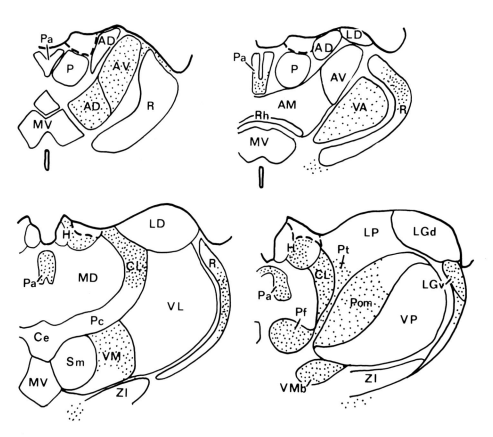

FIGURE 5.13.   Schematic map of the immunocytochemical localization of Leu- and Met-enkephalin-containing fibers in the thalamus of a rat. Redrawn from Sar *et al.* (1978).

The high concentrations in the intralaminar nuclei may reflect the close association of these nuclei with the globus pallidus (Chapter 12), which contains one of the highest concentrations of enkephalins in the brain. In view of the involvement of both intralaminar nuclei and globus pallidus in certain aspects of motor behavior, these high concentrations in them may relate to the effects of opiates on motility (Bloom *et al.*, 1978). If the high concentrations of opiate receptors and opioid peptides reported by some workers in the mediodorsal nucleus are verified, their concentrations there may reflect a role of this nucleus in affective aspects of pain appreciation (Snyder and Childers, 1979).

Initially, few of the known brain or brain–gut peptides were reported in significant amounts in the thalamus (Emson, 1979; Hökfelt *et al.*, 1980; Snyder, 1980). Recently, however, somatostatin has been localized immunocytochemically to neurons in the reticular nucleus that also contain GAD (Graybiel and

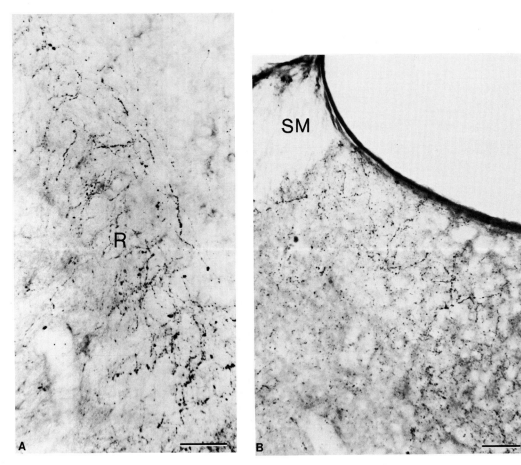

FIGURE 5.14. (A) Somatostatin-immunoreactive fibers in the reticular nucleus of a rat thalamus. (B) Somatostatin-immunoreactive fibers in the paraventricular nuclei of a rat. No somatostatin-positive cells occur in the rat thalamus. Bars = 50 μm.

Elde, 1983). Somatostatin immunoreactivity can also be demonstrated in fibers in the zona incerta and paraventricular nuclei (Fig. 5.14). Fibers with cholecystokinin-like (CCK) immunoreactivity have also been discovered in small numbers in the reticular, intralaminar, rhomboid, parataenial, and paraventricular nuclei (unpublished observations) and a few CCK-positive cells and fibers occur in the taste relay nucleus (Mantyh and Hunt, 1983). CCK receptors have been identified by the binding of a radioactive ligand in some of these same nuclei and in the ventral lateral geniculate nucleus (Zarbin *et al.*, 1983). Immunocytochemical staining of CCK fibers in the reticular nucleus appears to disappear after destruction of the cerebral cortex. CCK-immunoreactive fibers in the other nuclei seem to ascend to the thalamus from brain-stem levels (personal observations). Avian pancreatic polypeptide (neuropeptide Y) (Tatemoto *et al.*, 1982) is found in cells of the dorsal part of the ventral lateral geniculate nucleus that project to the suprachiasmatic nucleus (Card and Moore, 1982; Mantyh and Kemp, 1983) (Fig. 5.15). Neuropeptide Y immunoreactivity is particularly dense in fibers in the thalamic paraventricular nuclei, into which a continuous band of staining (Figs. 5.16–5.18) extends upwards from the paraventricular nucleus of the hy-

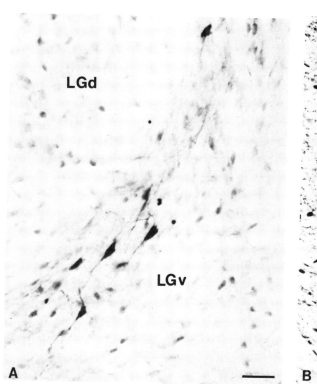
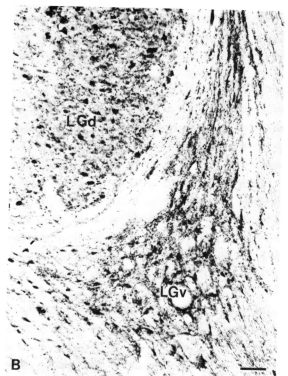

FIGURE 5.15.   (A) Neuropeptide Y-immunoreactive cells in the dorsomedial part of the ventral lateral geniculate nucleus of a rat. Such cells project to the suprachiasmatic nu- cleus of the hypothalamus. (B) GAD-immunoreactive cells in the ventral and dorsal lateral geniculate nuclei of a cat. Bars = 50 μm.

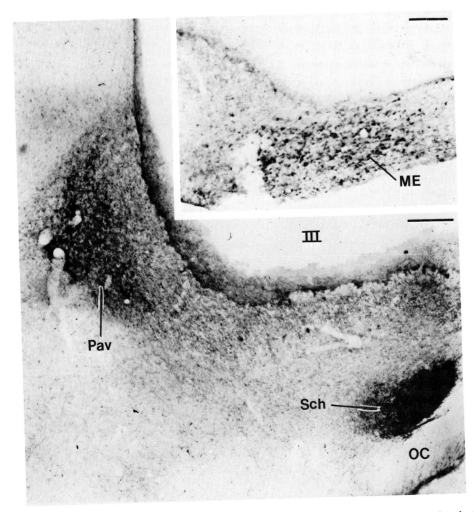

FIGURES 5.16–5.18. A series of sagittal and frontal sections showing the continuous band of neuropeptide Y-immunoreactive fibers ascending from cells in the vicinity of the median eminence (ME) and with dense ramifications in the suprachiasmatic (Sch) and the paraventricular nuclei (Pav) of the hypothalamus and in the anterior (Para) and posterior (Parp) paraventricular nuclei of the thalamus in a rat. OC: optic chiasm. Bars = 250 μm (16, 17A, C), 50 μm (17B), 500 μm (18). From unpublished work of S. H. C. Hendry and the author.

pothalamus and median eminence. The cells giving rise to these fibers appear to reside in the median eminence and arcuate nucleus (Fig. 5.18).

Among the other peptides, angiotensin II is said to be present in a few thalamic fibers (Fuxe *et al.*, 1976; Quinlan and Phillips, 1981). Substance P fibers, of unknown origin, are found in the ventral lateral geniculate nucleus (Mantyh and Kemp, 1983) but vasoactive intestinal polypeptide is said to be absent. Recently a relatively dense pathway containing neurotensin and arising in

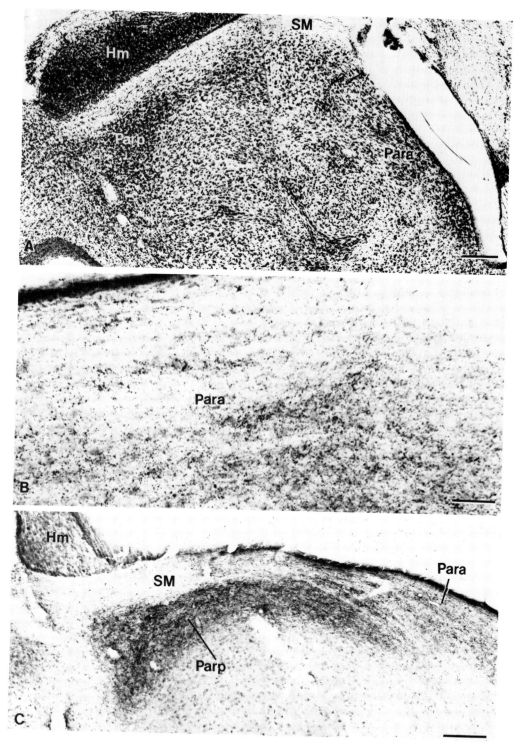

FIGURE 5.17

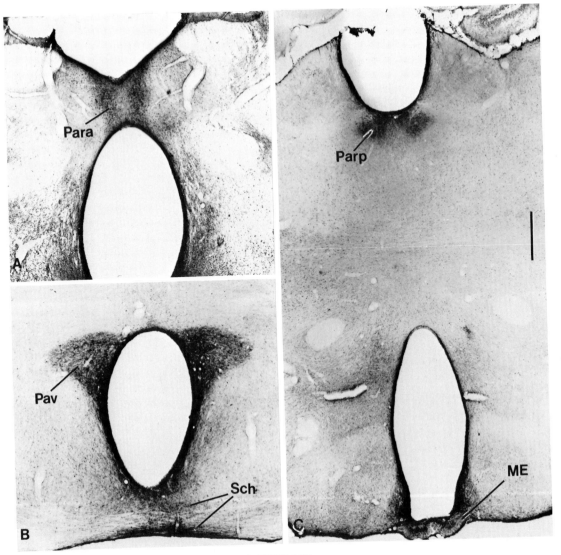

FIGURE 5.18

the endopiriform nucleus (the ventral part of the claustrum) has been shown to terminate in the central portion of the mediodorsal nucleus of the rat (Inagaki *et al.*, 1983). Other neurotensin-immunoreactive fibers enter the habenular nuclei (Uhl *et al.*, 1977), as do a few fibers positive for vasopressin and its neurophysin (Buijs, 1978; Sofroniew and Weindl, 1978) and others containing luteinizing hormone-releasing hormone (Barry, 1978; Silverman and Krey, 1978).

The mitochondrial enzymes succinic acid dehydrogenase and cytochrome oxidase can be demonstrated by relatively simple histochemical techniques and in the thalamus, as elsewhere, can be used to demonstrate concentrations of axon terminals and possibly recipient cells in particular nuclei. The concentration of each of the enzymes declines following chronic deafferentation or sensory deprivation. For example, removal of an eye or monocular visual deprivation in cats or monkeys leads to a reduction in cytochrome oxidase activity in the appropriate laminae of the dorsal lateral geniculate nucleus (Wong-Riley, 1979) (Fig. 5.19). Similarly, section of the trigeminal nerve during the first few days of life in rats leads to significant rearrangement of the distribution pattern in the ventral posterior thalamic nucleus (Killackey and Shinder, 1981) (Fig. 5.20).

As well as being valuable experimental tools, the enzymes can be used to demonstrate differences in the resting levels of activity among individual thalamic nuclei. Figure 5.21A, for example, a horizontal section through the thalamus of a monkey, indicates high levels of cytochrome oxidase activity in most of the ventral nuclei, in the reticular nucleus, and in the inferior nucleus of the pulvinar. Levels are much reduced in the intralaminar nuclei and in the anterior and lateral nuclei of the pulvinar; in the other thalamic nuclei present, levels are virtually zero. Other nuclei showing high activity are the medial and lateral geniculate nuclei (Fig. 5.19). These results suggest that those nuclei most directly related to the ascending sensory and motor pathways have higher levels of

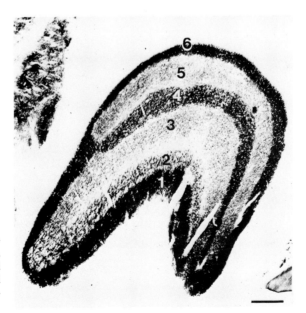

FIGURE 5.19.   Reduction of cytochrome oxidase activity in layers 2, 3, and 5 of a monkey's dorsal lateral geniculate nucleus following removal of the ipsilateral eye 6 days previously. Bar = 250 μm.

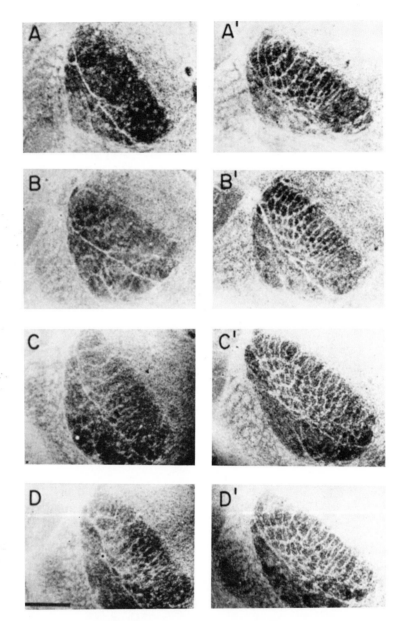

FIGURE 5.20.   Localization of succinic acid dehydrogenase activity, probably in axonal ramifications around cell clusters (Chapter 7) of the ventral posterior nuclei of a series of rats at 5 days of age. Right column is normal. Left column is contralateral to the side of the head on which the trigeminal nerve was divided at different postnatal ages from above down days 0 to 3. From Killackey and Shinder (1981).

FIGURE 5.21.   (A) Horizontal section through the thalamus of a monkey showing heavy cytochrome oxidase staining in the reticular nucleus, ventral nuclei, and inferior pulvinar nucleus, moderate staining in the intralaminar nuclei and the anterior and lateral pulvinar nuclei, but virtually no staining in other nuclei. (B) Frontal section through centre médian, VPM, VPL, VMb, and VPI nuclei of a monkey showing differences in density of cytochrome oxidase staining and staining of cell clusters in VPM and VPL nuclei. Bars = 1 mm (A), 500 μm (B). From Jones and Hendry (1984).

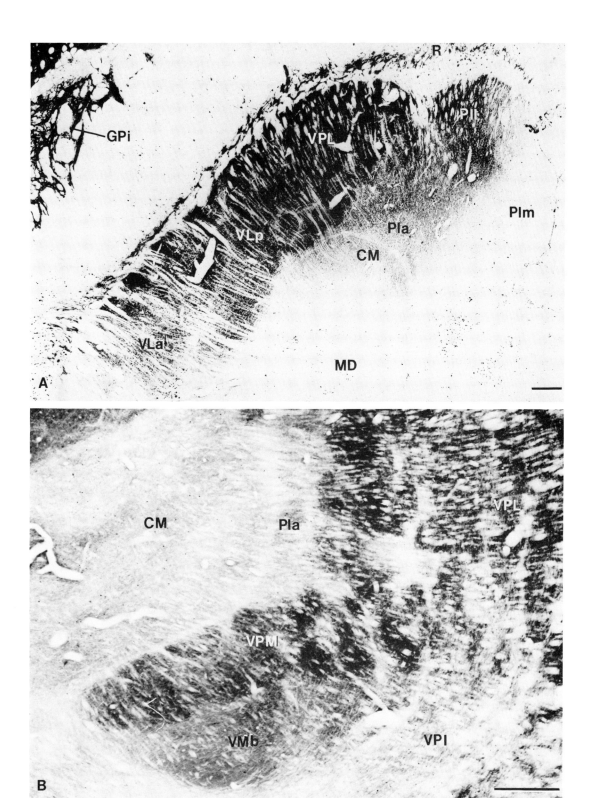

oxidative metabolism than those less directly related (Fig. 5.21B). Part of the explanation for the differences may lie in the generally larger size of axons and their terminals innervating the nuclei showing enhanced activity. But these nuclei are probably subjected to far higher resting levels of afferent input which would also tend to maintain a chronically high level of oxidative metabolism.

Man bestrebt sich, irgend eine Leistung als eine Funktion der
einzugehenden Bedingungen aufzufassen. . . .

Carl Ludwig (1858)

Development

Instead of diffuse, non-selective growth of nerve connections in brain development, we now have evidence for a highly ordered growth and patterning of brain pathways and fiber hookups, all strictly regulated with extreme precision through genetic control and a complex system of cytochemical affinities.

R. W. Sperry, *Science and Moral Priority* (1983)

Development of the Thalamus

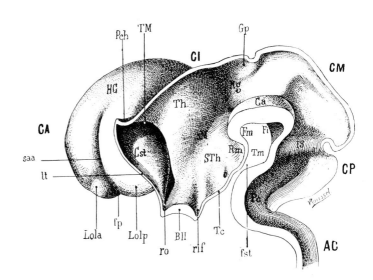

Copy of a drawing of His by one of Dejerine's illustrators, Roussel, showing a medial view of the brain of a human embryo 16.6 mm long and aged approximately 5 weeks. From Dejerine (1901).

6.1. Morphological Studies

6.1.1. Introduction

Descriptions of the development of the thalamus began soon after Nissl's (1889) original cytoarchitectonic parcellation. Wilhelm His (1893, 1904) himself was among the earliest workers in the field when he claimed that the hypothalamic sulcus (Fig. 6.1) on the medial wall of the human diencephalon represented a rostral extension of the sulcus limitans of the spinal cord and brain stem and thus divided the diencephalon into dorsal and ventral parts. The dorsal part, consisting of epithalamus, thalamus, and metathalamus* (the geniculate bodies), he thought was derived from the alar plate of the embryonic neural tube, and the ventral part, consisting of the hypothalamus, he thought was derived from the basal plate. His's point of view caused much argument partly because of disagreement about what constituted the hypothalamic sulcus in embryos but mainly on account of its implication that the diencephalon could be divided into alar and basal plates equivalent to the sensory and motor portions of the spinal cord and caudal brain stem (e.g., Kingsbury, 1920; Kuhlenbeck, 1948).

In English-speaking countries, at least, the fundamental plan of diencephalic development that was more widely accepted had its origins in the comparative anatomical studies of Herrick (1910, 1918) which gave us the terms *dorsal* and *ventral thalamus* and implied a separate developmental and evolutionary history for these and for the epithalamus and hypothalamus. Herrick's primary divisions were based upon his observations in amphibian embryos (Fig. 6.2): In the diencephalic wall he could detect a roof plate, a floor plate, and lateral plates. The lateral plates were grooved on their ventricular surfaces by longitudinally running dorsal, middle, and ventral diencephalic sulci which served to delimit four regions of cellular proliferation. These he called epithalamus, dorsal thalamus, ventral thalamus (sometimes giving it the alternative name subthalamus), and hypothalamus. This fundamental developmental subdivision of the diencephalon was confirmed for mammals by Droogleever-Fortuyn (1912) in his studies of the development of the rabbit thalamus (Fig. 6.3) and became widely accepted (e.g., Le Gros Clark, 1932a). The significance of these subdivisions was reaf-

---

* These terms were not introduced by His but date at least to Edinger (1885).

261

firmed by the demonstration of the clear connectional differences exhibited by their derivatives (Rose, 1942a). Herrick had earlier attempted to relate his subdivisions to different fiber connections but some of the connections that he regarded as basic have not stood the test of modern study.

There are two phases of thalamic growth: an early phase of cellular proliferation during which the epithalamus, dorsal thalamus, and ventral thalamus differentiate from one another, followed by a later phase during which the individual nuclei differentiate within the three larger subdivisions. Studies detailing the morphology of these two phases have been made on rabbits (Bianchi, 1909; Droogleever-Fortuyn, 1912; Rose, 1942a), rodents (Ströer, 1956; Niimi, et al., 1961; Coggeshall, 1964; Hyyppa, 1969; Keyser, 1972), man (Gilbert, 1934; Dekaban, 1954), and various other animals (Bergquist and Källén, 1954).

6.1.2. Early Development in the Rat

Of the studies that document the morphological development of the thalamus up to the time of differentiation of the individual nuclei, that by Coggeshall (1964) on the rat is possibly the most straightforward and easiest to follow. At 12 days of gestation, in the rostral part of the neural tube of the rat, the telencephalic and optic vesicles are just forming and the wall of the already folded diencephalon is divided by a vertical groove that divides it (Fig. 6.4) into a ventral or hypothalamic part which appears more continuous with the telencephalon, and a dorsal part which appears more directly continuous with the midbrain.

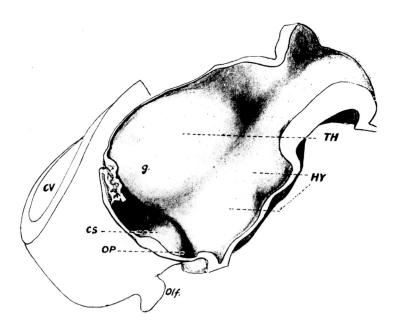

FIGURE 6.1.   Sagittal section of the forebrain in a human embryo of 28 mm, showing the thalamus (TH) and hypothalamus (HY) separated by the hypothalamic sulcus. CV: cerebral vesicle; CS: corpus striatum; Olf: olfactory bulb; OP: optic diverticulum. From Frazer (1931).

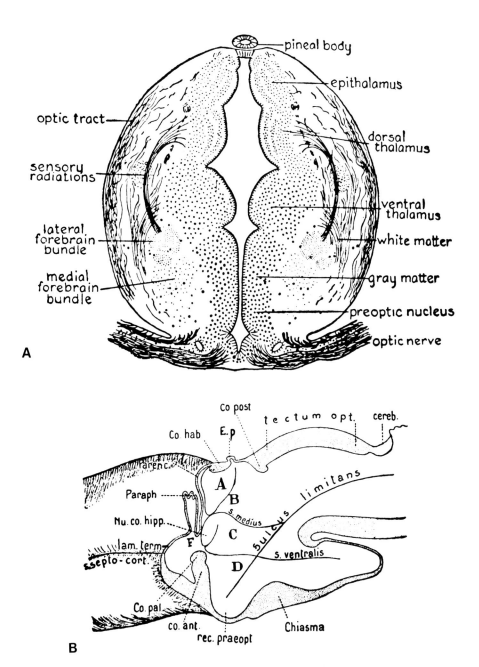

FIGURE 6.2. (A) Transverse section of the diencephalon of the salamander *Ambystoma* showing the (unlabeled) dorsal, middle, and ventral diencephalic sulci that delimit the epithalamus, dorsal thalamus, ventral thalamus, and hypothalamus. From Herrick (1926). Compare with Fig. 6.6. (B) Sagittal section of the diencephalon of *Ambystoma* showing the dorsal (unlabeled), middle (medius), and ventral (ventralis) diencephalic sulci and the supposed sulcus limitans. A: epithalamus; B: dorsal thalamus; C: ventral thalamus; D: hypothalamus. From Herrick (1933). Compare with Fig. 6.3.

This groove marks what Coggeshall called the anterior diencephalic fold. Caudal to it on day 13, appear middle and posterior diencephalic folds. The anterior and middle folds now delimit a wedge-shaped piece of diencephalic wall in which the dorsal and ventral thalamus will form. A second wedge-shaped piece between the middle and the posterior thalamic folds will become the epithalamus and pretectum. By the 14th day of gestation, the grooves marking each field have very clearly become relatively acellular regions, each associated with a developing major fiber tract: the anterior diencephalic fold with the external medullary lamina, the middle with the habenulopeduncular tract, and the posterior with the decussation and limbs of the posterior commissure (Figs. 6.4–6.6). This early delineation of the thalamus by developing fiber tracts was originally noted by Droogleever-Fortuyn (1912). A further developing fiber tract visible at this time lies on the dorsal surface of the part of the diencephalic wall anterior to the anterior diencephalic fold. It is the stria medullaris thalami (Figs. 6.4–6.6).

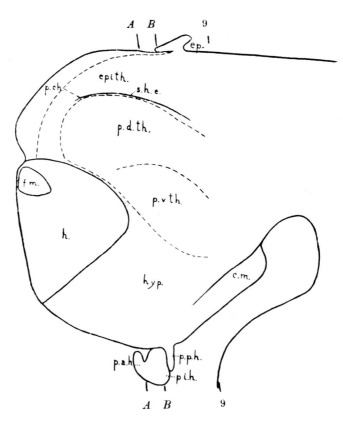

FIGURE 6.3. Sagittal section of the forebrain of a 15.5-mm rabbit embryo indicating the epithalamus (epith.), dorsal thalamus (p.d.th.), ventral thalamus (p.v.th.), and hypothalamus (hyp.) and their delimiting sulci. ep.: pineal; s.h.e.: external habenular sulcus; p.ch.: choroid plexus. From Droogleever-Fortuyn (1912).

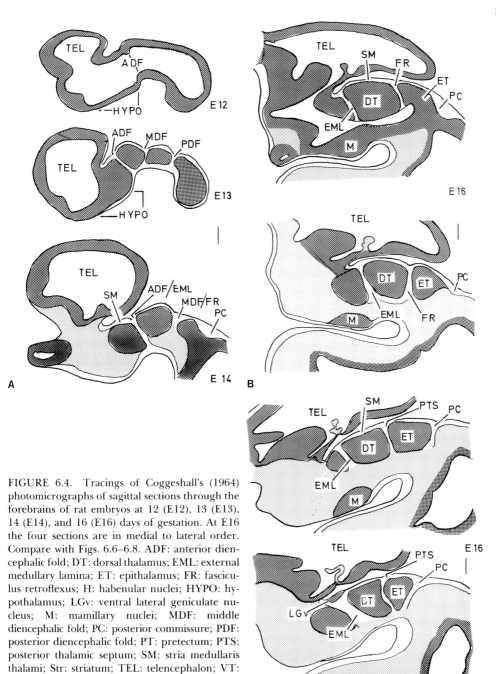

FIGURE 6.4. Tracings of Coggeshall's (1964) photomicrographs of sagittal sections through the forebrains of rat embryos at 12 (E12), 13 (E13), 14 (E14), and 16 (E16) days of gestation. At E16 the four sections are in medial to lateral order. Compare with Figs. 6.6–6.8. ADF: anterior diencephalic fold; DT: dorsal thalamus; EML: external medullary lamina; ET: epithalamus; FR: fasciculus retroflexus; H: habenular nuclei; HYPO: hypothalamus; LGv: ventral lateral geniculate nucleus; M: mamillary nuclei; MDF: middle diencephalic fold; PC: posterior commissure; PDF: posterior diencephalic fold; PT: pretectum; PTS: posterior thalamic septum; SM: stria medullaris thalami; Str: striatum; TEL: telencephalon; VT: ventral thalamus. Bars = 250 μm.

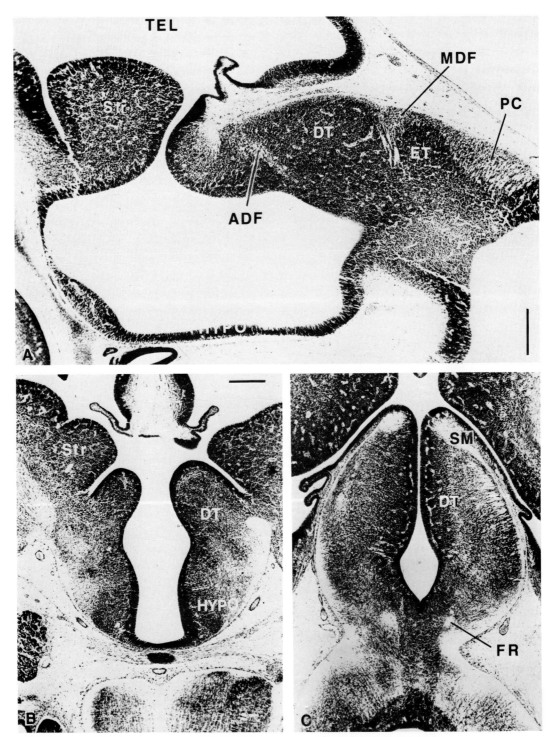

FIGURE 6.5.   Sagittal (A) and horizontal sections (B, C) through the diencephalons of rat fetuses at 13 days of gestation. Same abbreviations as Fig. 6.4. C is dorsal to B. Thionin stain. Bars = 250 μm. Preparations made in the author's laboratory by Dr. K. L. Valentino (Figs. 6.5–6.10).

As the cellular population of the developing thalamus increases, the three diencephalic folds disappear but the external medullary lamina, habenulopeduncular tract, and posterior commissure remain and serve to delimit the thalamus and epithalamus. Of the three tracts, the external medullary lamina is lengthened and comes to lie ventral as well as anterior to the thalamus by the 16th day of gestation. On this day, nuclear differentiation commences in the external medullary lamina with the appearance of a ventral lateral geniculate nucleus (Figs. 6.4–6.9) which at this stage lies dorsal and anterior, just beneath the overlying cerebral vesicle. The reticular nucleus and zona incerta appear in the remainder of the external medullary lamina by the next (17th) day but although cellular proliferation has largely ceased in the epithalamus and dorsal thalamus by this time (McAllister and Das, 1977; Section 6.2), no nuclear differentiation occurs in these regions until much later. The habenular nuclei are first to appear in the epithalamus on day 19 and the first dorsal thalamic nuclei appear on day 20 or 21 (Figs. 6.8–6.10).

Coggeshall notes that obvious fascicles of fibers can be seen joining the dorsal

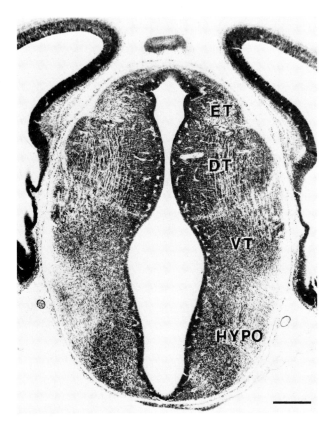

FIGURE 6.6. Transverse section of the diencephalon of a rat at 14 days of gestation. Same abbreviations as Fig. 6.4. Note similarity to Fig. 6.2A. Thionin stain. Bar = 250 μm.

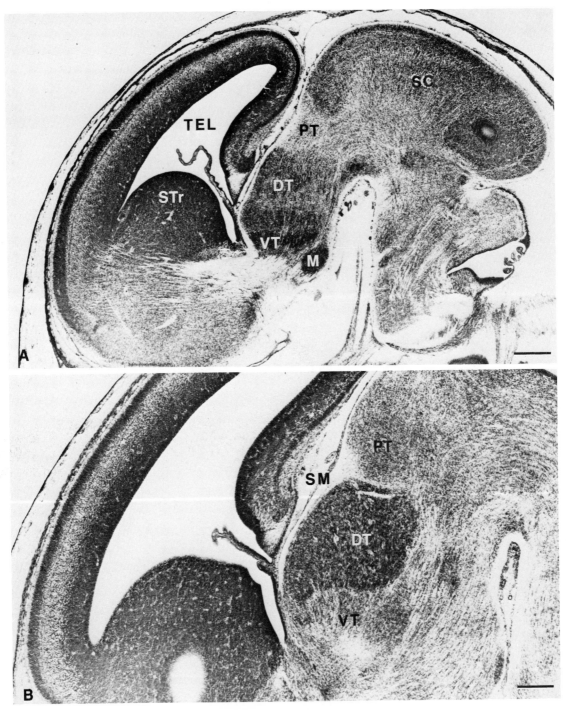

FIGURE 6.7. Sagittal sections in lateral (A) to medial (C) order through the forebrain of a rat fetus at 16 days of gestation. There is still virtually no nuclear differentiation in the dorsal thalamus. MTT: mamillotegmental tract. Thionin stain. Bars = 250 μm (B, C), 500 μm (A).

thalamus to the telecephalon via the internal capsule, by at least the 15th day of gestation but that the optic tract, medial lemniscus, and other ascending afferent pathways cannot be traced into the dorsal thalamus until rather later, the largest numbers of fiber fascicles not entering until about the 21st day. Thus, he felt he could correlate the onset of nuclear differentiation in the dorsal thalamus with the arrival of subcortical afferents: "This transforms the embryonic dorsal thalamus, a unit that is forming connections with the telencephalon alone, into the adult dorsal thalamus, a mosaic of units some of which connect the telencephalon synaptically with lower centers." He believed "it seems likely that optic, tactile, and other incoming pathways plug into circuits that are, to some extent, already formed." It is a nice metaphor but as we shall see (Section 6.3), it is unlikely that the thalamus has established connections with the cortex before subcortical afferents arrive.

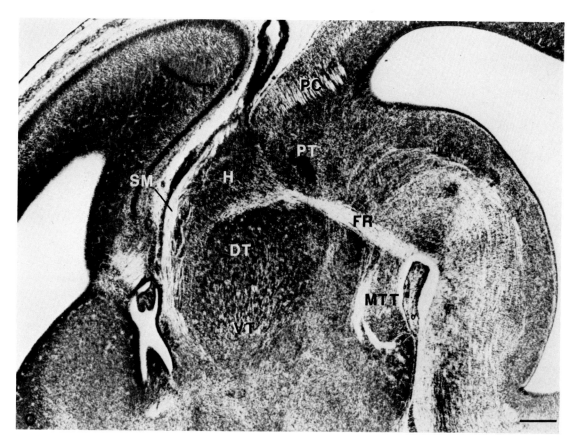

FIGURE 6.7. (*continued*)

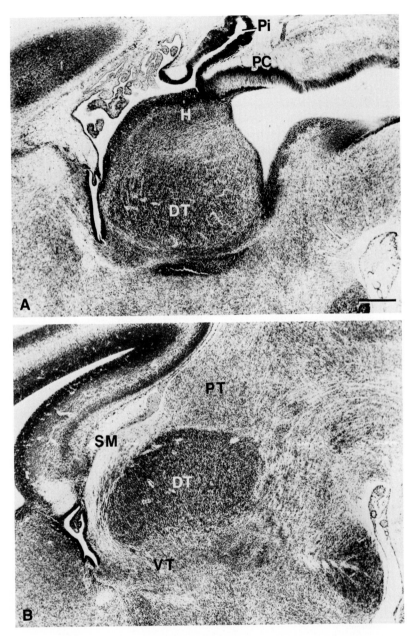

FIGURE 6.8.   Sagittal sections in medial (A) to lateral (D) order through the diencephalon of a rat fetus at 17 days of gestation. Nuclear differentiation is commencing in the epithalamus and ventral

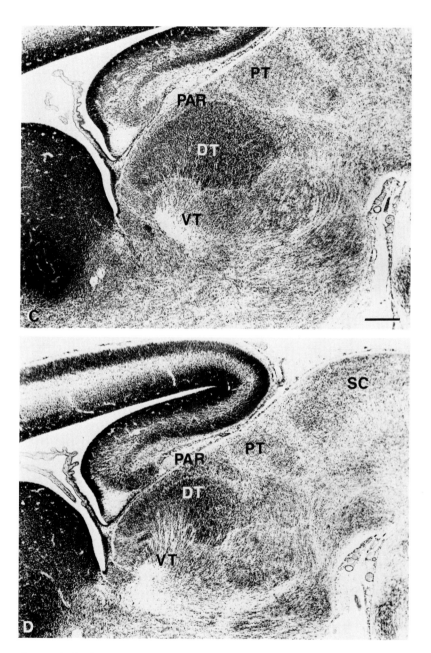

thalamus but not in the dorsal thalamus. Same abbreviations as Fig. 6.4. PAR: paraventricular nuclei;
Pi: pineal; SC: superior colliculus. Bars = 250 μm.

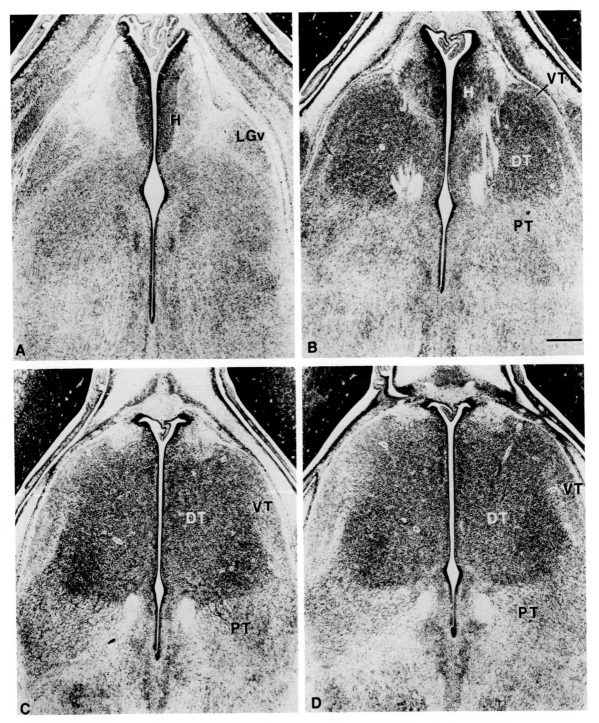

FIGURE 6.9. Horizontal sections in dorsal (A) to ventral (F) order through the diencephalon of a rat fetus at 17 days of gestation, showing continuing nuclear differentiation in epithalamus and ventral thalamus. Nuclear differentiation has not yet begun in the dorsal thalamus. Thionin stain. Bars = 500 μm.

The most comprehensive account of nuclear differentiation in the thalamus is that of Rose (1942a) in the rabbit. Using cross-sections to examine the earliest stages of thalamic development, whereas Coggeshall was later to use sagittal, Rose discerned the four regions of Herrick, each separated by a narrow acellular zone which was represented by a longitudinal groove on the ventricular wall. Rather than using Herrick's terms, however, Rose chose to name the regions according to his knowledge of the reactions of their derivatives to telencephalic ablation in the adult. Thus, the epithalamus was called "independent thalamus," the dorsal thalamus "dependent thalamus," and the ventral thalamus "connecting thalamic area." The precursor of the hypothalamus was called "hypothalamic plate," which developed further into a "dorsal hypothalamic area or subthalamus" and a ventral area giving rise to the rest of the hypothalamus and to the globus pallidus.

Each of the regions of Rose represents a localized thickening of the mantle zone occurring as the result of proliferation and migration of cells from the adjacent part of the ventricular zone (Fig. 6.9). The four regions become distinct as soon as proliferation gets well under way at the 16-mm stage of development (approximately 15 days of gestation in the rabbit). Proliferation, overall, ceases

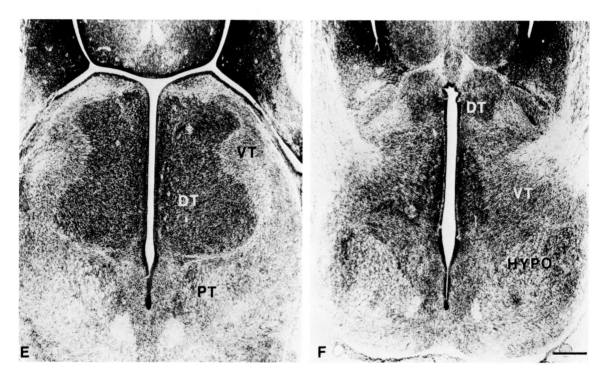

FIGURE 6.9.   (*continued*)

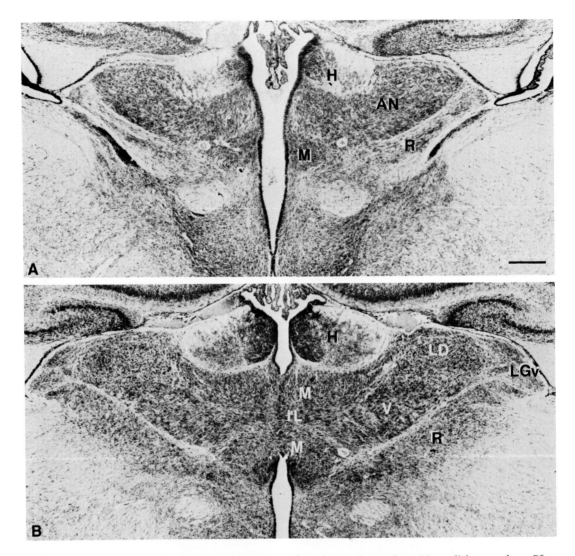

FIGURE 6.10. (A–D) Frontal sections in anterior (A) to posterior (D) order through the thalamus of a rat at 20 days of gestation. The major pronuclei and some of their nuclear derivatives can now be recognized in the dorsal thalamus. AN: anterior nuclei; H: habenular nuclei; IL: intralaminar nuclei; LD: lateral dorsal nucleus; LGd: dorsal lateral geniculate nucleus; LGv: ventral lateral geniculate nucleus; LP: lateral posterior nucleus; M: medial pronucleus; Pf: parafascicular nucleus; Po: posterior nuclei; PT: pretectum; R: reticular nucleus; V: ventral nuclei; VP: ventral posterior nucleus. (E) Frontal section through the thalamus of a rat at 6 days of age. Nuclear differentiation is now virtually complete. Thionin stain. Bars = 250 μm (A–D), 500 μm (E).

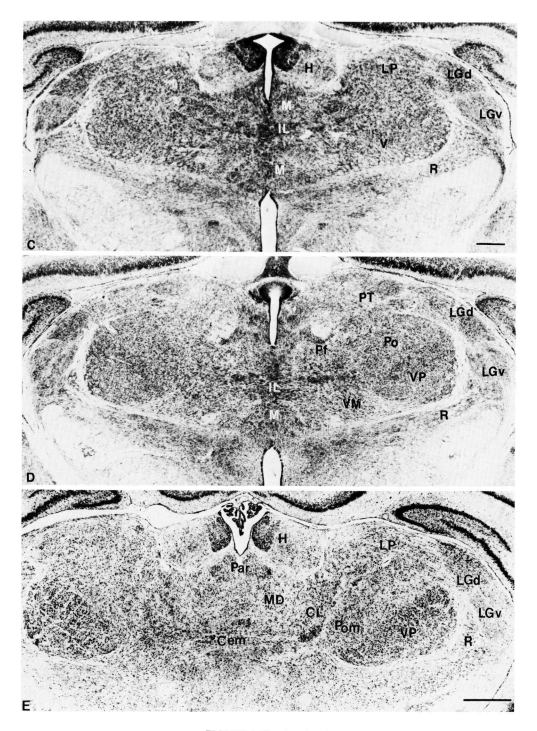

FIGURE 6.10. (*continued*)

in the diencephalic wall at about the 33-mm stage (approximately 18–19 days of gestation). Toward the end of this period, nuclear differentiation is complete in the epithalamus, has commenced in the ventral thalamus, but has not yet begun in the dorsal thalamus.

Nuclear differentiation begins in the epithalamus at the 18-mm stage (approximately 16–17 days) with a splitting off of a pretectal mass from the remainder of the epithalamus. Further differentiation commences in the pretectal mass and extends anteriorly in the epithalamus so that first the posterior paraventricular nucleus, then the habenular nuclei, and, finally, at 33 mm, the anterior paraventricular nucleus become recognizable. Later changes are changes in the relative dispositions of the nuclei only.

In the ventral thalamus, the precursors of the ventral lateral geniculate nucleus and reticular nucleus (including the zona incerta) are recognizable at 21 mm (approximately 17–18 days); they become very distinct from one another by 33 mm and reach their adult dispositions by 50 mm (approximately 20–21 days).

At 50 mm, differentiation is only just commencing in the adult thalamus with the appearance of five localized cellular aggregations or "pronuclei" (see Chapter 1). Rose distinguished medial, central, and dorsal pronuclei and pronuclei of the dorsal lateral and medial geniculate bodies. At 50 mm the latter two had already reached their definitive shape and position relative to other structures; further nuclear differentiation was occurring in the posterior part of the central pronucleus as the ventral nuclei formed; and the anterior part of the central pronucleus was starting to grow more or less horizontally through the medial pronucleus. By 65 mm (approximately 23 days), dorsal and ventral nuclei are visible in the medial geniculate body; the central pronucleus has become the three anterior nuclei, the anterior intralaminar nuclear group, and the ventral nuclear group; of the two components of the medial pronucleus, separated now by the intralaminar nuclei, the dorsal has become the parataenial, mediodorsal, and parafascicular nuclei and the ventral has become the medioventral nucleus; in the position of the dorsal pronucleus the lateral and posterior groups of nuclei are now seen. By 105 mm (approximately 27 days, 3 days before birth), all nuclei of the dorsal thalamus are clearly seen and have established their definitive dispositions in relation to the components of the epithalamus and ventral thalamus.

There have been no other descriptive studies as thorough as that of Rose on the development of the rabbit thalamus but it may be assumed that the temporal patterns of nuclear differentiation observed by Rose are typical of those of mammals generally (Fig. 6.10), always allowing, of course, for differences in the length of gestation. Rose emphasized one important principle of diencephalic development and a second is evident in his description. He noted that the ventricular zone of the neuroepithelium (called by him, following His, the *matrix layer*) was always "exhausted" before the commencement of differentiation of nuclear aggregations in the thickened intermediate ("mantle") layer. (By "exhaustion" he meant the reduction to a single layer of cells and the disappearance of mitotic figures.) That is, cellular proliferation and nuclear differentiation are events that are temporally separate. The second principle evident in Rose's work, though he did not emphasize it, is that nuclear differentiation commences posteriorly and laterally in the thalamus and proceeds anteriorly and medially. In Section 6.2, we shall see that these two principles have been reaffirmed in modern autoradiographic studies.

277

DEVELOPMENT OF
THE THALAMUS

6.1.4. The Pulvinar

In studying thalamic development in the rabbit, Rose did not have the opportunity of examining the origins of a pulvinar such as is present in primates. Early descriptions of the development of the human thalamus (Gilbert, 1934; Dekaban, 1954; Kahle, 1956) gave little reason for believing that the pulvinar was derived from anything other than the diencephalic wall which gave rise to the rest of the dorsal thalamic nuclei. In 1969, however, Rakic and Sidman discovered in human fetuses what appeared to be a massive influx of young neurons into the vicinity of the pulvinar from a proliferative region in the wall of the telencephalon (Fig. 6.11) (Rakic, 1974). They found that proliferation peaks in the diencephalic wall between 8 and 15 weeks in the human fetus and at the end of that period most thalamic nuclear groups except the pulvinar are large and separated by various fiber bundles and medullary laminae. (True cytologically distinct nuclear differentiation comes much later.) They were able to show that diencephalic proliferation had ceased after about 15 weeks by incubating diencephalic fragments from 18- to 22-week-old fetuses in [^{3}H]thymidine and noting little or no incorporation (Rakic and Sidman, 1968.)

Enlargement and differentiation of the pulvinar occur well after 15 weeks and it is thought that this is accounted for by an influx of differentiated but young (bipolar) neurons from the ganglionic eminence, a highly proliferative part of the wall of the lateral ventricle overlying the developing caudate nucleus (Fig. 6.11) and in the early stages of development closely adjacent to the thalamus. Incubation of 18- to 22-week-old fragments from the ganglionic eminence in [^{3}H]thymidine showed that 30% of the cells could be labeled after 1 hr. Moreover, beween 16 and 34 weeks streams of bipolar cells could be observed in sectioned material passing from the ganglionic eminence, under the region of the stria terminalis and into the dorsal part of the thalamus. These streams of cells were named *corpus gangliothalamicum*. It was thought that it is they which contribute to the late development of the pulvinar and it was speculated that they might also contribute cells to cortical areas with which the pulvinar nuclei are connected, for these cortical areas also seem to develop late in relation to the rest of the cortex (Rakic, 1977a).

The ganglionic eminence does not appear to contribute telencephalic cells to the pulvinar of other primates (Ogren and Rakic, 1981). All neurons of the rhesus monkey pulvinar are born from a localized region of the diencephalic ventricular and subventricular zone between E36 and E45 and reach their definitive site in about 3 days. Those of the inferior pulvinar nucleus are born first. The monkey pulvinar becomes separated from other thalamic nuclei on about E60 but nuclear differentiation in it is not complete until after E80. During this time there is considerable enlargement of the pulvinar but this is due to interstitial growth, not to the acquisition of telencephalic cells (Ogren and Rakic, 1981). There is no evidence to indicate that the thalamus of other mammals

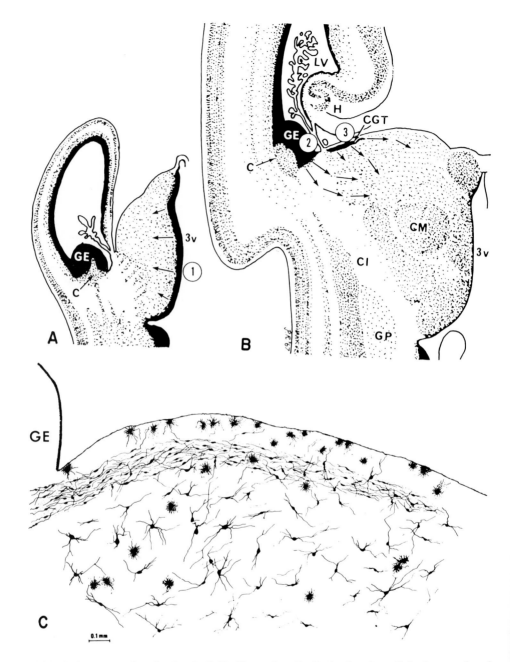

FIGURE 6.11. Drawings by Dr. P. Rakic illustrating (B, C) the invasion of the human dorsal thalamus by young neurons from a telencephalic proliferative zone, the ganglionic eminence (GE). In man these invading young neurons, seen as the corpus gangliothalamicum (CGT) in (B) and as a stream of bipolar cells in (C), will form the nuclei of the pulvinar. Other thalamic nuclei in man are formed (arrows in A) from the neuroepithelium lining the third ventricle (3v). From Sidman and Rakic (1974).

includes cells that have their origins in the telencephalon rather than in the diencephalon and no evidence that elements of midbrain origin contribute to the thalamus.

6.2. Studies of Cellular Proliferation and Migration

Some early authors had thought that localized aggregations of cells lining the diencephalic wall proliferated together and migrated into the mantle layer as a unit, differentiating into a single thalamic nucleus on arrival (e.g., Bergquist and Källén, 1954). The majority, however, had been impressed by the cessation of mitotic activity in the ventricular wall some time before the onset of nuclear differentiation in the intermediate (mantle) zone and did not commit themselves to a position on the mode of migration or subsequent differentiation of young neurons emigrating from the lining (Rose, 1942a; Dekaban, 1954; Ströer, 1956; Niimi *et al.*, 1961; Coggeshall, 1964).

The introduction of [^{3}H]thymidine autoradiography (Sauer, 1959; Sidman and Miale, 1959) led to a number of studies on the birthdates and subsequent histories of thalamic neurons. These include general studies on the mouse (Angevine, 1970), hamster (Keyser, 1972), and rat (McAllister and Das, 1977; Altman and Bayer, 1979a,b) as well as studies targeted at specific nuclei such as the anterior and medial in the rabbit (Fernández, 1969; Fernández and Bravo, 1973), and the lateral geniculate in the rat (Lund and Mustari, 1977; Brückner *et al.*, 1976) and monkey (Rakic, 1977b). Two major points emerge from these studies: all thalamic neurons are generated early and over a brief time span and there are gradients of neurogenesis operating across the thalamus in directions that ignore nuclear boundaries.

Virtually all thalamic neurons are born in mice between gestation days E10 and E16 (Angevine, 1970), and in the rat between E13 and E19 (McAllister and Das, 1977; Altman and Bayer, 1979a). Among individual nuclei, the time of origin of their cells may be quite brief; for example, most neurons of the rat dorsal lateral geniculate nucleus are born over a 2-day span (Brückner *et al.*, 1976; Lund and Mustari, 1977; McAllister and Das, 1977; Altman and Bayer, 1979a) (Fig. 6.12). However, it is more common for some neurons of most thalamic nuclei to be born over virtually the whole proliferative period. This results from the neurogenetic gradients mentioned above; in other words, young neurons migrating outwards from the ventricular zone come to rest in regions that are determined by their time of birth, not by nuclear boundaries. Three neurogenetic gradients have been described in virtually all the species studied. One operates from posterior to anterior so that the earliest born neurons accumulate progressively more anteriorly. A second operates from lateral to medial so that most neurons in the reticular nucleus, for example, are born before those of the mediodorsal nucleus (Fig. 6.13). The third gradient operates from ventral to dorsal but it is unclear whether this includes the epithalamus in which some neurons are born very early. An indication of the three gradients is given in

Fig. 6.13. Some of them may operate within individual nuclei as well as across the thalamus as a whole (Altman and Bayer, 1979b).

Angevine (1970) was impressed by the early origin of the magnocellular medial geniculate nucleus and of certain other large-celled nuclei in the mouse.

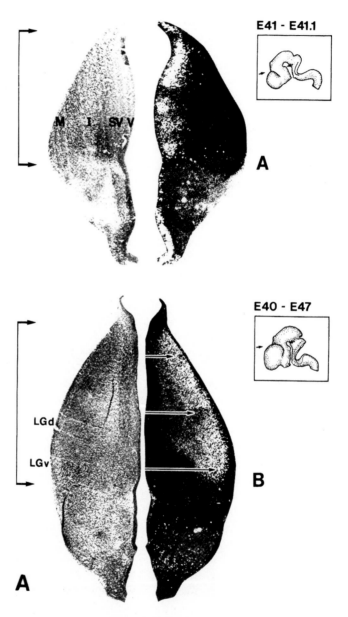

FIGURE 6.12. (A) Brightfield (left) and darkfield (right) photomicrographs of the two sides of the diencephalon in monkey fetuses at 41 (A) and 47 (B) days of gestation. Sections cut along plane indicated by arrows in insets. Animal in A received an injection of [³H]thymidine 1 hr before sacrifice and animal in B received a similar injection 1 week before sacrifice. Nuclei of young neurons that completed their last mitotic division close to time of injection are labeled in ventricular (V) and

He sought to correlate this with the earlier origins of layers V and VI in the cortex which he thought contained only projection neurons and not interneurons. He thus tentatively suggested that thalamic relay neurons might be born before interneurons. There is no evidence to support this hypothesis (McAllister and Das, 1977).

The most comprehensive description of the proliferation and migration of cells that will form a single thalamic nucleus is that of Rakic (1977b) on the dorsal lateral geniculate nucleus of the rhesus monkey (Fig. 6.14). By injecting [³H]thymidine into pregnant females at different times in the first quarter of the 165-day gestation period and killing the fetuses at different time intervals thereafter, he was able to provide a detailed developmental history of the dorsal lateral geniculate nucleus. Rakic found that virtually all geniculate neurons are born between E36 and E43 from a particular portion of the ventricular wall.

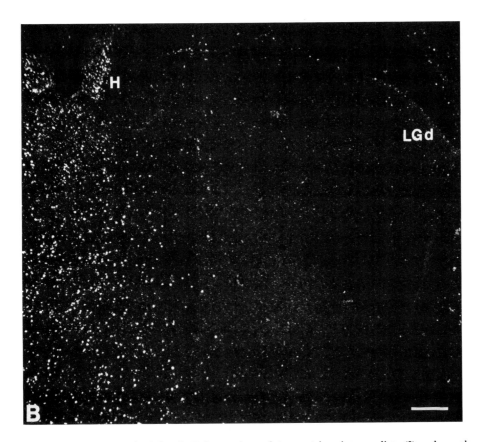

subventricular (SV) zones in A but in B have migrated (arrows) into intermediate (I) and mantle zones to form anlagen of dorsal (LGd) and ventral (LGv) lateral geniculate nuclei. From Ogren and Rakic (1981). (B) Autoradiograph showing [³H]thymidine-labeled nuclei in medial parts of thalamus in a 6-week-old rat that was administered the material at 16 days of gestation (cf. Fig. 6.13). H: habenula. Bar = 250 μm.

From his figures, this portion also contributes cells to nuclei adjacent to the lateral geniculate. Most proliferation occurs in the ventricular zone proper but there appears to be some mitotic activity in an underlying subventricular zone as well. At E35 Rakic estimates that the proliferative region contains 22,000 precursor cells that increase 10-fold by E41 and are then exhausted by E44. In the interim, they have produced the approximately 1.7 million nerve cells of the adult dorsal lateral geniculate nucleus. Possibly, the total produced is much greater, as he did not take any morphogenetic cell death into account.

Labeled young neurons migrate out from the proliferative region into radial fascicles. It is not yet known whether they are following radial glial fibers as has been postulated for cells migrating into the cerebral cortex (Rakic, 1972) but this seems possible because there are radially arranged glial processes in the diencephalon early in development (Fig. 6.15). On reaching the lateral surface of the diencephalon (a distance of 500–700 μm), the cells accumulate under the marginal zone from laterally, inwards. Fibers of the optic tract will arrive later in the overlying marginal zone. Successive generations of geniculate cells appear to migrate along the same fascicles and, thus, cells across the thickness of the

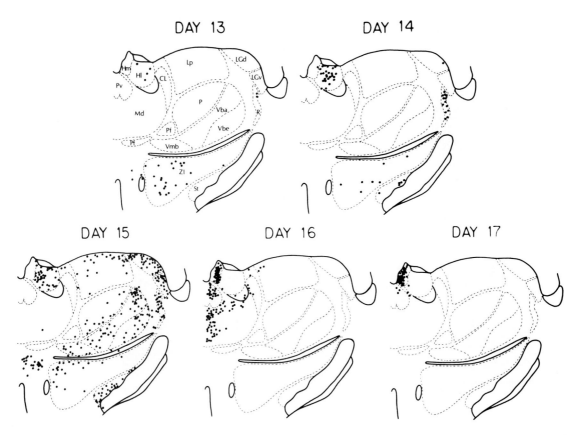

FIGURE 6.13. Sections through the thalami of postnatal rats that had been administered [³H]thymidine at different gestation ages (days 13–17), showing the lateral-to-medial gradient of neurogenesis in the dorsal thalamus. Dots indicate labeled cell nuclei. From McAllister and Das (1977).

dorsal lateral geniculate nucleus tend to line up in rows which Rakic feels may be the precursors of projection columns (see Chapter 3). In other words, the retinal coordinate system projected onto the dorsal lateral geniculate nucleus may have an analogous set of coordinates already present in the proliferating neuroepithelium.

The lateral-to-medial axis of cellular origins and arrival in the dorsal lateral geniculate nucleus becomes a ventral-to-dorsal one as the nucleus rotates ventrally during later development (Fig. 6.14), a feature originally emphasized by Le Gros Clark (1932a). This means that neurons of the magnocellular layers of the geniculate nucleus are born and reach the lateral aspect of the diencephalon before those of the parvocellular layers.

Although proliferation ceases by E44 and most neurons have reached the dorsal lateral geniculate nucleus soon after, cytological differentiation com-

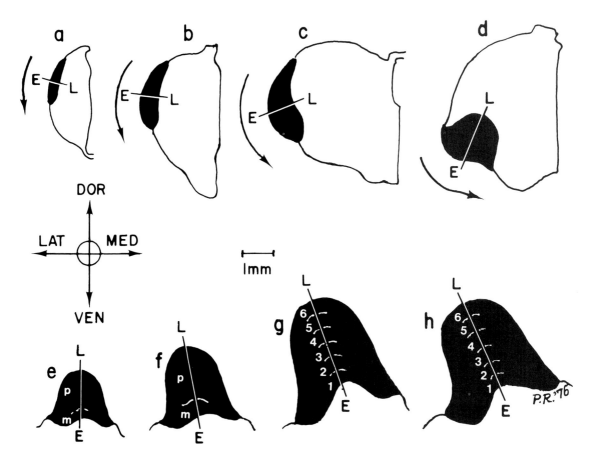

FIGURE 6.14.   The gradient of neurogenesis and the morphological rotation of the dorsal lateral geniculate nucleus of the rhesus monkey. a–h indicate the position of the nucleus at different fetal ages (a: E48; b: E58; c: E77; d: E84; e: E91; f: E97; g: E112) and postnatally (h). Neurons are generated and reach the nucleus along the gradient E to L, so that the cells of the magnocellular layers (1 and 2) are born before those of the parvocellular layers (3–6). The nucleus (black) rotates from a lateral to a ventral position (arrows) during development, and the magnocellular layers (m) become cytoarchitectonically distinct (e) before the parvocellular layers (p). From Rakic (1977b).

mences much later. The magnocellular layers separate off at about E90 and the parvocellular layers become distinct throughout the nucleus at E125–130, some 35–40 days prior to birth. Therefore, as pointed out earlier, there is a clear dissociation between proliferation, migration, and differentiation but a temporal correlation remains in the sense that the earliest born and earliest arriving cells differentiate first. Whether a similar correlation exists for other thalamic nuclei remains to be determined but the differentiation of ventral thalamic nuclei before dorsal and of posteriorly placed dorsal thalamic nuclei before anterior ones (Rose, 1942b) correlates well with the lateral-to-medial and posterior-to-anterior gradients of neurogenesis.

6.3. Neuronal Morphogenesis

There is no comprehensive account of neuronal morphogenesis in the mammalian thalamus. Rakic (1977b) has implied that young neurons migrating to the anlage of the monkey lateral geniculate nucleus are bipolar in form and comparable to migrating young neurons in other parts of the central nervous system. Rakic and Sidman (1969) have also published drawings of bipolar neurons migrating into the developing human pulvinar from the ganglionic eminence (Fig. 6.11; Section 6.2), and scattered throughout the literature there are other casual mentions of bipolar cells in the vicinity of various thalamic nuclei of infant animals (e.g., Garey and Saini, 1981). It is probable, therefore, that the initial phase of neuronal maturation in the thalamus involves a change from the bipolar to a multipolar form.

Very few Golgi studies have been carried out on neurons in the prenatal or very immature thalamus, so the exact details of early morphogenesis are not known. Morest (1969) first described in general terms the development of in-

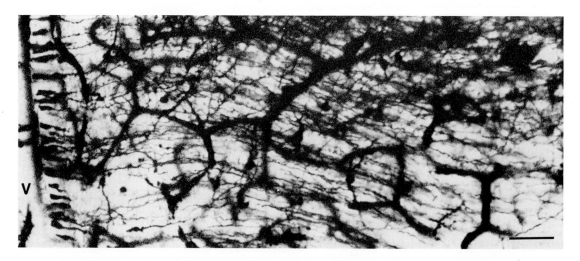

FIGURE 6.15. Golgi-impregnated radial glial cells with their cell bodies in the ventricular (V) lining, sending processes across the diencephalon in a rat fetus at 13 days of gestation. Large black profiles are blood vessels. Bar = 100 μm.

terneurons and relay neurons in the dorsal lateral geniculate nucleus and in the dorsal and ventral medial geniculate nuclei of postnatal cats, rabbits, and opossums. His results have generally been confirmed in the dorsal lateral geniculate nucleus of the rat and monkey by Parnavelas *et al.* (1977a,b) and by Garey and Saini (1981). Morest found that the enlargement of the cell body generally precedes the definitive expansion of the dendritic tree and that axons grow out and differentiate earlier than dendrites, as in many other central nervous regions. The dendritic maturation of relay neurons precedes that of interneurons by 1–2 weeks but in both cases is characterized by the appearance of terminal growth cones on the tips of the earliest formed dendrites, with other growth buds and filopodia forming in the vicinity of incipient branches. By the time of birth, or shortly after, most thalamic neurons have acquired dendritic morphologies and dendritic fields comparable to those of the adult in everything except size. Within the first few postnatal weeks there is a rapid increase in somal size and elongation of dendrites. In the early phases, many dendritic growth cones and filopodia are present but these later disappear. It seems likely that the disproportionate number of spines or other appendages shown on Golgi-impregnated neurons by Ramón y Cajal (1911) stem from his consistent use of neonatal animals in which these have not yet disappeared. As would be expected (Section 6.2) the neurons of some nuclei mature more rapidly than others: Morest finds that neurons in the ventral medial geniculate nucleus mature in advance of those in the dorsal medial geniculate nucleus and Parnavelas *et al.* (1977a,b) suggest that relay neurons mature earlier than interneurons.

A particularly significant study from the point of view of the differential maturation of individual classes of thalamic relay neurons is that of Friedlander (1982). In the dorsal lateral geniculate nuclei of kittens at 3–4 weeks of age, he was able to identify physiologically a small sample of X, Y, and W cells, to inject them intracellularly with horseradish peroxidase, and then to recover them morphologically. His results show that at this age W cells of the C laminae resemble W cells of the adult cat in terms of their physiological responses and morphology. Soma size, dendritic field size, and dendritic morphology are the same as in the adult. Y cells can also be identified physiologically in the A laminae and also resemble adult Y cells morphologically except that their somata are smaller and their dendritic fields are less extensive. X cells, on the other hand, are immature both physiologically and morphologically with very small somata and incomplete dendritic fields consisting of short dendrites often with filopodia and growth cones. At this stage all the principal cell types had observable axons, some with collaterals already in the perigeniculate nucleus. No interneurons were reported.

Although there have been few studies of the development of dendritic field organization in the thalamus, many have been devoted to the growth of cell somata, as seen in Nissl preparations. These studies have the advantage that the results are more readily quantifiable. Early neurons in a thalamic nucleus are small and consist of little more than a nucleus with minimal cytoplasm. Over the ensuing days, the cells grow in size, more and more cytoplasm appears, and nucleoli become more prominent than at early stages. In most mammals a considerable amount of cell growth continues postnatally. I can find no reports of quantitative studies of cell growth in the thalamus in the prenatal period so I have made a preliminary survey of cell sizes in several thalamic nuclei of fetal, early postnatal, and adult rats and cats in my collection (Table 6.1). These preliminary results indicate that there is on average a 50 to 100% increase in cell size from the time thalamic nuclei are becoming recognizable up to birth, and a further 100 to 200% increase between birth and maturity.

The postnatal maturation of thalamic cell size has been particularly well studied in the laminar dorsal lateral geniculate nucleus because it occurs more or less contemporaneously with the "critical period" when the neurons are unusually susceptible to visual deprivation or deafferentation (see Section 6.6). R. Kalil (1978) and Hickey (1980) in the common domestic cat and Robertson *et al.* (1980) in Siamese cats have demonstrated a period of extremely rapid postnatal growth in which mean cross-sectional area of the cells at birth reaches 95% of the mean adult area by 56 days postnatal. This represents an increase in size of some 300% (Fig. 6.16). Growth seems to be linear with the 150% increase occurring around 28 days postnatal. The period between 28 and 56 days is that during which geniculate neurons are most sensitive to the effects of visual deprivation or deafferentation (Section 6.6). R. Kalil (1978) considers that there is no difference in the rate of growth of cells in the binocular and monocular segments of lamina A. This contests earlier work of Garey *et al.* (1973) who felt that they could detect differential growth rates in the two segments. R. Kalil also reports that the largest cells of the laminar dorsal lateral geniculate nucleus continue to grow slowly beyond 56 days postnatal and well into adulthood. This has an interesting parallel with Hickey's (1977b) findings of differential rates of growth of cells in the parvocellular and magnocellular layers during the postnatal development of the human lateral geniculate nucleus. At birth, cells in the human nucleus are approximately 60% of their adult size. Then those in the parvocel-

TABLE 6.1. Mean Areas (μm^2) of Cells in Selected Thalamic Nuclei of Rats and Cats at Different Ages

		Nucleus				
	Age	AD	MD	VP	LGd	MGv
Rats	E18	78.5	50.3	28.3	78.5	50.3
	E19	78.5	50.3	38.5	95.0	76.5
	E20	88.6	63.6	58.7	103.9	86.5
	E21	95.0	76.5	95.0	113.1	95.0
	Birth	132.7	95.0	113.0	132.7	113.0
	P21	254.5	153.9	132.7	153.9	201.1
	Adult	263.5	168.7	153.9	168.7	254.5
Cats	E45	50.3	55.4	55.3	51.2	38.4
	E50	56.7	57.1	56.8	53.1	50.3
	E55	95.0	63.6	92.1	78.5	74.3
	E58	97.0	64.2	93.8	82.3	87.1
	Birth	188.7	76.5	122.7	103.1	95.0
	P21	258.6	104.2	132.7	114.2	101.2
	Adult	314.2	254.5	293.7	346.3	226.9

lular layers develop faster and achieve 95% of their adult size between birth and 6 months, with full adult size being reached by the end of the first year. Cells in the magnocellular layers develop slower and achieve 95% of the adult size only by the end of 12 months, with full adult size not being reached until the end of the second year.

The coincidence of the phases of most rapid geniculate cell growth with the critical period for deprivation effects on the cortex of both cats (Dews and Wiesel, 1970) and humans (see Hickey, 1977b) implies that there is a correlation between the two. Moreover, the apparent restriction of cells with Y-type responses (see Chapters 4 and 9) to the magnocellular layers of the primate dorsal lateral geniculate nucleus, the larger size of Y-type cells in the nucleus of the cat (Chapter 4), and the unusual susceptibility of Y-type cells in the cat and tree shrew to the effects of visual deprivation during the critical period (Section 6.6) imply that this susceptibility is in some way correlated with their protracted growth. This is still a little difficult to understand, however, in relation to Friedlander's (1982) report (see above), suggesting that Y-type cells, from the point of view of dendritic field architecture and physiological responses, are quite mature at the commencement of the critical period in kittens.

The only data on other nuclei, comparable to those on the growth of cells in the dorsal lateral geniculate nucleus, seem to be those which I have presented in Table 6.1. From these there appears to be a comparable rapid rate of cell growth in the early postnatal period. It is unclear, however, whether this is

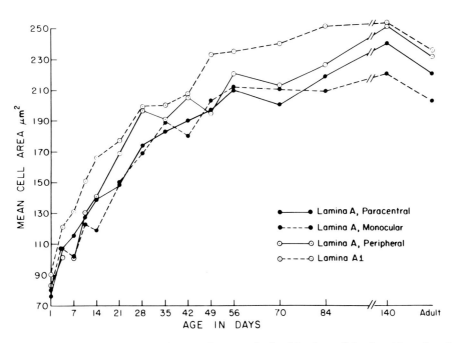

FIGURE 6.16. Mean cross-sectional areas of neurons in the A laminae of the dorsal lateral geniculate nucleus of cats at different postnatal ages. "Paracentral," "monocular," and "peripheral" indicate different parts of visual field representation in lamina A. Cells grow at same rate in each part of the laminae and are virtually fully grown at 56 days. From R. Kalil (1978).

associated with a similar critical period for the maturation of physiological properties. If the follicles of facial whiskers are removed or the trigeminal nerve cut in rodents within the first few days of life, cytoarchitectonic units fail to develop fully in the ventral posterior medial nucleus (Woolsey *et al.*, 1979; Killackey and Shinder, 1981; Section 6.5) and corresponding anatomical and physiological changes can be detected in the somatic sensory cortex (Van der Loos and Woolsey, 1973; Woolsey and Wann, 1976; Jeanmonod *et al.*, 1977; Killackey *et al.*, 1978; Killackey and Belford, 1979); the effect cannot be obtained after the first 5 days or so of life. This critical period in the rodent is also, from Table 6.1, a phase of rapid growth of cells in the ventral posterior nucleus, so here, too, rapid growth and sensitivity to peripheral manipulations may be correlated.

6.4. Formation of Connections

6.4.1. Afferent Innervation

The innervation of the thalamus is a relatively early event in development, and it is likely that some afferent fibers reach a thalamic nucleus even before that nucleus receives its full complement of cells from the ventricular zone. In the developing dorsal lateral geniculate nucleus of the cat, for example, optic fibers reach it first at E32 whereas neurons are still being added to the nucleus up to about E47 (Hickey and Cox, 1979; Shatz, 1981). Innervation by corticothalamic fibers, however, seems to occur after cell proliferation and migration are complete or at least toward the end of that period. Corticothalamic axons do not enter the rat thalamus before E19* (Schreyer and Jones, 1982; Jones *et al.*, 1982a) after the peak of neurogenesis has ended (McAllister and Das, 1977; Section 6.2) (Fig. 6.17). Though optic tract axons cross the rat lateral geniculate nucleus during the proliferative period at E16, they seem destined for the superior colliculus (Lund and Bunt, 1976) and in hamsters, although the axons arrive on the nucleus at postnatal day 0, they do not enter it in significant numbers for 3 more days (Frost *et al.*, 1979). Optic tract axons first enter the cat dorsal lateral geniculate nucleus (Shatz, 1983), 3 days after the end of the peak of neurogenesis for that structure (Hickey and Cox, 1979), though a few cells are still being born right up to the time of arrival of optic axons. In monkeys, optic tract fibers appear in the dorsal lateral geniculate nucleus at E54, 11 days after the end of neurogenesis (Rakic, 1977b), and corticothalamic axon terminations appear after E70 (Hendrickson and Rakic, 1977; Shatz and Rakic, 1981). It would seem, therefore, that afferent innervation follows on the heels of the end of proliferation and migration but this is long before the onset of cytological differentiation (Sections 4.3, 6.2, and 6.3).

---

* E18 in our publication but E19 in terms of McAllister and Das (1977) whose figures are quoted in Section 6.2.

It is not certain whether subcortical or corticothalamic afferents reach the thalamus first. In the case of the dorsal lateral geniculate nucleus of the monkey quoted above, optic tract axons commence innervating the nucleus many days before corticogeniculate synapses can be detected. In rats, optic tract fibers at least arrive in the vicinity of the geniculate 3–4 days before corticothalamic fibers enter other parts of the thalamus. Because the development of the occipital regions of the cortex lags behind that of other regions even in the rat, it is likely that corticothalamic fibers to the lateral geniculate nuclei reach them after E19 (see above). In cats, where optic tract axons innervate the lateral geniculate nucleus by E32 (Shatz, 1983) but in which corticothalamic axons cannot be traced to the nucleus before E48 (Anker, 1977), it is probable that subcortical innervation also precedes corticothalamic innervation. We do not know for sure, of course, that axons demonstrable in a nucleus light microscopically are necessarily forming synapses. Hendrickson and Rakic (1977) indicate that optic tract axons form synapses in the monkey lateral geniculate nucleus soon after arrival. On the other hand, in the lateral geniculate nucleus of the cat (Cragg, 1975) and rat (Karlsson, 1967), synaptogenesis is said to commence postnatally, a considerable time after the arrival of optic tract axons—at least 30 days after in the cat. There is, however, some reason to doubt this (Section 6.5).

Innervation of the thalamus by afferent fibers occurs long before innervation of the cortex by thalamic fibers. In rats with a gestation period of 22 days corticothalamic connections can be traced to the ventral thalamic nuclei by E19 (Schreyer and Jones, 1982; Jones *et al.*, 1982a) but thalamocortical axons from these nuclei do not invade the sensory motor regions until birth (Figs. 6.18 and

FIGURE 6.17. Anterogradely labeled corticofugal axons entering the thalamus (TH) from the internal capsule (IC) in an 18-day-old rat fetus that received an injection of horseradish peroxidase in the cortical plate 12 hr earlier. Bar = 100 μm. From Jones *et al.* (1982a).

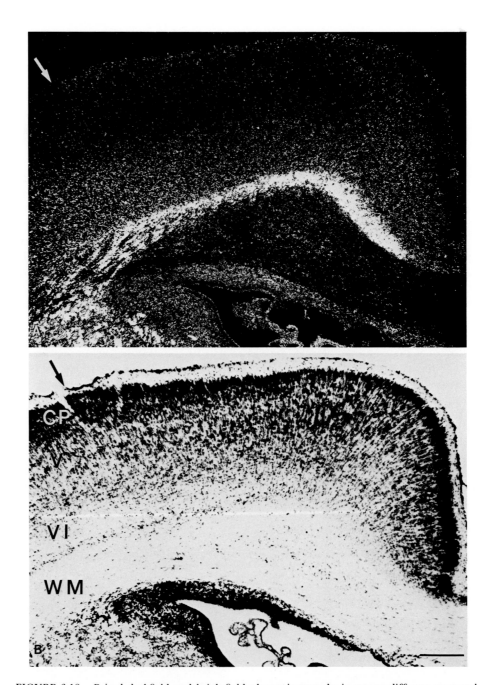

FIGURE 6.18. Paired darkfield and brightfield photomicrographs in rats at different postnatal ages, showing autoradiographically labeled thalamocortical axons entering the cerebral cortex. At 1 day of age (A, B), the axons are only just entering layer VI which, along with layer V, has become differentiated from the cortical plate (CP). On the third postnatal day (C, D), the fibers have penetrated the thickness of the cortex but do not ramify to any extent in the thin remaining cortical plate. On the fifth postnatal day (E, F), laminar concentrations of fiber ramifications are appearing in layers IV and VI. By the seventh postnatal day (G, H), columnar and laminar patterns of termination have become established in the somatic sensory cortex. All animals received injections of tritiated amino acids in the thalamus 18–24 hr before sacrifice. Bars = 200 μm (A, B), 100 μm (C–H). From Wise and Jones (1978).

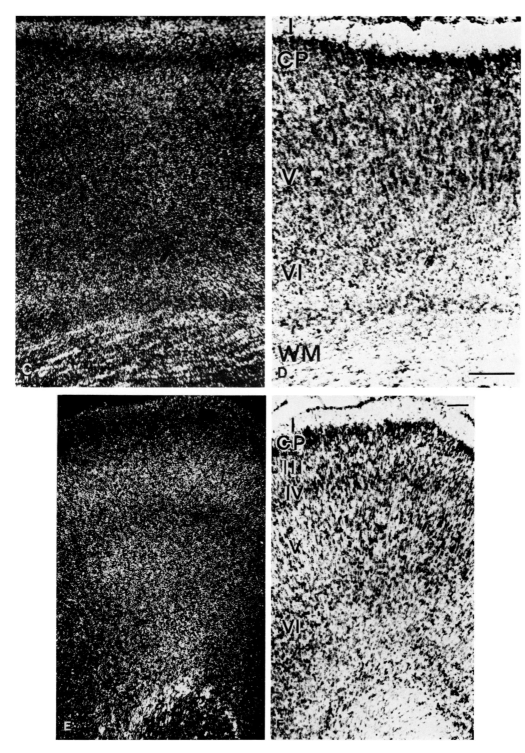

FIGURE 6.18. *(continued)*

6.19; Wise and Jones, 1978). In fetal cats with a gestation period of 63–65 days, optic tract axons commence innervating the dorsal lateral geniculate nucleus on E32 (Shatz, 1983) and corticothalamic fibers have entered the ventral nuclei by at least E50 (Wise *et al.*, 1977) (Fig. 6.20), yet innervation of the visual cortex by geniculate fibers commences only at about the time of birth (Anker and Cragg, 1974) and of the sensory motor regions by ventral nuclei fibers only at E57 (Wise *et al.*, 1977) (Figs. 6.20 and 6.21). In fetal monkeys, optic tract axons are in the dorsal lateral geniculate nucleus by E54 (Rakic, 1976, 1977b,c) and corticogeniculate axons apparently by E105 (Hendrickson and Rakic, 1977), but geniculocortical fibers do not innervate the striate cortex until about E124 (Rakic, 1976, 1977a–c). It is, therefore, not possible to accept Coggeshall's (1964; Section 6.1.2) suggestion that afferent fibers to the thalamus plug into already established circuits, particularly if this statement implies thalamocortical circuits. There is also reason to believe that corticothalamic connections may be formed after subcortical afferent connections. Nevertheless, the latter might be involved in some way in regulating the formation of connections of subcortical afferents with thalamic cells and/or of thalamic axons with the cortex.

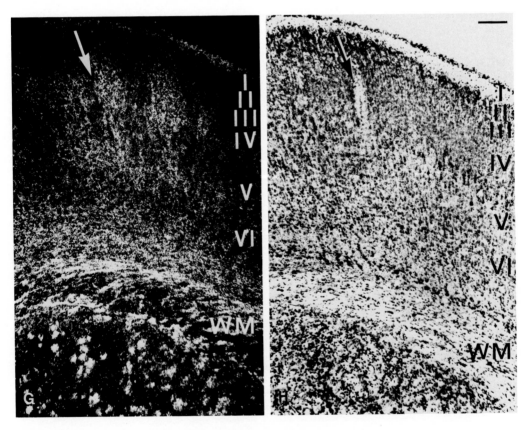

FIGURE 6.18. (*continued*)

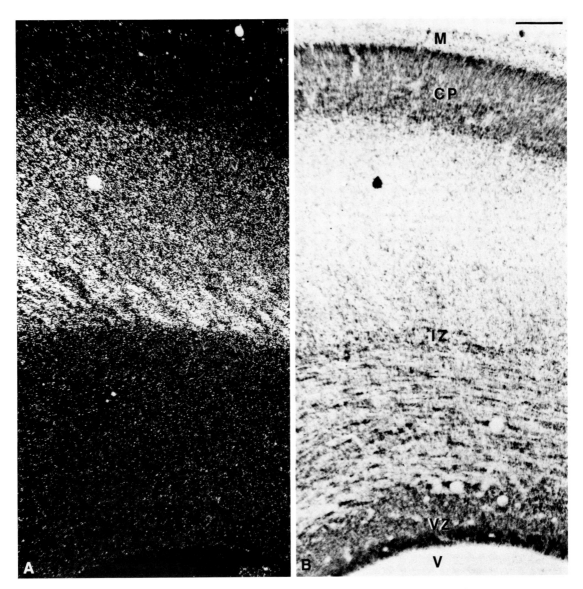

FIGURE 6.19. Darkfield and brightfield photomicrographs from the region of the developing sensory motor cortex in a cat fetus at 43 days of gestation. The thalamus was injected with tritiated amino acids 24 hr previously. Labeled thalamocortical fibers have accumulated in the outer part of the intermediate zone beneath the cortical plate (CP) but do not enter the cortex for approximately another 20 days. Many cells destined for the cerebral cortex are migrating through the waiting axons from the intermediate (IZ) and ventricular (VZ) zones over this period. Thionin stain. Bar = 100 μm. From Wise *et al.* (1977).

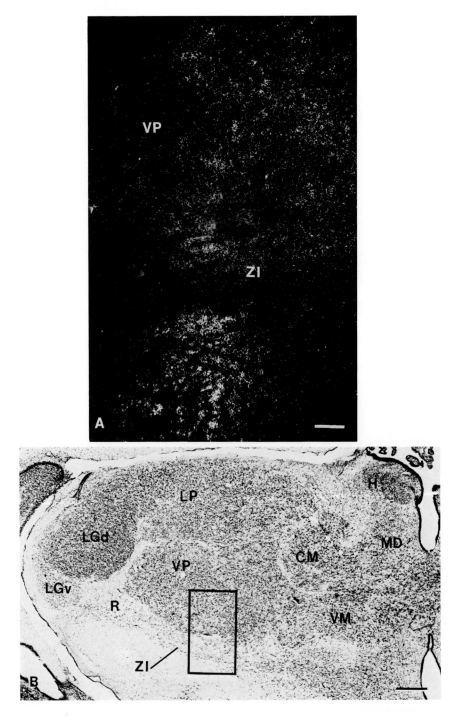

FIGURE 6.20. Autoradiographically labeled (A) cortical efferent axons descending in the cerebral peduncle (lower left) and distributing fibers to the zona incerta (ZI) and ventral posterior nucleus (VP) of the thalamus in a cat fetus at 55 days of gestation. The animal received an injection of tritiated amino acids in the presumptive sensory motor cortex 24 hr earlier. (A) is from box drawn in (B). Thionin stain. Bars = 100 μm (A), 500 μm (B). From Wise *et al.* (1977).

Although the innervation of the cortex is a late event in brain development, the growth of thalamic axons toward their cortical target occurs quite early and probably before the establishment of subcortical afferent connections in the thalamus. Thalamocortical fibers can be traced to the intermediate zone beneath the cortical plate of the cat sensory motor region at E40, some 25 days before birth (Wise *et al.*, 1977) (Fig. 6.19). Similarly, geniculocortical fibers in the monkey visual system reach the vicinity of the calcarine sulcus by E78, 87 days before birth (Rakic, 1976, 1977a–c) (Fig. 6.21). Commensurately early growth appears to occur in other species as well (Fig. 6.18). Yet the thalamic axons accumulate

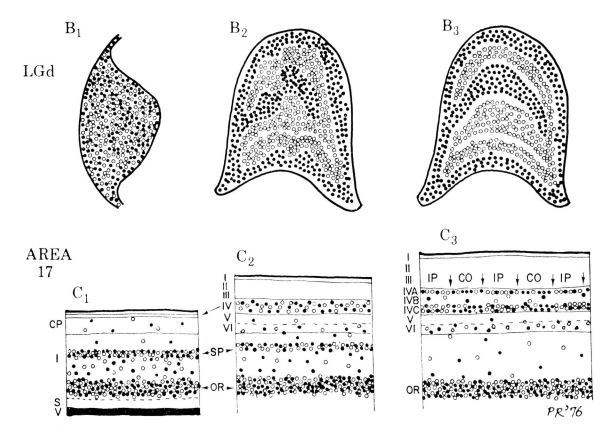

FIGURE 6.21.   (B1–3) Diagrams illustrating the progressive segregation of terminations of fibers projecting to the dorsal lateral geniculate nucleus from the ipsilateral and contralateral eyes, as demonstrated by autoradiography after injections of tritiated amino acids in the eyes of monkey fetuses at different ages. (B1: 78 days; B2: 124 days; B3: 144 days.) (C1–3) Diagrams showing progressive segregation of thalamocortical fiber ramifications in the striate cortex, as demonstrated by transneuronal autoradiography in the same brains. From Rakic (1977a).

for a protracted period in the intermediate zone beneath the cortical plate before making a second growth spurt that carries them into the cortex. Fibers enter the sensory motor areas of the cat at E55, the visual cortex of the monkey by E124, and the rat sensory motor and visual areas at birth (Wise *et al.*, 1977; Rakic, 1976, 1977c; Lund and Mustari, 1977; Wise and Jones, 1978). Over approximately the same waiting period, commissural axons arriving through the corpus callosum are also accumulating in the intermediate zone, yet they must receive a different signal to grow again, for they make their growth spurt into the cortex several days after the thalamocortical fibers—10 days in the case of the cat sensory motor regions and 3 days in the case of the same regions in the rat (Wise *et al.*, 1977; Wise and Jones, 1976, 1978). Destruction of callosal or thalamic fibers at birth in the rat is without effect on the subsequent developmental fate of the other fiber system (Wise and Jones, 1978).

The nature of the signal that causes accumulating thalamocortical fibers to make their second growth spurt into the cortex is unknown. Is it intrinsic to the axon? Is it a circulating messenger? Is it a local trophic factor released by the cortex at a particular point in time? Is it conditioned by the establishment of afferent synapses in the thalamus? There are no answers to these questions at the moment. In relation to the last, however, it may be noted that in congenitally anophthalmic mice a normal geniculocortical projection has been reported (Cullen *et al.* 1976; Godement *et al.*, 1979; Kaiserman-Abramoff *et al.*, 1980).

While the thalamocortical axons are accumulating in the intermediate zone beneath the cortical plate they are being traversed by young neurons migrating from the underlying ventricular zone (Fig. 6.19). These young neurons will form the granular and supragranular layers of the cortex in the case of the cat and rat. It is possible that these migrating cells could influence the subsequent fate of the thalamocortical axons as many of them will presumably be the target cells of the axons. Transient, synapselike structures have been reported in the intermediate zone which might represent a communication between axon and eventual target cell (Molliver *et al.*, 1973). But such contacts must be broken because cell migration to the cortex is completed by the time the first thalamic axons enter the cortex. Moreover, destruction of the precursor cells of the granular and supragranular layers by the administration of cytotoxic drugs at critically timed stages of development does not prevent thalamic axons from making their second growth spurt into the cortex and forming connections in an atypical site (Fig. 6.22; Jones *et al.*, 1982b; Yurkewicz *et al.*, 1984).

In normal animals the initial invasion of the cortex by thalamic afferents is somewhat diffuse (Figs. 6.18, 6.19, and 6.21). All layers of the cortex with the exception of the cortical plate proper appear colonized at first and the typical bi- or trilaminar pattern of specific connections only becomes resolved out of this in the ensuing few days. In rats, for example, thalamocortical fibers enter the sensory motor areas at birth but the two laminae of termination in layers III–IV and VI only become evident by day 6 postnatal (Wise and Jones, 1978). During this period, columnar zones of termination also become evident (Figs. 6.18 and 6.21). In the case of the monkey striate cortex, there is little evidence of segregation of ocular dominance columns at the outset but this begins to develop 3 weeks before birth (Rakic, 1976, 1977c). It is probable that synapses are formed as soon as thalamic fibers invade the cortex (Lund and Mustari,

1977; Jones, 1981b) (Fig. 6.23) but synaptogenesis appears to continue for many days and possibly weeks after that (Cragg, 1975; Kristt, 1978; Wise *et al.*, 1979; Juraska and Fifková, 1979). During this period, in the visual cortex the distribution of thalamic afferents can be modified by monocular deprivation (Hubel and Wiesel, 1977; Hubel *et al.*, 1977; LeVay *et al.*, 1978, 1980).

6.4.3. Relationships between Afferent Innervation and Cytoarchitectonic Differentiation

As pointed out in Section 6.4.1, innervation of the thalamus by afferent fibers does not appear to be the signal for nuclear differentiation to commence. The differentiation of a normal-appearing dorsal lateral geniculate nucleus in a strain of congenitally anophthalamic mice (Kaiserman-Abramoff *et al.*, 1980), also speaks against a dependence on innervation for early development. On the other hand, there is clearly some kind of relationship between the maturation of innervation patterns and cytoarchitectonic differentiation within a nucleus. Most data relating to this point are derived from studies of the development of the dorsal lateral geniculate nucleus, and the development of patterns of cy-

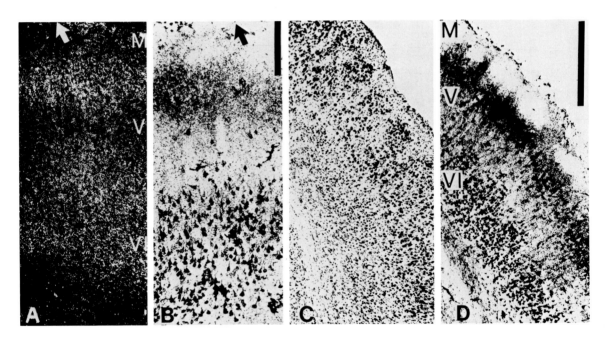

FIGURE 6.22. Distribution of thalamocortical fibers labeled autoradiographically (A) and by anterograde transport of horseradish peroxidase (B) in the sensory motor cortex of a rat in which the precursor cells of layers II–IV (C) were destroyed by injection of a cytotoxic drug during fetal life. Cells with corticothalamic projections are retrogradely labeled in layer VI in (B) and in (D) both corticothalamic cells (VI) and cells with axons in the pyramidal tract (V) are labeled following injections of horseradish peroxidase in both thalamus and cerebral peduncle. Thalamocortical axons which had not entered the cortex at the time of the cytotoxic perturbation have entered and are ramifying abnormally about layer V cells in (A), (B), and (D). Bars = 100 μm (A, B), 500 μm (C, D). From Jones *et al.* (1982b).

toarchitectonic lamination in this nucleus is associated with the progressive segregation of inputs from the two eyes.

6.4.3.1. Dorsal Lateral Geniculate Nucleus

In monkeys, retinal axons can be demonstrated by axonal transport methods in the lateral geniculate nucleus by E65 and fibers from the two eyes appear to overlap throughout the full extent of the anlage of the nucleus (Rakic, 1979). During the ensuing 80 days the left and right eye fibers gradually segregate out into their adult pattern of laminar distribution (see Chapter 9). At first the dorsal lateral geniculate nucleus is a homogeneous mass of cells (Fig. 6.21). Lamination commences on about E90 with the appearance of the magnocellular layers (Section 6.2) and continues in the parvocellular layers over a time frame that appears to be more or less contemporaneous with the segregation of left and right eye afferents and with the development of topographic order in corticogeniculate fiber terminations (Rakic, 1977b; Shatz and Rakic, 1981). It is probable that a similar course of development occurs in man in which the cellular lamination also occurs prenatally (Gilbert, 1934; Cooper, 1945; Moskowitz and Noback, 1962) and where the appearance of interlaminar spaces precedes that of cytological distinctiveness of the laminae (Hitchcock and Hickey, 1980).

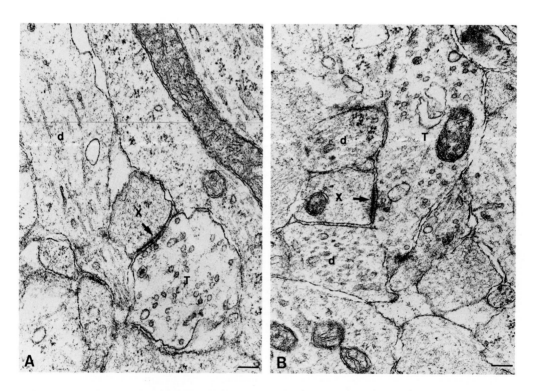

FIGURE 6.23. Axon terminals (T) thought to be from thalamocortical axons in the somatic sensory cortex of a 6-day-old rat ending on organelle-free processes (x) that are thought to be dendritic filopodia or early dendritic spines. Bars = 1 μm.

A very similar process occurs during the development of lamination of the dorsal lateral geniculate nucleus in carnivores. In ferrets (Linden *et al.*, 1981), which have a gestation period of 42 days, no laminae are visible in the nucleus at birth and fibers from the ipsi- and contralateral eyes are distributed to all parts of the nucleus, though the input from the contralateral eye is already denser. Over the next 2 days, ipsilateral fibers become more restricted in their distributions and contralateral fibers retreat from these areas, particularly from that which is destined to be the monocular segment. This early mapping out of monocular and binocular territories has been called "segmentation" by Linden *et al.* In the ensuing 2 weeks the overlap of ipsi- and contralateral fibers is progressively reduced elsewhere as the fibers retreat into the appropriate laminae. This is called "lamination" by Linden *et al.* The retinal axons commence segregating before cell layers become apparent but cellular lamination is complete a day or so before the complete segregation of left and right eye axons. The characteristic adult pattern of cell and fiber lamination is complete about 10 days before eye opening.

In cats (Shatz, 1983) there is a similar developmental sequence with the onset of fiber segregation preceding that of cellular lamination but it occurs largely prenatally and the two inputs may never have been totally overlapped. Ipsi- and contralateral fibers commence withdrawing from one another's territory by about E47 and their segregation is virtually complete by birth (E65). Cellular lamination, however, is only just commencing at birth and is not complete for another 10 days or so (R. Kalil, 1978; Guillery, 1979).

The early overlap and subsequent segregation ("segmentation") of ipsi- and contralateral retinal fibers is found in other species, such as rats (Bunt *et al.*, 1981) and hamsters (Frost *et al.*, 1979; So *et al.*, 1978) in which no cytological lamination develops. In these species, however, ipsilateral fibers never extend throughout the full volume of the nucleus. The earlier onset of fiber lamination in comparison with cellular lamination is also demonstrable in tree shrews (Brunso-Bechtold and Casagrande, 1981) in which ipsi- and contralateral fibers are more or less separated at birth but cellular lamination begins on day 5 postnatal. It is as well to remark that during the process of fiber and cellular lamination, axons innervating the ipsi-and conralateral eye laminae of the dorsal lateral geniculate nucleus are not only segregating themselves according to laminae but also according to projection lines. That is, fibers arriving from homonymous parts of the two retinae, probably at different times (see below), must arrange themselves so that their terminations are in precise register across all laminae. From the appearances of arriving fibers in the adult cat (Eysel and Wolfhard, 1983), it would appear that this fine sorting out occurs in the geniculate itself rather than in the optic tract.

6.4.3.2. Competition in the Dorsal Lateral Geniculate Nucleus

The nature of the influences that govern first segmentation and then lamination in the dorsal lateral geniculate nucleus is but little known. The idea of a direct competition for synaptic space seems reasonable to account for the segmentation process when it is considered that in at least opossums, rodents, and cats, contralateral fibers arrive first (Lund and Bunt, 1976; Cavalcante and

Rocha-Miranda, 1978; So *et al.*, 1978; Frost *et al.*, 1979; Bunt *et al.*, 1981; Shatz, 1983) and therefore may have a competitive advantage over ipsilateral fibers in the monocular segment. Some workers feel that the idea of a competition is confirmed by the apparent expansion of the ipsilateral field of termination in rodents in which the contralateral eye was enucleated at birth or congenitally absent (Lund *et al.*, 1973; Land *et al.*, 1976; Godement *et al.*, 1980). A more equal competition may normally be set up between ipsi- and contralateral fibers in the binocular segment of the nucleus and could account for the development of a more or less even pattern of lamination. The demonstration by Rakic (1981) that early (E63) prenatal removal of an eye from rhesus monkey fetuses not only leads to a failure of retraction of the remaining eye's fibers from inappropriate parts of the nucleus but also to a failure of cellular laminae to form, argues eloquently in favor of competition (see Section 4.3),* though death of a substantial number of cells that failed to be innervated by the removed eye could also account for some of the effect. On the other hand, Shatz (1983) points out that in the developing cat, ipsilateral retinal fibers never invade a significant part of the anlage of lamina A. This portion, as a consequence, represents a protected target for contralateral fibers and contralateral fibers innervating it may never be subjected to interlaminar competition.

In the tree shrew at birth, lamination of the dorsal lateral geniculate nucleus is well under way and retinogeniculate terminations from each eye have begun to segregate (Brunso-Bechtold and Casagrande, 1981, 1982; Brunso-Bechtold *et al.*, 1983). Corticogeniculate axons have not yet entered the nucleus, however, and the interlaminar zones in which they densely ramify in the adult have not yet formed. If the two eyes are removed at birth, the laminae of the nucleus fail to segregate completely because the interlaminar zones do not fully overlap (Brunso-Bechtold *et al.*, 1983). This may be due to the failure of corticogeniculate axons, in the absence of competition from retinogeniculate terminals, to concentrate their terminals at laminar borders. Organization of corticogeniculate terminations in columns corresponding to projection columns is, however, preserved.

6.4.3.3. Other Thalamic Nuclei

The association between afferent innervation and cytoarchitectonic delineation has not been worked out as thoroughly for other thalamic nuclei. The only potential correlations that can be drawn come from studies of the cytoarchitectonic development of the ventral posterior medial (VPM) nucleus of rats and mice. Here, the aggregations of cells that each represent a sinus hair on the face crystallize out of a more homogeneous mass of cells during the first 5 days of life but fail to delineate clearly if the trigeminal nerve is cut or whiskers removed at the periphery in the first 3 days (Chapter 5) (Woolsey *et al.*, 1979; Killackey and Shinder, 1981). Unfortunately, it is not known at what stage second-order trigeminal or corticothalamic afferents first enter the thalamus so the relationship between innervation and cytoarchitecture cannot be stated. Both sets of afferents have segregated out into terminal clusters associated with the

---

* When an enucleation is done at E91, some lamination is visible at 3 months postnatal but it is only comparable to the level of lamination attained at the time of enucleation.

cytoarchitectonic aggregates of the rat VPM by 4 days of age (Akers and Killackey, 1979). It would be interesting to determine whether there is an early overlap of inputs from the different sinus hairs followed by the restriction of each to an individual zone which later becomes a cytoarchitectonic entity. This would make the development of the VPM, in a sense, comparable to that of the dorsal lateral geniculate nucleus.

The cytoarchitecture of the rodent VPM can only be altered by vibrissal removal or nerve section up to day 3 postnatal whereas the somatic sensory cortex can still be affected up to day 5 postnatal (Woolsey *et al.*, 1979; Killackey and Shinder, 1981). This follows the ascending sequence of maturational changes in the rodent trigeminal system demonstrated by Belford and Killackey (1979). It also suggests that cytoarchitecture is influenced at certain periods by afferent connections, as in the dorsal lateral geniculate nucleus (Section 6.4.3.2). No evidence, such as changes in the sizes of cells within cellular aggregates neighboring those affected by the deafferentation, which might indicate competitive interactions similar to those in the geniculate, was reported.

6.4.3.4. Other Developmental Factors

The development of innervation and the correlative cytoarchitectonic delineation of the dorsal lateral geniculate nucleus cannot be regarded as a simple static matching of cells with inputs. The evidence for competitive interactions speaks against that. But other factors may also be involved in making the process an even more dynamic one. Young neurons are arriving at the dorsal lateral geniculate nucleus throughout the period of early innervation (Brückner *et al.*, 1976; Rakic, 1977b; Shatz, 1981; Hickey and Cox, 1979) and presumably must be sought out by arriving fibers. Then, too, retinal ganglion cells are being born throughout most of the early phase of innervation and so their axons (even those from the same eye) presumably arrive at different times and must be accommodated. In fetal cats (Walsh *et al.*, 1983) autoradiography after intra-amniotic injections [^{3}H]thymidine reveals that retinal ganglion cells are generated between E21 and E36 (or later). Small ganglion cells are produced throughout the period, medium-sized ganglion cells from E21 to E31, and large-sized ganglion cells from E26 to E30. But in all cases central retinal cells are produced first and peripheral last. Therefore, retinal ganglion cell development is characterized by three overlapping waves of neurogenesis spreading out from the center, each wave probably representing the generation of a particular functional class of cell (the W, X, and Y cells; Chapters 3, 4, and 9).

If we assume that the ganglion cells grow their axons toward the dorsal lateral geniculate nucleus soon after their birth, then from the above it follows that neither axons from the same part of the retina, nor those of a single functional class are arriving at the nucleus simultaneously. Hence, the nucleus and the arriving axons have to engage in some kind of shuffle that ensures that all fibers from the same part of the retina will meet up with one another and with their counterparts from the opposite retina in the same projection column. Moreover, within that projection column, axons of the several functional ganglion cell classes must be linked up with appropriate (and largely different) geniculate cells (Chapters 4 and 9). The axons must also cope with the fact that at least 50% of their number may die at the time innervation of the geniculate is at its peak, as the result of cell death in the retina (Ng and Stone, 1982; Stone *et al.*, 1982; Rakic and Riley, 1983). By analogy with other central nuclei (e.g., Hamburger, 1975; Cowan, 1973), there is probably a comparable degree of cell death in the geniculate itself which also must be accommodated to.

One can only assume that all of those factors that guide axons to their targets, that enable them to recognize appropriate target cells, that enable them to compete successfully for synaptic space with other axons, and that lead to stabilization of synapses once made, are brought into play in the setting up of this precise geometric and functional map. The nature of these factors is, of course, one of the biggest challenges to developmental neurobiology.

6.5. Maturation of Physiological Responsiveness

As we have seen in Section 6.4, the sensory thalamic nuclei in most mammals are innervated at birth and have established or are starting to establish connections with the cerebral cortex. Functional maturation would be expected to soon follow.

6.5.1. Dorsal Lateral Geniculate Nucleus

Work done on the maturation of single-unit responses in the dorsal lateral geniculate nucleus of kittens indicates that some kind of physiological response to optic nerve stimulation can be detected virtually as soon as the nucleus is innervated by optic nerve fibers. Kirkwood and Shatz (1982) observed postsynaptic responses to electrical stimulation of the optic nerves or tracts in *in vitro* preparations of the dorsal lateral geniculate nucleus as early as day 39 of gestation. This is only 7 days after the invasion of the nucleus by the first retinal fibers (Shatz, 1983). It suggests, contrary to the earlier electron microscopic study of Cragg (1975), that some kind of synapses are being formed soon after the arrival of afferent axons.

The earliest neuronal responses in the dorsal lateral geniculate nucleus are decidedly not facsimiles of those in the adult and further development of response patterns and of receptive field structure continue into the first 5 weeks of postnatal life at least. In the *in vitro* fetal preparation, Kirkwood and Shatz (1982) found an unusual number of geniculate cells receiving excitatory inputs from the two eyes—an exceptionally rare observation in the adult nucleus, except close to laminar borders (Chapter 9). Crossed inhibition only started appearing in late fetal and early postnatal preparations. These results indicate not only the later development of intrageniculate inhibition but also (Section 6.6) the making and breaking of synapses from an inappropriate eye, presumably as contralateral and ipsilateral axon terminations withdraw into their appropriate parts of the nucleus (Section 6.4).

The further maturation of physiological properties of geniculate neurons had been studied in postnatal kittens by Daniels *et al.* (1978). Their results (Table

6.2) also imply a considerable amount of synaptic remodeling as well as differential rates of maturation for different classes of relay neurons. Daniels *et al.* (1978) recorded extracellularly from single cells in the dorsal lateral geniculate nucleus of kittens between 6 and 40 days of age.* At the earliest time, retinotopy was already established in the nucleus. Moreover, as judged physiologically, X-type cell responses are even identifiable by 14 days of age (Norman *et al.*, 1977). In the first 2–3 weeks of postnatal life, however, most geniculate cells are relatively unresponsive to visual stimuli, hard to classify as X or Y type, have large receptive fields, usually without the classic center–surround antagonism (Chapters 4 and 9), without inhibitory zones, and they fatigue easily. A number even display "on–off"-center receptive fields, i.e., they respond to both switching on and switching off a visual stimulus applied to the receptive field center. Because a sample of optic tract axons recorded in the same study possessed only on- or off-center receptive fields, the on–off-center receptive fields were thought to indicate convergence of on-center and off-center optic tract axons on the same geniculate cell, something unusual in the adult nucleus. In the fourth week, after optic tract myelination is complete (Moore *et al.*, 1976), latencies of response to natural or electrical stimuli increase, receptive field size declines to approximately adult size, center–surround antagonism and inhibition appear, and the cells become exclusively on-center or off-center. A summary of all the changes in physiological responses is contained in Table 6.2.

These results suggest that there is still a considerable degree of reorganization occurring in the establishment of appropriate connections between retinal ganglion cells and geniculate neurons (see also Section 6.4) and that this is occurring around the time that visual experience is commencing. It would also appear that some X-type cells (and also a small, early maturing population of movement-selective or W-type cells in the C laminae) are able to acquire appropriate afferent connections quite early, whereas most X-type cells require some time and Y-type cells an even more protracted period. This protracted maturation of Y-type cells may be related to the greater sensitivity to early postnatal visual deprivation that they display in comparison with X-type cells (Section 6.6).

The physiological maturation of most retinal ganglion cells does not seem to precede that of dorsal lateral geniculate cells (Rusoff and Dubin, 1977), even though this might have been expected from the recordings of optic tract fibers by Daniels *et al.* (1978). Rusoff and Dubin (1977) found that the receptive fields of retinal ganglion cells situated outside the area centralis were very large up until the fourth or fifth week of life, only acquiring the small, adult-sized receptive field centers after the fourth week of postnatal life. Surround inhibition appeared even later. Despite this immaturity, however, adultlike X and Y characteristics could be identified as early as the fourth week of life. These findings suggest that retinal ganglion cell maturation also occurs in stages. From the preceding paragraph, it seems evident that this staged physiological development may be passed on to the developing dorsal lateral geniculate neurons.

In the medial interlaminar nucleus of kittens, Y-type cell properties also develop over the first few weeks of life (Wilson *et al.*, 1982). Latencies to stimulation of the optic chiasm are longer than in the adult, the cells have little

---

* Eye opening in kittens occurs after the end of the first week of life.

TABLE 6.2. Summary of Events in the Maturation of Physiological Responsiveness in Cats at Successive Postnatal Ages[a]

6–13 days	Basic visual topography is present
	Few cells per track; nonvisual cells in LGN; silent areas; few ipsilaterally driven cells
	Cells have zero to very low maintained rates
	Cells respond to full-field flash, but with long latencies (150–1000 msec)
	Cells fatigue with visual stimulation rates faster than 0.05 Hz
	Cells require high-frequency OX stimulation to respond
	Cells have large receptive fields; area summation sizes are often larger than mapped sizes
	No antagonistic surround responses; no surround inhibition
	Low-amplitude responses to flashing spots
	Large percentage of on–off cells
	Cells hard to classify as X or Y
	Sustained cells do not have onset transients
	Absence of inhibitory events in responses to flashing spots
	Low response rates to slowly moving stimuli; no response to rapid movement
	No response to hand-held or low-contrast targets
14–20 days	Increase in cells driven by the ipsilateral eye
	Decrease in nonvisual cells in LGN
	Sustained cells develop onset transients
	Some X cells have surround responses and surround inhibition
	Increase in peak response rates to moving targets
21–27 days	Increased maintained rates
	Less fatigue; more cells per track; few silent areas
	More variability in evoked and maintained rates
	Most cells can be classified as X or Y
	Decrease in receptive field sizes for X cells
	Most cells have surround responses
	X cells have surround inhibition
	Increase in amplitude of evoked response
	Cells show responses to hand-held and low-contrast targets
28–34 days	Decreased latencies to full-field flash (most are under 100 msec now)
	Decrease in receptive field sizes for Y cells
	Receptive field size now relates to field position
	Sustained/transient and X/Y proportions are adultlike
	On, off, on–off cell proportions are adultlike
35 days–adult	Evoked rates increase
	Peak firing rates to high-velocity stimuli increase
	Surround inhibition becomes stronger
	Inhibitory events in responses to flashing spots develop
	Periphery effects develop for Y cells

[a] Modified from Daniels *et al.* (1978).

spontaneous activity, and receptive fields are large with no opponent surrounds. Up until at least 8 weeks of age, the cells even display linear responses to a counterphased modulating grating and, thus, resemble X-type cells of the adult laminar dorsal lateral geniculate nucleus (Chapter 9). The capacity of the cells for spatial and temporal resolution is still developing at 16 weeks of age. In an earlier study, Kratz *et al.* (1978b) had shown that Y cells of the medial interlaminar nucleus in cats with one eye closed since birth, displayed similar response properties to those demonstrable in the neonatal kitten. Hence, it would appear that monocular deprivation of the medial interlaminar nucleus from birth freezes the Y cells therein at an early stage of development. In the visual cortex, many aspects of cellular organization such as receptive field size and ocular dominance are present within a week of birth (Hubel and Wiesel, 1970). On the other hand, certain properties such as orientation selectivity and binocular disparity only appear in visual cortex neurons between 3 and 5 weeks of age* (Pettigrew, 1972; Blakemore and van Sluyters, 1975; Buisseret and Imbert, 1976), correlating quite well with the acquisition of adult properties in the dorsal lateral geniculate nucleus.

Finally, the late appearance of inhibition in the dorsal lateral geniculate nucleus is supported, at least in monkeys, by the late dendritic maturation of interneurons (Garey and Saini, 1981), the appearance of presynaptic dendritic synapses well after other synapses (Hendrickson and Rakic, 1977), and the appearance of an adult ratio of symmetric to asymmetric synapses only at 60–70 days postnatal (Winfield *et al.,* 1976).

6.5.2.   Other Thalamic Nuclei

It is extremely disappointing to reflect that no other thalamic nucleus has been investigated for the development of cellular response properties with the same degree of finesse as the dorsal lateral geniculate nucleus. There would seem to be no inherent technical difficulty in undertaking correlative studies on the ventral posterior or medial geniculate nuclei. Moreover, adequate physiological parameters such as receptive field size, modality specificity, transient or sustained discharge properties, frequency tuning, and so on are established characteristics of the adult nuclei that could easily be explored in the developing nuclei.

A somatotopic organization can be detected by evoked potential or single-unit recording in the SI cortex of cats at birth (Rubel, 1971) and in rabbits and rats within 1–3 days of birth (Verley and Gaillard, 1978; Verley, 1977; Verley and Axelrad, 1975; Axelrad *et al.,* 1977). Columnar cortical properties in rats are present before 6 days of age (Armstrong-James, 1975). In kittens, apart from a longer latency of input, the SI cortex is remarkably adultlike (Rubel, 1971). These observations suggest that the ventral posterior nuclei are capable of relaying appropriate sensory information to the cortex at an early stage and it would be interesting to determine if the thalamic relay cells change with in-

* Some properties such as ocular dominance and direction selectivity are present even at the time of eye opening (Wiesel and Hubel, 1963a).

creasing age and experience along similar lines to those in the lateral geniculate nucleus (Section 6.5.1).

The medial geniculate nuclei could be a fruitful field of study on account of the changes that occur in the peripheral auditory apparatus during the first weeks of life. Electrical stimulation of the auditory nerve elicits evoked potentials in the auditory cortex of the cat at birth (Pujol and Marty, 1968), indicating that transmission occurs through the medial geniculate nucleus and that fibers have reached the cortex. However, the cortex appears very immature. Pujol and Marty (1968) could only evoke responses to tonal stimuli in the auditory cortex of 2- to 3-day-old kittens by using very high intensities and low-frequency tones (500–2500 Hz). At this age they could detect no tonotopicity in the auditory cortex. This immaturity of the thalamocortical pojection is no doubt related to the fact that the middle ear cavity only becomes excavated and filled with air toward the end of the third postnatal week. Prior to that, it is filled with mes-enchyme (Pujol and Hilding, 1973). By the end of the third week, auditory thresholds fall to adult levels and even in the first week there are significant changes in the responsiveness and frequency tuning of neurons in the cochlear nuclei and inferior colliculus (Pujol, 1972; Romand and Marty, 1975; Aitkin and Moore, 1975; Brugge *et al.*, 1978). The changes in the medial geniculate nucleus and auditory cortex that presumably accompany this maturation do not, how-ever, appear to have been charted.

Somatotopic organization of the facial whiskers is demonstrable by single-unit recording in the VPM nucleus of mice within 5 days of birth (Verley and Pidoux, 1981; Verley and Onnen, 1981). Projections from receptors related to the common fur of the face are not detectable until the second week of age. Though not described in any detail, receptive fields of cells related to the sinus hairs on the face initially appear larger than in the adult and the cells are less responsive. Reduction of receptive fields to a single vibrissa, the firing of bursts of discharges in response to a stimulus, and a reduction in latency of onset of the discharges occur toward the end of the second week of life. There is, thus, some similarity in the types of changes that occur in the VPM and the dorsal lateral geniculate nucleus as development progresses. More details, however, would be highly desirable.

6.6. Plasticity of Connections

It was pointed out in Section 6.5 that thalamic relay neurons, subsequent to their early innervation, undergo a relatively slow maturation process during which their adult physiological properties become fixed. This maturation occurs over approximately the same period during which the cells are growing rapidly (Section 6.4). As also pointed out in Section 6.5, the process of physiological maturation seems to involve the rearrangement or elimination of some early formed synapses and the late appearance of inhibitory synapses. The idea of competition between growing axons for synaptic space, with the stabilization of successful and the elimination of unsuccessful synaptic contacts, is currently very much in vogue (Changeux and Danchin, 1976; Purves and Lichtman, 1980).

Competitiveness of this kind is potentially one of the most important forces in the shaping of patterns of neural connectivity during development. If, as seems likely, it is a force that is at work in all parts of the nervous system, one might anticipate seeing evidence of it in the developing thalamus. There is a certain amount of evidence to indicate that competition between growing axons does affect the development of connectivity in the thalamus and, moreover, that this leads to a certain degree of plasticity that is subject to environmental influences.

6.6.1. Effects of Eye Removal and Visual Deprivation on the Developing Dorsal Lateral Geniculate Nucleus of the Cat

6.6.1.1. Geniculate Cell Growth

In the dorsal lateral geniculate nucleus of neonatal kittens, axons innervating one layer are kept out of inappropriate layers. The emphasis on an active process in this sentence is intentional for there are a large number of experiments that suggest an active process of competitive rejection is indeed at work. When a kitten is born, the segregation of ipsi- and contralateral retinal terminations is largely complete (R. Kalil, 1978; Shatz, 1983). However, if an eye is removed within the first 3 weeks of life, ganglion cell axons emanating from the remaining eye and terminating in the binocular segment (Chapter 9) of the nucleus will form sprouts that cross from their lamina of termination (A or A1) into adjacent parts of the denervated A1 or A lamina (Guillery, 1972; R. Kalil, 1972; Hickey, 1975, 1980; Robson *et al.*, 1978; Robson, 1981). There, the cells that manage to acquire this abnormal innervation are protected from the transneuronal atrophy that affects the other cells in the denervated lamina (Guillery, 1971c, 1972, 1973). This is clear evidence not only that innervation is essential for the survival and growth of cells in the binocular segment of the lateral geniculate nucleus but also that ipsilateral ganglion cell axons are capable of innervating contralateral cells and vice versa. Possibly the failure of translaminar sprouts to save all cells in a denervated lamina is related to the rate over which sprouts can grow during a critical time window during which denervated cells are retrievable.

If, instead of removing an eye, one eye is deprived of patterned visual input during the early postnatal period by suturing the eyelids closed or by installing a translucent contact lens, cells in certain parts of the visually deprived laminae of the dorsal lateral geniculate nucleus also fail to grow (Cook *et al.*, 1951; Wiesel and Hubel, 1963a; Kupfer and Palmer, 1964; Ganz *et al.*, 1968; Guillery and Stelzner, 1970; Guillery, 1972; Hickey *et al.*, 1977; Casagrande *et al.*, 1978; R. Kalil, 1980). The effect of monocular deprivation is rather less severe on the geniculate neurons than that of eye enucleation. Monocular eyelid suture at birth leads after 3 months to cells in the deprived A, A1, and magnocellular C laminae that are 40% smaller than those of a normal cat at the same age (Wiesel and Hubel, 1963b; Hickey, 1980; Leventhal and Hirsch, 1983a). This seems to represent a failure to grow rather than shrinkage, though some shrinkage of adult-sized cells occurs with eyelid suture at 3 months of age (Wiesel and Hubel, 1965a). Cells in the parvocellular C laminae (laminae C1–3) are less severely affected and cells in the undeprived A or A1 lamina may actually increase in size (Sherman and Wilson, 1975; Hickey *et al.*, 1977). It is a remarkable fact that cells in the part of lamina A that forms the monocular segment* of the dorsal lateral geniculate nucleus of the cat or dog are far less affected by this kind of visual deprivation than those in the binocular segment (Guillery and Stelzner, 1970; Guillery, 1972; Sherman and Wilson, 1975). This observation has profound implications for theories regarding the control of geniculate cell growth.

The differential effect of monocular deprivation on the binocular and monocular segments of the dorsal lateral geniculate nucleus has suggested that cell growth in the opposing laminae of the binocular segment is regulated by competitive interactions between the axons of these cells in the visual cortex. A further experiment (Fig. 6.24) speaks eloquently in favor of such a suggestion. If a small patch of retina is destroyed in the undeprived eye at the same time as the lid suture, the cells in the part of the deprived geniculate lamina lying opposite the consequent focus of severe transneuronal degeneration in the undeprived lamina are preserved from shrinkage, just like cells in the monocular segment (Guillery, 1972a,b).

These findings made by Guillery and his associates strongly reinforced an idea originally put forward on the basis of the effects of monocular deprivation on the developing visual cortex by Wiesel and Hubel (1965a). That is, during development of the binocular segment of the dorsal lateral geniculate nucleus there is normally intense competition between left eye and right eye driven cells for synaptic space in the cerebral cortex and that monocular deprivation throws the affected cells sufficiently off balance that their axons can no longer successfully compete for synaptic space in the cortex. They, thus, undergo what is, in effect, a "retrograde" atrophy. Cells in the monocular segment, not having to compete for synaptic space in parts of the cortex normally innervated by them alone, would be preserved. Cells acquiring innervation from translaminar sprouts after enucleation would compete successfully and those opposite a zone of severe transneuronal degeneration, though deprived, would compete more effectively with those in undegenerated parts of the other lamina. A deafferented monocular segment may be further protected by virtue of its invasion by axon sprouts from intact axons of the ipsilateral eye (Hickey, 1975; Polley and Guillery, 1980; Robson, 1981). It is doubtful, however, that significant sprouting occurs into a deprived monocular segment, for receptive fields and behavioral responses of cats to objects in the appropriate monocular part of the visual field are normal (Sherman *et al.*, 1972, 1974).

Sherman *et al.* (1974) sought single-cell responses in the visual cortex of cats subjected to the same experimental paradigm used by Guillery (1972), i.e., the creation of an artificial monocular segment within the monocularly deprived binocular segment by lesioning part of the retina in the undeprived eye. They found that the patch of cortex innervated by the part of the deprived lamina underlying the transneuronally degenerated part of the "undeprived" lamina

---

* Where the contralaterally innervated A lamina extends for some distance beyond the lateral edge of the ipsilaterally innervated A1 lamina (Chapter 9).

behaved very similar to that innervated by the deprived monocular segment (see also Wilson and Sherman, 1977). Cells responded to stimulation of the deprived eye and the cats oriented appropriately to stimuli presented to the deprived eye. Neither occurs in the case of deprivation without a contemporaneous lesion of the nondeprived retina (Sherman *et al.*, 1972; Wiesel and Hubel, 1963b, 1965b; Blakemore and van Sluyters, 1974; Kratz *et al.*, 1976; Shatz and Stryker, 1978).

These physiological and behavioral results, therefore, support the idea that lack of binocular competition within the cortex can render even deprived cells capable of effective synapse formation and/or maintenance and the establishment of behaviorally relevant innervation.

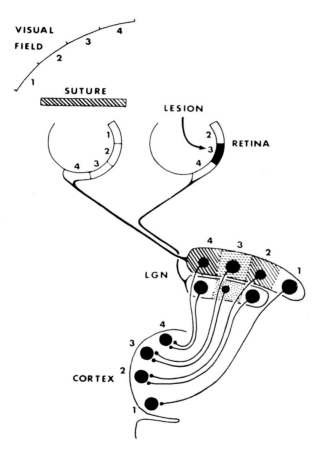

FIGURE 6.24. Guillery's experiment. A lesion destroys a portion of the retina in an otherwise normal eye of a cat. This causes a patch (3) of transneuronal atrophy in lamina A1 of the ipsilateral dorsal lateral geniculate nucleus. Atrophy of cells in this part of A1 preserves cells in part of lamina A opposite it from the shrinkage caused by deprivation of pattern vision due to suturing the lids of the other eye closed. The cells of the monocular segment of lamina A (1) are similarly preserved from the effects of visual deprivation. Protection from deprivation effects is thought to be due to lack of competition for synaptic space in the visual cortex. From Guillery (1972b).

Nature has created an experimental situation analogous to that devised by Guillery (1972b) and by Sherman *et al.* (1972). In the dorsal lateral geniculate nucleus of Siamese cats (Chapter 9), a portion of lamina A1 is innervated abnormally by fibers arising in the contralateral temporal retina. This is usually flanked by portions of lamina A1 that are innervated normally by the ipsilateral retina. When an infant Siamese cat is subjected to monocular visual deprivation (Guillery *et al.*, 1974; Guillery and Casagrande, 1977), it is only cells in these flanking regions, lying opposite a contralaterally innervated lamina A, that are severely affected (either by shrinkage if ipsilateral to the deprived eye or by hypertrophy if contralateral). The abnormal portion of lamina A1, lying opposite a lamina A which is, like itself, contralaterally innervated, is largely unaffected by ipsilateral deprivation. This effect would be expected in view of the theory that competition between left and right eye fibers in the cerebral cortex controls geniculate cell growth. In this case parts of laminae A and A1 are innervated by the same eye and neither gains a competitive advantage from monocular deprivation. As Guillery and Casagrande (1977) point out, however, the other parts of the Siamese cat's dorsal geniculate nucleus do not respond to monocular deprivation in a manner that could be predicted from the results of deprivation in normally pigmented cats. Lamina A is never affected by monocular deprivation as severely as the "normal" ipsilaterally innervated parts of lamina A1 and the part of lamina A in the monocular segment is not protected from the cell shrinkage that does occur.

The idea of a competition for synaptic space between left and right eye afferents to the visual cortex was first put forward by Wiesel and Hubel (1963b, 1965b) to account for their findings that most cells in the visual cortex receive binocular inputs but that these convert to monocular from the undeprived eye when the lids of the other eye are sutured closed before eye opening (Wiesel and Hubel, 1963b, 1965a,b; Ganz *et al.*, 1968). Their theory was based originally more on the idea of competition for synaptic space on single cortical cells than a competition for independent ocular dominance columns which had not then been demonstrated anatomically. It was in the same light that Guillery interpreted the results of his experiments. Recent work indicates that this viewpoint is not inappropriate because individual ganglion cell classes innervating the geniculate, or geniculate cell classes innervating the cortex may well compete with the one another for synaptic space on single cells, as well as for space in a lamina or in an ocular dominance column.

6.6.1.2. Competition at the Single-Cell Level

The first suggestion of a competition among cell types came in studies showing that some physiological classes of geniculate neurons could be more severely affected by neonatal visual deprivation than others. Sherman *et al.* (1972, 1975; Lin and Sherman, 1978; Friedlander *et al.*, 1982) discovered that the physiological class of cells termed Y-type cells (Chapters 4 and 9) could only rarely be identified in recordings made from a deprived A, A1, or C lamina in the binocular segment of cats subjected to unilateral suture of the eyelids in early life. Y cells, normally forming about 50–60% of a sample, were reduced to 10%. X-type cells were recorded in normal numbers (see also Hoffmann and Cynader, 1977; Hoffmann and Holländer, 1978; Eysel *et al.*, 1979; Sherman and Wilson, 1981; Geisert *et al.*, 1982). The selective loss of Y-type cells probably helps explain the impaired vision suffered by the cat when using the deprived eye to examine visual space represented by the binocular segment (Sherman, 1973, 1979; Dews and Wiesel, 1970; Chow and Stewart, 1972). By contrast, X-type cells were recorded in normal numbers in the deprived monocular segment, and the cats responded normally to visual stimuli in the part of visual space represented by it (Sherman *et al.*, 1972). The fact that large geniculate relay cells projecting to area 18 grow to only 50–60% of their normal size in monocularly deprived animals, whereas the smaller geniculate cells projecting to area 17 grow to 80% of their normal size (Garey and Blakemore, 1977; LeVay and Ferster, 1977), also fits with a more severe effect on Y cells.

A similar loss of physiologically definable Y-type cells has been reported in association with cell shrinkage of appropriate laminae in the lateral geniculate nucleus of the monocularly deprived tree shrew (Norton *et al.*, 1977).

Because in the cat the vast majority of geniculate neurons are innervated by X or Y retinal ganglion cells but rarely by both, these findings suggest that the axons of Y cells from a deprived retina are at a competitive disadvantage when establishing synaptic contacts in the binocular segment. This hypothesis is supported by the recent demonstration (Sur *et al.*, 1982) that Y axons injected with horseradish peroxidase have much reduced or even absent terminal fields in the dorsal lateral geniculate nucleus of monocularly deprived kittens. By contrast, the terminal fields of X axons are greatly expanded (Fig. 6.25). Coupled with this, the density of retinogeniculate synapses in the binocular segment seems to stay appoximately normal (Winfield and Powell, 1980; Winfield *et al.*, 1980).

These observations would suggest either that deprived Y-cell axons are unable to make or maintain synapses on their appropriate target cells in the face of competition from undeprived X-cell axons, or that they are unable to displace X-axon terminals from synapses that the latter might have already made. It has not yet been determined whether in the course of normal development X retinal axons only contact X-type geniculate cells and Y axons only Y-type cells or whether both types initially contact all cells. Therefore, both mechanisms of the Y-cell dropout are potentially possible.

Similar mechanisms are presumably operational in the medial interlaminar nucleus of the cat dorsal lateral geniculate nucleus. Here, early monocular deprivation also leads to a 30–35% reduction in cell size in parts of the nucleus (see Chapter 9) innervated by the deprived eye (Kratz *et al.*, 1978b), to a 90% reduction in the physiologically demonstrable Y-cell population (Kratz *et al.*, 1978a,b), to a reduction of Y-axon terminal ramifications, and to an increase of X-axon terminal ramifications (Sur *et al.*, 1982).

The large number of experiments that have been done on the dorsal lateral geniculate nucleus of monocularly deprived cats brings out the importance of competition in the normal development of the nucleus. What happens if both eye inputs suffer from reduced efficacy? Binocular deprivation deprives both monocular and binocular segments of Y-cell responses (Sherman *et al.*, 1972) and the receptive fields of striate cortex neurons fail to develop normally (Pettigrew, 1972). But cell size in the geniculate is nearly normal and in this it is comparable to the effect of rearing kittens in the dark (Sherman *et al.*, 1972; Hickey *et al.*, 1977; R. Kalil, 1978; Kratz *et al.*, 1979). Hence, competition is the key word in geniculate growth but normal physiological responsivity and the maturation of receptive field organization clearly depends upon normal visual experience—at least for Y cells in the geniculate and for the most forms of cells in the cortex (reviewed in Movshon and van Sluyters, 1981, and Sherman and Spear, 1982).

Further evidence for competition between retinal ganglion cell axons at the level of the single geniculate cell comes from the work of Archer *et al.* (1982). There workers have studied the effects of silencing action potentials in one optic nerve on the maturation of dorsal lateral geniculate neurons. To do this, tetrodotoxin was injected into one eye of kittens for the first 5 to 8 weeks of life.

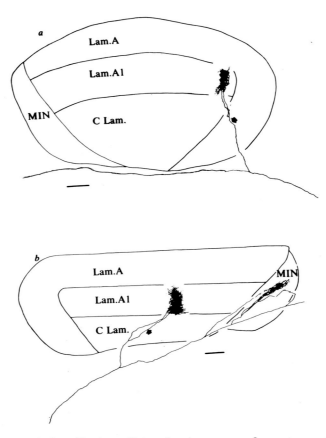

FIGURE 6.25.   Terminal ramifications of injected optic tract axons from cats reared with monocular eyelid suture. (Left) Upper axon is an X-type axon from the ipsilateral eye with a normal extent to its terminal ramifications; lower X-type axon has expanded terminal ramifications in both lamina

When the geniculate responses to visual stimuli were examined 5 to 7 days after cessation of the treatment and the return of action potentials in the fibers emanating from the treated eye, an abnormally high proportion of cells in the affected A or A1 lamina were excited by stimulation of either eye, and had abnormal "on–off" receptive fields. That is, they show transient discharges both when a visual stimulus applied to the centers of their receptive fields is turned on and when it is turned off. Very few geniculate cells in normal cats receive binocular excitatory inputs (Chapter 9) and none in the geniculate proper have receptive field centers showing "on–off" responses* (Cleland *et al.*, 1971). Normal cells have either "on" or "off" centers with an antagonistic surround (Chapter 9). To put it another way: normally developed geniculate cells receive direct excitatory inputs from only one eye and direct inputs from on- or off-center retinal ganglion cells, not from both. Half of a small sample of cells examined for X and Y properties in the study of Archer *et al.* showed convergence of X

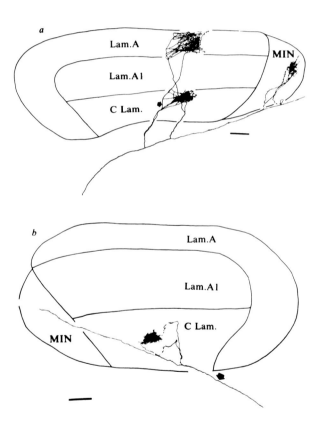

A1 and the medial interlaminar nucleus. (Right) Upper axon is a Y-type axon from the contralateral eye with apparently normal terminal ramifications. Lower Y-type axon has a much reduced terminal ramification. Bars = 250 μm. From Sur *et al.* (1982).

and Y ganglion cell inputs on the same cell. Most geniculate cells in normal animals receive only X or Y inputs, not both (Chapters 4 and 9). None of these effects seem to occur in cats reared in darkness or after binocular or monocular deprivation (Cleland *et al.*, 1971; Sherman, 1973; Kratz *et al.*, 1979; Mooney *et al.*, 1979). Hence, the effect of cessation of activity in one optic nerve seems to have been the promotion or maintenance of abnormal convergent inputs onto single cells.

Is it the promotion of new abnormal inputs, perhaps by sprouting of non-silenced axon terminals from adjacent cells, or is it the maintenance of inputs that are normally lost with visual experience that occurs in these deprivation experiments? Because of the findings of Kirkwood and Shatz (1982) (Section 6.5.1), there is reason to believe that it may be the maintenance of atypical inputs for these workers demonstrated bilateral excitatory inputs to a high proportion of geniculate neurons recorded extracellularly from *in vitro* preparations of the fetal cat thalamus between 43 and 59 days of gestation. After the 59th day and into the early postnatal period binocular excitation was still demonstrable in some neurons, but the more typical pattern of excitation by one eye and inhibition by the other (Chapter 9) was starting to emerge. These observations argue in favor of the elimination of synapses from ganglion cell axons of one eye during normal development and those of Archer *et al.* (1982), therefore, suggest their maintenance in the absence of impulses from the other eye.

Also in favor of the maintenance of inappropriate synapses in the absence of action potentials is the observation of Daniels *et al.* (1978) that a certain proportion of neurons in 2-week-old normal kittens display "on–off"-center receptive fields that later disappear. This argues for the early formation of synapses by both on- and off-center retinal ganglion cell axons on the same geniculate neuron and the later elimination of one of these types during normal development.

It is interesting that Y-type cells are more susceptible to the effects of deprivation than X-type cells in view of the seemingly later dendritic maturation of X-type cells (Friedlander, 1982). If, as everyone suspects, the competitive effect in the development of geniculate innervation is ruled from the cortex, then the axons of X cells are clearly able to establish effective innervation in the absence of dendritic maturation. The independent growth of axon and dendrites is becoming recognized elsewhere (e.g., Wise *et al.*, 1979; Jones, 1981b) so this is not surprising. On the other hand, one would feel that an immature dendritic field structure could reflect an afferent innervation in which synaptic numbers have not yet become established at their adult level. Therefore, such a cell might be predicted, *a priori*, to be more susceptible to deficient physiological inputs. The converse, however, seems to be the case. Perhaps the more rapid growth of the smaller, presumably X-type geniculate cells makes their axons more aggressive in seeking out synaptic contacts in the cerebral cortex, even in the face of a deficient afferent input.

---

* "On–off"-center receptive fields are regularly seen in the perigeniculate nucleus (Cleland *et al.*, 1971; Chapter 9).

There is no firm evidence that X-type cells are selectively influenced by experimental manipulations as Y-type cells are. There were suggestions that X-type geniculate cells may be more sensitive than Y-type cells to the effects of paralyzing the extraocular muscles of one eye even in adult cats (Brown and Salinger, 1975; Salinger *et al.*, 1977). However, this has not been supported by subsequent work (Winterkorn *et al.*, 1981).

6.6.1.3. Reversibility of the Effects of Visual Deprivation

The extent to which the effects of monocular deprivation are reversible in the cat have been investigated by a number of workers. If after monocular deprivation in infancy, the remaining undeprived eye is subsequently removed, then cells in the lateral geniculate nucleus return to their normal size in about 3 months and there is a substantial recovery of Y properties in the nucleus (Spear and Hickey, 1979; Geisert *et al.*, 1982). For reasons that have not yet been identified, driving of cells of the striate cortex by the deprived eye recovers long before the recovery of Y properties (Kratz *et al.*, 1976; Smith *et al.*, 1978; Spear *et al.*, 1980; Geisert *et al.*, 1982). Visual experience with the formerly deprived eye seems to be essential for the restoration of Y-type properties in the geniculate (Geisert *et al.*, 1982) but this is not essential for the restoration of driving of cortical cells by the deprived eye. This can be restored within a few hours of the enucleation of the normal eye (Kratz *et al.*, 1976; Smith *et al.*, 1978; Spear *et al.*, 1980). The rapid restoration of cortical driving suggests the presence of an element of cortical suppression during monocular deprivation but the slow recovery of Y-cell function may well depend upon the establishment of new connections in the lateral geniculate nucleus. Protracted visual experience seems essential for this and whatever the putative new connections are, they cannot be induced to form by simply closing the lids of the undeprived eye after opening the originally deprived one unless the cat is forced to use the latter (Hoffmann and Cynader, 1977; Sherman and Wilson, 1981; Geisert *et al.*, 1982).

6.6.2. Effects of Eye Removal and Visual Deprivation in Primates

It has been known for many years that removal of an eye in monkeys and other primates, including man, leads to transneuronal atrophy of the deafferented layers of the dorsal lateral geniculate nucleus (Minkowski, 1920; Le Gros Clark, 1932b; Glees and Le Gros Clark, 1941; Polyak, 1957). The effect is more severe in younger animals in which a high percentage of the deafferented cells may die (Hubel and Wiesel, 1977) although the amount of cell death is not clear. A severe shrinkage without cell loss is still detectable within 4 to 7 days of eye removal in a mature animal and reaches approximately a 40% reduction in cell size in 4 months (Glees and Le Gros Clark, 1941; Matthews, 1964). With the passage of many years, at least in man, considerable cell loss can occur (Goldby,

1957; Kupfer, 1965). In both monkeys and man, there is some evidence that cells in the contralaterally innervated laminae react to deafferentation before those in the ipsilaterally innervated laminae and it appears that the effect is more severe on the parvocellular than on the magnocellular laminae (Matthews *et al.*, 1960; Kupfer, 1965). No studies have reported differential effects upon the binocular and monocular segments of the nucleus, probably because the latter is so small.

Early studies on the effects of monocular or binocular visual deprivation in monkeys led to equivocal or contradictory results, possibly on account of differences in the ages of the animals used (Le Gros Clark, 1943; Chow, 1955). More recent studies have focused on monocular deprivation in infant monkeys with vision occluded from the first or second week of life and examined up to a year later (von Noorden, 1973; Headon and Powell, 1973, 1978; Hubel and Wiesel, 1977; Vital-Durand *et al.*, 1978). Severe transneuronal cell shrinkage occurs but, again, no difference was mentioned between the binocular and the monocular segments and whether translaminar sprouting occurs, as in the cat, has not been ascertained. Evidence for competitive interactions comparable to those postulated for the cat is, therefore, not yet forthcoming. Also potentially against the idea of competitive control of geniculate cell growth and physiological maturation is the lack of evidence for a differential effect of deprivation on X-type and Y-type geniculate neurons in monkeys (von Noorden and Middleditch, 1975). It is conceivable that the monkey dorsal lateral geniculate nucleus and the visual cortex where the putative competition occurs are too far advanced in development at birth for translaminar sprouting and differential effects on X- and Y-type cells to manifest themselves as they do in the cat. On the other hand, competition for space in layer IV of the visual cortex undoubtedly occurs between neighboring geniculate axons emanating from adjacent geniculate laminae during the early postnatal development of ocular dominance columns (Hubel and Wiesel, 1977; LeVay *et al.*, 1978). Therefore, one might anticipate seeing effects of this in the geniculate.

Headon *et al.* (1982) have reported that if eyelid closure is carried out from birth in monkeys, there is an initial hypertrophy of cells in the undeprived parvocellular laminae of the dorsal lateral geniculate nucleus. This represents an approximately 25% increase in cell size in comparison with the normal. It is maximal at 4–6 weeks of age before significant shrinkage occurs in the cells of the deprived parvocellular laminae. They suggest that it may be due to attempts by the undeprived cells to establish larger axonal domains than normal in the visual cortex. They also point out that the apparent shrinkage of deprived laminae when examined 4–6 weeks after eyelid closure at birth may be due primarily to the hypertrophy of the undeprived laminae.

The initial hypertrophy of cells in the undeprived laminae is succeeded by a gradual shrinkage amounting to approximately 10% in comparison with comparable cells in normal monkeys. The cells of the deprived laminae shrink by approximately 35% in comparison with the normal. This shrinkage of the undeprived laminae can also be induced by eyelid closure made several months after birth (Headon *et al.*, 1979, 1981a,b).

Monocular visual deprivation also causes failure of cell growth in the laminated dorsal lateral geniculate nucleus of the insectivore *Tupaia* (Casagrande *et al.*, 1974; Norton *et al.*, 1977) and in that of the prosimian primate *Galago* (Casagrande and Joseph, 1980; Sesma *et al.*, 1984).

Casagrande and Joseph (1980) find that monocular deprivation by eyelid suture in bush babies from the first week of life into the first year, leads to a failure of cell growth in the monocular as well as in the binocular segments of the dorsal lateral geniculate nucleus though the binocular segment is more severely affected and the monocular segment is not affected in all animals. As in cats, some recovery of the shrunken cells can be achieved by forcing the animals to use their deprived eye by "reverse suture" of the formerly open eye. Overall, neurons in the two small-celled layers (4 and 5) of the six-layered *Galago* dorsal lateral geniculate nucleus (see Chapter 9) are less affected by monocular deprivation than those in other layers. If the small-celled layers are preferentially populated by W cells as postulated by some authors (Chapter 9), this relative insensitivity to visual deprivation would make them comparable to the parvocellular C laminae of the cat dorsal lateral geniculate nucleus. These also contain many W cells and are relatively insensitive to visual deprivation (Hickey, 1980; Leventhal and Hirsch, 1983b; Section 6.6.1.3). Could it be that cortically projecting axons of parvocellular C-laminae cells in the cat, and of layers 4 and 5 in *Galago,* because they end in cortical layers different from those in which axons originating in the other layers end, are protected from competition and, therefore, their parent cells are protected from the effects of monocular deprivation?

6.6.3. Effects of Eye Removal and Visual Deprivation in Other Mammals

Anterograde transneuronal degeneration following removal of an eye has been demonstrated in the dorsal lateral geniculate nucleus in many mammals other than those already mentioned. They include: the goat (Minkowski, 1920), rat (Tsang, 1937), marsupial phalanger (Packer, 1941), sheep (Nichterlein and Goldby, 1944), rabbit (Cook *et al.*, 1951), guinea pig (Hess, 1957), and tree shrew (Glickstein, 1967). Monocular deprivation also affects geniculate cell growth in infant squirrels (Guillery and Kaas, 1974) as well as in the species earlier mentioned. Possibly, therefore, these are ubiquitous phenomena for the dorsal lateral geniculate nucleus but whether the principles of geniculate cell growth and maintenance that they bring out can be generalized to other thalamic nuclei remains to be determined.

6.6.4. Plasticity in the Development of Other Thalamic Nuclei

Does competition, either in the cortex (Section 6.6.1.1) or in the thalamus (Section 6.4.3.2) itself, govern the formation of connections and the growth of cells in thalamic nuclei other than the dorsal lateral geniculate nucleus? It is not hard to envisage competitive interactions ensuring the modality specificity of inputs to ventral posterior neurons as they seem to ensure the specificity of X or Y inputs to geniculate neurons (Section 6.6.1.2); or in setting up the lamellar-like tonotopic organization in the ventral medial geniculate nucleus; or in ensuring the predominance of excitatory drives from the contralateral ear on cells in the same nucleus (Chapter 8). At the moment, however, there are really no data to support or refute these possibilities.

As pointed out in Section 6.4.3.3, destruction of trigeminal nerve inputs in the neonatal period can lead to modifications in the cytoarchitectonic development of the VPM nucleus of rodents. There has been some physiological work to suggest that this is accompanied by changes in intranuclear connections. If the sinus hair follicles on the face of a mouse are destroyed shortly after birth, then single-unit responses induced by stimulation of the common fur of the buccal furry pad can be recorded throughout the VPM, as though afferent axons related to the furry pad had spread their terminations (Chapter 7) into the territory formerly innervated from the vibrissae (Verley and Onnen, 1981). The converse occurs in nude mice in which common fur is lacking from birth. Here, vibrissal responses can be detected in parts of the VPM normally reserved for inputs from the furry pad (Verley and Pidoux, 1981).

6.6.5. Formation of Aberrant Connections in the Thalamus

Axons growing toward the thalamus are presumably guided by those same factors that influence axon growth in other developing systems. These other factors probably represent a combination of innate programming, direction by axon–substrate interactions, and chemoaffinity between axons and target cells (Meyer and Sperry, 1976; Katz and Lasek, 1979; Constantine-Paton, 1979). The exact natures of these guidance mechanisms, however, remain to be determined and represent some of the most interesting areas in developmental neurobiology. Some measure of the specificity of connection formation in the thalamus can be gathered from the fact that few, if any, "mistakes" in afferent connections have been reported. Even in the Siamese cat (Chapter 9) in which a large proportion of fibers from the temporal retina decussate abnormally at the optic chiasm (Guillery, 1972; Shatz and Kliot, 1982), these fibers find their way appropriately to the dorsal lateral geniculate nucleus. There is no question, however, that anomalous connections can be induced to form experimentally, perhaps implying that axon and target matching may involve some competition between afferents or that a particular set of axons normally arrives during a developmental epoch when its appropriate cells are more ready to accept connections than cells in other nuclei.

Ablation of the dorsal lateral geniculate nucleus in newborn hamsters—at a time when retinogeniculate axons are just establishing connections (Section 6.4)—leads to invasion of inappropriate thalamic nuclei by the retinal axons, rather than the death of the axons (Schneider, 1973; Kalil and Schneider, 1975; Frost, 1981). Nuclei such as the lateral posterior, medial geniculate, and ventral posterior may, thus, come to receive geniculothalamic fiber terminations and invasion of these nuclei can be promoted by depriving them of some of their normal afferents (Schneider, 1973; Crain and Hall, 1980; So *et al.,* 1981, Frost, 1982) and by "pruning" the retinofugal axons by destroying their terminations in the superior colliculus (Schneider, 1973). A similar effect can be obtained by causing the dorsal lateral geniculate nucleus to degenerate by cortical ablation (Cunningham *et al.,* 1979). A retinotopic organization of the new connections may even be established (Frost, 1982) and the retinal connections with these unusual nuclei are capable of mediating transneuronal flow of intraocularly injected radioactive amino acids to the auditory and somatic sensory areas of the cerebral cortex (Frost, 1981).

There have been no successful experiments in which other afferent fiber systems such as the medial lemniscus and brachium of the inferior colliculus can be induced to grow into the dorsal lateral geniculate nucleus. Because all such experiments have been conducted on postnatal animals, this failure may suggest that at birth the lemniscal and brachial fibers are at too advanced a stage of development to undergo sprouting comparable to that demonstrated in geniculothalamic fibers.

Die einzelnen Thalamuskerne sind integrierende Bestandteile sehr
verschiedener Funktionssysteme. Die beschriebene Vielfältigkeit der
thalamischen Strukturen und ihrer Verbindungen kann daher nicht
überraschen. Mit einer globalen Betrachtung wird man nie zu einem
Verständnis der Funktionen des Thalamus gelangen. Dieses setzt vielmehr
eine *Analyse* der einzelnen in sich einheitlichen Teile des Thalamus voraus
und eine Analyse *der Elemente* dieser 'Kerne.' Erst wenn die Anordnung
der Elemente und ihre Verknüpfungen bekannt sind, wird man die
Ordnung im einzelnen verstehen können, nach denen die Thalamuskerne
arbeiten. Ordnung der Elemente ist aber die Voraussetzung für ein
angepaßtes und zweckentsprechendes biologisches Geschehen, ebenso wie
in den Apparaten der Technik.

Hassler (1959)

Individual Thalamic Nuclei

This ever-watchful sentinel of our corporeal frame—whose organ (through the medium of nerves of sensation) pervades the whole external surface of the body, including the intestinal canal—placed, as it is, to guard from external injury, this delicate machine—to keep in tune this harp of a thousand strings.

O. S. Fowler, Phrenology (1846)

The Ventral Nuclei

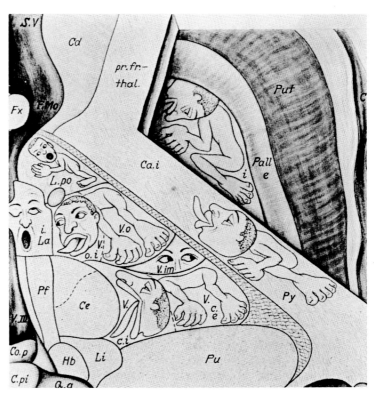

Hassler's (1962) sketch of multiple representations of the body form in the ventral nuclei of the human thalamus, as suggested by the results of stimulation in human subjects undergoing stereotaxic thalamotomy.

7.1. Introduction

In every mammal, with the exception of the monotremes, it is possible to identify at least three cellular components of the ventral nuclear group: the ventral lateral, ventral medial, and ventral posterior nuclei. In many species a ventral anterior nucleus can be distinguished as well and, in higher primates, the ventral lateral nucleus is clearly divided into at least two subnuclei. A ventral posterior inferior nucleus is also commonly identified.

7.2. Ventral Posterior Nucleus and the Somatic Sensory System

7.2.1. Description

The ventral posterior nucleus is one of the most clearly defined nuclei of the thalamus because of the large size and dense staining of its constituent cells and because of the lobulated appearance imposed on it by penetrating bundles of myelinated fibers (Figs. 7.1 and 7.2).

The nucleus is usually ovoid, stretching from the external medullary lamina to the internal medullary lamina, and with a tapering posterior pole in close proximity to the medial geniculate complex. In small species with relatively generalized brains such as the shrew, hedgehog, or opossum (Le Gros Clark, 1932a; Diamond *et al.*, 1970), the ventral posterior and medial geniculate nuclei may merge.

The ventral posterior nucleus is almost invariably divided by a sloping medullary lamina into two well-defined subnuclei: a ventral posteromedial nucleus (VPM) composed of smaller, relatively closely packed cells, and a ventral posterolateral (VPL) nucleus composed of larger cells, usually arranged in clusters. In some animals, such as the cat, the VPM extends posterior to the VPL. In others, such as monkeys, the VPL extends more posteriorly as a narrow tail. The trigeminal lemniscus terminates in the VPM and the medial lemniscus in the VPL. As a consequence, the relative sizes of the two nuclei vary according to the relative innervation density of the face and the rest of the body (Rose and

Mountcastle, 1959; Cabral and Johnson, 1971; Welker, 1974; Bombardieri *et al.*, 1975) (Fig. 7.3). In animals such as rodents in which the nose, mouth, and lips are the main tactile-receiving surfaces of the animal, the VPM is larger, and in the monotremes it dominates the whole ventral nuclear mass (Hines, 1929; Chapter 17). In other animals such as cats, the VPM and VPL are approximately the same size whereas in primates the VPL predominates on account of the larger representations of the hands and feet. Often the hand and foot representations in the VPL are divided by a further medullary lamina. Where the innervation of a part of the ventral posterior nucleus is extremely dense, probably reflecting a dense innervation of a particular body part, subsidiary focal aggregations of cells may be seen (Fig. 7.1). In the raccoon and slow loris, the VPL is multilobular, each lobule representing one of the densely innervated palmar or digital skin pads (Welker and Johnson, 1965; Krishnamurti *et al.*, 1972), and traces of similar lobulation are detectable in other primates such as the bush baby (Fig. 7.4). The VPM of rodents and some marsupials (Fig. 7.5) is broken up into elongated cell clusters (Van der Loos, 1976; Akers and Killackey, 1979; Woolsey *et al.*, 1979), each representing a facial sinus hair.

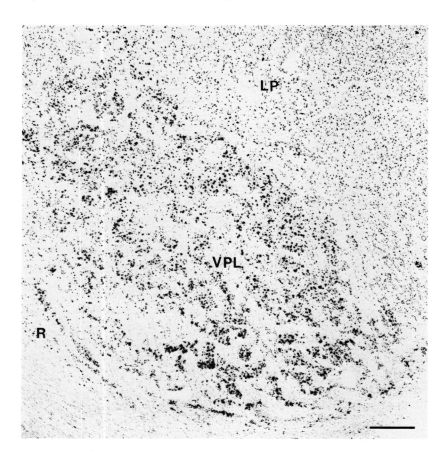

FIGURE 7.1. Lobulation of the ventral posterior nucleus in a harbor seal (*Phoca vitulina*). From a brain provided by Dr. R. W. Dykes. Bar = 1 mm.

7.2.2. Terminology

Apart from its characteristic cell structure, the ventral posterior nucleus can be distinguished by the fact that its borders confine the thalamic terminations of the medial and trigeminal lemnisci (Smith, 1975; Tracey *et al.*, 1980; Jones and Friedman, 1982) (Fig. 7.6). This excludes from the ventral posterior nucleus a small-celled region medial to the VPM that receives taste afferents. Though sometimes referred to as a pars parvocellularis of the VPM, for reasons discussed in Section 7.6.3, it seems preferable to regard this taste relay nucleus as a separate, basal ventral medial (VMb) nucleus.

The terms *ventral posterolateral* and *ventral posteromedial nuclei* were first introduced by Le Gros Clark (1930a) in his study of *Tarsius* but the nuclei had been recognized under different names from early times. von Monakow (1895) had called them *nucleus ventralis a* and *nucleus ventralis b*, through ventralis a probably also included parts of the ventral lateral complex. The VPL is largely the nucleus ventralis caudalis of Vogt (1909) and Friedemann (1911). The VPM had been termed the *semilunar nucleus* by Dejerine (1901) and Vogt (1909), the *saucer-shaped nucleus* by Flechsig (1886), and the *arcuate ventral nucleus* by Rioch (1929a). *Arcuate nucleus* had been used earlier by Münzer and Wiener (1902) for what is mainly the medioventral (reuniens nucleus). The VPL was referred to as the *external ventral nucleus* by Rioch (1929a). Rioch's terms still survive in many

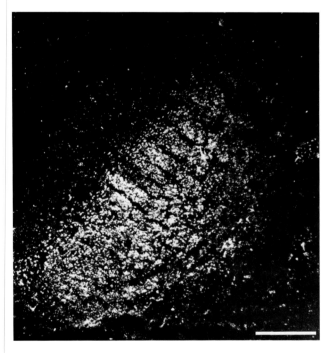

FIGURE 7.2. Darkfield photomicrograph showing retrograde labeling of "barreloids" in VPM nucleus and of cell clusters in VPL nucleus of a 2-day-old rat in which horseradish peroxidase was injected into the somatic sensory cortex. Bar = 250 μm.

studies on the ventral posterior nucleus. In man, Hassler (1959) identifies nuclei ventralis caudalis anterior externus and ventralis caudalis posterior externus which together appear comparable to the VPL, and nuclei ventralis caudalis anterior internus and ventralis caudalis posterior internus which together appear comparable to the VPM.

For many years the VPM or arcuate nucleus was held to include the small-celled medial part referred to above and probably first clearly identified by Friedemann (1911), though he included the nucleus VPI in it. In their early

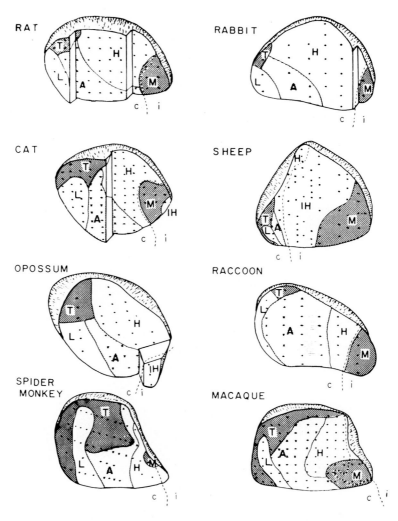

FIGURE 7.3. Relative proportions of VPL and VPM nuclei devoted to representations of body (T) and limbs (L, A), of contralateral head and face (H), and of ipsilateral facial structures (IH) and interior of mouth (M) in various mammals. Dashed line divides contralateral (c) from ipsilateral (i) representations. For rat the presence of an ipsilateral representation has now been questioned. From Cabral and Johnson (1971).

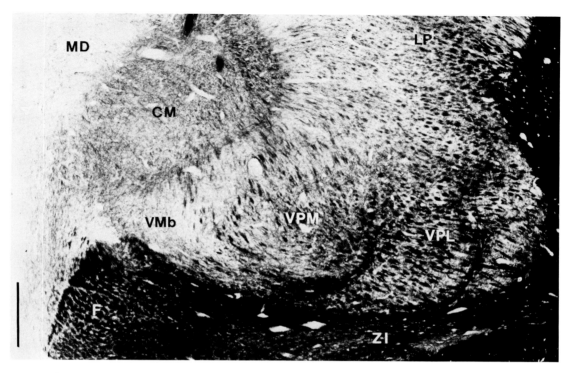

FIGURE 7.4. Lobulation of VPL nucleus in a bush baby (*Galago crassicaudatus*). Hematoxylin stain. Bar = 1 mm.

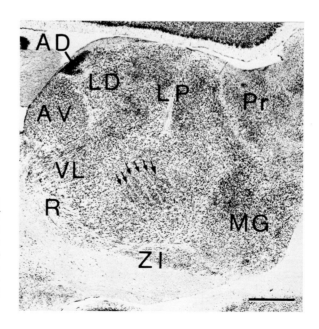

FIGURE 7.5. Oblique parasagittal section showing (arrows) six rows of cells ("barreloids") in VPM nucleus of a marsupial phalanger (*Trichosurus vulpecula*), each providing input to similar rows of cortical cells ("barrels") that represent the rows of mystacial vibrissae. Thionin stain. Bar = 1 mm.

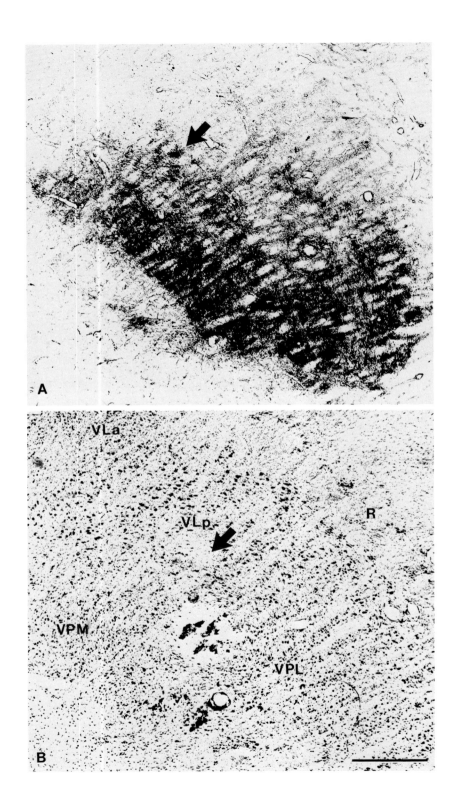

single-unit studies on the thalamus of the rabbit and cat, Rose and Mountcastle (1952, 1954, 1959) discovered that neuronal discharges could not be induced in this parvocellular component by natural somesthetic stimulation. The VPM and VPL, on the other hand, were readily activated by peripheral stimulation. Rose and Mountcastle, therefore, reintroduced M. Rose's (1935) term *ventrobasal complex* to refer to the parts of the ventral nucleus activated by somatic sensory stimulation. This was found to be coextensive with the VPL and VPM, exclusive of the parvocellular component. They placed the parvocellular component in a loosely defined ventral medial complex. Later the parvocellular nucleus was shown to be the thalamic relay for taste afferents (see Burton and Benjamin, 1971) but was not named separately from the other nuclei of the ventral medial complex. Therefore, Berman and Jones (see Jones and Leavitt, 1974) coined the term *basal ventral medial nucleus* for the parvocellular component.

7.2.3. Body Representation

Rose and Mountcastle's single-unit studies of the ventral posterior nucleus built upon the earlier evoked potential study of Mountcastle and Henneman (1949) in the monkey. Since that time, there have been numerous single-unit studies of the nucleus in a large number of species (e.g., Mountcastle and Henneman, 1952; Poggio and Mountcastle, 1960, 1963; Pubols and Pubols, 1966; L. M. Pubols, 1968; Cabral and Johnson, 1971; Welker, 1974; Emmers, 1965; Loe *et al.*, 1977; Dykes *et al.*, 1981; Jones and Friedman, 1982). All have served to show that the contralateral limbs, trunk, and tail are represented in the VPL and the head, face, and intraoral structures in the VPM. Each large body part is represented as a kind of curved lamella, with a lateral convexity and concentrically arranged, with neighboring lamellae representing neighboring body parts. The tail is represented most laterally (in the VPL) and the mouth and pharynx most medially (in the VPM). In-between are systematically represented: hindlimb, trunk, forelimb, head, and face (Fig. 3.17). These laminae, representing the large divisions of the body, can be demonstrated morphologically by labeling of lemniscal axons emanating from a part of one of the dorsal column nuclei (Fig. 3.17), by labeling of corticothalamic terminations emanating from one of the major subdivisions of the body representation in the somatic sensory cortex (Fig. 3.24), or by labeling of thalamocortical cells projecting to such a subdivision. Lateral lamellae tend to be more highly curved than medial ones; that for the tail, for example, curves around the posterior pole of the nucleus, at least in the monkey. The mapping is continuous. The lamellae do not appear to represent dermatomes, indicating that some reshuffling of afferents has already occurred at lower levels (Welker, 1974). The representation, though con-

FIGURE 7.6. Adjacent horizontal sections through ventral nuclei of the thalamus in a cynomolgus monkey showing anterogradely labeled medial lemniscal fibers outlining VPL nucleus (A). Arrows indicate lesion made just prior to killing animal along a horizontal microelectrode track at point of exit from zone of units with deep receptive fields in VPL nucleus. (A) Horseradish peroxidase labeling; (B) thionin stain. Bar = 1 mm. From Jones and Friedman (1982).

tinuous, is also distorted, reflecting the greater peripheral innervation density of the hand and foot and of areas in and around the mouth. The ventral half or more of the VPL contains the representations of the forepaw and hindpaw, that of the trunk and proximal parts of the limbs being compressed dorsally. And the greater part of the VPM is devoted to the representation of lips, tongue, and the inside of the mouth. Under most physiological conditions, even in conscious animals (Poggio and Mountcastle, 1963), only a representation of the contralateral side of the body can be detected, except for regions in and around the mouth. Neurons related to these regions are in the most medial part of the VPM and commonly have receptive fields that extend across the midline.

The VPM of the rat, mouse, and certain marsupials is divided into elongated cell clusters referred to as "barreloids" (Van der Loos, 1976; Haight and Neylon, 1978; Akers and Killackey, 1979; Woolsey *et al.*, 1979). Such configurations are identifiable in conventional Nissl stains and in material stained for certain oxidative enzymes (Belford and Killackey, 1978; Woolsey *et al.*, 1979) (Figs. 5.18 and 7.7).

The arrangement of the rodent "barreloids" has not been well described and is difficult to visualize in three dimensions. It appears that the row of barreloids representing a row of mystacial vibrissae is curved across the VPM from lateral to medial, with the caudal vibrissae of a row represented laterally and the rostral vibrissae medially. The most dorsal row of vibrissae is represented by the posterodorsal row of "barreloids" and the most ventral by the most anteroventral row (Waite, 1973a; Vahle-Hinz and Gottschaldt, 1983). Because of the tilted orientation, the rows can be visualized in both frontal and horizontal sections. The common hairs of the rat or mouse face appear to be represented in more anterior and ventral parts of the VPM (Verley and Onnen, 1981; Vahle-Hinz and Gottschaldt, 1983). In the cat they are represented in an interspersed manner with the representations of the sinus hairs (Vahle-Hinz and Gottschaldt, 1983).

Sinus hairs of the cat are represented in the VPM in rows comparable but not identical to those in the rat and mouse but no "barreloids" are detectable in the VPM of the cat (Vahle-Hinz and Gottschaldt, 1983) and the area of representation of one vibrissa is not as clearly demarcated from that of its neighbors as in the rat. There is a similar difference in the discreteness with which individual sinus hairs are represented in the cortex of the two species (Welker, 1971, 1976; Dykes *et al.*, 1977).

Though once reported (Emmers, 1965), there appears to be virtually no representation of ipsilateral intraoral and perioral regions in the rat VPM by contrast with the large ipsilateral representation in the cat and other species mentioned above (Waite, 1973a,b; Verley and Onnen, 1981; Vahle-Hinz and Gottschaldt, 1983). The large size of the ipsilateral representation in the cat may be reflected in a readily definable input from the ipsilateral trigeminal nuclear complex (Burton and Craig, 1979; Burton *et al.*, 1979). This has not been detected in rats (Lund and Webster, 1967b; Smith, 1973).

Most of the neurons residing in the ventral posterior nucleus are, in the words of Rose and Mountcastle, "modality and place specific." They have relatively small, well-localized receptive fields, and are specifically activated either by light tactile stimuli, by pressure, by movement of joints, or by manipulation of muscles and tendons (Rose and Mountcastle, 1954; Poggio and Mountcastle, 1960, 1963; Gordon and Manson, 1967; Harris, 1970; Loe *et al.*, 1977; Dykes *et al.*, 1981; Jones *et al.*, 1982c). In the conscious animal and in animals anesthetized with chloralose, for reasons that are not clear, much larger, often bilateral receptive fields with multiple kinds of sensory inputs may sometimes be detected (Baker, 1971; Jabbur *et al.*, 1972).

In anesthetized or sedated monkeys, the modality- and place-specific units are distributed in a way that not only suggests a finer level or organization in the lamellar representation plan but also implies that there may be more than one representation of the body in the nucleus. Microelectrodes introduced vertically from above, on entering the ventral posterior nucleus, record units only responding to stimulation of deep tissues for a distance of 300–500 μm, before passing into a larger central core in which units only respond to cutaneous stimuli (Figs. 3.18 and 7.8) (Poggio and Mountcastle, 1963; Loe *et al.*, 1977; Pollin and Albe-Fessard, 1979; Friedman and Jones, 1981; Jones and Friedman, 1982). There may be a further zone in which units respond only to tapping stimuli in the deepest part of the VPL or in the VPI (Dykes *et al.*, 1981). The thin strip

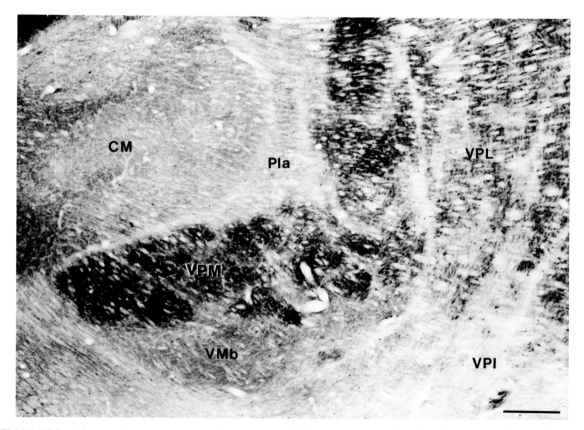

FIGURE 7.7. Clustered aggregations of cells in monkey VPM nucleus, outlined by histochemical staining for the mitochondrial enzyme cytochrome oxidase. Bar = 500 μm.

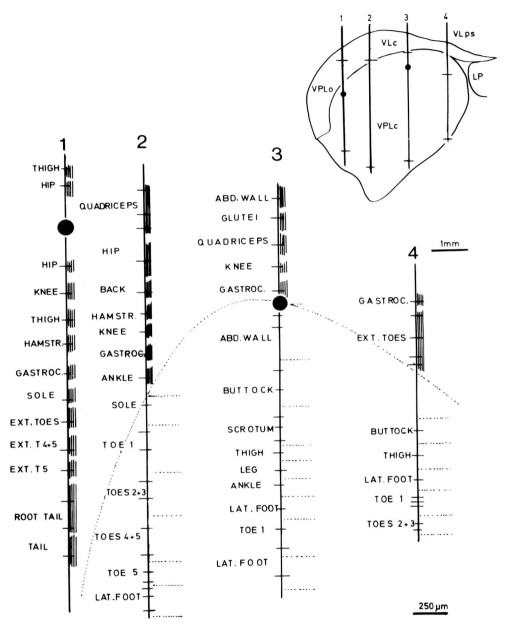

FIGURE 7.8. Vertical electrode tracks (1–4) penetrating VPL nucleus of a cynomolgus monkey in same parasagittal plane encounter units with receptive fields in deep tissues (hatching) in dorsal and anterior parts of the nucleus, and units with cutaneous receptive fields in the central part of the nucleus. Solid circles are lesions made at entry into (1) or exit from (4), region of deep responses. From Jones and Friedman (1982).

of deep responses on the dorsal surface of the ventral posterior nucleus extends down the full anterior surface of the nucleus and for a considerable extent down the posterior surface. This can be demonstrated either by a succession of vertically oriented microelectrodes or by microelectrodes that traverse the nucleus horizontally from behind and enter the deep shell region from the larger central cutaneous core (Friedman and Jones, 1981). A similar region responding to stimulation of muscle and joint afferents has been described with microelectrode recording in man, though it has been equated with the ventral intermediate nucleus of the ventral lateral complex (Hassler, 1959; Jasper and Bertrand, 1966; Goto et al., 1968). Both cutaneous and deep parts of the ventral posterior nucleus receive the terminations of the medial lemniscus (Jones and Friedman, 1982; Asanuma et al., 1983b), but the central cutaneous core is thought to receive inputs that have traversed the dorsal columns of the spinal cord, whereas the deep shell is thought to receive inputs mediated by the dorsolateral funiculus (Loe et al., 1977).

7.2.3.1. Finer Organization

The lamellae of the ventral posterior nucleus, as mentioned in Chapter 3, are to some extent descriptive conveniences and it is likely that dendrites or axons belonging to one overlap extensively into adjacent lamellae. However, those representing the hand and foot or thumb and face are often separated by a morphologically distinguishable fiber lamina, and when a microelectrode crosses the lamellae from lateral to medial, the receptive fields and modality properties of the units successively encountered can change over distances as small as 100 μm (Jones et al., 1982c) (Fig. 3.18).

A finer organization can be detected in each lamella by the use of electrophysiological techniques, coupled with anatomical studies of the distributions of medial lemniscal axons. As a microelectrode traverses the central cutaneous core of the monkey VPL or VPM horizontally from behind, it successively encounters, over distances of up to 800 μm, units with receptive fields on the same small area of the body surface (e.g., the tip of a finger) and sensitive to the same type of stimulus (e.g., touching the skin but not hair movement) (Friedman and Jones, 1981; Jones and Friedman, 1982; Jones et al., 1982c). Possibly if the electrode could follow the curvature of a lamella, it might remain within the same modality- and place-specific group of neurons over the full anteroposterior extent of the lamellae, though the elongated groups of cells need not be so long. Microelectrodes traversing the cutaneous core from lateral to medial or from dorsal to ventral, on the other hand, tend to cross from one place- and modality-specific cluster of units to another, within 100–300 μm of each other. The jumps in the shell of deeply responding units are less distinct, though clusters of cell responding to stimulation of the same group of muscles are common. At the junction of the deep and cutaneous regions, an occasional neuron may have two receptive fields, one deep and one cutaneous (Loe et al., 1977). Presumably this indicates a neuron that extends dendrites into the two regions.

The clustering of neurons with the same place and modality properties and the change in receptive field position from proximal to distal on a limb as an electrode penetrated dorsoventrally were first identified in the monkey ventral posterior nucleus by Poggio and Mountcastle (1963) and later confirmed by L. M. Pubols (1976) and Loe *et al.* (1977). The new findings (Jones *et al.*, 1982c) suggest that each lamella representing a large body part such as the hand or a finger is composed of multiple, narrow, elongated, rodlike configurations of neurons receiving input from the same small part of the body surface and from the same types of receptors. Some inkling of this pattern is evident in the paper of Loe *et al.* (1977) in which they reported that a single body part could be represented through much of the dorsoventral extent of the VPL but with modality-segregated neuronal clusters within it.

The rodlike pattern of organization can also be demonstrated anatomically by the curved, rodlike distribution of the bundle of lemniscal fibers emanating from a focal aggregation of cells representing the same body part in one of the dorsal column nuclei (Fig. 3.17B). The same rodlike configurations of label can also be seen when the full complement of lemniscal axon terminations are labeled (Fig. 3.17C). Individual rods labeled in this way rarely extend through the full anteroposterior dimension of a lamella in the monkey (Jones *et al.*, 1982c) and may be even shorter in cats (Berkley and Hand, 1978). Within a rod, individual lemniscal fibers run along it and can have concentrated terminal aggregations at several anteroposterior levels, though rarely throughout the full anteroposterior extent of the rod (Rainey and Jones, 1983; Jones, 1983b) (Figs. 3.28 and 7.9). This implies that the organizing principle is a bundle of lemniscal fibers of like place and modality properties, rather than the single lemniscal fiber.

The rods of afferent fiber distribution are duplicated by comparable rods of thalamocortical relay cells. A focal injection of tracer less than 1 mm in extent in a physiologically defined part of the somatic sensory cortex retrogradely labels cells in a curved rod extending through much of the anteroposterior dimension of the ventral posterior nucleus but over a very restricted extent in the mediolateral and dorsoventral dimensions (Jones *et al.*, 1979, 1982c; Jones, 1981a) (Fig. 3.25). In cats, rods labeled in this way appear to be shorter (Kosar and Hand, 1981). It seems probable that the "barreloids" of the rodent VPM and the clusters of cells detected in Nissl stains of the VPL or retrogradely labeled in the VPL of many species after large injections of tracer in the cortex are (e.g., Fig. 3.5A) similar to the rods selectively demonstrated in the experiments outlined here.

These observations suggest that a thalamic rod forms the basis of the place- and modality-specific input to a column of the somatic sensory cortex. The specificity of the input is determined by the bundling of lemniscal axons in the thalamus. It must be remembered, however, that "modality" as introduced by Mountcastle (1957) is used here in a very broad sense. It does not necessarily imply that the input to a thalamic rod and thus to a cortical column is receptor specific. The barreloids of the rodent VPM, receiving afferents related to several forms of receptor ending about the facial vibrissae (reviewed in Burgess and Perl, 1973), are clearly not receptor specific. There are hints, however, that some

FIGURE 7.9. Medial lemniscal axon and its terminals, anterogradely labeled by injection of horseradish peroxidase in medial lemniscus and ending in dorsal region of deep responses in the VPL of a cynomolgus monkey. Anterior to right. Bars = 1 mm and 200 μm. VPL cells drawn at same magnification as fiber. From Jones (1983b).

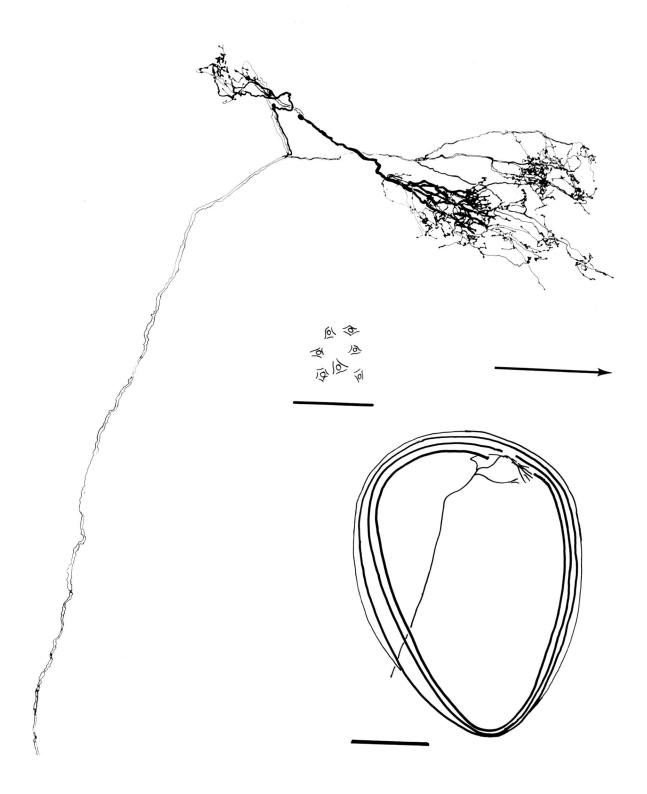

receptors, such as Meissner or Pacinian corpuscles, may have preferred routes through the thalamus to the cortex (McIntyre, 1962; Mountcastle *et al.*, 1969; Fisher *et al.*, 1983). It is conceivable, given the localized distribution of individual lemniscal fibers in the rods, that some cells in a rod and, therefore, some axons entering a column in the somatic sensory cortex may be more or less receptor specific. The obvious analogy here would be with the X and Y streams through the dorsal lateral geniculate nucleus (Chapter 9). X and Y axons from the same part of the retina innervate the same projection columns but end on different cells; these, though projecting to the same cortical column (or hypercolumn), remain largely separate.

Neurons situated in the thin shell that covers the anterior, dorsal, and partially the posterior surfaces of the ventral posterior nucleus respond to movement of joints, pulling on tendons, and to pressing on and kneading of muscles. The receptive fields are, as a consequence, hard to localize other than to a muscle group or a particular joint. Nevertheless, those responding to rotation of a joint can display discharge patterns that are joint specific (Mountcastle *et al.*, 1963; Yin and Williams, 1976). Units in the anterior part of the shell region respond to threshold stimulation of group I afferents in muscle nerves in both cats and monkeys (Mallart, 1968; Andersson *et al.*, 1966; Rosén, 1969a; Landgren and Silfvenius, 1970; Grant *et al.*, 1973; Maendly *et al.*, 1981). In view of this, and because many of the movement-related discharges in joint nerves may be derived from muscle receptors rather than from receptors in the joint capsule (Burgess and Clark, 1973; Clark and Burgess, 1975), it seems possible that many of the responses of neurons in other parts of the deep shell of the ventral posterior nucleus are mediated by muscle afferents rather than by joint afferents as previously supposed.

Neurons with receptive fields among the muscles and joints of the hands and feet are found in the anterodorsal deep shell as well as those with receptive fields situated more proximally on the limbs or body (Jones and Friedman, 1982). To some this may imply a separate full representation of the body in the shell, but the evidence on this point is incomplete.

7.2.4. Afferent Connections of the Ventral Posterior Nucleus

7.2.4.1. Lemniscal Inputs

The medial lemniscus, at its entry into the thalamus, consists of fibers arising in the contralateral dorsal column nuclei and certain of their satellite nuclei, as well as fibers from the contralateral trigeminal complex and from the contralateral lateral cervical nucleus.

a. Dorsal Column System. The cuneate and gracile nuclei project to the contralateral VPL (Fig. 7.10) in an organized manner that follows the overall somatotopy in each nucleus (Lund and Webster, 1967a; Boivie, 1971a; Hand and Van Winkle, 1977; and many others).

The cuneate and gracile nuclei of the several species that have been studied, each consist of two or more cytoarchitectonically distinct parts in which neurons have different kinds of receptive fields. A larger central and dorsal part of each nucleus consists of large, clustered cells with mainly cutaneous receptive fields while a ventral part, expanding caudally and rostrally, has a more mixed, unclustered cell population and responds to deep stimuli, including pressure and movement of joints (Kuypers and Tuerk, 1964; Gordon and Jukes, 1964; Winter, 1965; Millar and Basbaum, 1975). This region seems to be unique as being the principal site of termination of corticofugal fibers to the dorsal column nuclei (Kuypers and Tuerk, 1964). Rostral to the gracile nucleus is a more or less separate component, "nucleus Z" (Brodal and Pompeiano, 1957), in which group I muscle afferents from the hindlimb relay (Landgren and Silfvenius, 1971). These are second-order fibers that have arisen from cells lower in the spinal cord. Group I muscle afferents from the forelimb relay in a part of the cuneate nucleus, not obviously separated off from it (Rosén, 1969b). These appear to be primary afferent fibers. It is possible that all deep inputs to the dorsal column nuclei reach them via the dorsolateral funiculus, rather than with the cutaneous inputs via the dorsal columns (Whitsel *et al.*, 1969a).

Each of the receptive field classes in the three parts of the dorsal column nuclei project into the medial lemniscus and reach the contralateral VPL (Gordon and Seed, 1961; Gordon and Jukes, 1964; Rosén, 1969b; Landgren and Silfvenius, 1970). Though some neurons in the dorsal column nuclei receive convergent joint, muscle, and cutaneous inputs, few such cells project into the medial lemniscus (Tracey, 1980). Cells projecting to the thalamus are all large cells [more than 18 μm in diameter in cats (Berkley, 1975)] (Fig. 7.11). Smaller cells project elsewhere. Fibers arising in the different components of the dorsal column nuclei have different patterns of termination. The large cutaneous region projects to the central cutaneous core of the VPL with the clustered, or focal pattern of terminations already described in Section 7.2.3.1 (Groenewegen *et al.*, 1975; Hand and Van Winkle, 1977; Berkley and Hand, 1978; Tracey *et al.*, 1980). The ventrocaudal region of deep responses projects in a more diffuse manner to dorsal parts of the VPL (Hand and Van Winkle, 1977) in which "deep" units are also found (Section 7.2.3). Nucleus Z and presumably its cuneate equivalent send axons to terminate in the region on the anterior surface of the VPL in which group I responses are demonstrable (Grant *et al.*, 1973). Grant *et al.* consider that the terminals extend into the ventral lateral nucleus which seems unlikely in view of the data reviewed in Section 7.2.3.1. In this region there is significant overlap of the group I-related afferents with the terminations of fibers from the principal dorsal column nuclear components and the spinocervicolemniscal systems (Grant *et al.*, 1973). A few fibers from cells in the external cuneate nucleus also reach the dorsal region of the VPL (Boivie and Boman, 1981) but the significance of this projection is unknown. It could possibly represent a route whereby group I muscle afferents might reach cortical fields of SI other than area 3a (see Section 7.2.7).

b. Trigeminal System. The trigeminothalamic pathways have been a matter of some controversy, particularly revolving around the question of an ipsilateral route to the VPM from the principal sensory nucleus of the trigeminal complex. The contralateral pathway is not in doubt. Its axons arise from cells of all sizes (Burton and Craig, 1979) in the ventral two-thirds of the principal nucleus

(Torvik, 1957; Michail and Karamanlidis, 1970; Smith, 1975), decussate in the rostral pons and, joining the medial lemniscus, ascend with it to the thalamus. This pathway was demonstrated in early Marchi studies on rabbits (Wallenberg, 1895, 1896), man, monkeys (von Economo, 1911; Walker, 1938b), rat (Le Gros Clark, 1932a), cat (Russell, 1954), and other species. Modern studies with axonal degeneration (Mizuno, 1970; Smith, 1973, 1975) and autoradiographic techniques (Kruger *et al.*, 1977), confirm it and show that the fibers end with topography predictable from the single- and multiunit mapping studies (Section 7.2.3.1) in the VPM (Fig. 7.3).

The existence of an ipsilateral, dorsally running, trigeminothalamic pathway from the principal trigeminal nucleus has been argued about for some time. One might predict the existence of such a path in view of the ipsilateral components reported for receptive fields in and around the mouth among neurons in the most medial part of the VPM of most mammals (Section 7.2.3). Some of the early Marchi studies had reported such an ipsilateral pathway ascending dorsally in the brain-stem tegmentum, though these now appear to be fibers ascending from the reticular formation in the vicinity of the principal trigeminal nucleus to join the ipsi- and contralateral motor trigeminal nuclei (Smith, 1973).

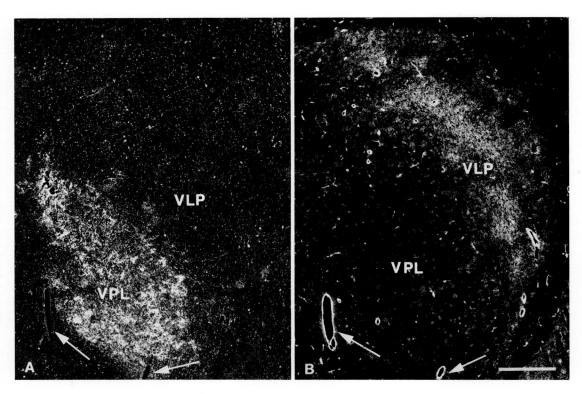

FIGURE 7.10. (A, B) Adjacent parasagittal sections through the thalamus of a cynomolgus monkey in which medial lemniscal terminations (A) were labeled by autoradiography af- ter injection of tritiated amino acids in dorsal column nuclei and cerebellar terminations (B) were labeled by degeneration after cerebellectomy. Arrows indicate same blood ves-

The ipsilateral trigeminal pathway to the medial part of the VPM has now been shown to ascend with the crossed trigeminal fibers in the medial lemniscus. However, there appear to be significant species differences for, though such a path can be demonstrated anatomically in species such as the goat, cat, and monkey (Karamanlidis and Voogd, 1970; Smith, 1975), there is little (Lund and Webster, 1967b) or no (Smith, 1973) evidence for its presence in the rat or rabbit (Karamanlidis *et al.*, 1978). This difference seems to be in line with the absence of significant ipsilateral receptive field components among VPM neurons related to the mouth in rats (Section 7.2.3.1). According to Smith (1975), in the monkey the uncrossed tract arises from cells in the dorsal one-third of the nucleus and ends in parts of the VPM lying dorsomedial to the part receiving the crossed fibers. In the cat, Torvik (1957) reported that large thalamic lesions caused retrograde degeneration of cells in the dorsal one-third of the ipsilateral principal trigeminal nucleus and this origin of the ipsilateral projection has been confirmed by retrograde labeling (Burton and Craig, 1979).

c. Spinocervical System. Fibers from the lateral cervical nucleus of the spinal cord after decussating in the first or second spinal cord segment travel in the

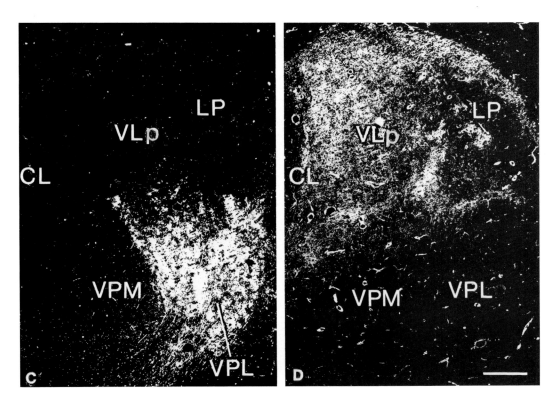

There is no overlap in terminations of the two sets of fibers. (C, D) Adjacent frontal sections through the thalamus of of another monkey in which the same procedure was carried out. Bars = 1 mm. Modified from Asanuma *et al.* (1983b).

medial lemniscus (Fig. 7.12). In the VPL, they appear to terminate with the other lemniscal fibers mainly in the deep shell region in primates (Boivie, 1980) and in an apparently equivalent region in cats (Landgren *et al.*, 1965) and possibly in rats (Lund and Webster, 1967a). In the cat, electrical stimulation of appropriate fiber pathways demonstrates the convergence of cervicothalamic and dorsal column–lemniscal afferents upon single, antidromically identified thalamocortical relay cells (Andersen *et al.*, 1966).

The lateral cervical nucleus of the spinal cord is now thought to be represented in most mammals (Rexed and Brodal, 1951; Mizuno *et al.*, 1967; Truex *et al.*, 1970), though it has been best studied in the cat. The principal afferents to the nucleus are second-order fibers arising from cells mainly at the neck of the dorsal horn in all segments of the spinal cord (Brodal and Rexed, 1953; Bryan *et al.*, 1973, 1974; Craig, 1978). These axons are rapidly conducting and ascend in the dorsolateral funiculus (Brodal and Rexed, 1953; Eccles *et al.*, 1960; Bryan *et al.*, 1974; Brown *et al.*, 1976). Probably most cells of the lateral cervical nucleus project, via the medial lemniscus, to the contralateral VPL (Morin and Catalano, 1955; Grant *et al.*, 1968; Boivie, 1970; Trevino, 1976; Blomquist *et al.*,

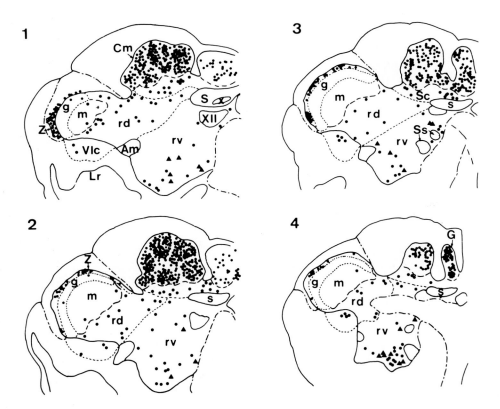

FIGURE 7.11.   Retrograde labeling of cells in the dorsal column (Cm, G) and spinal trigeminal nuclei after injection of horseradish peroxidase in the contralateral ventral posterior nuclei of a cat. Note that labeled cells in caudal nucleus of spinal trigeminal complex are mainly in marginal zone (Z) and in region deep to magnocellular zone (m). From Shigenaga *et al.* (1982).

1978; Craig and Burton, 1979), though a small population of putative inter-neurons has also been reported (Craig and Tapper, 1978) and 1–5% of the neurons have been said to project to the ipsilateral thalamus (Craig and Burton, 1979). In the VPL, cervicothalamic terminations are said, from physiological studies, to be primarily in its anterodorsal "deep" shell (Landgren *et al.*, 1965; Andersson *et al.*, 1966; Boivie, 1980), though anatomical studies suggest more

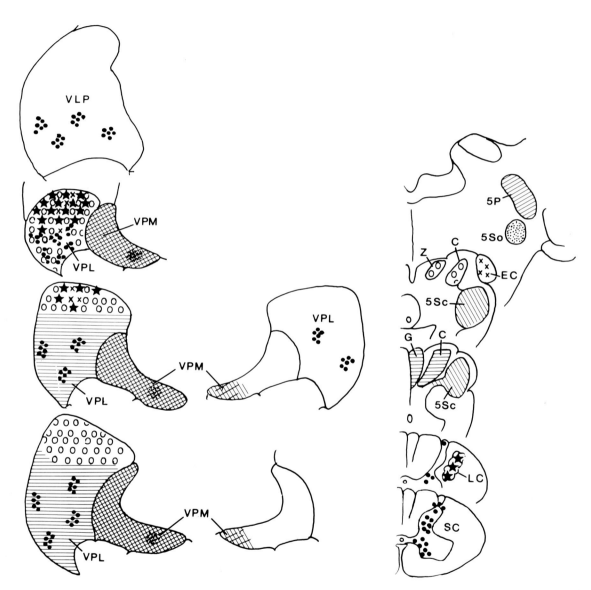

FIGURE 7.12. Semischematic diagram illustrating, on frontal sections, afferent pathways converging on ventral posterior nuclei and their regions of termination in monkeys. 5P: principal trigeminal nucleus; 5So: oral nucleus of spinal trigeminal complex; 5Sc: caudal nucleus of spinal trigeminal complex; LC: lateral cervical nucleus; SC: spinal cord; C: cuneate nucleus; G: gracile nucleus; Z: nucleus Z; EC: external cuneate nucleus.

widespread, though somatotopically organized terminations (Craig and Burton, 1979). This may reflect the body topography demonstrable in the lateral cervical nucleus itself.

The receptive fields of lateral cervical nucleus cells in the cat are probably representative of those projected onto the thalamus. Nearly all are cutaneous. Most (ca. 75%) have small receptive fields and respond to movement of hairs on the ipsilateral body surface but others receive inputs from high-threshold mechanoreceptors and may respond primarily to stimuli that approach a noxious level (Craig and Tapper, 1978). So far as they have been studied, the receptive fields of thalamic cells receiving spinocervicothalamic inputs resemble those of the low-threshold hair units in the nucleus [though a proportion may show convergence of dorsal column–lemniscal afferents as well (Andersen *et al.*, 1966)]. Muscle and joint afferents do not appear to be represented in the lateral cervical nucleus, and it seems that cells in the "deep shell" of the VPL that receive inputs from group I muscle afferents in the cat are not influenced by inputs from the spinocervicolemniscal system (Andersson *et al.*, 1966).

7.2.4.2. Spinothalamic Inputs

Spinothalamic afferents from the contralateral side of the spinal cord also terminate throughout the VPL (Getz, 1952; Kerr and Lippman, 1974; Boivie, 1979; Berkley, 1980; Asanuma *et al.*, 1983b; Burton and Craig, 1983; Mantyh, 1983). A presumably comparable set of fibers arising in the nucleus caudalis of the spinal trigeminal complex terminates in the VPM (Burton *et al.*, 1979). Spinothalamic axons were shown in one of the earliest studies conducted with the Nauta technique to have the same overall distribution as lemniscal fibers in the ventral posterior nucleus of monkeys (Mehler *et al.*, 1960). Then, in rats it was reported that spinothalamic axons terminated more anteriorly in the nucleus than those of the medial and trigeminal lemnisci (Lund and Webster, 1967a,b). In the cat, spinothalamic fibers were reported to terminate outside the ventral posterior nucleus altogether, in a part of the ventral lateral complex (Boivie, 1971b; Jones and Burton, 1974; Berkley, 1980). Earlier reports of degeneration in the ventral posterior nucleus after spinothalamic tractotomies in the cat were dismissed as being based upon involvement of the lateral cervical nucleus by cordotomies in the upper cervical segments. Reinvestigation of the spinothalamic system in all three species now reveals a far greater degree of uniformity than these earlier reports suggested (Fig. 7.13). Extensive overlap of both lemniscal and spinothalamic terminations is found in the ventral posterior nucleus, though the terminations of the spinothalamic system are very much heavier in monkeys than in the other species. In cats, for example, the spinothalamic terminations in the VPL appear confined to a few clusters in its ventral half (Burton and Craig, 1983). However, whereas lemniscal and cervicothalamic terminations are confined to the ventral posterior nucleus, spinothalamic terminations extend forwards into the ventral part of the adjoining ventral lateral complex in cats and monkeys (Boivie, 1979; Applebaum *et al.*, 1979; Tracey *et al.*, 1980; Berkley, 1980; Craig and Burton, 1981; Asanuma *et al.*, 1983b; Burton and Craig, 1983).

Early studies on the cells of origin of the spinothalamic tract did not specifically identify those projecting to the thalamus, for they identified the cells

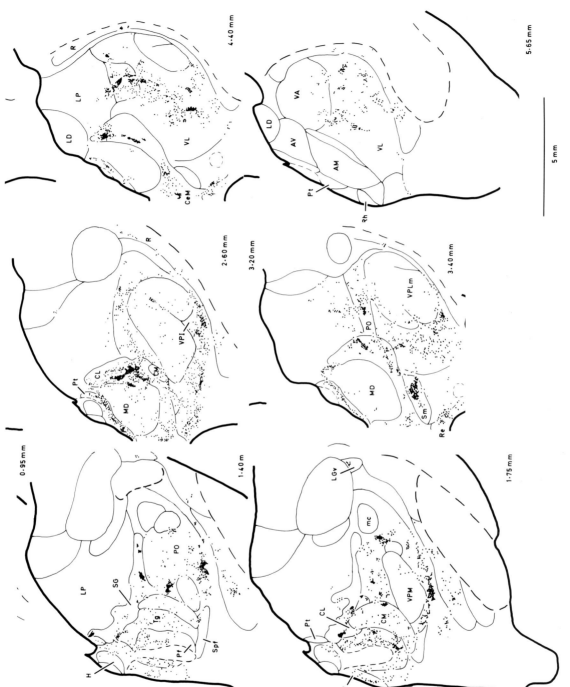

FIGURE 7.13. Distribution of spinothalamic fibers (dots) in the thalamus of the cat. Fibers anterogradely labeled by horseradish peroxidase injected in spinal cord. From Burton and Craig (1983).

by their chromatolytic reaction secondary to cutting the tract at higher spinal levels. Cells were reported in the ventral horn, in the base of the dorsal horn, and in the marginal zone (Foerster and Gagel, 1932; Cooper and Sherrington, 1940; Kuru, 1949; Morin *et al.*, 1951). Studies of spinothalamic tract cells antidromically activated from stimulation sites in the diencephalon found that most of the cells lay in the intermediate zone and ventral horn of cats (Dilly *et al.*, 1968; Trevino *et al.*, 1972) and throughout the dorsal horn, ventral horn, and intermediate zone of monkeys (Trevino *et al.*, 1973). The advent of cell labeling by retrograde axoplasmic transport enabled a far greater population of projecting cells to be reliably labeled from the thalamus. In cats, rats, and monkeys, most workers using retrograde labeling with horseradish peroxidase have reported spinothalamic tract cells projecting to the thalamus from all of the laminae of Rexed in the spinal cord, including the marginal zone and substantia gelatinosa but with particular concentrations in certain laminae (Fig. 7.14). Moreover, cells projecting to the VPL and to other parts of the thalamus are concentrated in different laminae and there are some species differences. In cats, most cells projecting to the VPL are situated in the deeper laminae (VI–VIII) of the ventral horn while those projecting to the intralaminar nuclei are found there and in the marginal layer (lamina I) (Trevino and Carstens, 1975; Carstens and Trevino, 1978). In monkeys the largest number of cells projecting to the VPL are in the marginal zone (lamina I) and in the neck of the dorsal horn (lamina V). More than 90% of these cells project to the contralateral VPL, except in the most caudal segments of the spinal cord where the number projecting ipsilaterally is higher. Cells projecting to medial parts of the monkey thalamus mostly occupy the intermediate zone and ventral horn (laminae VI–VIII) (Willis *et al.*, 1978, 1979). Some of these have branched axons to the VPL as well (Giesler *et al.*, 1981). In rats, spinothalamic cells projecting to lateral parts of the thalamus (including the VPL) are found in all laminae and their axons ascend in the ventrolateral funiculus whereas those projecting to medial parts of the thalamus arise mainly in the intermediate zone and ventral horn and their axons ascend in the ventral funiculus (Giesler *et al.*, 1979, 1981).

The cell types contributing to the spinothalamic tract have a variety of morphologies in retrogradely labeled material, and even in a single lamina of the spinal cord clearly include several of the classical anatomical types (Willis *et al.*, 1979). Few of these have been specifically related to the different receptive field types known to contribute to the spinothalamic tract.

In the spinal cord, spinothalamic tract cells responding to different types of peripheral stimulation tend to be concentrated in different laminae (Willis *et al.*, 1974, 1975; Applebaum *et al.*, 1975; Price *et al.*, 1976; Chung *et al.*, 1979). High-threshold neurons, i.e., neurons responding specifically to noxious mechanical and thermal stimuli, are especially concentrated in layer I (Christensen and Perl, 1970; Willis *et al.*, 1974; Price *et al.*, 1978; Kumazawa and Perl, 1978). Wide-dynamic-range neurons, i.e., neurons receiving inputs from sensitive mechanoreceptors and nociceptors, are concentrated at the neck of the dorsal horn (Wall, 1967; Willis *et al.*, 1974; Menétrey *et al.*, 1977). Low-threshold neurons, i.e., cells excited by low-threshold mechanoreceptors only, are found in all layers. Deep neurons, i.e., neurons receiving inputs from muscles, joints, and other subcutaneous tissues, are found in the deeper parts of the ventral horn (Wall,

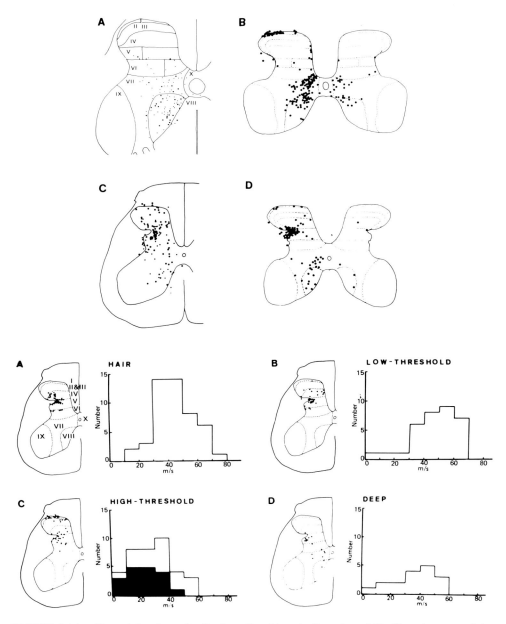

FIGURE 7.14. (Upper) Laminar distribution of antidromically activated (A, C) and retrogradely labeled (B, D) spinothalamic cells in the lumbosacral enlargement of the spinal cord of the cat (A, B) and monkey (C, D). From Willis and Coggeshall (1978). After Trevino *et al.* (1972, 1973) and Trevino and Carstens (1975). (Lower) Laminar distribution of identified spinothalamic tract cells with different types of receptive field and different latencies to antidromic stimulation in the spinal cord of the monkey. Many of the hair-movement and low-threshold cells had a wide dynamic range of responsiveness. From Willis and Coggeshall (1978).

1967; Pomeranz *et al.*, 1968; Willis *et al.*, 1974). Hence, from the distributions of the retrogradely labeled cells after injections of tracer in lateral or medial thalamus, one might expect the VPL to receive inputs primarily from wide dynamic range cells and low-threshold cells with fewer or no inputs from high-threshold and deep cells.

The receptive fields of spinothalamic tract cells projecting to the VPL in monkeys have been studied by Willis and his colleagues (Applebaum *et al.*, 1979; Kenshalo *et al.*, 1979; Giesler *et al.*, 1981; Gerhardt *et al.*, 1981). Forty-four cells in the lumbar segments of the spinal cord whose peripheral receptive fields were examined, were shown by antidromic activation to project to the VPL. Twenty of the cells had additional branches to the intralaminar nuclei but their characteristics were identical. Each of the cells had a localized excitatory receptive field on the side of the body ipsilateral to their location in the spinal cord (contralateral to the VPL). Electrical volleys in peripheral nerves indicated convergence of inputs to them from both cutaneous A and C fibers. The receptive fields varied in size but even the smallest were somewhat larger than those usually regarded as lemniscal. The smallest was restricted to the whole foot and the largest to the whole hindlimb. A few had even larger, discontinuous excitatory fields with contralateral components and all could be inhibited by stimuli applied to other parts of the body. The principal functional type contributing to the VPL input was the wide-dynamic-range cell, which receives inputs from both sensitive mechanoreceptors and nociceptors; high-threshold (i.e., nociceptor-specific) neurons made a smaller contribution. No low-threshold spinothalamic cells, which respond to sensitive mechanoreceptors, were found to project to the VPL and only one spinothalamic tract cell receiving inputs from deep tissues could be shown to do so.

The spinal trigeminal complex also projects to the VPM (Lund and Webster, 1967b; Tiwari and King, 1974; Ganchrow, 1978; Burton *et al.*, 1979; Hu *et al.*, 1981) and cells comparable in some ways to spinothalamic tract cells have been retrogradely labeled in the caudal subnucleus of the complex following injections of tracer in the contralateral VPM. However, in the cat at least, these are confined to the marginal layer and to the cells of the reticular formation just deep to the magnocellular layer (Hockfield and Gobel, 1978; Fukushima and Kerr, 1979; Burton *et al.*, 1979; Shigenaga *et al.*, 1983). A similar conclusion was reached in a study involving antidromic stimulation of trigeminothalamic fibers (Price *et al.*, 1976). Anterograde labeling of axons leaving the caudal nucleus of the cat and monkey shows that those projecting to the thalamus terminate throughout most of the VPM contralaterally and in the most medial part of the VPM ipsilaterally (Ganchrow, 1978; Burton *et al.*, 1979). This is the region in which neurons can be excited specifically by cooling of the tongue (Section 7.2.5.7). Furthermore, the marginal cells of the spinal trigeminal complex which give rise to the thalamic projection include cells that respond to innocuous cutaneous thermal stimuli or to noxious thermal and mechanical stimuli (Price *et al.*, 1976, 1979; Poulos *et al.*, 1979; Hu *et al.*, 1981). Such cells tend to have small receptive fields. Wide-dynamic-range neurons, with large receptive fields akin to those in the dorsal horn of the spinal cord, are found only in the magnocellular layer and, therefore, unlike those in the spinal cord, may not project to the thalamus.

A few trigeminothalamic cells are also demonstrated by retrograde labeling in the subnuclei oralis and interpolaris of the spinal trigeminal complex in cats (Burton and Craig, 1979). The functional significance of these cells is uncertain. The major projections of these two subnuclei are to the cerebellum (Cheek *et al.*, 1975).

7.2.4.3. Other Afferents

Electrical stimulation of splanchnic nerves leads to activation of neurons mainly in the trunk representation of the VPL (Patton and Amassian, 1951; McLeod, 1958). These afferents are in both the Aβ and the Aγ–δ ranges, and are thought to traverse both the dorsal and the ventrolateral columns of the spinal cord (see Section 7.2.4.2). Visceral afferents in the vagus appear to relay in the ventral medial nucleus (Section 7.6.3.4). Apart from the reticular nucleus of the thalamus (Jones, 1975c), and the cerebral cortical areas to which the VPL and VPM project (Jones and Powell, 1969a, 1970a), no other sources of afferents to the ventral posterior nucleus have been identified.

7.2.5. Receptive Fields of Ventral Posterior Neurons

Many of the data pertaining to the functional properties of ventral posterior neurons have been reviewed in the chapter on "integrative action in the thalamus." The concept of their place and modality specificity has been alluded to in the preceding section and the way in which these two, "lemniscal" characteristics are mapped in the VPL and VPM described in detail.

Poggio and Mountcastle's (1960, 1963) original definition of a lemniscal neuron depended upon certain properties that they called static and dynamic. The static properties included modality and place specificity: (1) Modality specificity was a rather broad qualitative measure based upon whether the neuron discharged in response to mechanical stimulation of a particular part of the body. Hence, skin units responded to movement of hairs and/or light stroking of the skin (42% of the ventral posterior neurons encountered in their study in unanesthetized monkeys). Deep units responded to indirect or direct stimulation of fasciae and periostea (32%) or to movement of joints (26%). (2) Place specificity meant that the neurons are "invariably related to circumscribed, continuous receptive fields on the contralateral side of the body." According to the authors, field size and shape do not change with time nor with experimental conditions such as end tidal CO_2 levels, levels of consciousness, or anesthetic state. Cutaneous fields, when measured in monkeys, varied from an average of 0.2 cm^2 on the fingers and toes to 20 cm^2 on proximal parts of the limbs and trunk. Baker (1971), working in the unparalyzed, unanesthetized cat, however, found that receptive fields of VPL neurons were overall much larger and only shrunk to these dimensions under the influence of barbiturates. This discrepancy has still not been resolved. Joint neurons, as studied by Poggio and Mountcastle, also had highly specific receptive fields, being excited by movement in one direction and with the greatest sensitivity occurring at a particular steady-state joint position (see Chapter 4).

The dynamic properties of lemniscal neurons included: (1) the recovery of excitability to a preexisting level following a stimulus, at a rate close to that of first-order afferents over the early phases of the recovery cycle but followed by a period of inhibition. The recovery time is strongly influenced by anesthetic state, and can be more than doubled in deeply narcotized animals. (2) The ability to follow repetitive stimuli applied to the receptive field at frequencies of 100–120 Hz. Beyond this frequency, the cells start to fail to respond on a one-to-one basis and even show a reduction in the peak response at 100 Hz. Ventrally placed VPL neurons with receptive fields on the digits of cats may be able to follow high-frequency stimuli more securely than dorsally placed neurons, with more proximal receptive fields (Iwamura and Inubushi, 1974b).

The work of Mountcastle and his colleagues on the ventral posterior nucleus had the obvious implication that different ventral posterior neurons should receive inputs from different types of peripheral somesthetic receptors. However, they made no attempt to fractionate their classification beyond the broad categories of "cutaneous" and "deep." Nor did they at that time make an attempt to classify ventral posterior neurons in terms of their discharge rates, adaptation rates, and so on. These are distinctions that have been attempted in more recent years (see Golovchinsky *et al.*, 1981).

The Mountcastle group also felt that it was the lemniscal properties that overwhelmingly dominate ventral posterior neurons (Poggio and Mountcastle, 1960, 1963). Though accepting that the spinothalamic tract terminated in the nucleus, they could find no neurons in it with receptive fields that they regarded as characteristic of the spinothalamic system. Hence, they felt that either spinothalamic activity was suppressed, even in their unanesthetized preparations, or that ventral posterior cells do not express typical spinothalamic properties. In commenting upon the observations of Perl and Whitlock (1961; Whitlock and Perl, 1959, 1961), to the effect that a significant number of VPL neurons could be activated by peripheral stimulation after section of the dorsal and lateral columns of the spinal cord in cats and monkeys, Poggio and Mountcastle noted that 69 of the 75 neurons reported had what they would regard as lemniscal properties. It has to be pointed out, however, that the remaining 6 did have large receptive fields and discharged in response to noxious stimuli.

Newer data, some of them already reviewed in Section 7.2.4.1, enable us to say that among the place- and modality-specific neurons of the lemniscal system in the ventral posterior nucleus, there are several types that show varying degrees of specificity of input from different peripheral receptors, though convergence is found. Differences in rates of adaptation of discharges to maintained stimuli are also coming to be emphasized. Moreover, cells with receptive fields more typical of the spinothalamic system are starting to be identified. The following is a broad-based classification derived from many sources and, consequently, may show some overlap between groups.

7.2.5.1. Neurons Receiving Group I Muscle Afferents

Neurons receiving group I muscle afferent inputs have receptive fields in one or more of a closely related group of muscles (e.g., flexors or extensors of the forearm but not both) and though their responses are on the whole faithful reflections of the dynamic properties of muscle spindle primary afferents, they clearly also receive cutaneous inputs. Inputs from spindle secondaries and Pacinian corpuscles also cannot be entirely ruled out (Andersson *et al.*, 1966; Maendly *et al.*, 1981; Chapter 4).

7.2.5.2. Neurons Responding to Movement of Joints

Neurons responding to movement of joints are difficult to evaluate on account of the different points of view adopted in the two major studies devoted to them (Chapter 4; Mountcastle *et al.*, 1963; Yin and Williams, 1976). Currently, one would have to say that two types may be present, each type receiving input during movement of one, or at most two, closely related joints; one, however, has tonic sensitivity and narrow excitatory joint angles, the other phasic sensitivity and is active over a wide range of joint angles. The nature of the peripheral receptors providing input to these two kinds of joint neurons remains to be determined.

7.2.5.3. Pressure Neurons and Neurons with Subcutaneous Receptive Fields

Pressure-sensitive units are obviously difficult to classify for, depending on the degree of pressure, anything from receptors in the dermis to those in muscles, joints, and interosseous membranes may be stimulated. Hence, units classified as "deep" may well include any of these classes. However, many workers would recognize as entities in the ventral posterior nucleus, typical "place and modality"-specific units with small receptive fields responding to rather light compression of the skin but silent when skin or hairs over the receptive field are lightly stroked. Most workers would also distinguish between these and units with deeper ("subcutaneous") receptive fields (Gordon and Manson, 1967; Tsumoto, 1974), the deepest of which are probably associated with Pacinian corpuscles (Dykes *et al.*, 1981).

VPL units with slowly adapting responses to a maintained light-pressure stimulus are found and these have receptive fields beneath either hairy or glabrous skin (Gordon and Manson, 1967; Tsumoto, 1974). As the depth of the receptive field increases, the units are innervated by lemniscal fibers with slower conduction velocities (Tsumoto, 1974). Slowly adapting VPL responses to light-pressure stimulation of glabrous skin are presumably mediated by inputs from Merkel cell receptors (e.g., Jänig *et al.*, 1968; Iggo and Ogawa, 1977; Ferrington and Rowe, 1980).

Rapidly adapting, light-pressure-sensitive units are also described in the VPL (Gordon and Manson, 1967). They probably receive their input from low-frequency vibration receptors of the Meissner type in glabrous skin (Talbot *et al.*, 1968). VPL neurons responding to Meissner corpuscle stimulation would be expected to respond in a phase-locked manner to stimuli of moderate amplitude at 20–50 Hz, because both the primary afferents and their cortical counterparts are capable of doing so (Mountcastle *et al.*, 1969). However, this feature has not been specifically examined in the thalamus.

Other rapidly adapting responses might be expected from VPL neurons receiving inputs from Pacinian corpuscles but this has not been studied except in the crudest manner. Such units, from their peripheral nerve counterparts, would be expected to be exquisitely sensitive to peripheral vibratory stimuli of very small amplitude at frequencies from 100 to 1000 Hz. Their cortical counterparts, however, are not capable of systematically following stimuli above 100 Hz in SI or above 300 Hz in SII (McIntyre, 1962; Mountcastle *et al.*, 1969; Fisher *et al.*, 1983). Therefore, the response in the VPL cannot be predicted and clearly needs to be investigated. Dykes *et al.* (1981) described multiunit responses to tapping of the wrist in the VPI of squirrel monkeys and attributed these to Pacinian inputs. Fisher *et al.* (1983) in the cat give indirect reasons for regarding Pacinian-sensitive cells as forming a large population in the VPL itself.

A further type of unit that may be classified in this section is that responding to movements of the claws in cats (Gordon and Manson, 1967; Tsumoto, 1974). These units can have particularly small receptive fields confined to a single claw but in some the receptive field can extend onto the adjacent glabrous skin pad as well, so that the classification of the unit as deep or subcutaneous is perhaps arguable. Claw units form from 4 to 9% of the units reported in studies of the VPL. Their discharges are slowly adapting.

In the VPM many units respond to light tapping of the teeth, usually with rapidly adapting discharges. The receptive fields of these "deep" units are quite small and can be confined to a single tooth, though receptive fields extending to two or three teeth are not uncommon (Poggio and Mountcastle, 1963; Bombardieri *et al.*, 1975). The teeth of the two jaws are represented systematically in different parts of the VPM. Presumably a mixed constellation of receptors in and around a tooth (Lisney, 1978) provides the input to the VPM neurons.

7.2.5.4. Cutaneous Neurons

There have been various attempts to classify cutaneous units in the VPL and VPM. It is clear that they form a major proportion of the responding neurons in the ventral posterior nucleus. There may be a fundamental division between units responding to hair movement and those responding only to light stroking of the skin but obviously, the latter is only readily studied in nonhairy skin, so the division in animal experiments is basically one between glabrous and hairy skin units. In the cat VPL, some units responding to light stroking of the glabrous foot pads may actually also extend their receptive field onto adjacent hairy skin so that deflection of adjacent hairs may also excite them (Gordon and Manson, 1967; Golovchinsky *et al.*, 1981). As usually described, each of the units with cutaneous receptive fields conforms to the standard lemniscal criterion, for their receptive fields are localized, commonly small, and always contralateral. Again typical of lemniscal units, the receptive fields of units related to the distal aspects of the limbs, are smaller than those on more proximal parts of the limbs and on the trunk (Poggio and Mountcastle, 1960, 1963). Tsumoto (1974) has reported that in the cat VPL, hair units with larger receptive fields on the trunk are innervated by more rapidly conducting lemniscal axons than those with smaller receptive fields on the limbs. For receptive fields on the forepaw digits and upper arm, the average conduction velocities reported were 22.9 and 37.2 m/sec, respectively, and for those with receptive fields on the trunk 53.1 m/sec, though the standard deviations were large. There was also some similar correlation with the conduction velocities of the VPL cells' thalamocortical axons, as judged by antidromic stimulation, though again the range of variability was large (11.8–57.1 m/sec).

Units in the VPL responding to deflection of the common body hairs are probably in the majority (Gordon and Manson, 1967; Golovchinsky *et al.*, 1981). They usually have rapidly adapting discharges (Fig. 7.15) and no direction selectivity. Light-touch units seem to be described as having either slowly or rapidly adapting discharges and no direction selectivity either but it is usually difficult to decide whether workers are describing a unique class of units or simply those referred to as light-pressure units in the preceding section. Until more comprehensive studies are made, Mountcastle's (1982) policy of referring to all such units as light-touch–pressure units may be desirable.

In the VPM, units receiving inputs from the mucous membranes of the lips, tongue, and mouth, in general behave as the light-touch units of glabrous skin. They have small localized receptive fields and usually rapidly adapting discharges. On the lips, nose, tongue, palate, and pharynx, close to the midline the receptive fields, though still small, can extend from the contralateral onto the ipsilateral side (Poggio and Mountcastle, 1963). There is some question, however, as to whether this is a feature of perioral receptive fields in rodents (Vahle-Hinz and Gottschaldt, 1983).

The hairs of the face are well represented in the VPM and particular attention has been paid to the representation of the large sinus hairs on the face which have a complex innervation (Andres, 1966) and play a particularly important role in the behavior of rodents and many other species. As mentioned in Section 7.2.3, the vibrissae and other sinus hairs of rodents are represented by individual aggregates of cells in the VPM. The receptive fields of the units within these aggregates have been studied by Waite (1973a,b), Verley and Onnen (1981), and Vahle-Hinz and Gottschaldt (1983). In the VPM of rodents, at least 75% of the neurons responding to sinus hair stimulation receive input from only one sinus hair. The units respond to bending of the relevant sinus hair with a tonic (i.e., slowly adapting) discharge. The frequency of discharge was thought by Waite (1973a) to be independent of the amplitude of hair bending, but Vahle-Hinz and Gottschaldt (1983) in a reexamination discovered that the frequency of discharge of tonic neurons could commonly covary with the degree of bending. Other neurons display phasic responses to hair bending and some such neurons can follow vibratory stimuli applied to the sinus hairs quite systematically and often on a one-to-one basis even up to frequencies of 700 Hz. However, many of the tonically discharging neurons can also show added phasic responses to a high-frequency vibratory stimulus superimposed upon the sustained bending of a hair. Hence, the distinction between tonic and phasic types is not completely clear and may reflect the experimental situation. At least 75% of the whisker-related neurons in the rat VPM display direction selectivity, i.e., they are particularly sensitive to movements of the whisker in one quadrant of the potential 360° through which the whisker can be moved (Waite, 1973b).

In the VPM of cats 70% of the neurons responding to sinus hair stimulation usually receive input from at least two and commonly from many more sinus hairs (Vahle-Hinz and Gottschaldt, 1983). Though their receptive fields are clearly larger than in rodents, their discharge properties are said to be rather similar, and both tonic and phasic types, some sensitive to vibratory stimuli, are present and most display some direction selectivity. The receptive fields of neurons responding to movement of the common hairs on the face and on the furry buccal pad near the mouth appear identical to those of hair neurons in the VPL.

In the VPM of the unanesthetized monkey, Hayward (1975) described differences in the distribution and in the discharge patterns of units responding to movements of the three different kinds of facial hairs. Neurons responding to movement of common hairs were situated in the dorsal half of the VPM, were rapidly adapting, and had slightly larger receptive fields than units responding to movements of circumoral vibrissae or of long facial whiskers which were situated in the ventral half of the VPM. Circumoral vibrissal units had slowly or rapidly adapting discharges and facial whisker units slowly adapting discharges, with direction selectivity.

7.2.5.5. Widefield Neurons

Most cutaneous neurons in the ventral posterior nucleus would fit into the category of place- and modality-specific neurons whose properties largely reflect those of the medial lemniscal system that seems to drive them (Rose and Mountcastle, 1959; Poggio and Mountcastle, 1960, 1963). There are, however, numerous reports scattered throughout the literature of particularly widefield ventral posterior neurons with receptive fields covering much of the body surface, both ipsi- and contralaterally and sometimes with discontinuous components. Such neurons have been described mainly in animals anesthetized with chloralose (Harris, 1970; Jabbur *et al.*, 1972; Berkley, 1973). Baker (1971), though finding relatively large cutaneous receptive fields, noted that all were contralateral and none discontinuous. Whether these represent the same type of neuron has not been ascertained.

Harris (1978a,b) made a population analysis of neurons responding to electrical stimulation of the contralateral forepaw in the VPL of chloralose-anesthetized cats. He reported that "early activity" (i.e., the earliest component of the evoked field potential occurring in the first 5 msec after stimulation) was due to activation of modality-specific neurons with small receptive fields confined to the contralateral forepaw. Later components of the potential, however (over the next 10–15 msec), he considered to be due to activation of "widefield neurons" with large bilateral receptive fields and modality convergence. His results thus suggest early, place- and modality-specific input via the medial lemniscus that respects the topographic representation in the nucleus and a later-arriving, bilateral input of uncertain origin that transcends the usually demonstrated somatotopy. One possibility is that this late bilateral input is mediated by the spinothalamic system, though Harris suggests that the dorsal column–lemniscal system itself may be involved. It is possible that widefield neurons are not thalamocortical relay cells, for Tsumoto and Nakamura (1974) were unable to antidromically activate such cells by stimulation of the cat sensory motor cortex.

7.2.5.6. Spinothalamic Tract Neurons

As reviewed in Section 7.2.4.2, there is ample evidence for the termination of spinothalamic tract axons in the VPL (Mehler *et al.*, 1960; Perl and Whitlock,1961; Mehler, 1969; Kerr and Lippman, 1974; Kerr, 1975; Carstens and Trevino, 1978; Willis *et al.*, 1979; Applebaum *et al.*, 1979; Giesler *et al.*, 1981; Burton and Craig, 1983). However, the evidence for a significant population of neurons in the VPL possessing receptive fields that could be attributed to the spinothalamic input has until recently been singularly small. Even in monkeys and cats in which all spinal pathways other than a single ventral quadrant were destroyed, Perl and Whitlock (1961) found that the majority of the receptive fields mapped for VPL neurons were small and contralateral and thus "lemniscal" in type. They did, however, note a small percentage of neurons with large fields even extending onto the ipsilateral side of the body. Such neurons commonly responded to stimuli that could be regarded as noxious in type and similar neurons have now been reported in unanesthetized monkeys by Pollin and Albe-Fessard (1979) and in anesthetized rats by Guilbaud *et al.* (1980). In cats, Gaze and Gordon (1954) also reported the presence of a small percentage of VPL neuronal responses from near-noxious stimulation of the ipsi- as well as of the contralateral side of the body and from stimulation of Aδ and even of C fibers in the ipsi- as well as the contralateral saphenous nerve. Guilbaud *et al.* also reported Aδ and C fiber inputs to the neurons they studied, from the sural nerve. Yokota and Matsumoto (1983a,b) report neurons responding to noxious stimuli in the VPM of the cat.

Studies by Honda *et al.* (1983) and Kniffki and Mizumura (1983) in the cat confirm all these observations, though, again, the population of neurons displaying the relevant properties is relatively small. Honda *et al.* dealt mainly with cells responding specifically to noxious stimulation of the skin. Kniffki and Mizumura dealt mainly with cells responding to noxious stimulation of muscle and tendon but their cells had additional cutaneous receptive fields from which most could be discharged by innocuous stimuli. In both cases the cells were driven by small myelinated afferents in the Aδ range. No additional C-fiber inputs were demonstrated. The responding cells in all these studies we may tentatively assume are "spinothalamic" in character but the reason for their seemingly low numbers still remains something of an enigma. A possible clue is furnished by the evidence from Willis and his colleagues about the receptive fields of cells in the spinal cord of monkeys that project their axons to the contralateral VPL (Applebaum *et al.*, 1979; Giesler *et al.*, 1981; Gerhardt *et al.*, 1981). They classified a sample of 44 projecting neurons into the types well known to occur in the spinothalamic system at spinal cord levels from past studies (Section 7.2.4.2). None were of the low-threshold type, i.e., of a type activated by stimulus intensities typical for lemniscal neurons; 16 were wide-dynamic-range neurons, i.e., normally excited both by sensitive mechanoreceptors and by nociceptors; 7 were high-threshold neurons, i.e., activated specifically by noxious heat or pinching the skin; 1 was a "deep" neuron responding only to joint movement and stimulation of muscles. The receptive fields of all these neurons were ipsilateral (contralateral to the VPL) and though of varying size, were generally larger than those of typical "lemniscal" neurons. All the projecting cells were shown to receive peripheral nerve inputs in both A and C ranges.

Casey and Morrow (1983) have also reported on nine wide-dynamic-range cells in the VPL of the awake squirrel monkey. These all had small receptive fields on the contralateral side of the body. Kenshalo *et al.* (1980) found that 73 of a sample of several thousand cells in the VPL of anethetized monkeys were excited by noxious mechanical stimuli or by noxious heat pulses at 43–50°C applied to small contralateral receptive fields. Ninety percent of the cells could be antidromically activated by stimulation of the SI cortex. The neurons appeared in an appropriate topographic sequence scattered among those responding to innocuous mechanical stimuli. The stimulus–response functions of these neurons were similar to those of spinothalamic tract neurons and both high-threshold nociceptive and wide-dynamic-range types could be identified (Fig. 7.14). Like spinothalamic tract neurons, an ascending series of noxious heat pulses leads to a burst of impulses which, though declining somewhat, are maintained at a relatively high rate for the duration of the stimulus. Lesions of the dorsolateral funiculus of the spinal cord did not affect the responses to noxious heat but lesions of the ipsilateral anterolateral quadrant did. This seems to provide conclusive evidence for a significant population of thalamic cells with spinothalamic-type receptive fields of both the high threshold and the wide dynamic range type. The extent to which the other spinothalamic inputs are reflected in the responses of ventral posterior neurons and the degree of convergence between spinothalamic and medial lemniscal inputs remain to be determined.

7.2.5.7. Thermal-Sensitive Neurons

The earliest studies on thalamic neurons responding to innocuous peripheral temperature changes were made in the tongue representation of the VPM

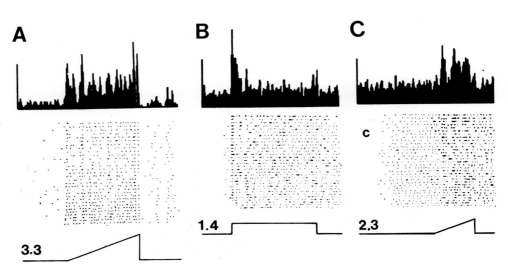

FIGURE 7.15.   (A) Rapidly adapting response of a neuron in the VPL nucleus of a cat to controlled mechanical stimulation at 3.3 mm/sec of guard hairs. (B, C) Slowly adapting responses of another neuron in VPL of a cat to a rectangular stimulus of 1.4 mm deflection (B), and to a ramp stimulus at 2.3 mm/sec (C). Upper figures are poststimulus time histograms summating responses to 32 stimuli shown individually in the rasters (middle). From Golovchinsky *et al.* (1981).

nucleus of the cat (Landgren, 1960; Emmers, 1966), rat (Benjamin, 1963), and monkey (Poulos and Benjamin, 1968). Here, some neurons were found to be specifically sensitive to thermal stimuli, though increasing their discharges only to cooling, not to warming of the tongue. The rate of discharge of such units increased during rapid cooling from any preadapted tongue temperature from 45 to 25°C. The lowest static discharge rates occurred at adapted temperatures above 45°C or below 15°C. Discharges were maximal between these, though each cell usually had a "best temperature" between 31 and 21°C and a function of the prevailing constant temperature. The cells, though coming in an appropriate topographic sequence in the tongue representation, did not respond to mechanical stimuli. Other tongue-related units, however, responded both to innocuous mechanical stimulation and to cooling, especially from relatively cool preadapted temperatures. In the early studies, both kinds of thermal-sensitive neurons in the VPM were found to have small (1–2 mm²) receptive fields that were usually contralateral, though some could extend across the midline. More recently, Auen *et al.* (1980) have restudied the two thermal-sensitive neuron types in the medial part of the cat VPM and have reported receptive fields of small and large size and mostly *ipsilaterally* situated. The reason for the discrepancy is not clear.

Neurons responding to cooling the glabrous skin of the contralateral hand and foot were discovered in the VPL of squirrel monkeys by Burton *et al.* (1970). But all cells were activated by innocuous mechanical stimulation as well as by cooling. No cells responded to warming the skin. All the cells with mixed thermal and mechanical responses had receptive fields, under anesthesia, of a small size comparable to lemniscal neurons.

In the rat VPL, many cells respond to changes in scrotal skin temperature. Some show convergence of thermoreceptor and mechanoreceptor inputs but many more are thermosensitive only and these have large, bilateral receptive fields (Hellon and Misra, 1973; Hellon and Mitchell, 1975; Schingnitz, 1981). Such cells change their firing rate in response to temperature changes over a relatively narrow range (1–2°C) with a threshold for a warm response at 33–40°C. The discharge rate appears to be proportional to the arithmetic sum of the differences in temperatures applied to the ipsi- and contralateral sides of the receptive field (Hellon and Mitchell, 1975). Jahns (1975) described this type of cell in about one-third of his sample of cells from the rat VPL indicating that their firing frequency, though falling with a drop in temperature, always maintained an even rate. Two-thirds of the cells in his sample, however, responded to cooling by changing a relatively even spike frequency to a bursting pattern. Such cells seemed to be located in parts of the VPL different from those in which the more typical warm neurons were situated.

7.2.5.8. Other Types

Cells responding to stimulation of vagally innervated visceral receptors and sometimes considered to be in the VPL or VPM are described with the account of the ventral medial complex (Section 7.6). Other visceral afferents traveling in the greater splanchnic nerves terminate in the VPL of the cat and baboon (Patton and Amassian, 1951; McLeod, 1958). The peripheral afferent fibers involved fall into both Aβ and Aγ–δ ranges (McLeod, 1958). The Aβ fibers include those from Pacinian corpuscles and their impulses probably reach the thalamus via the dorsal columns (Amassian, 1951a). The Aγ–δ fibers are probably from a variety of receptors, including visceral nociceptors, and their impulses reach the thalamus via the spinothalamic tracts (Downman and Evans, 1957; Hancock *et al.*, 1975). In a study of single units innervated by visceral afferents in the thalamus, McLeod (1958) found that most were activated by stimulation of the contralateral greater splanchnic nerve but a few were activated by stimulation of the ipsilateral nerve. Most of the units were confined to the trunk representation though a few were reported in the limb and tail representations in the VPL. Those innervated by Aβ fibers appeared distinct from those innervated by Aγ–δ fibers but both classes showed convergence of inputs from contralateral cutaneous receptors, usually on the trunk. It is likely that the convergence in the slower-conducting inputs occurs on the spinothalamic tract cells of origin of these fibers (Hancock *et al.*, 1975).

The cells innervated by the faster-conducting splanchnic afferents project to the trunk region of the SI cortex (Amassian, 1951a,b). The projection of those innervated by the slower-conducting splanchnic afferents is not known.

7.2.5.9. Lateral Inhibition

Poggio and Mountcastle (1963) noted that, like cutaneous lemniscal neurons elsewhere, neurons in the ventral posterior nucleus with cutaneous receptive fields need not be activated most powerfully by stimuli applied to the centers of their peripheral receptive fields but could have a functionally effective "center" eccentrically placed. They also pointed out that although cells are obviously excited by stimuli applied to their peripheral receptive fields, they might also be subjected to concomitant inhibitory influences from the same field. These influences might become obvious as a stimulus moved further away from the functional excitatory center. Despite this realization, Poggio and Mountcastle could demonstrate inhibition of spontaneous discharges by peripheral stimulation in only 5% of the cells that they studied in the VPL and VPM. This has always stood out as a major discrepancy between findings in the dorsal column nuclei and somatic sensory cortex in which lateral inhibition of most neurons is the norm (Mountcastle and Powell, 1959; Gordon and Jukes, 1962; Golovchinsky *et al.*, 1981).

In the cortex, for example, Mountcastle and Powell demonstrated that closely adjacent neurons responding to movement of a joint could show mutual interactions so that when one was excited by movement of the joint, the other was inhibited. In the thalamus, the only examples of this type of mutual inhibition came from the work of Tsumoto (1974). He showed afferent inhibition of certain hair units in the VPL of the cat. Inhibited neurons formed 2% of his total sample. The spontaneous and evoked discharges of such cells were characteristically suppressed by stimulation of hairs in an inhibitory receptive field which was the same as the (excitatory) receptive field of a neighboring VPL hair cell. Similar findings in regard to some VPL cells of the cat were reported by Iwamura and Inubushi (1974a,b). They applied paired, punctate electrical stimuli to various parts of the receptive fields of the cells and measured the effects on sizes of postsynaptic potentials, on the sizes of various components of the field potentials of the cells, as well as levels of poststimulus discharge. They determined that many cells could have both excitatory and inhibitory inputs from the centers of their receptive field. In many cases, the strongest inhibition, though coinciding with excitation, emanated from the receptive field center. In others, the inhibitory component of the receptive field was larger than and concentric with the excitatory center. Inhibitory effects appeared to be greater for neurons with small receptive fields on the digits.

The above observations seem to indicate that the lateral inhibition operating upon VPL neurons is mediated by intrinsic thalamic mechanisms for the inhibitory effects occur at latencies somewhat longer than the inhibitory ones. Though it seems justifiable to think of the effects being mediated by inhibitory thalamic interneurons, such neurons have not been specifically identified.

A slightly different form of afferent inhibition has been demonstrated by Gottschaldt *et al.* (1983a,b) in the VPM of the cat. They found that neurons responding with a slowly adapting discharge to trapezoidal displacements of sinus hairs on the face showed an additional phasic excitatory response to a superadded vibratory stimulus provided it was below 100 Hz. At higher frequencies of vibratory stimulation the cells were inhibited. Gottschaldt *et al.* interpret this to mean that the higher-frequency stimulation activated rapidly adapting, high-velocity-threshold receptors in the sinus hair follicles and that these led to inhibition of the thalamic units. Though we have no information about whether the different kinds of hair receptors have separate access to VPM cells, the effect is reminiscent of the inhibition of X cells in the dorsal lateral geniculate nucleus by Y cells, probably acting through an intermediate inhibitory thalamic interneuron (Chapter 4).

7.2.6. Corticothalamic Connections

Both the first and the second somatic sensory areas send corticothalamic fibers from layer VI cells to terminate in the VPL and VPM (Rinvik, 1968c, 1972; Jones and Powell, 1968, 1970a; Burton and Kopf, 1984; Jones *et al.*, 1979; Friedman *et al.*, 1983). It is probable that many of these layer VI neurons receive monosynaptic thalamocortical synapses (White, 1979). There is no evidence, despite earlier claims to the contrary (Ramón y Cajal, 1911), that the corticothalamic fibers are collaterals of pyramidal tract axons (Jones and Wise, 1977; Catsman-Berrevoets and Kuypers, 1981). En route, the corticothalamic fibers appear to give off collaterals in part of the reticular nucleus (Jones, 1975c). The projections from both SI and SII to the VPL and VPM are somatotopically organized so that inputs from the same part of the body representation in each cortical area converge in the VPL or VPM. For SI, areas 3a, 3b, 1, and 2 all contribute to the projection and return fibers to the parts of the ventral posterior nucleus from which they receive inputs. The corticothalamic fibers are thin and have the morphology described in Chapter 4, i.e., they have multiple en passant boutons and boutons on short side branches (Tömböl, 1967). They are distributed in a pattern that matches the rodlike configuration of thalamocortical relay cells with identical place and modality properties (Jones *et al.*, 1979).

As pointed out in Chapter 4, the functions of these corticothalamic connections have not been well worked out. Many workers have studied the effects of conditioning stimulation of the cortex on transmission of an afferent volley through the ventral posterior nucleus. A variety of effects have been reported: predominately inhibitory (Shimazu *et al.*, 1965), predominately facilitatory (Andersen *et al.*, 1967), powerful and lasting up to 300 msec (Shimazu *et al.*, 1965), or weak and best demonstrated when synaptic transmission is impaired by fatigue (Andersen *et al.*, 1967). From the effects of cooling the SI cortex, Burchfiel and Duffy (1974) concluded that the role of corticothalamic fibers was to provide a tonic inhibition of the ventral posterior relay neurons. If so, this effect must operate via an inhibitory interneuron for morphological and other evidence implies that corticothalamic fibers make excitatory synapses (Jones and Powell, 1969b). To date, however, there has been no study on corticothalamic functions in the ventral posterior nucleus as compelling as that by Lindström (1982) in the lateral geniculate nucleus (Chapter 4). Thus, the respective roles of inhibitory interneurons in the reticular nucleus and in the ventral posterior nucleus itself are unknown.

7.2.7. Cortical Projections

The cortical targets of the ventral posterior nucleus in (Fig 7.16) all species that have been examined are the first and second somatic sensory areas. Species in which this finding appears well documented include: monkeys (Jones and Powell, 1970a; Burton and Jones, 1976), cat (Jones and Powell, 1969a; Hand and Morrison, 1970; Morrison *et al.*, 1972; Spreafico *et al.*, 1981; Burton and Kopf, 1984), rat (Donaldson *et al.*, 1975; Wise and Jones, 1978; Donoghue *et al.*, 1979), and raccoon (Herron, 1982).

7.2.7.1. The First Somatic Sensory Area

The first somatic sensory area of primates (SI) was first identified cytoarchitectonically (Campbell, 1905; Brodman, 1909) and shown to consist of four architectonic fields, areas 3a, 3b, 1 and 2 (Vogt and Vogt, 1919; Fig. 7.16). These were only later shown to be coextensive with the first somatic sensory area (Powell and Mountcastle, 1959) which had been identified independently by surface evoked potentials (Woolsey *et al.*, 1942). SI soon became identified by the evoked potential method in other mammals (Fig. 7.19), though the correlation of this area with cytoarchitectonic fields comparable to those of primates has not always been evident.

Initially, the evoked potential method revealed only a single representation of the contralateral half-body surface in SI. Later, as single-unit studies began to be conducted on the area in monkeys, it was found that neurons in area 3a were driven primarily by stimuli applied to deep tissues of the body, neurons in areas 3b and 1 were driven by cutaneous stimuli and those in area 2 again by deep stimuli, mainly movements of joints (Powell and Mountcastle, 1959), though inputs from muscle afferents to area 2 have also been reported (Burchfiel and Duffy, 1972). Powell and Mountcastle interpreted their results in terms of a single half-body map but later work involving detailed multiunit mapping in monkeys, has indicated that separate half-body representations occur in areas 3b and 1 and that additional representations are likely to be present in areas 3a and 2 (Paul *et al.*, 1972; Merzenich *et al.*, 1978; Kaas *et al.*, 1979; Nelson *et al.*, 1980; Fig. 7.20). Area 3b was originally reported to receive only inputs from slowly adapting cutaneous receptors and area 1 from rapidly adapting cutaneous receptors (Kaas *et al.*, 1979). Now it appears that slowly adapting and rapidly adapting mechanoreceptors may project to different parts of area 3b (Sur *et al.*, 1981a). Similar indications of segregated inputs to different parts of SI have been forthcoming in cats (Fig. 7.20) (Felleman *et al.*, 1983; Sretevan and Dykes, 1982) but the situation is less clear in other species.

Early studies of the thalamic connectivity of the SI area were regional in their approach. von Monakow (1895) had very early pointed out that destruction of cortical areas posterior to the cruciate sulcus in cats led to atrophy of his nuclei ventralis a and b, the equivalents of the VPL and VPM. Probst (1900b,c) later noticed Marchi-stained degeneration in the coronal gyrus of the cat after

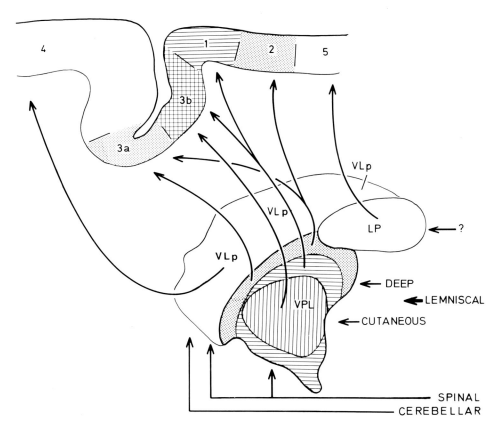

FIGURE 7.16. Schematic figure showing, on sagittal sections, the patterns of input–output connections of VPL and VLp nuclei. Modified from Jones and Friedman (1982).

destructive lesions of posterior parts of the ventral nucleus. Minkowski (1924), using cellular degeneration methods in monkeys, localized the projection of the ventral posterior nucleus to the postcentral gyrus. Thereafter, many studies reported projections to the postcentral gyrus or to its equivalent in nonprimates but it was not always made clear that the precentral gyrus is excluded from the projection zone (Polyak, 1932; Walker, 1934, 1938a; Le Gros Clark and Boggon, 1935). Virtually all authors from von Monakow on, however, had noted how degeneration moves medially in the ventral posterior nucleus as a lesion extends laterally in the first somatic area, an observation that reflects the topographic organization of the projection to different parts of the body representation. This observation has since been confirmed many times with many different techniques (e.g., Hand and Morrison, 1970; Jones and Powell, 1969a, 1970a; Saporta and Kruger, 1979; Jones *et al.*, 1979; Lin *et al.*, 1979; Nelson *et al.*, 1980).

The first author to pay attention to the possibility of differential projections to the cytoarchitectonic fields of the first somatic sensory area was Walker (1938a). On the basis of his retrograde degeneration studies, he concluded that area 3 received its input from anterior parts of the ventral posterior nucleus and area 1 from posterior parts. He considered that the thalamic input to area 2 came not from the ventral posterior but from the lateral posterior nucleus. Later, Le Gros Clark and Powell (1953) and Roberts and Akert (1963) noted that retrograde degeneration in the ventral posterior nucleus of monkeys was always severe after ablation of area 3 and mild after ablation of area 1 or area 2. But it became more severe if area 3 was ablated in conjunction with areas 1 and 2. This led them to believe that the principal thalamic input to areas 1 and 2 was via collateral branches of axons projecting to area 3. This interpretation was never satisfactorily reconciled with the observations of Powell and Mountcastle (1959) showing that the neurons in the various areas of the postcentral gyrus receive different kinds of inputs and that the receptive field properties of the cells are little different from those of individual cells in the ventral posterior nucleus (see also Paul *et al.*, 1972, and Sur *et al.*, 1981a).

The idea of a collateral or sustaining projection to areas 3, 1, and 2 is largely ruled out by the observation that injection of horseradish peroxidase in areas 1 and 2 six months after ablating area 3 leads to retrograde labeling of cells that remain unshrunken amidst the degenerated cells and intense gliosis resulting from the ablation of area 3 (Jones *et al.*, 1979). Moreover, different parts of the ventral posterior nucleus have since been shown to project to the separate fields. The current position seems to be (Fig. 7.16): the heart of the central cutaneous core of the nucleus projects only to the cutaneous-receiving area, area 3b; a thin surrounding region projects to area 3b and to area 1, a further cutaneous-receiving area; the anterior part of the deep shell projects to area 3a and the dorsal part to both areas 3a and 2, the deep recipient fields (Friedman and Jones, 1981; Jones and Friedman, 1982). Retrograde labeling with fluorescent dyes of different colors injected into various combinations of two fields in the SI shows that virtually no cells in regions projecting to two fields have branched axons ending in the two fields (Jones, 1983g) (Chapter 3).

The density of thalamocortical terminations in areas 1 and 2 in monkeys is remarkably light in comparison with that in areas 3a and 3b (Jones and Powell,

1970a; Jones, 1975b; Jones and Burton, 1976b). The parent axons entering and 3b but also they terminate only in the deep part of layer III (layer IIIB) in areas 1 and 2. In areas 3a and 3b they terminate in layers IIIB and IV (Jones, 1975b; Jones and Burton, 1976b). It is possible that the thin axons and their relatively small numbers play some role in determining that the retrograde degeneration occurring in the ventral posterior nucleus after ablations of areas 1 and 2 is always relatively insignificant. The differential retrograde labeling with fluorescent dyes suggests that the number of cells projecting to area 2, at least, is significantly smaller than that projecting to the other areas (Fig. 3.10).

Though areas 3a, 3b, 1, and 2 can be identified in cats (Hassler and Muhs-Clement, 1964), there is no differentiation of thalamocortical axons in regard to size, density, or layers of terminations as there is in monkeys (Jones and Powell, 1969a); the cytoarchitectonic segregation of neurons according to their receptive field properties also seems less clear, or at least the subject of conflicting reports (Dykes *et al.*, 1980; McKenna *et al.*, 1982; Felleman *et al.*, 1983) (Figs. 7.19, 7.20). Small cells at the perimeter of the VPL of the cat appear to project selectively to layer I of the cortex (Penny *et al.*, 1982). No comparable small-celled projection has yet been uncovered in monkeys.

In rats and mice, the first somatic sensory area can be less obviously divided into the four cytoarchitectonic fields found in other species, though attempts have been made to do so (e.g., Krieg, 1947; Caviness and Frost, 1980). The cortex is instead divided into highly granular patches, perhaps collectively equivalent to area 3b, with intervening and surrounding much less granular zones,

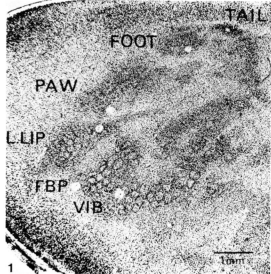
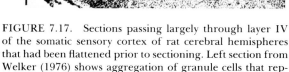
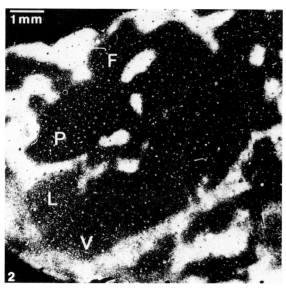

FIGURE 7.17. Sections passing largely through layer IV of the somatic sensory cortex of rat cerebral hemispheres that had been flattened prior to sectioning. Left section from Welker (1976) shows aggregation of granule cells that represent body parts indicated and that receive fibers from ventral posterior nucleus. Right section from Wise and Jones (1976) shows distribution of fibers of the corpus callosum to intervening and surrounding dysgranular zones.

perhaps equivalent to areas 1 and 2 (Woolsey and Van der Loos, 1970; Welker, 1976; Wise and Jones, 1976) (Fig. 7.17). These agranular zones are not readily responsive to peripheral stimulation in barbiturate-anesthetized animals (Welker, 1976). Thalamocortical fibers emanating from the ventral posterior nucleus of the rat terminate only in the granular zones (Wise and Jones, 1978) while the agranular zones receive the commissural inputs (Wise and Jones, 1976, 1978; Figs. 7.17, 7.18; Ivy *et al.*, 1979) (Fig. 7.19) and may receive thalamic fibers from other parts of the thalamus, possibly from the medial nucleus of the posterior complex (Pom) (Herkenham, 1980). The Pom receives spinothalamic fibers in the rat (Lund and Webster, 1967b) and a similar nucleus receiving spinal fibers in cats projects to area 5, posterior to area 2 (Tanji *et al.*, 1977) (Fig. 11.5). It is tempting to speculate that there may be a common plan in all this. That is, area 5 of the cat and the agranular zones of the rat might be equivalent to area 2 of the monkey, and the Pom of cats and rats might be equivalent to the dorsal part of the deep shell of the ventral posterior nucleus of monkeys. But there is as yet no evidence to support this.

The first somatic sensory area of marsupials and the hindlimb representation

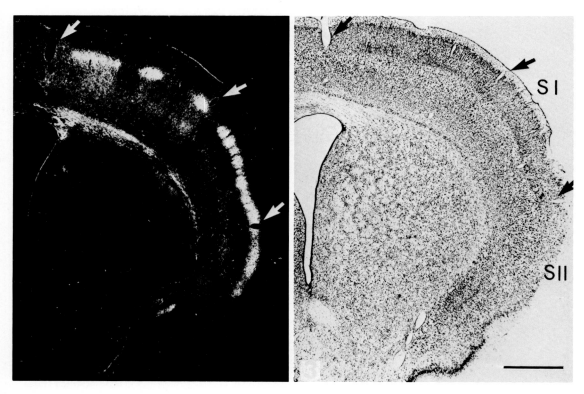

FIGURE 7.18. Adjacent frontal sections from a rat brain showing focal terminations of thalamic fibers from ventral posterior nucleus in granular aggregations in SI cortex and in a continuous band in SII cortex. Arrows indicate same blood vessels. Bar = 1 mm. From Wise and Jones (1976).

in that of the rat are regions of cortex in which functional areas overlap (see Chapter 3). They receive thalamocortical inputs both from the ventral posterior nucleus and from the part of the ventral lateral complex in which cerebellothalamic fibers terminate (Ebner, 1967; B. H. Pubols, 1968; Donoghue *et al.*, 1979; Ebner *et al.*, 1976). In association with this there is a complete superimposition of the first somatic sensory and motor cortex maps in marsupials (Lende, 1963, 1964) and a partial superimposition of the sensory and motor maps of the hindlimb in the rat (Hall and Lindholm, 1974). The superimposed parts can be

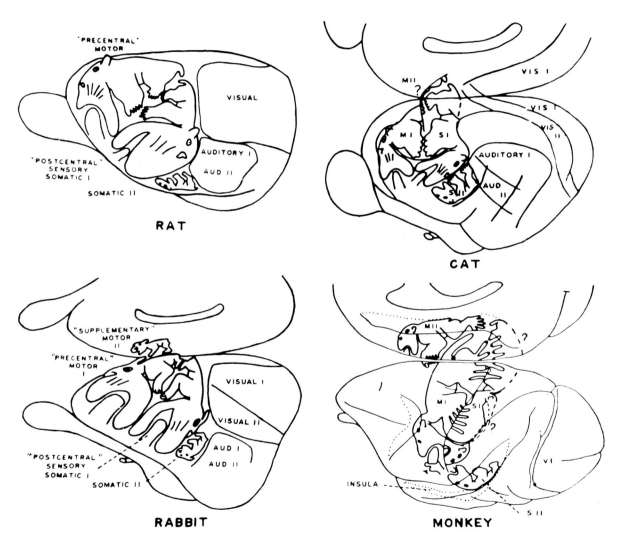

FIGURE 7.19. Evoked potential maps showing SI and SII and certain other areas of cortex in various species. From Woolsey (1958).

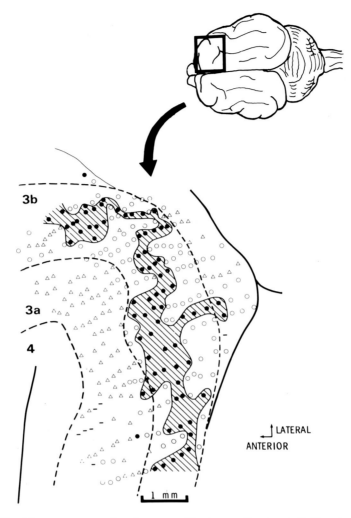

FIGURE 7.20. More recent maps based upon multiunit recording in cats (left) and monkeys (right) showing multiple representations in SI. On left, solid circles represent tracks in which slowly adapting units were identified, open circles rapidly adapting units, and triangles units with deep receptive

distinguished cytoarchitectonically by the overlap of the granular patch of layer IV that represents the hindlimb in the sensory component, on the giant layer V pyramidal cells that delineate motor cortex (Fig. 3.11) (Jones and Porter, 1980). Despite the complete superimposition in the rat, and the associated overlap of the cortical projections of the ventral posterior and ventral lateral nuclei, in neither case is there any overlap in the terminations of the principal afferent pathways to these two nuclei, the cerebellothalamic and medial lemniscal (Rockel *et al.*, 1972; Walsh and Ebner, 1973; Donoghue *et al.*, 1979). Possibly this betokens

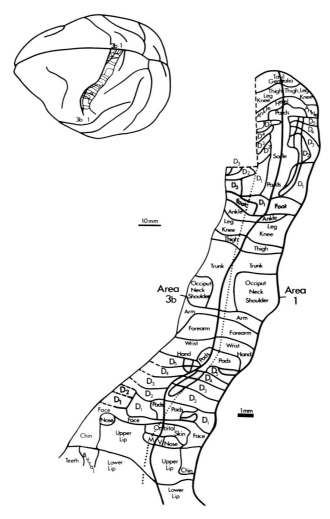

fields. On right, representation in area 3b was described as of slowly adapting and in area 1 as of rapidly adapting units. Left, from Sretavan and Dykes (1983); right, from Nelson *et al.* (1980).

a later state of evolutionary development for the sensory motor cortex than for the afferent pathways leading to it.

7.2.7.2. The Second Somatic Sensory Area

The second somatic sensory area was unknown to the earlier investigators of thalamocortical relations. First defined by Adrian (1940) and Woolsey (1943), it has now been identified in most of the commoner mammals with the exception of the monotremes (Bohringer and Rowe, 1977). In monkeys it occupies much of the parietal operculum, in carnivores anterior parts of the anterior ectosylvian gyrus, and in lissencephalic animals cortex close to the rhinal fissure.

The first intimation that the SII area received its thalamic input from the ventral posterior nucleus was made by Knighton (1950) who observed a second representation of the body in what he regarded as the posterior part of the ventral posterior nucleus of the cat and speculated that it might project to SII. Thereafter, Macchi *et al.* (1959) ablated most of the anterior ectosylvian gyrus in cats and reported retrograde degeneration in the posterior part of the ventral posterior nucleus. This result was discredited in a footnote to a single-unit investigation on the cat SII by Carreras and Andersson (1963) who, after examining brains with lesions of SI or SII, concluded that the results of Macchi *et al.* must have been due to undercutting of the first somatic sensory area. Earlier, Rose and Woolsey (1958), in analyzing the retrograde degeneration in the thalamus of cats resulting from very large and deep lesions of the auditory regions extending forward into the anterior ectosylvian gyrus, had concluded that SII and all the auditory fields received rather diffuse projections from the posterior complex of thalamic nuclei (see Chapter 11). SII was reaffirmed as the target of the ventral posterior nucleus by recording evoked responses in it and in SI following electrical stimulation of the nucleus (Guillery *et al.*, 1966). Subsequent work has shown that the anterior ectosylvian gyrus, which was regarded as synonymous with SII by Carreras and Andersson (1963) and Rose and Woolsey (1958), includes a number of areas additional to that containing the single body representation that defines SII (Haight, 1972; Robinson and Burton, 1980a–c; Burton *et al.*, 1982) (Fig. 7.21). Some of these areas do receive inputs from the posterior complex but others receive inputs from the medial geniculate complex and only a small anterior portion, containing neurons with receptive fields very similar to those in SI, is the target of the ventral posterior nucleus (Jones and Powell, 1969a; Heath and Jones, 1971a; Jones and Leavitt, 1973; Burton and Kopf, 1984). A comparable region has now been localized by single-unit mapping and by anatomical tracing in the brains of many species (Jones and Powell, 1969a, 1970a; Welker and Sinha, 1972; Johnson *et al.*, 1974; Welker *et al.*, 1976; Pubols, 1977; Herron, 1978; Wise and Jones, 1978; Donoghue *et al.*, 1979; Nelson *et al.*, 1979; Friedman *et al.*, 1980; Sur *et al.*, 1981a; Burton *et al.*, 1982). As well as receiving fibers from the ventral posterior nucleus, SII returns corticothalamic fibers to the nucleus in a way that follows the organization of the similar projection from SI (Jones and Powell, 1968; Friedman *et al.*, 1980; Burton and Kopf, 1984).

There has been some debate over whether single cells in the ventral posterior nucleus project by branched axons to both SI and SII (Chapter 3). Andersson *et al.* (1966) first suggested that single cells receiving group I muscle afferents in the ventral posterior nucleus might project by branched axons to the two areas because of similar latencies of antidromic activation from stimulation of SI and SII. This interpretation was later contested by Rosén (1969a). At about the same time, Andersen *et al.* (1966) reported antidromic invasion of a proportion of their sample of spinocervicolemniscal recipient cells in the VPL of the cat, from stimulation of both SI and SII. Rowe and Sessle (1968) also showed that single units in what they identified as the ventral posterior nucleus of the cat could be antidromically activated by stimulation of widespread areas of cortex including SI and SII. Then, Manson (1969), in a well-controlled antidromic stimulation study involving the collision technique, showed that a small percentage of units receiving cutaneous inputs in the ventral posterior nucleus sent branched axons to SI and SII. Most projected to SI alone and relatively few projected only to SII.

Jones (1975a) ablated SI of cats and 6 months later injected horseradish peroxidase in SII. It was shown that many cells, principally in the ventral part of the gliotic ventral posterior nucleus, were retrogradely labeled, implying a projection to SII. However, the cells were considerably shrunken, presumably because of destruction of an axon branch to SI (Fig. 7.22). A more recent study by Spreafico *et al.* (1981) shows double labeling of a few cells in the ventral posterior nucleus after injection of different fluorescent dyes in SI and SII. But most cells were labeled by one dye only, implying that many project to only one of the areas. A similar observation was made by Fisher *et al.* (1983) who injected different tracers into foci in SI and SII responding to stimulation of high-frequency cutaneous vibration receptors. Spreafico *et al.* interpret their results to indicate that SII receives its predominate projection from the posterior complex. But this is because their injections of "SII" involved the whole anterior ectosylvian gyrus, not simply the part already known to receive inputs from the ventral posterior nucleus. For consistency, it would be desirable to analyze such results in terms of the individual areas in the gyrus. Herron (1982), in a retrograde labeling study in the raccoon, has suggested that cells projecting to SII lie ventral to those projecting to SI. Fisher *et al.* (1983), however, found all their SI- and SII-projecting cells in the VPL.

A further problem in regard to SII is the clear evidence reported by many workers, for bilateral evoked responses and for units of the lemniscal type with bilaterally symmetrical receptive fields within it (Woolsey and Wang, 1945; Woolsey and Fairman, 1946; Whitsel *et al.*, 1969b; Pubols, 1977; Robinson, 1973; Robinson and Burton, 1980a). It seems unlikely, in view of the evidence against bilateral lemniscal responses in the ventral posterior nucleus, that the ipsilateral input to these SII neurons depends upon the thalamocortical route. Other routes that have been suggested are via the corpus callosum, either directly (Caminiti *et al.*, 1979) or indirectly via SI (Teitelbaum *et al.*, 1968; Friedman *et al.*, 1980). It has been claimed, however, that callosotomy does not entirely abolish bilateral responses in SII (Woolsey and Wang, 1945; Robinson, 1973) so this subject needs to be explored further.

7.2.7.3. Afferent Streams through the Ventral Posterior Nucleus

In Chapter 3, we referred to the relative independence of the X, Y, and W afferent streams as they flowed from the retina through the dorsal lateral geniculate nucleus en route to the cerebral cortex. This raises the question of whether the VPL–VPM complex of nuclei permits the separate relay of the dorsal column–lemniscal, spinocervical, and spinothalamic systems to the cerebral cortex and whether inputs from individual receptors or particular constellations of receptors may reach the cortex without interaction with inputs from others in the thalamus.

There are remarkably few data in this area and certainly none comparable to those which have been forthcoming on the X, Y, and W systems. Certain studies have dealt with the relative importance of the three pathways mentioned above in regard to the functions of the somatic sensory cortex. This may give us a few clues as to the extent of convergence in the ventral posterior nucleus by indicating the extent to which one route may compensate for the absence of the others.

Andersson (1962), in studying unit responses in the SII area of cats, reported that unit activity in response to natural peripheral stimulation disappeared after section of the dorsal columns and spinocervical tract in the spinal cord but not after section of one of these alone. Surface evoked potentials, however, could still be elicited after electrical stimulation of a limb. These were pre-

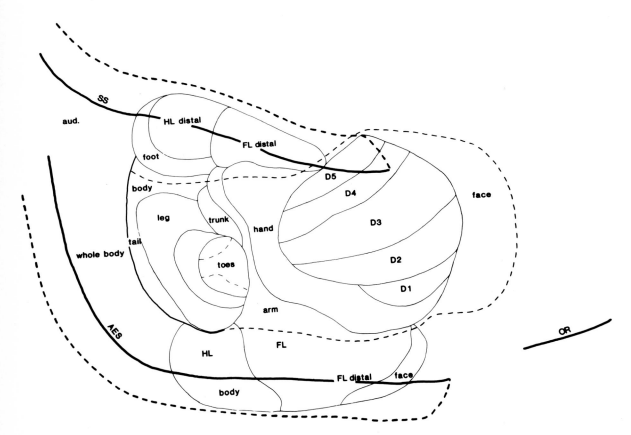

FIGURE 7.21. Position of SII in cat (left) and monkey (right) based upon distribution of fibers from ventral posterior nucleus and mapping of single-unit responses similar to those found in the ventral posterior nucleus. SS: suprasylvian

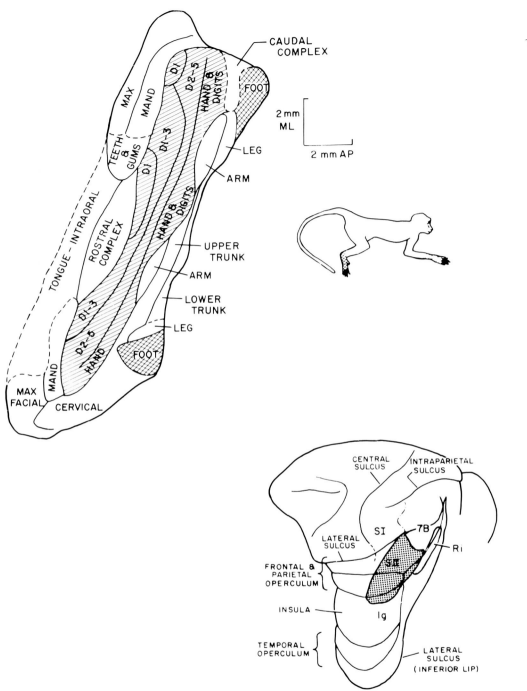

sulcus; AES: anterior ectosylvian sulcus; OR: orbital sulcus. Additional representation in banks of AES receives fibers from posterior nucleus of thalamus. Left, from Burton *et al.* (1982); right, from Robinson and Burton (1980a).

sumably mediated by the spinothalamic tracts. Andersson also noted that when units in SII were activated by the dorsal columns, lateral inhibition could be detected but not when the spinocervical tract was the route for activation. He, thus, suggested that the spinocervical tract was the most secure pathway to SII.

Comparable findings to these were obtained in SI as well as in SII of cats and dogs by Norrsell and Wolpaw (1966) and by Norrsell (1966), though only evoked potentials, not single units, were studied. In a later study in monkeys, however, Andersson *et al.* (1972) and Andersson and Norrsell (1973) concluded that significant, light-tactile input (as judged by evoked and single-unit responses) could reach SI when only the ventrolateral quadrants of the spinal cord were intact. The next step in these kinds of studies has not fully been taken, i.e., to study the extent of convergence of the three pathways on single units in the ventral posterior nucleus and their behavior after sequential lesions of the pathways in the spinal cord. Andersen *et al.* (1966) have reported convergence of dorsal column–lemniscal and spinocervical inputs on some neurons in the VPL of the cat, but not on others in the same part of the nucleus. In what is apparently

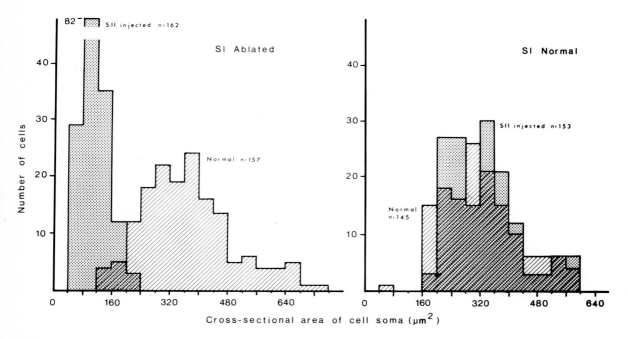

FIGURE 7.22.   Data from experiments in cats in which cells of the ventral posterior nucleus projecting to SII were retrogradely labeled with horseradish peroxidase in a normal animal (right) and in an animal (left) in which SI had been removed 6 months previously. Retrogradely labeled cells (stippling), in comparison with unlabeled cells in ventral posterior nucleus of contralateral side (hatching), are markedly shrunken on side on which SI was ablated, indicating that normally these cells have collateral axon branches to both areas. From Jones (1975a).

the only study of its type in unanesthetized monkeys, Pollin and Albe-Fessard (1979) found that section of the dorsolateral funiculus of the spinal cord at midthoracic levels in one monkey had no effect on the responses of cells in the contralateral VPL. This is surprising in view of the report of Dreyer *et al.* (1974) that section of the dorsolateral funiculus in monkeys leads to a lack of responsiveness of cells in areas 1 and 2 of the SI cortex to peripheral stimulation. It may reflect a failure to sample the more dorsal parts of the VPL in which the cells projecting to areas 1 and 2 are situated (Jones, 1983g). Pollin and Albe-Fessard also reported that section of one dorsal column and the corresponding dorsolateral funiculus at the same level caused a great reduction in the number of cells in the contralateral VPL that responded to stimulation of the hindlimb. Those that remained had receptive fields that resembled wide-dynamic-range cells of the spinal cord (Section 7.2.5.6). After 4 weeks, a number of cells responding to stimulation of the forelimb could be recorded in the hindlimb representation, but whether this represents an uncovering of a concealed spinothalamic input or sprouting of cuneate fasciculus axons to innervate the deafferented gracile nucleus is not clear. Further, more extensive studies of this kind would be highly desirable.

Behavioral studies tell us a little about the capacities of the individual somatic sensory pathways but nothing about their independence at the thalamic level. Tactile discrimination in monkeys (presumably a cortical function) is affected by dorsal column lesions but returns to normal eventually. Then, if lesions of the dorsal lateral funiculus are added to them (Vierck, 1973), the deficit returns and never fully recovers. Vierck (1974) has also shown that tactile movement detection is unaffected by combined dorsal column and spinocervical tract lesions. However, dorsal column lesions alone cause a lasting impairment in a monkey's ability to discriminate the direction of movement of a tactile stimulus. In their study of escape latencies exhibited by monkeys escaping from painful electrical stimulation of a leg, Vierck and Luck (1979) found that lesions of the dorsal columns or spinocervical tract had little effect, indicating little involvement in the conduction of information critical for pain perception. Lasting decreases in pain sensitivity, however, could be produced by lesions of the ventral and ventrolateral columns, provided they were bilateral.

If one turns to the question of whether separate classes of peripheral somesthetic receptors gain independent access to the cerebral cortex, he finds that, with a few exceptions, it can only be answered indirectly. This is because very little detailed work has been done on the relationships of individual thalamic cells to particular receptors or receptor groups. Hence, one cannot say what differences exist between their responses and the responses of the cortical cells that they innervate nor what differences they impose on the afferent inputs relayed through them.

It would appear, though, that at the level of analysis so far conducted, information transmitted from the ventral posterior nucleus to the SI cortex is a close facsimile of that entering the nucleus over its afferent pathways. The receptive fields of neurons in the ventral posterior nucleus and in the somatic sensory cortex and the specificity of their responses to particular kinds of peripheral stimulus are virtually identical (Sections 7.2.5 and 7.2.7); cells with different modality properties are segregated in different fields of SI and in different "rods" in the ventral posterior nucleus (Section 7.2.7); different parts of the ventral posterior nucleus project to different fields of SI (Section 7.2.7). Finally, few or no cells in the nucleus, even those in parts projecting to more than one cortical field, appear to have branched axons ending in more than one field (Sections 3.3.2. and 7.2.7).

We do have some information; it suggests that the inputs from certain receptor classes are relayed relatively unchanged to the somatic sensory cortex. For example, the discharges of VPL neurons responding to inputs in group I muscle afferents, accurately reflect changes in muscle length (Maendly *et al.*, 1981). Their responses are, on the whole, very comparable to those of muscle spindle primary afferents (Section 7.2.5.1). Moreover, the behavior of most of the neurons that they innervate in area 3a is also remarkably similar (Lucier *et al.*, 1975; Heath *et al.*, 1976; Hore *et al.*, 1976). Though convergence of cutaneous nerve inputs can be demonstrated by electrical stimulation on both cortical and thalamic group I neurons (Heath *et al.*, 1976; Maendly *et al.*, 1981), the overwhelming impression is of a secure and uncontaminated relay through the thalamus. Similar comments might well be applied when examining the similarities in the discharge patterns of cortical and thalamic neurons responding to movement of a joint or to light cutaneous stimulation (Mountcastle and Powell, 1959; Poggio and Mountcastle, 1963; Mountcastle *et al.*, 1963; Gardner and Costanzo, 1981), though in both cases much more diverse types of receptors are involved and have not been studied independently. It has been argued (Gardner and Costanzo, 1981) that cortical units showing rapidly adapting responses to movement of a joint receive inputs from slowly adapting receptors in muscles or joints whose tonic responses have been attenuated centrally. But whether this attenuation occurs in the thalamus or elsewhere has not been determined.

There has been no systematic attempt to study neurons in the ventral posterior nucleus that respond specifically to low- and high-frequency vibratory stimuli applied to the skin. The evidence, from studies on the cortex, suggests that inputs from low-frequency vibration receptors gain unimparied access to the cortex. Rapidly adapting neurons in SI that respond to low-frequency vibrations (5–10 Hz) of moderate amplitude applied to the glabrous skin of a finger, do so by discharging impulses in a phase-locked manner, closely resembling that of the primary afferents, and accurately reflecting the periodicity of the stimulus (Mountcastle *et al.*, 1969; Chapter 4). Most rapidly adapting neurons in SI and SII that respond to high-frequency vibrations of small amplitude can only follow systematically such stimuli up to 100 Hz for SI and 300 Hz for SII, whereas the primary (Pacinian) afferents can be phase-locked up to as much as 1000 Hz (Mountcastle *et al.*, 1969; Ferrington and Rowe, 1980). Possibly this failure occurs in the ventral posterior nucleus though there is some evidence that the arriving thalamocortical fibers bearing high-frequency messages and even the cells that they innervate monosynaptically can follow high-frequency stimuli quite well (Mountcastle *et al.*, 1969). The evidence, therefore, points to a fairly direct route for both kinds of mechanoreceptors through the thalamus to the cortex. Clearly, much further work needs to be done before we are in a position to say that this and the relay of other mechanoreceptor inputs mentioned above, is so independent as to be comparable to the X, Y, and W streams through the dorsal lateral geniculate nucleus, but the indications of similarity are obviously there.

7.3.   Ventral Posterior Inferior Nucleus

7.3.1.   Description

The region customarily referred to as the ventral posterior inferior nucleus (VPI) is really only distinct in the primate brain (Fig. 7.23). In monkeys it is a small-celled nucleus. It lies along the ventral surface of the ventral posterior nucleus on the ventral medullary lamina, extending posteriorly to the posterior pole of the ventral posterior nucleus and anteriorly invading the ventral aspect of the VLp nucleus. It is triangular in frontal section with its apex in the angle at which the VPL and VPM nuclei meet and its base on the ventral medullary lamina. The "nucleus" thus outlined, is the main portal of entry of fibers of the medial lemniscus and brachium conjunctivum into the ventral nuclei (Fig. 7.24). The greater part of its cell population is, therefore, glial. However, it does appear

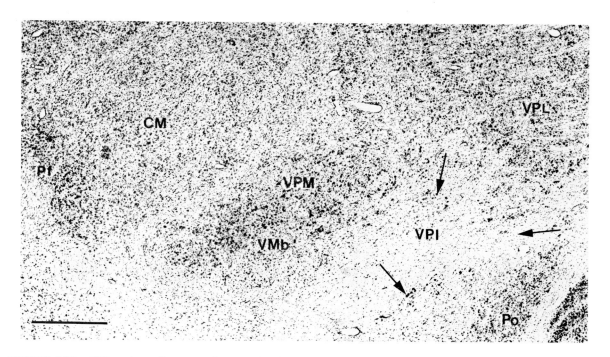

FIGURE 7.23.   VPI nucleus of a cynomolgus monkey in frontal section, showing (arrows) islands of VPL cells invading it. Thionin stain. Bar = 1 mm.

to contain a significant population of small, pale-staining neurons as well, especially anteriorly. In places it is deeply invaded by islands and fingers of the large, deeply staining cells of the VPL. I prefer to exclude these from the VPI because of their obvious continuity with VPL cells and because they project in a topographic sequence with the other VPL cells, to SI and SII (Fig. 7.23). Other workers have undoubtedly included them in the VPI and often, as a consequence, either have discovered a unique population of multiunit responses in the VPI (Dykes *et al.*, 1981) or have thought that the VPI projects to SI or SII (Friedman *et al.*, 1983).

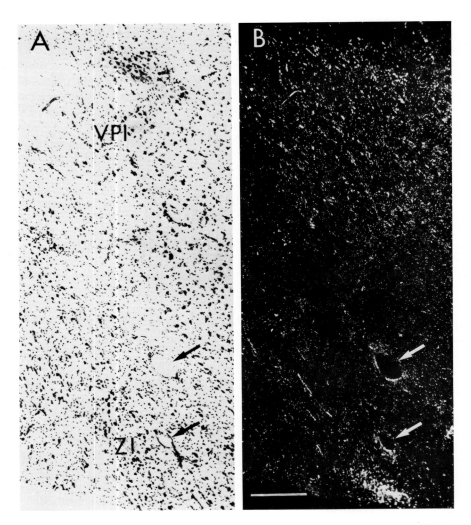

FIGURE 7.24. Autoradiographically labeled dentatothalamic fibers distributing a small patch of terminations in the zona incerta and then passing without termination through the VPI nucleus of a cynomolgus monkey. Sagittal section. Bar = 250 μm. From Asanuma *et al.* (1983b).

The VPI was identified in a guenon brain by Friedemann (1911) who included it in what he called the "pars parvicellularis of the caudal part of the ventral thalamic nucleus." This region also included, however, the basal ventral medial (VMb) nucleus of the present account. In Vogt's (1909) myeloarchitectonic account of the same monkey as Friedemann, the VPI region is labeled "radiation prelemniscale." Crouch (1934) seems to have introduced the term VPI in his brief description of the human brain and Olszewski (1952) adopted it in his atlas of the macaque monkey. He clearly distinguished the VPI from the VMb (which he called pars parvocellularis of the VPM). He did not include the large, invading cells of the VPL and, thus, his account is the same as that of the present one. It is also the same as that of Emmers and Akert (1963) in the squirrel monkey.

In man, Hassler (1959) returned to Friedemann's nomenclature, referring to the VPI as nucleus ventralis caudalis parvicellularis externa and the VMb as nucleus ventralis caudalis parvicellularis interna.

Attempts have been made to identify a VPI in other species, particularly the cat (Rinvik, 1968a). However, it seems doubtful to me that the region so labeled is anything other than the region of entry of the medial lemniscus.

7.3.3. Connections

There is relatively little information about the connectivity of the VPI. Neither lemniscal nor brachial conjunctival axon terminals can be clearly distinguished in it (e.g., Figs. 7.10 and 7.24; Tracey *et al.*, 1980; Asanuma *et al.*, 1983b). I believe that any terminal ramifications of the medial lemniscus apparently in it are invariably associated with the intruding clumps of VPL cells. Field potentials resulting from stimulation of the vestibular nerve have been described in VPI and adjacent nuclei (Deecke *et al.*, 1975) but because it has been shown subsequently that vestibulothalamic fibers actually terminate in the VLp (Lange *et al.*, 1979; Fig. 7.37), it seems likely that the field potentials in the VPI resulted from the fibers passing through it. A similar criticism (Fisher *et al.*, 1983) has been leveled against the suggestion of Dykes *et al.* (1981) that multiunit responses to stimulation of rapidly adapting, high-frequency vibration (Pacinian) receptors in the VPI of the squirrel monkey, represent Pacinian inputs specifically to VPI cells. It is not ruled out, however, that Dykes *et al.* were recording from the VPL cells that intrude on the VPI.

On the efferent side, there is indirect evidence that the VPI projects to the dysgranular field of the insular cortex (Roberts and Akert, 1963; Burton and Jones, 1976). A recent brief report suggesting that this actually represents a projection to SII (Friedman *et al.*, 1983) remains to be verified.

One must conclude that the VPI is only evident in primates but even here is not a well-defined entity. It is subject to much misinterpretation and we have few conclusive data about its connectivity or functions.

7.4. Ventral Lateral Complex and the Motor System

The ventral lateral complex occupies much of the ventral nuclear mass anterior to the ventral posterior nucleus. In small mammals such as rodents and even in the cat, it extends almost to the anterior pole of the ventral nuclear mass, separated only by a small ventral anterior nucleus. In the majority of mammals it extends for some distance posteriorly along the medial and/or dorsal surface of the ventral posterior nucleus, usually reaching the lateral posterior nucleus (Figs. 7.25 and 7.26). Though its cytoarchitecture is by no means identical in all species, it can usually be distinguished as a region of large, deeply staining cells that are more loosely packed than those in the ventral posterior nucleus. Along its medial border lie the intralaminar nuclei and the nuclei of the ventral medial complex. In primates parts of its dorsal surface reach to the dorsal surface of the thalamus (Fig. 7.25).

7.4.1. Description

In most nonprimates the ventral lateral complex is a homogeneous entity, no overt nuclear subdivisions being discernible. Even the distinction from the ventral anterior nucleus is questionable (Fig. 7.28). In primates it has been divided into many nuclear subdivisions, some of which, though justified cytoarchitectonically, have no differences in their connections and are named differently by different authors. After some years of maintaining a conservative approach to these divisions, I have been convinced that some rationalization of the primate terminology is needed and the present account endeavors to provide that.

On the basis of architecture and connections (Jones *et al.*, 1979; Asanuma *et al.*, 1983a–c), there are two major subdivisions of the ventral lateral complex in monkeys. The largest, the *posterior ventral lateral nucleus* (VLp), is made up of predominately large, deeply stained cells, quite widely dispersed in a clear neuropil and with fewer intervening small cells than in the ventral posterior nucleus. It lies immediately in front of the ventral posterior nucleus and here its cells are larger than elsewhere. The border between the two nuclei is sharp when seen in horizontal or sagittal sections (Figs. 7.25, 7.26); because the border lies in the frontal plane it is virtually impossible to see in frontal sections. This has led to many erroneous reports of the presence of a "transition zone" between the two nuclei. The VLp arches back over the dorsal surface of the ventral posterior nucleus, and insinuates itself as a narrow tail along the dorsolateral margin of the lateral posterior nucleus, reaching in that position to the ventricular surface of the thalamus at the level of the pulvinar. Anteriorly, a large protrusion extends along the lateral edge of the internal medullary lamina, between the lamina and the anterior ventral lateral nucleus (VLa). Smaller fingers of the VLp also insinuate themselves between the cell clusters that characterize the VLa (Fig. 7.27). VLp cells are largest ventrally adjacent to the VPL and become progressively smaller dorsally (Fig. 7.25).

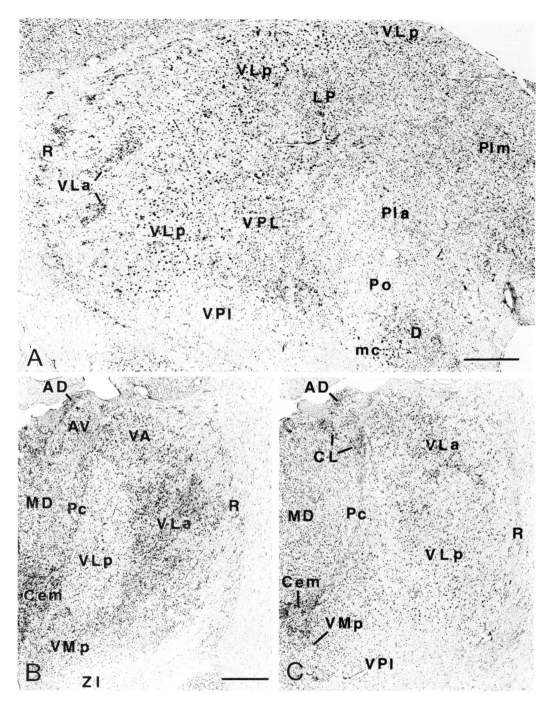

FIGURE 7.25.   Sagittal (A) and frontal (B, C) sections through VLp nucleus and adjacent nuclei in a cynomolgus (A) and a rhesus (B, C) monkey. Note how VLp nucleus extends posterodorsally over VPL nucleus and anteriorly medial to and between islands of VLa nucleus. Bars = 1 mm.

The VLa is composed of numerous large islands of small, closely packed cells lying anterior to the sloping anterior surface of the VLp and posterior to the ventral anterior nucleus. The islands tend to be separated by the fingers of less tightly packed cells that extend forward from the VLp.

7.4.2. Terminology

The VLp of the present account is virtually the same as the ventral intermediate nucleus of Vogt (1909) and Friedemann (1911) though they gave its

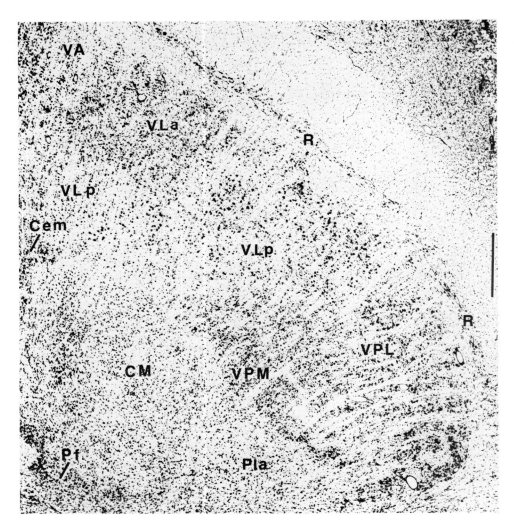

FIGURE 7.26. Horizontal section through the ventral nuclei of a cynomolgus monkey showing extension of VLp nucleus medial to VLa nucleus. Extension is "nucleus X" of Olszewski (1952) (cf. Fig. 7.25B). Bar = 1 mm. Modified from Asanuma *et al.* (1983a).

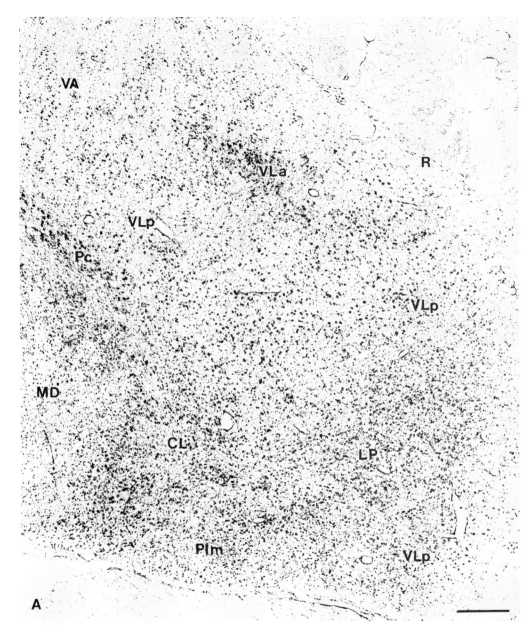

FIGURE 7.27. (A) Horizontal section through the dorsal parts of the ventral lateral nuclei showing extension of VLp cells posteriorly around the lateral posterior (LP) nucleus and anteriorly around and between the islands of the VLa nucleus. Bar = 1 mm. Modified from Asanuma *et al.* (1983a).

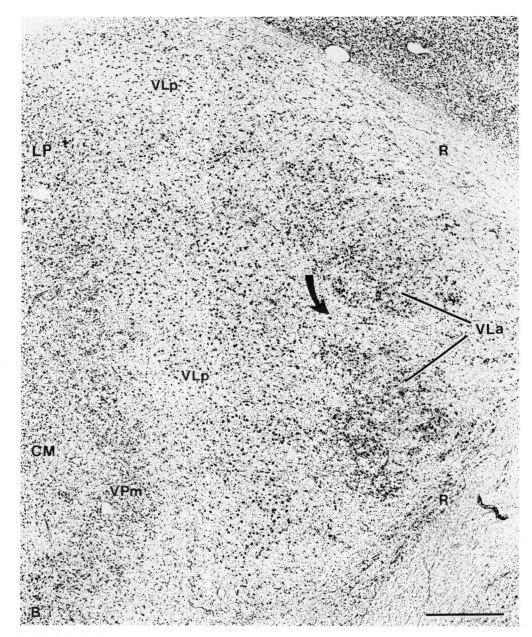

FIGURE 7.27. (B) Parasagittal section from a cynomolgus monkey showing (arrow) extensions of VLp nucleus between islands of VLa nucleus. Bar = 500 μm. Modified from Asanuma *et al.* (1983a). (C) Darkfield autoradiographs (A, B) from boxes indicated in C, showing labeled cerebellothalamic terminations in parts of the VLp medial to and between islands of VLa cells. Bars = 200 μm (A, B), 1 mm (C). Modified from Asanuma *et al.* (1983b).

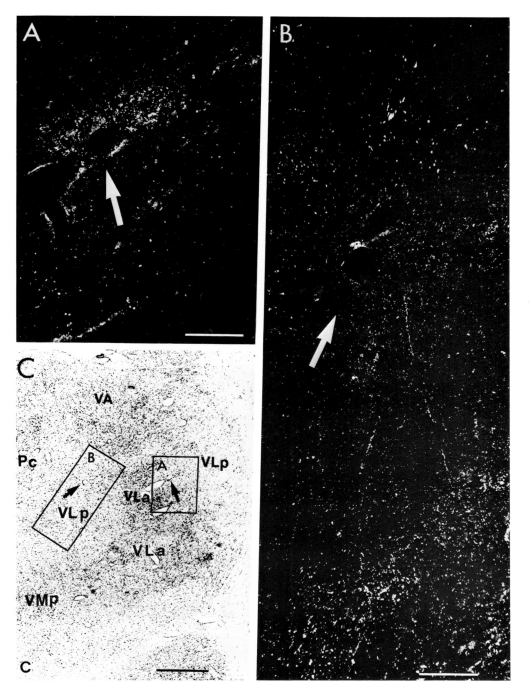

FIGURE 7.27.   (*continued*)

anterior and posterior extensions separate names or regarded them as transition zones [see also Hassler's (1959) nucleus ventralis intermedius externa and interna in man]. The VLp includes both ventral lateral and ventral intermediate nuclei of Walker (1938a). Olszewski (1952) referred to the large-celled ventral part lying anterior to the ventral posterior nucleus as the pars oralis of the ventral posterolateral nucleus (VPLo); he regarded the VPL of the present account (Section 7.2) as a pars caudalis (VPLc) of the same nucleus. Olszewski called the dorsal part of the VLp the caudal nucleus of the ventral lateral nucleus (VLc), its posterior tail the pars postrema of the ventral lateral nucleus (VLps), and its large anterior extension, nucleus "X." All four of these subdivisions have since been shown to receive the terminations of the cerebellothalamic system, which spreads from one division to the other without interruption (Asanuma *et al.*, 1983b,c). All the divisions also project to the same cortical area [area 4 (Jones *et al.*, 1979; Asanuma *et al.*, 1983a)]. On these grounds they form a common cerebellar relay nucleus to motor cortex and the cytoarchitectonic differences are thought to represent merely variants on a common theme, something Olszewski recognized when he drew interrupted rather than continuous lines between his divisions.

The VLa was identified as subdivision III of the lateral complex by Vogt (1909) and Friedemann (1911). It is the same as the pars oralis of the ventral lateral nucleus (VLo) of Olszewski (1952) and may be equivalent to the nucleus ventralis oralis anterior of Hassler (1959). No cerebellar terminations intrude on it, except for those in the fingers of the VLp that extend into it (Asanuma *et al.*, 1983b). The input to the VLa is from the internal segment of the globus pallidus and it projects to cortical areas anterior to area 4 (Kim *et al.*, 1976; Tracey *et al.*, 1980).

The medial ventral lateral nucleus (VLm) of Olszewksi (1952) is called the principal ventral medial nucleus (VMp) here for consistency with other mammals. It is the nucleus of termination of nigrothalamic afferents (Hendry *et al.*, 1979; Rinvik, 1975; Beckstead *et al.*, 1979; Carpenter and Peter, 1972; Carpenter *et al.*, 1976) and is described in Section 7.6.

7.4.3. Cerebellar Terminations and Body Representation

Axons of the brachium conjunctivum were conclusively demonstrated to terminate in the contralateral ventral lateral nuclei of the monkey by Vogt (1909) and in those of the dog by André-Thomas (1912). Since then the observation has been repeated many times in a large number of species (Batton *et al.*, 1977; Hendry *et al.*, 1979; Stanton, 1980; Kalil, 1981), though the descriptions of the precise origins and terminations of the axons have not always agreed.

It has been a common belief that cerebellothalamic fibers arise only in the dentate nucleus of the cerebellum, the interposed nucleus being thought to project principally to the red nucleus and the fastigial nucleus to the vestibular nuclei and reticular formation (André-Thomas, 1912; Mussen, 1927; Fulton, 1949). It is difficult to trace the exact origin of this belief, though it fitted in with a prevailing view that the cerebellar hemisphere, through the dentate nucleus, was related predominately to the cerebral cortex while the interposed and fastigial nuclei were concerned predominately with spinal cord and brain-stem mechanisms. Recent experimental evidence, however, indicates that all three deep cerebellar nuclei project upon the ventral lateral complex, the dentate and interpositus contralaterally and the fastigial bilaterally (Asanuma et al., 1983b,c). A further well-entrenched belief, that the red nucleus, in which some efferent fibers of the dentate and interposed nuclei terminate (Flumerfelt et al., 1973; Stanton, 1980; Asanuma et al., 1983b,c), also projects upon the ventral lateral complex, seems to have been disproved by the work of Edwards (1972) and of Hopkins and Lawrence (1975).

Most of the early workers who studied cerebellothalamic projections with the Marchi method concluded that the cerebellar fibers terminated anterior to the medial lemniscus but did not reach the anterior end of the ventral nuclear mass (Chapter 1) (Vogt, 1909; André-Thomas, 1912; Walker, 1934; Le Gros Clark, 1932a; Hassler, 1949b). In more recent times, it had come to be reported that the cerebellar terminations were far more extensive than that, extending posteriorly into the ventral posterior nucleus and anteriorly into the ventral anterior nucleus (Mehler, 1971; Chan-Palay, 1977; Berkley, 1980). The most recent work, however, localizes the cerebellothalamic projection to what is called here the posterior ventral lateral nucleus (VLp) of monkeys (Percheron, 1977; Stanton, 1980; Tracey et al., 1980; Kalil, 1981; Asanuma et al., 1983b,c) and to a comparable region in other animals (Angaut, 1970; Hendry et al., 1979) (Figs. 7.10 and 7.28). Slight discrepancies in the most recent accounts stem largely from the confused terminology of the region. For example, Kalil (1981), in noting extension of the cerebellothalamic terminations along the fingers of the VLp that intrude into the VLa, identifies a projection to the latter nucleus. It is very clear, however, that no cerebellar efferents terminate in the islands of small, tightly packed cells that characterize the VLa (Asanuma et al., 1983b,c).

In cats the cerebellothalamic projection terminates in what is usually called the ventral lateral nucleus but extends into parts of a region sometimes called the ventral anterior nucleus (Hendry et al., 1979). Close examination of the architecture or cortical connectivity of the two, however, reveals no justifiable reason for distinguishing separate nuclei in the region of cerebellar terminations. We prefer, therefore, to define the boundaries of a posterior ventral lateral nucleus of the cat in terms of the endings of the cerebellar fibers. The same definition seems reasonable in other species as well (Rockel et al., 1972; Walsh and Ebner, 1973; Faull and Carman, 1968, 1978; Asanuma et al., 1983a).

As the finer organization of the cerebellothalamic projection system has been more thoroughly characterized in monkeys, the remainder of our account will concentrate on the monkey (Thach and Jones, 1979; Asanuma et al., 1983b,c). All three deep cerebellar nuclei terminate throughout the full extent of the VLp. The terminations of fastigiothalamic fibers are less dense than those of the fibers from the other two cerebellar nuclei. The ipsilateral fastigiothalamic projection is sparsest of all. These ipsilateral fibers enter the ipsilateral thalamus after traversing the posterior commissure, commissure of the superior colliculus, or internal medullary lamina of the thalamus, i.e., they have crossed twice, once on leaving the cerebellum and once on approaching the thalamus (Asanuma *et al.*, 1983b).

There is a precise topography in the organization of the thalamic projection from all three deep cerebellar nuclei (Fig. 7.29). Anterior parts of each nucleus project to lateral parts of the VLp and posterior parts of each nucleus project medially in the VLp, extending into the part called nucleus X by Olszewski (1952). Lateral parts of each cerebellar nucleus project dorsally in the VLp and

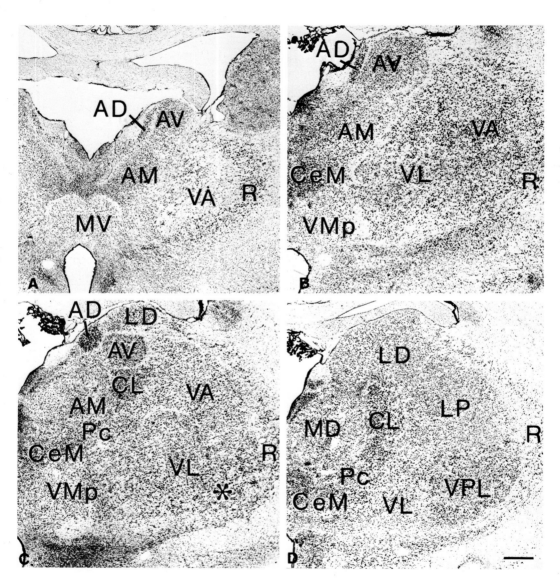

FIGURE 7.28. Paired brightfield and darkfield photomicrographs showing autoradiographically labeled cerebellothalamic terminations in VL nucleus of a cat. Bar = 1 mm. From Hendry *et al.* (1979).

medial parts ventrally (Fig. 7.29). In other experiments it has been discovered that lateral parts of the VLp project upon medial parts of the motor cortex; medial parts of the VLp, including the nucleus X component, project upon lateral parts of the motor cortex; and intervening parts project in-between (Walker, 1938a; Strick, 1976a; Jones *et al.*, 1979; Asanuma *et al.*, 1983a). Hence, because of the known body somatotopy in area 4, we may assume a similar somatotopy in the VLp and, by extrapolation, also in each of the deep cerebellar nuclei (Fig. 7.29).

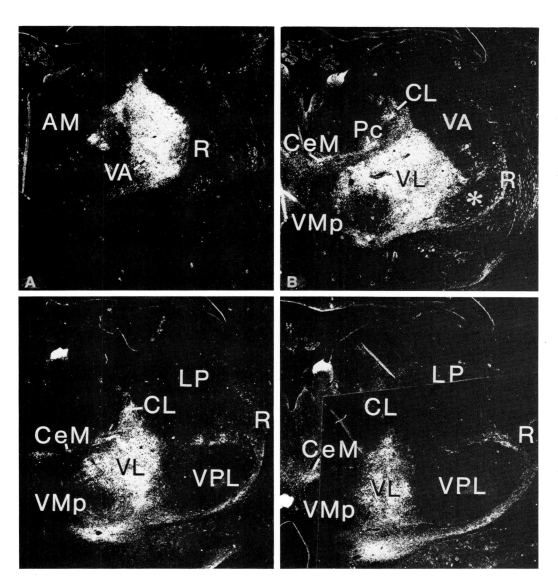

FIGURE 7.28. (*continued*)

There are relatively few conclusive data for body somatotopy in the deep cerebellar nuclei. From electrophysiological studies of various kinds, some workers have concluded that there is no discrete representation of individual muscle groups or movements (Bantli and Bloedel, 1976); others have suggested a single representation spanning all three nuclei (Goldberger and Crowdon, 1973; Massion and Rispal-Padel, 1972; Schultz *et al.*, 1976; Snider and Stowell, 1944a,b). There is now, however, some preliminary evidence from single-unit studies of neurons in the dentate nuclei of conscious monkeys trained to perform a task

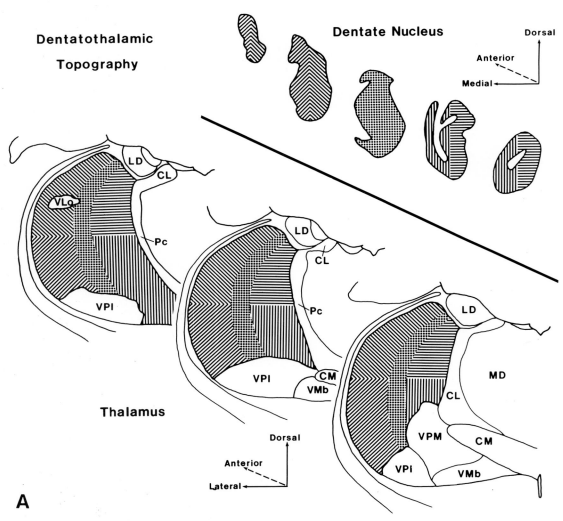

A

FIGURE 7.29. Topography in the cerebellothalamic projections. The three deep cerebellar nuclei all project in a similar topographic sequence upon the VLp nucleus. Knowing that medial parts of the VLp project to the face representation in the motor cortex, intermediate parts to the fore-limb representation, and lateral parts to the hindlimb representation, suggests that in each deep cerebellar nucleus, face is represented posteriorly, hindlimb anteriorly, and forelimb in-between. From Asanuma *et al.* (1983c).

with different muscle groups, indicating that neurons whose discharge is modulated during activity of one muscle group lie in a different part of the dentate nucleus from those modulated during activity of another muscle group (Perry and Thach, 1979).

The indirect anatomical evidence for a somatotopic representation in the VLp of the thalamus is supported by the results of single-unit studies in conscious animals, which indicate that neurons whose discharge alters in relation to movements of the forelimb, lie in the middle of the mediolateral extent of the nucleus

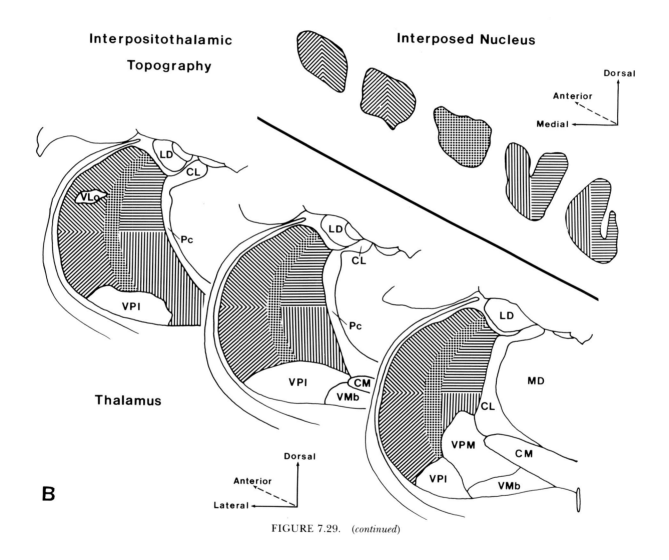

FIGURE 7.29. (*continued*)

(Strick, 1976b; Macpherson *et al.*, 1980). Neurons in this region project to the forelimb representation in area 4 (Strick, 1976a; Jones *et al.*, 1979) and, from the cerebellothalamic topography, we may predict the position of the representation of the forelimb in the deep cerebellar nuclei (Fig. 7.29).

As deduced from the topographic ordering of the thalamocortical projection from the VLp to area 4, the representation of a large body part such as the hand occupies a lamella comparable to that in the ventral posterior nucleus (Walker, 1938a). Unlike the lamellae in the ventral posterior nucleus, which are more or less vertical, those in the VLp curve backwards over the dorsal surface of the ventral posterior nucleus (Fig. 7.25) as they follow the curved contours of the VLp.

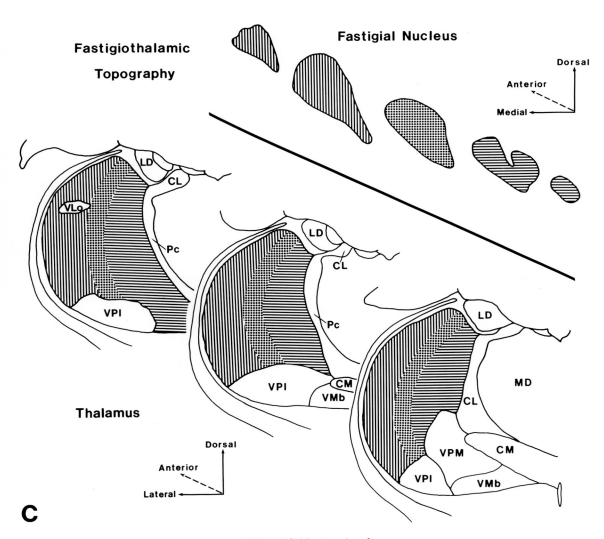

FIGURE 7.29. (*continued*)

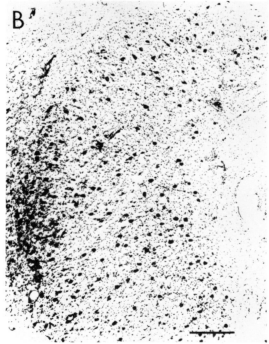

FIGURE 7.30. (A) Rodlike termination of autoradiographically labeled dentatothalamic fibers in dorsal part of VLp nucleus of a monkey. Sagittal section. (B) Injection of tritiated amino acids in lateral part of contralateral dentate nucleus. Frontal section. Bars = 250 μm. From Asanuma *et al.* (1983c).

Within a lamella of the VLp, there is a finer fractionation of the cerebellothalamic terminations, very similar to that of lemniscal terminations in the ventral posterior nucleus (Thach and Jones, 1979; Kalil, 1981; Asanuma *et al.*, 1983c). Fibers emanating from a small focus of cells in the dentate nucleus end as a single, elongated, rodlike zone of terminations that extend anteroposteriorly through the VLp. Injections of tracer that affect larger parts of the dentate nucleus give rise to multiple rods of terminations (Fig. 7.30). The intervening gaps are to a large extent filled by the terminations of fastigiothalamic, interpositothalamic, and spinothalamic fibers (Fig. 7.31), though the terminations of fastigiothalamic and interpositothalamic fibers are more focal than rodlike (Fig. 7.32).

At the present time, we have little inkling as to the significance of these alternating domains of afferent terminations. It is by no means certain that

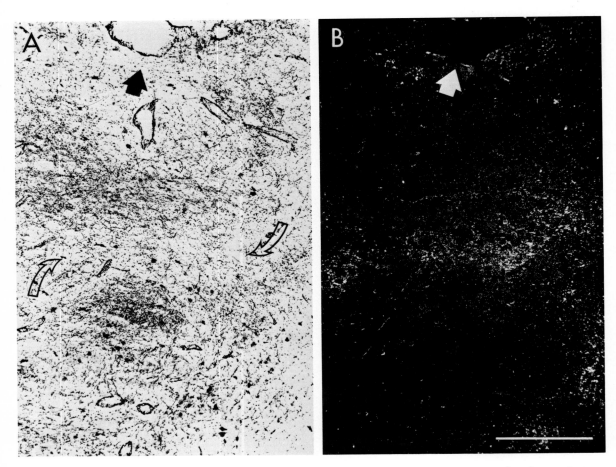

FIGURE 7.31. Adjacent sagittal sections showing alternation of degenerating dentatothalamic fiber terminations (A) and autoradiographically labeled interpositothalamic terminations (B) in a monkey in which the contralateral dentate nucleus was lesioned and the interposed nucleus on the same side was injected with tritiated amino acids. Bar = 500 μm. From Asanuma *et al.* (1983c).

adjacent zones of dentatothalamic and interpositothalamic terminations end on different VLp cells. The length and radiating nature of the dendrites of these large cells make such a possibility rather unlikely and there is a limited amount of data from intracellular recordings in the VLp of the cat to support the idea of dentate and interposital inputs to single thalamic cells (Shinoda *et al.*, 1982). But the sizes of the zones of termination make it likely that some cells should be more profoundly influenced by one input than by another. Hence, because of the focal or columnar nature of the inputs to the motor cortex, it is likely that some cerebellar columns in the motor cortex may be more influenced by one cerebellar nucleus than by others.

7.4.4. Cortical Target of the VLp

The cortical target of the ventral lateral complex has been known for many years to include the motor cortex. This information is evident in the early work of von Monakow (1895) and his observations were confirmed many times with the retrograde degeneration method (Minkowski, 1924; Walker, 1938a; Le Gros Clark and Boggon, 1935; Waller and Barris, 1937). Despite the well-established

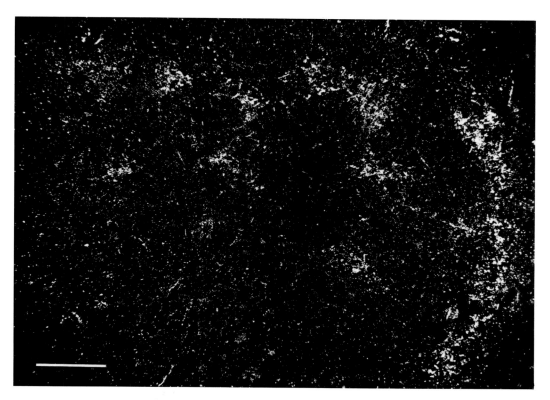

FIGURE 7.32. Frontal section showing focal clusters of autoradiographically labeled interpositothalamic fiber terminations in VLp nucleus of a cynomolgus monkey. Bar = 300 μm. From Asanuma *et al.* (1983c).

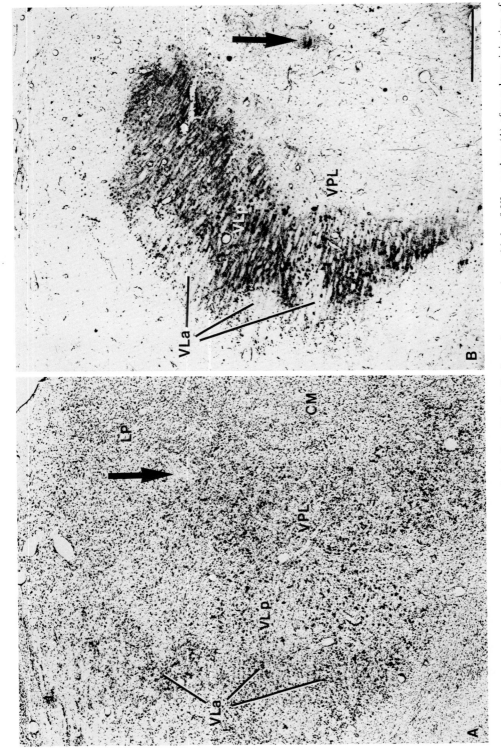

FIGURE 7.33. Alternate sections showing retrograde labeling (B) of cells throughout ventral two-thirds of VLp nucleus (A) after a large injection of horseradish peroxidase in the motor cortex of a monkey. Note how labeled cells extend around and between islands of the VLa. Lesions on same electrode track indicated by arrows. Bars = 1 mm. After Jones and Friedman (1982).

nature of the information, exact details of the distribution of thalamocortical fibers from the ventral lateral complex have been surprisingly lacking. Recent studies indicate that in monkeys, the whole VLp projects only to area 4 of the cortex; i.e., to an area 4 defined cytoarchitectonically as the gigantocellular field and excluding the adjacent area 6aα of Vogt and Vogt (1919), but including area 4a of von Bonin and Bailey (1947) [i.e., area FA of von Economo and Koskinas (1925)] (Figs. 7.33 and 7.34) (Jones *et al.*, 1979; Friedman and Jones, 1981; Asanuma *et al.*, 1983a). Though area 6aα was included in the primary motor cortex by Woolsey (1958) in his surface stimulation studies, there is mounting evidence from anatomical (Muakkassa and Strick, 1979) and physiological (Weinrich and Wise, 1982) studies that it should be regarded as a separate field.

We have not been able to find any evidence of a projection to area 4 from the VLa. Strick (1976a) showed a few retrogradely labeled cells in this [the VLo nucleus of Olszewski (1952)] in his drawings illustrating the results of his experiments involving injections of tracer in the arm area of the motor cortex. However, because the intrusion of the fingers of the VLp among the islands of

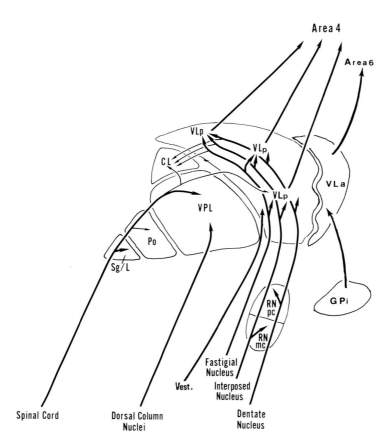

FIGURE 7.34. Scheme of afferent and efferent connections of VLp nucleus in monkeys. RNpc, RNmc: parvocellular and magnocellular parts of red nucleus; Vest: vestibular nuclei. Modified from Asanuma *et al.* (1983b).

the VLa was not recognized at that time, it seems likely that these few cells lay in the fingers of the VLp. Alternatively, the spread of the injection in the cortex may have been underestimated, for the small cells in the islands of the VLa become retrogradely labeled if area 6aα is involved (Asanuma *et al.*, 1983c).

In cats the cortical projection of the ventral lateral complex has usually been studied in relation to the complex as a whole, rather than specifically to the zone receiving cerebellar terminations. Such studies have indicated projections to areas 6 and 4 of the cortex (Strick, 1970, 1973). When the separation of the complex into cerebellar and noncerebellar recipient zones is taken into account, the limited evidence currently available suggests that the cerebellar recipient zone projects only to area 4 (Hendry *et al.*, 1979). Comparable results have been obtained in raccoons (Sakai, 1982) and rats (Donoghue *et al.*, 1979), though in rats a distinction between areas 4 and 6 is not particularly easy to make.

7.4.5. Other Sources of Afferents to the VLp and the Motor Cortex

7.4.5.1. Spinothalamic

As mentioned in Section 7.2.4.2, the spinothalamic tract extends its terminations from the ventral posterior nucleus into the VLp. The terminations are relatively sparse, alternate with cerebellothalamic terminations, and appear to be confined to the ventral, larger-celled region of the VLp in monkeys (Asanuma *et al.*, 1983b,c) and to a region of the ventral lateral nucleus immediately anterior to the ventral posterior nucleus in cats and other species (Lund and Webster, 1967b; Boivie, 1971b; Jones and Burton, 1974; Applebaum *et al.*, 1979; Burton and Craig, 1983; Asanuma *et al.*, 1980). The region of terminations is included in the part of the ventral lateral complex projecting to the motor cortex (Hendry *et al.*, 1979; Larsen and Asanuma, 1979; Asanuma *et al.*, 1983a–c). In these regions in monkeys and cats, particularly short-latency (3–8 msec) responses to electrical stimulation of peripheral nerves have been reported (Horne and Tracey, 1979; Lemon and van der Burg, 1979; Larsen and Asanuma, 1979). In conscious monkeys, neurons with cutaneous receptive fields have been identified in ventral parts of the VLp (Horne and Porter, 1980). Neurons in more dorsal parts of the VLp do not have readily identifiable peripheral receptive fields.

The relay of afferent information at short latency through the ventral part of the VLp has suggested to some that this could provide the basis for the short-latency somatic sensory input to the motor cortex (Horne and Tracey, 1979; Lemon and van der Burg, 1979; Asanuma *et al.*, 1980). Many neurons in the motor cortex of cats and monkeys have peripheral receptive fields similar to those in the somatic sensory cortex (Albe-Fessard and Liebeskind, 1966; Rosén and Asanuma, 1972; Conrad *et al.*, 1975; Porter and Rack, 1976) and evoked potentials or unit discharges can be elicited at latencies as short as 5–10 msec in the motor cortex following electrical stimulation of peripheral nerves (Brinkman *et al.*, 1978; Lemon and Porter, 1976; Asanuma *et al.*, 1979b, 1980).

The spinothalamic input to the VLp is an obvious candidate for mediating the short-latency input to area 4 and the receptive fields of spinothalamic tract cells would be appropriate for cells that provided the inputs to the motor cortex cells. Unfortunately, there is a considerable problem associated with this point of view: the receptive fields of motor cortex neurons disappear in monkeys after sectioning the dorsal columns of the spinal cord (Brinkman et al., 1978) and the evoked potential in area 4 is concomitantly reduced (Asanuma et al., 1979b). This finding was originally taken to imply terminations of the medial lemniscus in the VLp but this has not been confirmed (Tracey et al., 1980; Kalil, 1981; Asanuma et al., 1983b), and it is clear that the lemniscal relay, the ventral posterior nucleus, does not send branched projections to both area 4 and the somatic sensory cortex (Jones et al., 1979). An alternative view might be that the cerebellar projection to the VLp could provide the somatic sensory input to area 4. Here too, there are problems: first, evoked responses in the motor cortex survive ablation of the cerebellum (Malis et al., 1953); second, it is often reported that relatively few neurons in the deep cerebellar nuclei have detectable somatic sensory receptive fields and that if they respond to natural or electrical stimuli, do so rather weakly and at long latencies [12–16 msec or much longer (Eccles et al., 1977a,b; Harvey et al., 1979)]. More recently, however, shorter-latency responses have been reported in the nucleus interpositus of cats (Shinoda, 1983). It has also been shown that although the earliest responses to a sensory perturbation of a manipulandum in awake monkeys are in the motor cortex (15 msec), rather than in the interposed nucleus (18 msec), the median response latency was shorter in the interposed nucleus (44 msec) than in the motor cortex (57 msec) (Thach, 1978). Hence, the interposed nucleus could still play a role in influencing the slower-responding motor cortex cells (Asanuma et al., 1983c).

An alternative route for short-latency somatic sensory input to reach the motor cortex has been suggested to be formed by corticocortical fibers linking the somatic sensory cortex to the motor cortex (Wiesendanger et al., 1976). This, too, presents problems for though ablation of the postcentral gyrus may cause a reduction in the amplitude of an evoked potential in the motor cortex, it does not abolish it completely (Malis et al., 1953; Asanuma et al., 1980). There is currently no ready explanation for what seems at first sight a relatively straightforward observation.

Even when the route responsible for short-latency somatic sensory input to the motor cortex is finally elucidated, it is unclear what role this plays in the behaving animal. It is established that the discharges of some motor cortex neurons are profoundly affected by sensory perturbations occurring prior to or during a movement (Evarts and Fromm, 1977). But, though losing their peripheral receptive fields after destruction of the dorsal columns of the spinal cord, the discharges of many motor cortex neurons firing in relation to a movement are unaffected (Brinkman et al., 1978). In the ventral lateral complex of the thalamus in conscious monkeys, neurons discharging in relation to movement, if affected at all by peripheral perturbations that lead to sensory feedback, are affected only at long latencies (Strick, 1976b; Macpherson et al., 1980). By contrast, the movement-related discharge occurs particularly early in relation to the movement (Fig. 7.35). This has led some investigators to view the cerebellothalamic input as providing an initiating signal to the motor cortex prior to

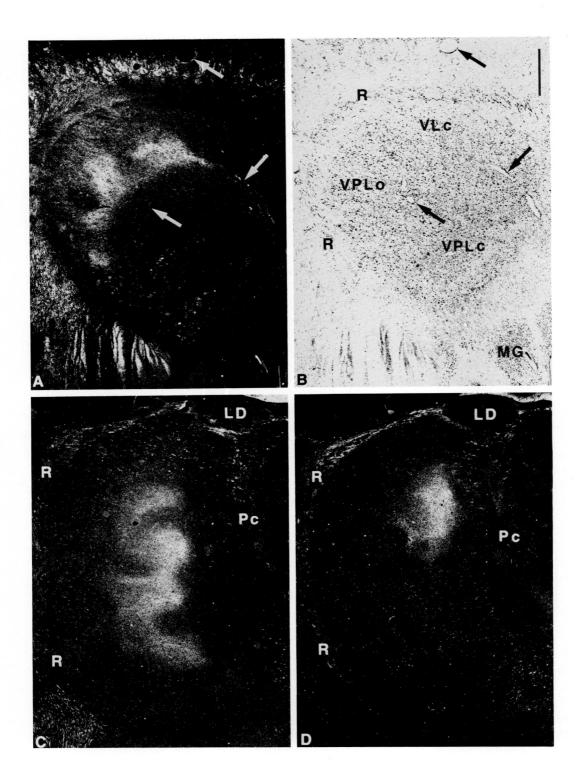

certain kinds of movement (Evarts and Thach, 1969; Allen and Tsukahara, 1974) instead of simply forming one of several possible sensory feedback routes from the periphery.

7.4.5.2. Dual Representation in the Motor Cortex

The relay of spinothalamic inputs through the ventral aspect of the VLp is also of interest in terms of the differential distribution of peripheral receptive fields in the motor cortex. In monkeys, neurons in posterior parts of area 4 of both the hand and foot representation have receptive fields that are cutaneous in character while those in anterior parts of area 4 mostly have receptive fields in deep tissues (Strick and Preston, 1978a,b; Tanji and Wise, 1981). It appears that there is a similar differential distribution of receptive field types in the motor cortex of the cat (Pappas and Strick, 1981). Though only small patches of cortex have been mapped in each case, these results have been taken to imply a separate representation of the body in each part of the motor cortex, formerly thought to contain a single representation such as is commonly drawn in "homuncular fashion" (e.g., Fig. 7.36).

In monkeys at least, it is clear that ventral parts of the VLp project to posterior parts of area 4 (Jones *et al.*, 1979; Tracey *et al.*, 1980), so that the spinothalamic input to this region might form the basis for the cutaneous receptive fields. Dorsal parts of the VLp project to anterior parts of area 4 but these dorsal parts receive only cerebellar inputs and only a few of the neurons there in conscious monkeys have obvious peripheral receptive fields (Horne and Porter, 1980; Macpherson *et al.*, 1980). It seems possible that the VLp may well be subdivided in a manner rather similar to the ventral posterior nucleus with different parts projecting to different functional areas of the cortex, and that within the single rather broad representation in the VLp, more than one body representation may lie hidden.

7.4.5.3. Vestibulothalamic Projection

The ventral, larger-celled part of the VLp is also the site of termination of a rather sparse vestibulothalamic projection in monkeys (Lange *et al.*, 1979; Asanuma *et al.*, 1983b). A vestibulothalamic projection in the cat has not yet been so accurately delimited (Raymond *et al.*, 1976; Maciewicz *et al.*, 1982). In both species the origin of the projections is a rather modest number of cells in the ventral part of the vestibular nuclei, in a region where the spinal, lateral and medial vestibular nuclei come together (Condé and Condé, 1978; Kotchabhakdi

---

FIGURE 7.35. (A, B) Darkfield and brightfield photomicrographs from the same sagittal section of a cynomolgus monkey, showing autoradiographic labeling of corticothalamic fiber terminations in VLp nucleus. (C, D) Frontal sections from different cynomolgus monkey brains showing labeling of corticothalamic fiber terminations in ventral part of VLp (C) after injection of tritiated amino acids in posterior part of area 4, and in dorsal part of VLp (D) after injection in anterior parts. Bar = 1 mm. Modified from Jones *et al.* (1979).

et al., 1980; Tracey et al., 1980; Maciewicz et al., 1982) and more or less coextensive with the site of termination of the relatively small spinovestibular projection (Mehler, 1969).

Studies of the vestibulothalamic pathways have been carried out intermittently for many years. Evoked potential and single-unit studies in cats had originally localized the thalamic vestibular relay in the vicinity of the magnocellular medial geniculate nucleus of the cat (Mickle and Ades, 1954) and its cortical target within the anterior suprasylvian sulcus (Walzl and Mountcastle, 1949; Mickle and Ades, 1952). But the consensus of anatomical opinion following Tarlov's (1969) careful degeneration study in primates was that no direct vestibulothalamic pathway existed. Tarlov pointed out that previous reports of

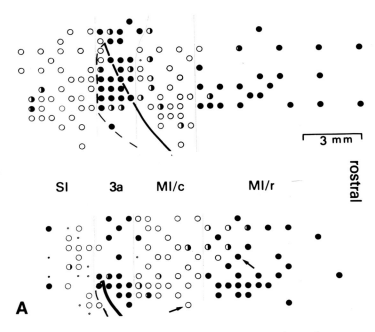

FIGURE 7.36. (A) Submodality segregation in the foot representations of sensory motor cortex of two cynomolgus monkeys. Heavy lines indicate central sulcus and dashed lines cytoarchitectonic borders. Solid circles indicate microelectrode penetrations that encountered units with receptive fields in deep tissues; open circles indicate penetrations that encountered units with cutaneous receptive fields; half-filled circles indicate penetrations that encountered units with mixed, deep, or cutaneous receptive fields. SI indicates mainly area 3b; MI/c and MI/r the "cutaneous" and "deep" subdivisions of area 4. From Tanji and Wise (1981). (B) A and B show data from two movement-related cells in the ventral lateral nucleus of a conscious monkey. Movement trials are aligned at onset of movement (arrows). Individual trials are in raster form with each dot representing an action potential. Histograms with data from multiple trials are at top of each composite figure. With cell A there is consistent excitation prior to flexion only. Responses of cell A are shown in situations in which signal to move was a movement (line step) in a line on an oscilloscope screen viewed by the monkey. Cell B responses are shown in situations in which cue to move was a visual signal, a line step, or a twisting force (torque) applied to the manipulandum grasped by the animal. Cell discharges primarily in relation to flexion. "On" and "off" refer to movements when monkey aligns his visual tracking signal with a line that is displaced and when he realigns signal after line returns to original position on cessation of perturbation. From Macpherson et al. (1980).

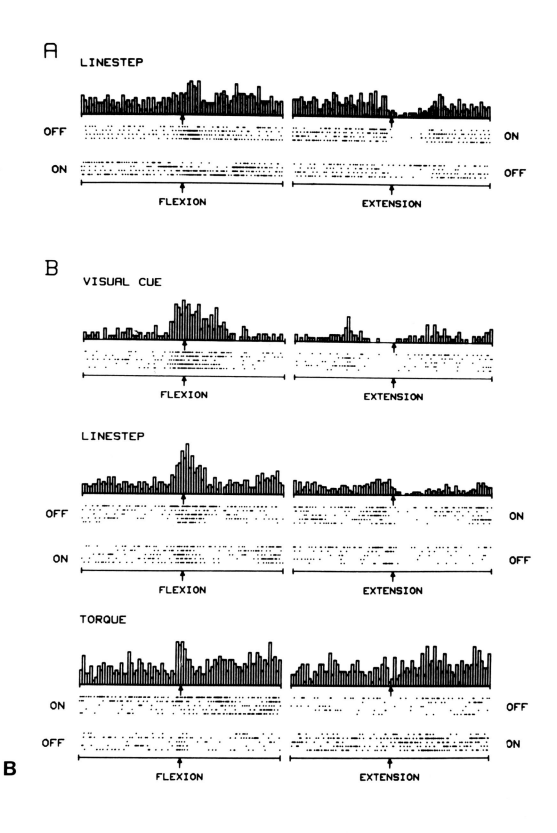

positive findings in degeneration studies were probably based upon incidental damage to adjoining structures such as the superior cerebellar peduncle and trigeminal nuclei.

Field potential and single-unit studies subsequent to the study of Tarlov continued to show short-latency responses to stimulation of the vestibular nerve (either directly or by rotation of an animal) in several thalamic nuclei of monkeys and in various parts of the cerebral cortex of both cats and monkeys (Sans *et al.*, 1970; Landgren *et al.*, 1967; Schwarz and Fredrickson, 1971; Schwarz *et al.*, 1973; Liedgren *et al.*, 1976; Büttner *et al.*, 1977; Deecke *et al.*, 1975; Büttner and Henn, 1976; Lange *et al.*, 1979; Blum *et al.*, 1979). The relevant thalamic

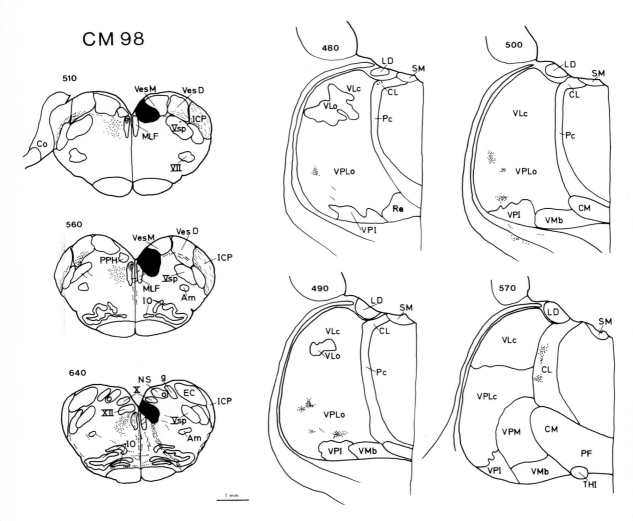

FIGURE 7.37. Distribution of autoradiographically labeled vestibulothalamic fibers following injection of tritiated amino acids in contralateral medial vestibular and prepositus hypoglossi nuclei. From Asanuma *et al.* (1983b).

nuclei were variously reported as the ventral posterior, ventral lateral, and ventral posterior inferior. The exact termination of primary vestibulothalamic fibers was localized anatomically to the ventral, larger-celled part of the VLp by the use of the autoradiographic method by Lange *et al.* (1979) and was subsequently confirmed by Asanuma *et al.* (1983b) (Fig. 7.37). Presumably the lack of sensitivity of the axonal degeneration method had previously prevented Tarlov from identifying the rather sparse projection.

Though the vestibular fibers terminate in the relay to the motor cortex, the vestibular projection field in the cerebral cortex has been localized within the somatic sensory or parietal cortex (Fredrickson *et al.*, 1966), in parts of areas 2 and 3a of monkeys (Odkvist *et al.*, 1974; Schwarz *et al.*, 1973), and close to the second somatic sensory area in cats (Landgren *et al.*, 1967). It is possible that the responses elicited in these areas by vestibular nerve stimulation are mediated by corticocortical fibers from area 4 or by other thalamic nuclei such as the VPI (Burton and Jones, 1976) though the source of vestibular inputs to the latter suggested in field potential studies (Deecke *et al.*, 1975) has not been confirmed anatomically.

7.4.6. VLa Nucleus and Pallidal Inputs

It is a traditional belief that two other motor-related pathways—the pallidothalamic and nigrothalamic—converge on the same thalamic territory as the cerebellothalamic pathway (Nauta and Mehler, 1966; Kemp and Powell, 1971; Mehler and Nauta, 1974). It is difficult to trace the origins of this belief. Vogt (1909) and Vogt and Vogt (1920) had early remarked upon the apparent segregation of the pallidal and cerebellar terminations in monkeys and man. Wilson (1914) made no clear statement and Walker (1938a) devoted little attention to the pallidal input. Hassler (1949b,c, 1959) in man reemphasized the Vogts' point of view. When single-unit investigations of the ventral lateral complex were made in cats, the number of neurons receiving convergent inputs from two or three of these sources was exceedingly small (Purpura *et al.*, 1966; Uno *et al.*, 1970; Purpura, 1972) though the few neurons demonstrating convergent inputs tended to be emphasized, and continue to be so (Chevalier and Deniau, 1982). Part of the reason for the belief in convergence undoubtedly stemmed from the loose use of the term "ventral lateral" nucleus, as though it were a homogeneous entity, and the inclusion in it of parts of other nuclear complexes. Added to this were the descriptions, from axonal degeneration studies of more extensive terminations of cerebellothalamic and pallidothalamic axons (Mehler, 1971; Mehler and Nauta, 1974; Nauta and Mehler, 1966) than are now known to exist (Percheron, 1977; Tracey *et al.*, 1980; Kalil, 1981; Asanuma *et al.*, 1983a–c). The present position seems to be that fibers from the internal segment of the globus pallidus terminate only in the VLa of monkeys (Kuo and Carpenter, 1973; Tracey *et al.*, 1980; De Vito and Anderson, 1982; Parent and De Bellefeuille, 1982) and in comparable anteromedial parts of the ventral lateral–ventral anterior complex in cats (Hendry *et al.*, 1979) (Figs. 7.38 and 7.39). In both species their termi-

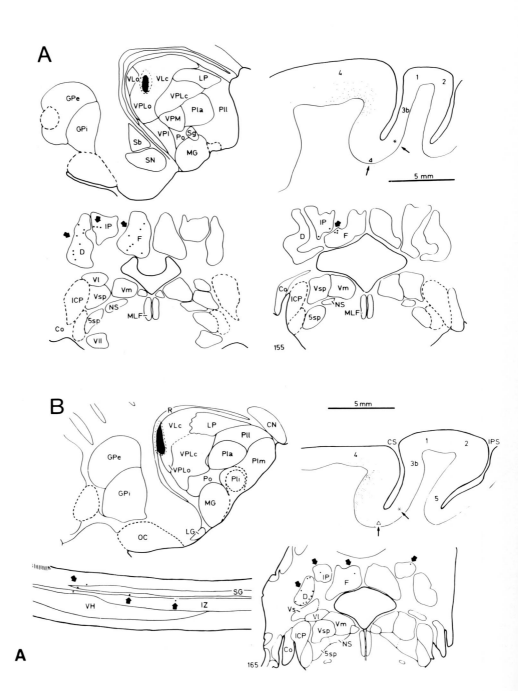

nations appear to be independent of those of axons from the deep cerebellar nuclei. The terminal nucleus or region of the pallidal fibers, unlike that of the cerebellar inputs, does not project to area 4 but to parts of area 6 anterior to area 4 (as defined on p. 395). Its projection probably includes the supplementary motor area. Neurons of the supplementary motor area discharge earlier in relation to movement onset than those in area 4 (Tanji *et al.*, 1980) and the supplementary motor area in man shows enhanced blood flow even while thinking about a movement (Roland *et al.*, 1980). It has, thus, been considered to play a role in planning movement stratagems and it is interesting to speculate that the pallidothalamic input to it may also be involved.

In cats, rats, and monkeys, axons from the pars reticulata of the substantia nigra terminate more medially in the principal ventral medial nucleus [called

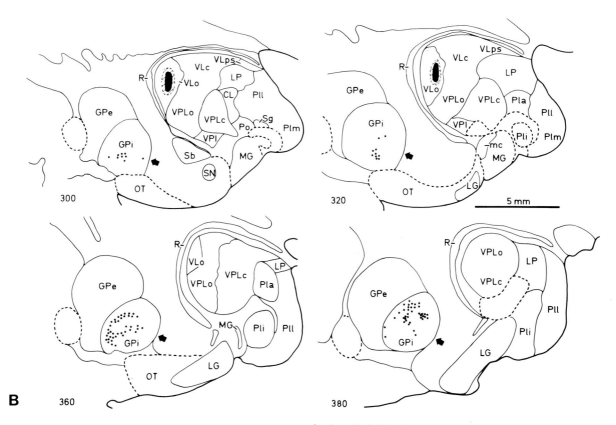

FIGURE 7.38.   (A) Sagittal sections of thalamus and cortex and frontal sections of the deep cerebellar nuclei in two different experiments in monkeys in which horseradish peroxidase was injected in dorsal (A, VLc) and ventral (B, VPLo) parts of the VLp nucleus. Retrogradely labeled corticothalamic cells appear in anterior parts of area 4 after the dorsal injection and in posterior parts after the ventral in-jection. Both injections cause retrograde labeling of cells in the contralateral deep cerebellar nuclei. But only the ventral injection leads to retrograde labeling of spinal cord cells (B). (B) Retrograde labeling of cells in the internal segment of the monkey globus pallidus after injection of horseradish peroxidase in the ipsilateral VLa nucleus. Both from Tracey *et al.* (1980).

the pars medialis of the ventral lateral nucleus by Olszewski (1952)] (Faull and Carman, 1968; Carpenter *et al.*, 1976; Beckstead *et al.*, 1979; Hendry *et al.*, 1979) (Figs. 7.40 and 7.41). This nucleus projects to cortical areas on the medial surface of the cerebral hemisphere that are closely related to the limbic system (Beckstead, 1976; Hendry *et al.*, 1979). There is little or no overlap of nigrothalamic territories into the terminal territories of the cerebellar and pallidal inputs in monkeys and, though there may be slight overlap with the pallidal terminations in the other species, the inputs are definitely concentrated in independent regions (i.e., the VMp and VL). These regions are not necessarily as clearly distinguishable cytoarchitectonically as they are in monkeys. The present viewpoint about the motor-related pathways must be, therefore, that they each terminate in well-circumscribed components of the thalamus which though they may not be cytoarchitectonically separable in nonprimates, have different input–output connections so that convergence of the cerebellar, pallidal, and nigral efferent pathways in the thalamus and cortex is minimal or absent. This would seem to

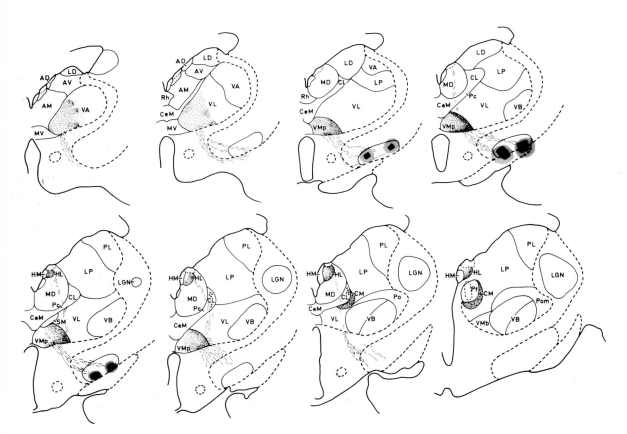

FIGURE 7.39. Autoradiographic demonstration of terminations of pallidothalamic fibers in the cat. From Hendry *et al.* (1979). Compare with Figs. 7.40 and 7.41.

have profound implications for theories and experiments on the functions of the thalamus and cortex in motor behavior.

7.4.7. Cerebellar Inputs to Parietal Cortex?

Sasaki *et al.* (1972, 1975) have repeatedly demonstrated that short-latency evoked potentials can be elicited in the anterior parietal cortex following electrical stimulation of either side of the cerebellum. They have concluded that the effect emanates from the deep cerebellar nuclei and is relayed through the ventral anterior nucleus because recruiting responses can be elicited in the anterior parietal cortex (areas 5 and 7) by electrical stimulation in these two regions (Sasaki *et al.*, 1972, 1975, 1976). An alternative explanation of these results might have been that the effect is mediated by the intralaminar complex which has a heavy projection to areas 5 and 7 (Hendry *et al.*, 1979; Macchi *et al.*, 1977) and stimulation of which elicits recruiting responses in those areas (Jasper, 1960). However, there is now a small amount of evidence that some cells lying among the terminations of the cerebellothalamic fibers in the ventral lateral complex of the cat project to anterior parietal areas 5 and 7 (Hendry *et al.*, 1979). These cells lie in a portion of the ventral lateral complex that is separate from that projecting to area 4. There are no comparable data yet available in monkeys. Retrogradely labeled neurons reported in the magnocellular division of the ventral anterior nuclei by Divac *et al.* (1977b) do not appear to lie within the zone of cerebellothalamic terminations (Stanton, 1980; Kalil, 1981; Asanuma *et al.*, 1983b), though this was once described (Mehler, 1971).

7.5. Ventral Anterior Nucleus

7.5.1. Description

A ventral anterior nucleus, though often described, is scarcely distinguishable in small mammals. When labeled, it is usually identified as a region of relatively large cells at the extreme anterior end of the ventral nuclear complex. The cells are usually less densely packed than those in the ventral lateral complex. When the afferent connections of the region are taken into account, it seems clear that the architectonic separation is fairly arbitrary though there does appear to be a small region that is devoid of pallidal, cerebellar, and substantia nigral inputs (Figs. 7.28, 7.39, and 7.41). By analogy with monkeys, this region probably warrants the name *ventral anterior nucleus,* though it is doubtful that it is only this region that is labeled in descriptive accounts.

In monkeys a ventral anterior nucleus is a clearly defined cytoarchitectonic entity. It lies between the reticular nucleus anteriorly and the VLa posteriorly (Fig. 7.42) and extends from the external to the internal medullary lamina. Two parts are commonly recognized: an anterolateral smaller-celled, principal division (VAp) and a posteromedial magnocellular division (VAmc), surrounding the mamillothalamic tract and, medially, not clearly distinguishable from the central lateral nucleus (Fig. 7.42). In myelin-stained preparations, penetration of the nucleus by bundles of fibers of the anterior thalamic radiation gives it a lobulated appearance. In man the ventral anterior nucleus, called nucleus latero-polaris by Hassler (1959), is particularly distinct and divided into as many as five subnuclei.

7.5.2. Terminology

Most early writers on the thalamus, even of primates, did not identify a ventral anterior nucleus as an entity separate from the ventral lateral complex. In 1934 Aronson and Papez in the rhesus monkey and Grünthal in several species recognized a separate anterior component of the ventral nuclei, calling it respectively the nucleus ventralis, pars anterior or nucleus ventralis oralis. It appears to be the same as the anteromedial ventral nucleus earlier described in the tree shrew by Le Gros Clark (1929a).

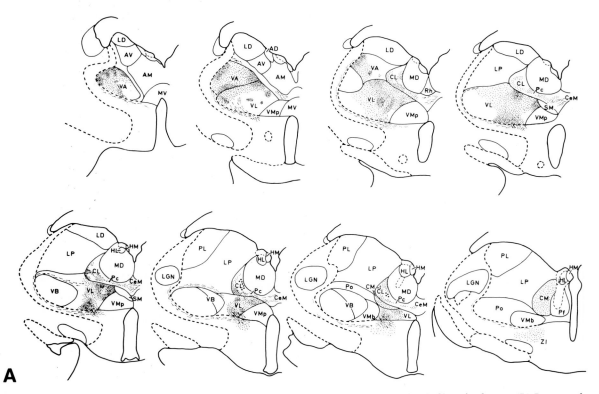

A

FIGURE 7.40. (A) Autoradiographic demonstration of terminations of cerebellothalamic fibers in the cat. (B) Retrograde labeling of thalamocortical cells projecting upon area 4 of the cat, following a small injection of horseradish peroxidase in the forelimb representation. Both from Hendry *et al.* (1979).

Walker (1935, 1938a), in particular, drew attention to the nucleus when he claimed that in the monkey and chimpanzee, it remained "almost completely intact following hemidecortication." He believed that it only showed severe retrograde degeneration if the striatum was involved. In later years this comment sometimes came to be taken as confirmatory evidence for a postulated functional similarity between the ventral anterior and the intralaminar nuclei. Subsequent to Walker's remarks, Waller (1940b) in the cat and Rose and Woolsey (1943) claimed that a region comparable to the ventral anterior nucleus did show severe atrophy after destruction of the motor cortex in cats or after hemidecortication in rabbits, implying a cortical projection. It is now customary to identify a cortically projecting ventral anterior nucleus in most animals, though, as pointed out in the preceding section, this is not particularly easy on cytoarchitectonic grounds in subprimates.

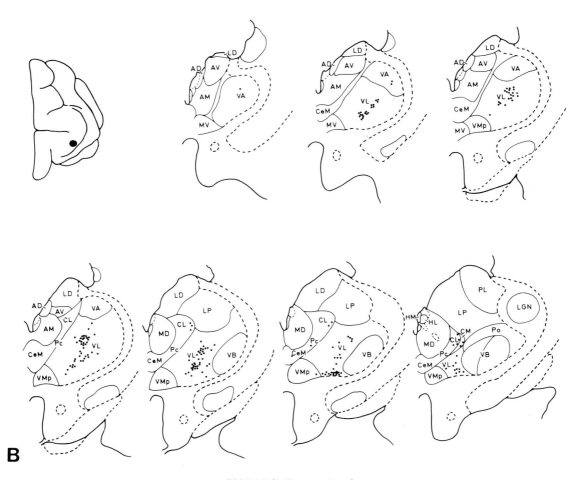

FIGURE 7.40. (*continued*)

7.5.3. Connections

A critical reader of the literature on the connections of the thalamus would probably be led to conclude that the input–output connections of the ventral anterior nucleus are still unknown. Occasionally a few retrogradely labeled cells are reported in it after injections of tracer in various parts of the cerebral cortex in the striatum. However, given the difficulty of identifying the nucleus accurately in nonprimates and the small number of labeled cells, it is difficult to accept the validity of many of these accounts. The only cortical areas that seem to be consistently involved in the various reports that have emerged seem to be those of the anterior parietal regions (Robertson, 1977; Divac *et al.*, 1977b; Hendry *et al.*, 1979). This is somewhat surprising in view of the usually accepted relationship of the ventral nuclei with the central gyri and the premotor cortex. Probably the soundest judgement one can make at the moment about the cortical connectivity of the ventral anterior nucleus is that the subject requires more work, particularly in the monkey in which the nucleus is large and readily identified.

The afferent connections of the ventral anterior nucleus are equally uncertain. Though cerebellar, pallidal, and substantia nigral fibers have sometimes

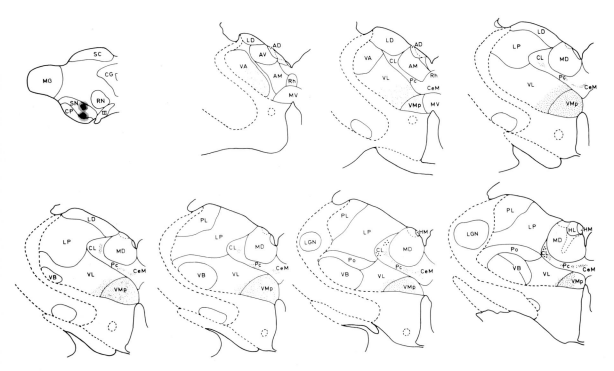

FIGURE 7.41. Autoradiographic demonstration of terminations of nigrothalamic fibers in the cat. From Hendry *et al.* (1979).

been reported ending in it, the ventral anterior nucleus is not described as the principal site of termination so the positive finding of a few terminations, again, probably arises from the difficulty of accurately delimiting the nucleus. There seems to me to be no clear connectional difference between the principal and the magnocellular parts as currently described in the literature but their cytoarchitecture is so different that connectional differences would be anticipated. The magnocellular is probably part of the central medial nucleus (Fig. 7.42).

Nauta and Whitlock (1954) described axonal degeneration in the ventral anterior nucleus of cats following destruction of the intralaminar nuclei and particularly of the centre médian. This observation was taken as confirmation of the viewpoint that nonspecific projections from the ventral anterior nucleus to the cortex, like those postulated from the intralaminar and certain other thalamic nuclei, could be involved in the cortical "recruiting response" (Jasper, 1960). The observation of Nauta and Whitlock does not seem to have been duplicated with modern tracing techniques. It is, therefore, uncertain whether the degeneration reported represented degeneration of terminals or simply of fibers traversing the nucleus en route to the cortex or striatum.

Sasaki *et al.* (1975, 1976) report that electrical stimulation of the cerebellum of the cat leads to evoked potentials and recruiting responses in the parietal cortex and consider that this effect is mediated by the ventral anterior nucleus, and a few retrogradely labeled cells in the thalamus after injections of tracer in the anterior parietal cortex lie within the zone of cerebellar terminations in the cat (Hendry *et al.*, 1979). However, in the monkey thalamus from which the definition of the ventral anterior nucleus derives, the nucleus lies outside the cerebellar terminal region (Stanton, 1980; Kalil, 1981; Asanuma *et al.*, 1983b).

7.6. Ventral Medial Complex

7.6.1. Description

The ventral medial complex is a group of nuclei, some of which have ill-defined borders, mainly extending along the medial borders of the ventral lateral complex and ventral posterior nucleus and across the midline beneath the internal medullary lamina. By contrast with many of the other nuclear complexes referred to in this chapter, its several component nuclei are probably best identified in nonprimates, though part of the reason for their apparent omission from standard atlases of primate thalami stems from their inclusion in other nuclear groups.

Three nuclei make up the ventral medial complex: a basal ventral medial nucleus (VMb), a submedial nucleus (Sm), and a principal ventral medial nucleus (VMp) (Figs. 7.43 and 7.44). The VMb is readily distinguished in virtually all mammals as a small-celled region closely associated with the ventromedial edge of the ventral posterior medial nucleus (VPM) in cats and monkeys, though in the rat and other rodents commonly separated from the VPM by an extension

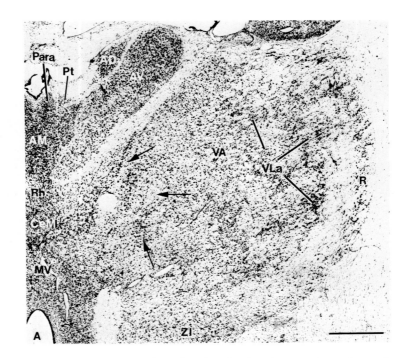

FIGURE 7.42. (A) Thionin-stained frontal section showing the ventral anterior nucleus in a cynomolgus monkey at level at which islands of VLa nucleus start to encroach on it. Arrows indicate large cells called magnocellular VA nucleus by Olszewski (1952) which may belong to central medial nucleus. (B) Thionin-stained frontal section through the principal ventral medial nucleus of a rhesus monkey. (C) Thionin-stained frontal section through the basal ventral medial nucleus of a cynomolgus monkey. Bars = 1 mm (A), 500 μm (B, C).

of the ventral part of the external medullary lamina. The nucleus is especially large in ungulates (Solnitzky, 1938; Rose, 1942b) and monotremes (Fig. 7.48).

The Sm is a small nucleus containing a few, relatively widely dispersed, small cells, usually lying on the dorsal or anterodorsal aspect of the VMp. In cats it is at the anterior pole of the centre médian nucleus and is enclosed in a sling of the internal medullary lamina (Figs. 7.44, 12.2, 13.1, 13.2, and 16.1). In rats it surrounds the mamillothalamic tract as this enters the thalamus (Figs 7.43 and 13.2). An equivalent Sm has not usually been identified in monkeys, except in the squirrel monkey by Emmers and Akert (1963).

The VMp is a region of loosely packed, medium-sized cells with ill-defined lateral and anterior borders. As a consequence, the whole nucleus or parts of it are frequently included in the ventral lateral or ventral anterior nucleus. It extends along the medial edges of the ventral lateral and basal ventral medial nuclei where it is especially large in some bats and insectivores (Le Gros Clark, 1932a; Campbell and Ryzen, 1953). Medially, posterior to the medioventral (reuniens) nucleus, the two VMp nuclei fuse across the midline beneath the internal medullary lamina. The region of fusion is sometimes called the inter-

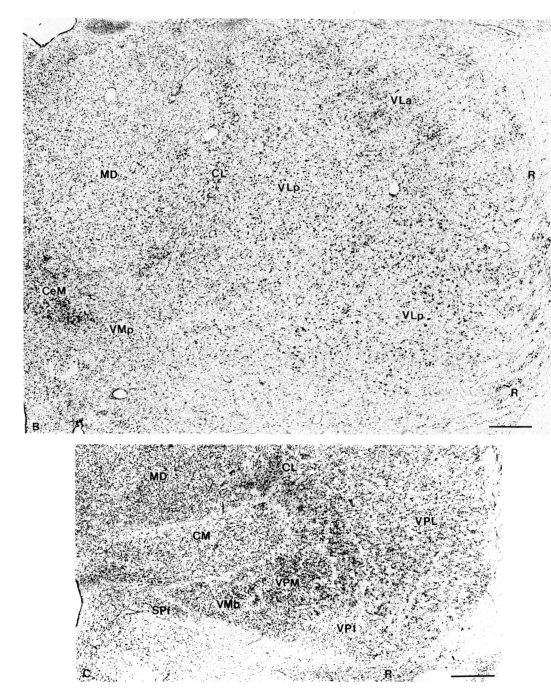

FIGURE 7.42. (*continued*)

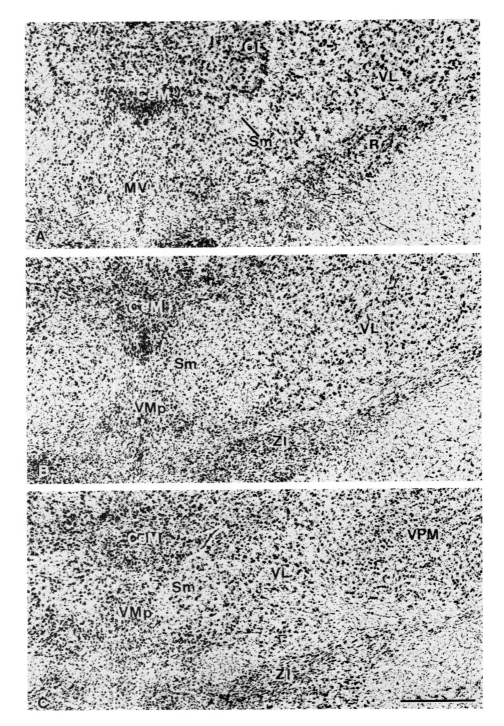

FIGURE 7.43.   The three nuclei (Sm, VMb, VMp) of the ventral medial complex in the rat.

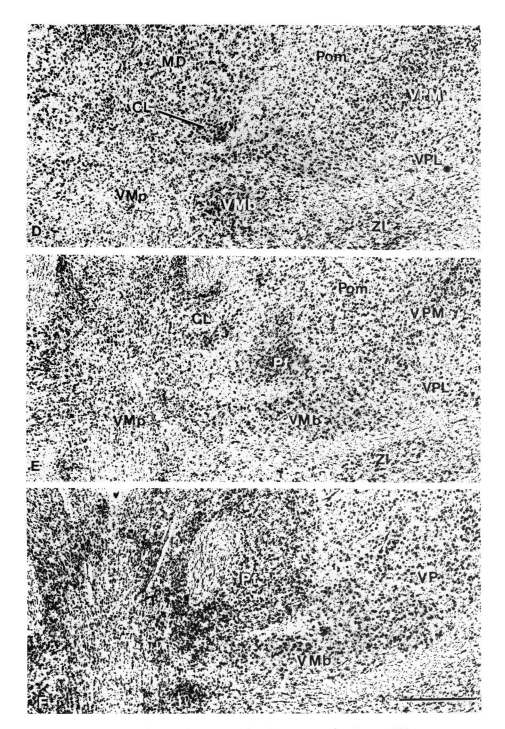

Thionin-stained frontal sections. A is most anterior, F most posterior. Bars = 500 μm.

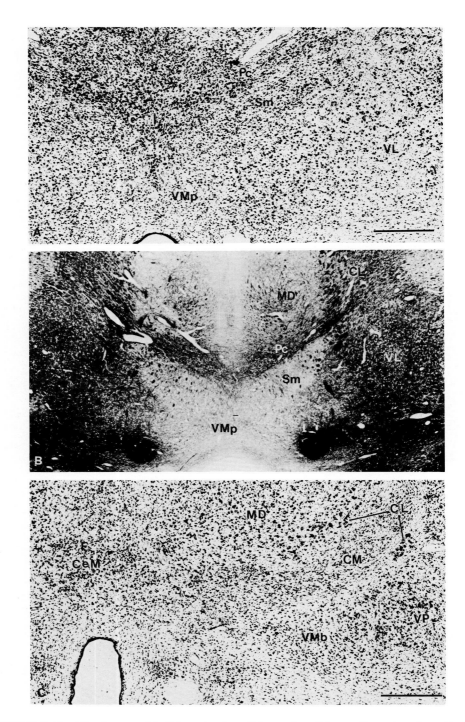

FIGURE 7.44.   The three nuclei of the ventral medial complex of the cat. Frontal sections. (A, C) Thionin stain; (B) hematoxylin stain. Bars = 1 mm.

ventral nucleus. Anteriorly, the VMp expands laterally in front of the ventral lateral complex. In monkeys this part has been commonly referred to as the medial division of the ventral lateral nucleus (VLm) (Olszewski, 1952). In man it has been called nucleus ventralis oralis medialis (Hassler, 1959).

7.6.2. Terminology

The systematization of the ventral medial nucleus terminology used here was first made by A. L. Berman and the author in 1973, though the atlas of the cat brain in which this delineation was drawn up was not published until 1982. In the intervening period, the terminology had crept into relatively common use, though it is not universal and many older terms are still in vogue. Their persistence, like that of any terminology, is not unjustified, as it reflects past usage. Our new description was an attempt at a consensus viewpoint applicable to all mammals and reflecting fiber connections.

The term *ventral medial complex* derives from the work of M. Rose (1935) on the rabbit thalamus in which he recognized three nuclei (VM 1–3) equivalent to the three named here. The ventral medial nucleus of Rioch (1929a) in the cat referred only to the VMp. Variants of the term that referred more or less to the VMp can also be found in earlier works. Among these are the *medioventral nucleus* of Le Gros Clark (1929a, 1930a, 1932a) and Crouch (1934) in monkeys. *Submedial nucleus* (Noyau sous mediale) is Vogt's (1909) term, though it referred mainly to the reuniens nucleus (i.e., to the medioventral nucleus of the present account). Rioch (1929a) in the cat may have been the first to use *submedial nucleus* in the present sense. The submedial nucleus of Walker (1937, 1938a) in the monkey is the central medial nucleus. In the rat, our submedial nucleus was called *nucleus gelatinosus* by Krieg (1944), a name still in common use.

In Vogt's account of the monkey as well as in those of Friedemann (1911) and Pines (1927), most of the VMp is referred to as the lateral division of the oral part of the ventral nucleus (Vtl α and β). Our VMp in the monkey is the same as the medial division of the ventral lateral nucleus (VLm) of Olszewski (1952). Walker (1938a) included it in his medial ventral nucleus.

Gurdjian (1927) seems to have introduced *interventral nucleus* in his study of the rat thalamus. It was adopted by Rioch (1929a) soon after for the same region in the cat. The *nucleus ventralis commissurae mediae* of Friedemann (1911) and some later authors may include parts of the VMp but it mostly referred to the reuniens nucleus.

Basal ventral medial nucleus (VMb) was coined by Berman and Jones as a new term for what had been recognized previously usually as a parvocellular division of the VPM. *Pars parvicellularis der caudalen Parte des ventralen Thalamuskernes* was first used by Friedemann (1911) in the monkey brain although he appears to have included much of the VPI as well as the VMb in it, as did Pines (1927) and Hassler (1959). As a pars parvocellularis of the arcuate (VPM) nucleus, it seems to have entered the English literature in the study of Aronson and Papez (1934). It is clearly shown but not labeled in Rioch's (1929a) earlier work on the cat.

Gurdjian (1927) in the rat also clearly illustrated it but called it the *medioventral nucleus.*

M. Rose's (1935) concept of a ventral medial complex of nuclei was reintroduced in 1952 by J. E. Rose and Mountcastle in order to distinguish the region from what they called the *ventrobasal complex.* The ventrobasal complex is equivalent to our VPL and VPM and was identified by Rose and Mountcastle as the part of the rabbit and cat thalamus activated by somatic sensory stimuli. Later this definition was extended to other species (Mountcastle and Henneman, 1952; Rose and Mountcastle, 1959). Our VMb was found to be unaffected by somatic sensory stimulation and, though earlier included by Rose (1942b) as a pars medialis of the ventrobasal complex, he and Mountcastle now included it in the ventral medial complex.

Subsequently, second-order taste afferents were found to relay in the VMb but not in the rest of the ventral medial complex (Blomquist *et al.*, 1962; Emmers *et al.*, 1962; Emmers, 1964, 1966) and lesions largely confined to it in rats led to hypogeusia (Ables and Benjamin, 1960; Oakley and Pfaffmann, 1962). Thereafter, regrettably, "ventral medial complex or nucleus" appeared in many publications as synonymous with the taste relay nucleus alone. The confusion caused by this seemed to compound that already in the literature, so we felt it necessary to introduce the new term, *basal ventral medial nucleus,* for the taste relay nucleus and to refer to the other parts of the ventral medial complex by the names already mentioned.

7.6.3. Connections

7.6.3.1. VMp and the Substantia Nigra

The VMp is the principal thalamic target of axons arising from cells in the pars reticularis of the ipsilateral substantia nigra (Faull and Carman, 1968; Carpenter *et al.*, 1976; Faull and Mehler, 1978; Beckstead *et al.*, 1979; Hendry *et al.*, 1979) (Fig. 7.41). Many of these nigrothalamic cells send branched axons to both the thalamus and the superior colliculus (Bentivoglio *et al.*, 1979; Anderson and Yoshida, 1980; Parent *et al.*, 1983). This input is, therefore, not dopaminergic, but it has been suggested that it is GABAergic as biochemical markers for GABA transmission in tissue samples from the general region of the VMp decline after destruction of the substantia nigra (DiChiara *et al.*, 1979). There is also physiological evidence that thalamic neurons in the vicinity of the VMp may be inhibited by stimulation of the substantia nigra (Ueki *et al.*, 1977; Deniau *et al.*, 1978). The only other nigral input to the thalamus is formed by a small zone of terminations in the posterior part of the central lateral nucleus (Hendry *et al.*, 1979).

The cortical target of the VMp cells appears to be relatively diffuse and unconstrained by the architectonic borders of the cortical fields in the region of projection. As identified in anterograde tracing experiments in the rat and cat, the axons of VMp cells terminate in layer I of frontal and medial cortical areas (Krettek and Price, 1977b; Herkenham, 1979; Glenn *et al.*, 1982). In the cat the projection is less extensive on the medial surface than in the rat (Glenn *et al.*, 1982). The type of influence that the substantia nigra, via the VMp, may exert over these fields is currently unknown.

7.6.3.2. Sm and the Spinal Cord

The connections of the rather insignificant-looking Sm have not been the subject of intensive study. In rats there is evidence from retrograde tracing studies that it projects upon the cortex in the vicinity of the frontal pole (Jones and Leavitt, 1974; Price and Slotnik, 1983). In cats the equivalent region may be deep in the presylvian sulcus (Craig *et al.*, 1982). Its afferent connections have not been thoroughly explored. In rats, cats, and monkeys, there is a rather modest input arising from the spinal cord and caudal medulla (Craig and Burton, 1981; Mantyh, 1983) (Fig. 7.13). In the cat the fibers arise from cells in the

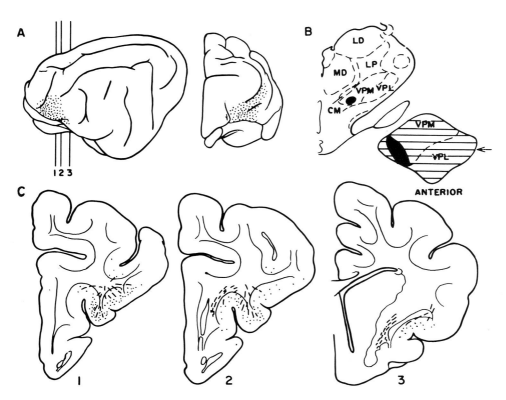

FIGURE 7.45.   Distribution of degenerating thalamocortical fibers (stippling) in the two taste representations of the cat cortex after destruction of VMb nucleus. From Jones and Powell (1969a).

marginal layer of the dorsal horn of the spinal cord and of the caudal nucleus of the spinal trigeminal complex. This observation has led Craig and Burton (1981) to elevate the Sm to the status of a thalamic pain center. Further work is clearly called for.

7.6.3.3. VMb and the Taste Pathway

Central taste afferents terminate in the VMb (Figs. 7.46, 7.47). The representation of the ipsilateral side of the tongue is greater than the contralateral in cats and monkeys, but both sides are equally represented in rats (Blomquist *et al.*, 1962; Emmers *et al.*, 1962; Emmers, 1964, 1966; Ganchrow and Erickson, 1972). The representation is contiguous with and only overlaps partly the representation of other (mechanoreceptive and thermal) tongue afferents in the VPM (Landgren, 1960; Blomquist *et al.*, 1962; Emmers, 1966; Burton and Benjamin, 1971). The gustatory neurons do not apparently respond to tactile or thermal stimuli as well, but in rats ipsi- and contralateral taste inputs converge on the same cells (Ganchrow and Erickson, 1972).

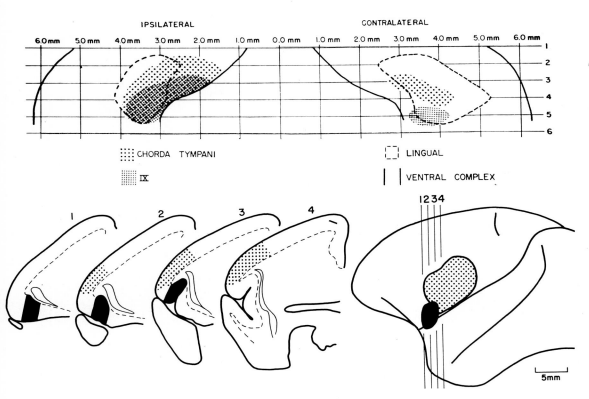

FIGURE 7.46.   (Upper) Projections of two of the taste nerves (chorda tympani, IX) in VMb nucleus of the squirrel monkey. (Lower) Dual projection of taste afferents to SI and to opercular cortex in the squirrel monkey. Both from Burton and Benjamin (1971).

Benjamin and Akert (1959) first described cellular degeneration in the VMb of rats following ablation of a part of the cerebral cortex shown to be involved in taste discrimination behavior. This has subsequently been confirmed in other animals with similar methodology and by evoked potentials and single-unit recording in rats (Roberts and Akert, 1963; Benjamin *et al.*, 1968; Benjamin and Burton, 1968; Ganchrow and Erickson, 1972; Norgren and Wolf, 1975). Benjamin and Burton (1968) localized a taste nerve projection to the tongue representation in the first somatic sensory area of the squirrel monkey and a similar region in the cat contains single units responding specifically to taste solutions (Landgren, 1957; Cohen *et al.*, 1957). In cats, degeneration of thalamocortical axons following lesions of the VMb (Jones and Powell, 1969a; Ruderman *et al.*, 1972) has been traced to one focus within the representation of the tongue in the first somatic sensory area and to a second focus, identified by single unit-recording as a taste area (Burton and Earls, 1969) in the banks of the presylvian sulcus (Fig. 7.45). A dual taste representation receiving inputs from the VMb,

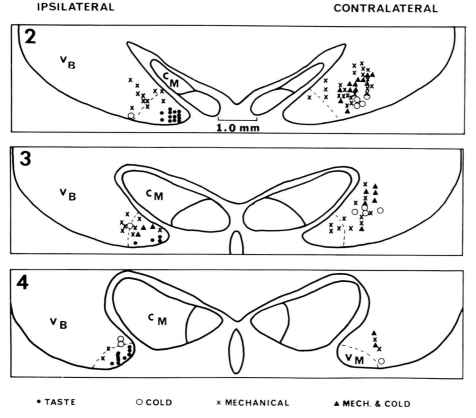

FIGURE 7.47. Localization of taste units, and units responding to cooling and/or mechanical stimulation of tongue of squirrel monkey. VM is basal ventral medial nucleus, VB ventral posterior nucleus. From Burton and Benjamin (1971).

is present in comparable areas in monkeys (Benjamin and Burton, 1968; Burton and Jones, 1976) (Fig. 7.46). In monkeys, in order to cause severe retrograde degeneration of cells in the VMb, it is necessary to ablate both representations, suggesting the possibility of a collateral projection (Roberts and Akert, 1963; Benjamin and Burton, 1968). In rats there is reason to believe that the VMb is divided into two parts, a dorsal part, projecting to the gustatory cortex, and a ventral part, projecting to an adjacent region of the insular cortex (Krettek and Price, 1977b). Some cytoarchitectonic gradation in the rodent nucleus tends to support this.

The nature of the taste projection to the VMb, classically assumed to arise in the nucleus of the solitary tract (von Economo, 1911), has been reinvestigated in recent years. According to Norgren and Leonard (1971, 1973), Ricardo and Koh (1978), and Norgren (1976), the nucleus of the solitary tract distributes relatively few ascending fibers to the VMb. Instead, they consider that its major outflow is to a further region of known gustatory representation in the vicinity of the parabrachial nucleus of the pons. This nucleus can be shown to project heavily and bilaterally to the VMb (Saper and Loewy, 1980; Beckstead *et al.*, 1980; Block and Schwartzbaum, 1983). The parabrachial neurons projecting to the VMb are said also to send axon branches to other more rostral and ventral forebrain areas such as the amygdala (Norgren, 1976; Voshart and Van der Kooy, 1981). The solitary tract nucleus–parabrachial nucleus–VMb axis is relatively rich in fibers containing neuroactive peptides and there appear to be a few peptide-containing cells in the VMb itself (Mantyh and Hunt, 1983).

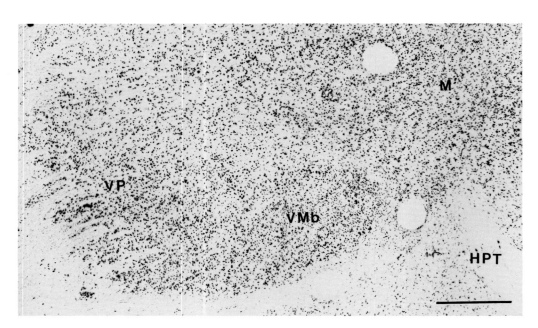

FIGURE 7.48. Large, presumptive basal ventral medial nucleus in a monotreme, the echidna. Frontal section, thionin stain. From a brain provided by Dr. D. J. Tracey. Bar = 1 mm.

There is an older literature, based mainly upon clinicopathological observations, relating the region of the ventral medial complex to visceral sensation (see Penfield and Rasmussen, 1950).

In an early study in the cat, Dell (1952) demonstrated that evoked potentials could be recorded in a region medial to the VPM following electrical stimulation of the vagus nerve. The thalamic representation of the vagus does not seem to have been pursued since that time except for a recent study of Rogers *et al.* (1979) in which it was reported that numerous cells in or near the VPM of the rat were responsive to stimulation of hepatic sodium and osmoreceptors. The afferent fibers innervating (or, perhaps forming) these receptors are known to travel in the hepatic branch of the vagus nerve. Moreover, the right vagus nerve is essential for the failure of reversal of a rat's taste preference for saline that normally occurs when sodium chloride is injected into the hepatoportal circulation. The Nissl photograph presented by Rogers *et al.* is, unfortunately, rather poor and does not clearly illustrate the sites from which they recorded responsive cells in the thalamus. The region illustrated could be the VMp or the Sm, but it is unlikely to be the VMb or VPM.

There is in souls a sympathy with sounds;
And, as the mind is pitch'd the ear is pleased
With melting airs, or martial, brisk, or grave;
Some chord in unison with what we hear,
Is touch'd within us, and the heart replies.

W. Cowper, *The Winter Walk at Noon*

8

The Medial Geniculate Complex

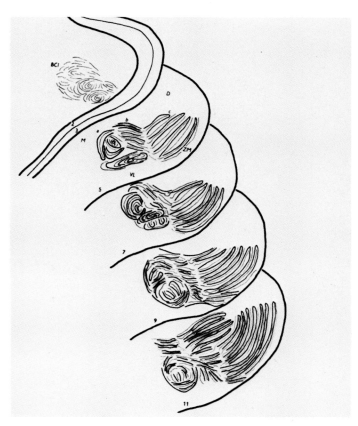

Transverse reconstructions of Golgi preparations showing fibrodendritic layers of the ventral medial geniculate nucleus in a 15-day-old cat. From Morest (1965a).

8.1. Description

The medial geniculate body is a complex of three major nuclei (Figs. 8.1–8.3): ventral or principal, medial or magnocellular, and dorsal or posterior. The ventral nucleus lies lateral to the brachium of the inferior colliculus as this traverses the complex. It consists usually of medium-sized, tightly packed, and darkly staining cells that have a tendency to form dorsoventral rows. The medial or magnocellular nucleus lies medial to the brachium, in the angle between the brachium and the medial lemniscus. Usually, it contains relatively few cells, the most obvious of which are deeply staining, angular, and large, thus accounting for the name, but populations of smaller cells of the adjacent posterior complex tend to invade it and some workers recognize intrinsic small cells as well. The dorsal nucleus caps the ventral nucleus and expands posteriorly so as to fill the whole cross-sectional area of the medial geniculate body at its posterior pole (Figs. 8.2B and 8.3A). It usually consists of small- to medium-sized cells that are paler staining and less densely packed than those in the ventral medial geniculate nucleus.

The three medial geniculate nuclei are clearly seen in most mammals. However, in small animals with fairly generalized brains, such as shrews, hedgehogs, and marsupials (Le Gros Clark, 1933; Bodian, 1939; Erickson *et al.*, 1967; Rockel *et al.*, 1972; Aitkin and Gates, 1983) the separation of dorsal and ventral nuclei may be indistinct. Conversely, in monkeys, not only are the three major nuclear divisions very distinct but the dorsal nucleus can be clearly subdivided further (Jordan, 1973; Burton and Jones, 1976; Fitzpatrick and Imig, 1978) (Fig. 8.3). Similar, though less distinct, divisions of the dorsal nucleus can also be detected in animals such as the cat (Fig. 8.2) (Ramón y Cajal, 1911; Morest, 1964; Berman and Jones, 1982) and tree shrew (Casseday *et al.*, 1976; Oliver and Hall, 1978a,b) and in prosimians (Figs. 8.4A,B). It is my impression that the dorsal nucleus is particularly large in animals such as bats, porpoises, and seals which are known or thought to use echolocation (Fig. 8.4C). In animals with relatively generalized brains, the magnocellular nucleus may not contain particularly large cells and, here, the term *medial* or *internal nucleus* is probably more appropriate (Fig. 8.1).

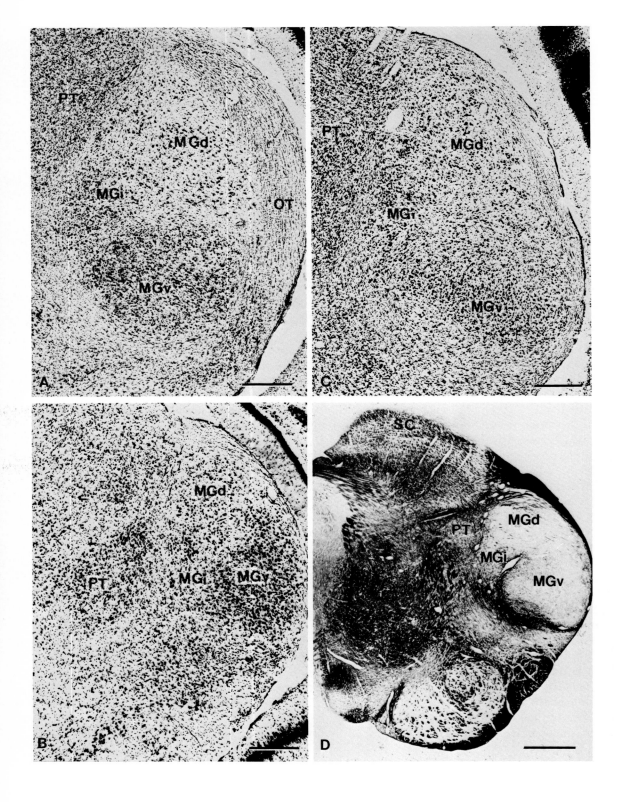

The three major nuclei of the medial geniculate complex were recognized in the rabbit and cat by von Monakow (1895), Ramón y Cajal (1911), and Nissl (1913) and by certain other early workers (Berman and Jones, 1982). Ramón y Cajal described in the cat a superior lobe, corresponding to the dorsal nucleus, and an inferior lobe which he further divided into a pars ovoidea (corresponding to the ventral nucleus) and an internal segment (corresponding to the magnocellular nucleus). Ramón y Cajal also recognized in Golgi preparations certain finer divisions of both the ventral and the dorsal nucleus, namely a marginal zone of loosely packed cells on the anterior aspect of the ventral nucleus and superficial and deep components of the dorsal nucleus. Sychowa (1962) in a myeloarchitectonic study of the dog and Morest (1964) in a Golgi study of the cat subsequently redescribed most of these finer divisions of Ramón y Cajal. Morest divides the ventral nucleus into a pars lateralis and a pars ovoidea, plus marginal and ventrolateral nuclei; he divides the dorsal nucleus into dorsal, superficial dorsal, and deep dorsal nuclei; he divides the medial or magnocellular nucleus into large-celled and small-celled parts (see also Winer and Morest, 1983). As will emerge in Section 8.6, there is some connectional basis for many of these subdivisions. A suprageniculate nucleus is described by some authors as part of the medial geniculate body. In some cases, it is likely that this represents part of the dorsal nucleus. In others, it is clearly the same as the suprageniculate nucleus that is described in Chapter 11 as part of the posterior complex of nuclei. Commonly a part of the lateral posterior nucleus invades the dorsal part of the medial geniculate body (Figs. 8.2C and 11.7) and this, too, may be misinterpreted as part of the dorsal nucleus.

The three medial geniculate nuclei of von Monakow and Ramón y Cajal were often described in other mammals, including primates (Figs. 8.3 and 8.4), and were usually given names derivative of those of Ramón y Cajal (Friedemann, 1911; Nissl, 1913; Müller, 1921; Pines, 1927; Hornet, 1933; M. Rose, 1935; J. E. Rose, 1942b). Many authors, however, did not detect a difference between the dorsal and the ventral nucleus and the practice became established of dividing the medial geniculate complex into only two parts: a magnocellular, medial, or internal nucleus and a principal, lateral, or external nucleus (e.g., Gurdjian, 1927; Rioch, 1929a; Walker, 1938a; Walker and Fulton, 1938; Bodian, 1939; Rose and Woolsey, 1949b, 1958; Olszewski, 1952; Jaspar and Ajmone-Marsan, 1954; Kruger, 1959; Rockel *et al.*, 1972). In some species such as the rat and opossum, this probably reflects the genuine difficulty of detecting cytoarchitectonic differences between the dorsal and the ventral nucleus. In others the failure to define a dorsal medial geniculate nucleus may have stemmed from the belief, expressed by Le Gros Clark (1933), that the superior lobe (dorsal nucleus) of Ramón y Cajal was a posterior protrusion of the pulvinar (cf. Fig. 8.1).

FIGURE 8.1. Frontal sections showing the dorsal (MGd), ventral (MGv), and internal (MGi) or magnocellular nuclei of the medial geniculate complex in a rabbit (A), rat (B), and guinea pig (C, D). Thionin (A–C) and hematoxylin (D) stains. Bars = 500 μm.

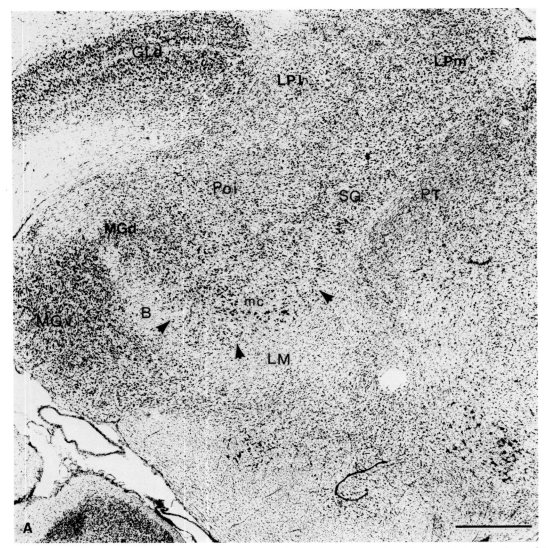

FIGURE 8.2. (A) Frontal section near the anterior pole of the medial geniculate complex of a cat. Dorsal nucleus (MGd) is only just visible. Region called ventrolateral nucleus by Morest (1964) lies ventrolateral to MGv. Arrowheads outline posterior extension of medial nucleus of posterior complex. Thionin stain. Bar = 1 mm. From Jones and Burton (1974). (B) Horizontal section showing the three major nuclei (MGd, MGv, and mc) of the medial geniculate complex in a cat. Arrows indicate portion of ventral lateral nucleus of thalamus in which spinothalamic fibers end. Thionin stain. Bar = 1 mm. From Jones and Burton (1974). (C) Frontal section near the middle of the medial geniculate complex of a cat showing invasion of the complex by part of the lateral posterior lateral nucleus (LPl). Thionin stain. Bar = 1 mm. Compare with Fig. 11.7.

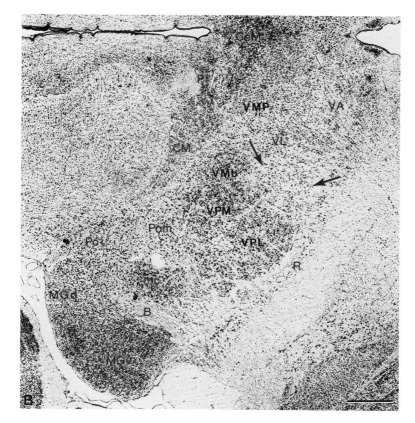

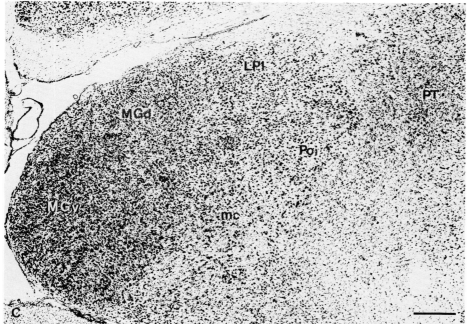

FIGURE 8.2.   (*continued*)

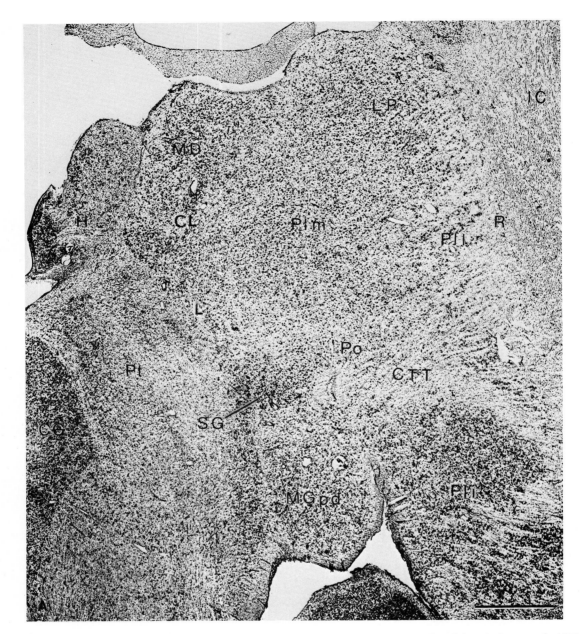

FIGURE 8.3. (A) Frontal section through the posterior pole of the medial geniculate complex of a rhesus monkey, showing the posterodorsal nucleus (MGpd) and adjoining nuclei. Thionin stain. Bar = 1 mm. (B) Frontal section anterior to (A) showing commencement of ventral nucleus (MGv) of medial geniculate complex. (C) Frontal section anterior to (B) showing commencement of anterodorsal (MGad) and magnocellular nuclei (mc) of medial geniculate complex. PPN: peripenduncular nucleus of midbrain. (D, E) Frontal sections near anterior pole of medial geniculate nucleus in a rhesus monkey (D) and a squirrel monkey (E) showing anterodorsal (MGad), ventral (MGv), and magnocellular (mc) nucleus and adjacent nuclei. Bars = 500 μm. All from Burton and Jones (1976).

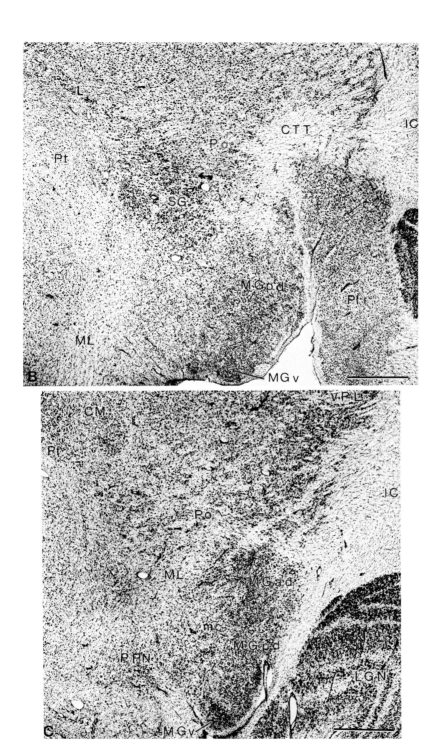

FIGURE 8.3. (*continued*)

With the modern studies of connectivity of the medial geniculate complex in the cat, tree shrew, and monkey, has come a better understanding of its three nuclei (Diamond *et al.,* 1969; Oliver and Hall, 1975, 1978a,b; Burton and Jones, 1976; Casseday *et al.,* 1976; Fitzpatrick and Imig, 1978; Winer *et al.,* 1977; Andersen *et al.,* 1980a,b). The ventral nucleus is established as the recipient of the most direct ascending auditory pathway, as the nucleus with the most clear-cut cochleotopic representation, and as the nucleus that projects upon the primary auditory cortex (area AI; Sections 8.3 and 8.6). Sometimes, therefore, it has been called the *principal nucleus* though, from what has been said above, there seems to be good reason for avoiding this term. The subdivisions of the dorsal nucleus seem to have less clearly defined brain-stem inputs and form the relays to auditory (and perhaps to certain nonauditory) areas around AI. The magnocellular nucleus is a diffusely projecting nucleus (Chapter 3).

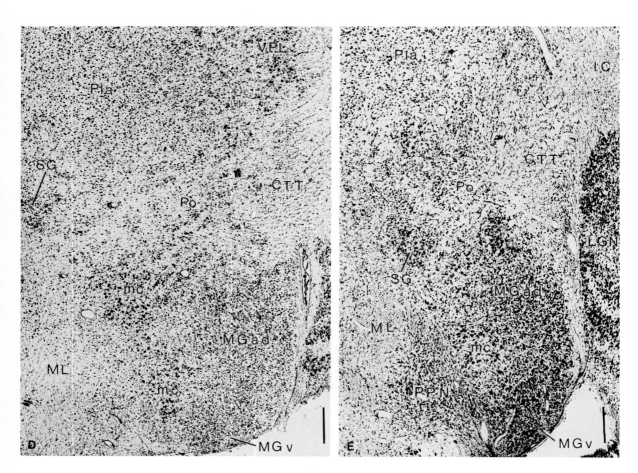

FIGURE 8.3. (*continued*)

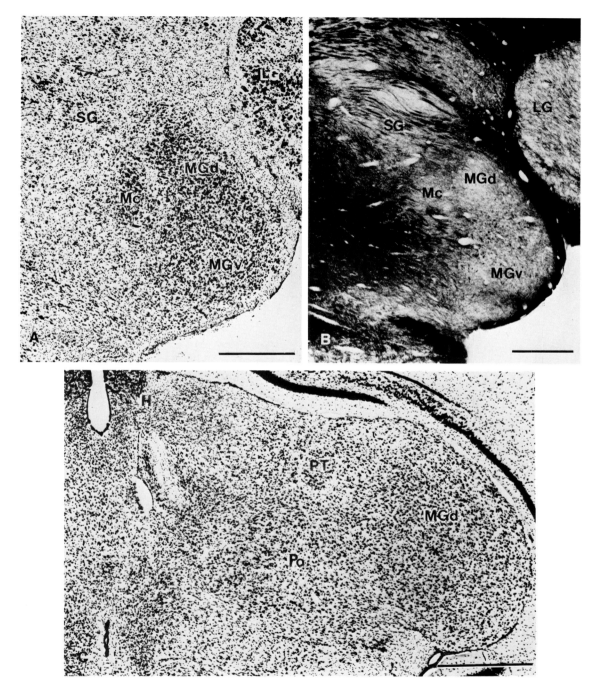

FIGURE 8.4. (A, B) Frontal sections near the middle of the medial geniculate complex of a bush baby showing the three major subdivisions. (C) Frontal section through the enlarged dorsal nuclei of the medial geniculate complex in a moustache bat (*Pteronotus parnelli*). (A) Hematoxylin stain; (B, C) thionin stain. Bars = 1 mm (A, B), 500 μm (C).

8.3. Subcortical Connections

8.3.1. Ventral Nucleus

The principal ascending input to the ventral nucleus arises in the central nucleus of the inferior colliculus of the ipsilateral side (Figs. 8.5 and 8.6). A weaker input arises from the contralateral central nucleus (Woollard and Harp-

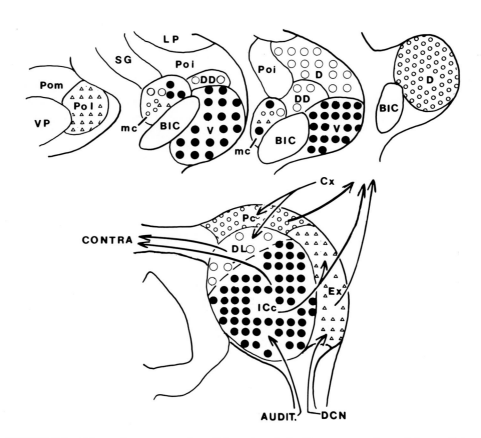

FIGURE 8.5. Schematic representation of the subcortical afferent connections of the medial geniculate complex in the cat. Central nucleus of inferior colliculus (ICc) receives inputs from most brain-stem auditory centers (AUDIT.) and projects bilaterally to ventral medial geniculate nucleus (V); external nucleus of inferior colliculus (Ex) receives inputs from dorsal column nuclei (DCN) and collaterals of central nucleus efferents, and projects bilaterally to lateral region (Pol) of the posterior nucleus. Pericentral nucleus of inferior colliculus (Pc) projects to posterior part of dorsal (D) medial geniculate nucleus. Cx indicates cortical efferents which terminate in pericentral nucleus and dorsal (DL) division of central nucleus. Dorsal division may project (bilaterally) to anterior part of dorsal (D) and to deep dorsal (DD) medial geniculate nuclei. All nuclei of inferior colliculus project to magnocellular (mc) medial geniculate nucleus. Drawn from data in Kudo and Niimi (1978) and Andersen *et al.* (1980a).

man, 1940; Moore and Goldberg, 1960, 1963; Morest, 1965b; Jones and Rockel, 1971; Casseday *et al.*, 1976; Kudo and Niimi, 1978; Oliver and Hall, 1978a; Andersen *et al.*, 1980a; Calford and Aitkin, 1983). There appears to be little firm evidence for afferent fibers arising from lower brain-stem auditory centers. Fibers enter the ventral nucleus in serial order from the brachium of the inferior colliculus. At their region of entry the fibers follow a somewhat helical course, leading Morest to call this region of the ventral nucleus, as seen in Golgi preparations, the *pars ovoidea*. The fibers ascend more or less vertically in the rest of the nucleus following the contours of the gently curving lamellae of cells that are situated there (Morest, 1965a; Chapter 3). Morest calls this region the *pars lateralis*. Fibers conveying impulses related to low-pitched sounds terminate laterally, those conveying impulses related to higher-pitched sounds terminate progressively more medially (Andersen *et al.*, 1980a), thus setting up the tonotopic organization demonstrable in the nucleus (Aitkin and Webster, 1972; Gross *et al.*, 1974; Calford and Webster, 1981; Chapter 3). Injections of small amounts of anterograde tracer in the central nucleus of the inferior colliculus tend to give labeling in a series of bursts at different levels in a lamella (Andersen *et al.*, 1980a; Fig. 3.27), suggesting that, as with the medial lemniscal fibers in the ventral posterior nucleus (Chapter 3), individual collicular afferents terminate at selected levels along the dorsoventral extent of a lamella (and see Section 8.6). This was not detected by Morest (1964, 1965a) in his Golgi studies which originally established the lamellar organization of the ventral nucleus.

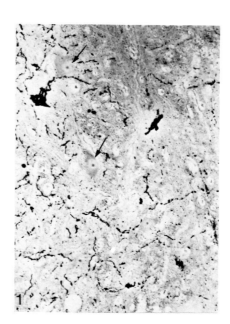

FIGURE 8.6. Degenerating spinothalamic fibers in a cat terminating in relation to the posterior extension of the medial region of the posterior nucleus (see Fig. 8.2B), and avoiding the large cells (arrows) of the magnocellular medial geniculate nucleus. ×400. From Jones and Burton (1974).

8.3.2. Dorsal Nucleus

In his studies of afferents to the medial geniculate complex of the cat, Morest (1965b) concluded that the dorsal nucleus received brain-stem afferents, not from the inferior colliculus, but via a pathway of undetermined origin situated deep in the midbrain tegmentum. Conceivably, such a pathway comes directly from the cochlear nuclei (Strominger *et al.*, 1977). More recent work with anterograde and retrograde tracing methods appears to have established the pericentral nucleus of the inferior colliculus (Fig. 8.5) and possibly adjacent regions as the major sources of fibers to the dorsal nuclei (Rockel *et al.*, 1972; Casseday *et al.*, 1976; Oliver and Hall, 1978a; Kudo and Niimi, 1978, 1980; Andersen *et al.*, 1980a; Calford and Aitkin, 1983). Whether these fibers have differential projections to the several subnuclei of the dorsal nucleus seems to me to have not been satisfactorily determined. It appears that the pericentral nucleus of the inferior colliculus may project only to posterior parts of the dorsal nucleus in the cat and tree shrew (Casseday *et al.*, 1976; Kudo and Niimi, 1978; Andersen *et al.*, 1980a; Calford and Aitkin, 1983). The other nucleus of the inferior colliculus, the external nucleus, has not been determined as the origin of inputs to other parts of the dorsal nucleus but may project to the lateral nucleus (PoI) of the posterior complex (Kudo and Niimi, 1978).

8.3.3. Magnocellular Nucleus

The magnocellular medial geniculate nucleus has often been claimed to be a region in which ascending auditory, medial lemniscal, spinothalamic, and vestibular fibers converge. There is some physiological evidence for such convergence at the single-cell level, though the responses of most cells in the nucleus are strictly auditory in character (Poggio and Mountcastle, 1960; Aitkin, 1973). It is very difficult to authenticate many of the anatomical reports of convergence because the nucleus is situated in a region through which many afferent systems enter the thalamus and fibers of passage can be readily mistaken for terminations. I have not been able to convince myself of medial lemniscal terminations in it in material from my own laboratory through they have been reported by others (Lund and Webster, 1967a; Boivie, 1971a; Hand and Van Winkle, 1977; Berkley, 1980). Spinothalamic termination (Nauta and Kuypers, 1958; Mehler, 1966a, 1969; Boivie, 1971b; Kerr, 1975; Berkley, 1980) in both cat and monkey, I consider to be around the smaller cells of the posterior complex or suprageniculate nucleus that intrude on the magnocellular nucleus, rather than around its large cells (Jones and Burton, 1974; Berman and Jones, 1982; Asanuma *et al.*, 1983b; Fig. 8.6 and Chapter 11). I can find no convincing anatomical report of afferents from the vestibular nuclei terminating in it. Only the long-described afferent inputs from the inferior colliculus (Fig. 8.5) (Woollard and Harpman, 1940; Moore and Goldberg, 1960, 1963; Casseday *et al.*, 1976; Oliver and Hall, 1978a; Andersen *et al.*, 1980a) and deep layers of the superior colliculus (Graham, 1977) seem clearly established. Recently Calford and Aitkin (1983) have suggested that the external nucleus of the inferior colliculus and parts of the central nucleus may be a source of afferents from this structure.

439

THE MEDIAL
GENICULATE
COMPLEX

8.4. Receptive Fields and Other Properties of Medial Geniculate Neurons

8.4.1. Ventral Nucleus

A cell in the ventral medial geniculate nucleus will respond to quite wide ranges of pure tone stimuli delivered to one or both ears; but at intensities of sound pressure close to threshold, the cell responds only to a certain "best frequency" which is characteristic for that cell (Rose *et al.*, 1966; Aitkin and Webster, 1972; see Section 3.4.3). Using this method of analysis in the cat (Aitkin and Webster, 1972; Calford and Webster, 1981), it can be shown that throughout most of the ventral division, cells with similar best frequencies tend to be arranged in the more or less sagittally oriented lamellae stretching through the antero-posterior extent of the ventral division as described in Chapter 3. The lamellae are arranged sequentially so that a microelectrode advanced from lateral to medial across the ventral nucleus encounters cells whose best frequencies change systematically from low to high except in the pars ovoidea where the coiling of the lamellae makes tonotopicity difficult to demonstrate (Fig. 3.16). Because the lamellae appear to curve around the posterior surface of the ventral nucleus, a dorsoposterior to ventroanterior penetration also encounters a systematic sequence of best frequencies from low to high. At most levels of the nucleus, an electrode penetrating dorsoventrally tends to encounter only cells with closely similar best frequencies because it tends to stay within a lamella. The ventral division, thus, appears to contain a complete tonotopic representation of the audible frequency range based upon the regular lamellarlike organization of the dendritic fields of the cells in the ventral division and of the afferent axons which run along them (Morest, 1964, 1965a; Jones and Rockel, 1971). For convenience, we call these lamellae *isofrequency lamellae*.

The majority of cells responding to tonal stimuli in the ventral nucleus respond transiently with on responses, are sharply tuned to frequency, and are influenced by stimuli applied to either ear (Galambos *et al.*, 1952; Adrian *et al.*, 1966; Aitkin and Webster, 1972; Calford and Webster, 1981; Calford, 1983; Chapter 3.4.3 and Fig. 4.28). The effects mediated by the two ears may both be excitatory (EE cells), or stimulation of one ear may cause excitation and of the other inhibition (EI cells). Pure tone stimuli applied to the two ears within a few hundred microseconds of each other usually have interacting effects. However, in all cases of binaural interaction the effect mediated by the contralateral ear leads that mediated by the ipsilateral ear (Aitkin and Webster, 1972). This probably reflects the fact that the most direct pathway to the ventral nucleus is the crossed one from the contralateral cochlear nuclei. The cochlear nuclei of one side project to the central nucleus of the inferior colliculus of the opposite side and not to that of the ipsilateral side (Osen, 1972; Goldberg and Moore, 1967; Roth *et al.*, 1978). Inputs to the medial geniculate nucleus from the ipsilateral ear must, therefore, involve more synaptic interruptions.

8.4.2. Dorsal Nucleus and Magnocellular Nucleus

The dorsal nucleus and the magnocellular nucleus have been less well studied electrophysically (Chapter 3). Tonotopicity has not been described in either nucleus. Many cells encountered in the dorsal division (Aitkin and Webster, 1972; Aitkin, 1973; Calford and Webster, 1981) do not respond to auditory stimuli and those that do, respond very irregularly, at long latency and usually to a broad range of frequencies. In the posterior part of the dorsal nucleus, a few typical EE cells have been encountered (Calford and Webster, 1981) and there are slight differences between cells there and those in the deep dorsal nucleus (Calford, 1983). Many cells of the magnocellular nucleus do respond to pure tone stimuli and show evidence of binaural interaction with the leading effect being mediated by the contralateral ear (Aitkin, 1973; Calford, 1983). However, they usually respond to very much wider ranges of frequency than cells in the ventral division, i.e., they have very broad tuning curves (Fig. 4.28) and their responses tend to be tonic. Some cells in the magnocellular nucleus that respond to auditory stimuli may also respond to tactile and vibratory stimuli (Poggio and Mountcastle, 1960). However, such cells appear to be more common in the anterior part of the nucleus where it merges with the medial region of the posterior group and the remainder of the nucleus appears to be more purely auditory in function (Poggio and Mountcastle, 1960).

The evidence for convergence of dorsal column–lemniscal and auditory inputs (from the central nucleus) in the external nucleus of the inferior colliculus (Aitkin *et al.*, 1975, 1978) suggests that similar interactions between the two sensory systems might be detected in nuclei of the medial geniculate complex to which the external nucleus might project. To date there have been no reports of such interactions other than in the magnocellular nucleus.

8.5. Corticothalamic Connections

Diamond *et al.* (1969), using axonal degeneration techniques in the cat, first reported that corticothalamic fibers descending from the various auditory cortical fields (Woolsey, 1964; Fig. 8.7) were differentially distributed among the ventral nucleus and the various subdivisions of the dorsal nucleus. All fields, however, projected to the magnocellular nucleus. Since that time, the delineation of some of the auditory fields in the cat cortex has been modified as the result of newer mapping studies (Merzenich *et al.*, 1975; Knight, 1977; Fig. 8.8). The current position in the cat (Fig. 8.9) seems to be that the AI field projects heavily to the ventral nucleus, the anterior auditory field to the deep dorsal nucleus, the AII field to the posterior part of the dorsal nucleus (Andersen *et al.*, 1980b); the posterior auditory and suprasylvian fringe fields appear to project to deeper parts of the dorsal nuclei (Diamond *et al.,* 1969; Sousa-Pinto, 1973; Pontes *et al.,* 1975). All fields project on the magnocellular nucleus (Andersen *et al.,* 1980b; Pontes *et al.,* 1975). In monkeys and tree shrews, the AI field projects to the ventral nucleus and various surrounding auditory fields appear to project differentially upon the anterodorsal and posterodorsal nucleus (see Fig. 8.12). All fields project to the magnocellular nucleus (Fitzpatrick and Imig, 1978; Oliver and Hall, 1978b).

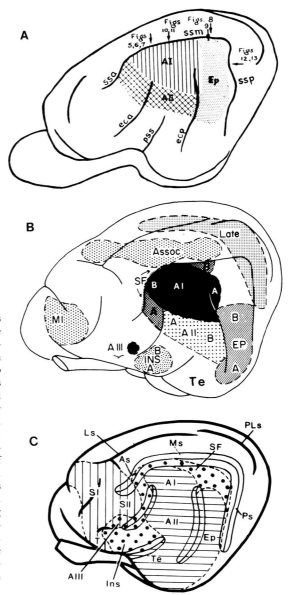

FIGURE 8.7.   Auditory cortical fields as delineated cytoarchitectonically (A) by Rose (1949) and with the evoked potential technique (B) by Woolsey (1964). A and B in each field indicate the regions thought to represent the apical (A) and basal (B) turns of the cochlea. AI, AII: first and second auditory fields; AIII: region of auditory evoked potentials earlier described by A. Tunturi; Ep: ectosylvian auditory field; Ins: insular field; SF: suprasylvian fringe auditory field; Assoc and Late: regions of long-latency auditory evoked responses; MI: motor cortex; Te: added to Woolsey's figure is temporal area of Diamond et al. (1958). (C) Schematic diagram indicating thalamocortical connections as determined with the Nauta technique in cats. Horizontal hatching: medial geniculate nucleus; vertical hatching: ventral posterior nucleus; dots: posterior complex. From Heath and Jones (1971b).

Generally speaking, the corticothalamic connections from a particular cortical field return to the nucleus of the medial geniculate complex from which they receive thalamocortical inputs (Pontes *et al.*, 1975; Casseday *et al.*, 1976; Oliver and Hall, 1978b). In the cat, however, the rigidity of this organization has been questioned by Andersen *et al.* (1980b). From experiments in which small injections of tritiated amino acids were made in the AI, AII, or anterior auditory fields, they demonstrated strong connections to one principal nucleus and to the magnocellular nucleus as would be expected, but they also found significant, though less strong connections to adjoining nuclei. That is, AI projected to the deep dorsal as well as to the ventral and magnocellular nuclei, the anterior auditory field projected to the lateral division of the posterior nucleus as well as the deep dorsal and magnocellular nuclei. The AII field, however, had no projections outside the posterior dorsal and magnocellular nuclei. Andersen *et al.* insist that their results are not caused by spread of injections from AI to the anterior auditory field, though the borders between these fields can be quite variable (Merzenich *et al.*, 1975; Knight, 1977). Their findings suggest a rather more diffuse organization in the relationship between auditory cortex and auditory thalamus than has previously been contemplated and need to be confirmed.

8.6. Thalamocortical Connections

8.6.1. Background

The projection of the medial geniculate complex to the auditory koniocortex was established with the retrograde degeneration method many years ago (Walker, 1938a; Waller, 1940a) but there were no indications in these early studies of differential projections from the various nuclei of the complex. In 1949 Rose subdivided the auditory region of the cat cortex into a central, moderately granular area, coincident with the first auditory or AI field as delineated by Woolsey and Walzl (1942) with the evoked potential method, and a number of surrounding areas with different cytoarchitectonic characteristics (Fig. 8.7A). These were

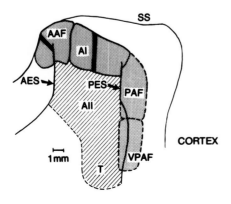

FIGURE 8.8. Auditory cortical fields in the cat as delimited by multiunit recording in recent studies. AI and AII fields are similar to AI and AII of Woolsey (Fig. 8.7); anterior part of suprasylvian fringe area is now regarded as a separate anterior auditory field (AAF) with a complete representation of the cochlea. Posterior ectosylvian area is divided into at least two fields [posterior auditory field (PAF) and ventral posterior auditory field (VPAF)] with complete representations. Black lines indicate isofrequency bands. From Andersen *et al.* (1980b).

later mapped with the evoked potential method by Woolsey and his co-workers, further subdivided, and most of them demonstrated to contain complete and independent representations of the cochlea (Fig. 8.7B; Woolsey, 1964). Rose and Woolsey (1949a) were able to show that destruction of the AI area resulted in retrograde degeneration in the anterior part of the medial geniculate complex, in a region corresponding to the ventral nucleus. They also found that lesions of different parts of AI led to degeneration in different parts of the nucleus in a manner that implied a cochlear representation in the nucleus and a cochleotopic projection on the cortex. Isolated destruction of each of the other fields, however, resulted in very little degeneration in the medial geniculate complex.

Diamond *et al.* (1958) were the first to demonstrate an independent projection of the dorsal medial geniculate nucleus when they showed that destruction of the so-called temporal field lying ventral to the second auditory area (AII, Figs. 8.7B,C) led to retrograde degeneration at the posterior pole of the medial geniculate body. The degeneration became especially severe if the insular field and the AII field were destroyed as well. They interpreted their results to indicate direct projections from the posterior pole to the temporal region and collateral projections to the insular and AII fields. It is interesting to note that, unlike all

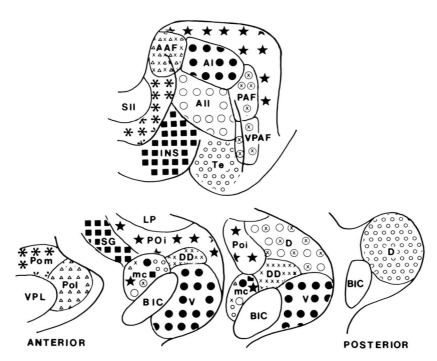

FIGURE 8.9. Schematic representation of thalamocortical (and of reciprocal corticothalamic) projections in the auditory system of the cat. The same symbol indicates the cortical field and the nucleus of the medial geniculate or posterior complex with which it is interconnected. Magnocellular medial geniculate nucleus (mc) has diffuse connections. This figure presents a consensus viewpoint of the thalamocortical connections derived from Andersen *et al.* (1980b), Bentivoglio *et al.* (1983), and Winer *et al.* (1977), none of whom agree on all counts except for AI and MGv. See Figs. 8.5 and 11.8.

the other cortical fields related to the medial geniculate nucleus, the temporal field has never been shown to contain a representation of the cochlea or to be responsive to cochlear stimulation. Rose and Woolsey (1958) in the same year reported observations that can be interpreted in the same way as those of Diamond *et al.*, though they were based on extremely deep cortical lesions that caused degeneration in many thalamic nuclei. Rose and Woolsey, however, were able to make a strong case for extremely widespread cortical projections from the magnocellular nucleus. Destruction of the AI or more ventral auditory fields alone, resulted in little or no retrograde degeneration in the magnocellular nucleus but the degeneration grew more and more severe as fields additional to AI, including those such as SII outside the auditory regions, became involved. This, Rose and Woolsey felt, was evidence for widespread, probably collateral, projections from the magnocellular nucleus. From these observations grew their concept of essential and sustaining projections (Chapter 1).

In the era of anatomical studies involving axonal degeneration, Diamond *et al.* (1969) argued from the differential distribution of corticothalamic fibers in the medial geniculate complex that the ventral medial geniculate nucleus would project to AI, the various subnuclei of the dorsal nucleus to independent fields around AI, and the magnocellular nucleus to all fields. The more direct method of lesioning the medial geniculate complex and studying the distribution of axonal degeneration in the cortex did not lead to particularly clear-cut results in the cat (Wilson and Cragg, 1969; Heath and Jones, 1971a,b; Niimi and Naito, 1974). In monkeys, Mesulam and Pandya (1973) indicated projections from anterior parts of the medial geniculate complex to area AI and from posterior and medial parts to adjacent fields.

8.6.2. Recent Studies

In recent years when more sensitive techniques have been used, a slight controversy appears to have arisen over the nature of the geniculocortical projection (Figs. 8.9, 8.10, and 8.11). Some workers in the cat, tree shrew, and monkey have found that the ventral nucleus projects only to AI and have favored the view that the various subdivisions of the dorsal nucleus project independently to separate cortical fields around AI (Sousa-Pinto, 1973; Burton and Jones, 1976; Casseday *et al.*, 1976; Oliver and Hall, 1978b). Only the magnocellular nucleus is considered to project widely and diffusely. Against these results, Winer *et al.* (1977) have reported that the subnuclei of the dorsal nucleus may project upon more than one auditory cortical field in the cat. Their interpretation is based upon experiments in which they found retrograde labeling of cells in several nuclei of the medial geniculate complex after injections of tracer aimed at the individual auditory cortical fields. The labeling of cells in the magnocellular nucleus was consistent with other reports of widespread thalamocortical projections from this nucleus. But they also interpreted their results to indicate similarly widespread projections from several of the subdivisions of the dorsal nucleus and even from the ventral nucleus. This viewpoint was supported by Andersen *et al.* (1980b) in their study of corticothalamic connections. Nevertheless, when one looks at the results of Winer *et al.*, it is evident that an injection apparently

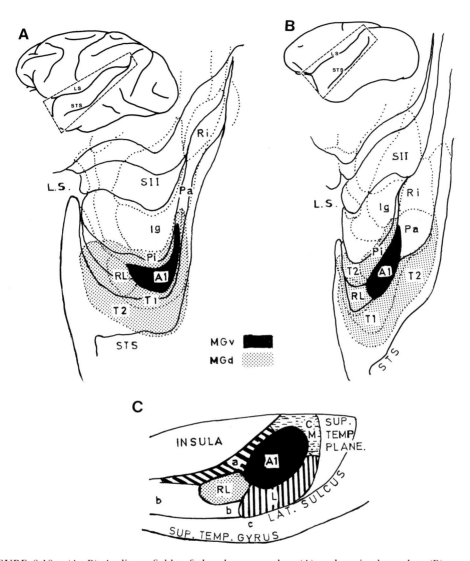

FIGURE 8.10. (A, B) Auditory fields of the rhesus monkey (A) and squirrel monkey (B), as delimited by cytoarchitecture and thalamocortical connections. Lateral sulcus (LS) is reconstructed as though opened out to expose insula. Symbols in boxes indicate nuclei of medial geniculate complex projecting to each field, along with projections from other nuclei to adjoining fields. A1: first auditory field; Pi: parainsular (AII) auditory field; Ig: Granular insular field; Pa: posterior auditory field; Ri: retroinsular field; RL: rostrolateral auditory field; SII: second somatic sensory area; STS: superior temporal sulcus; T1, T2: first and second temporal auditory fields. Modified from Burton and Jones (1976). (C) Auditory fields (A1, CM, RL, L, a) of the rhesus monkey, as delimited by multiunit recording. Redrawn from Merzenich and Brugge (1973).

centered in one or other of the auditory fields invariably led to a major focus of retrograde labeling in only one nucleus other than the magnocellular; fewer, less concentrated cells were labeled in other nuclei or subnuclei. An alternative interpretation of the results of Winer *et al.,* in terms of the old nucleus-to-field concept, would have the ventral nucleus projecting to AI, the deep dorsal nucleus to AII, the posterior pole of the dorsal nucleus to the temporal field, and the other components of the dorsal nucleus to the two posterior auditory (posterior ectosylvian) (Fig. 8.9) fields. After injection of one of these fields, label in nuclei other than the primary nucleus or the magnocellular nucleus could well be due to spread of an injection into white matter or into adjacent fields. It seems to me that the thalamocortical connections of the cat auditory system warrant fur-

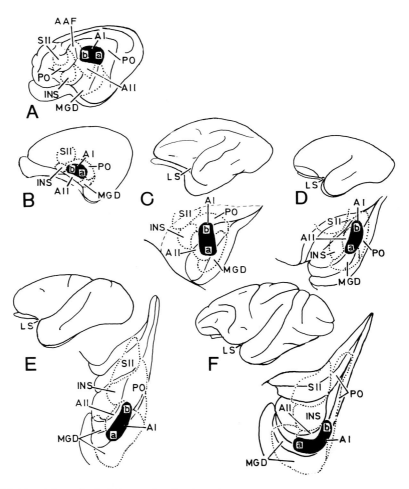

FIGURE 8.11. A hypothetical figure showing how downward and forward growth of a temporal lobe would lead to rotation of the AI field of a brain like that of a cat (A), rodent or insectivore (B), so that apex (a) to basal (b) representation of cochlea would come to be reversed, as it is in the higher primate brain. (C) Lemur; (D) marmoset; (E) squirrel monkey; (F) macaque monkey.

ther study in order to resolve the uncertainty posed by these differing interpretations.

No data are available from the study of Winer *et al.* on the thalamic inputs
to the anterior auditory field of the cat. Studies of corticothalamic connections,
however, relate it to the superficial dorsal (Diamond *et al.*, 1969) or deep dorsal
(Andersen *et al.*, 1980b) nucleus. The insular field is the target of the suprageniculate nucleus of the posterior complex (Jones and Leavitt, 1973; Winer *et
al.*, 1977). The origins of the cochlear-related inputs to the insular field are,
thus, uncertain, for the subcortical inputs to the suprageniculate nucleus arise
in the superior colliculus (Chapter 11).

In both New World and Old World monkeys, the downward and forward
growth of the temporal lobe has led to a rotation of the auditory cortical fields
in comparison with those of the cat (Figs. 8.10 and 8.11). The representation of
the basal turn of the cochlea in the most granular field (AI) now lies posteromedially instead of anteriorly as it does in the cat; the representation of the
apical turn lies anterolaterally instead of posteriorly (Woolsey, 1964; Merzenich
and Brugge, 1973). The AI field receives input from the ventral medial geniculate nucleus (Burton and Jones, 1976) (Fig. 8.10). The AII field has come to
lie medial instead of ventral to AI (Woolsey, 1964) and the apparent equivalents
of the several other auditory fields of the cat lie posterior, lateral, and anterior
to AI. The anterior of these fields, at the tip of the supratemporal plane (called
RL by Merzenich and Brugge, 1973), receives its thalamocortical input from the
most posterior part of the dorsal medial geniculate nucleus (Burton and Jones,
1976; see also Fitzpatrick and Imig, 1978, and Mesulam and Pandya, 1973). It
thus seems equivalent to the temporal field of the cat. AII and the other fields,
like their apparent equivalents in the cat, are the targets of the other subdivisions
of the dorsal nucleus, though details of differential projections have not been
worked out (Burton and Jones, 1976).

The tree shrew auditory cortex (Fig. 8.12) resembles a situation intermediate
between the cat and the rotated cortex of the monkey. It consists of a highly
granular AI field that receives thalamic input from the ventral medial geniculate
nucleus, and up to five surrounding (mainly dorsal, posterior, and ventral) fields
receiving inputs from the dorsal nuclei (Casseday *et al.*, 1976; Oliver and Hall,
1978b). Oliver and Hall consider that each of the surrounding fields receives
inputs from a separate subdivision of the dorsal nucleus.

In the monkey and tree shrew, as in the cat, the magnocellular medial
geniculate nucleus projects diffusely to all the auditory fields, to adjacent insular,
frontoparietal, and temporal fields and perhaps even beyond (Burton and Jones,
1976; Oliver and Hall, 1978b). I have seen an occasional retrogradely labeled
cell in it after injections of tracer in the postcentral gyrus and superior parietal
lobule. One feature distinguishes the projection of the magnocellular nucleus
from that of the ventral and dorsal nuclei in the monkey and tree shrew. Where
thalamocortical axons from the ventral and dorsal nuclei terminate in the middle
layers of the cortex (layers IV and/or III) those from the magnocellular nucleus
terminate elsewhere. In the monkey, magnocellular fibers end in layer I (Jones
and Burton, 1976b), while in the tree shrew they are reported to end predominately in layer VI (Oliver and Hall, 1978b).

Differential projections from the magnocellular nucleus to layer I and from the equivalents of the ventral and dorsal nuclei to the middle layers have also been reported in the rat (Ryugo and Killackey, 1974). There has been no comparable published study in the cat but I have noted degenerating layer I terminations after magnocellular nucleus lesions in old cat material of Heath and Jones (1971a; Fig. 8.13).

8.6.3. Tonotopic Organization and Functional Segregation in the Thalamocortical Projection from the Ventral Nucleus

Merzenich *et al.* (1982) were able to confirm the original observations of Rose and Woolsey (1949a; Section 8.6.1) on the tonotopicity of the thalamic projection to AI in the cat. By injecting horseradish peroxidase into parts of AI

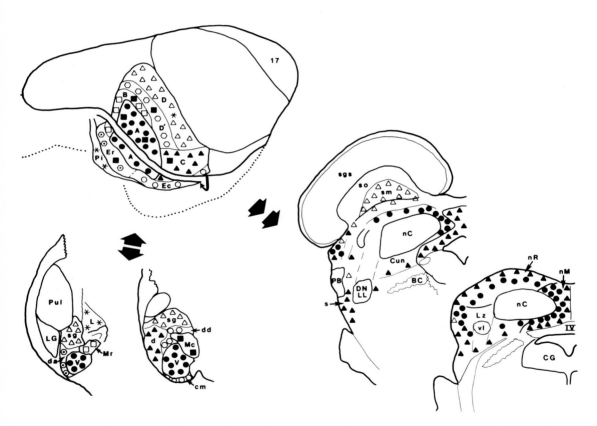

FIGURE 8.12. Auditory cortical fields in the tree shrew (top) defined in terms of their interconnections with the ventral (V), dorsal (sg, d, dd), and magnocellular (Mc) nuclei of the medial geniculate complex (bottom left), and their corticofugal connections to subnuclei of the inferior colliculus. Pi, Ec, Er indicate transitional and perirhinal areas which receive from nuclei flanking the medial geniculate complex. From Oliver and Hall (1978b).

representing a small range of frequencies, they observed retrograde labeling in lamellar configurations in the ventral nucleus that clearly correspond to the anatomical and physiological "isofrequency lamellae" found there (Morest, 1965a; Aitkin and Webster, 1972; Section 8.4 and Chapter 3). In AI of the cat (Merzenich *et al.*, 1975; Reale and Imig, 1980) a single frequency is represented as a line running mediolaterally across this cortical field (Fig. 8.14). The results of Merzenich *et al.* (1982) imply, therefore, that an isofrequency lamella in the ventral medial geniculate nucleus is projected onto a mediolaterally oriented isofrequency band in AI. High frequencies are represented anteriorly in the cat AI and low frequencies posteriorly (Woolsey and Walzl, 1942; Merzenich *et al.*, 1975) and, in keeping with the observations of Aitkin and Webster (1972) on tonotopicity in the ventral medial geniculate nucleus, medial lamellae project anteriorly in AI and lateral lamellae posteriorly (Andersen *et al.*, 1980b; Merzenich *et al.*, 1982).

Each isofrequency band in AI of the cat cortex does not contain a homogeneous population of neurons. Neurons excited by both ears (EE neurons) and

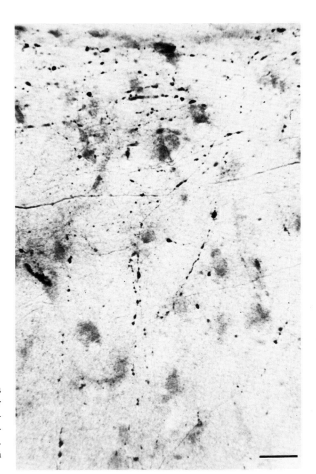

FIGURE 8.13. Degenerating axons ramifying in layer I of the anterior auditory field of a cat following a lesion in the magnocellular medial geniculate nucleus. Bar = 100 μm. From material illustrated in Heath and Jones (1971a).

neurons exicted by one ear and inhibited by the other (EI neurons) (see Section 8.4 and Chapter 4) are segregated in alternating regions. The groups of EE and EI neurons, particularly in regions of high-frequency representation, line up with those in neighboring isofrequency bands so as to form a series of alternating EE and EI bands that run across AI orthogonal to the isofrequency bands (Imig and Adrian, 1977; Imig and Brugge, 1978; Middlebrooks *et al.,* 1980; Imig and Reale, 1981) (Fig. 8.15).

Middlebrooks and Zook (1983) (Fig. 8.15) using single- and multiunit mapping in the high-frequency representation of AI have consistently identified a ventral pair of EI and EE bands running continuously anteroposteriorly across AI. The EI band is the more ventral and at the border with AII. In the middle part of AI the bands are more variable and discontinuous, while dorsally in AI the bands are not found; there, units have broader tuning curves, and binaural neuronal responses are less distinct.

Middlebrooks and Zook (1983) attempted to inject small quantities of horse-radish peroxidase into single EI or EE bands in order to detect any differential

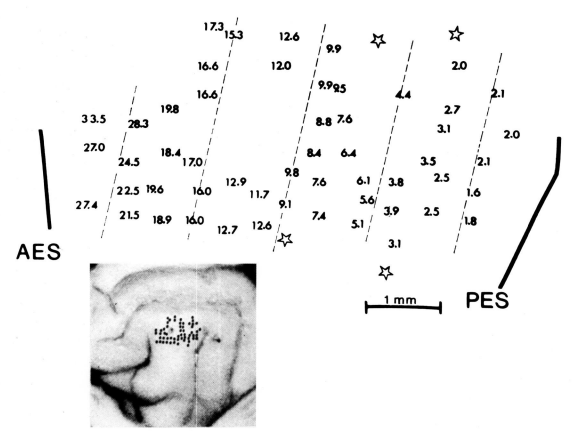

FIGURE 8.14.   Surface map of cat first auditory area showing best frequencies (in kHz) of units recorded in perpendicular electrode tracks at points indicated. Note progression from high (base of cochlea) to low (apex of cochlea) in anteroposterior dimension and mediolateral isofrequency bands. From Merzenich *et al.* (1975).

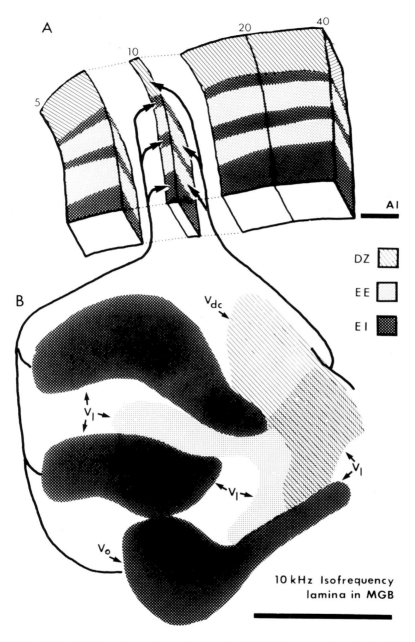

FIGURE 8.15.   EE and EI bands running orthogonal to the isofrequency lines in the AI auditory cortex (A) of the cat. Anterior is to right. (B) A parasagittal section of the ventral medial geniculate nucleus showing an isofrequency lamella and its EE and EI bands postulated from cortical connectivity. DZ indicates an uncertain dorsal zone. From Middlebrooks and Zook (1983).

inputs from the ventral medial geniculate nucleus. They discovered that probably all EI bands receive inputs from thalamic neurons lying in three separate parts of each isofrequency lamella in the ventral nucleus. An injection of tracer in the ventral EI band at, say, the representation of 18 kHz thus labels three long columns of cells extending through the anteroposterior dimension of the appropriate isofrequency lamella. The most ventral of these columns is in the part of the lamella that coils medially in the pars ovoidea of Morest (1965a). EI projecting cells in neighboring isofrequency lamellae of the ventral medial geniculate nucleus are aligned, as in the cortex, to form bands stretching orthogonally across the isofrequency lamellae (Fig. 8.15).

Injections of tracer in cortical EE bands led to retrograde labeling of two anteroposterior columns of cells that alternate with those labeled after injections of an EI band. One lies in the dorsalmost part of an isofrequency lamella and the other between the dorsal and the middle EI projecting columns (Fig. 8.15). Again, EE-related cells in adjacent isofrequency lamellae are aligned across the lamellae. The dorsal region of AI probably also receives input from the dorsalmost EE column.

The observations of Middlebrooks and Zook (1983) imply that there should be a segregation of EE and EI responding cells in different parts of each isofrequency lamella in the ventral and medial geniculate nucleus. In the only relevant study, Calford and Webster (1981) noted that EE and EI units were often found in separate clusters in the ventral nucleus but they did not report any systematic shift of the kind to be expected from the findings of Middlebrooks and Zook. The study of Middlebrooks and Zook not only implies a variation in functional properties along the dorsoventral dimension of the isofrequency lamellae in the ventral medial geniculate nucleus but also indicates an anteroposteriorly elongated, rodlike arrangement of functional classes of cells and of thalamocortical relay cells. In both respects the pattern of organization bears a remarkable resemblance to the modality-specific rods of thalamocortical relay cells in the ventral posterior nucleus (Jones et al., 1982; Chapters 3 and 7). One would anticipate that individual axons entering an isofrequency lamella in the ventral medial geniculate nucleus would, like medial lemniscal fibers entering the ventral posterior nucleus, turn horizontally and end along part of the length of an EI or EE rod.

9

Lateral Geniculate Nucleus

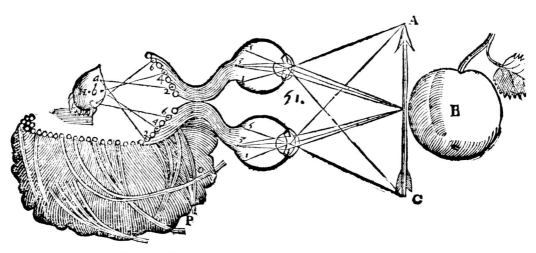

A figure from Descartes's *L'Homme* (1664) showing the projection of an object onto the retina and thenceforth into the lateral ventricles and pineal body.

9.1. Description

The lateral geniculate nucleus is remarkably stable across mammals and can be recognized even in those such as moles in which the eyes are congenitally absent or greatly reduced in size. It is stricly a nuclear complex, rather than a single nucleus, for in virtually every mammal two or more components can be recognized. One of these, the dorsal lateral geniculate nucleus, is developmentally a derivative of the dorsal thalamus and projects to the cerebral cortex. In a number of species it is laminated and may contain a number of subsidiary nuclei. The other component, the ventral lateral geniculate nucleus, is developmentally a part of the ventral thalamus and does not project to the cortex. In primates, seemingly on account of the descent and rotation of the dorsal lateral geniculate nucleus during development (Le Gros Clark, 1932b) (Fig. 9.1), the ventral nucleus has come to lie dorsal to the dorsal nucleus; in this position it is often referred to as the pregeniculate nucleus.

9.1.1. Dorsal Lateral Geniculate Nucleus

Because of the rather variable structure of the dorsal lateral geniculate nucleus, it is not possible to give a generalized account that is relevant to all mammals. Among the commoner experimental animals and their relatives, three principal forms can be arbitrarily recognized: these are the forms typical of rodents, carnivores, and primates.

9.1.1.1. The Dorsal Lateral Geniculate Nucleus of Rodents

The nucleus of rodents and many other small mammals might arbitrarily be called the more generalized form. It is usually small, dorsolaterally situated on the surface of the thalamus, and not clearly separated from the underlying lateral nuclei by a medial ramus of the optic tract; it can be quite flattened (Fig. 9.2). In Nissl preparations, the nucleus contains a fairly homogenous cell population that is not overtly laminated and which can closely resemble the cells of the underlying lateral complex.

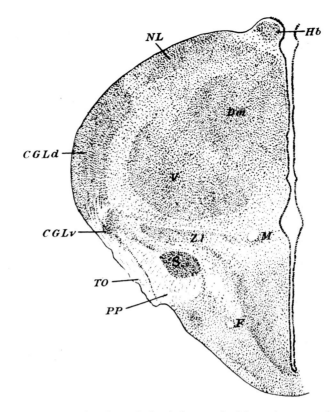

FIGURE 9.1. Transverse section through the thalamus of a 92-mm human embryo showing the early position of the anlage of the pregeniculate or ventral lateral geniculate nucleus (CGLv) before rotation of the dorsal lateral geniculate nucleus (CGLd) forces it into a position different from that in lower primates and nonprimates. From Le Gros Clark (1932b).

A hidden alternation of the left eye and right eye inputs may be revealed by selective staining of retinal axon terminations (Fig. 9.3). This shows an outer and an inner layer of terminations of fibers emanating from the contralateral eye and a much shorter central layer of terminations arising from fibers of the ipsilateral eye (Giolli and Guthrie, 1969; Hayhow *et al.*, 1962; Lund, 1965; Montero *et al.*, 1968; Cunningham and Lund, 1971; Creel and Giolli, 1972; Lund *et al.*, 1974; Wise and Lund, 1976; Takahashi *et al.*, 1977). In highly visual rodents, such as squirrels, an incipient triple lamination can often be detected with cell stains and becomes unmasked by labeling of retinal axon terminations (Tigges, 1970; Kaas *et al.*, 1972a). The layers so revealed consist of three longer layers that receive fibers from the contralateral eye and two shorter intervening layers that receive fibers from the ipsilateral eye (Fig. 9.3). It has become customary

FIGURE 9.2. Frontal sections through the dorsal lateral geniculate nucleus of a rat (A), gray squirrel (B), and guinea pig (C, D). Thionin stain (A–C) and hematoxylin stain (D). Bars = 250 μm.

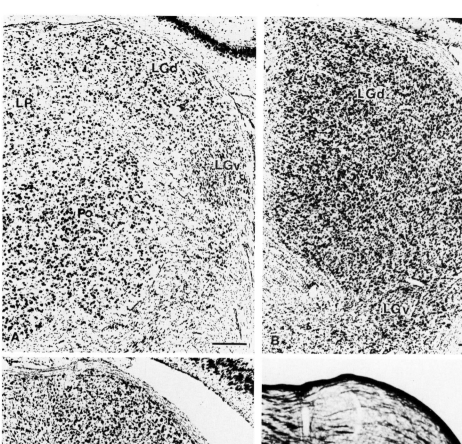

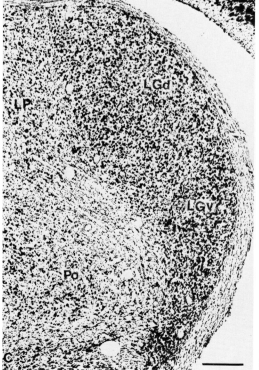

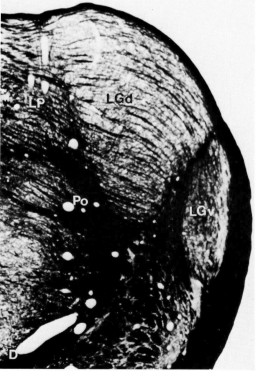

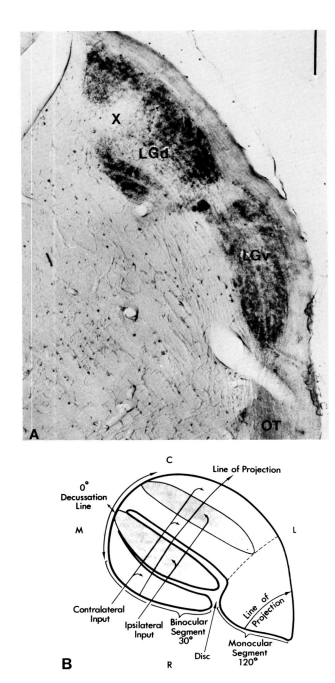

FIGURE 9.3. (A) Anterograde labeling with horseradish peroxidase of afferent fibers arising in the contralateral eye and terminating in two major bands in the dorsal lateral geniculate nucleus of a rat. The small unlabeled (X) zone deep to the outer band of label is the site of termination of fibers from the ipsilateral retina. Bar = 250 μm. (B) Schematic figure from Kaas *et al.* (1972) showing the laminar pattern revealed in the squirrel by staining of the retinogeniculate projections. Three laminae are innervated by the contralateral retina and two laminae by the ipsilateral. The parts of the contralaterally innervated laminae not matched by ipsilaterally innervated laminae constitute the "monocular segment." C: caudal; R: rostral; L: lateral; M: medial.

to refer to the zones of alternating retinal inputs in these and other species as
the binocular region or segment of the dorsal lateral geniculate nucleus and the
zone containing only parts of the contralaterally innervated laminae, the monocular region or segment.

459

LATERAL
GENICULATE
NUCLEUS

9.1.1.2. The Dorsal Lateral Geniculate Nucleus of Carnivores

A second distinctive form of the lateral geniculate nucleus is characteristic
of the carnivore brain and has been described in the cat, dog, and other carnivores (Thuma, 1928; Rioch, 1929a; Minkowski, 1920; Guillery, 1970; Hickey
and Guillery, 1974; Sanderson, 1974). The nucleus is large, dorsolaterally or
laterally situated, and over much of its extent clearly separated from the rest of
the dorsal thalamus by a medial ramus of the optic tract; it is often S shaped
when viewed from the side, with an upturned, posterior tail (Fig. 9.4); the cell
population is clearly mixed, with both large and smaller forms easily recognized.
Two components are usually present: a larger lateral, *laminar dorsal lateral geniculate nucleus* and a smaller, nonlaminated, *medial interlaminar nucleus* (Fig. 9.4).
The laminar nucleus has three overt laminae separated by interlaminar fiber
plexuses which can be invaded by displaced cells. The most dorsal and the middle
laminae are composed of large and small- to medium-sized cells and in the cat
are referred to respectively as lamina A and lamina A1. Lamina A, which receives
the terminations of axons from the contralateral retina, extends more laterally
than lamina A1, which receives the terminations of fibers from the ipsilateral
retina. Laminae A and A1 can each be divided into two cellular leaflets by a thin
fiber plexus in certain carnivores such as the mink and weasel (Sanderson, 1974;
Guillery *et al.*, 1974; Guillery and Oberdorfer, 1977). They are incompletely
divided in others such as the ferret and certain procyonids, but not divided in
the cat, dog, fox, other procyonids, or in the seal (Fig. 9.5). The third and deepest
lamina, adjacent to the optic tract is now recognized in the cat and other carnivores to contain several sublaminae. Collectively they may be called the "C
laminae." (There is no lamina B in modern terminology.) The most dorsal of
these (lamina C) is composed of large, deeply staining cells similar to those in
lamina A1. Progressively deeper to it are laminae C1, C2, and C3, all composed
of small cells and not always separable from one another in Nissl stains alone,
except perhaps in the fox. They are brought out, however, by labeling of retinal
fiber terminations: laminae C and C2 receive fibers from the contralateral eye
and C1 from the ipsilateral eye; C3 may not receive retinal axon terminations
(Guillery, 1970; Hickey and Guillery, 1974). Lamina C2 is particularly thick in
the mink whereas lamina C dominates in the seal (Fig. 9.5).

The *medial interlaminar nucleus* of the carnivore dorsal lateral geniculate
nucleus is composed of medium to large, moderately deeply staining cells, which
tend to be vertically oriented but are not laminated. The medial interlaminar
nucleus occupies approximately the middle one-third of the anteroposterior
extent of the dorsal lateral geniculate nucleus in most carnivores; it tends to be
comma shaped in frontal section, with its lateral tail invading or undershooting
the C laminae of the laminar nucleus (Fig. 9.4). A vertical sheet of cells in the
lateral edge of the pulvinar nucleus of the cat, receives retinal axons (Berman
and Jones, 1977) and has sometimes been regarded as an extension of the medial
interlaminar nucleus, the "geniculate wing" (Guillery *et al.*, 1980).

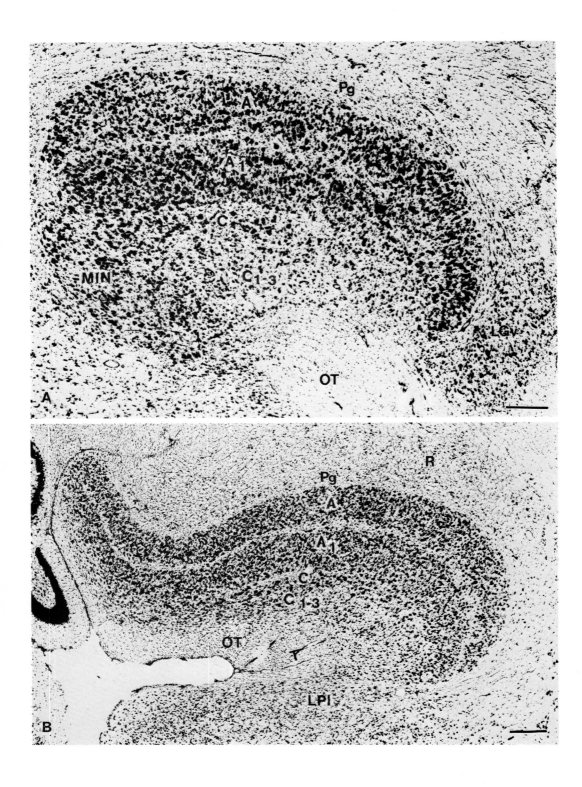

The medial interlaminar nucleus was identified and named by Thuma (1928) in the cat thalamus. Earlier it had been illustrated by Winkler and Potter (1914) who regarded it as part of the ventral lateral geniculate nucleus. Thuma is also responsible for naming laminae A and A1 in the laminar lateral geniculate nucleus. He referred to lamina C, however, as a central interlaminar nucleus and to laminae C1–3 collectively as a single lamina B. Though Thuma's terminology for the laminated dorsal lateral geniculate nucleus survived until Guillery's revision of 1970, the medial interlaminar nucleus was largely ignored until Hayhow (1958) once more drew attention to it and pointed out that it received input from the two eyes. This observation has been repeatedly confirmed (Seneviratne and Whitteridge, 1962; Laties and Sprague, 1966; Stone and Hansen, 1966; Garey and Powell, 1968; Kinston *et al.*, 1969; Guillery, 1970; Kratz *et al.*, 1978a; Guillery *et al.*, 1980; Rowe and Dreher, 1982) and the presence of the nucleus and its binocular input has been detected in a variety of other carnivores (e.g., Sanderson, 1974; Hickey and Guillery, 1974).

The anatomical studies have indicated that the ipsilateral eye projects to a relatively small central core of the medial interlaminar nucleus while the contralateral eye projects to a rather larger dorsal, medial, and ventral shell enclosing the ipsilaterally innervated core (Fig. 9.6). Guillery *et al.* (1980) and Rowe and Dreher (1982; Fig. 9.6) have shown that the input to the lateralmost component of the contralaterally innervated region is derived from the 20° or so of *temporal* retina adjacent to the vertical meridian. A majority of the retinal inputs, as judged from the responses of cells in the medial interlaminar nucleus, are of the Y type (Mason, 1975; Palmer *et al.*, 1975; Dreher and Sefton, 1978; Kratz *et al.*, 1978a) though other types have also been described. These will be mentioned in a later section.

An equivalent of the medial interlaminar nucleus of the carnivore dorsal lateral geniculate nucleus has not been positively identified in all other species. Campos-Ortega and Hayhow (1971) consider that the intermediate cell group of Minkowski (1922), essentially a deep component of the pregeniculate nucleus, could be its equivalent in monkeys. Additional representations of the visual field or retina-innervated regions lying medial or deep to the dorsal lateral geniculate nucleus have been equated with the medial interlaminar nucleus in the rat, rabbit, and ungulates (Choudhury and Whitteridge, 1965; Montero *et al.*, 1968; Cummings and de Lahunta, 1969; Campos-Ortega, 1970). In all these species it appears to correspond to what Rose (1942b) called the *pars geniculata pulvinaris*.

The Perigeniculate Nucleus. A thin layer of scattered small cells lying in the external medullary lamina and closely apposed to the dorsal, lateral, and anterior surfaces of the dorsal lateral geniculate nucleus in the cat was called the *substantia grisea perigeniculata* by Rioch (1929a). It was left unnoticed for many years until Sanderson (1971) drew attention to the facts that it could be activated by visual

---

FIGURE 9.4. Frontal (A) and sagittal (B) sections through the dorsal lateral geniculate nucleus of cats showing the various laminae (A, A1, C, C1–3) of the laminar nucleus, the medial interlaminar nucleus (MIN), and the perigeniculate nucleus (Pg). Thionin stain. Bars = 250 μm.

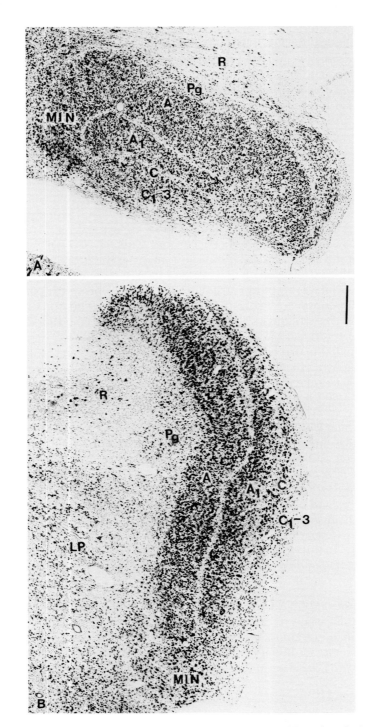

FIGURE 9.5. Frontal (A) and sagittal (B) sections through the dorsal lateral geniculate nucleus of a seal (*Phoca vitulina*) and sagittal section through that of a raccoon (C). Seal brain kindly provided by Dr. R. W. Dykes. Thionin stain. Bars = 1 mm (A, B), 500 μm (C).

stimuli and showed a retinotopic organization, and that its cells had properties that did not resemble those of the underlying dorsal lateral geniculate nucleus. Szentágothai (1963) had rather vaguely suggested that it might be an extension of the reticular nucleus of the thalamus and this has come to be generally accepted (Chapter 15). It appears that it may form a concentration of inhibitory interneurons for the dorsal lateral geniculate nucleus (Chapters 5 and 15) and that it is innervated by collaterals of thalamocortical and corticothalamic axons passing between the dorsal lateral geniculate nucleus and the visual cortex (see Chapters 3 and 4).

A perigeniculate nucleus has been identified in representatives of several carnivore families. It is particularly large in procyonids, mustelids (Sanderson, 1974), and in at least one species of seal (Fig. 9.5).

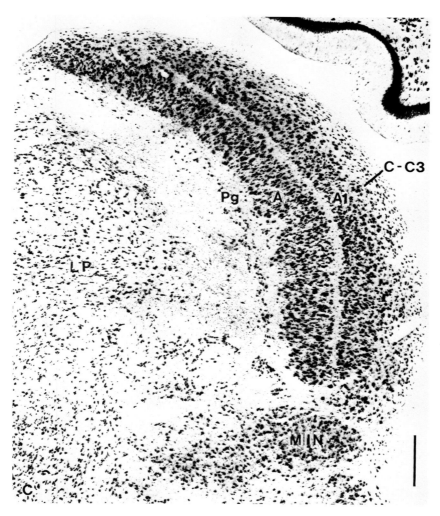

FIGURE 9.5   (*continued*)

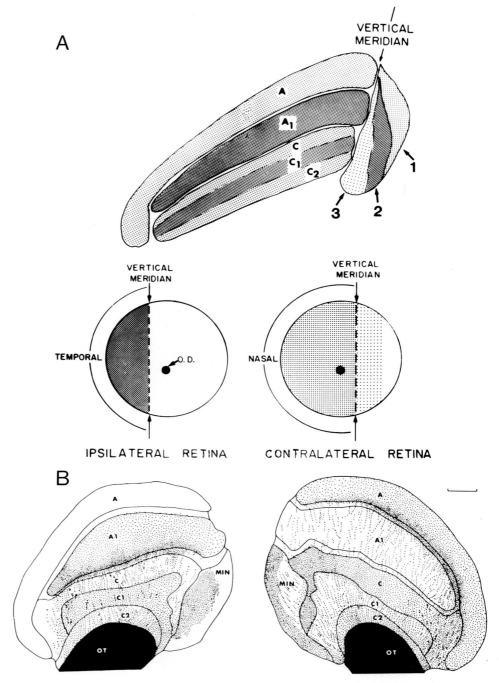

FIGURE 9.6. (A) Representation of the visual field in the laminar and medial interlaminar nuclei of the dorsal lateral geniculate complex of the cat. O.D.: optic disc. From Rowe and Dreher (1982). (B) Distribution of retinogeniculate axons arising in the ipsilateral (left) and contralateral (right) eyes of the cat. From Kaas *et al.* (1972) after Guillery (1970).

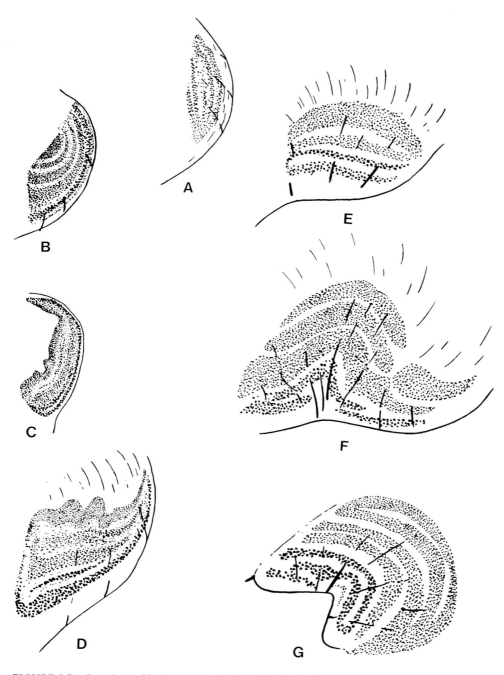

FIGURE 9.7.   Drawings of laminar organization of the dorsal lateral geniculate nucleus in the tree shrew (A) and in various primates. (B) Mouse lemur; (C) Coquerell's dwarf lemur; (D) *Lemur catta;* (E) orangutan; (F) man; (G) cercopithecine monkey. Lateral to the right in each case. From Le Gros Clark (1932b).

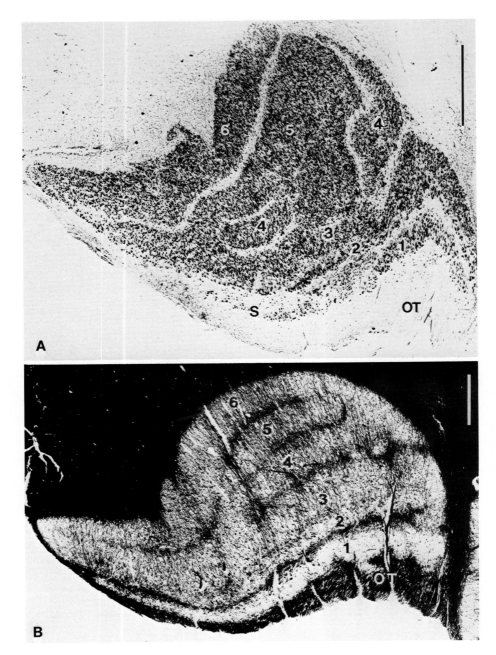

FIGURE 9.8. Gallocyanin (A) and hematoxylin (B) stained frontal sections through the human dorsal lateral geniculate nucleus, lateral to left. (A) is from lateral edge of nucleus and shows fusion of layers 3 and 5. OT: optic tract. (C) Frontal section through the dorsal lateral geniculate nucleus of *Macaca mulatta;* lateral to right. Thionin stain. Bars = 1 mm (A, B), 500 μm (C).

An equivalent of the perigeniculate nucleus has been identified in ungulates (Rose, 1942b) but not in other species.

9.1.1.3. The Dorsal Lateral Geniculate Nucleus of Primates

The dorsal lateral geniculate nucleus of primates presents a third characteristic form of the nucleus. It is relatively very large, laterally or ventrolaterally situated, and frequently totally isolated from the rest of the dorsal thalamus by being enveloped in the optic tract; it forms an elongated lateral bulge on the thalamus in prosimians; it is usually in the shape of an inverted U in monkeys, apes, and man. The U shape is said to be less marked in the orangutan and the U appears opened out in man (Figs. 9.7 and 9.8). The optic tract and branches of the posterior cerebral artery enter the nucleus at the hilus and the optic radiations leave from the dorsal convexity.

The cell population is mixed and in many primates, irrespective of taxonomic status, the nucleus is very overtly laminated (Le Gros Clark, 1932b, 1941a,b; Chacko, 1948, 1954a,c; Hassler, 1966, Kaas *et al.*, 1978). The clearest laminar

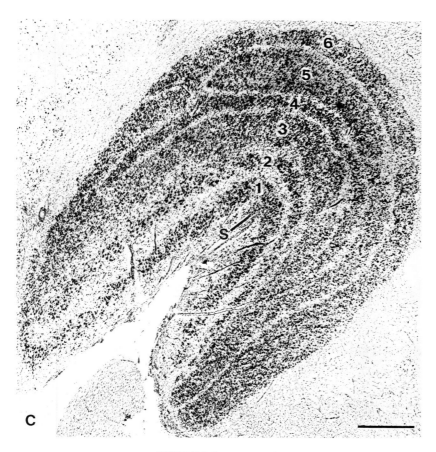

FIGURE 9.8. (*continued*)

distinction is between magnocellular and parvocellular layers. The magnocellular laminae, usually two (laminae 1 and 2 in man), are relatively thin and composed of large, deeply staining cells (Figs. 9.7 and 9.8). In man, apes, and monkeys, the magnocellular layers lie ventrally—close to the pial surface of the nucleus. The external layer is innervated by the contralateral eye and the internal by the ipsilateral eye (Fig. 9.9). Dorsal or dosolateral to the magnocellular laminae are the parvocellular laminae, composed of paler staining, small to medium-sized cells. Usually four of these (laminae 3–6) are recognized in man, chimpanzee, and Old World monkeys (Figs. 9.7–9.9). Laminae 4 and 6 are innvervated by the contralateral eye and laminae 3 and 5 by the ipsilateral eye (Fig. 9.9). In some Old World monkeys such as the guenons, however, the four parvocellular layers are ill-defined in Nissl-stained preparations, whereas in others, such as macaques and baboons, some parts of the laminae are reduplicated to give an appearance of more than four in parts of the nucleus (Le Gros Clark, 1932b; Chacko, 1948, 1955a; Balado and Franke, 1937; Hassler, 1959; Hickey, 1977a; Tigges *et al.*, 1977a) (see Fig. 9.19). At the anterior and posterior ends of the nucleus, however, the parvocellular laminae always fuse, giving an appearance of less than four. The greatest degree of laminar variability may be in the human brain (Hickey and Guillery, 1979)

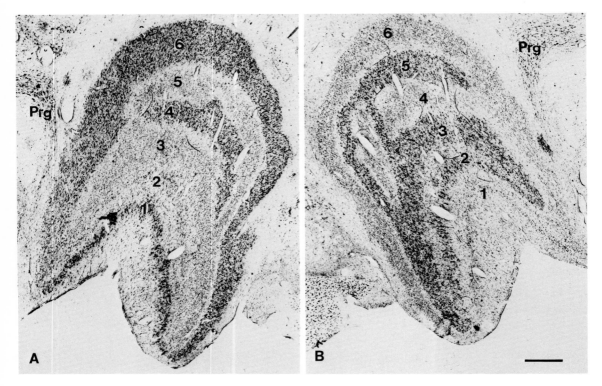

FIGURE 9.9. Transneuronal degeneration of the dorsal lateral geniculate nucleus of a rhesus monkey subsequent to removal of one eye several weeks previously. (A) Ipsilateral and (B) contralateral to eye removal. Prg: pregeniculate nucleus. Thionin stain. Bar = 500 μm. From a preparation of Dr. W. M. Cowan.

In many New World monkeys (e.g., the squirrel monkey and spider monkey), the parvocellular region is not laminated in Nissl preparations (Chacko, 1954b) (Fig. 9.10) but a hidden, seemingly quadripartite lamination, identical to that in Old World monkeys and in man, is revealed by the segregated terminations of axons from each eye (Jones, 1964; Doty *et al.,* 1966; Jacobs, 1969; Kaas *et al.,* 1972; Tigges and O'Steen, 1974). In some New World monkeys, (e.g., the marmosets, the capuchin monkey, and the owl monkey), the parvocellular mass may show varying degrees of splitting in Nissl preparations, though usually into only two layers. A similar pattern in seen in the orangutan and gibbon (apes) and in *Tarsius,* lemurs, and lorises (prosimians) (Solnitzky and Harman, 1946; Chacko, 1948, 1954c, 1955b; Hassler, 1966; Campos-Ortega and Oliver, 1968; Kanagasuntheram *et al.,* 1968, 1969; Kanagasuntheram and Wong, 1968; Kanagasuntheram and Krishnamurti, 1970)

Kaas *et al.* (1972, 1978) have proposed that a lamination pattern consisting of two ventral (or lateral) magnocellular and two dorsal (or medial) parvocellular laminae represents the fundamental anthropoid plan. Their idea, which has similarities to earlier suggestions of Minkowski (1920) and Walls (1953), is based upon a consideration of descriptions in the literature of a number of different species of monkey and ape, careful reconstructions from serially sectioned monkey and human brains, the distribution of ipsi- and contralateral retinogeniculate axons, and the systematic mapping of the visual fields in the dorsal lateral geniculate nucleus. In the owl monkey, with four geniculate layers, fibers from the ipsilateral eye terminate in the ventralmost magnocellular and in the dorsalmost parvocellular laminae. Fibers from the contalateral eye terminate in the inner

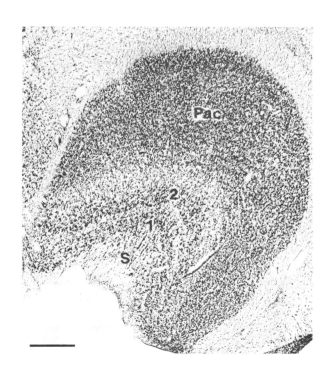

FIGURE 9.10. Frontal section through the dorsal lateral geniculate nucleus of a squirrel monkey (*Saimiri sciureus*) showing two magnocellular layers (1, 2), the S layer (S), and an unlaminated parvocellular mass (Pac). Thionin stain. Bar = 500 μm. Lateral to right.

magno- and parvocellular laminae. Each of the four layers, from receptive field mapping, contains a full representation of the appropriate visual field. In forms such as Old World monkeys, certain apes, and man, in which four parvocellular layers are customarily described, Kaas's group points out that the layers are incomplete, show considerable degrees of fusion, and, when mapped electrophysiologically, may not contain full representations of the visual field of the relevant eye. It is their belief that the two fundamental parvocellular laminae, one with ipsi- and the other with contralateral input, split and interweave with one another, especially in the region of representation of central vision, thus giving an appearance of four (or even more) laminae with alternating ipsi- and contralateral retinal inputs. At the margins of the nucleus, patterns of lamination and retinal axon distribution closely resemble those of the owl monkey, which Kaas's group takes as its prototype (Fig. 9.11). Given this new information, it might seem appropriate to dispense with the current system of numbering the parvocellular layers 3–6 (which dates to Le Gros Clark, 1932b), in favor of something that more accurately reflects their arrangement. Kaas *et al.* (1978) and their followers have adopted the policy of referring to the old layers 3 and 5 together as the internal parvocellular lamina and the old layers 4 and 6 together as the external parvocellular lamina (even though it lies deep within the brain). This has some merit. Yet, on the other hand, a new nomenclature designed to clear up a discrepancy in one aspect of the thalamus almost invariably leads to confusion in another. Hence, so long as it is recognized that layers 3 and 5 and 4 and 6 go together as components of single laminae and are only identifiable

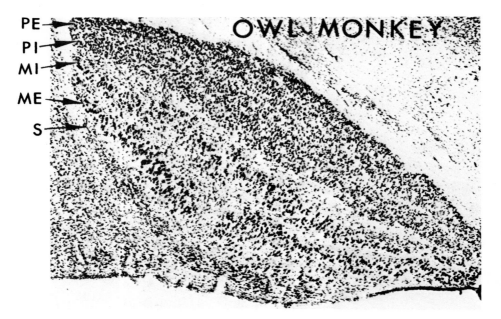

FIGURE 9.11. Sagittal section through the dorsal lateral geniculate nucleus of an owl monkey (*Aotus trivirgatus*) showing the S lamina, the external (ME) and internal (MI) magnocellular laminae, and the internal (PI) and external (PE) parvocellular laminae. PI corresponds to layers 3 and 5 and PE to layers 4 and 6 of Old World primates. From Kaas *et al.* (1972).

in the middle parts of the lateral geniculate nucleus, little harm will probably accrue.

In most prosimians, in which the dorsal lateral geniculate nucleus is often more laterally placed and usually not folded on itself as in anthropoids, two layers corresponding to the magnocellular layers lie external to three or more parvocellular laminae (Le Gros Clark, 1932b; Chacko, 1954a; Hassler, 1966; Ionescu and Hassler, 1968; Kanagasuntheram *et al.*, 1969; Campos-Ortega and Hayhow, 1971; Kaas *et al.*, 1978; Fitpatrick *et al.*, 1980) (Fig. 9.12). Of these, however, the central one or two are usually thinner and composed of cells smaller than in the more typical external and internal two (Hassler, 1966; Kaas *et al.*, 1978). The magnocellular layers receive alternating contra- and ipsilateral retinal inputs, the external parvocellular layer ipsi-, and the internal parvocellular layer contralateral retinal inputs. The central small-celled layer or layers usually receive a mixture of ipsi- and contralateral inputs. From the work of Kaas, Dia-

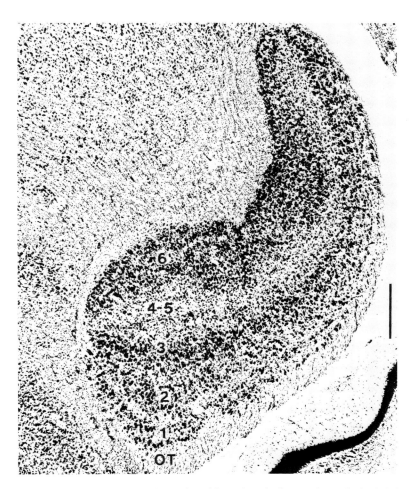

FIGURE 9.12. Frontal section through the dorsal lateral geniculate nucleus of a bush baby (*Galago crassicaudatus*). Lateral to right. Layers 4 and 5 are thought to be comparable to cells in the interlaminar spaces of other primate lateral geniculate nuclei. Thionin stain. Bar = 500 μm.

mond, and their co-workers, to be discussed in Section 9.3.2, it would appear that the thinner, central layers should be regarded as intercalated layers, akin to the less overt but nevertheless identifiable cells situated in the fibrous plexuses between the laminae of the dorsal lateral geniculate nucleus of anthropoids (Figs. 9.7–9.10). If this is so, then the fundamental primate plan is preserved.

Interlaminar cells receiving retinal inputs have been described between the parvocellular layers and between parvo- and magnocellular layers in several species of monkeys (Kaas *et al.*, 1978; Hendrickson *et al.*, 1978). In some species, such as the owl monkey, one as least of these intercalated laminae is particularly thick (Jones, 1966; Kaas *et al.*, 1978), and resembles the central small-celled laminae of prosimians.

A further additional row of small cells of greater or lesser thickness is often detected in the optic tract external to the magnocellular laminae in monkeys, apes, and man and may also be present in prosimians (Le Gros Clark, 1941b; Kanagasuntheram *et al.*, 1969; Giolli and Tigges, 1970; Tigges and Tigges, 1970; Kaas *et al.*, 1972, 1978; Tigges *et al.*, 1977a). This row of cells has been referred to as the S lamina and is itself often divided into two concealed subsidiary laminae, one receiving ipsi- and the other contralateral retinal fibers.* The S laminae and the intercalated cell groups in the interlaminar plexuses have input–output relations rather different from those of the principal laminae and have a special significance that will be taken up in Section 9.3.2.

9.1.1.4. The Dorsal Lateral Geniculate Nucleus of Other Species

The dorsal lateral geniculate nucleus of many species of insectivores resembles the nucleus of the rat (Le Gros Clark, 1928; Campbell, 1969). In some such as the hedgehog and the mole the nucleus is reduced to a narrow strip of cells scarcely distinguishable from the rest of the thalamus. With the possible exception of the essentially sightless mole, however (Lund and Lund, 1965), there is an outer zone of contralateral retinal terminations and an inner, smaller zone of ipsilateral retinal terminations (Campell *et al.*, 1967). By contrast, the tree shrew, *Tupaia* (Fig. 9.13), possesses a lateral geniculate nucleus that resembles in size and complexity of lamination that of many of the lower primates (Le Gros Clark, 1929a, 1932b; Glickstein, 1967; Diamond *et al.*, 1970). Such features led Le Gros Clark (1962) to classify *Tupaia* among the primates.

Le Gros Clark (1932b) recognized five laminae to the tree shrew dorsal lateral geniculate nucleus; outer and inner magnocellular laminae and three intervening parvocellular laminae. More recently, Diamond and his associates and others have identified six layers which, rather confusingly, they number from deep (1) to superficial (6), rather than from superficial to deep as has become customary in primates (Campbell *et al.*, 1967; Glickstein, 1967; Harting *et al.*, 1973). Layers *2, 3, 4,* and *6* receive the terminations of contralateral retinal fibers and layers *1* and *5* those of ipsilateral retinal fibers. Some scattered cells also appear between some of the laminae, as described in Section 9.3.2.

---

* In a paper published after this chapter was written, D. Fitzpatrick *et al.* (1983) referred to most of the S layers of the squirrel monkey as intercalated layers and restricted their use of the term "S layer" to islands of large cells which they consider to be displaced parts of layer 2 of the dorsal lateral geniculate nucleus.

Variability in the dorsal lateral geniculate nucleus, similar to that of the insectivores, is to be found among the bats. Small, echo-locating bats such as *Myotis* and *Pteronotus* (Fig. 9.14) have small unlaminated nuclei with a predominately contralateral retinal input while large, non-echo-locating bats such as the fruit bat, *Pteropus,* have large, partially laminated nuclei with three layers of contralateral retinal terminations and a small, mainly dorsally placed zone of ipsilateral terminations somewhat similar to that of the rat (Cotter and Pierson Pentney, 1979).

The dorsal lateral geniculate nucleus of ungulates has a pattern of organization quite different from that of carnivores and primates, though it shows some similarities with the nucleus of the rabbit (Rose, 1942b). Rose in 1942 described a series of laminae in the nucleus of the sheep and pig, though more recent workers in these and other species have tended to question the ease with which these laminae can be discerned in Nissl preparations (Nichterlein and

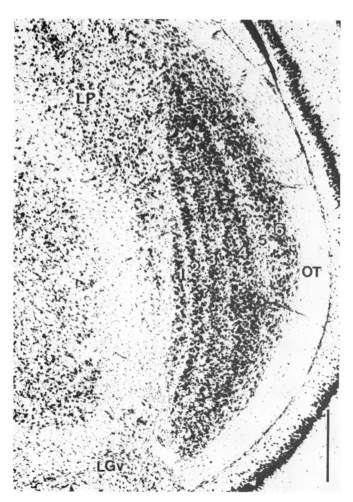

FIGURE 9.13.   Nissl-stained frontal section through the dorsal lateral geniculate nucleus of a tree shrew (*Tupaia glis*). From a preparation of Dr. M. Conley. Bar = 500 μm.

Goldby, 1944; Campos-Ortega, 1970; Karamanlidis and Magras, 1972, 1974). The nucleus tends to be laterally placed and elongated dorsoventrally (Fig. 9.15). Covering its posterolateral surface is a thick and very extensive region of cells of mixed size (called the α layer by Rose). It curves ventrally under the deeper elements from which it is separated by a thin, incomplete fibrous layer (layer β of Rose). Deepest of all is a small region of large cells (the γ layer) which can be incompletely split into layers γ1 and γ2 (Fig. 9.15). A dorsomedially situated pars geniculata pulvinaris of Rose is the probable equivalent of the feline medial interlaminar nucleus (Campos-Ortega, 1970). The greater part of the nucleus is filled by terminations of fibers from the contralateral eye. These fill regions equivalent to layers α and γ of Rose with the exception of a small dorsomedial region more or less equivalent to Rose's layer γ2; this region receives fibers from the ipsilateral eye (Minkowski, 1920; Nichterlein and Goldby, 1944; Campos-Ortega, 1970; Karamanlidis and Magras, 1972, 1974). It is interesting that this binocular segment of the nucleus shows some degree of laminar separation comparable to the cat. The medial interlaminar nucleus, similarly, has a large contralateral input and a small ipsilateral input (Fig. 9.15)

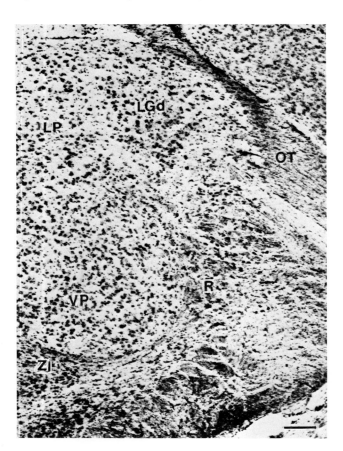

FIGURE 9.14. Frontal section through the dorsal lateral geniculate nucleus of a moustache bat (*Pteronotus parnelli*). Luxol fast blue and thionin stain. Bar = 100 μm.

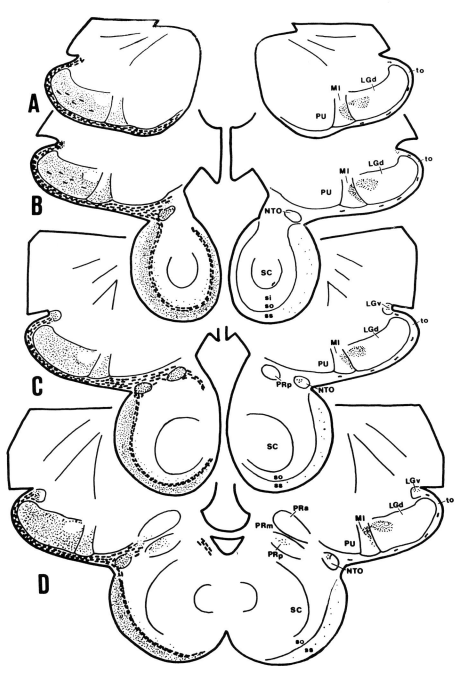

FIGURE 9.15. Distribution of ipsilateral (right) and contralateral (left) retinogeniculate fibers in horizontal sections from the thalamus of a horse. MI: medial interlaminar nucleus. From Karamanlidis and Magras (1974).

As Sanderson and his colleagues have clearly pointed out, there is no pre-dictability about cellular lamination or retinal fiber distribution even in different species belonging to the same family (Pearson *et al.,* 1976; Sanderson and Pearson, 1977; Sanderson *et al.,* 1979). The dorsal lateral geniculate of the marsupials, for example, appears to consist of two fundamental cellular masses: an outer usually more compact cellular region (the α segment) and a looser inner region (the β segment). But the number of subsidiary laminae in the α segment can range from none in the American opossum, *Didelphis* (Fig. 9.16) with complete superimposition of ipsi- and contralateral retinal fiber terminations (Bodian, 1937; Lent *et al.,* 1976; Royce *et al.,* 1976), to four in the phalanger, *Trichosurus* (Fig. 9.16), with clear alternation of ipsi- and contralateral retinal terminations in what are presumed to be the binocular regions of the nucleus (Packer, 1941; Hayhow, 1967; Rockel *et al.,* 1972; Sanderson *et al.,* 1980). Even among members of the same family these patterns can vary enormously. The South American opossum, *Marmosa,* though in the same family as *Didelphis,* displays two cellular laminae in its α segment and a considerable degree of alternation of ipsi- and contralateral retinal inputs in both α and β segments (Royce *et al.,* 1976). Of two members of a single family of Australian marsupials, *Pseudocheirus* and *Petaurus,* the former shows no cellular lamination of its α segment but pronounced alternation of its ipsi- and contralateral retinal inputs in the binocular part of the nucleus (Pearson *et al.,* 1976). The latter species displays four cellular laminae in the α segment (Johnson and Marsh, 1969) but the pattern of retinal inputs has not yet been determined. In the marsupials, and possibly in other mammals as well, a fairly rigid stratification of input from one eye is no guarantee that inputs from the other eye will be comparably stratified. The α segment of the dorsal lateral geniculate nucleus of *Dasyurus* is divided into two cellular layers and though in the binocular part of the nucleus the contralateral retinal axon terminations are clearly stratified in three laminar zones, the ipsilateral inputs spread fairly diffusely over three of these and the intervening spaces (Sanderson and Pearson, 1977).

The smallest mammalian lateral geniculate complex may be that of the monotremes (Campbell and Hayhow, 1971, 1972) (Chapter 17). A posterolat-erally situated group of cells called LGNa by Campbell and Hayhow consists of a small-celled lateral part receiving few optic axons and a large-celled medial part receiving many. An anterodorsally situated small group of cells called LGNb by Campbell and Hayhow, also receives optic axons. LGNa seems to have some similarities to a ventral lateral geniculate nucleus and LGNb, therefore, may be equivalent to a dorsal. LGNa was misidentified by Hines (1929) as the nucleus of the stria terminalis.

9.2.   Visual Field Representation

Each of the laminae of a dorsal lateral geniculate nucleus receiving fibers from one or other retina contains a systematic representation of the appropriate half visual field. The proportion of the half field represented in a particular

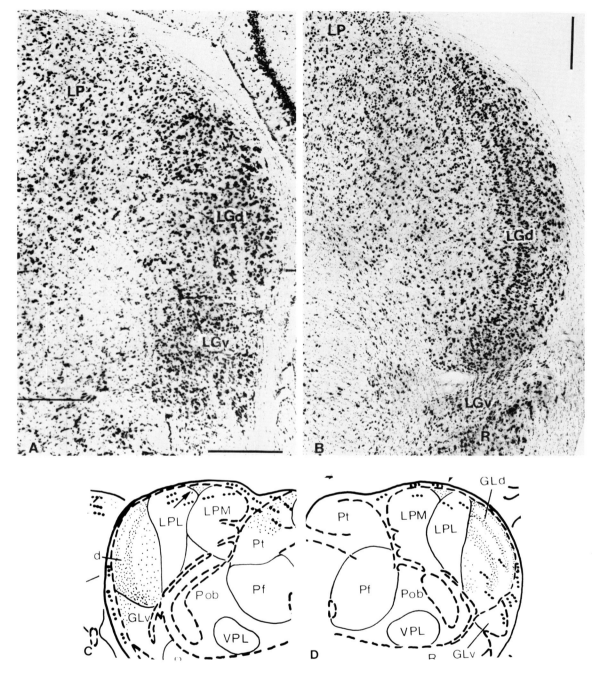

FIGURE 9.16.   Frontal sections through the dorsal lateral geniculate nucleus in (A) a polyprotodont marsupial, the opossum (*Didelphis virginiana*); (B) a diprotodont marsupial, the brush-tailed possum or phalanger (*Trichosurus vulpecula*).

Thionin stain. Bars = 500 μm. Drawings, from Rockel *et al.* (1972), show distribution of retinogeniculate fibers in the latter species: ipsilateral, right; contralateral, left.

lamina is obviously directly related to the extent of the ipsi- or contralateral retina providing input to that lamina. With the possible exception of man, the contralateral nasal half retina usually contributes far more fibers to the lateral geniculate nucleus than does the temporal half retina of the same side. As is well known, the greater the contralateral nasal retinal input, the greater the extent of the monocular temporal visual field. In man, this is reduced to a rather insignificant peripheral crescent (Fig. 9.17), but in many animals, especially those with laterally directed eyes, the monocular temporal field is far greater than the binocular. This organization is grossly manifested in the sizes of relevant laminae in the dorsal lateral geniculate nucleus, whether the lamination is overt or concealed. Where the contralateral nasal and ipsilateral temporal inputs are approximately equal, as in man, matched pairs of laminae receiving contra- and ipsilateral inputs are of approximately equal horizontal extent. Where the contralateral input far exceeds the ipsilateral, the ipsilaterally innervated laminae are proportionately shorter. In the cat, for example, lamina A extends far beyond the lateral margin of lamina A1 and the unpaired lateral part of lamina A represents the monocular portion of the temporal visual field. In animals with concealed lamination, such as the rat, the ipsilaterally innervated region is usually much shorter than the two adjoining contralaterally innervated regions (Fig. 9.3). Armed with this information, when confronted with a pattern of lamination such as exhibited by the dorsal lateral geniculate of a sheep or horse (Fig. 9.15), one could reasonably predict that layer $\gamma 1$ is ipsilaterally innervated and layers α and $\gamma 2$ contralaterally innervated.

The principle just outlined does not necessarily imply that all laminae innervated by the same eye in multilaminated dorsal lateral geniculate nuclei will be of equal length. The two magnocellular laminae of the monkey nucleus are clearly shorter than the parvocellular, and the contralaterally innervated C laminae of the cat are shorter than lamina A. But the magnocellular laminae of the monkey, receiving approximately equal ipsi- and contralateral inputs, are of approximately equal extent, whereas in the cat, laminae C and C2, receiving contralateral inputs, are more extensive than lamina C1 which receives a reduced ipsilateral input; hence, the use of the expression "matched pairs" in the preceding paragraph.

Each lamina of the lateral geniculate nucleus can be considered a flat sheet on which the whole or some part of the contralateral half visual field is laid out, depending upon whether it is innervated from the contra- or ipsilateral retina. The vertical meridian of the visual field, and along it the center of gaze, are customarily represented along one border of a lamina, the horizontal meridian across the lamina, and the periphery around its other borders. In the cat, for example (Bishop *et al.*, 1962; Seneviratne and Whitteridge, 1962; Stone and Hansen, 1966; Laties and Sprague, 1966; Garey and Powell, 1968; Kaas *et al.*, 1972), the vertical meridian is represented along the medial borders of the A and C laminae and the periphery around their anterior, posterior, and lateral borders. The horizontal meridian crosses more or less from medial to lateral; the representation of the superior quadrant of the half field is contained posteriorly in the geniculate and of the inferior quadrant anteriorly. The representation of central vision is enlarged in comparison with that of the periphery

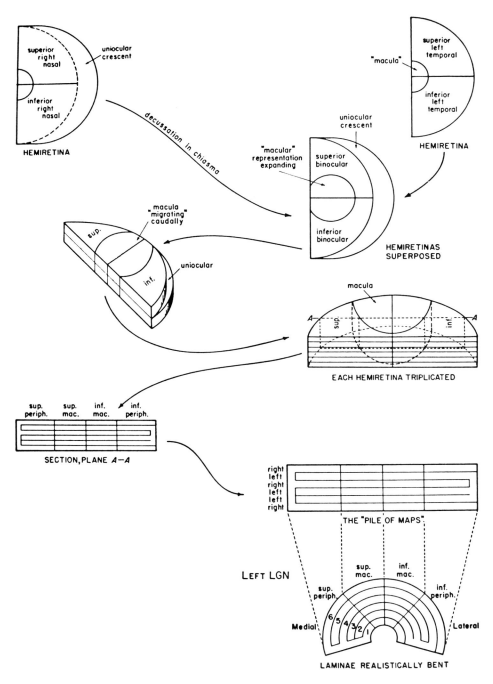

FIGURE 9.17. Walls's (1953) scheme of the organization of the human dorsal lateral geniculate nucleus as a stack of folded, visuotopic maps.

and the first 5° of visual angle takes up the medial one-quarter of the nucleus (Fig. 3.19). The rest of lamina A1 represents visual angles from 5 to 45° whereas the wider A lamina represents visual angles to 90°. The optic disc is usually "represented" as an acellular discontinuity in the contralaterally innervated lamina A (Fig. 3.19). Similar discontinuities can often be detected in several laminae of the monkey nucleus (Malpeli and Baker, 1975) and in other species (Kaas *et al.*, 1973).

Microelectrode mapping indicates that the vertical meridian of the half visual field is represented along the lateral edge of the medial interlaminar nucleus of the cat so that a reversal in receptive field trends occurs as an electrode, crossing the A and C laminae from lateral to medial, enters the medial interlaminar nucleus. The periphery of the visual field is represented medially, the lower quadrant anteriorly, and the upper quadrant posteriorly in the medial interlaminar nucleus (Kinston *et al.*, 1969; Sanderson, 1971; Kratz *et al.*, 1978a).

In higher primates the horizontal maps of the visual half field are folded on themselves about the hilus of the dorsal lateral geniculate nucleus. Until recently (Section 9.1.1.3) it was customary to think of six maps of the contralateral half field: one generated from the ipsilateral retina in laminae 2, 3, and 5 and one generated from the contralateral retina in laminae 1, 4, and 6. Now, in view of the work of Kaas *et al.* (1972) mentioned above, only four complete maps are recognized. In the geniculate as a whole, the vertical meridian is represented across the posterior surface, the periphery around the anterior margin, the horizontal meridian from anterior to posterior with the upper field quadrant represented laterally and the lower field quadrant medially. The area centralis is represented posteriorly and the central 20° of vision occupies most of the posterior half of the nucleus (Kaas *et al.*, 1972; Malpeli and Baker, 1975).

This pattern of visuotopic organization in the primate lateral geniculate nucleus was worked out primarily by observing the distribution of isolated patches of Marchi-stained degeneration or of transneuronal cellular atrophy ensuing from focal lesions of different portions of the retina (Brouwer and Zeeman, 1926; Le Gros Clark and Penman, 1934; Polyak, 1957). Several of the early investigators remarked that only the magnocellular laminae stretched uninterruptedly through the anteroposterior extent of the nucleus. The four parvocellular laminae that they recognized were only fully evident in middle parts of the nucleus. Anteriorly all four parvocellular laminae tend to fuse and posteriorly only two are detectable. Nevertheless, interpretations of experimental results and textbook accounts were presented in terms of six layers and six visuotopic maps. Wall's (1953) figures of the human dorsal lateral geniculate nucleus as a set of six stacked and bent half discs are particularly illustrative of the established dogma (Fig. 9.17).

As pointed out in the preceding section, modern mapping studies that made use of microelectrode recording (Kaas *et al.*, 1976, 1978; Malpeli and Baker, 1975) have led to a new intepretation that stresses the existence of only two parvocellular laminae the splitting and interdigitation of which in the middle of the geniculate, leads to the appearance of four or more laminae. When this is taken into account and the mapping data carefully collated, only two complete parvocellular laminae and two visuotopic representations emerge: one generated from the ipsilateral retina and the other from the contralateral retina, the reduplicated portions of the parvocellular laminae representing only central vision (Malpeli and Baker, 1975) (Fig. 9.18). Of some interest is the fact that in an albino human, the splitting of the parvocellular laminae occurred in a manner that would be commensurate with an abnormal contralateral retinal input to layers 2, 3, and 5 (Guillery *et al.*, 1975).

Despite the reduplication of parts of the two fundamental parvocellular laminae, the pattern of visuotopic organization in them is consonant with the pattern outlined above. Where two parts of the same parvocellular laminae lie separated from one another by a portion of the other parvocellular lamina, they appear to receive fibers from the same part of the retina and to represent the same point in visual space. The reason for believing this lies in the older literature; if a tiny lesion is made in the retina, the ensuing axonal or transneuronal cellular degeneration occurs in a line extending across all the appropriate ipsi- or contralaterally innervated laminae at the same level. Similarly, a focal lesion or injection of tracer in part of the representation in the visual cortex leads to axonal or cellular labeling in a line extending across all laminae at the same level (Fig. 9.19). Hence, if a microelectrode is driven perpendicularly across all geniculate laminae, irrespective of whether the parvocellular laminae are split or not at that level, the units successively encountered in each lamina or partial lamina will have receptive fields in the same region of visual space. Such a linear arrangement of receptive fields across the laminae is referred to as a projection line or column (Sanderson, 1971). The functional significance of the duplication of parts of the visual field representation in the parvocellular laminae, nevertheless, continues to elude us.

The evidence just alluded to, for the representation of homonymous points in the visual fields of the two eyes being in register across all laminae of the primate dorsal lateral geniculate nucleus, also extends to other species such as the cat (Bishop *et al.*, 1962; Stone and Hansen, 1966; Garey and Powell, 1968; Kinston *et al.*, 1969; Sanderson, 1971). The underlying basis of this was pointed out by Tello (1904). In his Golgi studies of the cat lateral geniculate nucleus, he noted that the optic tract axons terminated perpendicular to the laminae in formations he thought resembled cypress trees (but see Mason and Robson, 1979, and Chapter 4). Orientation across the thickness of the laminae has subsequently been confirmed for several species (Bowling and Michael, 1980, 1984; Michael and Bowling, 1982; Sur and Sherman, 1982b). Axons from homonymous points in the two retinae must obviously align themselves across alternate laminae in order to generate the projection lines that extend across the thickness of all laminae. In the cat, some but not all optic tract axons can branch to terminate in two contra- or two ipsilaterally innervated laminae (Sur and Sherman, 1982b) but in monkeys single axons appear to innervate one lamina only (Michael and Bowling, 1982).

In animals such as rats and mice with a small ipsilateral retinal input, a coherent map of visual space, as projected through the contralateral eye, can be readily detected in the dorsal lateral geniculate nucleus by microelectrode map-

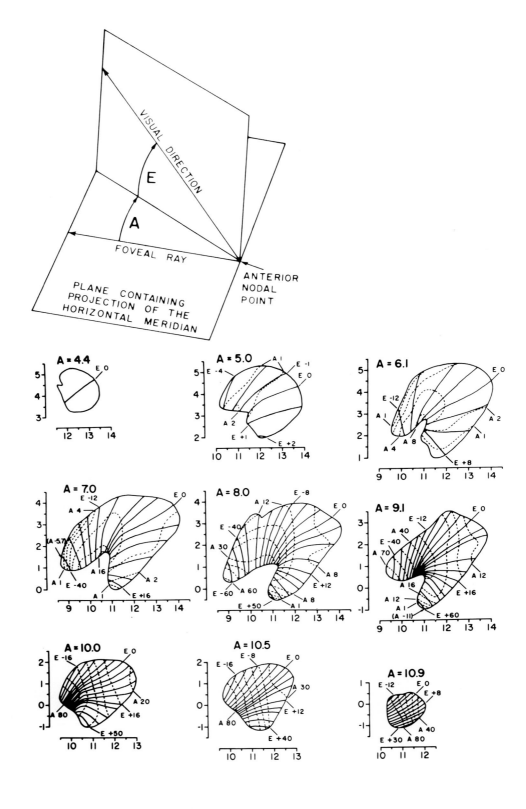

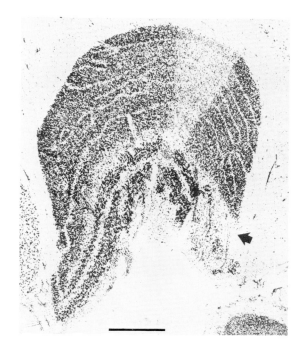

FIGURE 9.19. Retrograde degeneration spanning all laminae of the dorsal lateral geniculate nucleus of a baboon that had sustained a small traumatic lesion of the striate cortex. The column of degeneration is comparable to a projection column (Fig. 3.20). Arrow indicates a small patch of additional degeneration. Bar = 1 mm. From Kaas *et al.* (1972).

ping (Montero *et al.*, 1968; Reese and Jeffery, 1983) or by anatomical means (Montero and Guillery, 1968; Lund *et al.*, 1974). In the region of ipsilateral terminations, a recent mapping study in the rat indicates that a portion of far, upper nasal visual space is mapped. This part of the representation, at 50–90° eccentricity, appears to be absent from the representation mapped via the contralateral eye (Reese and Jeffery, 1983). In other rodents such as squirrels in which the ipsilateral retinal input is larger and bilaminar, it has been implied that projection lines passing through adjacent portions of ipsilaterally and contralaterally innervated laminae represent conjugate points in the 30° of overlap in the visual fields of the two eyes (Kaas *et al.*, 1972). However, no data have been presented to support this position.

In the Siamese cat the retinogeniculate projection is abnormal and the dorsal lateral geniculate nucleus shows corresponding cytoarchitectonic disruptions (Guillery, 1969c; Guillery and Kaas, 1971; Hubel and Wiesel, 1971). The principal defect in this animal is misrouting of optic tract axons at the optic chiasm so that a proportion of the fibers from the temporal retina send their axons to the contralateral, rather than the ipsilateral lateral geniculate nucleus. In the dorsal lateral geniculate nucleus, the normally contralaterally innervated laminae (A and C) are intact but, of the normally ipsilaterally innervated laminae (A1

FIGURE 9.18. Lower figure shows isoelevation (solid lines) and isoazimuth (broken lines) curves drawn on frontal sections of the dorsal lateral geniculate nucleus of a rhesus monkey. Upper figure indicates how visual direction is defined by azimuth (A) and elevation (E). Plane of horizontal meridian is perpendicular to plane of visual direction. From Malpeli and Baker (1975).

and C1), lamina A1 in particular is broken up into clumps, some of which are ipsilaterally innervated and others contralaterally innervated. Those that are contralaterally innervated tend to invade and fuse with laminae A and C (Fig. 9.20). The abnormal contralateral projection arises from a strip of temporal retina up to 20° temporal to the vertical meridian. Within it, in the abnormal lamina A1, a retinotopic sequence is preserved but because the projection goes to the contralateral rather than the ipsilateral lamina A, the abnormal clumps of lamina A1 receive a mirror image of the normal representation (Guillery and Kaas, 1971). This is associated with a modification of the geniculocortical projection that can take different forms in different strains of Siamese cats (Hubel and Wiesel, 1971; Kaas and Guillery, 1973; Shatz, 1977). It is assumed that the characteristic strabismus of many Siamese cats is related to the deformation of the visual field projection.

The reduced ipsilateral and enhanced contralateral innervation of laminae

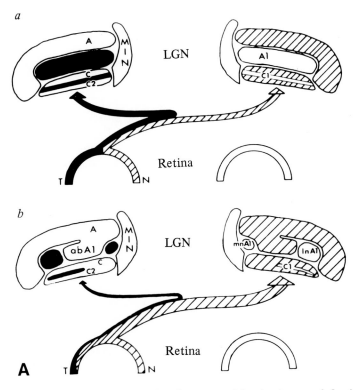

FIGURE 9.20.   (A) Schematic figure illustrating the nature of the visual system defect in the Siamese cat (b); (a) is arrangement in common domestic cat. Fibers from a large portion of the temporal retina decussate abnormally at the optic chiasm and, thus, parts of lamina A1 of the dorsal lateral geniculate nucleus receive a contralateral innervation, with consequent disruption of the laminar cytoarchitecture. From Shatz and Kliot (1982). (B, C) Photomicrographs of frontal sections through

A1 and C1 appear to be associated with the albino trait, for which the Siamese cat is homozygous, and the ipsilateral projection is even further reduced in tyrosinase-negative albino cats (Creel *et al.*, 1982). Albinism has now been discovered to be associated with similarly abnormal retinogeniculate projections in other species including man. Albino rats (Lund, 1965) and rabbits (Giolli and

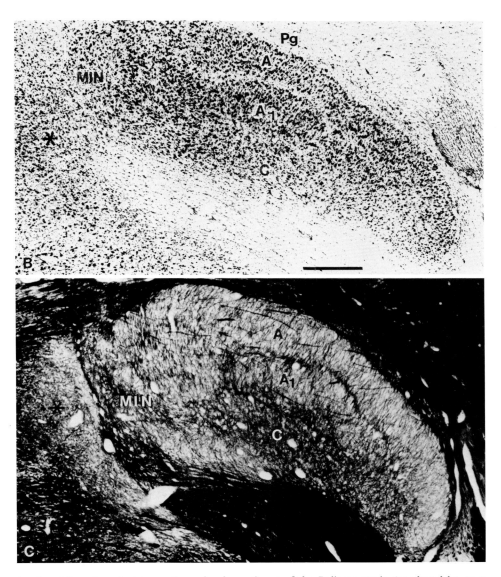

the dorsal lateral geniculate nucleus of a domestic cat of the Balinese strain (produced by cross-breeding between the Siamese and the chinchilla strains). Lamina A1 appears foreshortened because the parts abnormally innervated by the contralateral eye tend to fuse with lamina A. There appears to be some disruption of the C laminae as well. (B) Thionin stain; (C) hematoxylin stain. Bar = 500 μm.

Creel, 1973) show a reduced ipsilateral retinogeniculate input, though, in the relatively homogeneous dorsal lateral geniculate nucleus, this cannot be correlated with architectonic changes. Albino members of the carnivore order other than cats, such as ferrets, mink, and a tiger, all display changes in lamination and altered retinogeniculate connections comparable to those of the Siamese cat (Guillery, 1971a; Guillery and Kaas, 1973; Sanderson *et al.*, 1974; Guillery *et al.*, 1971). In one example of albinism in man (Guillery *et al.*, 1975), the multilaminar cytoarchitectonic pattern was disrupted in a manner that could be correlated with a reduced ipsilateral and added contralateral input to the traditional laminae 2, 3, and 5.

9.3. Afferent Connections to the Dorsal Lateral Geniculate Nucleus

The principal afferent drive to the dorsal lateral geniculate nucleus is obviously the retina and much has already been described of the differential laminar distributions of axons from the ipsi- and contralateral eyes. Other sources of afferent input that can be predicted from the general organizational pattern of all dorsal thalamic nuclei (Chapter 3) are the appropriate areas of the cerebral cortex and the thalamic reticular nucleus. To these we may add afferents from the so-called perigeniculate nucleus (Rioch, 1929a) which has only received attention in the cat and which many workers regard as part of the reticular nucleus, closely applied as a thin strip to the dorsolateral surface of lamina A. A thin sheet of cells comparable to the perigeniculate nucleus has been described in some other animals such as the sheep (Rose, 1942b). Other sources of input to the dorsal lateral geniculate nucleus are the superior colliculus and pretectum. Other fibers are said to come from the brain-stem reticular formation.

9.3.1. Retinal Afferents

The separation of ipsi- and contralaterally projecting retinal ganglion cells in the primate retina is particularly sharp. Along the vertical meridian there is a strip no more than 1° wide along which contra- and ipsilaterally projecting ganglion cells mingle (Stone *et al.*, 1973; Bunt *et al.*, 1977). In the cat all X-type (see below) retinal ganglion cells of both nasal and temporal retina also overlap by no more than 1° and other ganglion cell types of the temporal retina are similarly organized, i.e., ipsilaterally projecting cells extend no more than 1° into the nasal retina. However, some contralaterally projecting cells extend across the vertical meridian for up to 3° into the temporal retina (Stone and Fukuda, 1974; Kirk *et al.*, 1976a,b). This means that a small proportion of the contralaterally innervated laminae of the dorsal lateral geniculate nucleus may be receiving some input from the ipsilateral retina. In rats and mice a large temporal region, approximately one-quarter the surface area of the retina, projects bilaterally (Lund *et al.*, 1974; Dräger and Olsen, 1980), though the number of ipsilaterally projecting cells in it is quite small (Cowey and Perry, 1979; Jeffery et al., 1981).

The retinal fibers innervating the dorsal lateral geniculate nucleus have been the subject of a great deal of investigation in the last decade on account of the ease with which different functional classes can be distinguished. These have been the subject of several recent reviews (Rodieck, 1979; Sherman, 1979; Stone et al., 1979). Enroth-Cugell and Robson (1966) were the first to point out that retinal ganglion cells in the cat could be characterized in terms of their capacity for spatial summation of a visual stimulus applied to their receptive fields. One category of cells, which they called X cells, was capable of responding in a linear fashion to a sine wave grating going in and out of phase (a contrast reversal test). That is, for the center–surround-type receptive field of such a cell the grating could be placed in a "null position" so that the sum of excitatory and inhibitory influences acting on the cell summed to zero and the cell ceased discharging. When the stimulus was applied elsewhere in the receptive field, the cell responded at a frequency close to the fundamental temporal frequency of the change of phase and changed its response as the temporal frequency of the grating changed (Fig. 9.21). A second category of center–surround-type cells, termed Y cells, were not capable of linear summation. That is, no null position could be found in their receptive fields and their responses were usually twice that of the temporal frequency of the change in phase of the grating (Fig. 9.21). Both on-center and off-center ganglion cells can be classified as X or Y types. In recording from the optic tract, Enroth-Cugell and Robson encountered 20% X-cell axons and 80% Y-cell axons.

The X- and Y-type retinal ganglion cells later came to be recognized as having other distinguishing characteristics. Though the spatial frequency test remains, in the eyes of the purist, the only true and satisfactory distinction, X cells have often been termed "brisk-sustained" neurons because of the nature of their discharges in response to a stimulus located in the center of their receptive fields. Y cells by contrast are referred to as "brisk-transient" neurons. X cells have more slowly conducting axons: their conduction velocities are 18–25 m/sec compared with 30–40 m/sec for Y cells in the cat (Cleland et al., 1971; Fukada, 1971). X cells tend to respond best to small, slower-moving targets, to higher spatial frequencies, but lower temporal frequencies than Y cells (Ikeda and Wright, 1972; Cleland et al., 1973; Cleland and Levick, 1974; Bullier and Norton, 1979; So and Shapley, 1981).

Subsequent to the discovery of the X and Y categories of retinal ganglion cells a further set of retinal ganglion cells was discovered in the cat and called W cells. These had large receptive fields, sluggish responses, and slowly conducting axons and appeared to outnumber significantly the X and Y cells (Stone and Hoffman, 1972; Cleland and Levick, 1974; Stone and Fukuda, 1974; Cleland et al., 1975a,b; Kirk et al., 1975). More recently, it has become evident that the W cells cannot constitute a single functional class (e.g., Sur and Sherman, 1982a). It would also be wrong to leave this section without pointing out that several varieties of functional retinal ganglion cell types undoubtedly exist outside the X, Y, and W classification (see Rodieck, 1979). Some of these provide color-coded inputs to the lateral geniculate nucleus (see Section 9.3.1.6). However,

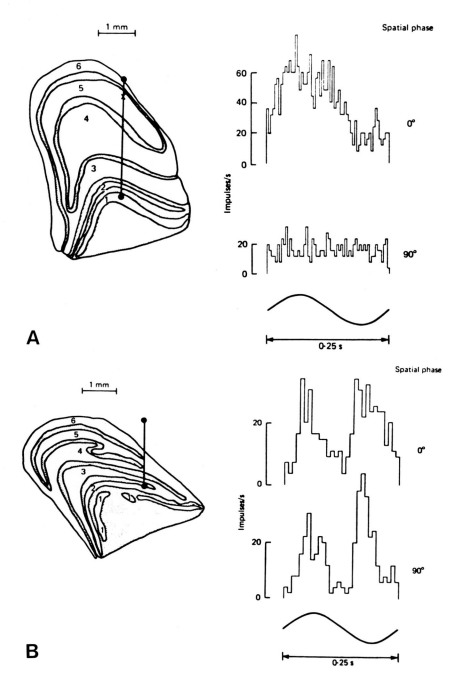

FIGURE 9.21. (A) Location (x) of an X-type neuron along an electrode track through the dorsal lateral geniculate nucleus of a cynomolgus monkey (*Macaca fascicularis*). On the right are shown the responses of the cell to a contrast reversal grating (temporal modulation signal at bottom), as a function of spatial phase, 90° being the null position at which on and off influences on cell sum to zero. (B) Location (lower dot) of a Y-type neuron in one of the magnocellular layers of *M. fascicularis*. At right, it is seen that the cell responds to a grating at approximately twice the modulation frequency and largely independent of spatial phase. Both from Kaplan and Shapley (1982).

far less work has been devoted to these than to X, Y, and W cells and especially not to their central projections.

9.3.1.1. Morphology and Central Distributions of X, Y, and W Ganglion Cells

These three functional classes of retinal ganglion cells correlate quite well with different morphological classes of retinal ganglion cells in the cat. There are at least three classes of morphologically distinguishable retinal ganglion cells (Boycott and Wässle, 1974; Rowe and Stone, 1977; Leventhal *et al.*, 1980b; Hughes, 1981) and each of these has now been directly correlated with the X, Y, or W classes by intracellular recording and injection (Peichel and Wässle, 1981; Wässle *et al.*, 1981a,b; Saito, 1983). α cells have large somata and medium-sized dendritic fields, form 4–7% of the ganglion cell population, and are evenly distributed across the retina. There is a one-to-one correlation between the number of Y cells recorded in a given patch of cat retina and the number of retinal ganglion cells with large somata in that patch (Cleland *et al.*, 1975a,b).

β cells have small somata and small dendritic fields, form a higher percentage of the ganglion cell population, and are especially concentrated in parts of the retina representing central vision. They are thought to correspond to X cells (see below).

γ cells have small somata but large dendritic fields, form approximately 50% of the ganglion cell population, and are fairly evenly distributed across the retina with a slight increase near the central retina. They are thought to correspond largely to W cells (Rowe and Stone, 1977), though some putative W cells may have medium-sized somata (Stone and Clarke, 1980; Leventhal *et al.*, 1980b; Rowe and Dreher, 1982). Saito (1983) has correlated W cells with a subclass of γ cells, referred to by Boycott and Wässle (1974) as δ class.

The central projections of X, Y, and W cells were determined in the cat some years ago by electrophysiological methods. X cells were shown to project to the dorsal lateral geniculate nucleus but not to the superior colliculus and Y cells were shown to project by branched axons to both sites (Cleland *et al.*, 1971; Hoffmann *et al.*, 1972; Hoffman, 1973; Singer and Bedworth, 1973; Cleland and Levick, 1974; Fukuda and Stone, 1974). In the laminar dorsal lateral geniculate nucleus, X- and Y-cell axons end in laminae A and A1 and in lamina C (Wilson *et al.*, 1976; Mitzdorf and Singer, 1977). W cells were later shown to project to both the superior colliculus and the dorsal lateral geniculate nucleus (Wilson and Stone, 1975; Wilson *et al.*, 1976; Cleland *et al.*, 1975a,b, 1976) where they end only in the C laminae, especially laminae C1 and C2. W cells constitute the only type of retinal ganglion cell axon terminating in the ventral lateral geniculate nucleus (Spear *et al.*, 1977). Y cells form the principal input to the medial interlaminar nucleus (Mason, 1975; Palmer *et al.*, 1975; Dreher and Sefton, 1979; Kratz *et al.*, 1978a,b). These, too, appear to be branches of Y-cell axons projecting to the superior colliculus and laminar dorsal lateral geniculate nucleus (Dreher and Sefton, 1979; Bowling and Michael, 1980, 1984; Sur and Sherman, 1982b).

Anatomical tracing studies have provided a good deal of confirmation for the differential distribution of the axons of different classes of retinal ganglion cells. Kelly and Gilbert (1975), for example, found that injection of horseradish peroxidase into the A laminae of the dorsal lateral geniculate nucleus led to retrograde labeling of only retinal ganglion cells with middle- and large-sized somata, i.e., somata corresponding to α and β but not to γ cells. This has now been confirmed by Leventhal (1982), Leventhal *et al.* (1980b), and Rowe and Dreher (1982). Leventhal *et al.* (1980b), and Rowe and Dreher (1982) have extended the observations by showing labeling of α, β, and γ cells after injections of the C laminae. Injections of the medial interlaminar nucleus label α and several varieties of γ cells but not β cells. Injection of horseradish peroxidase into the superior colliculus led, in the hands of Kelly and Gilbert, to retrograde labeling of small- and large-sized somata but not of middle-sized somata. That is, the somata of β (or X) cells were not labeled. Illing and Wässle (1981) now consider that all α cells are labeled from the tectum, together with 10% of β cells and 50% of γ cells. By contrast, from the thalamus, all α and β and 50% of the γ cells are labeled, making 77% of the retinal cells project to the thalamus.

The newest data in the cat come from the intraaxonal injection of single, physiologically characterized retinal ganglion cell axons into the optic tract (Bowl-

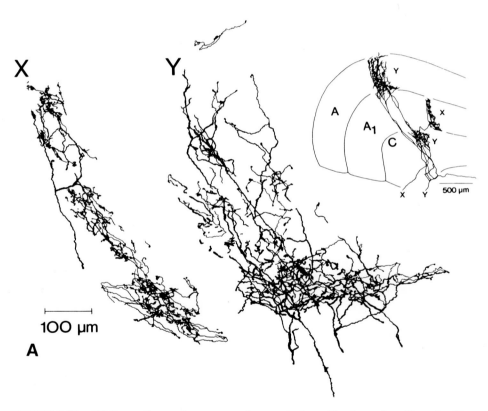

FIGURE 9.22. (A) Large figure shows parts of a terminal ramifications of an X- and a Y-type retinogeniculate axon injected in the optic tract of a cat. Reconstruction at top right shows narrow terminal field of X-type axon from ipsilateral eye in lamina A1 and wider terminal fields of the Y-type axon from the contralateral eye in laminae A and C. (B) Terminal ramifications of injected Y

ing and Michael, 1980, 1984; Sur and Sherman, 1982b) (Fig. 9.22). To date, only the X and Y categories have been distinguished. All X-cell axons innervate only lamina A or A1 in narrow zones and may have a thin branch with relatively few terminals in the medial interlaminar nucleus. Y-cell axons branch to innervate lamina A and C (if contralateral) or lamina A1 alone (if ipsilateral). They have broad zones of termination with more boutons than X axons and most have added dense terminations in the medial interlaminar nucleus. The results, though understandably derived from a relatively small sample, suggest that the A laminae are X and Y innervated, lamina C is Y innervated, and presumably laminae C1 and C2 are W innervated. They also indicate (see below) the preponderance of Y inputs to the medial interlaminar nucleus.

9.3.1.2. X, Y, and W Cells in Other Species

Retinal ganglion cells of three sizes and with different axon conduction velocities have been described in the rat (Bunt *et al.,* 1974; Fukuda, 1977; Perry, 1979) and provisionally correlated with the X, Y, and W cells of the cat. Retinal ganglion cells with X- and Y-like properties have also been described in monkeys (Gouras, 1969; Schiller and Malpeli, 1978; DeMonasterio, 1978) and it was found

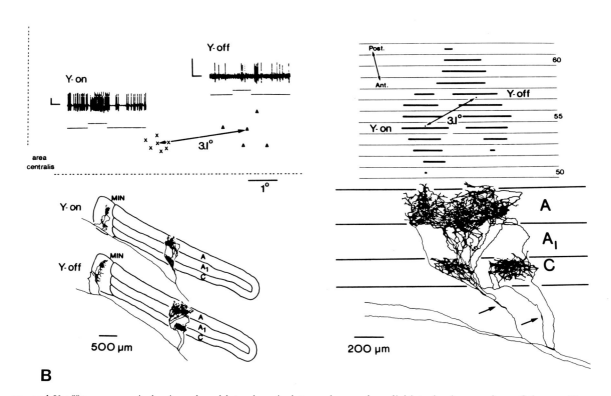

B

on- and Y off-type axons in laminar dorsal lateral geniculate nucleus and medial interlaminar nucleus of the cat. Upper right is a reconstructed surface view of the ramifications of the axons in lamina A. Upper left shows responses to pulses of light and positions of receptive fields which were 3.1° apart. Both from Bowling and Michael (1984).

that many of the X cells could not be antidromically activated from the superior colliculus (Schiller and Malpeli, 1978). Morphological equivalents of X and Y retinal ganglion cells have been detected by retrograde labeling following injections of tracer in the parvo- or magnocellular layers of the dorsal lateral geniculate nucleus of monkeys (Bunt et al., 1975). Neither appears to correspond closely to the types described in Golgi preparations by Polyak (1957). Those called B cells and apparently comparable to β cells of cats project to the parvocellular layers. Those called A cells and comparable to the α cells of cats project to the magnocellular layers (Leventhal et al., 1981). W-type cells have also been reported in monkeys though they have not yet been shown to project to the dorsal lateral geniculate nucleus (Leventhal et al., 1981).

In the rabbit, ganglion cells with concentrically arranged receptive fields fall into classes that resemble X and Y cells in terms of briskness of discharge and axon conduction velocity but the distinction between the two is far less clear than in the cat and both types send axons to the superior colliculus (Molotchnikoff and Lessard, 1980; Vaney et al., 1981; Lo, 1981, 1983).

In opossums, three size groups of retinal ganglion cells can be retrogradely labeled from injections of tracer in the lateral geniculate nucleus. All of these, and a fourth group of the smallest cells, could be labeled from the superior colliculus (Rapaport and Wilson, 1983).

9.3.1.3. Receptive Fields of Lateral Geniculate Neurons

The majority of neurons in the A, A1, and C laminae of the cat lateral geniculate nucleus and in all laminae of the monkey geniculate have the center–surround-type receptive field organization, very similar to that of retinal ganglion cells (Hubel and Wiesel, 1961; Bishop et al., 1962; Kozak et al., 1965; Wiesel and Hubel, 1966). On-center and off-center receptive fields appear to be present in approximately equal numbers. Fuller details of these spatially selective types of receptive fields are given in Chapter 4.

In studies carried out subsequent to the original work of Hubel and Wiesel (1961), the ubiquity of concentric receptive fields in the lateral geniculate nucleus has not been questioned and it stands as one of the basic tenets of geniculate physiology.

Several workers had suggested that the center–surround antagonism of lateral geniculate cells is likely to be produced by an interaction between on and off retinal channels converging on individual geniculate cells (e.g., Singer and Creutzfeldt, 1970). Schiller (1982), however, has given reasons for doubting this by showing that in both parvo- and magnocellular layers of the monkey lateral geniculate nucleus, infusion of DL-2-amino-4-phosphorobutyric acid eliminates all responses of on-center cells, including the off-response from their receptive field surrounds, while not affecting the responses of off-center cells. The infused material appears to act selectively on on-cells in the retina, and the results suggest that the effect of their blockade is relayed to lateral geniculate cells whose receptive fields are independent of innervation by off-cells of the retina. The partial or complete segregation of on- and off-center neurons in different parvocellular laminae of the monkey and tree shrew geniculate (Schiller and Malpeli, 1978; Conway et al., 1980; Conway and Schiller, 1983) and in different sublaminae of the mink and ferret geniculate A and A1 laminae (LeVay and McConnell, 1982; Stryker and Zahs, 1983) also suggests that these two afferent streams are relatively independent. In the mink they may even project to separate cortical columns (McConnell and LeVay, 1983).

9.3.1.4. The X, Y, and W Streams through the Dorsal Lateral Geniculate Nucleus

Neurons in the dorsal lateral geniculate nucleus appear to acquire most of their receptive field properties from optic tract fibers (Hubel and Wiesel, 1961; Bullier and Norton, 1979; So and Shapley, 1981; Chapter 4). Hence, in addition to the now-classic center–surround organization, many cells in the nucleus display characteristics that have come to be termed X-like, Y-like, and W-like. As would be expected from the preceding section, most work on these cell types has been done on the cat: X and Y cells are found in the A and A1 laminae, W cells in the C laminae (C, C1, and C2), and Y cells are also found in lamina C and in the medial interlaminar nucleus (Cleland *et al.*, 1971; Fukuda and Saito, 1972; Hoffmann *et al.*, 1972; Fukuda and Stone, 1974; Palmer *et al.*, 1975; Mason, 1975, 1976; Wilson and Stone, 1975; Wilson *et al.*, 1976; Cleland *et al.*, 1975a,b, 1976; Mitzdorf and Singer, 1977; Dreher and Sefton, 1979; Kratz *et al.*, 1978a,c; Lehmkuhle *et al.*, 1980; Sur and Sherman, 1982a). There has been a little work on X- and Y-type cells in other species but the rather mixed collection of ganglion cell types that constitute the W class of the cat have not yet been equated with a functional class in primates.

a. X and Y Cells in the Cat. The receptive field properties of X, Y, and W cells in the cat dorsal lateral geniculate nucleus do not differ significantly from those of the corresponding retinal ganglion cells, indicating not only a tendency toward laminar segregation but also that there is little convergence of the afferent streams at the single-cell level even in the A laminae. This has tended to be borne out by intra- and extracellular recording though debate still goes on about the extent to which X and Y inputs may converge on inhibitory interneurons in the A laminae (Singer and Bedworth, 1973; Eysel, 1976; Dubin and Cleland, 1977; Mooney *et al.*, 1979; Lindström, 1982; see also Chapter 4).

The differential distribution of α and β ganglion cells in the retina is reflected in the distribution of X and Y cells in the A and A1 laminae. Y cells are found throughout all parts of the laminae, whereas X cells are more common in regions representing central vision (i.e., medially). Taking advantage of this fact, LeVay and Ferster (1977) were able to show that Y cells probably correspond to large neurons without lamellated inclusions in their cytoplasm and X cells to medium-sized neurons with the inclusions. These in turn correspond respectively to Guillery's (1966) class 1 and class 2 cells, based on Golgi staining (see Chapter 4) and seem to be comparable to the small sample of physiologically characterized X and Y cells that have been recovered histologically after intracellular injection (Friedlander *et al.*, 1981; Stanford *et al.*, 1981, 1983). It has now become common to regard the larger cells of the A laminae, even when identified only in Nissl stains, as Y cells. When characterized electrophysiologically, X cells form approximately 40% of a sample and Y cells 50–60% (Sherman *et al.*, 1972). The

Y cells achieve receptive field maturity later than X cells (Daniels *et al.*, 1978) and in binocular parts of the nucleus are especially sensitive to visual deprivation during early life (Sherman *et al.*, 1972, 1975; LeVay and Ferster, 1977; Garey and Blakemore, 1977).

W cells have been examined less thoroughly in the cat lateral geniculate nucleus and there is as yet little positive evidence for a comparable class of cells in other species. In the cat W cells appear to be confined to laminae C, C1, and C2 (Wilson *et al.*, 1976; Sur and Sherman, 1982a; Stanford *et al.*, 1983) and based upon their response properties, probably do not conform to a single class but are linked by the common property of being innervated by slowly conducting retinal afferents (Sur and Sherman, 1982a). When identified by intracellular injection of a marker subsequent to physiological characterization (Stanford *et al.*, 1981, 1983), W cells have the morphological characteristics of a small (type 4) class of relay neurons originally identified in the C laminae by Guillery (1966) (see also Hitchcock and Hickey, 1983). It also seems likely from morphological studies that the W input to the C laminae is derived from γ-type retinal ganglion cells of more than one somal size (Leventhal *et al.*, 1980a,b; Rowe and Dreher, 1982).

The cells of the cat medial interlaminar nucleus are approximately 30% larger than the mean of those in the A laminae, which tends to confirm their Y character. The Y cells of the developing medial interlaminar nucleus are also affected by visual deprivation in much the same way as those in the A laminae (see Chapter 6).

When studied physiologically Y cells represent approximately 90% of all cells identified in the medial interlaminar nucleus (Kratz *et al.*, 1978a,c). They are distinguished from Y cells in the A laminae by relatively minor differences such as somewhat smaller receptive field centers, slightly shorter latencies to stimulation of the optic chiasm, and an apparent lack of inhibition from their nondominant eye (Kratz *et al.*, 1978a). It is likely that all are innervated by branches of Y-type optic tract axons destined for the laminated part of the dorsal lateral geniculate nucleus and the superior colliculus (Hoffmann, 1973; Singer and Bedworth, 1973; Fukuda and Stone, 1974; Dreher and Sefton, 1979) (Fig. 9.22 and Chapter 3). The branching of Y axons to innervate these sites has been morphologically confirmed (Bowling and Michael, 1980; Sur and Sherman, 1982b). Most of those ending in the medial interlaminar nucleus of the mink are of rather coarse caliber, presumably reflecting their Y character (Guillery and Oberdorfer, 1977). Some, however, are relatively fine and Guillery and Oberdorfer suggest that they may innervate a small group of cells with non-Y-type properties located near the ventrolateral edge of the nucleus in the cat (Dreher and Sefton, 1979). Some cells in this position possess lamellated bodies like X cells in the laminar lateral geniculate nucleus (LeVay and Ferster, 1977). It has also been reported that the medial interlaminar nucleus receives a substantial W input (Dreher and Sefton, 1979) which seems to be derived from retinal ganglion cells of the γ type but with medium- rather than small-sized somata (Rowe and Dreher, 1982). These correspond to the so-called W1 cells of Stone and Clarke (1980).

The differences in the conduction velocities of retinal ganglion cell axons of the X, Y, and W classes are reflected in the conduction velocities of the axons projecting to the cerebral cortex from geniculate X-, Y-, and W-like cells (Fig.

9.23) (Cleland *et al.*, 1971, 1976; Hoffmann and Stone, 1971; Stone and Hoffman, 1971; Hoffmann *et al.*, 1972; Fukuda and Saito, 1972; Wilson and Stone, 1975; Wilson *et al.*, 1976; Bullier and Henry, 1979). Y-like cells are the fastest and those in the A laminae and in lamina C project to areas 17 and 18 (Hoffmann *et al.*, 1972; Stone and Dreher, 1973; LeVay and Ferster, 1977), though it is thought that independent Y cells project to each area (LeVay and Ferster, 1977). Y cells in the medial interlaminar nucleus project to area 18 and to certain of the lateral suprasylvian areas (Burrows and Hayhow, 1971; Rosenquist *et al.*, 1974; Maciewicz, 1975; Gilbert and Kelly, 1975; LeVay and Ferster, 1977). Holländer and Vanegas (1977) indicate a projection to area 19 as well. The ventrolateral region of small cells in the medial interlaminar nucleus is said to project only to area 17 (LeVay and Ferster, 1977).

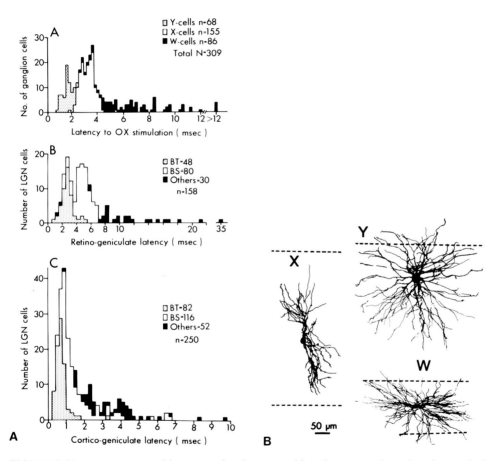

FIGURE 9.23. (A) Frequency histograms showing (A) antidromic response latencies of cat retinal ganglion cells following electrical stimulation of the optic chiasm; (B) response latencies of cat dorsal lateral geniculate cells to electrical stimulation of the appropriate parts of the retina; (C) antidromic response latencies of cat dorsal lateral geniculate cells to electrical stimulation of the visual cortex. BT (brisk-transient) and BS (brisk-sustained) are equivalent respectively to Y-type and X-type cells. From Bishop (1983) after Rowe and Stone (1977; A) and Cleland *et al.* (1976; B, C). (B) Injected X, Y, and W neurons in the cat dorsal lateral geniculate nucleus. From Stanford *et al.* (1981).

X cells in the A laminae of the cat project only to area 17 (Stone and Dreher, 1973; LeVay and Ferster, 1977). In area 17 individual thalamocortical axons, interpreted as belonging to X or Y cells on the basis of their diameters and laminae of termination, each have collateral branches to the upper part of layer VI and then a complementary pattern of terminations in more superficial layers. Those thought to be X-cell axons terminate in the deepest part of layer IV while those thought to be Y-cell axons end in the more superficial parts of layer IV (Ferster and LeVay, 1978). In the cortex there is physiological evidence that neurons receiving thalamocortical inputs may be influenced primarily by X or Y but not by both afferent types (Bullier and Henry, 1979; Gilbert and Wiesel, 1979; see also: Leventhal, 1979; Leventhal et al., 1980b; Dreher et al., 1980). This is perhaps determined by the different levels of termination of X and Y axons.

b. X and Y Cells in Other Species. X- and Y-type cells have been reported in the lateral geniculate nucleus of the tree shrew (Sherman *et al.,* 1975; Conway *et al.,* 1980; Conway and Schiller, 1983), bush baby (Norton and Casagrande, 1982), owl monkey (Sherman *et al.,* 1976), and macaque monkey (Schiller and

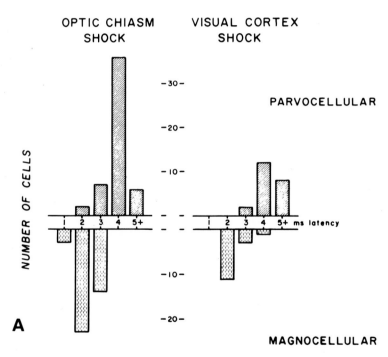

FIGURE 9.24. (A) Response latencies of cells in the magno- and parvocellular layers of the monkey dorsal lateral geniculate nucleus to electrical stimulation of optic chiasm or visual cortex. (B) Multiunit histograms showing distribution of on- and off-type cells in the different layers (1–6) of the monkey dorsal lateral geniculate nucleus. Two histograms (a, b) are shown for each lamina. Stimuli were small spots of red, green, or white light applied to centers of receptive fields. Both from Schiller and Malpeli (1978).

Malpeli, 1978; Dreher *et al.,* 1976). Fast and slow thalamocortical cells have been reported in the rat (Fukuda, 1973). In monkeys it was reported that X cells occur only in the parvocelluar laminae (3–6), while Y cells constitute some 90% of the population in the magnocellular laminae (1 and 2) and are innervated by more rapidly conducting retinal axons than X cells (Fig. 9.24). The thalamo-cortical axons of the X and Y cells in the geniculate are of comparably different conduction velocities (Marrocco and Brown, 1975). The magno- and parvocel-lular layers project to different levels of the striate cortex (Chapter 3). The axons arising in the parvocellular layers terminate primarily in the deeper half of layer IV (layer IVcβ) with a smaller more superficial region of terminations in the upper part of layer IV (layer IVa) bordering layer III and with occasional fibers to layer I. By contrast, axons arising from the magnocellular layers terminate just above the deeper terminations of fibers from the parvocellular layers (in

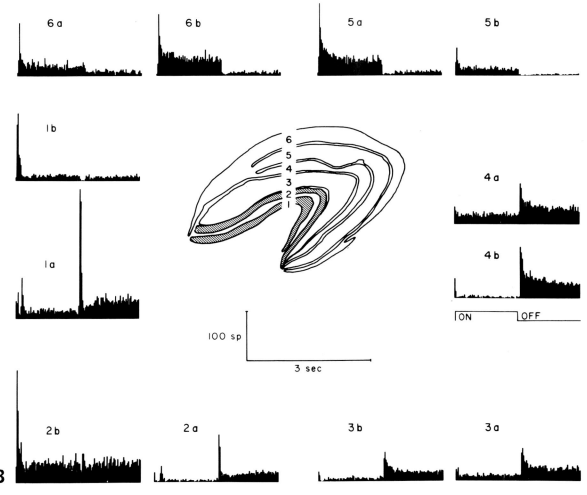

FIGURE 9.24. (*continued*)

layer IVcα) (Hubel and Wiesel, 1972; Glendenning *et al.*, 1976; Kaas *et al.*, 1976; Hendrickson *et al.*, 1978) (Chapter 3). These observations tended to reinforce the belief in the independent relay of X and Y properties to the cortex and even suggested that the axons of X and Y cells might influence separate populations of cortical neurons.

In the bush baby, Norton and Casagrande (1982) report that the two magnocellular layers (1 and 2) contain mainly Y cells, the two parvocellular layers (3 and 6) mainly X cells, and the two intercalated layers (4 and 5) W-like cells.

Some doubt has recently been cast upon the rigidity of separation of the X and Y streams in the laminae of the monkey dorsal lateral geniculate nucleus, by Kaplan and Shapley (1982). These workers found that 99% of the cells they recorded from in the parvocellular layers were indeed X-type cells, but they encountered both X- and Y-type cells in the magnocellular layers. There, X cells formed 75% of their sample and Y cells 25%. They noted that X- and Y-type cells of the magnocellular layers had much higher contrast sensitivities than X cells in the parvocellular laminae. That is, magnocellular layer cells reached a criterion level of firing with lower levels of contrast between the bars of a sine wave grating stimulus than did parvocellular layer cells. On these grounds, Kaplan and Shapley consider that X-type cells of the magnocellular layers should be regarded as a group separate from those of the parvocellular layers. The discrepancy between their results and those of previous workers seems to rest upon the criteria used and the range of tests performed in order to distinguish X and Y cells. Kaplan and Shapley based their identification of X cells on the capacity of the cells for linear spatial summation as in the original study of Enroth-Cugell and Robson (1966). The other workers, however, tended to regard the magnocellular layer cells as Y cells because of their high velocity sensitivity, fast retinal inputs, and brisk-transient discharges. Doubtless disagreements of this kind will continue until a single set of criteria, acceptable to all, can be worked out.

c. W Cells.   W cells in the C laminae of the cat lateral geniculate nucleus, though still displaying properties indicative of their not belonging to a single functional class, are said to be a little more homogeneous than in the retina (Sur and Sherman, 1982b). Types with linear and nonlinear responses have been particularly well documented by Sur and Sherman (1982a).

The W cells of the C laminae have slowly conducting axons (Wilson and Stone, 1975; Cleland *et al.*, 1975a,b, 1976; Wilson *et al.*, 1976) which probably correspond to a thin class of thalamocortical axon identified morphologically by Ferster and LeVay (1978). Many C-laminae cells, as retrogradely labeled from the cortex, are small and the W cells, therefore, probably correspond to the class 4 cell described in Golgi preparations by Guillery (1966). Those labeled in layer C, however, are much larger and probably correspond to the Y cells of that layer (Wilson *et al.*, 1976; Holländer and Vanegas, 1977). Of the small number of W cells identified physiologically and injected intracellularly in the C laminae, most were small and had the morphology of class 4 neurons (Friedlander *et al.*, 1981; Stanford *et al.*, 1983). As yet, there is no evidence for W-like cells in the monkey lateral geniculate nucleus. The C laminae of the cat project widely and rather diffusely over the cortex, to areas 17, 18, 19 and possibly to all areas of the lateral suprasylvian region (Gilbert and Kelly, 1975; LeVay and Gilbert, 1976; Raczkowski and Rosenquist, 1983). In the cortex the fibers, as traced auto-radiographically, terminate in the outer part of layer I and in layer IV. In area 17 they overlap the terminations of both X- and Y-cell axons. It has been suggested that lamina C projects primarily to area 18 while laminae C1–C3 project to the other areas (Holländer and Vanegas, 1977).

9.3.1.5. Binocularity in the Dorsal Lateral Geniculate Nucleus

Since the earliest studies of transneuronal degeneration following eye removal, it has not been contested that individual laminae of the dorsal lateral geniculate nucleus in cats and monkeys are dominated by inputs from a single eye (Minkowski, 1920; Le Gros Clark and Penman, 1934; Barris, 1935; Cook *et al.*, 1951; Cohn, 1956; Polyak, 1957). There has been a considerable amount of discussion, however, about the extent to which cells in or between the laminae may be subjected to binocular influences.

Some of the earliest single-unit studies on the cat lateral geniculate nucleus demonstrated a small proportion of cells that were excited by stimuli applied to either eye (Erulkar and Fillenz, 1960; Bishop *et al.*, 1959b, 1962; Kozak *et al.*, 1965), though this could not be demonstrated by all authors (Hubel and Wiesel, 1961). It has been pointed out that virtually all of these binocularly driven cells lay at the borders of adjacent laminae or in the interlaminar plexuses. The contemporary studies carried out with axonal degeneration methods on the distribution of optic tract fibers appeared to show an overlap in the terminal distribution of left and right eye fibers in the interlaminar plexuses (Hayhow, 1958; Laties and Sprague, 1966; Garey and Powell, 1968). These plexuses, either with or without lamina C, were customarily referred to, following Thuma (1928), as a "central interlaminar nucleus." The left and right eye overlap was thought to be sufficient to account for binocularity of some cells in the cat and, by extrapolation, it was often assumed that there should be similar overlap and binocularity in monkeys.

In his reevaluation of the distribution of optic tract fibers as seen in the cat by axonal degeneration methods, Guillery (1970) failed to find any evidence of overlap in the interlaminar plexuses and gave reasons for regarding the earlier positive results as being based upon inadequate histology or misinterpretation of the picture of degeneration. Neither Guillery (1969a,b) nor later workers (Robson and Mason, 1979) could detect optic tact terminals in the interlaminar plexuses electron microscopically. Nevertheless, as Guillery (1966) and others (Famiglietti and Peters, 1972) have shown, the two or three classes of larger cells in the geniculate laminae can have dendrites that extend from one lamina to the next (see Fig. 4.6). Moreover, cells in the interlaminar plexuses and seemingly displaced from the adjacent A, A1, or C laminae can extend dendrites into the two adjacent laminae. These seem to offer an adequate explanation for the presence of a population of binocularly excited cells at laminar interfaces.

A different type of binocular interaction is implied by the observation that a substantial majority of the cells throughout laminae A, A1, or C of the cat dorsal lateral geniculate nucleus are subjected to inhibitory influences from the eye that does not provide the direct retinal input to that lamina. In 1966 Suzuki and Kato recorded intracellularly from neurons in the lateral geniculate nucleus of cats while electrically stimulating the optic nerves. They found that for a large majority of cells, EPSPs were engendered only by stimulation of one nerve (ipsi- or contralateral) but that IPSPs could be engendered by stimulation of either nerve and at the same latency. A later study by Suzuki and Takahashi (1970) reported similar findings though in a smaller percentage of cells. The work has been confirmed again recently by Lindström (1982) who has pointed out that because the two IPSPs do not facilitate or occlude one another when both optic nerves are stimulated, they must be mediated by separate populations of inhibitory interneurons. Extracellular studies have provided further details of the inhibitory effect from an eye that is inappropriate for a particular lamina (Singer, 1970; Sanderson *et al.*, 1971; Rodieck and Dreher, 1979). In these experiments the spontaneous activity (and to some extent the stimulus-evoked discharges) of a unit innervated in lamina A or C by the contralateral eye or in lamina A1 by the ipsilateral eye could be inhibited (at a latency of 30–80 msec) by sweeping a slit of light across a homonymous (or corresponding) receptive field in the other eye. This eye came to be called the nondominant eye. The inhibitory receptive fields in the nondominant eye were customarily several times larger than those in the dominant eye. The effect was seen with both on-center and off-center units innervated by the dominant eye and the great majority of geniculate cells had such binocular receptive fields. In a small number of cases, however, the nondominant eye was reported to exert an excitatory effect. Rodieck and Dreher (1979) compared the visual suppression from the nondominant eye in the lateral geniculate nucleus of cats and monkeys. In the cat they found that cells in all laminae of the laminar nucleus could be suppressed in this way but especially X cells with dominant input from the ipsilateral eye. However, in the medial interlaminar nucleus, the members of the large Y cell population are reportedly excited by one eye only and lack an inhibitory input from a nondominant eye (Kratz *et al.*, 1978a). In monkeys, X-like cells showed no nondominant eye suppression; only Y-like cells in the magnocellular layers were affected in that way.

The idea that there might be a substructure in the usually inhibitory receptive field of the nondominant eye was taken up by Schmielau and Singer (1977). They discovered that in the majority of A or A1 layer cells of the cat, stimulation of the receptive field center in the nondominant eye actually led to a facilitation of the response of a cell to stimulation of its dominant eye. Stimulation of the receptive field surround in the nondominant eye, on the other hand, led to inhibition. These effects occurred irrespective of whether a cell had an on-center or off-center receptive field and seemed to affect exclusively the initial on or off responses of the cell rather than any later tonic discharge. The facilitatory effect had a latency of 30–180 msec and lasted 60–600 msec; the inhibitory effect had a latency of 50–80 msec and lasted 60–230 msec.

The long latencies of these responses suggest either rather complicated intrageniculate mechanisms or mechanisms, such as a thalamocorticothalamic loop, that operate beyond the geniculate nucleus. In their study, Sanderson *et al.* (1971) reported that ablation of the visual cortex had little or no effect on binocular interactions in the dorsal lateral geniculate nucleus and felt, therefore, that intrageniculate mechanisms were preemptive. In their later study, Schmielau and Singer (1977) found that inactivation of corticothalamic fibers by cooling the cortex abolished the facilitatory effect from the receptive field center of the nondominant eye. From this region, only inhibitory responses can then be elicited. The former inhibitory effect from the receptive field surround in the nondominant eye is also rendered less effective but is not abolished under these conditions. Hence, binocular facilitation is thought to be conveyed by this cortical loop while binocular inhibition is thought to be mediated probably by this in cooperation with intrageniculate mechanisms. This cooperativity forms an essential component of Singer's (1977) theory of the action of corticogeniculate fibers in stereopsis, as reviewed in Chapter 4.

Because much of the inhibitory effect of stimulating a nondominant eye is not based on a corticothalamic loop, it must be mediated by intrageniculate mechanisms. Because it is binocular, interlaminar routes must be operational and because it involves corresponding receptive fields, the effect probably operates along the projection columns that span all laminae of the geniculate and represent a part of visual space (see Chapter 3).

In attempting to discover a route for interlaminar and, therefore, binocular inhibitory effects in the cat lateral geniculate nucleus, attention has naturally focused on the long-axoned interneuron described by Tömböl (1969) (Chapter 4). Although all of the classes of larger cells that have been described in the cat lateral geniculate nucleus have dendrites that can often cross from one lamina to another (Guillery, 1966; Famiglietti and Peters, 1972), any binocular effect mediated by this route would presumably only affect cells situated close to laminar borders or in the interlaminar plexuses.* It is only in these positions that the relatively small number of cells that are excited equally from the two eyes are found (Erulkar and Fillenz, 1960; Bishop *et al.*, 1962; Kozak *et al.*, 1965). The long-axoned interneuron, on the other hand, though having its soma in one lamina, sends its axon to an adjacent lamina and, furthermore, has fine structural characteristics that suggest it may well be inhibitory. To date, however, no parameters which would serve to characterize physiologically the long-axoned interneuron have been established.

9.3.1.6. Color in the Dorsal Lateral Geniculate Nucleus

Le Gros Clark (1940, 1942, 1949) put forward an idea for the geniculate transfer of color information by speculating that matched pairs of layers could receive axons of retinal ganglion cells sensitive to only one range of wavelengths: red, green, or blue. His evidence was rather slender and, for his attempt, he was subjected to a sustained, savage, though nonetheless witty attack by Walls (1953). This served to discredit the concept and must have daunted many a would-be theorist from proposing a new one.

De Valois *et al.* (1958) seem to have been the first to report that many single neurons in the lateral geniculate nucleus of macaque monkeys are sensitive to monochromatic light. Those showing sensitivity to wavelengths with a peak at 510 nm were presumed to receive rod connections whereas others with peaks

---

* Friedlander *et al.* (1981), in their intracellular injection study, found only the Y cells identified had dendrites that crossed interlaminar borders (Fig. 4.6).

of sensitivity at 440, 550, 590, and 620 nm were presumed to receive different cone inputs. Color-sensitive cells in layers 5 and 6 had only on responses to light of the specific wavelength but some in layers 3 and 4 could be excited by light of one wavelength and inhibited by that of another. Hubel and Wiesel (1960), in recording from single optic nerve fibers in the spider monkey, noted that a proportion of these fibers showed similar opponent-color responses. They then went on to seek similar color-coded receptive fields in the lateral geniculate nucleus of light-adapted monkeys (Wiesel and Hubel, 1966). In the parvocellular layers, the commonest type (77%) of color-sensitive receptive field was the classical concentric type already described, the center and surround having different spectral sensitivities (Fig. 9.25A). That is, the cells with such fields showed opponent-color responses to two different sets of wavelengths and no response to intermediate wavelengths. The commonest arrangement was a red on or off center and an antagonistic green off or on surround. These cells had particularly small receptive fields, the centers measuring no more than 2 mins of visual angle near the fovea.

A smaller number of cells in the parvocellular layers did not have the center–surround arrangement and gave opponent-color responses over all regions of their receptive fields with the neutral point being independent of stimulus geometry. Another small group, found in both parvo- and magnocellular layers, had concentric fields but the center and surround were sensitive to the identical wavelength. A third minor group only in the magnocellular layers had an on center and a large surround whose sensitivity was displaced toward the red with respect to the center, the surround usually predominating over the center in both red and white light.

Wiesel and Hubel (1966) made some attempt, in dark-adapted monkeys, to

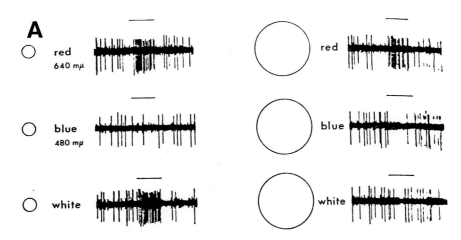

FIGURE 9.25. (A) Responses of a cell in one of the parvocellular layers of the monkey dorsal lateral geniculate nucleus to white and monochromatic light. The left column shows responses to stimulation of the receptive field center with a ½° spot. The right column shows responses to stimulation of the center and opponent surround of the receptive field. The cell is shown to be red on center with a blue opponent ("off") surround. From Wiesel and Hubel (1966). (B) Types of opponent-color cells found at various levels of the visual system in several different species. From Daw (1972).

determine to what extent color-coded geniculate cells representing parts of the retina away from the fovea received both rod and cone connections. They concluded that there was evidence both for and against rod connections, but that where a cell had both, the rod and cone inputs to the center or to the surround were always of the same sign (i.e., on or off).

Brief descriptions of opponent- and other color-coded cells in the monkey lateral geniculate nucleus were given by De Valois (1972), Daw (1972), Marrocco and Brown (1975), and Padmos and Norren (1975). Daw specifically mentions that he and Pearlman were unable to discover "double-opponent"-color cells. These had been described by Hubel and Wiesel (1968) (also Michael, 1978, 1981) in the monkey striate cortex and by Michael (1971, 1973) in the ground squirrel lateral geniculate nucleus. Such cells are characterized by a concentric receptive field in which the opponent colors for the center are the reverse of those for the periphery. For example, such a field would have a red on center and a red off surround but a green off center and a green on surround (Fig. 9.25B). Such cells are optimally influenced, therefore, by simultaneous stimulation with two colors. The presence of cells with receptive fields of this type would imply con-

B OPPONENT COLOR CELLS WITH CENTER–SURROUND ARRANGEMENT

Wiesel + Hubel, 1966 (type I)
Michael, 1968 (class 3)

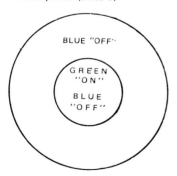

Wagner, MacNichol + Wolbarsht, 1963
Michael, 1968 (class 2)
Kaneko, 1972

OPPONENT COLOR CELLS WITH NO CENTER–SURROUND ARRANGEMENT

Wiesel + Hubel, 1966 (type II)
Michael, 1963 (class I)
Norton, Spekreijse, Wolbarsht, Wagner, 1968

DOUBLE OPPONENT CELLS

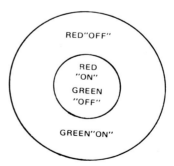

Daw, 1968
Hubel + Wiesel, 1968
Michael, 1970
Pearlman + Daw, 1970

FIGURE 9.25. *(continued)*

vergence of opponent-color cells with the more common single, center–surround receptive field. Hence, in monkeys, this type of convergence appears to occur at cortical levels only. In the ground squirrel, however, although double-opponent-color cells are not found in the optic nerve, they form the major class of color-sensitive cells in the lateral geniculate nucleus (Michael, 1968, 1973), suggesting convergence in this animal in the geniculate itself.

Dreher *et al.* (1976) in two species of macaque, other than the rhesus monkey, confirmed Wiesel and Hubel's (1966) observations on the laminar distribution of the four types of color-coded cells. In addition, they showed that the predominate opponent-color-type cell in the parvocellular layers could be classified as an X-type cell in terms of its response to spatial stimuli such as moving gratings and its latency to stimulation of the optic chiasm. Of the less common color-coded cell classes, all those encountered in the parvocellular layers could also be classified as X type and those in the magnocellular layers as Y type (see also Hicks *et al.*, 1983). Michael (1983) has recently briefly reported that different categories of color-opponent cells may be localized in different parvocellular laminae of the monkey.

Cats had generally been considered to have a rather limited ability to discriminate hue. Daw and Pearlman (1970) trained cats to distinguish red from cyan and orange from cyan, both in the mesopic and photopic levels of illumination, i.e., independent of whether rods were or were not saturated. Cat color discrimination was, therefore, proven to depend on more than one kind of cone though the colors needed to be particularly bright. In the same study they found that the vast majority of single units recorded throughout all layers of the laminar dorsal lateral geniculate nucleus were sensitive to a single wavelength in the green and with a peak at 556 nm. That is, all received input from a single type of cone. Four cells out of 434, however, found in the C laminae had concentric color-opponent receptive fields with a blue on center and a red and green off surround. At least two of these cells were double opponent cells that were excited by green and inhibited by blue light in their receptive field surrounds. There is thus some evidence for the relay of information from more than one type of cone through the cat lateral geniculate nucleus.

There have been no studies of the cat lateral geniculate nucleus specifically directed at color mechanisms since that time. Recordings from cat retinal ganglion cells suggest that X and Y cells are not color-opponent (see Rodieck, 1979), thus supporting the finding of Daw and Pearlman of no color-opponent cells in the major geniculate layers in which their axons terminate.

Schiller and Malpeli (1978) also attempted to correlate X- and Y-like properties of monkey geniculate neurons with color sensitivity. Their findings in this respect are essentially the same as those of Dreher *et al.* (1976). That is, cells in the parvocellular layers were predominately color-opponent and X-like, while those in the magnocellular layers have broad-band properties and are Y-like. Among the cells of the parvocellular layers, however, they found that those with red, green, or white on-center receptive fields were mainly in layers 5 and 6 and those with red, green, or white off-center receptive fields were mainly in layers 3 and 4. All blue-sensitive cells had blue on centers and were found almost exclusively in layers 3 and 4.

As Schiller and Malpeli conclude, there is indeed evidence that "the three pairs of geniculate laminae are channels carrying different sorts of information." Perhaps, after all, Le Gros Clark did have some inkling of a fundamental pattern and was not, as Walls caustically put it, attempting to play the cerebral piano with a broom. Here, the analysis of color mechanisms in the lateral geniculate nucleus appears to have ceased. It is interesting to speculate, however, that neurons in the interlaminar zones of monkeys might be predominately color coded, if only for the reason that they project to layer III of the striate area (Section 9.4.2 and Chapter 3) where they end in patches that are rich in cytochrome oxidase and GAD activity (Hendrickson *et al.*, 1981; Horton and Hubel, 1981). Such patches contain cells that have poor orientation selectivity but which are color coded (Livingstone and Hubel, 1982).

9.3.2. Tectogeniculate Afferents

The afferent fibers forming the tectal input from the superior colliculus arise in the superficial part of the stratum griseum superficiale of the colliculus (Casagrande *et al.*, 1972; Robson and Hall, 1977; Harting *et al.*, 1978; Albano *et al.*, 1979; Graham and Casagrande, 1980; Weber *et al.*, 1983). They are distributed in relatively small numbers to various parts of the dorsal lateral geniculate nucleus, depending on the species. In the cat the tectogeniculate fibers terminate primarily in lamina C3 with perhaps some overlap innto lamina C2 (Graham, 1977; Torrealba *et al.*, 1981) (Fig. 9.26). Although those arising from a particular part of the retinal representation in the superior colliculus terminate opposite the comparable part of the (contralateral) representation in lamina C2, the degree of overlap with the terminations of retinal fibers in C2 is relatively small (Graham, 1977; Torrealba *et al.*, 1981). In other species the tectogeniculate projection commonly terminates among retinal fiber terminations in interlaminar regions and in subsidiary laminae such as the S layer of monkeys (Fig. 9.26), though tectogeniculate fibers may innervate certain principal layers as well (Benevento and Fallon, 1975a; Robson and Hall, 1976; Harting *et al.*, 1978, 1980; Fitzpatrick *et al.*, 1980). In the squirrel, layer 3c is the site of termination; this is innervated by the contralateral retina (Robson and Hall, 1976). In the tree shrew, layer 3, the interlaminar regions between layers 4 and 5, and a region deep to layer 1 receive tectal fibers. Layer 3 receives contralateral retinal inputs, the other regions probably both ipsi- and contralateral fibers (Laemle, 1968; Albano *et al.*, 1979; Fitzpatrick *et al.*, 1980). In the bush baby (a prosimian) the tectal fibers end in layers 4 and 5, in the interlaminar regions between layers 1 and 2, between layers 2 and 3, and in a region between layer 1 and the optic tract. Layer 4 receives ipsilateral and layer 5 contralateral retinal inputs; the other regions are probably innervated by both retinae (Laemle and Noback, 1970; Fitzpatrick *et al.*, 1980). In monkeys the tectal fibers terminate in the S layers and in the interlaminar regions between the more ventral layers, all of which are probably innervated by both retinae (Kaas *et al.*, 1978; Harting *et al.*, 1978, 1980).

Though this may seem to be yet another bewildering array of connectional differences in the visual thalamus, one common factor seems to bind them together. As pointed out by Diamond and his associates (Fitzpatrick *et al.*, 1980; Raczkowski and Diamond, 1981) all of the regions receiving tectogeniculate fibers are small-celled regions and all project to superficial (I–III) layers of the cerebral cortex, often over wide areas (Leventhal *et al.*, 1979; Carey *et al.*, 1979; Fitzpatrick

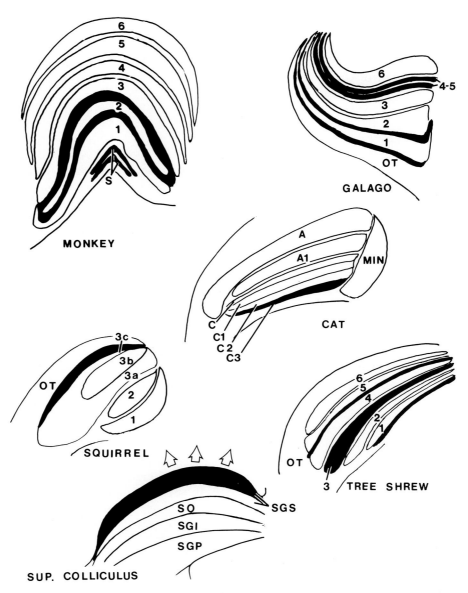

FIGURE 9.26. Schematic representation of the origin and distribution of tectal fibers innervating the dorsal lateral geniculate nucleus in the cat, tree shrew, bush baby, squirrel, and monkey. Fibers arise from cells in the superficial part of the stratum griseum superficiale (SGS) and terminate in interlaminar zones, in subsidiary laminae such as the S laminae of the monkey, and/or in small-celled layers such as layers 4 and 5 of the bush baby.

et al., 1980). In the cat, as pointed out below, the deeper C laminae (C1, C2) receive the W class of retinal ganglion cell axons whose functional characteristics are different from the X and Y classes that innervate the other laminae. Hence, Diamond's group suggests that a comparable pattern of organization may exist in other species; i.e., certain small-celled layers, interlaminar regions, and supernumerary laminae such as the S layer of monkeys, may represent the equivalents of the feline deep C laminae and receive W-type inputs. As yet, there are little or no physiological data to support this interesting and potentially useful generalization. Of two units recorded in the S laminae of fascicularis monkeys, one had X-type and the other Y-type properties (Kaplan and Shapley, 1982) and units recorded in the smallest-celled layer of the tree shrew did not have W-like properties (Conway and Schiller, 1983).

As pointed out by Raczkowski and Diamond (1981), there appears to be a divergence of the W and Y streams at the level of the superior colliculus. In the cat, W-like cells are found in a more superficial tier of the stratum griseum superficiale (Stone *et al.*, 1979) and Y-like cells in a deeper tier (Hoffmann, 1973). This is supported by the observation of Itoh *et al.* (1981) that the smallest (γ) retinal ganglion cells are retrogradely labeled from injections of tracer in the superficial tier while the largest (α) cells are retrogradely labeled from deeper injections. Furthermore, in many species it is the superficial tier of the stratum griseum superficiale that projects to the C laminae or interlaminar regions of the dorsal lateral geniculate nucleus while the deeper tier projects to the pulvinar or lateral posterior nucleus (Harting *et al.*, 1973; Jacobson and Trojanowski, 1975; Glendenning *et al.*, 1975; Robson and Hall, 1977; Albano *et al.*, 1979; Kawamura *et al.*, 1980; Raczkowski and Diamond, 1981). The possibility exists, therefore, for the superior colliculus to mediate different effects upon the extrastriate cortex, a W effect through the C laminae or their equivalents in noncarnivores and a Y effect through the lateral posterior–pulvinar complex. All layers of the cat dorsal lateral geniculate nucleus may also receive fibers from the nucleus of the optic tract of the pretectum (Graybiel and Berson, 1980a).

9.3.3. Functional Attributes of the X, Y, and W Systems

The highly specialized characteristics of X, Y, and W cells and their seeming independence of one another from retina, through geniculate and/or superior colliculus to cerebral cortex, have suggested to many workers that they form three parallel streams with different functional roles to play in vision (e.g., Stone *et al.*, 1979; Lennie, 1980a,b; Barlow, 1981).

Both X and Y cells obviously possess properties that make them suitable for participating in the highly discriminative aspects of vision. Both appear to have high resolving power; X cells have smaller receptive field centers and can resolve higher spatial frequencies than Y cells but Y cells are more sensitive to rapidly moving stimuli. It has often been predicted, therefore, that X cells are involved in aspects of vision involving high spatial resolution whereas Y cells should be involved in movement discrimination. In view of this, it seems significant that X cells tend to concentrate near the central retina whereas Y cells are found in all parts, including the periphery. Furthermore, the streaming through the dorsal lateral geniculate nucleus takes both X and Y inputs to areas 17 and 18 in the cat and to area 17 in the monkey. When these areas or their homologs in other species are removed, animals suffer a great reduction in visual acuity usually demonstrated by a loss of fine pattern discrimination (Snyder and Diamond, 1968; Schneider, 1969; Weiskrantz, 1972; Weiskrantz *et al.*, 1974; Sprague *et al.*, 1977). Yet in these animals a considerable capacity for gross pattern and form detection and visual orientation usually remains (Snyder and Diamond, 1968; Humphrey, 1974; Sprague *et al.*, 1977). Nonprimates usually show more residual capacity in this regard but even in humans object detection may remain (Weiskrantz *et al.*, 1974).

This residual capacity for pattern and form as opposed to fine discriminative vision has been thought likely to depend on the W-cell input to the extrastriate visual areas, coming in the cat through the C laminae of the laminar dorsal lateral geniculate nucleus, though some Y-cell input may reach the extrastriate areas through the medial interlaminar nucleus. It seems appropriate, then, that if the extrastriate areas are removed, leaving the geniculate input to striate cortex (and, in the cat, to area 18) intact, visual acuity surives but overall pattern and form detection appear compromised (Sprague *et al.*, 1970, 1977; Berlucchi *et al.*, 1972). The comparable route for a W-cell input to the extrastriate areas of animals other than the cat is not clear because W-cell inputs have not been demonstrated to the dorsal lateral geniculate nucleus in species other than the cat. One potential route is indirect: via the superior colliculus and its projection to the lateral posterior–pulvinar complex of the thalamus, which, in turn, projects upon the extrastriate areas and is thought to help maintain an animal's capacity for orientation toward a visual stimulus in the absence of a striate cortex (Snyder and Diamond, 1968; Schneider, 1969). It is conceivable, however, that some diffuse projections arising in the monkey lateral geniculate nucleus may spread beyond the striate areas and, thus, provide a remaining geniculocortical route in the absence of a striate cortex. But reports of these seem less than adequately documented and have been contested (Yukie and Iwai, 1981; Benevento and Yoshida, 1981; see Ogren and Hendrickson, 1976).

Given the very large number of visually related fields in the cerebral cortex, it would be unduly optimistic to think that the whole of vision can be explained in terms of two or three parallel retinothalamocortical pathways. There are as many as 13 retinotopically organized visual areas in the cat cortex (Palmer *et al.*, 1978; Tusa and Palmer, 1980) and perhaps almost as many in monkeys (Allman and Kaas, 1971; Zeki, 1978; Van Essen *et al.*, 1982). Some of these have been reported to display an over- or underrepresentation of parts of the visual field (Hubel and Wiesel, 1965; Palmer *et al.*, 1978; Tusa and Palmer, 1980); there are reports of neurons with special classes of receptive field confined to some (Spear and Baumann, 1975, 1979; Zeki, 1978); each appears to receive an overall pattern of thalamocortical, corticocortical, and callosal connections somewhat different from the others (see Section 9.4). Hence, it seems inevitable that each will play a different functional role in the whole visual process. Probably the functions of each of them will turn out to depend upon the relative balance of X, Y, W, other retinal and nonretinal inputs, coming over a variety of both direct and indirect routes, rather than upon any one of these.

9.3.4. Afferents from the Brain-Stem Reticular Formation

It seems unquestionable that the thalamus is influenced by the reticular formation of the brain stem, for there are numerous reports of transmission through the thalamus being modified by stimulation of the reticular formation (Singer, 1977; Burke and Cole, 1978). Though the pathways mediating the reticulothalamic effect are anything but well documented anatomically (see Mehler, 1966a), several kinds of data force us to take them seriously. Because much of the information relevant to this topic comes from studies on the dorsal lateral geniculate nucleus, it seems appropriate to review the subject here.

Perhaps the most widely studied of the reticulothalamic effects is the "PGO wave" phenomenon demonstrable in the cat dorsal lateral geniculate nucleus at the onset of and during the phase of sleep characterized by a desynchronized EEG and rapid eye movements (REM or paradoxical sleep) (Jouvet, 1962; Bizzi and Brooks, 1963; Maffei *et al.*, 1965; Bizzi, 1966; Brooks, 1967; Malcolm *et al.*, 1970; McIlwain, 1972; Bartlett *et al.*, 1973; Hobson, 1984) (Fig. 9.27). The PGO waves are large, stereotyped, negative field potentials recordable from the pontine reticular formation, the dorsal lateral geniculate nucleus, and the occipital

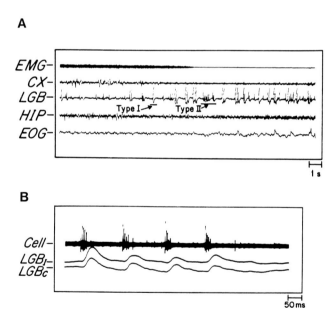

FIGURE 9.27. (A) PGO waves in the dorsal lateral geniculate nucleus (LGB) of a cat during the transition from synchronized (NREM) to desynchronized (REM) sleep. EMG indicates atonia as REM sleep becomes established; CX shows the desynchronized EEG; HIP indicates the accompanying theta rhythm in the hippocampus; EOG shows the rapid eye movements. Type I are the early PGO waves which become clustered and of reduced amplitude (type II) as the signs of REM sleep become more prominent. (B) Clustered firing (cell) of a cell in the peribrachial region of the pons. Each burst precedes the onset of PGO waves in the ipsilateral (LGB_I) and contralateral (LGB_c) dorsal lateral geniculate nucleus by 10–20 msec. Both from Hobson (1984).

cortex and often from other parts of the thalamus and cortex as well. They appear singly at the onset of REM sleep and in bursts during the eye movements once REM sleep is established. The waves in the pons precede those in the cortex and geniculate by 10 msec or more. They are lateralized in that they appear first and remain longest on the side toward which the eye movements occur. They do not depend on visual input as they still appear after eye removal (Sakakura and Iwamura, 1967).

During sleep, presumably accompanied by PGO waves in the dorsal lateral geniculate nucleus, the excitability of geniculate relay neurons to peripheral stimuli is enhanced (Fig. 9.28), probably by the mechanisms alluded to in Chapter 4.

PGO waves are generated by the discharge and progressive recruitment of large numbers of reticular cells throughout the length of the pons but especially near the pontomesencephalic junction (Morrison and Pompeiano, 1966; Laurent *et al.*, 1974; Doty *et al.*, 1973). These cells, which may be cholinergic, are probably under the influence of noradrenergic and serotoninergic inputs from the locus coeruleus and dorsal nucleus of the midbrain raphe (Jacobs and Jones, 1978; Hobson, 1984). They probably project directly to oculomotor and vestibular nuclei, and are thought to affect the lateral geniculate nucleus and cortex both directly and via connections to neurons in the "parabrachial region" (around the superior cerebellar peduncle). The parabrachial region is then thought to project directly to the geniculate and cortex. Along with similar direct projections from the locus coeruleus and dorsal raphe, it is thought to mediate the PGO effect. Hobson's idea is that the aminergic projections are normally inhibitory (as suggested by Foote *et al.*, 1974) and that a reduction in discharges from the locus coeruleus and dorsal raphe would not only initiate the discharges of the pontine generator cells but also enhance the excitability of geniculate and cortical neurons, making them more responsive to the excitatory influences from the parabrachial region. Because the generator neurons also cause activation of oculomotor neurons, it has been suggested that the effect at the level of the lateral geniculate nucleus is in the nature of a corollary discharge associated with the command to change eye position (see Singer, 1977; Hobson, 1984). This has been contested as occurring, however, in alert animals performing saccadic eye movements (Büttner and Fuchs, 1973; Duffy and Burchfiel, 1975).

The evidence that PGO waves in the geniculate and cortex are generated by brain-stem mechanisms is strong because similar changes in excitability can be readily produced by reticular stimulation [especially of the midbrain (Wilson *et al.*, 1973; Bartlett *et al.*, 1973; Doty *et al.*, 1973; Cole *et al.*, 1980)] and because prepontine transection of the brain-stem abolishes transfer from the pontine generator to the geniculate and cortex. The lesion studies also show that neither activation of the cortex nor its laterality depends on the lateral geniculate nucleus, as both lateral geniculate nuclei can be destroyed without abolishing the transmission of the pontine excitation to the cortex (e.g., Bizzi and Brooks, 1963; Cohen and Feldman, 1968; Malcolm *et al.*, 1970; Orban *et al.*, 1972; Singer, 1973a; Wilson *et al.*, 1973; Singer and Bedworth, 1974; Doty *et al.*, 1973).

Noradrenergic pathways from the locus coeruleus to the vicinity of the lateral geniculate complex have been demonstrated autoradiographically and immunocytochemically (Swanson and Hartman, 1975; Jones and Moore, 1977; Kromer and Moore, 1980), though the nature of the terminations of these fibers has never been established and the ventral lateral geniculate nucleus receives a far denser innervation than the dorsal.

The anatomical evidence for direct pathways from the raphe, the parabrachial region, and the reticular formation to the lateral geniculate nucleus seems to me to be less strong. Retrogradely labeled cells have certainly been demonstrated in these sites after injections of tracers in the vicinity of the dorsal lateral

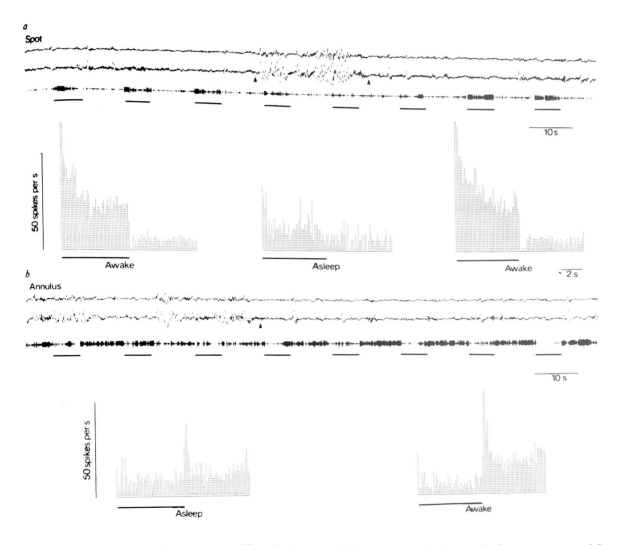

FIGURE 9.28.   Responses of an on-center cell from lamina A of the dorsal lateral geniculate nucleus of a drowsy cat which alternately wakes and sleeps. The upper two traces in (a) and (b) are EEGs recorded from anterior and posterior head regions. Sleeping is indicated by the slow waves in the middle of each trace in (a) and in the left half of each trace in (b). The third trace records spikes from the geniculate cell. The histograms indicate spike frequency averaged for several responses. In (a) a spot is flashed on the receptive field center: responses of the cell are weaker during slow-wave sleep. In (b) an annulus is flashed on the receptive field surround: the suppression of firing and the off response are weaker during slow-wave sleep. From Livingstone and Hubel (1981).

geniculate nucleus in cats and rats (Leger *et al.*, 1975; Gilbert and Kelly, 1975; Ahlsén and Lo, 1982; Mackay-Sim *et al.*, 1983). However, I feel that few steps have been taken to control for involvement of fibers of passage passing around the geniculate en route to other forebrain sites. A similar reservation applies to studies involving electrical stimulation in the lateral geniculate nucleus and the recording of antidromic potentials in the brain-stem sites (Sakai and Jouvet, 1980; Leger *et al.*, 1975; Ahlsén and Lo, 1982). Finally, where anterogradely transported tracers have been injected into the parabrachial region, dorsal raphe, and other parts of the pontine reticular formation, projections have been demonstrated to the intralaminar and parts of the ventral thalamic nuclei but not to the lateral geniculate nuclei (Edwards and De Olmos, 1976; Saper and Loewy, 1980; McKellar and Loewy, 1982; see Chapter 3). Stimulation of the locus coeruleus in cats leads to depression of the activity of putative interneurons and enhancement of the excitability of relay neurons in the lateral geniculate nucleus (Nakai and Takaori, 1974; Rogawski and Aghajanian, 1982), an effect inhibited by α-adrenergic blockers. The increased excitability seems to be the reverse of the effect predicted above during PGO wave induction. Noradrenaline when iontophoresed onto geniculate neurons has a weak depressant effect and serotonin a powerful one (Phillis *et al.*, 1967; Tebēcis and DiMaria, 1972; Rogawski and Aghajanian, 1980). These effects are also inconsistent with the theory, though Doty *et al.* (1973) have shown that the enhanced excitability of neurons in the lateral geniculate nucleus of squirrel monkeys ensuing from stimulation of the mesencephalic reticular formation may be preceded by a brief period of inhibition. Acetylcholine, which is thought to be involved somewhere in the pontine generator circuit (Jouvet, 1972), does excite geniculate neurons quite powerfully (Tebēcis and DiMaria, 1972) and cholinergic blockers tend to depress the effects of reticular formation stimulation on geniculate neurons (Phillis *et al.*, 1967).

Reticular formation influences have been reported upon transmission through the intralaminar thalamic nuclei (Chapter 12) (Purpura *et al.*, 1966; Mancia *et al.*, 1971; Steriade, 1981; Glenn and Steriade, 1982). The intralaminar nuclei have been described mainly in anatomical studies as receiving inputs from various parts of the reticular formation including the parabrachial region and locus coeruleus (Nauta and Kuypers, 1958; Conrad *et al.*, 1974; Pickel *et al.*, 1974; Edwards and De Olmos, 1976; Graybiel, 1977; Jones and Moore, 1977; Saper and Loewy, 1980; Kromer and Moore, 1980; McKellar and Loewy, 1982; Steriade and Glenn, 1982). However, because they lack projections to other dorsal thalamic nuclei, the intralaminar nuclei are unlikely candidates for mediating the reticular effect directly through the dorsal lateral geniculate or other nuclei. A circuit through intralaminar nuclei and cerebral cortex, and back to the thalamic relay nuclei is a possibility on account of the diffuse cortical projection from the intralaminar nuclei (Chapter 12). Efferent cells of the cat visual cortex can be directly activated by stimulation of the reticular formation (Singer *et al.*, 1976) and cooling of the visual cortex renders less effective the facilitation of geniculate transmission normally induced by stimulation of the reticular formation (Schmielau and Singer, 1977).

A further indirect route whereby the reticular formation could influence transmission in the dorsal lateral geniculate nucleus or in thalamic nuclei generally, is via the reticular nucleus (or the perigeniculate nucleus in the case of the cat lateral geniculate nucleus). As pointed out in Chapters 3, 4, and 15, there is evidence for inhibitory effects being exerted by these nuclei on the underlying dorsal thalamic nuclei. Both receive corticofugal fibers and could participate in a reticulothalamocorticothalamic loop. Whether the reticular and perigeniculate nuclei are activated directly from the reticular formation or indirectly from the intralaminar nuclei, however, remains an open question. The reticular nucleus certainly receives many afferent fibers from the intralaminar nuclei in which reticulothalamic axons are said to end. Neurons of the reticular nucleus are inhibited by stimulation of the reticular formation (Fukuda and Iwama, 1971; Yingling and Skinner, 1976) and are less active during arousal than during slow-wave (non-REM) sleep (Mukhametov *et al.*, 1970a,b; Lamarre *et al.*, 1971). Finding a direct anatomical pathway from the reticular formation to the reticular nucleus, however, has not been easy. In my opinion, the only well-documented reports are those indicating a pathway arising in the dorsal aspect of the rostral midbrain reticular formation (Edwards and De Olmos, 1976) or deeper part of the pretectum (Berman, 1977). Neither of these regions has been implicated in the ascending reticular effect, though it is, of course, possible that the axons of pontine generator cells or of subsequent cells in an ascending chain, may relay there.

9.3.5. Corticogeniculate Afferents

Fibers from the visual cortex return to the dorsal and ventral lateral geniculate nuclei and to certain other nuclei of the dorsal thalamus. They also give collateral branches to the perigeniculate and reticular nuclei. Their general properties and synaptic organization have been described in Chapter 4. Here, we shall be concerned only with the pattern of organization that they display in their projections upon the dorsal lateral geniculate nucleus. As might be anticipated from the diversity of structure exhibited by this nucleus, there are significant species differences. Perhaps the only features held in common across species are that area 17 always contributes a heavy corticogeniculate projection and that it arises from the modified pyramidal cells of layer VI.

Corticogeniculate axons in all species with a laminated nucleus ramify densely in the interlaminar spaces between the cell laminae and appear to terminate primarily in parts of the laminae abutting on the spaces (Guillery, 1967; Holländer, 1974; Sanderson and Kaas, 1974; Updyke, 1975; Lin and Kaas, 1977; Brunso-Bechtold *et al.*, 1983).

In rodents and other species with dorsal lateral geniculate nuclei of relatively simple form, all corticogeniculate fibers probably arise in area 17 and terminate throughout the length, breadth, and depth of the nucleus (Nauta and Bucher, 1954; Montero and Guillery, 1968; Benevento and Ebner, 1970; Rockel *et al.*, 1972; Robson and Hall, 1975; Giolli *et al.*, 1978; Holländer *et al.*, 1979). On occasion, the density of terminations has been described as being lighter in the deepest part of the nucleus. The projection is topographically organized so that the fibers arising from a small punctate focus in area 17 will distribute to a visuotopically related part of the nucleus, ending in a column stretching across both contralaterally and ipsilaterally (where present) innervated regions.

Whether the "belt" regions of cortex surrounding area 17 in rodents and in species with comparable geniculate organization contribute corticothalamic fibers to the dorsal lateral geniculate nucleus seems to be undecided.

In the cat the pattern of corticogeniculate projection is rather complicated and has been argued over for many years. The present position seems to be (Updyke, 1975, 1977, 1983): area 17 sends fiber to all laminae of the laminar nucleus, and the density of termination appears equal in all of them; it has an additional, somewhat sparser projection to the medial interlaminar nucleus; area 18 sends fibers to all the laminae and to the medial interlaminar nucleus but the density of terminations is greater in the interlaminar zones, in lamina C, and in the medial interlaminar nucleus than in laminae A and A1; area 19 sends fibers primarily to laminae C1–C3 and the medial interlaminar nucleus; areas beyond area 19 may project to the deeper C laminae (Updyke, 1981a). The projections are retinotopically organized (Garey et al., 1968) and those from the three areas end in register. The corticothalamic cells, thus, seem to project axons primarily to those parts of the dorsal lateral geniculate from which the cortical area in which they lie receives afferent connections, maintaining the rule of reciprocity alluded to in Chapter 4. It should be noted, however, that the literature on corticothalamic connections in the cat is enormous and often contradictory. What has been given above seems to represent the current status.

In monkeys, area 17 seems confirmed as the major source of corticothalamic fibers to the dorsal lateral geniculate nucleus (Holländer, 1974; Spatz and Erdmann, 1974; Holländer and Martinez-Milan, 1975; Lund et al., 1975). All layers of the dorsal lateral geniculate nucleus, including the interlaminar zones, appear to receive corticogeniculate projections (Spatz and Erdmann, 1974; Benevento and Fallon, 1975b; Holländer and Martinez-Milan, 1975; Ogren and Hendrickson, 1976; Lin and Kaas, 1977; Graham et al., 1979; Graham, 1982). The projection is retinotopically organized, following the retinal projection lines through the nucleus (Holländer and Martinez-Milan, 1975; Graham, 1982). Fibers to the magnocellular layers arise in the deepest part of layer VI of the striate cortex and those to the parvocellular layers arise in more superficial parts of layer VI (Lund et al., 1975).

There have also been a few reports of sparse projections from area 18 and other extrastriate areas (such as the middle temporal area) to the magnocellular layers and the S layers (Wong-Riley, 1976; Lin and Kaas, 1977).

In lower primates and in the insectivore *Tupaia*, the corticogeniculate projection is similar to that described in monkeys (Harting and Noback, 1971; Tigges et al., 1973; Casagrande, 1974; Symonds and Kaas, 1978).

9.4. Geniculocortical Connections

It is a remarkable fact that despite the enormous amount of effort that has gone into analyzing the visual system over the last few years, the precise details of the geniculocortical projection in two of the commonest laboratory animals, the monkey and cat, have still not been completely elucidated. New data are being added continually and gradually a consensus is being arrived at but it has to be admitted that there are still many reports not in complete harmony with others.

9.4.1. The Cat

Studies carried out with cellular and axonal degeneration techniques in the 1960s claimed to show projections from the dorsal lateral geniculate nucleus not only to the striate area (area 17) but to rather widespread extrastriate areas as well (Glickstein *et al.,* 1967; Wilson and Cragg, 1967; Niimi and Sprague, 1970). Gradually there was a shift of opinion in favor of a more limited projection upon areas 17 and 18 and possibly the lateral suprasylvian area of Clare and Bishop (1954), in which short-latency responses to optic tract stimulation had been reported in 1961 by Vastola (Wilson and Cragg, 1967; Heath and Jones, 1971b; Burrows and Hayhow, 1971). Yet, even so, the projection of the lateral geniculate nucleus upon two or three cytoarchitectonic fields containing a complete visuotopic representation of the contralateral half field, seemed strangely at odds with the geniculocortical projection to a single cortical field that appeared to exist in other animals.

Attempts were made, from careful analyses of patterns of retrograde cellular and anterograde axonal degeneration, to work out the differential projections of the laminar and medial interlaminar nuclei of the dorsal lateral geniculate complex. Garey and Powell (1967) decided that the retrograde degeneration occurring in the lateral geniculate nucleus as a result of selective cortical lesions indicated that small and medium cells of the laminar nucleus projected to area 17, large cells to areas 17 and 18, and the medial interlaminar nucleus primarily to area 19. It was implied that the lateral suprasylvian area could have received inputs from all parts of the nucleus. In a carefully controlled study based upon small, stereotaxically placed lesions in the various parts of the lateral geniculate complex and cortex, Burrows and Hayhow (1971) decided that the laminar nucleus projected to areas 17 and 18 and possibly the lateral suprasylvian area, and that the main cortical target of the medial interlaminar nucleus was area 18.

Little more was added until the application of the newer anatomical tracing techniques in the middle 1970s. In one of the first autoradiographic studies on thalamocortical projections, Rosenquist *et al.* (1974) determined that the laminar lateral geniculate nucleus sent axons to areas 17 and 18 and the medial interlaminar nucleus to areas 18, 19 and the lateral suprasylvian area. One year later, Gilbert and Kelly (1975) applied the newly introduced horseradish peroxidase tracing method and found that the laminar nucleus projects to areas 17, 18, 19, and the lateral suprasylvian area but that the medial interlaminar nucleus projects only to area 18 (see also Maciewicz, 1975).

Gradually the conflicts in these observations have been worked out (Fig. 9.29). The first major contribution was made by LeVay and Gilbert (1976) when

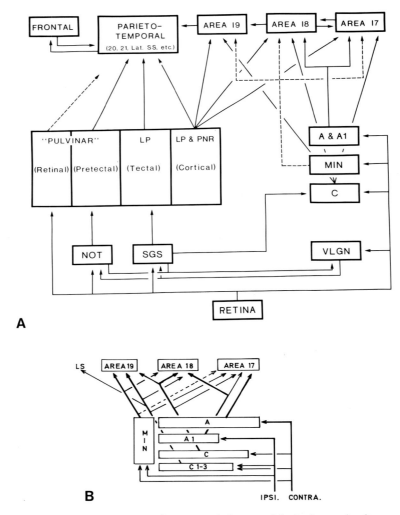

A

B   IPSI. CONTRA.

FIGURE 9.29.   (A) A general scheme of thalamocortical connectivity in the cat visual system. NOT: nucleus of optic tract; SGS: superficial layer of superior colliculus; VLGN: ventral lateral geniculate nucleus. (B) Cortical projections of the different components of the cat dorsal lateral geniculate nucleus, based mainly upon Holländer and Vanegas (1977). Thin or interrupted lines indicate a minor projection. Note that areas 17 and 18 appear to receive contralateral retinal inputs (CONTRA) independently through layers A and C but ipsilateral retinal input (IPSI) passes to both via lamina A1. LS indicates cortical areas of lateral suprasylvian region.

they injected small amounts of radioactive amino acids in individual laminae and showed autoradiographically that the A laminae project to layers III, IV, and VI of areas 17 and 18 where they form the well-known ocular dominance columns (LeVay *et al.*, 1978; Shatz and Stryker, 1978). The C laminae project to layers I and IV of areas 17, 18, 19, and several other areas of the lateral suprasylvian gyrus as well. In layer IV of areas 17 and 18 their terminals lie on each side of the two bands formed by the endings of X and Y axons arising in the A laminae (LeVay and Gilbert, 1976; Ferster and LeVay, 1978). This differential laminar pattern has recently been confirmed by the retrograde labeling method (Leventhal, 1979). Area 19 and many of the areas outside areas 17 and 18 have been shown to receive thalamic inputs also from components of the lateral posterior complex (Gilbert and Kelly, 1975). So far as is known these other inputs terminate in layer IV.

The differential projections of the laminar and medial interlaminar nuclei were confirmed by LeVay and Ferster (1977). As pointed out in a preceding section, these workers also showed morphological distinctions between X and Y cells and indicated that although X cells projected only to area 17, Y cells projected to areas 17 and 18. Nevertheless, they felt that individual Y cells did not have branched axons to both areas. In this, their results did not support the earlier studies of Garey and Powell (1967) and of Stone and Dreher (1973). However, the more recent work of Geisert (1980) who injected different tracers into areas 17 and 18 of the same animal suggests that a small proportion of cells in the A laminae (19%) may indeed have axon branches to both areas. In the C laminae, he found 50% of the projecting cells had branched axons. It was also LeVay and Ferster who first noted a small population of presumed morphological equivalents of X cells projecting to area 17 among the larger population of presumed Y cells in the medial interlaminar nucleus and both presumed X and Y cells projecting to area 17 in the large-celled lamina C. The presence of a small population of presumed X cells in the medial interlaminar nucleus projecting to area 17 was later confirmed by Meyer and Albus (1981).

In a retrograde labeling study that appeared in the same year as that of LeVay and Ferster but which has received relatively less attention, Holländer and Vanegas (1977) presented a slightly different view of the thalamocortical projection pattern. Following small injections of horseradish peroxidase in area 17 they noted that most retrograde labelled cells were in lamina A and the upper part of lamina A1. Comparable injections in area 18 led to labeling of cells mainly in laminae A1 and C. Injections in area 19 led to labeling mainly in the deeper C laminae and in the medial interlaminar nucleus. This pattern they discern, therefore, stresses localized concentrations of cells mostly projecting to a single area but not totally segregated in individual subnuclei. Perhaps this is the explanation for many discrepancies in the literature of thalamocortical connections in the cat visual system. As Rodieck (1979) points out, the organization elucidated by Holländer and Vanegas (1977) implies that area 17 receives its contralateral eye input from lamina A and its ipsilateral eye input from lamina A1, while area 18 receives its ipsilateral eye input from lamina A1, and its contralateral eye input primarily from lamina C (Fig. 9.29). It is as if the contralateral eye inputs to area 17 and 18 have acquired separate lateral geniculate laminae but that their ipsilateral eye inputs have failed to segregate.

The lateral suprasylvian area still remains something of an enigma and at present it would appear that all of its constituent areas receive input from the C laminae of the laminar dorsal lateral geniculate nucleus (Raczkowski and Rosenquist, 1980; Dreher et al., 1980; Tong et al., 1982) while two (the AMLS and PMLS areas) also receive fibers from the medial interlaminar nucleus.

9.4.2. The Monkey

From the earliest retrograde degeneration studies, the dorsal lateral geniculate nucleus of New World monkeys, Old World monkeys, apes, and man has been thought to project only upon area 17 of the cerebral cortex (Minkowski, 1920; Walker, 1936, 1938a,d; Walker and Fulton, 1938; Polyak, 1957). The pattern of retrograde degeneration in the lateral geniculate nucleus ensuing from small lesions of the striate area had by the 1920s been used to map out the visuotopic representation in the geniculate (Fig. 9.19).

It was some years after the introduction of modern axon degeneration methods before the monkey thalamocortical projection came to be studied but when studies were made they supported the older findings (Wilson and Cragg, 1967; Hubel and Wiesel, 1969, 1972; Garey and Powell, 1971). Axoplasmic transport methods, when they first began to be applied, also demonstrated a projection only to the striate area (Wiesel *et al.*, 1974; Hubel *et al.*, 1977; Kaas *et al.*, 1976; Tigges *et al.*, 1977b; Hendrickson *et al.*, 1978).

Emerging from a series of studies from several laboratories are remarkable new data on the pattern of ocular dominance columns in area 17 and on the differential levels of termination of axons arising in different laminae of the monkey dorsal lateral geniculate nucleus (see Hubel and Wiesel, 1977). In their first study utilizing the axon degeneration method, Hubel and Wiesel (1972) noted that axons arising from the parvocellular layers of the geniculate terminated in a thick band in layer IVcβ and in a thinner band in layer IVa of the striate area (Chapter 3). By contrast, axons from the magnocellular layers terminated just above the deeper terminations of fibers from the parvocellular layers, in layer IVcα. Lesions of parvo- but not of magnocellular layers also tended to give some degeneration in layer I. These results have been confirmed many times and in addition it has been shown that there is a small band of terminations probably arising from both magno- and parvocellular layers of the geniculate at the junction of layers V and VI of the cortex (Wiesel *et al.*, 1974; Glendenning *et al.*, 1976; Hubel *et al.*, 1977; Kaas *et al.*, 1976; Hendrickson *et al.*, 1978). The S laminae and the interlaminar regions are thought also to project to area 17 (Wong-Riley, 1974; Ogren and Hendrickson, 1976). These subsidiary regions outside and between the main laminae appear to project to layer III or other more superficial layers (Fitzpatrick *et al.*, 1983; Weber *et al.*, 1983). This therefore resembles the situation in the tree shrew and in prosimian primates (see Section 9.4.3).

The question of whether the monkey dorsal lateral geniculate nucleus projects outside area 17 has started to come up lately. For some time projections from the geniculate to area 18 have been alternately postulated (Wong-Riley, 1976) and denied (Benevento and Rezak, 1975; Ogren and Hendrickson, 1976; Hendrickson *et al.*, 1978). Those who have denied it have felt that the layer I projection arises in the pulvinar. On the other hand, there are now at least two anatomical reports of inputs from the lateral geniculate nucleus to extrastriate areas in the monkey (Benevento and Yoshida, 1981; Yukie and Iwai, 1981). One of these claims that cells retrogradely labeled in the lateral geniculate nucleus after injections of horseradish peroxidase involving the prestriate cortex lie primarily in the interlaminar regions (Yukie and Iwai, 1981). Norden and Kaas (1978) detected significant retrograde labeling in these regions only after large injections of area 17. Conceivably, as suggested by Diamond (see Section 9.3.2) here is the homolog of the cat C-laminae and W-cell projection system.

9.4.2.1. Ocular Dominance Columns

When examined with microelectrodes, cells throughout the vertical depth of the visual cortex at any one spot show a preference for all being driven by the same eye, left or right, and in layer IVc are driven solely by the eye preferred by the cells lying above and below them (Hubel and Wiesel, 1962, 1977). Such vertical arrays of neurons, each 300–500 μm wide in the horizontal dimension, are referred to as ocular dominance columns. Across the surface of the monkey visual cortex those related to each eye alternate. Their basis lies in the alternation of thalamocortical axon teminations arising from lateral geniculate laminae representing the left and right eyes (Fig. 9.30). Hubel and Wiesel (1969) first demonstrated this by making electrolytic lesions in individual laminae of the monkey dorsal lateral geniculate nucleus and staining the ensuing axonal degeneration in the striate area. In layer IVa and IVc, irrespective of whether the geniculate lamina lesioned was parvo- or magnocellular, the degeneration was in patches each about 300–500 μm wide and separated by a gap of equal width containing few or no degenerating axon terminations. Lesions involving two laminae, representing left and right eyes, led to a filling in of the gaps of degeneration. This confirmed the alternation of laminar and thus of left and right eye inputs to layer IV. Careful reconstruction of the patterns of degeneration from serial sections showed that each left eye or right eye patch was, in fact, part of an elongated strip of fairly constant width extending for a considerable distance through the horizontal dimension of layer IV. Each eye's input is represented by many such strips that alternate with strips of equal dimensions receiving input from the other eye. By the use of transneuronal autoradiography, it has now become possible to obtain more complete anatomical images of the strips than could be gained from the lesion material (Wiesel et al., 1974; Hubel et al., 1977). In this method a tritiated amino acid such as proline and/or a tritiated sugar such as fucose are injected in rather large amounts into an eye, and after 2–3 weeks labeled axoplasmically transported material crosses the geniculocortical synapses related to that eye and reaches the terminals of geniculocortical axons in the visual cortex. This serves to outline the layer IVa and layer IVc components of the ocular dominance strips related to the injected eye throughout their full extent. It also shows up similar though thinner strips in layer VI, immediately below each ocular dominance strip in layer IVc. The strips in layer IVc can now also be outlined in normal brains by a reduced silver stain apparently for myelinated axons (LeVay et al., 1975). Histochemical staining for cytochrome oxidase in monkeys subjected to enucleation of one eye (Fig. 9.30) will also reveal the ocular dominance strips in layer IVc as well as the additional strips and patches in layers III, IVa, and VI.

The ocular dominance strips, when labeled or stained in their entirety, approach the representation of the vertical meridian of the visual half field at

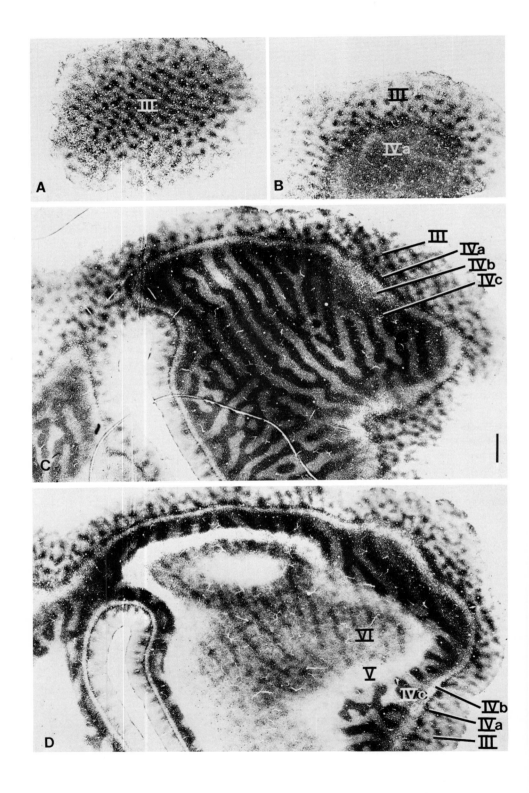

right angles (Fig. 9.31). Those related to the same eye can divide and reunite and, moreover, are capable of expanding into the territory of the other eye's strips if this eye is deprived of visual input during a critical phase of development (see Chapter 6). In the representation of the monocular segment of the contralateral temporal visual field, all thalamocortical input is from the contralateral eye and, thus, alternating strips are not seen (Fig. 9.32).

The ocular dominance strips in the normal, visually experienced animal are particularly rigidly organized, never varying in thickness across the cortex. They are, thus, in a sense independent of the projection of the visual field onto the cortex, for the number of degrees of visual field angle represented in 1 mm of the striate cortex (the reciprocal of the magnification factor) (Daniel and Whitteridge, 1961) (Chapter 3) increases markedly toward the representation of the periphery. It also appears that there is considerable overlap in the receptive fields of individual neurons in the visual cortex so that as much as 1 mm of cortex must be traversed horizontally before neurons are encountered whose receptive fields do not overlap those 1 mm away (Hubel and Wiesel, 1974). At the borders of an ocular dominance strip, on the other hand, there is a sharp cutoff between neurons related to the left or right eye (Hubel and Wiesel, 1974). Hubel and Wiesel refer to each pair of left and right eye ocular dominance strips, as a "hypercolumn." These have a combined width approaching 1 mm and, as pointed out by Hubel and Wiesel, contain all of the circuitry necessary to analyze a single small region of the binocular visual field.

The columnarity of ocular dominance throughout the thickness of the cortex is thought to be carried from the ocular dominance strips of layer IV by the vertical axons of cells upon which the thalamic axons end, making synaptic contacts with cells above and below the layers of thalamic terminations (e.g., Gilbert and Wiesel, 1979). The vertical flow of activity following stimulation of one eye is made particularly clear in [^{14}C]-2-deoxy-D-glucose autoradiographs (Kennedy *et al.,* 1975) (Fig. 9.32). Here, the glucose analog, given before or during a period of eye stimulation or after removal of one eye, accumulates in the cortical neurons activated by the stimulated eye.

9.4.2.2. Monkeys without Ocular Dominance Strips

Ocular dominance strips have been reported from physiological studies of Old World macaque monkeys and of a single species of New World monkey, the spider monkey *(Ateles)* (Hubel and Wiesel, 1968). The anatomical correlates

---

FIGURE 9.30. Sections cut more or less parallel to the surface of the striate cortex in a macaque monkey from which one eye had been removed 2 weeks before sacrifice. The sections were reacted for cytochrome oxidase activity. (A) and (B) show rows of punctate periodicities in layer III; (C) shows heavily stained ocular dominance strips related to the remaining eye and pale strips related to the removed eye in layer IVc; surrounding these can be seen the unstained layer IVb and the stained layer IVa; (D), mainly through layer VI, shows alternating rows of thick and thin periodicities related to the intact and removed eye, respectively. All cytochrome oxidase-stained layers and patches receive terminations of thalamocortical fibers (see Fig. 3.8A). Periodicities in layers III and VI are aligned with centers of ocular dominance strips in layers IVa and IVc. Bar = 1 mm.

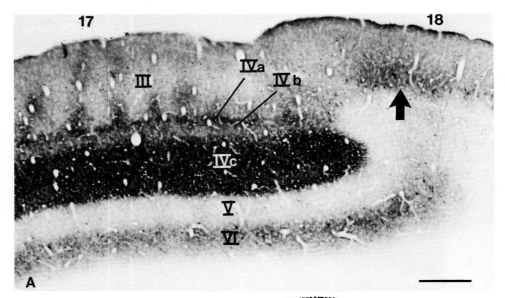

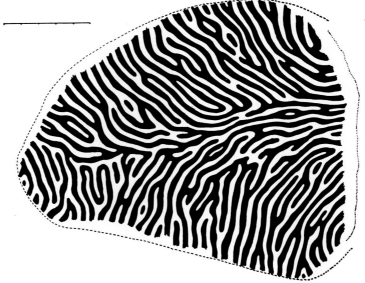

FIGURE 9.31. (A) Cytochrome oxidase-positive periodicities in a conventional section through areas 17 and 18 of a monkey with both eyes intact. Note larger stained and unstained regions in area 18. (B) A reconstruction of the layer IVc ocular dominance strips; black strips are related to one eye and white to the other. All strips intersect the representation of the vertical meridian of the visual field at right angles (to right). Bars = 500 μm (A), 5 mm (B). From Hubel and Freeman (1977).

FIGURE 9.32. [^{14}C]-2-deoxy-D-glucose autoradiographs from a normal monkey (above), from a monkey from which one eye had been removed (bottom) and from a monkey in which both eyes had been occluded (middle). Ocular dominance columns are evident in the lowest figure except in the representation of the peripheral, monocular part of the temporal visual field (arrow). Elsewhere the columns related to the seeing eye extend virtually through all layers of the cortex, indicating the vertical flow of intracortical activity. From Kennedy *et al.* (1976).

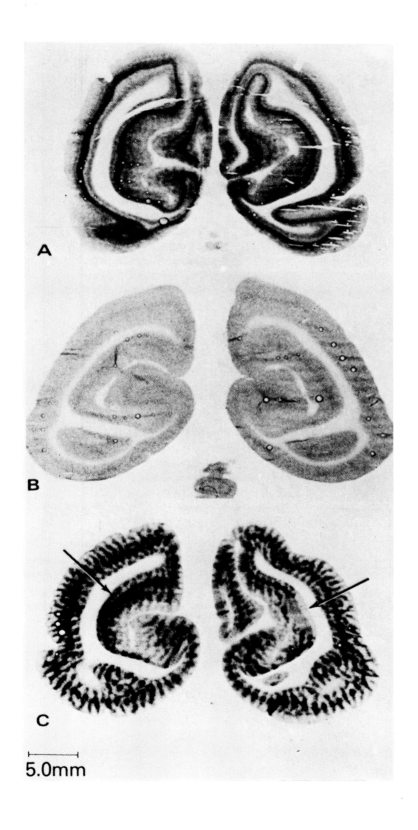

5.0mm

of the columns have been demonstrated by transneuronal autoradiography in both genera, and in other Old World monkeys and apes (Kaas *et al.*, 1976; Tigges *et al.*, 1977a; Hendrickson *et al.*, 1978; Florence and Casagrande, 1978) and by direct labeling of thalamocortical axons arising from single geniculate laminae in macaques (Hendrickson *et al.*, 1978). Ocular dominance strips, however, cannot be demonstrated anatomically in species of New World monkeys other than *Ateles* (e.g., the squirrel monkey, capuchin monkey, and owl monkey), though homogeneous laminae of labeled axon terminations can be demonstrated by transsynaptic autoradiography in layers IVa, IVc, and VI (Hubel *et al.*, 1977; Kaas *et al.*, 1976; Tigges *et al.*, 1977b; Hendrickson *et al.*, 1978). From the physiological point of view, ocular dominance columns have also not been described in New World monkeys other than *Ateles*. For a time, it appeared that the absence of segregated transneuronal labeling in the visual cortex of these other New World monkeys following an injection of isotope in one eye could have resulted from the lack of interlaminar plexuses in the dorsal lateral geniculate nucleus and the consequent spread of material leaking from ganglion cell axon terminals in laminae related to the injected eye into those related to the other eye. This explanation, however, has generally been discounted because transneuronal labeling of ocular dominance columns occurs in genera of Old World monkeys such as *Erythrocebus* and *Cercopithecus* in which the parvocellular geniculate layers are not well differentiated. Conversely, failure to label the thalamocortical input in ocular dominance strips still occurs in the New World owl monkey *(Aotus)* in which lamination of the geniculate is evident (Hendrickson *et al.*, 1978).

The presence of ocular dominance strips in at least one genus of New World monkeys makes unlikely a simple evolutionary explanation of their absence in others based upon early isolation of New World and Old World genera. The appearance of the strips in some but not other representatives of quite unrelated orders such as the cat (Shatz *et al.*, 1977; Shatz and Stryker, 1978; LeVay *et al.*, 1978) suggests that clues to the basis for their presence may be found in behavioral attributes but no plausible ideas have yet been advanced. As pointed out by Hendrickson *et al.* (1978) and others, these clues are unlikely to be found in the morphology of the retina, for some animals with anatomically demonstrable ocular dominance columns have mainly cone retinae and obvious foveae while others have mainly rod retinae and no foveae.

9.4.3. Geniculocortical Projections in Other Species

On the whole, geniculocortical projections in other species resemble those of the monkey more than they do those of the cat (Fig. 9.33).

Possibly the most convenient way to think of the visual cortex of the lower primates, of insectivores, rodents and lagomorphs, marsupials, and possibly of other orders as well, is in terms of the "core and belt" concept advanced by Diamond and his associates (Hall and Diamond, 1968; Diamond and Hall, 1969; Diamond *et al.*, 1970; Kaas *et al.*, 1970; Diamond, 1979). Here, the equivalent of the striate area, containing a full representation of the contralateral half visual field, is regarded as the "core" and receives a heavy geniculocortical input ending

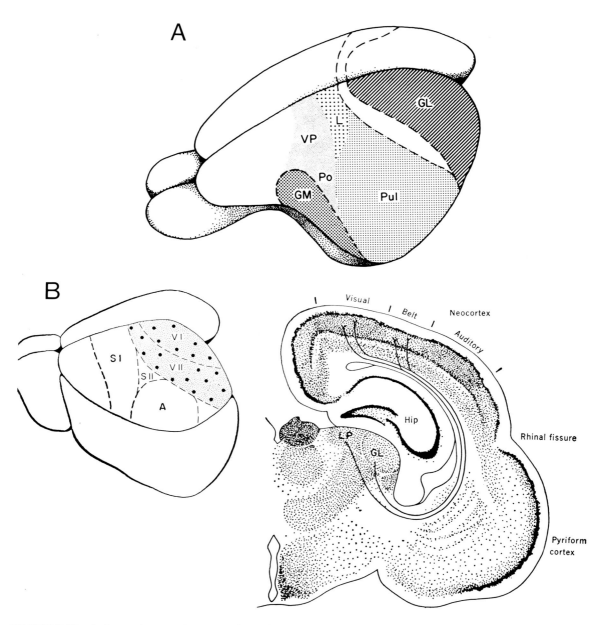

FIGURE 9.33.   Striate and extrastriate cortex in two insectivores, the tree shrew (A) and hedgehog (B). Peristriate belt is white in (A) and labeled "belt" and VII in (B). Auditory cortex also indicated by A or GM; temporal cortex indicated by Pul, parietal by L, and somatic sensory cortex by SI, SII, VP, and Po. (A) from Diamond *et al.* (1970); (B) from Diamond and Hall (1969).

predominately in layer IV. Around this, especially anteriorly, laterally, and posteriorly, is a cytoarchitectonically distinct "belt" containing at least one complete visuotopic representation, usually reversing at the border with the core. Other representations of the half visual field, sometimes associated with cytoarchitectonically definable cortical fields, can lie outside the core and belt, commonly anteriorly and laterally (Diamond *et al.,* 1970; Allman and Kaas, 1971, 1974a,b, 1976; Kaas *et al.,* 1972a,b). These have generally been shown to receive their thalamocortical inputs from nuclei of the lateral posterior–pulvinar complex, but controversy has often revolved around the question of whether the "belt" region receives its thalamic input from the lateral posterior–pulvinar complex or from the dorsal lateral geniculate nucleus itself and there are reports based upon different techniques favoring either position (Diamond and Utley, 1963; Rose and Malis, 1965; Hall and Diamond, 1968; Diamond *et al.,* 1970; Benevento and Ebner, 1971; Dräger, 1974; Hubel, 1975; Ribak and Peters, 1975; Hughes, 1977; Karamanlidis and Giolli, 1977; Weber *et al.,* 1977; Gould *et al.,* 1978; Dürsteler *et al.,* 1979; Karamanlidis *et al.,* 1979; Holländer and Hälbig, 1980; Haight *et al.,* 1980; Coleman and Clerici, 1980, 1981; Towns *et al.,* 1982). Geniculocortical fibers in rodents terminate in layers I, IV, and VI. Differential origins of the fibers terminating at each level are not known.

One nonprimate that again deserves mention is the tree shrew which shows a pattern of laminar projection to the cortex having many parallels with that of monkeys, though its cortex lacks ocular dominance columns (Hubel, 1975; Casagrande and Harting, 1975; Humphrey *et al.,* 1977). Despite this lack, there is a very clear dissociation of inputs derived from the different laminae of the

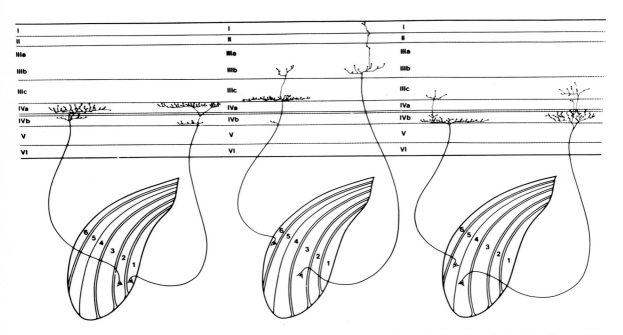

FIGURE 9.34. Schematic figure indicating the differential levels of intracortical termination of geniculostriate fibers arising in the different layers of the tree shrew dorsal lateral geniculate nucleus. From Conley *et al.* (1984).

lateral geniculate nucleus. Fibers emanating from geniculate laminae innervated by the contralateral eye (2, 3, 4, and 6) terminate throughout the thickness of layer IV; those emanating from the laminae innervated by the ipsilateral eye (1 and 5) terminate at the superficial and deep margins of layer IV, leaving a zone in between that is therefore innervated only from the contralateral eye (Fig. 9.34). Fibers arising from cells in the interlaminar zones of the geniculate terminate in layer III and in layer I; there is an additional band of termination, probably derived from all laminae, in layer VI (Hubel, 1975; Casagrande and Harting, 1975; Carey *et al.*, 1979; Fitzpatrick *et al.*, 1980). There are suggestions of a comparable pattern of laminar segregation in the gray squirrel (Harting and Huerta, 1983).

9.5. Ventral Lateral Geniculate Nucleus

The ventral lateral geniculate nucleus of nonprimates and its primate analog, the pregeniculate nucleus, are not parts of the dorsal thalamus and are described in Chapter 15 with other components of the ventral thalamus.

Where like a pillow on a bed,
A pregnant banke swel'd up, to rest
The violets reclining head,
Sat we too, one anothers best.
Our hands were firmely cimented
With a fast balme, whence did spring,
Our eye-beames twisted, and did thred
Our eyes, upon one double string;
So to entergraft our hands, as yet
Was all the means to make us one,
And pictures in our eyes to get
Was all our propagation.
                    John Donne, *The Extasie*

Lateral Posterior and Pulvinar Nuclei

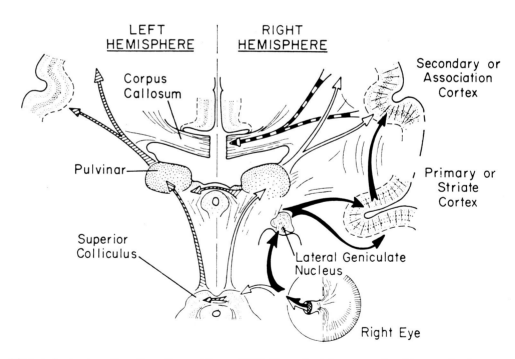

The two visual systems, from Trevarthen and Sperry (1973). Note callosal, interpretectal, and intertectal pathways linking the systems of the two sides.

10.1. Description

The lateral posterior and pulvinar nuclei can be regarded as parts of a larger nuclear complex. One reason for bringing the two together is that undoubtedly equivalent nuclei have been regarded as part of the lateral posterior nucleus in some species and as parts of a pulvinar in others. The cat, for example, is customarily said to have a lateral posterior nucleus divided into three or more subnuclei and a single pulvinar nucleus (e.g., Berman and Jones, 1982). Monkeys, on the other hand, are usually said·to have a single lateral posterior nucleus and at least four pulvinar nuclei (e.g., Olszewski, 1952). Apart from this idiosyncrasy, however, the lateral posterior and pulvinar nuclei are closely associated topographically and have many morphological and connectional similarities.

10.1.1. Higher Primates

10.1.1.1. Lateral Posterior Nucleus

In monkeys, apes, and man, the lateral posterior nucleus is relatively small and consists of a homogeneous population of moderately densely packed, pale staining, medium-sized cells (Figs. 10.1 and 10.2). The nucleus lies on the dorsal surface of the ventral posterior nucleus against the external medullary lamina, overlain by the backwards-curving tails of the posterior ventral lateral (VLp) and lateral dorsal (LD) nuclei. Posterior to these tails, it comes to the dorsal surface of the thalamus and then ends by merging with the medial and lateral pulvinar nuclei.

10.1.1.2. Pulvinar Nuclei

Distinct medial, lateral, inferior, and anterior pulvinar nuclei are found within the large posterior protrusion of the lateral thalamic mass called "the pulvinar" in man, apes, and most monkeys. The medial pulvinar nucleus is dorsomedially situated, forms the dorsal and most of the posterior surface of the pulvinar mass, and is the region of the thalamus which, apart from the intervening limitans–suprageniculate nucleus, fuses with the anterior portion of the midbrain (Fig. 10.2). Its cells are relatively small, dispersed, and pale staining.

531

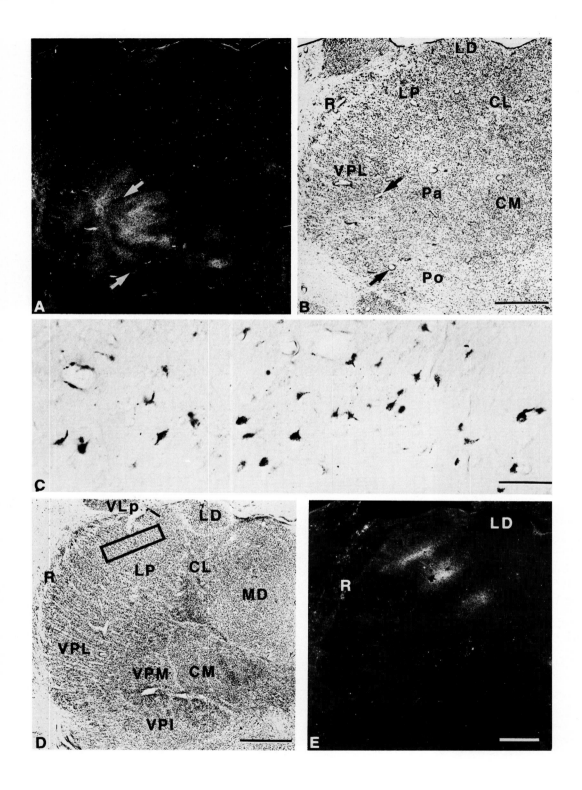

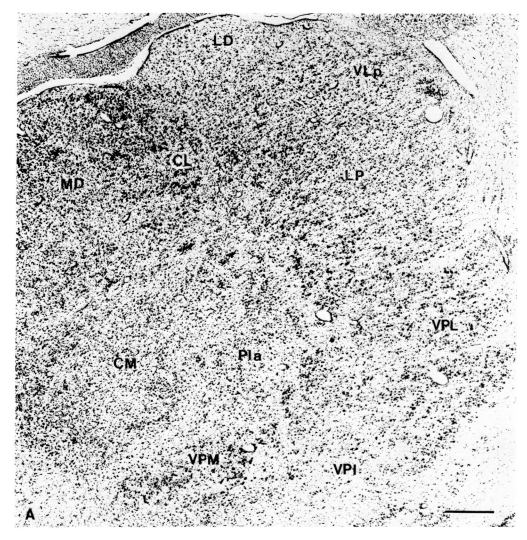

FIGURE 10.2. A series of thionin-stained frontal sections in anterior–posterior order through the thalamus of a rhesus monkey showing the lateral posterior and anterior pulvinar nuclei (A); the anterior pulvinar nucleus (B); the anterior, lateral, and inferior pulvinar nuclei (C); the medial, lateral, and inferior pulvinar nuclei (D); and the medial and lateral pulvinar nuclei (E). Bars = 500 μm. CTT: corticotectal tract. (B, D) From Burton and Jones (1976).

FIGURE 10.1. (A, B) Darkfield and brightfield photomicrographs of the same section, showing distribution of autoradiographically labeled corticothalamic fibers in the VPL and anterior pulvinar (Pa) nuclei of a cynomolgus monkey, following spread of an injection of tritiated amino acids from area 2 into the anterior part of area 5. (C) Retrogradely labeled cells from part of the lateral posterior nucleus outlined by box in (D), following injection of horseradish peroxidase in posterior part of area 5 in a rhesus monkey. (E) Anterograde labeling of corticothalamic fiber ramifications in lateral posterior nucleus of a cynomolgus monkey that received an injection of tritiated amino acids in the posterior part of area 5. Bars = 1 mm (A, B, D, E), 100 μm (C). All from Jones et al. (1979).

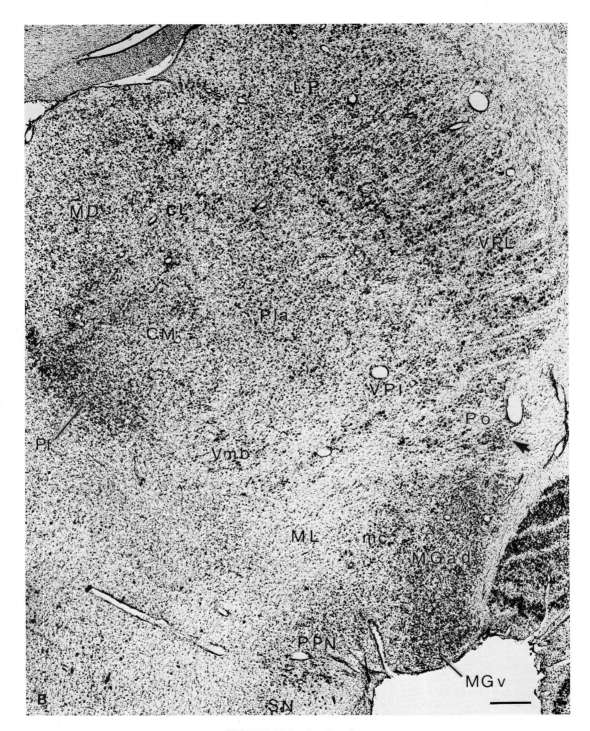

FIGURE 10.2. *(continued)*

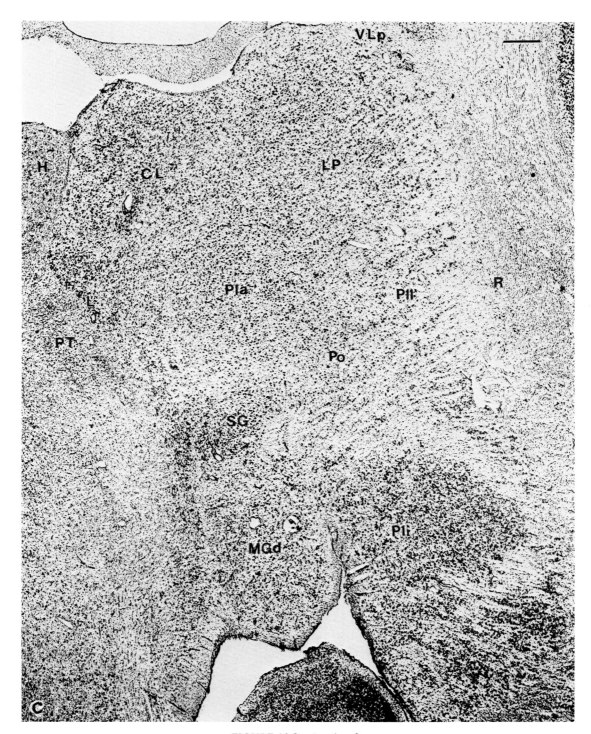

FIGURE 10.2.   (*continued*)

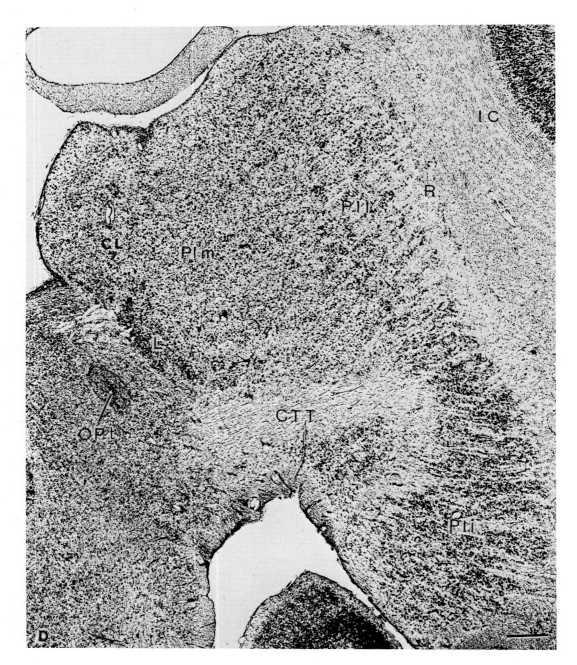

FIGURE 10.2. *(continued)*

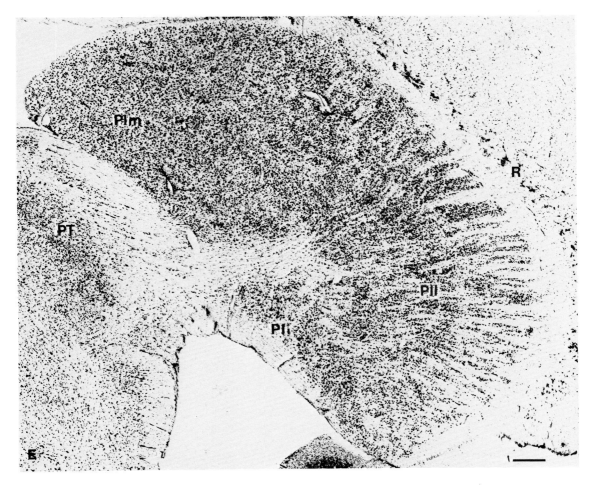

FIGURE 10.2. (*continued*)

The lateral pulvinar nucleus is dorsolaterally situated, adjacent to the external medullary lamina and posterior to the ventral posterior and lateral posterior nuclei. Its cells are similar in size to those of the medial pulvinar nucleus but are more darkly staining and broken up into linear arrays by the bundles of fibers of the so-called corticotectal tract that traverse the nucleus (Fig. 10.2).

The inferior pulvinar nucleus hangs from the ventral aspect of the lateral pulvinar nucleus, forming a conspicuous mass just lateral to the posterior part of the medial geniculate complex; it insinuates itself anteriorly between the medial and lateral geniculate nuclei (Fig. 10.2). Its cells are the darkest staining and most closely packed of all the pulvinar nuclei.

The anterior pulvinar nucleus extends from the medial pulvinar and posterior nuclei, anteriorly between the centre médian and ventral posterior nuclei. Over much of its extent it is separated from the ventral posterior medial (VPM) nucleus by a thin fiber lamina. Its cells are the smallest, palest staining, and most widely dispersed of all the pulvinar nuclei (Figs. 10.1 and 10.2).

10.1.2. Lower Primates

A lateral posterior nucleus comparable to that in monkeys is readily detectable in prosimian brains. Extending posteriorly from the lateral posterior nucleus between the lateral geniculate nucleus and the midbrain, is a large, bulbous protrusion referred to as the pulvinar (Fig. 10.3). It has the same relative disposition with regard to the lateral ventricle, medial medullary lamina, limitans–suprageniculate and posterior nuclei, as the pulvinar mass in higher primates (Section 10.1.1). However, it has a fairly homogeneous architecture and it is difficult to detect finer nuclear subdivisions within it (Fig. 10.3). Diamond and his school (Glendenning *et al.*, 1975; Carey *et al.*, 1979; Fitzpatrick *et al.*, 1980; Symonds and Kaas, 1978; Raczkowski and Diamond, 1981) have divided the pulvinar in the prosimians into dorsal ("superior") and ventral ("inferior") nuclei primarily on the basis of the visuotopically organized connections that these dorsal and ventral regions have with the superior colliculus and cerebral cortex (Sections 10.2.1 and 10.2.2). Armed with this knowledge, it may be possible to visualize two divisions in well-stained Nissl sections: the dorsal division has slightly more dispersed cells than the ventral (Fig. 10.3) and they are separated by an incomplete fiber lamina.

10.1.3. Cat

10.1.3.1. Lateral Posterior Complex

The lateral posterior complex of the cat is large and dominates the dorsal half of the lateral thalamic mass (Fig. 10.4). The complex commences near the anterior pole of the ventral posterior nucleus and extends posteriorly between the ventral posterior and lateral dorsal nuclei and dorsally between the central lateral and pulvinar nuclei, to reach the dorsal surface of the thalamus; it then invades the dorsal aspect of the medial geniculate body between the pretectum, dorsal lateral geniculate nucleus, and dorsal medial geniculate nucleus. This large mass of cells can be divided, though not without difficulty, into three constituent nuclei which are approximately in the form of parallel slabs running in horizontal section from anterolateral to posteromedial and in frontal section from dorsomedial to ventrolateral (Berman and Jones, 1982). The *lateral posterior medial nucleus* (LPm) is the anteromedial slab and forms most of the anterior pole of the lateral posterior complex. It consists of small pale-staining, closely packed cells and stains heavily for acetylcholinesterase.

The *lateral posterior intermediate nucleus* (LPi) lies lateral and posterior to the LPm and reaches to the posterior pole of the thalamus adjacent to the anterior pretectal nucleus. It is composed of medium-sized, moderate-staining, dispersed cells and stains moderately for acetylcholinesterase.

The *lateral posterior lateral nucleus* (LPl) is lateral and posterior to the LPi and extends along the medial border of the pulvinar nucleus, reaching from the medial interlaminar nucleus of the dorsal lateral geniculate complex, pos-

FIGURE 10.3. The dorsal and ventral pulvinar nuclei of a bush baby (*Galago crassicaudatus*). (A) Thionin stain; (B) hematoxylin stain. Bar = 500 μm.

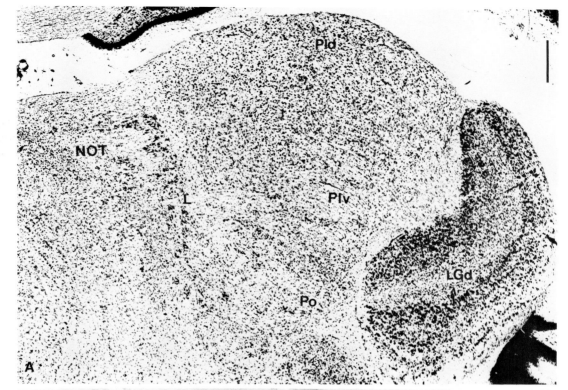

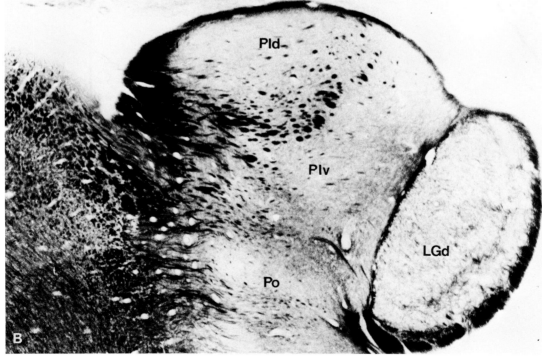

teriorly into the dorsal aspect of the medial geniculate body. Its cells are medium-sized, moderately stained, and relatively densely packed. It stains weakly for acetylcholinesterase but is more uniformly myelinated than the LPm, LPi, or pulvinar nuclei.

10.1.3.2. Pulvinar Nucleus

The pulvinar nucleus of the cat is a single homogeneous entity composed of small- to medium-sized cells and streaked by the fibers of the optic tract and dorsal thalamic radiation that traverse it. It forms a fourth slab, in parallel with

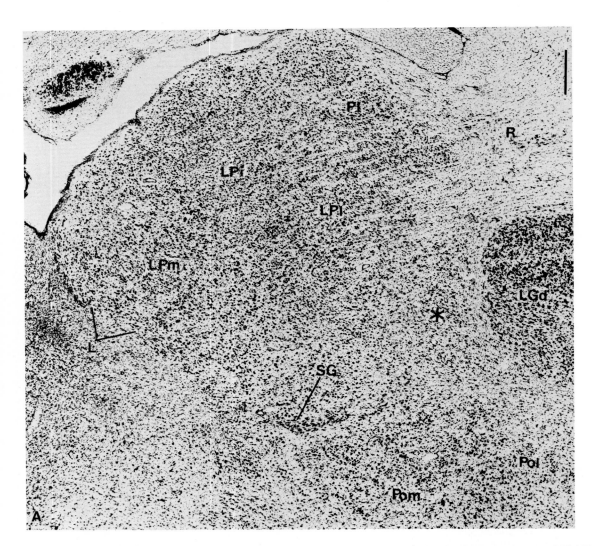

FIGURE 10.4.   Thionin-stained (A) and hematoxylin-stained (B) frontal sections showing the three components (LPl, LPi, and LPm) of the lateral posterior nucleus and the pulvinar nucleus (Pl) in a cat. Asterisk indicates part of LPl nucleus formerly called "posterior nucleus of Rioch." Bars = 500 μm.

and dorsolateral to the slabs formed by the three lateral posterior nuclei and for most of its extent lying dorsomedial to the dorsal lateral geniculate nucleus. It stains heavily for acetylcholinesterase.

10.1.4. Rodents and Other Species

In most rodents, the lateral posterior nucleus is a single homogeneous entity in the dorsal aspect of the thalamus, intercalated between the dorsal lateral geniculate nucleus, medial region of the posterior nucleus (Pom), and the central lateral nucleus. It extends posteriorly between the pretectum and the dorsal medial geniculate nucleus. It stains well for acetylcholinesterase anteriorly but poorly posteriorly. No pulvinar nucleus can be detected (Fig. 10.5A).

In lagomorphs and ungulates, the appearance is rather similar to that in rodents though there is some evidence for a separation of the lateral posterior nucleus into dorsal and ventral parts. No pulvinar nucleus is seen.

In insectivores, such as *Tupaia* (Fig. 10.6), there is a large lateral posterior nucleus lying along the medial surface of the highly developed, dorsal lateral

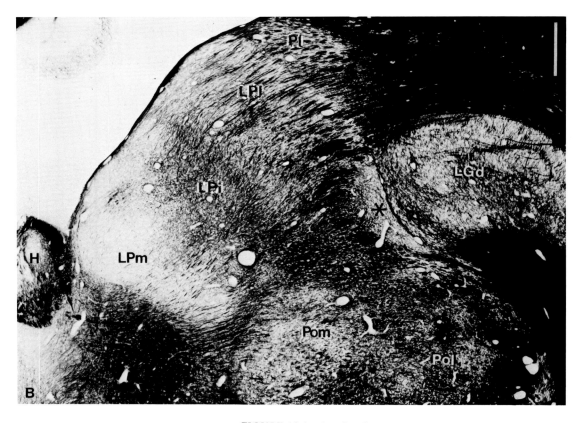

FIGURE 10.4. (*continued*)

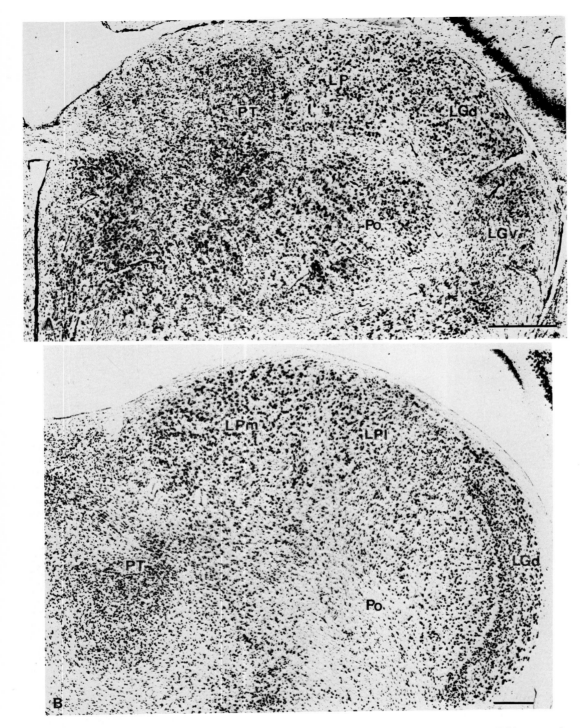

FIGURE 10.5. Frontal sections showing the lateral posterior nuclei of a rat (A), a marsupial phalanger (*Trichosurus vulpecula*) (B), and a guinea pig (C). Thionin stain. Bars = 500 μm.

geniculate nucleus. The cells of the lateral posterior nucleus are small, pale-staining, and dispersed and the nucleus is incompletely divided into dorsal and ventral divisions, not unlike those of the lower primate pulvinar.

In marsupials, some rodents, and some bats, the lateral posterior nucleus is relatively large and quite distinct, and is commonly incompletely divided into medial and lateral subnuclei (Figs. 10.5B,C).

10.2. Terminology

10.2.1. Lateral Posterior Nucleus

Lateral posterior nucleus ("lateralen hinteren Kern") is Nissl's (1889, 1913) term, applied in his study of the rabbit thalamus to a region equivalent to that which is recognized today. The latinized form of the name appears to have first entered the literature in Münzer and Wiener's (1902) description of the rabbit thalamus, though they included part of the nucleus in the medial geniculate body. Lateralis posterior was applied to a more accurately delineated nucleus in

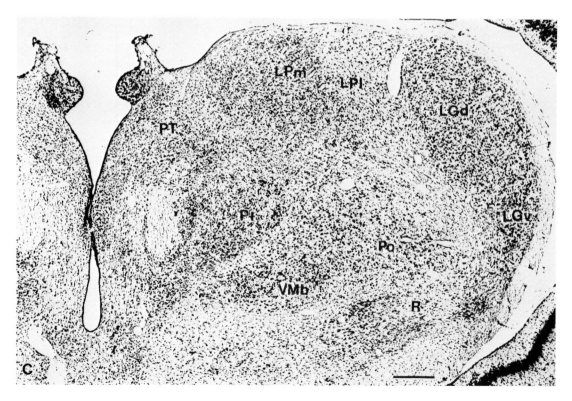

FIGURE 10.5. (*continued*)

the rabbit brain by D'Hollander (1913). From his work it appears to have been adopted for equivalent nuclei in the rat by Gurdjian (1927), in the cat by Rioch (1929a), and in lower primates and insectivores by Le Gros Clark (1932a). Le Gros Clark (1928, 1929a,b, 1930a) and other authors had previously used the term *nucleus lateralis b*, applied by von Monakow (1895) to the cat lateral posterior nucleus.

In the first descriptions of the monkey thalamus, Vogt (1909) and Friedemann (1911) also adopted von Monakow's term for the lateral posterior nucleus but called it *lateral ventral nucleus* (as opposed to the lateral dorsal nucleus), and subdivided it into several parts. Crouch (1934) continued this practice in monkeys but Walker (1937, 1938a) felt that the differences were insignificant and the habit of subdividing the monkey lateral posterior nucleus ended there. In man, Hassler (1959), who calls the lateral posterior nucleus *nucleus dorso-intermedius*, continues to recognize several incomplete subdivisions.

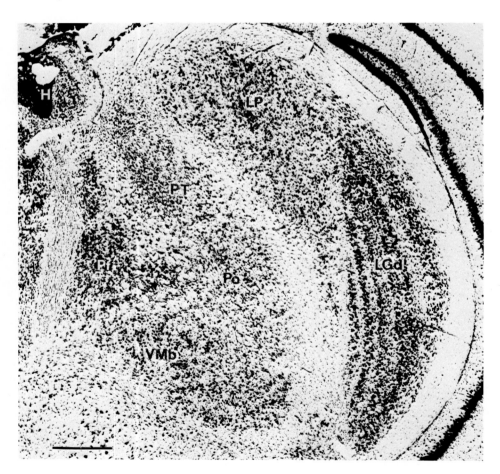

FIGURE 10.6.   Thionin-stained frontal section showing the lateral posterior nucleus of a tree shrew (*Tupaia glis*). From a preparation of Dr. M. Conley. Bar = 500 μm.

The lateral posterior nucleus of the cat and dog was divided into several parts by Rioch (1929a) who clearly recognized some of the regional variations in cytoarchitecture described in Section 10.1.3. Some hints of these subdivisions can be seen in the older atlas of the cat by Winkler and Potter (1914). Rioch's nucleus lateralis posterior intermedius corresponds in part to the present LPm, but also includes elements of the present LPl. His nucleis lateralis posterior dorsalis and lateralis posterior medialis are primarily the present LPi. Rioch's nucleus lateralis posterior ventralis is the Pom region of the posterior nucleus. Rioch's (1929a, 1931) nucleus posterior, originally illustrated by von Monakow (1895), is now regarded as part of the LPl (Section 10.3.1). Part of the lateral posterior nucleus was originally included in the pretectum by Rioch (1929a) though he later rescinded this position (1931).

Ingram *et al.* (1932) made a bipartite division of the cat lateral posterior nucleus but subsequently the nucleus came to be regarded as a single entity (Jimenez-Castellanos, 1949; Jasper and Ajmone-Marsan, 1954). It was only when the connections of the nucleus started to be examined that the tripartite division given in the present account started to become clear and the finer architectonic parcellation thus acquired some relevance (Graybiel, 1970, 1974b, 1977; Graham, 1977; N. Berman and Jones, 1977; Updyke, 1977, 1983; A. L. Berman and Jones, 1982). To this has now been added the correspondence of the subnuclei of the cat lateral posterior nucleus with different patterns of acetylcholinesterase staining made by Graybiel and Berson (1980b) who relate this to subnuclei named according to Rioch rather than to those to the present account which are derived from Updyke (1977) and Berman and Jones (1982).

Comparing the subdivided lateral posterior nucleus and single pulvinar nucleus of the cat with the single lateral posterior nucleus and subdivided pulvinar of the primate, one can readily imagine that one or more nuclei comparable to those of the monkey pulvinar may lie buried in the cat lateral posterior nucleus. Niimi and Kuwahara (1973) have attempted to resolve the discrepancy between the terminology of carnivores and primates by applying terms derived from the pulvinar nuclei of monkeys to what they regard as the equivalent nuclei in the cat (see Fig. 10.8). My own preference is to wait until more data are available on the connections of the nuclei of the primate pulvinar before changing the cat nomenclature.

Though the lateral posterior nucleus of some small mammals such as marsupials and some rodents (e.g., the guinea pig, Fig. 10.5C) is sometimes divisible into medial and lateral parts (Goldby, 1941; Oswaldo-Cruz and Rocha-Miranda, 1967), there is little evidence to enable a correlation to be made between these divisions and those of the carnivore lateral posterior nucleus or primate pulvinar.

10.2.2. Pulvinar

Pulvinar is Burdach's (1822) term, though it is not clear from his writings whether he regarded it as a thalamic nucleus comparable to the anterior, medial, and lateral nuclei that he contemporaneously described, or simply as the protrusion from the posterior surface of the human thalamus. In 1897, van Gehuchten still did not regard it as a nucleus in his account of the human thalamus. By 1901, however, Dejerine was referring to the pulvinar as a fourth thalamic nucleus.

The name did not appear in Nissl's (1889, 1913) first description of thalamic cytoarchitecture in the rabbit and the first use in animals appears to be by von Monakow (1895) and Probst (1903) in their accounts of experimental work on the brains of cats and dogs. The same nucleus was later designated *pulvinar* in the cat and dog by Rioch (1929a) and the name remains.

After Probst, the next usage of *pulvinar* in animals seems to have come in the accounts of the monkey thalamus given by Vogt (1909) and Friedemann (1911). Vogt, in myelin preparations, recognized medial and lateral regions of the pulvinar, the medial divided into four nuclei and the lateral into five, all of which she designated by Greek letters. Most of them were later identified in Nissl preparations from monkeys by Friedemann and in Nissl and myelin preparations from lemurs by Pines (1927). In these subdivisions can be detected the four pulvinar nuclei that we recognize today. The identification of medial, lateral, and inferior pulvinar nuclei in the monkey dates to Crouch (1934) who seems to have included the anterior pulvinar nucleus in the medial pulvinar nucleus. Crouch's divisions were accepted by Walker (1937, 1938a–c) for the monkey and chimpanzee. The anterior pulvinar nucleus can be seen in Walker's drawings of the monkey thalamus but he does not label it. Aronson and Papez (1934) call it the *lateral posterior nucleus*. Their subdivisions of the rest of the pulvinar bear little relationship to those of other workers.

Publication of the atlas of the rhesus monkey brain by Olszewski (1952) reaffirmed the identity of medial, lateral, and inferior pulvinar nuclei and Olszewski also clearly delineated what he called the *oral pulvinar nucleus*. This is in the region previously referred to as nucleus P,mu,beta by Friedemann (1911) and Pines (1927) and as an oral prolongation of the human pulvinar by Vogt and Vogt (1941). The term *oral pulvinar nucleus* was changed in the interests of terminological consistency to *anterior pulvinar nucleus* by Burton and Jones (1976). The pulvinar of New World monkeys appears virtually identical to that of the macaque (Emmers and Akert, 1963). Hassler's (1959) account of the human pulvinar is also basically the same as Olszewski's (1952) though Hassler subdivides most of the nuclei several times.

Application of the term *pulvinar* to the thalami of lower primates started with the work of Pines (1927) on a species of lemur. In this animal, Pines felt that he could detect most of the subdivisions of the pulvinar earlier described by Vogt (1909) and Friedemann (1911). Le Gros Clark (1930a) in *Tarsius* recognized that the configuration of the posterior part of the lateral nuclear region was not greatly different from that described by Rioch (1929a) in the cat. Indeed, Le Gros Clark mentions four parallel slabs in *Tarsius* very similar to those described in the cat LP in the present account. In the insectivore *Tupaia*, with a very small pulvinar nucleus, Le Gros Clark (1929a) saw an intermediate stage between *Tarsius* and the cat. From observations such as these, he was led to believe (1932a) that a pulvinar nucleus could be detected in most mammals provided that the investigator could rid himself of the belief that a pulvinar was a unique, higher primate characteristic. Le Gros Clark's point of view was adopted by a number of other workers, e.g., Papez (1932), and it became more or less accepted that the lateral posterior nucleus of animals such as the rat incorporated elements of the pulvinar. It is considerations such as this that led Rose (1942b) to refer to the whole lateral posterior complex of the sheep, rabbit, and other species as pulvinar.*

Since Rose's paper, application of the term *pulvinar* to nonprimates other than the cat has declined though one still sees it used in the tree shrew, *Tupaia* (Carey *et al.*, 1979), and squirrel (Robson and Hall, 1977). On the whole, *lateral posterior nucleus* has become the preferred term for the posterior part of the lateral nuclear mass. *Pulvinar nucleus* remains as an idiosyncratic term for a part of the carnivore thalamus that is probably equivalent to only one of the several pulvinar nuclei of primates, the others being represented by components of the lateral posterior complex (Section 10.2.1). There seems little reason, therefore, for reintroducing the term in other nonprimates.

There has been some revival of interest in the pulvinar of prosimian primates in recent years as the result of studies by Diamond and his associates on prosimians and on the tree shrew, *Tupaia,* which though an insectivore, has a lateral posterior–pulvinar region somewhat similar to that of the prosimians, as mentioned above (Harting *et al.*, 1973; Glendenning *et al.*, 1975; Raczkowski and Diamond, 1978, 1981; Carey *et al.*, 1979; Fitzpatrick *et al.*, 1980; see also Symonds and Kaas, 1978). Primarily on the basis of afferent connections, these workers define a dorsal (their superior) and a ventral (their inferior) pulvinar nucleus in the prosimians and a pulvinar nucleus (situated dorsally) and a lateral intermediate nucleus (situated ventrally) in the tree shrew. Carey *et al.* confess that the naming across species may not be consistent and that the tupaiid pulvinar may be more appropriately considered the lateral posterior nucleus and the lateral intermediate nucleus the pulvinar. Symonds and Kaas (1978) have discerned finer connectional subdivisions of the two pulvinar nuclei of the bush baby that do not seem matched by clear cytoarchitectonic parcellations. In these, they see certain similarities to the pattern of connections in the pulvinar nuclei of a New World monkey, *Aotus,* and have thus divided the pulvinar of this monkey into dorsal and ventral divisions comparable to those of the prosimian, rather than in the manner that has become more conventional for the anthropoid pulvinar.

10.3. Connections

In this section, unlike in other chapters, I have chosen to consider afferent and efferent connectivity together, partly because the connections in many instances provide the basis for subdividing the nuclei and partly because a consideration of the inputs and outputs together brings more potential functional relevance to the nuclear parcellations.

---

* Rose's pars geniculata pulvinaris is the medial interlaminar nucleus of the lateral geniculate complex (Chapter 9).

10.3.1.   Lateral Posterior and Pulvinar Nuclei in the Cat

It seems appropriate to commence this discussion of afferent connectivity with the cat as this species has been the most thoroughly studied and in the studies that have been conducted on it certain organizational principles have emerged that may provide a basis for looking at other less well-studied species.

10.3.1.1.   Background

Early studies conducted with the retrograde degeneration technique told us little about the cortical relationships of the cat lateral posterior nucleus, other than that it was connected with the lateral and suprasylvian gyri (Waller and Barris, 1937). During the middle years of the era of the Nauta technique, the cat lateral posterior complex occupied a rather unusual position, for the primary visual cortical areas, areas 17 and 18, were shown to project to it as well as to the dorsal lateral geniculate nucleus (Beresford, 1961; Altman, 1962; Guillery, 1967; Garey *et al.*, 1968). Because it was then believed that only the dorsal lateral geniculate nucleus furnished a thalamocortical input to areas 17 and 18 in the cat, the corticothalamic projection to the lateral posterior complex seemed to break the rule of thalamocortical and corticothalamic reciprocity (Chapter 3). This seeming discrepancy only became resolved a decade or more later when the lateral posterior complex was shown to furnish a "nonspecific" input to areas 17 and 18 and several other cortical areas (Section 10.3.1.3).

Unraveling of some of the complexities of the cat lateral posterior pulvinar complex commenced in the later years of the Nauta era at the hands of Graybiel (1970, 1972, 1974b; Graybiel and Nauta, 1971). She was able to show a tripartite division of the nuclei on the basis of afferent connections. The lateralmost division, consisting of the present pulvinar nucleus, received afferents from the pretectum; a division medial to that (approximately equivalent to our LPl) from the visual cortex; and a division medial to that again (equivalent to our LPi) from the superior colliculus (Fig. 10.7). These fibers are now known to arise only in the superficial layers of the colliculus (Graham, 1977; Itoh *et al.*, 1979; Graybiel and Berson, 1980b). A fourth even more medial and anterior region (equivalent to our LPm) was evident at that time but its afferent connections were not clearly established (Fig. 10.7). There was some evidence for inputs from the deep layers of the superior colliculus (Jones, 1974); and it is now stated to receive inputs from a pretectal–tectal junctional region (Graybiel and Berson, 1980b). The efferent connections of the four parts of the lateral posterior–pulvinar complex were, in retrospect, not clearly elucidated but there was sufficient evidence to suggest that they projected upon different cortical fields outside areas 17 and 18 and in the middle and posterior suprasylvian gyri (Heath and Jones, 1971b; Graybiel, 1974b; Jones, 1974). Even now the exact cortical targets of the various subnuclei are by no means clear, though the basic quadripartite division on the basis of afferent connections, originally defined by Graybiel, has been confirmed many times (Kawamura *et al.*, 1974; Graham, 1977; Berman, 1977; Itoh, 1977; Berson and Graybiel, 1978; R. T. Robertson *et al.*, 1980; Hughes, 1980).

After Graybiel's, the next major contribution to the analysis of the cat lateral

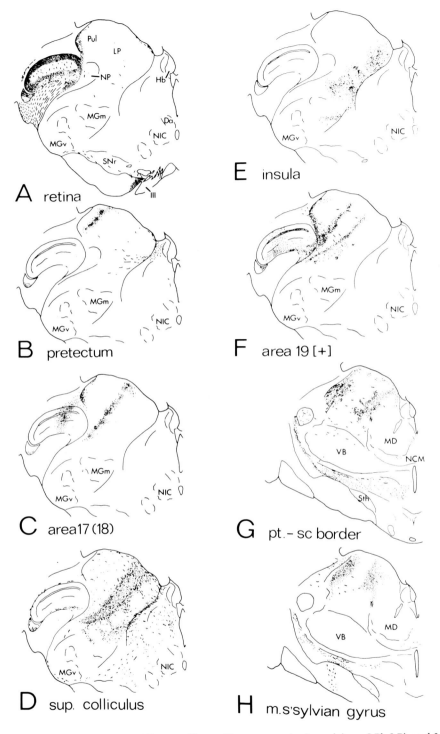

A retina

B pretectum

C area17 (18)

D sup. colliculus

E insula

F area 19 [+]

G pt.- sc border

H m.s'sylvian gyrus

FIGURE 10.7. Distribution of different afferent fiber systems in the pulvinar, LPl, LPi, and LPm nuclei of the cat. From Berson and Graybiel (1978).

posterior–pulvinar complex came from Updyke (1977). Using the autoradiographic method to trace corticothalamic connections from areas 17, 18, and 19 of the cat cortex, he was able to confirm the existence of the visuotopic organization in these connections first demonstrated by Garey *et al.* (1968) with the Nauta method. He showed that areas 17, 18, and 19 projected visuotopically to the LPl and area 19, again visuotopically, to the pulvinar nucleus, with a reversal at the border between the nuclei (Figs. 10.7 and 10.8). From the distribution of labeled corticothalamic terminations after injections of tritiated amino acids in different parts of the visual field representation in the three cortical areas, Updyke outlined a representation in the LPl and pulvinar nuclei in which the upper quadrant of the contralateral visual field is represented posterodorsally, the lower quadrant anteroventrally, and the vertical meridian along the LPl/LPi border and at the lateral margin of the pulvinar nucleus. The representation of the periphery occurs at the border between the LPl and pulvinar nuclei (see also Raczkowski and Rosenquist, 1981). In the course of these studies Updyke was able to confirm the existence of two other divisions of the lateral posterior complex comparable to those made by Graybiel (essentially our LPi and LPm,

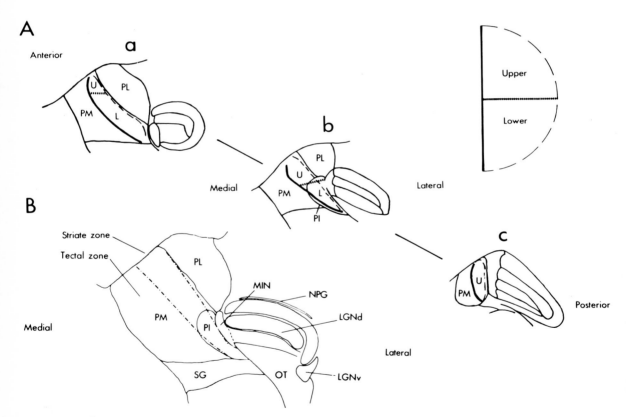

FIGURE 10.8.   Mason's (1981) reconstruction of the visual field representation in the striate cortex-recipient, LPl nucleus of the cat. Mason used the terminology of Niimi and Kuwahara (1973) in naming the subdivisions of the LP–pulvinar complex. In the LPl nucleus, thick line is representation of vertical meridian of visual half field, interrupted line periphery, dotted line horizontal meridian, U and L upper and lower fields.

TABLE 10.1. Distribution of Receptive Field Types among Visually Responsive
Neurons Recorded in the LP–Pulvinar Complex of the Cat[a,b]

RF type	Retinal zone	Striate zone	Tectal zone	PL
Concentric (X)	3	—	—	—
Concentric (Y)	14	—	—	—
Concentric (W)	37	—	—	—
Diffuse	2	2	14	15
Movement-sensitive	—	15	45	23
Direction-sensitive	3	76	19	27
Orientation-sensitive	—	—	—	4
Unclassified	10	37	46	39

[a] From Mason (1981).
[b] Total sample = 431 cells.

though with some variation anteriorly), and to show that the posterior nucleus of Rioch (1929a) and von Monakow (1895) was essentially a continuation of the LPl. In the LPi, a further visuotopic organization, reversed with respect to that in the LPl, can be demonstrated by the corticothalamic projection from certain visual areas in the suprasylvian sulcus (Updyke, 1981a,b).

Further confirmation of a quadripartite division of the lateral posterior–pulvinar complex of the cat has come from the differential distribution of acetylthiocholinesterase staining in the cat thalamus (Graybiel and Berson, 1980b; Section 10.1.3). The histochemical stain appears to pick out selectively the LPi and pulvinar nuclei and certain parts of the LPm.* The LPl is essentially unstained.

The exact origins of the collicular and pretectal inputs have also been established by the use of anterograde and retrograde tracing techniques. Fibers from the superior colliculus providing input to the LPi appear to arise primarily from cells in the deeper part of the stratum griseum superficiale (Kawamura *et al.*, 1980), a major site of retinal axon terminations. The additional observations that larger retinal ganglion cells project primarily to the deeper part of the stratum griseum superficiale (Itoh *et al.*, 1981) and that cells with Y-type properties are found in this part of the stratum (McIlwain and Lufkin, 1976; Hoffmann, 1973) have suggested to some that the collicular input to the LPi may be part of the Y-type central visual system, though this has not been confirmed in one study of the cat lateral posterior–pulvinar complex (Mason, 1981; Table 10.1).

The pretectal nucleus that furnishes that major, perhaps the sole, afferent input to the pulvinar nucleus is the nucleus of the optic tract, including its extension the olivary pretectal nucleus (Berman, 1977). These are the major pretectal nuclei in which retinal afferents terminate (Berman, 1977). The type of retinal ganglion cell that projects to the pretectum is not clear at present. However, the pulvinar nucleus, like the LPi, is also only one synapse removed from the retina.

---

* In Graybiel and Berson's terminology, LPM is our LPi, and LI (lateral intermediate nucleus) is our LPm plus the anterior part of the Pom nucleus. LPl is the same in both terminologies.

10.3.1.2.   Present Position

     Though the division of the lateral posterior–pulvinar complex of the cat into four nuclei each dominated by a different afferent input (pulvinar: pretectal; LPl: visual cortical; LPi: superficial superior collicular; LPm: deep collicular or pretectal) was inherently satisfying, a number of additional facts suggested that things were not that simple. First, came evidence for afferent connections from additional sources such as the retina and, second, came evidence for cortical connections well outside the conventionally accepted visual association areas (Figs. 10.7, 10.9, and 10.10).

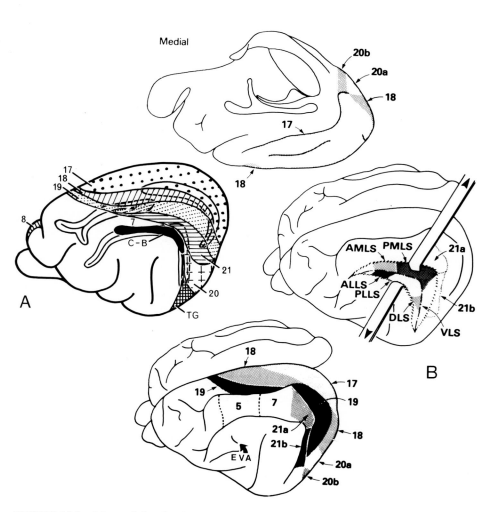

FIGURE 10.9.   Maps of the visual areas in the cortex of the cat. (A) Cytoarchitectonic and connectional map from Heath and Jones (1971b). C-B: lateral suprasylvian or Clare–Bishop area. (B) Multiunit maps from Symonds *et al.* (1981) with addition of ectosylvian visual area (EVA) of Mücke *et al.* (1982).

A direct retinal input to parts of the lateral posterior complex lying adjacent to the dorsal lateral geniculate nucleus of the cat had been suggested from axonal degeneration material by Hedreen in 1970. The connection was shown autoradiographically to be bilateral and to occupy a thin slab at the lateral edge of the pulvinar nucleus (Berman and Jones, 1977). The region has subsequently been called the *geniculate wing* by Guillery *et al.* (1980). Retinal ganglion cells whose axons form this ʿrojection appear from retrograde labeling and receptive field mapping to be mainly of the W type (Leventhal *et al.*, 1980a; Mason, 1981). Other afferents from the pretectum, additional to those terminating in the pulvinar nucleus, appear to reach anterior parts of the LPm (Robertson *et al.*, 1983). I have not been able to duplicate the results of Itoh *et al.* (1979) suggesting a cerebellar input to parts of the pulvinar nucleus.

The widespread origins of corticothalamic fibers to the lateral posterior–pulvinar complex of the cat have been difficult to reconcile with a straightforward quadripartite division. As well as arising from the obvious visual areas 17, 18, and 19 and from the several visual areas in the middle suprasylvian sulcus, these appear to originate in parts of areas 5 and 7 at the anterior end of the middle suprasylvian gyrus and in the temporal regions as well (Kawamura *et al.*, 1974; Updyke, 1977, 1981a, 1983; Robertson and Cunningham, 1981; Olson and Graybiel, 1981; Tong *et al.*, 1982; Mücke *et al.*, 1982). The terminations of the axons derived from many of these areas can occupy parts of the

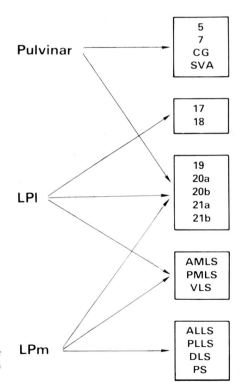

FIGURE 10.10. Thalamocortical connectivity of the various visual areas illustrated in Fig. 10.9. From Raczkowski and Rosenquist (1983).

lateral posterior–pulvinar complex seemingly independent of the four major nuclear subdivisions and overlap, seemingly without a corresponding visuotopic order, the orderly projections of areas 17, 18, and 19 to the LPl or the pulvinar nucleus. In an effort to make order out of apparent chaos, Updyke (1983) has adopted a shell and core pattern of organization for the complex. He sees the pulvinar, LPl, LPi, and LPm nuclei as the central core with particular patterns of input–output connections; around them, particularly anteriorly, dorsally, and ventromedially, and even intruding into the four elements of the core, is the shell in which all the other subcortical and cortical afferents terminate. Updyke's viewpoint has much to commend it as it could explain the apparent imposition of certain sets of connections on parts of a series of coherent visuotopic maps formed by other major connections and could also explain the morphological heterogeneity of the four primary nuclei of the lateral posterior–pulvinar complex.

10.3.1.3. Cortical Connections

It may be useful at this point to summarize the current understanding of the cortical connectivity of the cat lateral posterior–pulvinar complex. Before doing this, it is necessary for the reader to realize that there are a very large number of cortical areas in the cat brain that respond at short latency to visual stimuli and contain complete or partial representations of the visual field (Palmer *et al.*, 1978; Tusa *et al.*, 1979; Symonds *et al.*, 1981) (Figs. 10.9 and 10.10). Areas 17 and 18 are the primary fields lying along the lateral gyrus; area 19 closely related to them is mainly in the lateral sulcus. Four or more areas (AMLS, PMLS, ALLS, PLLS) lie in the banks of the middle suprasylvian sulcus; two (DLS, VLS) in the banks of the posterior suprasylvian gyrus and at least one (EVA) in the banks of the anterior ectosylvian sulcus. These show various patterns of interconnection and can only inspire admiration for those who painstakingly mapped them and bewilderment as to what they might be doing for the cat in terms of visual behavior. Probably in this multitude, however, lies one of the reasons for the comparable multiplicity of subdivisions and connectional patterns of the lateral posterior–pulvinar complex. Further working them out should keep some anatomists in business for years.

Before returning to our summary (Figs. 10.9 and 10.10) we should recall that the dorsal lateral geniculate nucleus provides the major thalamic input to areas 17, 18, and 19 and a probably more diffuse input to two visual areas (AMLS and PMLS) in the medial bank of the middle suprasylvian sulcus and to two (20a and 21a) in the middle and posterior suprasylvian gyri (LeVay and Gilbert, 1976; Geisert, 1980; Hughes, 1980; Marcotte and Updyke, 1981; Niimi *et al.*, 1981b; Raczkowski and Rosenquist, 1980).

a. Pulvinar Nucleus. This projects to visual area 19, to two visual areas (AMLS and PMLS) in the medial bank of the middle suprasylvian sulcus, to four visual areas (20a, 20b, 21a, 21b) in the posterior and middle suprasylvian gyri, to a visual area in the splenial sulcus, and to an apparent nonvisual area toward the anterior end of the cingulate gyrus (Raczkowski and Rosenquist, 1983; Marcotte and Updyke, 1981; Symonds *et al.*, 1981; Updyke, 1983). Virtually all of these areas return fibers to the pulvinar nucleus (Kawamura *et al.*, 1974; Updyke, 1977, 1981a,b, 1983; Robertson and Cunningham, 1981). The pulvinar nucleus receives a small direct afferent connection from the retina (Berman and Jones, 1977; Guillery *et al.*, 1980) and a heavy input via the nucleus of the optic tract of the pretectum (Berman, 1977) which is only one synapse removed from the retina. It, thus, stands in a position to furnish a fairly direct visual input to many of the cortical visual areas additional to that furnished by the dorsal lateral geniculate nucleus.

b. LPl.   This projects (Figs. 10.9 and 10.10) to visual areas 17, 18, and 19, to six of the areas (AMLS, PMLS, 20a, 20b, 21a, 21b) receiving input from the pulvinar, and to two additional areas (VLS and PS) in the posterior suprasylvian sulcus (Raczkowski and Rosenquist, 1983; Hughes, 1980; Marcotte and Updyke, 1981; Niimi *et al.*, 1981b; Symonds *et al.*, 1981; Updyke, 1981a). The LPl seems to receive fibers back from most of these areas (Updyke, 1977, 1981a, 1983; Graybiel and Berson, 1980b). It is not possible to find any convincing evidence for a subcortical input to the LPl though a "weak" collicular input has been claimed (Graybiel and Berson, 1980b). If it lacks a subcortical input, it is the only dorsal thalamic nucleus in the cat that does so, which is rather suprising. The nucleus appears to be dominated by the visuotopically organized corticothalamic input from areas 17, 18, and 19 and might, therefore, represent a group of thalamic cells whose function is to monitor visual cortical activity uncontaminated by a direct view of the external world. One cannot help but find such an idea a little hard to digest.

*c. LPi.**   This projects (Figs. 10.9 and 10.10) to two visual areas (ALLS and PLLS) in the lateral bank of the middle suprasylvian sulcus, to two areas (DLS and PS) in the posterior suprasylvian sulcus, and to only two of the areas (20b and 21b) in the posterior and middle suprasylvian gyri (Raczkowski and Rosenquist, 1983; Hughes, 1980; Marcotte and Updyke, 1981; Symonds *et al.*, 1981; Updyke, 1981a,b, 1983). It seems to receive fibers back from all of them (Updyke, 1977, 1981a, 1982, 1983). The primary subcortical input to the LPi is from the superior colliculus (Graybiel, 1972), particularly from the deeper tier of the stratum griseum superficiale (Kawamura *et al.*, 1980; Caldwell and Mize, 1981). The stratum griseum superficiale is a major site of retinotectal and corticotectal terminations (Garey *et al.*, 1968; Graybiel, 1975, 1976; Harting and Guillery, 1976). The retinal input to the deep tier is mainly derived from Y-type retinal ganglion cells, by contrast with the superficial tier whose retinal input is derived from W cells (Hoffmann, 1973; McIlwain and Lufkin, 1976; Stone *et al.*, 1979; Itoh *et al.*, 1981). The LPi, therefore, might provide a route for Y-cell inputs to gain access to those cortical areas that stand off-line from the dorsal lateral geniculate nucleus (Fig. 10.10), though Mason (1981) found few Y-type cells in his study of the cat lateral posterior complex and these were all in the retinal recipient zone of the pulvinar nucleus (Table 10.1).

* This is called by Updyke (1977) the "interjacent" LP nucleus.

d. LPm. This projects (Figs. 10.9 and 10.10) to the ectosylvian visual area (EVA) in the banks of the anterior ectosylvian sulcus and also to areas 7 and 5b of the middle suprasylvian gyrus and to temporal regions. It receives fibers back from all of them (Olson and Graybiel, 1981; Mücke *et al.*, 1982; Updyke, 1983). Different parts of the LPm may project upon each of these fields (Updyke, 1983). The subcortical input to the LPm is not clear. It has variously been reported as arising in the deep layers of the superior colliculus (Jones, 1974), pretectum (Robertson, 1983), or at the border between the two (Graybiel and Berson, 1980b), so it is unclear what type of input it furnishes to its cortical targets.

10.3.1.4. Laminar Terminations in Cortex

Considering all the energy that has been expended on unraveling the circuitry that is distilled into Fig. 10.10, it is surprising that so little attention has been devoted to the question of whether these diverse thalamocortical pathways, many of which converge on the same cortical areas, terminate in the same or different cortical layers or to the question of whether they end diffusely or in a focused manner that reflects the visuotopic organization demonstrable in the pulvinar, LPl, and LPi nuclei. In primates and in certain nonprimates other than the cat, thalamocortical fibers arising in parts of the lateral posterior nucleus or pulvinar terminate in layer IV of the extrastriate visual areas but mainly in layer I in the striate area (Sections 10.3.2 and 10.3.3).

In the cat (Fig. 10.15) the LPl projection to the cortex is also characterized by a heavy concentration of terminal ramifications in layer I of virtually all the cortical visual areas, including areas 17 and 18. There appear to be some terminal ramifications in the vicinity of layer IV in areas 17 and 18 (mainly in layer III and at the deep border of layer IV) but by far the heaviest layer IV projection occurs in area 19 and in the visual areas of the middle suprasylvian sulcus (Rosenquist *et al.*, 1974; Miller *et al.*, 1980; Symonds *et al.*, 1981). Hence, in these areas other than 17 and 18 the LPl appears to furnish the "specific" input (Chapter 3) but to these areas, and to areas 17 and 18, it also appears to furnish a nonspecific, layer I input. The relative diffuseness of the layer I input, however, has not been determined. This is perhaps reflected in the fact that corticothalamic fibers to the LPl arising in areas 17 and 18 come from layer V cells whereas those to the dorsal lateral geniculate nucleus come from layer VI cells (Gilbert and Kelly, 1975). The layer I and layer IV projections from the LPl to areas 17 and 18 appear to overlap those of the C laminae of the lateral geniculate nucleus (Fig. 10.15) (Miller *et al.*, 1980). If Y-type inputs should prove to relay through the LPl, then the convergence of LPl projections upon layers in which W-type inputs (relayed through the C laminae of the dorsal lateral geniculate nucleus) terminate, may represent the first major point of interaction between these two functional streams in the cat visual system.

Coming to terms with the enormously complicated connectivity of the cat lateral posterior–pulvinar complex still represents a major challenge and one doubts that the full story has yet been told. It seems inevitable that the laminar distribution and relative density of its terminations will determine its effect on the various visual areas in which its inputs converge with those of the dorsal lateral geniculate nucleus. Until we know the relative weightings of all these inputs, however, it is difficult to make any kinds of predictions regarding their functional attributes.

10.3.2. Lower Primates and a Tree Shrew

The afferent and efferent connectivity of the pulvinar nuclei has been particularly well studied in the bush baby, *Galago*, and in the tree shrew, *Tupaia*, by Diamond and his followers. It will be recalled that the lemuroid, *Galago*, possesses a dorsal and a ventral pulvinar nucleus and a relatively small, homogeneous lateral posterior nucleus, and that the insectivore, *Tupaia*, shows some similarities to this (Section 10.1). Fortunately, the pattern of connections, at least as now known, is simple in comparison with the cat, and though increasing complexities are being described, fundamental similarities can still be identified.

Glendenning *et al.* (1975) first emphasized the division of the *Galago* pulvinar into dorsal and ventral nuclei (their superior and inferior divisions). They showed that both projected to layer IV of several extrastriate visual areas and determined that the superior colliculus provided afferents to only the posterior part of the ventral nucleus. Later work involving injections of retrograde tracers in the pulvinar nuclei has indicated that, as in the cat and several other species (Section 10.3.1), the collicular afferents arise from the deeper tier of the stratum griseum superficiale which may be a main site of termination of Y-type retinal ganglion cell axons (Raczkowski and Diamond, 1981). Other work on the thalamocortical connections of the two pulvinar nuclei in the *Galago* has shown that, in addition to projections to the several extrastriate areas, both nuclei furnish a projection to the superficial layers (possibly only layer I) of area 17 (Carey *et al.*, 1979). As in the cat, certain subsidiary laminae of the dorsal lateral geniculate nucleus also project to superficial layers of the extrastriate areas (Fitzpatrick *et al.*, 1980); the main geniculate laminae project only to area 17. Corticothalamic connections returning to the two pulvinar nuclei of the *Galago* arise (Fig. 10.11) in area 17 as well as in the extrastriate areas (Symonds and Kaas, 1978; Wall *et al.*, 1982). One at least of these areas also projects to the lateral posterior nucleus. As in the cat (Section 10.3.1) and monkey (Section 10.3.3), the corticothalamic connections from area 17 originate primarily from layer V cells and those from the several extrastriate areas primarily from layer VI cells (Raczkowski and Diamond, 1981). This pattern seems to follow that described in Chapter 3 in which corticothalamic connections returning to the thalamic nucleus that provides layer IV input arise in layer VI whereas those to the nucleus providing "nonspecific" layer I input arise in layer V.

In studies of the corticothalamic connectivity of the *Galago* pulvinar nuclei, Symonds and Kaas (1978) reported that fibers from area 17 did not terminate in the part of the ventral nucleus receiving fibers from the superior colliculus (Figs. 10.11A, B). This observation bears comparison with the finding that the LPi of the cat in which collicular afferents end is independent of corticothalamic fibers from area 17 (Section 10.3.1.1). From the work on the cat, one might expect the region of termination of collicular afferents to receive corticothalamic fibers from cortical areas equivalent to those in the suprasylvian sulcus of the cat (Section 10.3.1). One area, the middle temporal area, thought to be comparable to one at least of the suprasylvian areas, does not, however, contribute fibers to the collicular receiving subnucleus (Wall *et al.*, 1982). There is a similar pattern of connections in the insectivore, *Tupaia*, in which a region identified as "pulvinar" and projecting to extrastriate visual areas receives input from the superior colliculus (Harting *et al.*, 1973; Albano *et al.*, 1979), but another area, identified as "lateral intermediate nucleus," is devoid of collicular inputs, receives pretectal inputs (Weber and Harting, 1980), and is connected with superficial layers of area 17 (Carey *et al.*, 1979). It should be noted, however, that the restriction of superior collicular inputs to only part of the *Galago* pulvinar complex was not confirmed by Raczkowski and Diamond (1981) who described retrograde labeling in the superior colliculus (Fig. 10.11C) after injections of tracer in all parts of both dorsal and ventral pulvinar nuclei.

The corticothalamic (Symonds and Kaas, 1978; Wall *et al.*, 1982) and thalamocortical connections (Raczkowski and Diamond, 1981) seem to define two visuotopic representations in adjacent parts of the dorsal and ventral pulvinar nuclei of the *Galago* such that the vertical meridian of the visual field is represented along their common border. In this, there is also some similarity to the visuotopically organized cortical projections to and from the adjacent LPl and

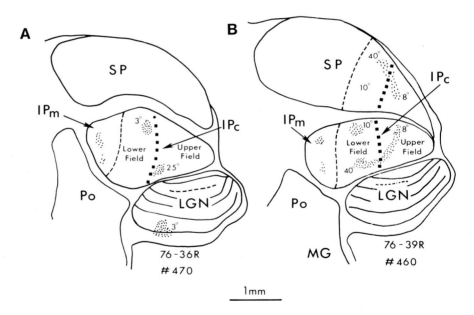

FIGURE 10.11. (A, B) Location of autoradiographically labeled fiber ramifications (dots) in the dorsal (SP) and ventral (IP$_c$, IP$_m$) divisions of the pulvinar in two experiments (A, B) in the prosimian bush baby following injections of tritiated amino acids at two points in the striate cortex. In (A), injections were approximately in parts of cortex representing 3° and 25° from center of gaze. In (B), injections were approximately at 8° and 40°. Label appears in regions of the two pulvinar nuclei

pulvinar nuclei of the cat (Section 10.3.1). Other parts of the dorsal pulvinar nucleus are free of connections with the visual cortical areas. In the ventral pulvinar nucleus, there are three zones: the one visuotopically connected with the visual cortical areas is large and central; the one receiving afferents from the superior colliculus is posterior; a third is medial and also receives visual corticothalamic fibers but seemingly without the precise visuotopic organization suggested for the central part of the ventral pulvinar nucleus (Symonds *et al.*, 1978). Again, it is possible to discern certain parallels between the visually connected and visually unconnected parts of the two pulvinar nuclei of the *Galago* and components of the lateral posterior–pulvinar complex of the cat. Exact parallels, however, are hard to draw, especially in view of the conflicting data over the sites of termination of superior collicular afferents in the pulvinar of the *Galago*.

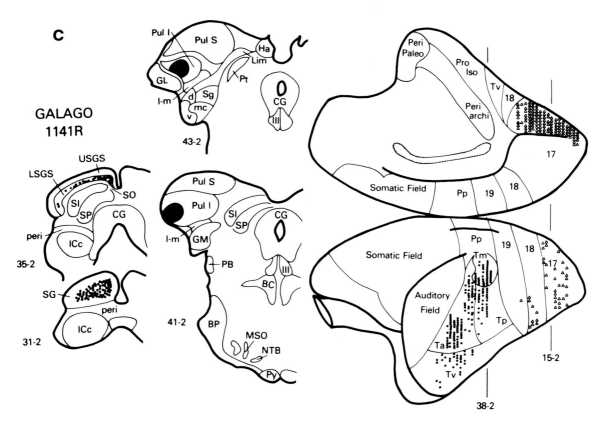

corresponding to the appropriate part of the representation. Heavy interrupted line indicates representation of horizontal meridian. From Symonds and Kaas (1978). (C) Retrograde labeling of cells in striate and parietotemporal cortex and in deep stratum of stratum griseum superficiale of superior colliculus following injection of horseradish peroxidase in ventral pulvinar (Pul I) nucleus of a bush baby. Modified from Raczkowski and Diamond (1981).

10.3.3. Monkeys

10.3.3.1. Introduction

Most workers studying the pulvinar complex of New World and Old World monkeys have adopted the broad parcellation of the complex into medial, lateral, inferior, and anterior nuclei as popularized by Olszewski (1952) in the Old World, macaque monkey and by Emmers and Akert (1963) in the New World, squirrel monkey. Apart from size, there is very little difference between the pulvinar in the two groups of primates and nuclear parcellation is equally clear (or not so clear depending upon one's point of view). One group of investigators, however, influenced by the work of their school on the prosimian pulvinar, has described the pulvinar nuclei of the New World, owl monkey *(Aotus)* in terms of superior and inferior divisions (Section 10.3.2) (Lin *et al.*, 1974; Lin and Kaas, 1977, 1979, 1980; Graham *et al.*, 1979). Their work deals only with the more posterior parts and the anterior pulvinar nucleus is largely ignored; but their superior division includes the medial and lateral pulvinar nucleus of others and their inferior division is coextensive with the inferior pulvinar nucleus of others.

There is now a large literature on the connectivity of the various nuclei of the monkey pulvinar. It is difficult to distill it into one frame of reference, for the various workers in the field have commonly adopted different points of view, particularly in regard to the naming of the nuclei and to the weight they have placed upon foci of axon terminations as bases for making finer subdivisions. What follows is an overview that attempts to bring the main features of the connectivity together without obscuring basic features in a welter of detail (Fig. 10.12).

It is clear that not all parts of the monkey pulvinar are visual in function. The anterior pulvinar nucleus and the medial two-thirds or more of the medial pulvinar nucleus appear to have no afferent connections from visually related brain-stem nuclei and do not project to any of the six or so cortical areas which have been shown to contain a map of all or some part of the visual field in the monkey cortex (Allman and Kaas, 1971, 1974a,b; Allman *et al.*, 1973; Baker *et al.*, 1981). The lateral posterior nucleus should also be included with the anterior and medial pulvinar nucleus as none of its connections are immediately visual in character.

10.3.3.2. Medial Pulvinar Nucleus, Anterior Pulvinar Nucleus, Lateral Posterior Nucleus

Apart from corticothalamic connections which reciprocate the thalamocortical connections, the afferent connections of these nuclei are, to my mind, not clearly established.

The medial pulvinar nucleus is clearly established as projecting to the cortex of the superior temporal gyrus (Burton and Jones, 1976) (Figs. 3.12 and 3.19). The cortical area to which it projects stretches from the temporal pole up the exposed surface of the superior temporal gyrus and through the anterior bank of the superior temporal sulcus, between the cortical territories of the dorsal medial geniculate nuclei and of the inferior and lateral pulvinar nuclei. There is also evidence for a small projection to part of the lateral nucleus of the

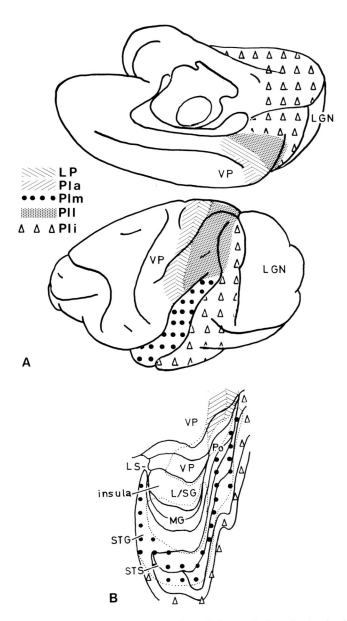

FIGURE 10.12. Large figure (A) shows an outline of the cortical territories in the Old World monkey brain that receives inputs from the components of the LP–pulvinar complex. Inputs from certain of these nuclei to layer I of the striate cortex (LGN) and questionable projection of medial pulvinar nucleus to frontal eye field are not illustrated. Anterior pulvinar territory abuts on area 2 of the somatic sensory cortex, which receives its input from the ventral posterior nucleus (VP). Small figure (B) adapted from Burton and Jones (1976) shows an opened-out view of lateral (LS) and superior temporal (STS) sulci in the rhesus monkey; thalamic nuclei providing input to each field are indicated as in (A). Note that territory of medial pulvinar nucleus extends along superior temporal gyrus (STG) to tip of temporal lobe.

amygdala (Jones and Burton, 1976a). In my opinion, reports of corticothalamic and thalamocortical connections of the medial pulvinar nucleus with the frontal eye field (Trojanowski and Jacobson, 1974, 1976; Bos and Benevento, 1975; Barbas and Mesulam, 1981; Künzle and Akert, 1977) probably reflect connections with the rather diffuse part of the central lateral nucleus that extends into the medial pulvinar nucleus by wrapping around the posterior pole of the mediodorsal nucleus (Figs. 13.2C and 13.3).

The anterior pulvinar and lateral posterior nuclei project to areas 5 and 7 of the superior and inferior parietal lobules (Fig. 10.12) (Burton and Jones, 1976; Baleydier and Maugiére, 1977; DeVito, 1978; Kasdon and Jacobson, 1978; Pearson *et al.*, 1978). The anterior pulvinar nucleus is more closely associated with parts of these areas lying adjacent to the first somatic sensory cortex on the surface of the hemisphere and adjacent to the second somatic sensory area in the lateral sulcus, whereas the lateral posterior nucleus is more closely related to slightly more posteriorly situated parts (Burton and Jones, 1976; Jones *et al.*, 1979) (Figs. 10.1 and 10.12). These posteriorly situated parts include area 7 (Burton and Jones, 1976) but may also include three of the visually mapped areas [the dorsomedial, medial, and posterior parietal areas (Graham *et al.*, 1979)] (Fig. 10.14).

10.3.3.3. Inferior and Lateral Pulvinar Nuclei

Large, but not all parts of the inferior pulvinar nucleus receive fibers from the ipsilateral superior colliculus (Benevento and Fallon, 1975a; Trojanowski and Jacobson, 1975; Partlow *et al.*, 1977; Harting *et al.*, 1980; Benevento and Standage, 1983), particularly from cells situated deep in the stratum griseum superficiale (Trojanowski and Jacobson, 1975; Marrocco *et al.*, 1981; Huerta and Harting, 1983). This is a relatively slow pathway, the latency to antidromic stimulation being 5–10 msec (Marrocco *et al.*, 1981). The terminations in the inferior pulvinar nucleus extend in part into the lateral pulvinar nucleus and may encroach on the lateral edge of the medial pulvinar nucleus, forming three main foci of input (Benevento and Standage, 1983). Projections to more or less the same foci emanate in the parts of the pretectum that receive retinal afferents (Benevento and Standage, 1983). There are also grounds for believing that a small direct retinopulvinar connection comparable to that in the cat terminates in the most lateral part of the inferior pulvinar nucleus in at least some monkeys (Campos-Ortega *et al.*, 1972; Hendrickson *et al.*, 1970).

On the basis of single-unit mapping (Allman *et al.*, 1972; Gattass *et al.*, 1978a,b; Bender, 1981), a complete representation of the contralateral visual half field has been demonstrated in the inferior pulvinar nucleus with a second in approximately the ventral half of the lateral pulvinar nucleus (Fig. 10.13). Other parts of the lateral pulvinar nucleus are also visually responsive but have not yet been mapped. There is some question as to the precision with which the two complete maps coincide with architectonic borders. Bender (1981), adopting Olszewski's (1952) drawings of the borders of the lateral and inferior pulvinar nuclei, considers that the representation in the inferior pulvinar nucleus fills that nucleus and that the central 2–3° of the representation extends into the lateral pulvinar nucleus. The border between the two representations, he correlates with a line of myelinated fibers first described by Vogt (1909). The second map fills only the ventral half or so of the lateral pulvinar nucleus and covers the dorsal, lateral, and posterior aspects of the map in the inferior pulvinar nucleus (Fig. 10.13). The two maps are such that the vertical meridian of the visual field is represented along the border between the two, with the representation of the upper field quadrant situated ventrally and that of the lower field quadrant situated dorsally. The representation of the horizontal meridian extends continuously across the map in the inferior nucleus but in the lateral nucleus, after 2–3°, it splits and the two parts extend up the lateral border of that nucleus (Fig. 10.13).

Charting of corticothalamic connections from area 17 of the cortex to the two nuclei (Spatz and Erdmann, 1974; Holländer, 1974b) follows the visuotopic map in the inferior pulvinar nucleus but only the central part of the representation in the lateral pulvinar nucleus (Ungerleider *et al.*, 1983). All parts of both maps, however, appear to receive corticothalamic fibers from the several extrastriate visual cortical areas (Benevento and Davis, 1977; Graham *et al.*, 1979; Lin and Kaas, 1979; Lund *et al.*, 1981; Graham, 1982) (Fig. 10.14). The extent to which the two maps and the corticothalamic connections to them are superimposed on the terminal regions of superior collicular afferents is argued about. It has been suggested that in both the lateral and the inferior pulvinar nucleus, there are regions in which the terminations of corticothalamic fibers from the visual areas and of afferent fibers from the superior colliculus do not superimpose (Benevento and Fallon, 1975a; Partlow *et al.*, 1977; Lin and Kaas, 1979, 1980; Harting *et al.*, 1980). Once again, it is possible to see some resemblance to the lateral posterior–pulvinar complex of the cat, with a dual visuotopic map, receiving corticothalamic connections from the striate and extrastriate areas, and a part dominated by tectal inputs. However, more data are clearly needed on the exact distributions of the various inputs and on their relationships to the two visuotopic maps before too close a parallel can be drawn.

Cortical projections from the lateral and inferior pulvinar nuclei have been demonstrated to the striate cortex and to most of the visuotopically mapped extrastriate visual areas (Benevento and Rezak, 1975, 1976; Ogren and Hendrickson, 1976, 1977; Trojanowski and Jacobson, 1976, 1977; Wong-Riley, 1977; Curcio and Harting, 1978; Rezak and Benevento, 1979; Lund *et al.*, 1981; Benevento and Standage, 1983) in some of which they may end in strips resembling ocular dominance strips (Ogren and Hendrickson, 1977; Curcio and Harting, 1978). The relationships of these inputs to periodicities in cytochrome oxidase staining in area 18 have not yet been established (Tootell *et al.*, 1983). Both nuclei project to area 17 but I have been able to find no reliable data on whether they have differential projections to the six or so extrastriate visual areas. From the distributions of corticothalamic fibers from the extrastriate areas (Lin and Kaas, 1979; Graham *et al.*, 1979; Graham, 1982) one would anticipate that some parts of both nuclei should project to all six areas but that other parts might project to certain areas only.

In area 17, pulvinar afferents clearly terminate in layer I whereas in the extrastriate areas they terminate in layer IV and the adjacent part of layer III

(Benevento and Rezak, 1975; Orgren and Hendrickson, 1976, 1977; Lund *et al.,* 1981). In keeping with the general principle that diffuse (i.e., layer I) projections are usually reciprocated by corticothalamic projections arising in layer V (Chapter 4 and Sections 10.3.1 and 10.3.2), it is this layer that gives rise to projections from the striate area to the pulvinar nuclei, whereas in the extrastriate areas, such connections arise from layer VI (Lund *et al.,* 1975, 1981; Ogren and Hendrickson, 1977; Trojanowski and Jacobson, 1977).

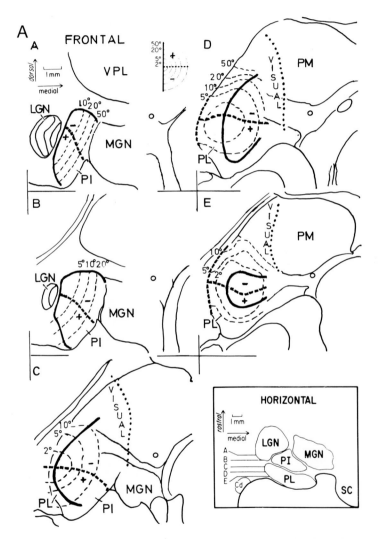

FIGURE 10.13. Tracings and reproductions of figures of Bender (1981) showing representation of the visual field projected onto frontal (A) and sagittal (B) sections of the macaque monkey inferior and lateral pulvinar.

10.3.4. Other Species

There are relatively few data on the connections and general organization of the lateral posterior nucleus in other species (Fig. 10.16). It is known in general terms that the superior colliculus projects upon parts of the lateral posterior ("pulvinar nucleus") in rodents (Robson and Hall, 1977), in insectivores other than *Tupaia* (Diamond and Utley, 1963; Hall and Diamond, 1968; Diamond and Hall, 1969; Abplanalp, 1970; Harting *et al.,* 1972; Hall and Ebner, 1970a; Gould *et al.,* 1978), and in marsupials (Benevento and Ebner, 1970, 1971; Rockel *et al.,* 1972; Coleman *et al.,* 1977). It is also evident that area 17 returns fibers to a part of the lateral posterior nucleus in these and other species (Nauta and Bucher, 1954; Abplanalp, 1970, 1971; Hughes, 1977; Holländer *et al.,* 1979), and that the lateral posterior nucleus projects upon both area 17 and a surrounding extrastriate visual belt (Fig. 10.16), in both of which the projections overlap those of the dorsal lateral geniculate nucleus (Hughes, 1977; Karamanlidis and Giolli, 1977; Robson and Hall, 1977; Gould *et al.,* 1978; Olavarria, 1979; Dürsteler *et al.,* 1979; Herkenham, 1980; Chapter 9). There is also a

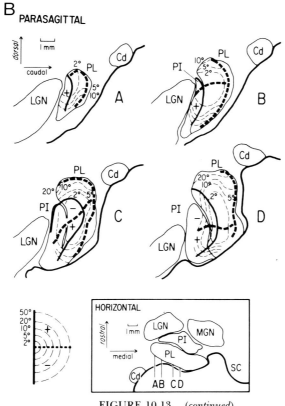

FIGURE 10.13.   (*continued*)

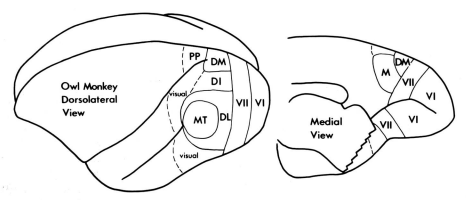

FIGURE 10.14. The multiple areas containing representations of the visual field, as determined by multiunit mapping in the owl monkey. From Graham *et al.* (1979).

dissociation of layer I and layer IV inputs so that lateral posterior fibers ending in area 17 terminate in layer I but in the surrounding cortical areas they terminate in the middle layers (Hughes, 1977; Robson and Hall, 1977; Gould *et al.*, 1978; Olavarria, 1979; Herkenham, 1980). There are obviously many parallels in the pattern of organization between these species and that seen in the cat and primate brain but there is insufficient information yet for a truly common plan to be discerned.

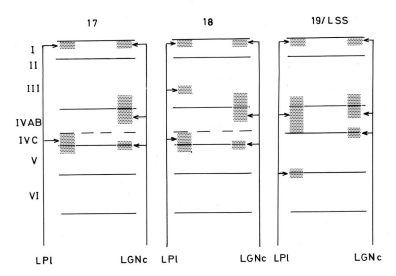

FIGURE 10.15. Schematic representation of the layers of termination in areas 17, 18, and 19 and the lateral suprasylvian areas (LSS), of fibers arising in the LPl nucleus and in the C laminae of the dorsal lateral geniculate nuclei of the cat. Drawn from sources quoted in the text.

10.3.5. Summing Up

567

LATERAL
POSTERIOR
AND PULVINAR
NUCLEI

There is evidently no single organizational plan of the lateral posterior–pulvinar complex across all species. The lack of such a set of principles may well reflect the lack of data from some species as well as the increasing parcellation into subnuclei in carnivores and primates. The biggest problem appears to be in deciding which subdivisions have the same patterns of connections from species to species. Yet there are some features that all mammals, so far as they have been studied, hold in common. It is clear that the superior

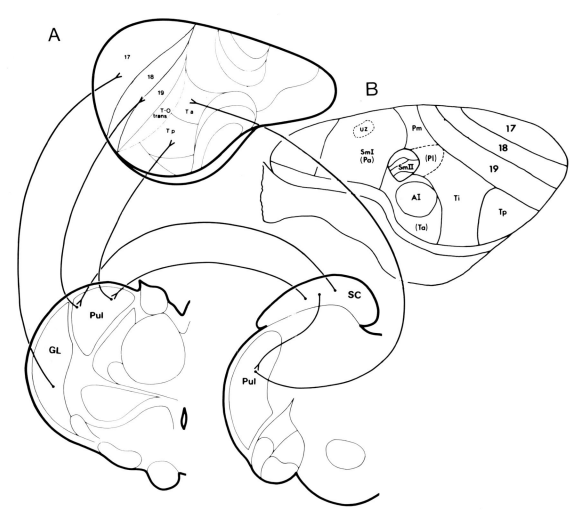

FIGURE 10.16. (A) Striate and adjoining cortical areas in the tree shrew and the thalamic nuclei providing input to them. From Casseday *et al.* (1976). (B) Visual (17–19), temporal (Ti, Tp), parietal (Pm, Pl), auditory (AI, Ta), somatic sensory (SmI, SmII), and frontal (presumably motor) areas of the cerebral cortex in the gray squirrel. From Nelson *et al.* (1979).

colliculus and pretectum are the major sources of noncortical input and that it is the parts of these structures on direct line with the retina that furnish much of this input. The collicular afferents arise primarily in deeper parts of the stratum griseum superficiale which has led to the suggestion that the retinocolliculopulvinar (or lateral posterior) system is of the Y type (see Chapter 4). This would, in turn, imply an involvement of the relevant part of the pulvinar or lateral posterior nucleus and its cortical target(s) in visually guided behavior other than that necessitating fine-grain discriminations which seems to be more a feature of the X-type retinal ganglion cell system and its central projections (Chapter 9). There is firm evidence in at least one species, the cat, for a direct retinal input to the lateral part of the lateral posterior–pulvinar complex. This seems to be furnished by retinal ganglion cells mainly of the W class (Mason, 1981).

Collicular, pretectal, and direct retinal afferent terminations do not fill the whole lateral posterior–pulvinar complex, though from the projection of known visual areas of the cerebral cortex (including the striate area) and the visuotopic maps in the lateral posterior–pulvinar complex, it is assumed that several additional parts of the complex are also visual in function. The cortical territory to which the visually related parts of the complex project includes the primary visual cortex (i.e., the striate area in most species) as well as every other extrastriate visual area in which a visuotopic map has been detected. There is a division in these projections in the sense that the relevant parts of the lateral posterior–pulvinar complex project to layer I of the primary visual cortex but to the middle layers of the extrastriate areas (Fig. 10.15). In some of these extrastriate areas, a layer I projection is furnished by the dorsal lateral geniculate nucleus (Chapter 9). There is, thus, a complicated, interrelated mixture of presumed diffuse, layer I connections and of presumed topographically organized, layer III–IV connections involving the dorsal lateral geniculate nucleus, parts of the lateral posterior–pulvinar complex, and virtually the whole visually responsive cortex.

Other parts of the lateral posterior–pulvinar complex are less obviously visual in terms of their connections. Their cortical territory includes most of the parietal areas 5 and 7, parts of which are involved in visual attention and visually guided behavior (Lynch *et al.*, 1977; Mountcastle *et al.*, 1981). The nature of the subcortical inputs to these other parts of the lateral posterior–pulvinar complex is not clear. It is possible that such inputs are derived primarily from deeper layers of the superior colliculus and parts of the pretectum that are not directly connected with the retina. Inputs from these structures which seem to be involved in some aspects of visually guided behavior, might well be appropriate for the kinds of visually guided behaviors apparently mediated by area 7.

10.4. Functions

The lateral posterior–pulvinar complex is obviously part of a visual pathway that ascends from the superior colliculus and pretectum and, bypassing the lateral geniculate nucleus, reaches virtually all the visually related areas of the cerebral cortex. What is its function? How does it stand in relation to the geniculostriate pathway? What are the functional roles of those parts of the primate pulvinar that are less obviously wired into this massive extrageniculate visual system?

The idea of a relationship between the pulvinar and vision had been in the minds of comparative anatomists for many years, on account of the parallel enlargement of pulvinar and temporal lobe in primates (Le Gros Clark, 1962). This idea crystallized in modern times in Schneider's (1969) captivating behaviorial work in which he formally enunciated the idea of two parallel visual systems, one geniculostriate and the other extrageniculostriate. Schneider's plan was reinforced by the previous and subsequent anatomical studies of Diamond and his collaborators (e.g., Snyder and Diamond, 1968; Diamond and Hall, 1969; Section 10.3). At the time of publication of Schneider's paper in *Science* in 1969, it had become clear that there was a strong tectothalamic pathway (Altman and Carpenter, 1961; Tarlov and Moore, 1966) that could be thought of as carrying retinal information via the superior colliculus to the lateral posterior–pulvinar complex and thereafter to the extrastriate cortex. In retrospect (see Section 10.3), the anatomy of this pathway was then known only imperfectly but the principle was sufficiently clear to provide a basis for Schneider's experiments. On ablating the primary visual cortex of hamsters and, thus, presumably rendering the geniculocortical system inoperative, he found that the animals could still be trained to orient toward visual stimuli but were unable to learn those visual discriminations of which hamsters are normally capable. By contrast, if the superior colliculus were ablated, thus presumably deactivating the pathway via the lateral posterior nucleus to nonprimary visual cortex, he found that the animals were virtually incapable of tracking or making orienting movements toward appropriate visual stimuli. However, if they stumbled up against the stimuli, they could learn to discriminate the appropriate one. Schneider's results, obtained in his thesis work of 1966, were to some extent replicated for the cortex in the hedgehog and tree shrew by Hall and Diamond (1968) and Snyder and Diamond (1968). In the tree shrew, however, it was necessary to ablate both areas 17 and 18 to affect visual pattern discrimination and some forms of pattern discrimination were, in fact, affected by removal of the target of the pulvinar nucleus alone (Killackey and Diamond, 1971; Killackey *et al.*, 1972; Ware *et al.*, 1974). It has become clear since then that in most animals all discriminative capacity is not lost after striate cortex lesions alone. Some pattern and form discrimination remains or can be regained with training in rats (Hughes, 1977; Spear and Barbas, 1975), cats (Doty, 1971; Spear and Braun, 1969; Sprague, 1972; Sprague *et al.*, 1977, 1981), squirrels (Levey *et al.*, 1973), rabbits (Moore and Murphy, 1976), and even in primates including man (Humphrey, 1974; Weiskrantz, 1972, 1974; Weiskrantz *et al.*, 1974). In a later study, Casagrande and Diamond (1974) determined that ablation of the superficial layers of the superior colliculus in the tree shrew, the layers that project to the lateral posterior–pulvinar complex, actually leads to the defect of pattern discrimination and similar results have been reported by Berlucchi *et al.* (1972) in the cat. By contrast, in the tree shrew at least, destruction of the deeper layers which project to other parts of the thalamus (Chapter 11) caused defective orientation toward visual stimuli. Schneider's later work (1970) also indicates that the deeper layers of the colliculus in hamsters must be removed to maintain the visual orientation and tracking defect. There is recovery if only the superficial layers are lesioned.

Though the dissociation of discriminative vision and visual orientation appears no longer complete for the two systems, the concept of two visual systems, presumably interacting by interconnections at the cortical level, has nevertheless become one of the fundamental tenets in the study of visually guided behavior. Obviously, from what has just been said, it is not clear that the collicular input to the lateral posterior–pulvinar complex is primarily involved in the phenomenon of visual orientation and tracking. Moreover, from what has been said in Section 10.3, the anatomy of the system is more complicated than originally assumed, with multiple subsets of connections, some arising in the pretectum as well as in the superior colliculus. Furthermore, the extrastriate visual areas, particularly in the monkey (Dubner and Zeki, 1971; Baizer *et al.*, 1977; Zeki, 1973, 1978; Van Essen and Zeki, 1978) but also in the cat (Spear and Baumann, 1975), display receptive field specializations among their neurons which suggest involvement in aspects of visual behavior more complex than simply orientation of the body and eyes toward a stimulus. Depth-specific cells, for example, are particularly characteristic of area 18 in the monkey (Poggio and Fischer, 1977), and directionally selective cells are particularly concentrated in an area in the superior temporal sulcus suggesting it is a movement detection area (Dubner and Zeki, 1971; Maunsell and Van Essen, 1983a,b). This would be appropriate for an area dominated by Y-type inputs coming via the superior colliculus (see Section 10.3). Other specialized functions have been suggested for other areas including mechanisms of color vision (Zeki, 1978).

Coupled with this evidence suggesting functional diversity in the extrageniculate visual system, there also has to be taken into account the fact that the lateral posterior–pulvinar system clearly provides a direct, albeit layer I, input to the primary visual cortex; also, in many mammals, certain cell groupings of the dorsal lateral geniculate nucleus itself provide a comparable layer I input to the extrastriate areas. Hence, both striate and extrastriate areas are subject to influences from the same thalamic nuclei, though clearly the different laminar patterns of terminations and the origins of the thalamocortical axons from different cell groups could make these influences quite distinct in the various areas.

Neuronal responses to visual stimuli have been recorded in the pulvinar of monkeys for at least a decade (Mathers and Rapisardi, 1973; Gattass *et al.*, 1978a,b, 1979; Lin and Kaas, 1979; Allman *et al.*, 1972) and stimulation of the pulvinar in man or of the lateral posterior nucleus in the cat modifies the wave pattern of the electrocorticogram recorded over the visual cortex in response to a subsequent visual stimulus (Brown and Marco, 1967; Crighel *et al.*, 1974). Bender (1982) studied the visual responsiveness of neurons in the visual field representation in the inferior pulvinar nucleus of paralyzed monkeys under nitrous oxide. He found that most units possessed receptive fields very comparable to those of cells in the striate cortex, rather than to those in the superior colliculus. Though receptive fields were 3 to 5 times larger than in the striate cortex, they were still relatively small, 1 to 5° in diameter. Most cells responded to stationary or moving slits of light. One-third showed no selectivity for the orientation of a stimulus but the remainder were orientation selective and, of these, some were direction selective as well. These are features that are typical of the striate cortex, though orientation tuning was said to be broader in pulvinar neurons than in cortical neurons. Moreover, cells projecting from the monkey superior colliculus to the inferior pulvinar are not direction selective (Marrocco *et al.*, 1981). Bender, therefore, believes that the principal drive to inferior pulvinar neurons may well come from the striate cortex. He does not rule out the extrastriate areas as influences over the pulvinar neurons because orientation selectivity and direction selectivity are characteristic features of neurons in many of these areas as well and may be particularly evident in some. However, he points out that neuronal selectivity for depth is a feature of at least one extra-striate area (area 18) but that depth-selective neurons are not found in the inferior pulvinar nucleus. The superior colliculus obviously cannot be ruled out also as an influence, for it sends fibers to the visual representation in the inferior pulvinar and the rather long latency of response of inferior pulvinar neurons to light flashes also suggests this: nonoriented cells require an average of 44 msec to reach 75% of their peak firing rate and orientation-selective cells require 60 msec. The pulvinar neurons may also be subject to other, as yet undefined influences, for they have prolonged (up to 100 msec) afterdischarges that are quite unusual in geniculate or striate cortex cells. Moreover, they show considerable variability in their responses to repeated visual stimuli, a variability that can be correlated with the level of arousal, as indicated by the EEG; they may also be influenced in a rather nonspecific way by other sensory stimuli (see also Gattass *et al.*, 1978b). The origins of these influences are not clear.

Most earlier studies of receptive field organization in the lateral posterior–pulvinar complex of the cat had yielded less specific results than those of Bender on the monkey. In the visuotopically organized pretecto-recipient, tecto-recipient, and striate cortex-recipient nuclei (pulvinar, LPi, and LPl), units tend to have large, diffuse receptive fields (Kinston *et al.*, 1969; Godfraind *et al.*, 1972; Veraart *et al.*, 1972) but a few are orientation selective or selective for moving stimuli and some have direction selectivity, especially in the pulvinar nucleus (Mason, 1978, 1981; Chalupa *et al.*, 1983). The striate-recipient zone cells appear to display a predilection for moving and changing visual noise ("texture"). The only concentric receptive fields were found in the retinal-recipient zone (Mason, 1981). The LPM was virtually unresponsive (Table 10.1). It is possible that with more detailed study, proportions of direction- and orientation-selective cells similar to those in the monkey inferior pulvinar nucleus might be encountered. Stimulation in the lateral posterior–pulvinar complex of the cat can affect the character of a subsequently evoked potential in the visual cortex (Brown and Marco, 1967), perhaps suggesting that, via its projections to layer I, the lateral posterior–pulvinar complex may be able to control lvels of excitability in the striate cortex. Steriade *et al.* (1977a,b) have suggested that the cat lateral posterior nucleus may itself be under the control of the brain-stem reticular formation.

If the visual parts of the lateral posterior–pulvinar complex are particularly concerned with the analysis of moving stimuli, then one might anticipate some defects relevant to this after destruction of the nuclei. To date, relatively little has been determined from ablations in monkeys or from pulvinotomies in man. Conventional testing of monkeys for acquisition of a visual discrimination after pulvinar lesions showed no defect in several studies (Chow, 1954; Thompson and Meyers, 1971; Mishkin, 1972), and only when stimuli to be discriminated were presented tachistoscopically for 10 msec (Chalupa *et al.*, 1976) did a defect emerge. During the acquisition of a similarly presented visual discrimination in normal monkeys, average evoked potentials recorded in the inferior pulvinar nucleus undergo significant changes in latency, amplitude, and wave form (Gould *et al.*, 1974). It is possible that the defect demonstrated by tachistoscopic stimulation is one of oculomotor control, for monkeys subjected to the more conventional discrimination situation, after lesions of the inferior and lateral pulvinar nuclei, show prolonged fixation on the stimulus and far fewer saccades to blank portions of the visual field (Ungerleider and Christensen, 1977, 1979). Pennyman *et al.* (1980) have shown some neurons with discharges related to saccadic eye movements in the two nuclei of normal monkeys.

Some parts of the superior temporal projection area of the medial pulvinar nucleus are polysensory in character (Bruce *et al.*, 1981), perhaps reflecting reports of multisensory inputs to single neurons in parts of the monkey and cat lateral posterior–pulvinar system (Mathers and Rapisardi, 1973; Avancini *et al.*, 1983).

Other hints of potential nonvisual functions of the less overtly visual parts of the human pulvinar have come from stimulating and lesioning the region in man. Ojemann and his collaborators (Ojemann and Fedio, 1968; Ojemann *et al.*, 1968) have found that bipolar stimulation, particularly of the anterodorsal part of the pulvinar, during a word-naming task, led to a failure in object naming and a striking reduction in short-term verbal recall. The effect was asymmetric and seen only when the left pulvinar of right-handed persons was stimulated. Presumably this reflects a close association between the part of the pulvinar stimulated (the medial nucleus?) and the angular and supramarginal gyri of the dominant hemisphere. Seemingly in confirmation, Ciemens (1970) reported a case in which a lesion of the left lateral posterior and medial pulvinar nuclei was associated with prolonged aphasia.

Some neurosurgeons have advocated the use of pulvinotomy for the relief of hypertonias and dyskinesia (Cooper *et al.*, 1973; Gillingham *et al.*, 1977). It is difficult to see the anatomical or physiological basis for the positive results claimed.

There appears to be little in the clinical literature that enables a correlation to be made between the functions of the lateral posterior and anterior pulvinar nuclei and their cortical targets: the superior and inferior parietal lobules. As is well known, these areas are concerned with extrapersonal space and the body image in man (De Renzi, 1982; Hyvarinen, 1982). Electrophysiological studies in awake, behaving monkeys have demonstrated neurons with functional properties associated with attention to objects in space, with movements of the eyes toward stimuli of interest, and with projection of the limbs into extrapersonal space (Lynch *et al.*, 1977; Yin and Mountcastle, 1977; Rolls *et al.*, 1979; Sakata *et al.*, 1980; Mountcastle *et al.*, 1975, 1981, 1983; Bushnell *et al.*, 1981). There are no comparable electrophysiological data on the lateral posterior–pulvinar complex and it seems to me that this might well be a fertile area in which to commence studies, for the nature of the subcortical drive to the neurons in the parietal cortex is virtually unknown (Lynch, 1980).

. . . one can consider the ventrobasal complex and the geniculate bodies, together with their essential cortical projection areas as the essential cores of the somatic, visual and auditory systems. Surrounding them at both thalamic and cortical levels, and fusing between them, are areas which are less specific with regard to local sign and modality served, in which complex sensory interactions occur, and which are thought to subserve sensory integrative actions.

                                            Poggio and Mountcastle (1960)

The Posterior Complex of Nuclei

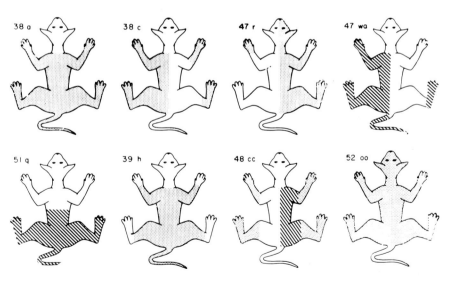

Receptive fields of neurons in the posterior nuclear complex of cats. Stippling: mechanical stimuli; hatching: noxious stimuli. From Poggio and Mountcastle (1960).

11.1. Description

11.1.1. Primates

The group of nuclei forming the posterior complex of the thalamus are probably best appreciated in sections of a higher primate brain such as one of the New World or Old World monkeys (Fig. 11.1, and see figures in Chapters 8 and 10). Similar appearances are found in apes and man (Kanagasuntheram *et al.,* 1969; Hassler, 1959). Two principal aggregations of cells are involved: the limitans–suprageniculate nucleus, consisting of large, deeply staining, close-packed cells, and the posterior nucleus, consisting of smaller, pale-staining, dispersed cells. The limitans part of the limitans–suprageniculate nucleus commences at the posterior end of the habenular complex, crosses the posterior pole of the mediodorsal nucleus where it fuses with the similar cells of the central lateral and parafascicular nuclei, and runs as a thin line of cells down the medial medullary lamina in the posteroventral aspect of the medial pulvinar nucleus. The medial medullary lamina separates it from the anterior pretectal nuclei. At some distance dorsomedial to the medial geniculate complex, the limitans part expands considerably as the suprageniculate part of the common nucleus. This suprageniculate part continues the oblique line of the limitans part down into the magnocellular medial geniculate nucleus with which it fuses; the magnocellular medial geniculate cells are sufficiently larger, however, that the two nuclei are readily distinguishable (Chapter 8; Figs. 11.1 and 11.2). The dorsal part of the suprageniculate element lies in the ventral aspect of the medial and anterior pulvinar nuclei.

The posterior nucleus, with its smaller, dispersed cells, commences at the dorsolateral aspect of the suprageniculate nucleus and expands laterally across the anterodorsal aspect of the medial geniculate complex, and anteriorly ventral to the ventral posterior lateral (VPL) nucleus, eventually becoming continuous with the ventral posterior inferior (VPI) nucleus (Fig. 11.1). Its borders with the anterodorsal medial geniculate nucleus and with the VPI are hard to define. It is traversed by fibers of the medial lemniscus as they enter the thalamus and invaded by islands of posteriorly situated cells of the VPL (Fig. 11.2B). The part subjacent to the VPL, we term the medial region of the posterior nucleus and the part adjacent to the anterodorsal medial geniculate nucleus, the lateral region

(Burton and Jones, 1976). Boivie (1978, 1979) in the rhesus monkey has described a portion of the medial region extending anteriorly on the dorsal aspect of the VPM. I consider this to be mainly the medullary lamina separating the VPM from the anterior pulvinar nucleus (see Figs. 7.7 and 10.2A).

In lower primates such as lemurs and lorises (Fig. 11.3; Kanagasuntheram *et al.*, 1968; Simmons, 1980), and *Tarsius* (Le Gros Clark, 1930a), the same

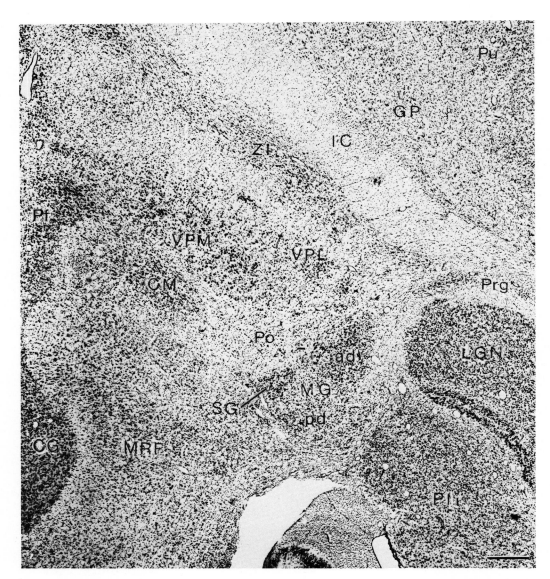

FIGURE 11.1.   Thionin-stained horizontal section through the thalamus of a squirrel monkey showing the suprageniculate and posterior nuclei and adjacent nuclei. Bar = 500 μm. From Burton and Jones (1976).

components of the posterior complex are readily detectable but the continuity between the limitans and the suprageniculate components of the limitans–suprageniculate nucleus is commonly broken and the cells more dispersed. The medial medullary lamina is often broken and the nucleus limitans and nucleus of the optic tract of the pretectum tend to fuse. The posterior nucleus, though distinguishable from the ventral posterior and magnocellular medial geniculate nuclei, is far less easily dissociated from the dorsal medial geniculate nuclei or from the overlying pulvinar or lateral posterior nuclei and is not mentioned in some descriptions.

11.1.2. Carnivores

In the cat, dog, and other carnivores such as raccoons and seals (Figs. 11.4, 11.5, and 11.7, and also Figs. 8.2 and 10.4) the component nuclei of the posterior complex, though present and having the same relative dispositions, have far less well-defined boundaries than in primates. The limitans component of the limitans–suprageniculate nucleus lies along the medial medullary lamina in the medial aspect of the lateral posterior nucleus but is reduced to such a thin line of cells that it is indistinguishable in places. Elsewhere, it may form a few broken clumps of cells. The suprageniculate component is also made up of loose clusters of cells and only their denser staining makes them distinguishable from the lateral posterior and posterior nuclei. The suprageniculate cells, nevertheless, abut on the magnocellular medial geniculate nucleus, as in primates. The clumps of cells representing the limitans–suprageniculate nucleus are particularly clear in material stained for acetylthiocholinesterase (Graybiel and Berson, 1980b). The posterior nucleus is relatively large but has ill-defined boundaries. One component, the intermediate region (Poi), expands laterally from the suprageniculate element of the limitans–suprageniculate nucleus, beneath the lateral posterior nucleus and well into the dorsal nuclei of the medial geniculate complex (Fig. 11.7 and see Chapters 8 and 10). From the Poi a medial region (Pom) expands anteriorly toward the posterior pole of the ventral posterior nucleus and then continues well anteriorly on the dorsal aspect of the ventral posterior nucleus and ventral to the lateral posterior nucleus. A few cells of the Pom also expand posteriorly among the cells of the magnocellular medial geniculate nucleus. Also from the Poi, a lateral region (Pol) expands anteriorly around the lateral aspect of the posterior pole of the VPL (Figs. 11.4 and 11.7) in close contact with the external medullary lamina.

11.1.3. Rodents and Other Small Mammals

In rodents, tree shrews, the rabbit, and marsupials, I find it difficult to identify deeply staining cells along the medial medullary lamina that might be regarded as a limitans–suprageniculate nucleus (Fig. 11.6). A region of small, loosely packed cells invading the dorsal aspect of the medial geniculate complex, though sometimes called suprageniculate nucleus (Lund and Webster, 1967a,b)

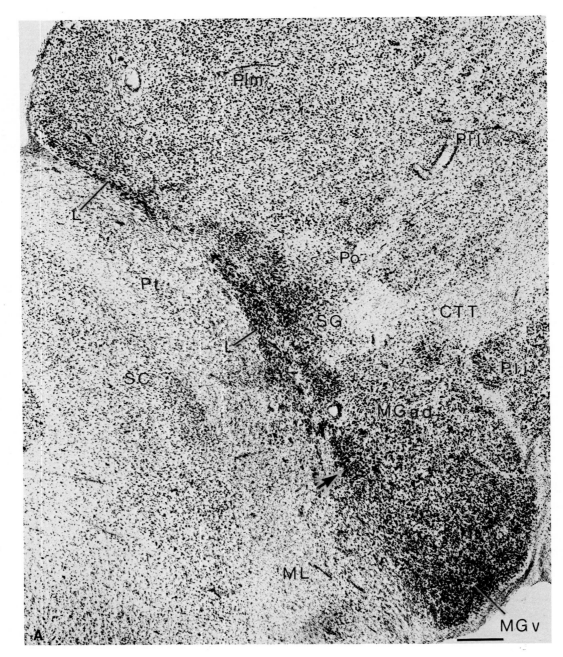

FIGURE 11.2. Thionin-stained frontal sections through the thalamus of a rhesus monkey showing the components (L, SG, Po) of the posterior nuclear complex and adjacent nuclei. Bars = 500 μm. (A) is posterior to (B). Arrow in (A) indicates

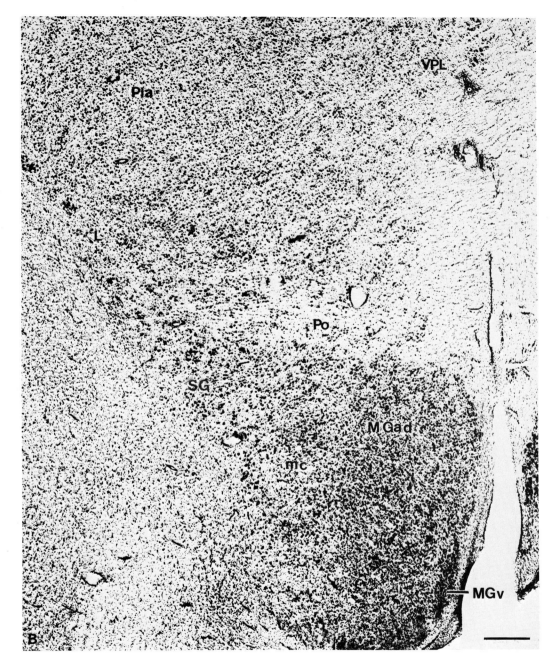

fusion of suprageniculate nucleus with medial geniculate complex. (A) from Burton and Jones (1976). See Chapters 8 and 10 for other illustrations of monkey posterior complex.

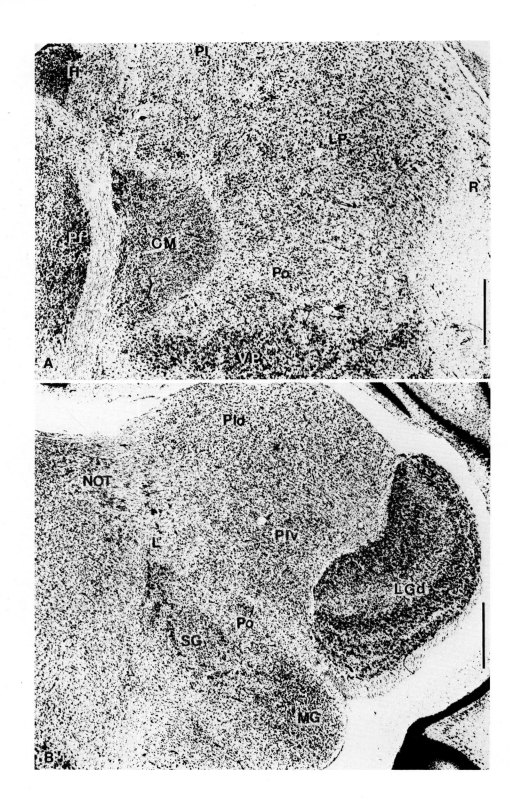

with or without a limitans nucleus (Oliver and Hall, 1978a,b) and sometimes called a suprageniculate part of the posterior complex (Rockel *et al.*, 1972), is to my mind more comparable to the Poi region of cats and may also include elements of the dorsal medial geniculate nuclei (Section 11.1.2). A diffuse region, continuous with this lies at the posterior pole of the ventral posterior nucleus (Fig. 11.6) and may well be comparable to parts of the Pom and Pol regions of cats. The Pom region, however, as it extends anteriorly over the dorsal surface of the ventral posterior nucleus, becomes remarkably well defined—much more so than in the cat—and appears more like a conventional nucleus than any other part of the posterior complex (Figs. 11.6A,C).

11.2. Terminology

11.2.1. Primates

The names *limitans* and *suprageniculate nuclei* were used by Friedemann (1911) in his cytoarchitectonic studies of the monkey brain and the regions delimited by him correspond to those identified under the same names by later workers (Olszewski, 1952; Emmers and Akert, 1963). *Limitans* appears to have been coined by Friedemann but *suprageniculate* is Münzer and Wiener's (1902) term for the dorsal part of the medial geniculate complex in the rabbit. Vogt (1909) identified our posterior nucleus in myelin preparations as a region traversed by the medial lemniscus and Friedemann (1911) and Pines (1927) in Nissl preparations called it a transitional region (region of Übergangzellen). Some workers on primates bring the limitans–suprageniculate and posterior nuclei together as a single group while others separate them. Olszewski (1952) in the rhesus monkey has separate limitans and suprageniculate nuclei but includes the posterior nucleus in the latter. Hassler (1959), following Friedemann, includes the suprageniculate nucleus in a large nucleus limitans but separates the posterior nucleus off as a nucleus limitans portae. Emmers and Akert (1963) in the squirrel monkey have separate limitans, suprageniculate, and posterior nuclei. Aronson and Papez (1934) mistakenly called the nucleus limitans, the nucleus of the optic tract. My own preference (Burton and Jones, 1976) is to link the limitans and suprageniculate elements into a single limitans–suprageniculate nucleus because of their identical structure and connections, and to divide the posterior nucleus into regions which though similar in structure have different connections (Section 11.3.1).

Le Gros Clark (1932a) felt that the posterior enlargement of the pulvinar in higher primates forced the nucleus of the optic tract of the pretectum downwards and medially to fuse with the nucleus limitans. To anyone familiar with

---

FIGURE 11.3. Thionin-stained frontal sections through the posterior complex of the prosimian, bush baby (*Galago crassicaudatus*). (A) is anterior to (B). Note close association of nucleus limitans (L) with pretectal nucleus of optic tract (NOT). Bars = 1 mm.

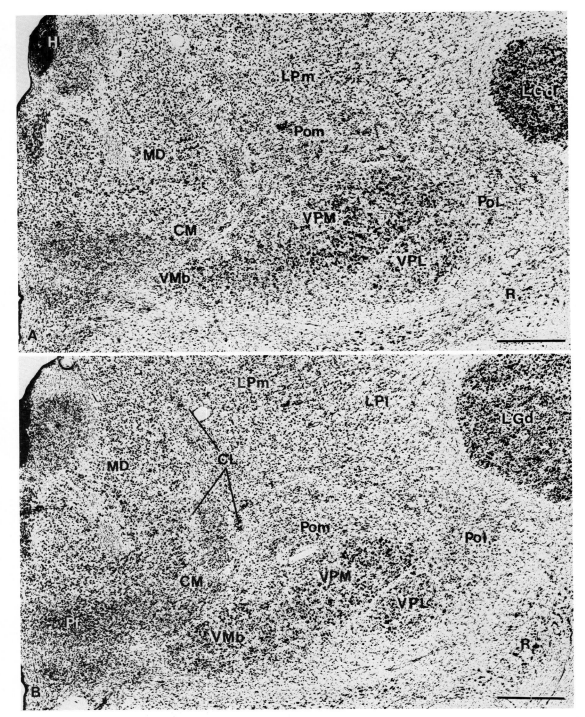

FIGURE 11.4. Thionin-stained frontal sections through the posterior complex of a cat. (A) is anterior, (C) posterior. Note extensions of Pom and Pol divisions of posterior nucleus dorsal and lateral to ventral posterior nucleus (A–C) and islands of cells of suprageniculate (SG) nucleus (C). For intermediate division of posterior nucleus, see Figs. 11.7 and 8.2A,C. Bars = 1 mm.

the cat brain in which the two nuclei are widely separated, this may seem unlikely. Inspection of a lorisoid or lemuroid brain, however, illustrates the difficulty of distinguishing the two (Fig. 11.3B).

11.2.2. Carnivores

Limitans and suprageniculate nuclei were identified in the cat by most early workers (Rioch, 1929a; Ingram *et al.*, 1932) but the present posterior nucleus was not separately identified. Some more recent workers on the cat still include most of the posterior nucleus in the suprageniculate nucleus (Niimi and Kuwahara, 1973).

The term *posterior complex* or *posterior group* of nuclei appears to have been first used in the present sense by J. E. Rose (1942b) in his account of the thalamus of the sheep but it quickly became applied to the cat. The term derives from the posterior thalamic nucleus described by M. Rose (1935) in the rabbit. In his initial description of the posterior complex, J. E. Rose (1942b) included in it: most of the lateral posterior nucleus, plus an ill-defined area extending posteriorly and ventrally from the lateral posterior nucleus and pulvinar and between

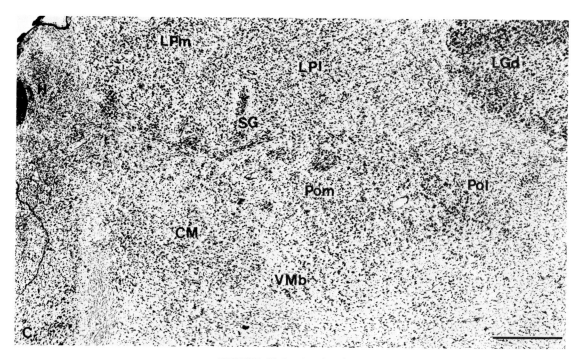

FIGURE 11.4. (*continued*)

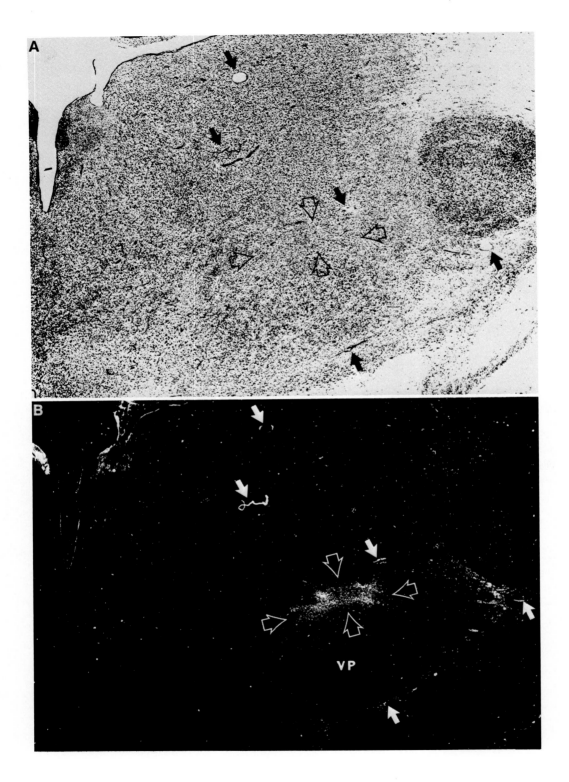

the medial and lateral geniculate nuclei, to the pretectal area. Subsequently, Rose and Woolsey (1958) described in the cat what they called a "pulvinar–posterior system." This had a dorsal part, consisting essentially of the lateral posterior nucleus and pulvinar, and a ventral part made up of heterogeneous cell types and with ill-defined boundaries. This ventral part embraced the posterior pole of the ventral posterior nucleus and extended medially across the dorsal half of the medial geniculate complex. It included what others had called suprageniculate nucleus, nucleus limitans, posterior nucleus of Rioch, ventral parts of the lateral posterior nucleus, and the magnocellular and dorsal nuclei of the medial geniculate complex. Rose and Woolsey used the term *pulvinar–posterior system* operationally, to include those regions of the thalamus in which cellular degeneration occurred after large lesions of the auditory and adjacent regions of the cerebral cortex. From these experiments, they concluded that the ventral portion of the pulvinar–posterior system, which they called the *posterior group*, probably formed a part of the thalamic auditory system. Subsequently, Poggio and Mountcastle (1960) investigated this posterior group electrophysiologically and defined its boundaries in somewhat more detail. In terms of the present account, they included the limitans–suprageniculate nucleus, the three regions of the posterior nucleus, and the magnocellular medial geniculate nucleus in their "posterior group." Most succeeding accounts of the cat followed Poggio and Mountcastle's delineation.

Investigations of the connections of the posterior group of the cat served to define it further (Moore and Goldberg, 1963; Mehler, 1966b; De Vito, 1967; Jones, 1967; Rinvik, 1968a,c,d; Boivie, 1971a,b; Jones and Powell, 1968; Diamond *et al.*, 1969; Graybiel, 1970; Heath, 1970; Heath and Jones, 1971a; Robertson and Rinvik, 1973; Jones and Leavitt, 1973; Jones and Burton, 1974). The Pol region was referred to as a lateral division of the posterior group by Moore and Goldberg (1963) when they showed that it received afferent fibers from the inferior colliculus, whereas other parts of the complex (except for the magnocellular medial geniculate nucleus) did not. The Pol was also found to be distinct in receiving fibers from the AI and AII fields of the auditory cortex (Diamond *et al.*, 1969).

The Pol has usually been regarded as the anterior portion of the medial geniculate nucleus (Rioch, 1929a; Ingram *et al.*, 1932; Jimenez-Castellanos, 1949) or as a part of the ventral posterior nucleus (Jasper and Ajmone-Marsan, 1954). The remainder of the posterior group, including the limitans–suprageniculate nucleus, was called the medial division by Moore and Goldberg, but as it was later shown to include three regions with different connections, it was subdivided further (Jones and Powell, 1968; Diamond *et al.*, 1969; Heath and Jones, 1971a). The *medial region* (Pom) as redefined by Mehler (1966b), Jones and Powell (1968),

FIGURE 11.5. Brightfield (A) and darkfield (B) photomicrographs of the same section showing autoradiographic labeling of corticothalamic fiber ramifications in Pom division of posterior nucleus of a cat following injection of tritiated amino acids in area 5a of the cerebral cortex. Bar = 1 mm. From Tanji *et al.* (1977).

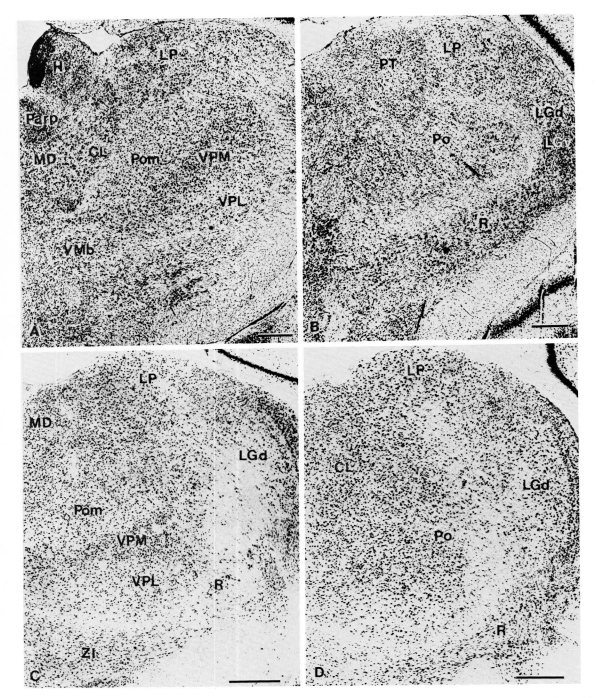

FIGURE 11.6.   Thionin-stained frontal sections through the posterior nucleus of a rat (A, B) and a marsupial phalanger (C, D). Note the prominent medial division (Pom) dorso-medial to the ventral posterior nucleus. Bars = 500 μm (A, B), 1 mm (C, D).

Jones and Burton (1974), and in a slightly modified fashion by Rinvik (1968a), is the part receiving afferent fibers from the spinal cord, and from what was then thought to be the somatic sensory cortex (see Section 11.2.1). The extension of the Pom back along the magnocellular medial geniculate nucleus has usually been included as a part of the magnocellular medial geniculate nucleus (Rioch, 1929a; Jasper and Ajmone-Marsan, 1954; Morest, 1964; Mehler, 1966b) or has been labeled the *parageniculate nucleus* (Nauta and Kuypers, 1958, after Lewandowsky, 1904) but because it receives spinal fibers, it is regarded as a part of the Pom (Jones and Burton, 1974).

The part of the Pom lying dorsal to the ventral posterior nucleus has usually been included as a part of the overlying lateral posterior nucleus and Rioch (1929a) referred to it as the ventral division of the lateral posterior nucleus. Ingram *et al.* (1932) labeled it as a region of transition between the ventral posterior and centre médian nuclei.

The *intermediate region* (Poi) of the posterior nucleus was named by Diamond *et al.* (1969) because, unlike the Pol, Pom, and the limitans–suprageniculate nucleus, it received afferent fibers descending from fields posterior ectosylvian

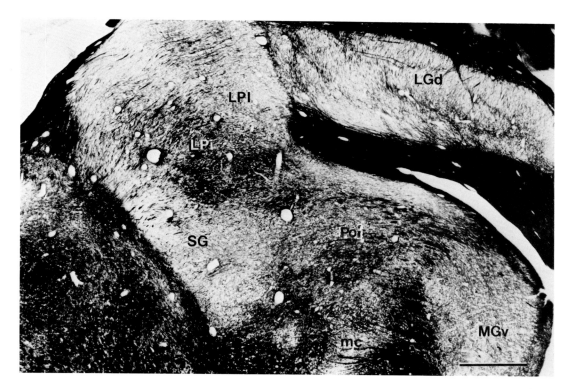

FIGURE 11.7.   Hematoxylin-stained frontal section through the suprageniculate nucleus (SG) and the intermediate (Poi) division of the posterior nucleus of a cat. Extension of LPl nucleus into dorsal aspect of medial geniculate body is clearly evident in this photomicrograph (cf. Fig. 8.2C). Bar = 1 mm.

(EP) and temporal of the auditory cortex and from dorsal and posterior parts of the suprasylvian gyrus (Diamond *et al.,* 1969; Heath and Jones, 1971a,b). Its ascending afferent connections were then uncertain and they remain so. The *limitans–suprageniculate nucleus* became evident as a separate entity in the cat when Heath and Jones (1971a,b) showed that it is connected primarily to the insular cortex and Graybiel (1970) indicated that it is one of the principal thalamic targets of the superior colliculus.

The magnocellular medial geniculate nucleus, though often considered a part of the posterior group, was sufficiently distinct to be regarded as an independent entity by workers in the field, though the suprageniculate nucleus was sometimes erroneously included in it (Rinvik, 1968a). The posterior nucleus of Rioch, though once regarded as a part of the posterior group in the cat (Rose and Woolsey, 1958; Poggio and Mountcastle, 1960), is clearly part of the lateral posterior nucleus (Chapter 10).

11.2.3. Rodents and Other Small Mammals

The use of the terms *posterior nucleus* or *posterior complex* in mammals other than primates or carnivores is confused because of the historical precedent, created by Nissl (1889, 1913), of terming the pretectal region the *nucleus posterior of the thalamus.* This practice was adopted by Gurdjian (1927) and by Le Gros Clark in his early work (Le Gros Clark, 1929a) though he later abandoned it (Le Gros Clark, 1930a). *Nucleus posterior thalami* for the pretectum continued to be used by some authors and it still survives in some atlases (e.g., Snider and Niemer, 1961; König and Klippel, 1963).

The parts of the posterior complex in the vicinity of the medial geniculate complex were not identified by Gurdjian (1927) in his study of the rat. He delineated the Pom region, terming it *nucleus ventralis pars dorsomedialis.* In the much later work on the rat of Lund and Webster (1967a,b), the Pom became the *posterolateral complex,* and the pars lateralis and the remaining parts of the posterior complex and dorsal medial geniculate nuclei the *suprageniculate nucleus.*

In the rabbit, M. Rose's (1935) posterior nucleus corresponds to dorsal parts of the posterior nucleus of the present account but other parts are included by him in the medial geniculate nucleus and the anterior Pom region is regarded as a division (Vbα) of the ventral posterior nucleus. Posterior and suprageniculate nuclei were later detected in the rabbit by J. E. Rose (1942b).

Rockel *et al.* (1972) in studying the afferent connectivity of the thalamus in the marsupial phalanger, attempted to correlate regions with comparable connections with those in the posterior complex of the cat. Lateral, intermediate, and medial regions were identified but a limitans–suprageniculate nucleus separate from the intermediate region could not be distinguished. The extension of this terminology to the rat was made by Jones and Leavitt (1974). The lateral and intermediate regions have received little further study in the rat but the Pom region, probably because it is so distinct, has become well established in the rat. In the opossum, it has been referred to as *nucleus C* (Oswaldo-Cruz and Rocha-Miranda, 1967) or as a central intralaminar nucleus (Killackey and Ebner, 1973).

11.3. Connections

11.3.1. Afferent Connections

11.3.1.1. Ascending

The several components of the posterior complex can be distinguished by their afferent connections though these have only been thoroughly studied in the cat (Fig. 11.8). There is sufficient information from other species, however, to suggest a comparable organization in all mammals.

The limitans–suprageniculate nucleus is the target of the intermediate or deep layers of the superior colliculus. This has been confirmed in cats (Graybiel, 1970; Graham, 1977; Calford and Aitkin, 1983), tree shrews (Oliver and Hall, 1978a,b), and monkeys (Benevento and Fallon, 1975a; Partlow *et al.*, 1977; Harting *et al.*, 1980). Calford and Aitkin (1983) suggest it may also receive fibers from the interstitial nucleus of the brachium of the inferior colliculus.

Much of the medial region (Pom) of the posterior nucleus is taken up with terminations of spinothalamic fibers (Mehler, 1966c, 1969; Boivie, 1979; Berkley, 1980; Asanuma *et al.*, 1983b). In monkeys, and in man, these terminations commence in the ventral aspect of the suprageniculate nucleus (Fig. 11.9) and were once mistakenly believed to be in the magnocellular medial geniculate nucleus. The terminations extend forwards toward the posterior pole of the ventral posterior nucleus. Boivie (1979) believes that they continue anteriorly dorsal to the ventral posterior nucleus but this has not been confirmed (Mantyh, 1983; Asanuma *et al.*, 1983b). In cats, spinothalamic terminations commence among the small cells of the Pom that invade the magnocellular medial geniculate nucleus (Fig. 8.6) and continue forwards throughout most of the Pom (Boivie, 1971b; Jones and Burton, 1974; Berkley, 1980). In rats, marsupials, and other small mammals, the spinothalamic terminations also commence in the vicinity of the internal (magnocellular) medial geniculate nucleus and continue forward into the Pom though they do not seem to fill the Pom (Lund and Webster, 1967b; Mehler, 1969; Rockel *et al.*, 1972). Retrograde tracing studies in the cat suggest that spinothalamic axons innervating the Pom region may arise from cells in laminae of the spinal cord somewhat different from those giving rise to axons ending in other thalamic nuclei (Carstens and Trevino, 1978). Dorsal column–lemniscal and cervicothalamic fibers have also been described as contributing fibers to the Pom region (Berkley and Hand, 1978; Boivie, 1970, 1971a, 1978). It is difficult to evaluate these reports because fibers of these systems traverse the Pom en route to the ventral posterior nucleus.

The lateral region of the posterior nucleus (Pol) in the cat and monkey and a comparable region in other animals is the recipient of fibers from the inferior colliculus (Moore and Goldberg, 1963; Tarlov and Moore, 1966; Rockel *et al.*, 1972; Moore *et al.*, 1977), particularly from its external nucleus (Kudo and Niimi, 1980). The intermediate region of the posterior nucleus (Poi) in the cat has not been adequately investigated for afferent connections.

There is no anatomical evidence to support earlier physiological reports (e.g., Mickle and Ades, 1954) of inputs from the vestibular system to the vicinity of the posterior complex.

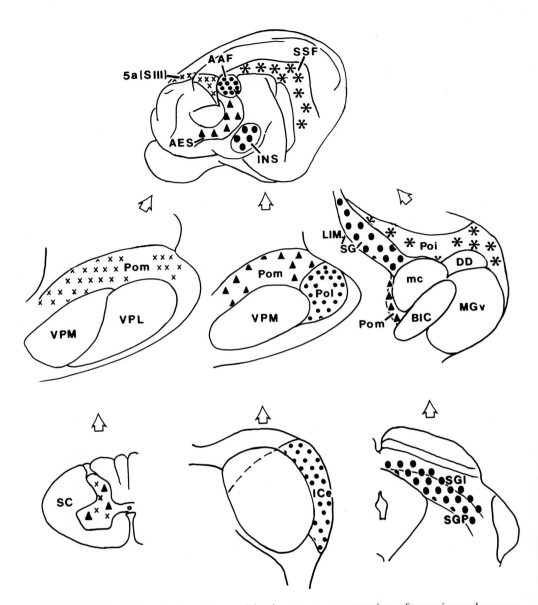

FIGURE 11.8. Schematic figure summarizing input–output connections of posterior nuclear complex in the cat. Known sources of inputs are spinal cord (SC), external nucleus of inferior colliculus (ICe), and deeper layers of superior colliculus (SGI, SGP). Cortical areas receiving inputs from posterior complex form a band which commences in insular cortex (INS), extends up banks of anterior ectosylvian sulcus (AES), across anterior auditory field (AAF) and third somatic sensory area (area 5a or SIII), then in suprasylvian fringe (SSF) along ventral bank of middle suprasylvian sulcus and onto posterior ectosylvian gyrus.

11.3.1.2. Corticothalamic

The components of the posterior complex appear to receive corticothalamic fibers from regions of cortex to which they project, though the amounts of data available are small. The limitans–suprageniculate nucleus clearly receives fibers from the insular cortex of the cat (Heath and Jones, 1971a; Graybiel and Berson, 1980b). There is no comparable work in monkeys but because the limitans–suprageniculate nucleus projects to the granular field of the insular cortex (Burton and Jones, 1976; Section 11.3.2), one may anticipate that this part of the primate insula will return fibers to the nucleus. There are no relevant data from other mammals.

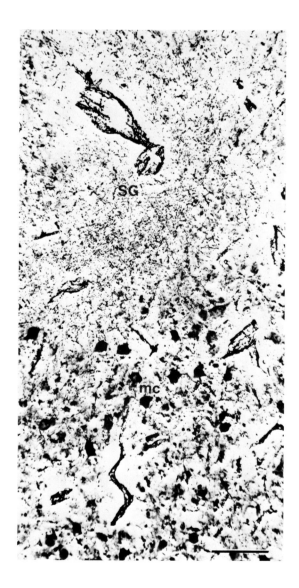

FIGURE 11.9. Degenerating spinothalamic tract fibers ramifying in suprageniculate nucleus (SG) of a cynomolgus monkey, rather than around large cells of magnocellular medial geniculate nucleus (mc). From Asanuma *et al.* (1983b).

The part of the Pom region lying dorsal to the ventral posterior nucleus in the cat receives fibers from the anterior part of parietal area 5 (Tanji *et al.*, 1977). This corticothalamic projection was originally thought to emanate from the somatic sensory cortex adjacent to area 5 (Jones and Powell, 1968; Rinvik, 1968c,d) (Fig. 11.5). The earlier reports were probably based upon interruption of fibers from area 5 when lesions were placed in the somatic sensory cortex. I have never been able to label the Pom region of monkeys from injections of tracer in or near the somatic sensory cortex. Commonly, however, such injections when they spread to area 5 result in labeling of the anterior pulvinar nucleus (Jones *et al.*, 1979; Friedman *et al.*, 1980; Chapter 10). One wonders, therefore, whether the anterior pulvinar nucleus, a nucleus with no obvious equivalent in the cat brain, should be considered comparable to this part of the Pom. Against this, however, is the evidence that the anterior pulvinar nucleus of the monkey, unlike the part of the Pom lying dorsal to the ventral posterior nucleus in the cat, does not obviously receive spinothalamic terminations (Section 11.3.1.1). I do not think that the Pom of the cat can be regarded as equivalent to the dorsal part of the monkey ventral posterior nucleus. In the rat the first somatic sensory area (SI) clearly does project to both the ventral posterior nucleus and the Pom (Wise and Jones, 1977a; Fig. 11.10). There is reason to suspect, however, that the two corticothalamic projections may arise in different cytoarchitectonic elements of the SI (Donoghue *et al.*, 1979).

The Pol and Poi regions appear to be the targets of different fields in and around the auditory region of the cat cortex (Diamond *et al.*, 1969) (Fig. 8.9). The details of these corticothalamic connections have not been thoroughly elucidated. It appears that the Pol region receives corticothalamic fibers mainly from the anterior auditory field (Andersen *et al.*, 1980b). The Poi region may be more closely connected with the fields in the lateral bank of the suprasylvian sulcus and/or in the posterior ectosylvian gyrus (e.g., Diamond *et al.*, 1969;

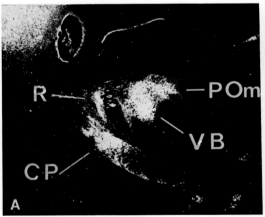
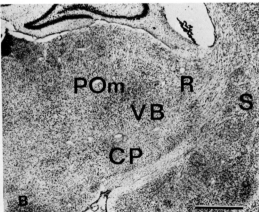
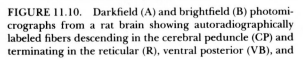

FIGURE 11.10. Darkfield (A) and brightfield (B) photomicrographs from a rat brain showing autoradiographically labeled fibers descending in the cerebral peduncle (CP) and terminating in the reticular (R), ventral posterior (VB), and medial division of the posterior nucleus (POM) following an injection (A, top left) of tritiated amino acids in the first somatic sensory area. S: striatum. Bar = 500 μm. From Wise and Jones (1977a).

Updyke, 1981a). It is doubtful, however, that the Poi region has been clearly
distinguished from the dorsal nuclei of the medial geniculate complex in some
of the studies reporting corticothalamic connections to the Poi region.

593

THE POSTERIOR
COMPLEX
OF NUCLEI

In primates or rodents, there are no studies of the corticothalamic connections of the posterior complex comparable in detail to those in cats.

11.3.2. Efferent Connections

11.3.2.1. Background

Until quite recently, the only significant studies of the thalamocortical connections of the posterior complex were carried out on the cat. Work on the cat commenced with the study of Rose and Woolsey (1958). Prior to that time, the posterior nucleus, probably because of its ill-defined character, had been ignored. The limitans–suprageniculate nucleus had been found not to undergo retrograde degeneration after cortical lesions in monkeys and it was thought that it might be a part of the intralaminar complex and to project to the striatum (Le Gros Clark, 1936a; Le Gros Clark and Northfield, 1937; Walker, 1938a).

Rose and Woolsey's investigation arose out of a study in which Meyer and Woolsey (1952) had attempted to define the cortical areas necessary for the ability to make discriminations between different frequencies of sound.* In the course of this study, they ablated various large areas of the auditory cortex and Rose and Woolsey later analyzed the distribution of retrograde degeneration in the thalamus. The degeneration was always extensive and in all cases affected the lateral geniculate, pulvinar, and lateral posterior nuclei in addition to the medial geniculate and posterior complexes, because of disruption of fibers running in the white matter deep to the auditory areas. However, a subtractive analysis from case to case permitted conclusions to be drawn about the projections of the posterior complex.

Their basic finding was that parts of the posterior complex that we would now regard as the Poi and Pol underwent severe retrograde degeneration when cortical removals were sufficiently large to cause almost total atrophy of the dorsal and ventral medial geniculate nuclei and degeneration of the posterior part of the magnocellular medial geniculate nucleus. Such cortical ablations had to include virtually all the auditory fields then known (Fig. 8.7). When a lesion expanded forwards through and beneath the anterior ectosylvian sulcus so as to involve the second somatic sensory area (SII) and adjoining fields such as the insular cortex and the region now known as the anterior auditory field, the parts of the posterior complex equivalent to our limitans–suprageniculate nucleus and the Pom region underwent total atrophy, along with the anterior part of the magnocellular medial geniculate nucleus. Destruction of the SII area alone, however, was without effect on the posterior complex and destruction of the auditory cortex minus the SII region was without effect on anterior parts of the

---

* It has subsequently been shown that the cortex is not essential for frequency discrimination, cats being able to do this provided the auditory midbrain is intact (Goldberg and Neff, 1961). The cortex is, however, necessary for most aspects of sound localization (e.g., Whitfield *et al.*, 1972; Heffner, 1978).

posterior complex. Rose and Woolsey (1958) argued from these findings that anterior parts of the posterior complex (mainly the Pom) would project in a sustaining fashion upon both the second somatic sensory area and some of the auditory fields. Similarly, because other parts of the posterior complex and the magnocellular medial geniculate nucleus only underwent severe retrograde degeneration when several fields of the auditory cortex were removed together, they argued for widespread sustaining projections from these nuclei also.

No further contributions were made on the subject of the cortical projections of the posterior complex for several years. In 1966 Guillery *et al.* cast doubt upon the idea that anterior parts projected to SII by showing that electrical stimulation in the thalamic ventral posterior nucleus of the cat could elicit focal evoked potentials in SII as well as in SI (Chapter 7). The projection of the ventral posterior nucleus to both SI and SII was confirmed anatomically with axonal degeneration methods a few years later (Jones and Powell, 1969a; Hand and Morrison, 1970). In 1971, by making lesions in the posterior complex with electrodes introduced into the thalamus horizontally from behind, Heath (1970) and Heath and Jones (1971a) were able to avoid the problem of interference with fibers of passage and show that large parts of the posterior complex as a whole projected upon an extensive band of cortex (Fig. 11.11): starting in the insular area, this extended upwards through the banks of the anterior ectosylvian sulcus between SII and the auditory fields; it then crossed the upper part of the anterior ectosylvian gyrus to enter the suprasylvian sulcus, in the ventral bank of which it extended over the AI auditory field and back into the posterior ectosylvian gyrus. The projection to the banks of the anterior ectosylvian sulcus was detected independently by Hand and Morrison (1970).

It is easy to see from the localization of this band of cortex how Rose and Woolsey's (1958) lesions, extending forward from the auditory areas, would have damaged the insular cortex and the cortex in the anterior ectosylvian sulcus and how large posteriorly placed lesions would have undercut cortex in the suprasylvian sulcus. The Pol region and parts of the Poi region were clearly not involved in the lesions made by Heath and Jones (1971a) so the experiments were inconclusive about their cortical targets. Moreover, the results seemed to indicate that the magnocellular medial geniculate nucleus was also targeted on the same cortical band as the limitans–suprageniculate nucleus and Pom regions which were destroyed. The magnocellular nucleus has since been shown to have more widespread cortical projections, as Rose and Woolsey (1958) originally suggested (Chapter 8), but the parts of the posterior complex are obviously focused on more limited regions.

11.3.2.2. Recent Studies

a. Cat. In one of the earliest studies made using the retrograde transport of horseradish peroxidase, Jones and Leavitt (1973) confirmed the projection of the limitans–suprageniculate nucleus to the insular cortex and suggested that other parts of the posterior nucleus projected upon different parts of the band of cortex originally delimited by Heath and Jones (1971a). The absence of retrograde labeling in the posterior complex after injections of tracer in the SI and SII cortex cast doubt upon the observations of Rowe and Sessle (1968) and

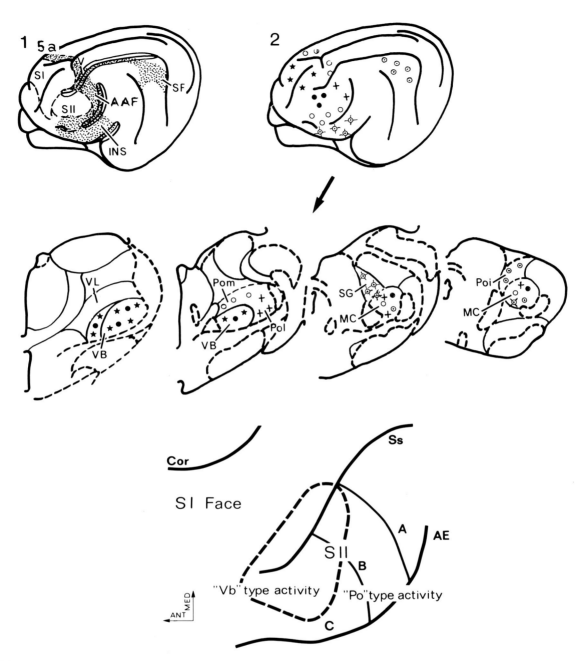

FIGURE 11.11. (Top) Figure modified from Jones and Leavitt (1973) showing (1) band of projection (stippling) of posterior nuclear complex in cat and (2) parts of posterior complex containing retrogradely labeled cells after injections of horseradish peroxidase in different parts of the band. Magnocellular medial geniculate nucleus (MC) contains labeled cells after injections of any part of band as well as of adjacent (unmarked) auditory areas. Ventral posterior nucleus only contains labeled cells after injections of first (SI) or second (SII) somatic sensory areas. (Bottom) Drawing from Haight (1972) showing older subdivision (A, B, C) of anterior ectosylvian gyrus and further division, corresponding to projection areas of ventral posterior and Pom nuclei, in which microelectrode recordings reveal units with receptive fields typical of the two projecting nuclei.

Curry (1972b) that neurons in the posterior complex could be antidromically activated by electrical stimulation of these cortical areas. For some years very little was added to the anatomical findings until the detailed studies of thalamocortical connectivity of the third somatic sensory area (SIII) of the cat cortex by Tanji *et al.* (1977) and of the cat auditory cortex by Winer *et al.* (1977), complemented by the study of corticothalamic connectivity made by Andersen *et al.* (1980a) (see Chapter 8).

SIII, the physiologically definable correlate of the architectonic area 5a of the cortex, is reciprocally connected with the part of the Pom that expands forwards dorsal to the ventral posterior nucleus (Figs. 11.5 and 11.11). Previous anatomical reports of projections to this part of the Pom from the SI and SII cortex (Jones and Powell, 1968; Rinvik, 1968c,d) undoubtedly stemmed from interruption of fibers from area 5a which adjoins both of them. It is possible that the more posterior parts of the Pom are also connected with area 5a. Rowe and Sessle (1968) reported a few cells in the vicinity of the magnocellular medial geniculate nucleus that were antidromically activated from the SIII area. The reports of Winer *et al.* and of Andersen *et al.* are not particularly clear on the cortical relations of the Poi and Pol but, looking at their figures, it appears to me that heavy and consistent retrograde labeling was obtained in the limitans–suprageniculate nucleus after injections of the insular cortex, in the Poi region after large injections of the posterior ectosylvian gyrus and of the Pol region after injections close to the suprasylvian sulcus. The recent report of Bentivoglio *et al.* (1983) is in general agreement with this interpretation. The data, however, are insufficient to determine whether the three regions of the posterior nucleus project to single auditory fields or diffusely to several (Fig. 11.8).

b. Monkeys. The cytoarchitecture and cortical projections of the posterior complex of rhesus monkeys and squirrel monkeys were examined by Burton and Jones (1976) who concluded that the cortical target of the region as a whole could, like in the cat, be regarded as a band starting on the insula and curving between and around SII and the first auditory field (AI) (Fig. 11.12). The limitans–suprageniculate nucleus clearly projects to the granular insular field, a medial part of the posterior nucleus, equivalent to that receiving spinothalamic afferents, projects to the retroinsular field (Ri) between SII and AI, and a lateral, presumed auditory part projects to the postauditory field adjacent to Ri and posterior to AI (Fig. 11.12). There was no evidence for a differentially projecting region equivalent to the Poi region of the cat.

c. Rat and Other Animals. Injections of retrograde tracer into the vicinity of the SI area of rats, retrogradely label cells both in the ventral posterior nucleus and in the anterior part of the Pom region that lies dorsal to it (Jones and Leavitt, 1974; Donoghue *et al.*, 1979). It is difficult to be sure that this represents a projection to an area adjacent to SI or to a particular cytoarchitectonic component of SI. SI of the rat is divided into granular regions that form the basis of the body map, and intervening and surrounding dysgranular regions that are relatively unresponsive to somesthetic stimuli (Welker, 1976) (Fig. 7.17). Only the granular regions have been shown by anterograde labeling to receive an

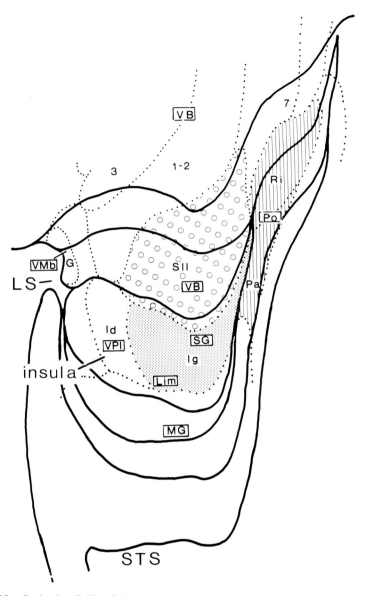

FIGURE 11.12. Projection fields of the posterior nucleus complex in the rhesus monkey. Lateral sulcus (LS) is drawn as though opened out to reveal insula. STS indicates superior temporal sulcus. Granular insular area (Ig) receives fibers from suprageniculate–limitans (SG-Lim) nucleus; retro-insular (Ri) and postauditory (Pa) fields receive inputs from posterior nucleus (Po). Adjacent somatic sensory areas (3, 1–2, SII) receive fibers from ventral posterior nucleus (VB), auditory fields from medial geniculate nucleus (MG), gustatory area (G) from basal ventral medial nucleus (VMb), and dysgranular insular area (Id) probably from ventral posterior inferior nucleus (VPI). After Jones and Burton (1976b).

input from the ventral posterior nucleus (Wise and Jones, 1978). It is possible, therefore, that the dysgranular zones [which are the commissurally connected parts of SI in the rat (Wise and Jones, 1976)] receive their thalamic inputs from the Pom region. This has not, to my knowledge, been investigated.

Cortical connectivity of other parts of the rat posterior complex has not been examined.

Anterograde labeling of thalamocortical projections in the tree shrew (Oliver and Hall, 1978a,b) (Fig. 8.12) indicates that the posterior complex in that insectivore projects to a region anteroventral to AI and perhaps corresponding to the insula, and to a second region enclosing the anterior, dorsal, and posterior parts of AI and other auditory fields. In this pattern, there is some resemblance to the curved band of projection detected in the cat and monkey.

11.4. Physiology of the Posterior Complex

11.4.1. Single-Unit Studies in Animals

In recording from single units throughout a large part of the posterior complex of barbiturate-anesthetized cats, Poggio and Mountcastle (1960) remarked on the great difference between their receptive field properties and those of the adjoining ventral posterior nucleus. The majority of ventral posterior neurons (Chapter 7) possess small, contralateral receptive fields and respond to only one type of peripheral stimulus. That is, they are place and modality specific. By contrast, 60% of the units recorded in the posterior complex by Poggio and Mountcastle responded not to gentle mechanical stimulation of the body but to cuts, pinpricks, and other stimuli that threatened destruction of tissue. Many of the remaining neurons could be activated by gentle mechanical stimulation of skin or hairs but all neurons had extremely large receptive fields covering much of the contralateral, ipsilateral, and even both sides of the body. There was no somatotopy in the posterior complex. Moreover, many cells were polymodal, responding to noxious and gentle mechanical stimulation applied to different parts of the body and often to auditory and vibratory stimuli as well. The parts of the posterior complex recorded from by Poggio and Mountcastle included the suprageniculate, Pom, Pol, and Poi regions, as well as the magnocellular medial geniculate nucleus (Fig. 11.13). They remarked that somatic sensory neurons were more common anteriorly and polymodal neurons elsewhere, except in more posterior parts of the magnocellular medial geniculate nucleus in which neurons were more purely auditory in character. They also illustrate one completely auditory track through the Pol region (Fig. 11.13). Responses to vibration were thought to be mediated by the vestibular system, for Mickle and Ades (1954) had previously reported vestibular evoked potentials in the vicinity of the magnocellular geniculate nucleus.

Poggio and Mountcastle reasoned that the neurons of the posterior complex behaved in a manner expressive of the functional properties of the spinothalamic system, though not exclusively so. They concluded that the posterior complex as a whole and its cortical projection (thought at that time to be primarily to the

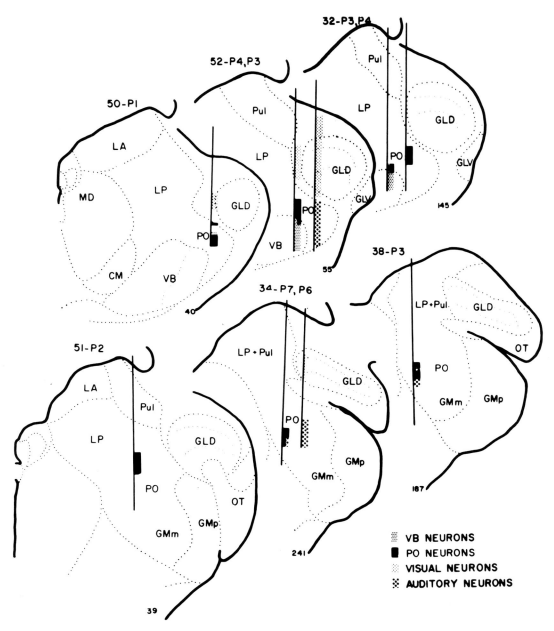

FIGURE 11.13. Outline drawings of sections through the posterior complex of a cat on which nine electrode penetrations are reconstructed and location of single units activated by stimuli indicated are marked. Note lack of penetrations through heart of Pom region and appearance of a pure auditory track through the Pol region. From Poggio and Mountcastle (1960).

SII region) were concerned in some special way with pain sensibility. This contrasted with the ventral posterior nucleus whose properties and cortical projection (thought to be only to SI) they reasoned made it suitable for relaying the discriminative forms of somatic sensibility.

Essentially similar results to those of Poggio and Mountcastle (1960) were reported from studies of the posterior complex in the unanesthetized squirrel monkey by Casey (1966) and in anesthetized monkeys and cats by Whitlock and Perl (1961), Perl and Whitlock (1961), Calma (1965), Rowe and Sessle (1968), and Berkley (1973). Casey reported that anesthetic could have a powerful effect upon the responses of the cells: even a light level of anesthesia could convert multimodal cells to apparently purely nociceptive cells. Perl and Whitlock and Whitlock and Perl noted that unit responses in the posterior complex survived destruction of all spinal pathways except those in the ipsilateral ventrolateral quadrant, indicating a predominance of spinothalamic inputs to the posterior complex. Rowe and Sessle confirmed this by showing that reversible cooling of the dorsal column nuclei had little effect on the posterior complex cells. Berkley did note a few cells in the Pom responsive to stimulation in the vicinity of the dorsal column nuclei and attributed these responses to a small dorsal column–lemniscal input to the Pom reported by Boivie (1971a).

Probably the most detailed physiological study of the somatic sensory part of the posterior complex is that of Curry (1972a,b; Curry and Gordon, 1972). This study focused on the Pom region which had by then been identified by anatomical connections, and, though including the magnocellular medial geniculate nucleus, it excluded the Poi and Pol regions and the limitans–suprageniculate nucleus. Selective electrical stimulation of the spinal cord suggested, on the basis of the size and latencies of gross potentials elicited in the Pom region, that the contralateral dorsal columns and dorsolateral funiculus (spinocervical tract), rather than the spinothalamic tract, provided the major input to the Pom. However, all responses were significantly smaller in amplitude than those in the ventral posterior nucleus.

By contrast with the observations of Poggio and Mountcastle (1960), Curry (1972a) reported that most single units recorded in the Pom responded to innocuous somesthetic stimuli, particularly to movement of hairs. Receptive fields were, however, large and sometimes bilateral with dissociated inhibitory receptive fields and the cells were relatively unresponsive. A few cells did respond to noxious stimulation, a few to tapping the skin, a few to auditory stimulation, and approximately 20% had convergent somatic and auditory input. Because of the paucity of cells responding to noxious stimuli and the suggestive evidence of their main input being the dorsal columns, Curry felt that the Pom region could not be regarded as a pain relay. Given the recent data on the projection of high-threshold and wide-dynamic-range spinothalamic tract neurons to the ventral posterior nucleus (Giesler et al., 1981; Honda et al., 1983; Chapter 7), it is probably necessary to agree in part with Curry. On the other hand, the cells have some resemblance to those spinothalamic tract cells that project to the intralaminar nuclei (Giesler et al., 1981; Chapter 12) so that some role in pain transmission is not completely ruled out. As to the discrepancy between Curry's results and those of Poggio and Mountcastle (1960), I believe that it may be attributed to the fact (illustrated in their figures) that Curry was recording more ventral and medial than Poggio and Mountcastle. He may, thus, have been dealing with a more purely somatic population of neurons, though from the anatomy, one would expect these to receive a predominately spinothalamic input.

None of the studies on the physiology of posterior complex cells in the cat have explored the anterior extension of the Pom (where it lies dorsal to the ventral posterior nucleus). In the cortical field to which it projects, area 5a or the third somatic sensory area (Darian-Smith *et al.*, 1966; Dykes *et al.*, 1977; Tanji *et al.*, 1977), unit responses are more typical of those of the lemniscal system than of those so far reported in the posterior complex (Dykes *et al.*, 1977; Felleman *et al.*, 1983). This may support Boivie's (1971a) contention that fibers from the dorsal column nuclei to the posterior complex end primarily in this anterior extension of the Pom.

Newer physiological data on the other cortical areas to which the subregions of the posterior complex project furnish little evidence to implicate the complex in any easily identifiable aspect of pain sensibility. An early single-unit study of Carreras and Andersson (1963) on the anterior ectosylvian gyrus of the cat provided evidence for the presence of a significant number of neurons with properties akin to those reported by Poggio and Mountcastle (1960) in the posterior complex. These were initially considered to be in the SII cortical area but more recent anatomical and physiological parcellations of the anterior ectosylvian gyrus would place them outside the borders of SII as defined by the projection of the ventral posterior nucleus (Heath and Jones, 1971a,b; Haight, 1972; Burton *et al.*, 1982) (Figs. 11.8, 11.1, and 11.14). They do lie, however, within the target cortex of the Pom region and bilateral lesions of this cortex in cats may under some circumstances increase the threshold for escape from electric shock (Berkley and Parmer, 1974).

In monkeys the Pom projection cortex [the retroinsular area (Burton and Jones, 1976)] (Figs. 11.12 and 11.15) does not contain any neurons uniquely sensitive to noxious stimuli though it does contain neurons with large receptive fields somewhat similar to those of the Pom region of the cat. The Pol projection cortex in the monkey (the postauditory area) is purely auditory in character (Robinson and Burton, 1980a–c). The Pol and Poi projection cortex in cats would also be predicted to be auditory.

The cortical target of the limitans–suprageniculate nucleus is the granular insular area in monkeys (Burton and Jones, 1976) and the so-called insular area in the cat (Heath and Jones, 1971a; Jones and Leavitt, 1973) (Figs. 11.11, 11.12, and 11.15). In this area in the cat, neurons have been reported to show convergence of visual, auditory, and somesthetic inputs (Loe and Benevento, 1969) as might be expected from the target of the limitans–suprageniculate nucleus which receives its inputs from the deeper layers of the superior colliculus (Section 11.3.1.1). In the monkey, however, most neurons in the granular insular area are reported to be driven by innocuous somatic stimuli and to have large receptive fields (Robinson and Burton, 1980c).

11.4.2. Lesions in Man

None of the recent studies on the posterior complex provide much evidence for its major involvement in a central pain pathway, and the work reviewed in Chapter 7 makes it likely that the ventral posterior nucleus would at the moment

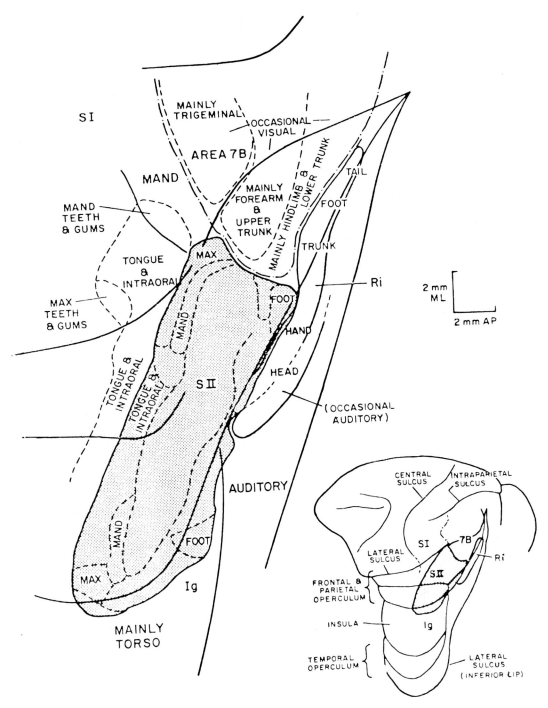

FIGURE 11.14. Topographic map of second somatic sensory area (SII) and nature of stimuli eliciting single- or multiunit responses in adjoining regions of the cynomolgus monkey brain. Retroinsular area (Ri) has predominately somatic sensory responses. From Robinson and Burton (1980b). Compare with Figs. 11.12, 7.21, and 8.10.

be more fruitful ground for the study of thalamic pain mechanisms. Nevertheless, a number of tantalizing observations in human patients still lead one to associate some aspects of pain with the posterior complex, just as one does with the intralaminar system (Chapter 12). Hassler (1960) has reported that stimulation of the posterior nucleus (his nucleus limitans portae) in man elicits pain and that large lesions in this region alleviate painful conditions. It is possible, of course, that these effects are caused by stimulation or destruction of spinothalamic fibers that pass through the posterior nucleus en route to other thalamic nuclei but this has not been determined.

Electrical stimulation of the presumed SII area and the adjacent, probable target cortex of the posterior complex does not elicit reports of pain in man (White and Sweet, 1969; Lende *et al.*, 1971). However, deep undercutting or excision of parts of the parietotemporal operculum has alleviated painful conditions (Talairach *et al.*, 1949; Lende *et al.*, 1971) and Biemond (1956) has described two cases in which very small infarctions in the opercular–insular region were associated with hyperpathia or hemi-hypalgesia. It is easy to dismiss reports of this kind on the grounds that damage to white matter disconnects many different cortical areas from one another and from the thalamus. The effects may have nothing to do with the posterior complex or its cortical target but they cannot be dismissed out of hand.

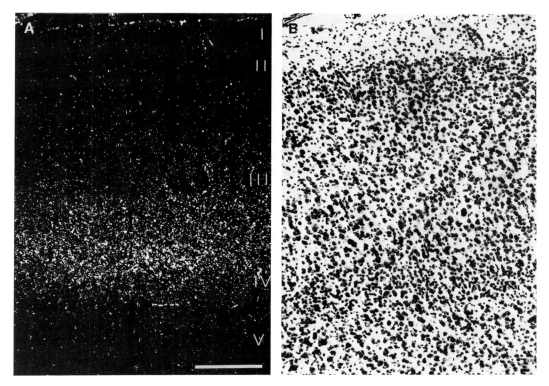

FIGURE 11.15. Paired brightfield and darkfield photomicrographs showing layers of termination of thalamocortical fibers in granular field of monkey insular cortex. Bar = 250 μm. From Jones and Burton (1976a).

11.4.3. Conclusion

One is obliged to conclude that there is little about the posterior complex or its cortical target that argues strongly in favor of their involvement in the direct transmission of information about the quality, intensity, or localization of a painful stimulus. All current evidence would favor the viewpoint that these are relayed through the ventral posterior nucleus. On the other hand, an admittedly rather small subdivision (the Pom region) of the posterior complex receives spinothalamic afferents and its neurons respond to noxious stimuli in a manner reminiscent of those of another spinothalamic target, the central lateral nucleus of the rostral intralaminar group (Chapter 12). There is much circumstantial evidence to support the notion that the intralaminar nuclei and their widespread cortical projections are implicated in some way in raising the level of cortical excitability in relation to a peripheral stimulus and perhaps in controlling the affective response to such a stimulus (Chapter 12). Perhaps in the Pom region of the posterior complex and its projection to the parietal operculum or its equivalent in animals, we should seek an involvement in other aspects of the pain response, say something more typical of parietal lobe function such as pain symbolism.

One thing is clear, however, and that is that the posterior complex cannot be thought of as a single entity, perhaps engaged in some as yet ill-defined aspect of pain sensibility. Rather, it is a group of several nuclei each with distinct input–output connections. The only thing they have in common is their proximity in the thalamus and their projections to cortical fields in and around the insula. The limitans–suprageniculate nucleus is evidently multimodal, perhaps with a preponderance of visual influences from its connections with the superior colliculus; the Pol and Poi regions are presumably auditory from their connections and the Pom region is evidently somatic sensory. But none appear to have the specificity of organization typical of relays in the major sensory pathways. If there is some general physiological characteristic that provides them with some commonality of function, it has thus far eluded us.

The low tide and nadir of hope from 2 a.m. to 4.
Magical Euphoria wells from 4 a.m. to 6—the
thalamic "All Clear."
          Cyril Connolly, *The Unquiet Grave* (1981)

The Intralaminar Nuclei

Surface slow waves elicited in the cerebral cortex from repetitive stimulation at 7 per second of the intralaminar nuclei. Note sequential augmentation and decline (recruiting). From Jasper (1949).

12.1. Definition

The obvious cell masses within the internal medullary lamina are the central medial, paracentral, and central lateral nuclei. The central lateral nucleus may be subdivided further, especially in primates (Figs. 12.1 and 12.2). The central medial nucleus is a single nucleus in the part of the lamina that crosses the midline. With it at the midline are found additional cell groups, sometimes referred to as a single rhomboid or central nucleus or subdivided further into subsidiary nuclei with other names. Collectively, the nuclei so far mentioned form the rostral group of intralaminar nuclei (Berman and Jones, 1982) or the central commissural system (Rose, 1942b).

Toward the posterior end of the internal medullary lamina and contained within a splitting of the lamina, is the centre médian nucleus, which is large in primates, of medium size in some mammals but only distinguished with difficulty or not at all in small mammals such as rodents, lagomorphs, and marsupials (Figs. 12.3–12.5). Posteromedial to the centre médian nucleus and extending to the medial medullary lamina and posterior surface of the thalamus, is the parafascicular nucleus (Figs. 12.3–12.5). The centre médian and parafascicular nuclei constitute the caudal group of intralaminar nuclei (Le Gros Clark, 1932a; Berman and Jones, 1982) or the postmedial group of nuclei (Rose, 1942b).

12.2. Rostral Group of Intralaminar Nuclei

12.2.1. Description

The *central medial, paracentral, central lateral,* and *rhomboid* nuclei form the rostral group of intralaminar nuclei. The first three are not easy to separate from one another and, collectively, form an extensive, thin sheet stretching through much of the anteroposterior extent of the thalamus. The rhomboid nucleus is in the midline and less clearly associated with the internal medullary lamina but extends winglike groups of cells into the lamina anteriorly (Fig. 12.1A).

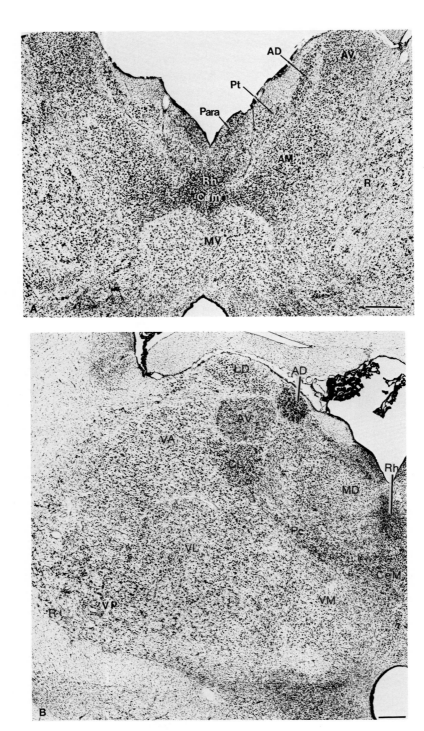

FIGURE 12.1. Frontal sections in anterior to posterior order, showing the rostral group of intra-laminar nuclei in a cat. Arrows in (D) indicate posterior, large-celled component of central lateral nucleus. Thionin stain. Bars = 500 μm.

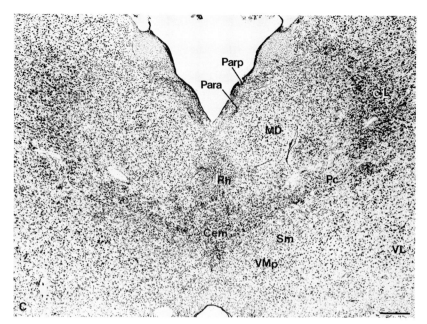

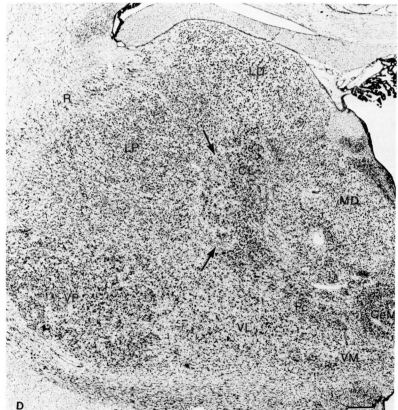

FIGURE 12.1. *(continued)*

12.2.1.1. Central Medial Nucleus

The central medial nucleus is an unpaired structure situated across the midline and joining together the paracentral and central lateral nuclei of the two sides (Figs. 12.1–12.3). It contains large, deeply staining cells which at the anterior pole of the nucleus merge with the interanteromedial part of the two anteromedial nuclei and with the rhomboid nucleus. The posterior pole of the central medial nucleus lies between the caudal group of intralaminar nuclei immediately anterior to the mesencephalic central gray matter. The parafascicular nucleus can often be traced in continuity with it.

Over the anterior half of its extent the central medial nucleus lies between the rhomboid nucleus dorsally and the fibrous capsule of the medioventral nucleus ventrally. This fibrous capsule is the continuation of the internal medullary lamina across the midline. Posteriorly, as the rhomboid nucleus thins out, the central medial nucleus can lie directly adjacent to the mediodorsal nucleus. Its lateral parts extend out into the internal medullary laminae and become continuous with the paracentral nucleus of each side.

The central medial nucleus is evident in all species and there is usually little difficulty in identifying it (Figs. 12.1, 12.6, and 12.7).

12.2.1.2. Rhomboid Nucleus

The rhomboid ("central") nucleus is a single midline mass of cells rather smaller than those of the central medial nucleus. It has lateral winglike extensions

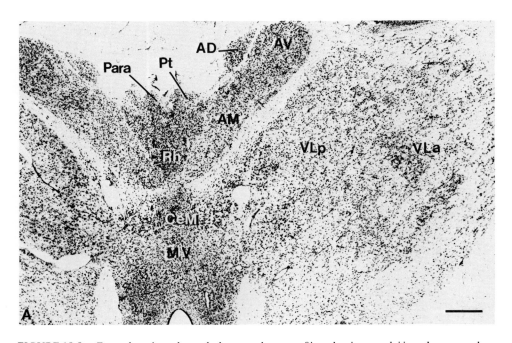

FIGURE 12.2. Frontal sections through the rostral group of intralaminar nuclei in a rhesus monkey (A, B) and a raccoon (C). Thionin stain. Bars = 500 μm.

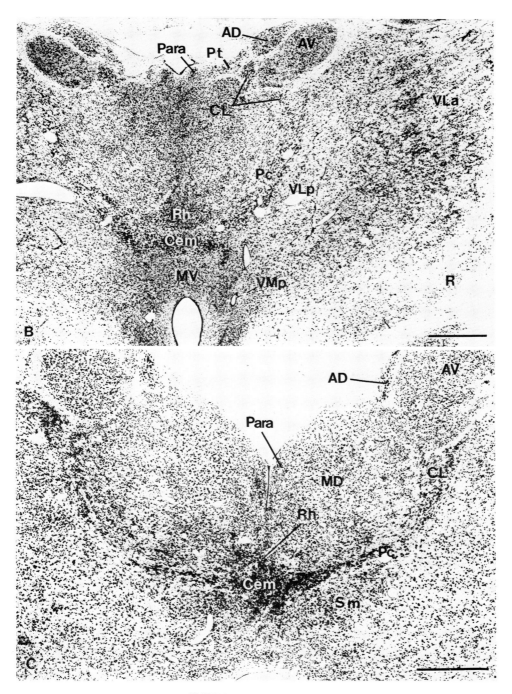

FIGURE 12.2. (*continued*)

composed of smaller, elongated cells extending into the internal medullary lamina of each side and this may be the most obvious part in some species such as rodents (Fig. 12.6). The central mass of the nucleus is quite large in many mammals and extends from the posterior end of the interanteromedial component of the anteromedial nuclei (with which it fuses) to the anterior poles or even the middles of the mediodorsal nuclei. Over most of its extent, the central mass lies between the two anterior paraventricular nuclei dorsally and the central

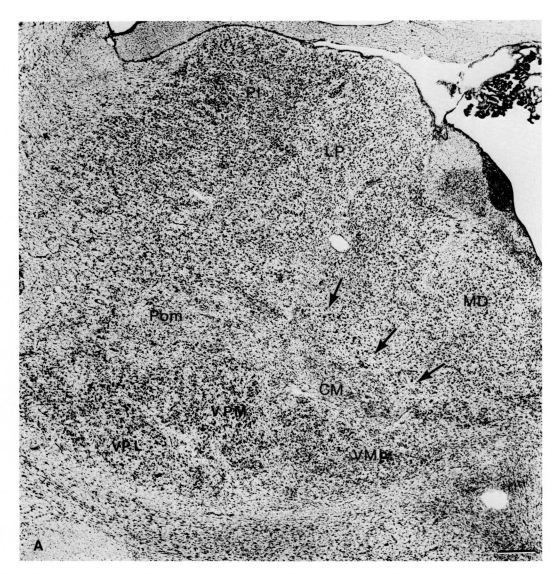

FIGURE 12.3.   Frontal sections through the centre médian and parafascicular nuclei of a cat. Arrows indicate large-celled component of central lateral nucleus. Thionin stain. Bars = 500 μm.

medial nucleus ventrally; posteriorly, it intervenes between the two mediodorsal nuclei. *Central nucleus* is, thus, an obvious alternative name. In its dorsal part in primates, it approaches the medially compressed dorsal part of the central lateral nucleus (Fig. 12.2) so that each mediodorsal nucleus may appear as though completely surrounded by intralaminar nuclei.

12.2.1.3. Paracentral Nucleus

The cells of the paracentral nucleus are usually of approximately the same size and staining intensity as the cells of the central medial nucleus. Therefore, the boundary between the two nuclei is determined mainly by the fact that the cells in the paracentral nucleus are more flattened (Figs. 12.1–12.6).

The paracentral nucleus occupies the anterior and medial portions of the internal medullary lamina. It is extremely thin in primates and its cells can form no more than a series of disconnected clusters in regions where the internal medullary lamina is at its thinnest. Over most of its extent the nucleus lies between the mediodorsal nucleus and the ventral nuclei. At its dorsal and posterior edges, it gives place to the central lateral nucleus which in most species expands to form a rather thicker component of the intralaminar system.

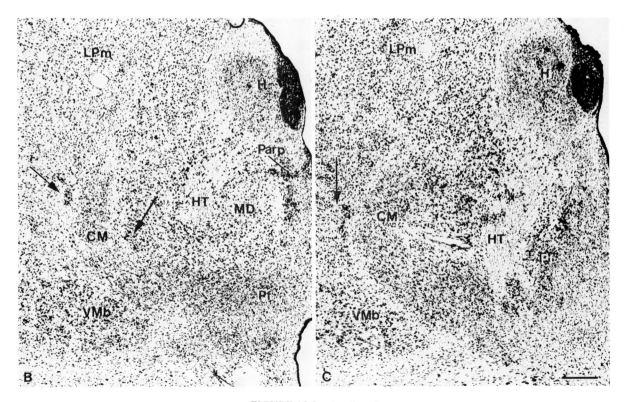

FIGURE 12.3. (*continued*)

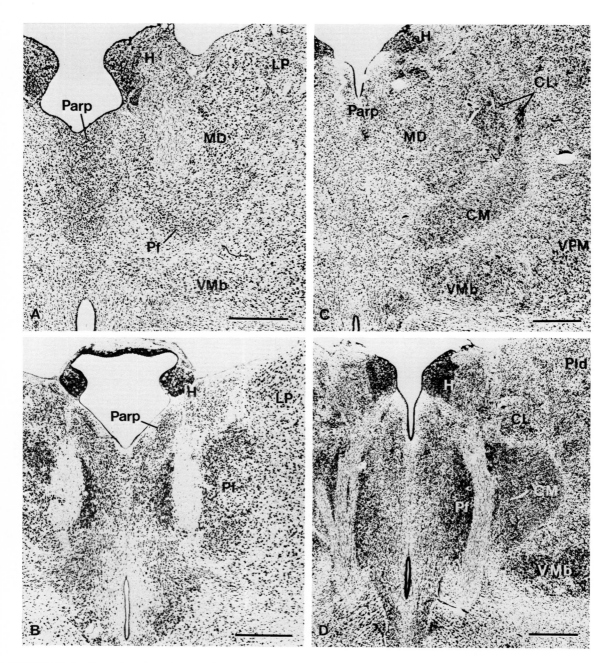

FIGURE 12.4.   Frontal sections through the parafascicular nucleus in a guinea pig (A), marsupial phalanger (B), and bush baby (C, D). Thionin stain. Bars = 1 mm.

12.2.1.4. Central Lateral Nucleus

This nucleus consists for the most part of cells of similar size, shape, and staining properties to those of the paracentral nucleus (Figs. 12.1–12.6). The two nuclei can often be distinguished from one another in the anterior and middle parts of their anteroposterior extent, for the central lateral nucleus forms a large dorsolateral expansion which is only tenuously connected to the paracentral nucleus. The dorsolateral expansion of the central lateral nucleus occurs where the internal medullary lamina splits to enclose the lateral dorsal and anteroventral nuclei and this part of the central lateral nucleus is, thus, Y shaped in frontal sections of many animals. In rabbits, it is even further expanded into a large triangular mass, sometimes named separately.

In primates the dorsal part of the central lateral nucleus is compressed medially by the enlarged lateral nuclear mass of nuclei and forms a small triangular region between the parataenial and lateral dorsal nuclei (Figs. 12.2A,B) and may approach the midline adjacent to the paraventricular nuclei. This part has been called the *superior central lateral nucleus* (Olszewski, 1952). Posteriorly, the cells of the central lateral nucleus become larger and even more deeply stained, but because the internal medullary lamina becomes less distinct the boundaries between the central lateral, lateral posterior, and mediodorsal nuclei become less obvious. Therefore, in its most posterior part, the central lateral nucleus consists of scattered, large cells lying at the lateral and ventrolateral edges of the mediodorsal nucleus. These large cells of the central lateral nucleus persist to levels more posterior than the central medial or paracentral nuclei. In rodents, they form isolated clusters that are probably usually included in the mediodorsal nucleus or regarded as isolated islands of the parafascicular nucleus (Fig. 12.4). In the cat, the large cells are intimately related to the centre médian nucleus, forming a small strip or cluster enclosed in the medially directed concavity of the centre médian nucleus (Figs. 12.1D and 12.3C). In primates, in my opinion, the large, deeply staining cells, sometimes called a paralamellar or densocellular component of the mediodorsal nucleus, are these posterior cells of the central lateral nucleus (Figs. 12.5, 12.7, 12.8, and 11.2). They extend around the posterior surface of the mediodorsal nucleus so as to fuse with the parafascicular nucleus and nucleus limitans (Fig. 12.8), a fact recognized by Grünthal (1934) many years ago.

12.3. Caudal Group of Intralaminar Nuclei

12.3.1. Centre Médian Nucleus

The centre médian nucleus, when present, is usually composed of a homogeneous population of small, closely packed cells. It is enclosed by a splitting of the posterior part of the internal medullary lamina and in an animal such as the cat, it is slung beneath the internal medullary lamina, intercalated mainly between the central lateral nucleus and the ventral posterior medial nucleus (Fig.

12.3). In primates the nucleus is greatly expanded and assumes a rounded shape in cross-section (Fig. 12.5). The nucleus is not distinguishable in lagomorphs, rodents, marsupials, bats, shrews, and certain other species with small brains, though some atlases of the rat label posterior parts of the internal medullary lamina centre médian (Albe-Fessard *et al.*, 1966). It is small in ungulates and carnivores, detectable in the tree shrew and in lower primates (Chapter 2), but reaches its maximum size in monkeys, apes, and man. The centre médian nu-

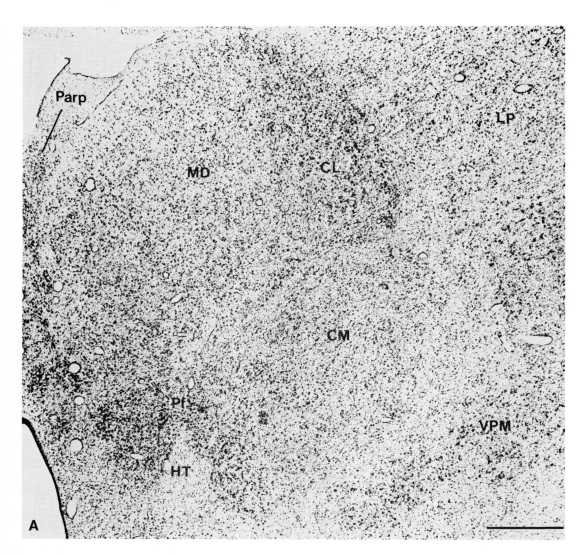

FIGURE 12.5.   Frontal (A) and sagittal (B) sections showing the centre médian and parafascicular nuclei in cynomolgus monkeys. Note large CL nucleus in (A) which has sometimes been called a paralamellar part of the mediodorsal nucleus. Thionin stain. Bars = 1 mm.

cleus, when present, is separated from the mediodorsal nucleus by the large-celled posterior extension of the central lateral nucleus and these large cells have often been included in the centre médian nucleus in animals such as the cat. In man, in this region, Vogt and Vogt (1941) and Hassler (1959) identify an extensive magnocellular division of the centre médian nucleus which Niimi *et al.* (1960) and Mehler (1966a) argue should be regarded as equivalent to the posterior part of the central lateral nucleus. At about the level of the habenulopeduncular tract, the parafascicular nucleus abuts upon the centre médian nucleus. The two nuclei are fairly readily distinguished because the parafascicular cells stain more intensely and are more tightly packed but there is usually a region of transitional cell types joining the ventral aspects of the two and in this region the distinction is less clear. Anteriorly, the ventral aspects of the two centre médian nuclei approach the midline beneath the central medial nucleus but the two nuclei do not usually fuse across the midline. The wide posterior

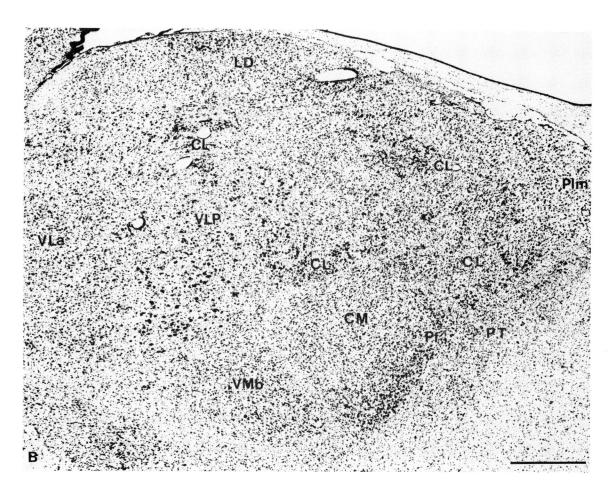

FIGURE 12.5. (*continued*)

surface of the centre médian nucleus abuts directly upon the pretectum of the midbrain, separated only by the thin medial medullary lamina (Fig. 8.2B).

12.3.2. Parafascicular Nucleus

This nucleus is a densely aggregated mass of moderately large, deeply staining cells lying posterior or posteroventral to the mediodorsal nucleus and pierced by the habenulopeduncular tract (Figs. 12.3–12.5). Anteriorly, it can usually be seen to be continuous with those cells of the central lateral nucleus that extend backwards in the lateral edge of the mediodorsal nucleus. Laterally, it merges with a centre médian nucleus, when present, and medially it lies adjacent to the anterior end of the periaqueductal gray matter.

In frontal sections the junction between the thalamus and the midbrain in this region is often difficult to define and there is the suggestion of a transitional region joining the posterior parts of the parafascicular, centre médian, mediodorsal, and lateral habenular nuclei to the periaqueductal gray matter, the pretectal nuclei, and the nucleus of the posterior commissure. This difficulty can be resolved by reference to sections cut in the other two standard planes in which the medial medullary lamina which forms the line of junction between the thalamus and the midbrain is usually quite distinct (e.g., Fig. 8.2).

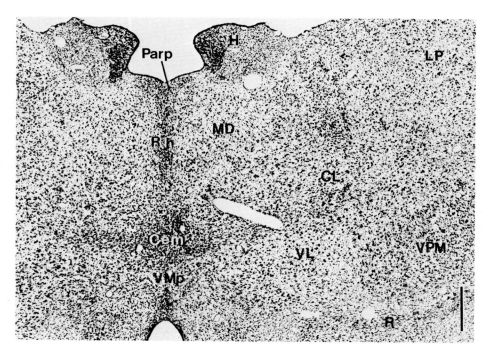

FIGURE 12.6. Frontal section showing the posterior part of the rhomboid nucleus forming the so-called "intermediodorsal nucleus" in a guinea pig. Thionin stain. Bar = 500 μm.

12.4. Terminology

Many of the nuclei described as intralaminar nuclei have also been referred to as nuclei of the *midline*. The term *Mittelliniekern* was first used by Nissl (1889, 1913) although Dejerine (1901) had also noticed that within the human thalamus, close to the third ventricle, there was a "central gray substance" separate from the other nuclei. In the rabbit, Nissl (1889, 1913) described the Mittelliniekern as having a lateral winglike expansion (approximately corresponding to the paracentral nucleus) which passed out to join a magnocellular nucleus corresponding to our central lateral nucleus, or at least to its anterior expanded part. The magnocellular nucleus and nucleus of the midline in the rabbit were referred to respectively as the *falciform* and *commissural nuclei* by Ramón y Cajal (1911).

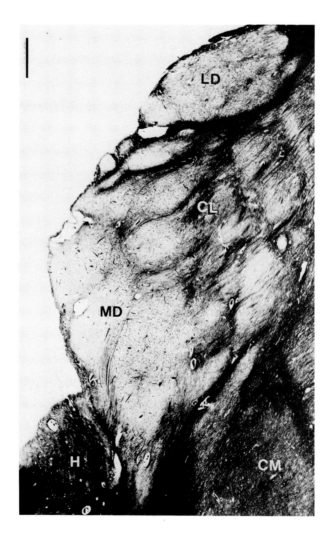

FIGURE 12.7. An oblique section, through the human thalamus, approximately midway between frontal and horizontal, showing the part of the internal medullary lamina containing cells of the central lateral nucleus, that extends around the posterior surface of the mediodorsal nucleus and which is often regarded as part of the medial pulvinar nucleus. Hematoxylin stain. Bar = 1 mm.

Rhomboid nucleus is also Ramón y Cajal's term, first used in reference to the guinea pig. Rose's (1942b) central commissural system, consisting of the rhomboid, interanterodorsal, interoanteromedial, central (medial), paracentral, and central lateral nuclei, derives from Nissl's Mittelliniekern. The so-called magnocellular nucleus of the rabbit, named separately by Nissl, has served to cause a great deal of confusion since Nissl's time because it is so large that it has often not been recognized as the expanded anterior part of the central lateral nucleus. Some authors, not seeing its equivalence to the central lateral nucleus, have argued, without justification, that it is a unique feature of the rabbit brain. M. Rose (1935) called it *nucleus angularis.*

The terms *central medial, paracentral,* and *central lateral nuclei* derive from the work of Gurdjian (1927; *paracentral nucleus*) and Rioch (1929a; *central medial, central lateral*) and were quickly adopted by others (e.g., Le Gros Clark, 1932a; Ingram *et al.,* 1932; Aronson and Papez, 1934; Walker, 1935). *Nucleus centralis* is a common alternative for *central medial nucleus,* either alone (Gurdjian, 1927;

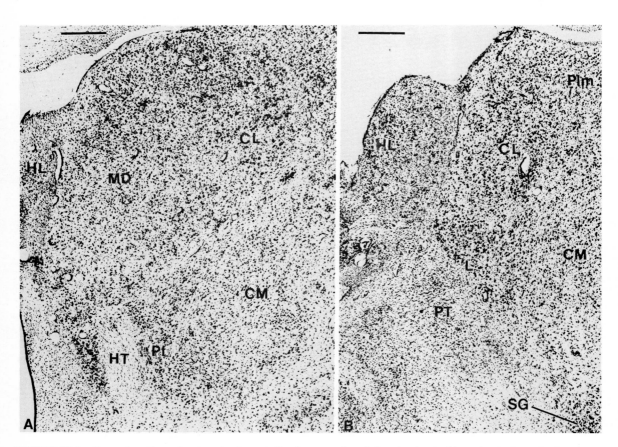

FIGURE 12.8. Frontal sections showing the large cells of the posterior part of the central lateral nucleus, which are sometimes referred to as a paralamellar or densocellular part of the mediodorsal nucleus. The expanded part of the internal medullary lamina in this region is shown in the human thalamus in Fig. 12.7. (A) Squirrel monkey; (B) rhesus monkey. Thionin stain. Bars = 500 μm.

Olszewski in the monkey refers to the posterior part of the central medial nucleus as nucleus centralis inferior, divides most of the rhomboid nucleus into nuclei centralis, centralis latocellularis superior, and centralis intermedius, but includes anterior parts of both central medial and rhomboid nuclei in a common nucleus centralis densocellularis. In nonprimates, posterior parts of the rhomboid nucleus have been called an *intermediodorsal nucleus* (Gurdjian, 1927; Rioch, 1929a). The rhomboid nucleus of the rabbit was called *nucleus alatus* by M. Rose (1935), a name that is still sometimes seen.

I have followed the precedent set by Berman and Jones (1982) in referring to all the midline elements other than the central medial nucleus as *rhomboid nucleus,* to avoid confusion across species. To avoid further confusion with the centre médian nucleus (see below), I have avoided the term *central nucleus* altogether.

The names of the individual members of the rostral group of intralaminar nuclei derive from work in nonprimates, but the term *intralaminar* itself and the names of the two members of the caudal group of intralaminar nuclei derive from primates. The *centre médian* (or *centrum medianum*) is, of course, Luys's term, introduced in 1865 as the result of Luys's researches on the human brain. *Nucleus centrum medianum* is neither good Latin nor correct in the sense in which Luys envisaged this "center" operating (see Chapter 1). For consistency, however, the practice has grown up of referring to it as a nucleus. Le Gros Clark (1930a) may have been the first to adopt this usage in his study of *Tarsius.* Vogt (1909) and Friedemann (1911) had used *corpus Luysii.* Le Gros Clark was soon to be followed by Rioch (1931) who had not detected the nucleus in his earlier study of the cat (1929a). Since then, the name has become universal for those animals that possess the nucleus. In man, however, some have referred to the nucleus as the *central nucleus* (e.g., Hassler, 1959), inviting confusion with the rhomboid nucleus or other parts of the central commissural system (Section 12.4). Vogt and Vogt (1941), more in keeping with Luys, simply call it *Griseum centrale.*

Parafascicular nucleus is Vogt's (1909) and Friedemann's (1911) term, derived from their work on the monkey brain. Prior to that, in other animals it had generally been regarded as a posterior component of the medial nuclear group (Nissl, 1889, 1913; von Monakow, 1895), a practice maintained by M. Rose as late as 1935. Otherwise, however, it was almost universally adopted, probably the first author to do so being D'Hollander (1913), followed by Gurdjian (1927), Pines (1927), and Rioch (1929a). The old idea still survives, however, in J. E. Rose's (1942b) use of the collective term *post medial group* for the parafascicular and centre médian nuclei.

Intralaminar nuclei seems to have been Le Gros Clark's (1932a) abbreviation of the term *nuclei of the internal medullary lamina,* also first used by Vogt (1909) and Friedemann (1911) and applied to the rabbit as *nucleus lamellaris* by D'Hollander (1913). Le Gros Clark (1932a), from his comparative studies, regarded the parafascicular nucleus as having differentiated from the caudal part of the paracentral nucleus and then, the centre médian in those animals which possess such a nucleus was said by him to differentiate in turn from the caudal part of the parafascicular. This suggestion was later supported by Kuhlenbeck (1954) and others. Le Gros Clark, therefore, was probably the first to formally associate the parafascicular and centre médian nuclei with the intralaminar system. The expression *caudal group of intralaminar nuclei* is that of Berman and Jones (1982). As the centre médian and parafascicular nuclei, like the rostral intralaminar group of nuclei, clearly have a close relationship with the striatum (Vogt and Vogt, 1941; Powell and Cowan, 1967; Mehler, 1966a), there is justification for supporting Le Gros Clark's viewpoint. Le Gros Clark also considered, as did Walker (1938a), that the nucleus limitans and possibly the suprageniculate nucleus, mainly because they showed similar undramatic cellular changes following cortical ablations, might also belong to the intralaminar system. The continuity between central lateral, parafascicular, and limitans nuclei at the posterior pole of the thalami of monkeys, chimpanzees, and man is clearly evident in the photomicrographs of Grünthal (1934) who referred to the three as *nucleus circularis* because of the way they encircled the mediodorsal nucleus. There is, however, little recent evidence that the limitans and suprageniculate nuclei project to the striatum (Burton and Jones, 1976).

As the years have gone by, the equivalence of the various intralaminar nuclei between primates and nonprimates has come to be accepted. Apart from the debatable question of whether a centre médian nucleus can be distinguished in the thalami of rodents and other small mammals, the only major contention is over the posterior part of the central lateral nucleus in primates. As pointed out in Chapter 13, much of this nucleus has been regarded as a pars paralamellaris or densocellularis of the mediodorsal nucleus, even though the original users of these terms (Friedemann, 1911; Olszewski, 1952) recognized the association with the internal medullary lamina and were hesitant about including the nucleus in the mediodorsal nucleus. The comparable large cells at the posterior end of the cat central lateral nucleus (Fig. 12.3) have had a similar history (Rinvik, 1968a, 1972; Mehler, 1966a; Jones and Burton, 1974). Walker's (1938a) insistence that the paralamellar cells underwent degeneration after cortical ablations whereas the other intralaminar nuclei did not, was probably the single major influence in causing the paralamellar cells of the monkey to be included in the mediodorsal, rather than in the central lateral nucleus. Because these cells along the edge of the primate mediodorsal nucleus, like their counterparts in the cat, receive spinothalamic afferents (Mehler, 1966a,b; Jones and Burton, 1974; Ganchrow, 1978; Burton and Craig, 1983; Mantyh, 1983), I concur with Mehler (1966a) that they should be regarded as cells of the central lateral nucleus. Whether they do or do not undergo degeneration after cortical lesions now seems immaterial (see Chapter 3 and Section 12.5).

The status of the dorsomedial, magnocellular part of the centre médian nucleus outlined in man by Vogt and Vogt (1941) and Hassler (1959) is still uncertain. Niimi *et al.* (1960), from their comparative anatomical observations in a number of species, suggested that the human magnocellular part corresponded to the lateral part of the parafascicular nucleus of small mammals. To me, this seems to mean the part that extends forwards to join the central lateral nucleus. Niimi *et al.* concluded that only the parvocellular part of the human centre médian nucleus corresponds to the nucleus of that name in animals such as the cat. Mehler *et al.* (1960) and Mehler (1966b) concurred with this viewpoint, on the grounds that only the small-celled part undergoes the disproportionate

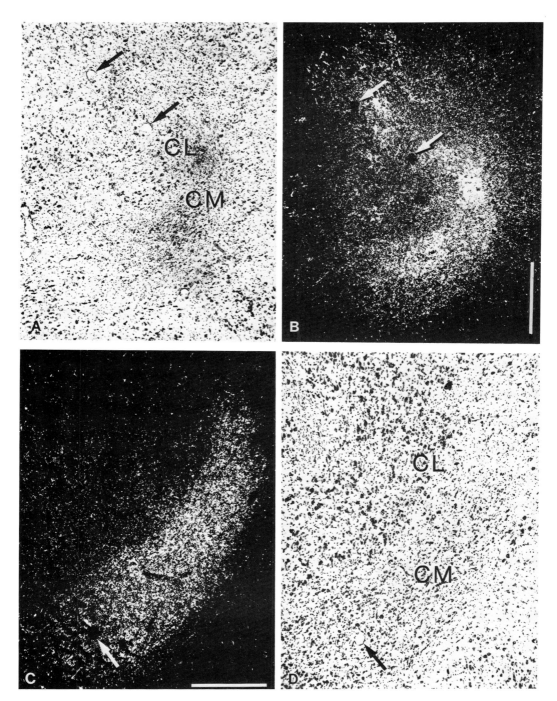

FIGURE 12.9. Autoradiographic labeling of fiber terminations from the deep cerebellar nuclei (A, B) and from the entopeduncular nucleus (C, D) around the small cells of the centre médian nucleus in the cat. Cerebellar terminations but not pallidal terminations extend into other intralaminar nuclei of which the central lateral (CL) is shown. Bars = 500 μm. From Hendry *et al.* (1979).

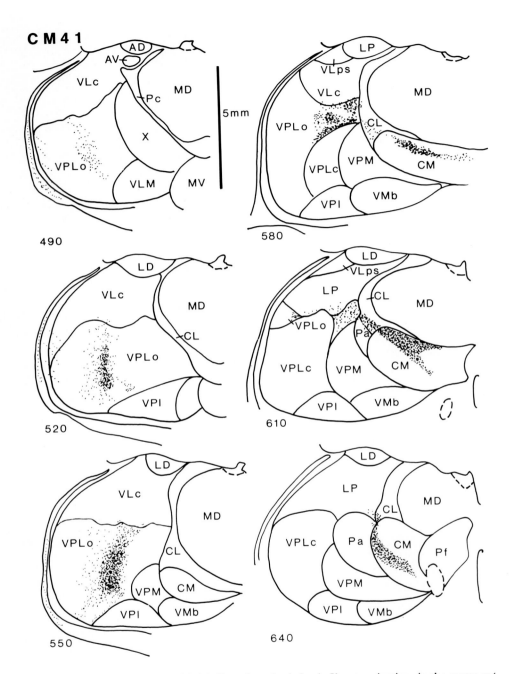

FIGURE 12.10.   Autoradiographic labeling of corticothalamic fiber terminations in the centre médian and central lateral nuclei of a monkey in which tritiated amino acids were injected in arm representation of area 4. Nuclei labeled X, VPLo, VLc, and VLps are all parts of VLp nucleus of present work (Chapter 2); VPLc is VPL nucleus. From Jones *et al.* (1979).

increase in size typical of the primate nucleus and that only the small-celled part in both cats and monkeys receives fibers from the internal segment of the globus pallidus and from the motor cortex [see also Kuypers (1966) in the chimpanzee]. In examining material from my own laboratory (Hendry *et al.*, 1979; Jones *et al.*, 1979; Mink, Thach, and Jones, unpublished observations), I am led to concur with Mehler (1966b), except in regard to the terminations of corticothalamic fibers from the motor cortex. These seem to extend their terminations into the central lateral nucleus and may in the centre médian itself delimit a part of the nucleus different from that receiving pallidal inputs (Fig. 12.9).

The other area of some confusion is at the midline among the nuclei of the rostral intralaminar group. Confusion over the rhomboid, central medial and a "central" nucleus has previously been mentioned. However, parts of the inter-anteromedial, parataenial, and paraventricular nuclei continue to be mislabeled intralaminar nuclei. Equally, labeling of cells or axons in the dorsal components of the central lateral nucleus is often mistakenly referred to these other non-intralaminar nuclei.

12.5. Connections

12.5.1. Efferent Connections

12.5.1.1. Background

The midline and intralaminar nuclei were said by Nissl (1913) to undergo cellular degeneration following destruction of the cerebral cortex and although it has been generally accepted that the nuclei show some cellular reaction in these circumstances, the degree and significance of this reaction and whether it indicated a thalamocortical projection have been a matter of debate. Walker (1935, 1936, 1938a) and Le Gros Clark and Boggon (1935) noted minimal or absent changes in the nuclei following hemidecortications in monkeys; on the basis of a small number of inconclusive experiments and certain antecedent work, Walker (1938a) considered that they probably sent their efferent axons not to the cortex but to the globus pallidus. In studies of human pathological material, Le Gros Clark and Russell (1939), Vogt and Vogt (1941), Freeman and Watts (1947), McLardy (1948), Hassler (1948), and Simma (1950) defined the target of the centre médian in man as the putamen (Chapter 3). The connections of the rostral group of intralaminar nuclei, however, continued to be debated. Rose and Woolsey (1943) in their study of thalamocortical relations in the rabbit, rejected the idea that the intralaminar nuclei projected primarily on the striatum. They described degeneration in the intralaminar group after cortical ablations as being as intense as in the main nuclei of the thalamus and held that the intralaminar and related "commissural" nuclei probably projected to olfactorily related cortex at the base of the telencephalon.

Droogleever-Fortuyn (1953), Droogleever-Fortuyn and Stefens (1951), Powell and Cowan (1954, 1956, 1967), and Cowan and Powell (1955) contested this point of view and considered that the striatum in experimental animals was the target of both rostral and caudal groups of intralaminar nuclei. Although lesions of the cerebral cortex caused some pallor and shrinkage of cells in the nuclei, complete degeneration (with atrophy and cell loss) occurred only after lesions involving the striatum. The position of the degeneration in the intralaminar groups varied depending upon the site of the striatal lesion, as originally reported by Vogt and Vogt (1941) in the centre médian nucleus alone. Vogt and Vogt (1941) had found in man that lateral parts of the centre médian nucleus degenerated after pathological destruction of the putamen and medial parts after destruction of the caudate nucleus. They detected no degeneration in the parafascicular nucleus and made no mention of any in other intralaminar nuclei. Hassler (1959), reexamining the Vogts' series, related the parafascicular nucleus to the vicinity of the nucleus accumbens. Powell and Cowan (1956) in monkeys (Fig. 12.11A) reported that only the centre médian and parafascicular nuclei projected to the putamen, while the rostral group of intralaminar nuclei projected to the caudate nucleus. There was a topography in the projection such that destruction of anterior or posterior parts of the putamen led to degeneration in anterior or posterior parts of the centre médian nucleus and destruction of anteromedial parts of the putamen led to degeneration in the parafascicular nucleus. Of the rostral group, the central medial nucleus degenerated after lesions in anterior parts of the head of the caudate nucleus and the central lateral and paracentral nuclei degenerated after more posteriorly placed lesions of the head of the caudate nucleus. The parataenial nucleus was considered to show degeneration as the result of destruction of the nucleus accumbens. In nonprimates without a centre médian nucleus, the parafascicular nucleus degenerates after lesions of the putamen and the rostral group of intralaminar nuclei after lesions of the head of the caudate nucleus (Gerebtzoff, 1940b; Droogleever-Fortuyn and Stefens, 1951; Powell and Cowan, 1954; Cowan and Powell, 1955). The rhomboid nucleus of the rabbit was said to degenerate after lesions restricted to cortex over the nucleus accumbens rather than after involvement of the nucleus accumbens itself (Cowan and Powell, 1955). Again, the parataenial nucleus was related to the nucleus accumbens. In cats, which do possess a centre médian nucleus, evidence for its projection to the putamen came from studies using anterograde axonal degeneration (Nauta and Whitlock, 1954).

All of the above findings suggest a highly ordered projection of the intralaminar nuclei upon the striatum. This has largely been confirmed with newer techniques (Sato *et al.,* 1979), though the terminations of axons from an individual intralaminar nucleus in the striatum appear somewhat more widespread than suggested by the retrograde degeneration studies (Krettek and Price, 1977b; K. Kalil, 1978; Royce, 1978a).

At the time much of the retrograde degeneration work was being done, it was becoming established from anterograde degeneration studies that the cerebral cortex sends axons to the intralaminar nuclei (e.g., Auer, 1956; Petras, 1965; Powell and Cowan, 1967; Rinvik, 1968b–d; DeVito, 1969). Therefore, a possible explanation (Powell and Cowan, 1967) for the cellular changes reported in the intralaminar nuclei after cortical lesions is that they are transneuronal in character and caused by deafferentation of the intralaminar cells rather than by section of their efferent axons. This might account for the lack of severity of the retrograde effect reported by some observers after cortical lesions.

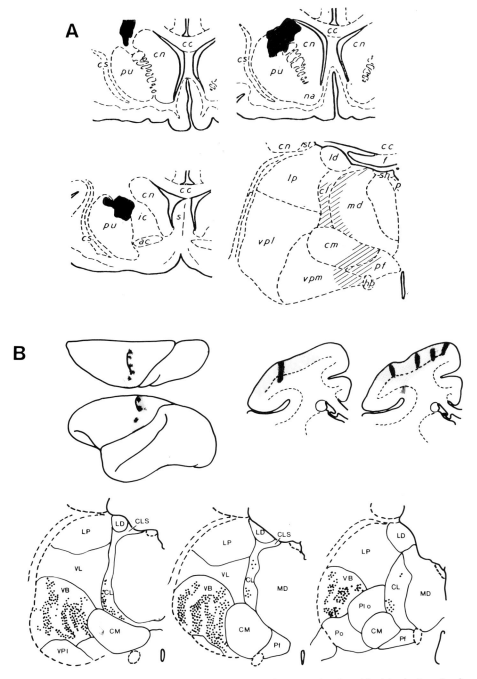

FIGURE 12.11.   (A) Distribution of retrograde cellular degeneration (hatching) in the intralaminar nuclei of a monkey following destruction of parts of the caudate nucleus and putamen. Frontal sections. From Powell and Cowan (1956). (B) Distribution of retrogradely labeled cells in the intralaminar nuclei of a squirrel monkey following injection of horseradish peroxidase in the somatic sensory cortex. Frontal sections. From Jones and Leavitt (1974); cf. Fig. 3.5.

Murray (1966), however, in a retrograde degeneration study in the cat, returned to the position of Rose and Woolsey. She considered that the cellular changes occurring after cortical lesions in the central medial, paracentral, and central lateral nuclei were severe with definite cell loss, rather than mild. She regarded the severe changes occuring after striatal lesions in the experiments of Powell and Cowan (1956) as being due to interference with the internal capsule. Murray's studies were later reinforced by similar findings made quanitatively in the cat by Macchi *et al.* (1975). According to Murray, the central medial and paracentral nuclei are related primarily to the anterior limbic and cingulate cortex and the central lateral nucleus to the parietal (suprasylvian) cortex. The results of Macchi *et al.* (1975) are similar (Fig. 12.12). This organization is in keeping with the pattern of corticofugal projections passing from the cortex to the intralaminar nuclei (Powell and Cowan, 1967). There were few studies of the intralaminar nuclei with techniques for demonstrating anterograde axonal degeneration, rather than retrograde cellular changes. Two at least, however, did succeed in showing projections from the centre médian and adjacent nuclei to the putamen in the cat and monkey (Nauta and Whitlock, 1954; Mehler, 1966a).

12.5.1.2. Recent Studies

More recent studies using methods based upon anterograde and retrograde axonal transport have clearly established that the intralaminar nuclei, including the rhomboid, centre médian, and parafascicular nuclei of the cat and other mammals, project heavily upon the striatum (Jones and Leavitt, 1974; Nauta *et al.*, 1974; Kuypers *et al.*, 1974; K. Kalil, 1978; Royce, 1978a,b; Herkenham and Pert, 1981; Krettek and Price, 1977b; Sato *et al.*, 1979; Macchi *et al.*, 1984) but also lightly and diffusely upon widespread areas of the cerebral cortex (Fig. 12.11B and Chapter 3). The projections seem concentrated in frontal, medial, and dorsolateral cortex but a few cells can be retrogradely labeled from virtually

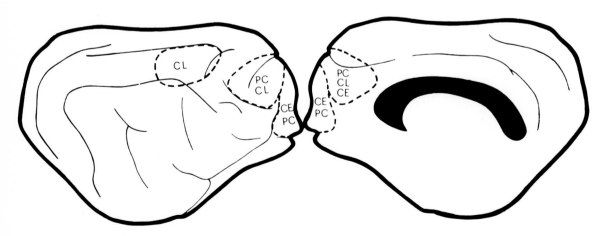

FIGURE 12.12. Cortical zones in which the majority of the fibers from the rostral group of intralaminar nuclei appear to terminate in the cat. Based on retrograde degeneration studies. From Macchi *et al.* (1975).

every cortical area, including that of the piriform lobule and occipital pole (Jones and Leavitt, 1974; Jones, 1975a; Herkenham, 1980; Strick, 1975; Macchi *et al.*, 1977; Kievet and Kuypers, 1975; Mizuno *et al.*, 1975; Itoh and Mizuno, 1977; Kennedy and Baleydier, 1977; Haberly and Price, 1978; Bentivoglio *et al.*, 1981). Herkenham (1980) hints that a few intralaminar fibers may reach every cortical area. Detailed physiological study gives no evidence for collateral projections from the same intralaminar neuron to both striatum and cortex (Steriade and Glenn, 1982).

As pointed out in Chapter 3, each of the intralaminar nuclei, though projecting rather widely upon the cortex, appears to have a particular region upon which most of its cortical projection is focused (Macchi *et al.*, 1975, 1977, 1984). Hence, the intralaminar nucleus-to-cortex projection cannot be regarded as totally diffuse or nonspecific. Bentivoglio *et al.* (1981, 1983), in experiments involving the injection of two different retrogradely transported tracers at points separated by several millimeters in the cat or rat cortex, found that few intralaminar cells were double labeled. Steriade and Glenn (1982) showed that of more than 200 intralaminar neurons antidromically activated by stimulation of the cortex, only one could be discharged by stimuli applied to the pericruciate and suprasylvian gyri. These findings also suggest some degree of precision in the projection with individual cells projecting to relatively localized regions.

The topography of the intralaminar projection upon the cortex (Section 12.5.1.1) is such that it conforms to the pattern of "nonspecific" thalamocortical connections (Fig. 12.13B) demonstrated following electrical stimulation of the intralaminar or midline nuclei by Morison and Dempsey (1942), Jasper (1949, 1960), Starzl and Magoun (1951), Hanberry and Jasper (1953), and others. These connections were thought to mediate the cortical "recruiting response" and the relevant axons were believed by Lorente de Nó (1949) to pass from the intralaminar nuclei to layer I of the cerebral cortex (see Section 12.6.1).

Fibers thought to arise in the intralaminar nuclei were traced by autoradiography to layer I of several areas of the cat cortex (Jones, 1975a). In the rat, however, Herkenham's (1980) short report suggests that intralaminar fibers terminate only in layer VI and that the layer I projection arises from nuclei such as the medioventral and ventromedial (see also Glenn *et al.*, 1982). This is clearly an area that deserves some anatomical attention in order to resolve the discrepancy.

12.5.2. Afferent Connections

12.5.2.1. Corticothalamic Connections

The intralaminar nuclei receive fibers in an organized manner from the cerebral cortex. The cells of origin (Fig. 4.31) are smaller pyramidal cells of layer V (Catsman-Berrevoets and Kuypers, 1978). Some of the cells may have branched axons to the thalamus and striatum (Royce, 1983). The organization of the projection has been demonstrated in the primate and cat but there are indications of a comparable topography in other species (e.g., Petras, 1969; Kuypers, 1966; Powell and Cowan, 1967; Künzle, 1978; Goldman, 1979; Wise and Jones, 1977a; Beckstead, 1979; Jones *et al.*, 1979; Hendry *et al.*, 1979; Künzle and Akert, 1977; De Vito,

1967, 1969; Berman and Payne, 1982; Royce, 1983; Macchi *et al.*, 1984). In general, the prefrontal granular cortex and the limbic cortex on the medial surface of the hemisphere send fibers to the paracentral and central medial nuclei, the premotor area (area 6) to the central lateral and parafascicular nuclei, and the motor cortex (area 4) to the centre médian nucleus and/or adjacent central lateral nucleus. The somatic sensory and anterior parietal areas send fibers to the posterior

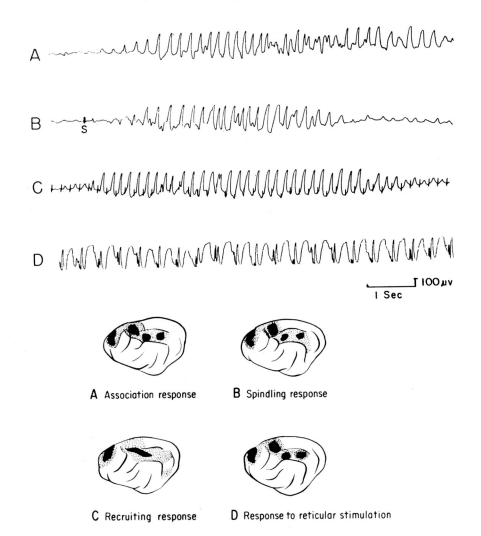

FIGURE 12.13. (Upper) Cortical slow-wave activity generated in the anterior suprasylvian gyrus of barbiturate-anesthetized cats by stimulation of the central medial intralaminar nucleus. A: spontaneous spindle burst; B: spindle burst tripped by a 1-msec shock to the nucleus; C: waxing and waning of recruiting response elicited by repetitive stimulation at 5 per second; D: spike and wave response to stimulation at 2–5 per second. From Jasper (1960). (Lower) Cortical foci in which recruiting responses, polysensory responses, spontaneous barbiturate spindles, and responses to stimulation of the midbrain reticular formation, are best recorded. From Thompson (1967); cf. Fig. 12.12.

part of the central lateral nucleus. Few or no fibers reach the intralaminar nuclei from more posteriorly and laterally situated areas of cortex such as the visual and temporal areas.

12.5.2.2. Subcortical Afferents

Although the details have not been firmly established, it has been known for a long time that the intralaminar nuclei receive afferent fibers arising in the basal ganglia, cerebellum, and spinal cord. The centre médian nucleus receives a heavy projection from the entopeduncular nucleus (internal segment of the globus pallidus) (Fig. 12.9) (Nauta and Mehler, 1961, 1966; Mehler, 1966a; Hendry *et al.*, 1979; Kim *et al.*, 1976; Larsen and McBride, 1979; Van der Kooy and Carter, 1981; Parent and DeBellefeuille, 1983) and many internal pallidal cells may have branched axons to both the ventral lateral thalamic nucleus and the centre médian nucleus in monkeys (Parent and DeBellefeuille, 1983); the central lateral nucleus and adjoining parts of the parafascicular nucleus are major targets of the deep cerebellar nuclei (Faull and Mehler, 1978; Asanuma *et al.*, 1983b) (Figs. 7.28 and 7.40). A projection from the spinal cord and caudal nucleus of the spinal trigeminal nucleus to the posterior parts of the central lateral nucleus was originally reported as small (Mehler, 1966b, 1969; Mehler *et al.*, 1960; Boivie, 1971b; Jones and Burton, 1974), and to involve only the large-celled component (including the so-called paralamellar part of the mediodorsal nucleus). Newer work with more sensitive anterograde tracers, however, suggests that the terminations of spinothalamic fibers may extend more anteriorly in the rostral group of intralaminar nuclei (Craig and Burton, 1981; Mantyh, 1983; Burton and Craig, 1983) (Fig. 7.13). Fibers to the intralaminar nuclei arise mainly in deep ventral parts of the spinal cord gray matter (Willis *et al.*, 1979; Trevino, 1976; Giesler *et al.*, 1981). Some appear to be branches of those directed toward the ventralposterior nucleus (Giesler *et al.*, 1981).

The little definite evidence available regarding the afferent connections of the rhomboid nucleus indicates inputs from the nucleus cuneiformis of the midbrain reticular formation (Edwards and De Olmos, 1976).

Other afferent fibers to the intralaminar nuclei have been reported from parts of the reticular formation of the medulla oblongata (Nauta and Kuypers, 1958; Loewy *et al.*, 1981; McKellar and Loewy, 1982) and pons (Graybiel, 1977; Robertson and Feiner, 1982), parabrachial nucleus of the pons (Saper and Loewy, 1980), nucleus cuneiformis of the midbrain reticular formation (Edwards and deOlmos, 1976; Graybiel, 1977), deep layers of the superior colliculus (Graham, 1977; Partlow *et al.*, 1977; Harting *et al.*, 1980; Graham and Berman, 1981), pretectum (Benevento *et al.*, 1977; Berman, 1977; Graham and Berman, 1981), and substantia nigra (Carpenter *et al.*, 1976; Beckstead *et al.*, 1979; Hendry *et al.*, 1979; Faull and Mehler, 1978). None of this work, as presented, suggested particularly heavy terminations. The fibers from most of the above sites are shown rather diffusely throughout the central lateral, paracentral, and central medial nuclei. It is hard to identify any localized concentrations of terminals. Possibly, all of these various afferents converge on the same cells though one has the impression that some at least are concentrated in different regions. Inputs from the substantia nigra in the monkey, for example, appear mainly concentrated in the part of the central lateral nucleus commonly referred to as the paralamellar part of the mediodorsal nucleus. It is also possible that fibers from some of the regions listed above are not even terminating among the cells of the intralaminar nuclei but merely traveling in the internal medullary lamina to other nuclei.

The sources of afferent fibers to the parafascicular nuclei are not completely clear. There are reports in rodents and marsupials that spinothalamic fibers from the upper part of the spinal cord terminate in it (Lund and Webster, 1967a,b; Mehler, 1969; Rockel *et al.*, 1972) but such connections have not been demonstrated in other species. Probably in these small mammals they represent spinothalamic connections to the posterior part of the central lateral nucleus, as seen in other species. The only other reports of subcortical afferents to the parafascicular nucleus are those indicating inputs from the periaqueductal gray matter of the midbrain (Hamilton, 1973; Edwards and deOlmos, 1976; Ruda, 1976).

12.6.   Functions

In one sense, the intralaminar nuclei can be regarded as diffusely innervated and to project diffusely upon widespread areas of the cortex and striatum. In another sense, different parts receive different sets of afferent inputs and project with a fine-grain topography upon the striatum and have zones of regional dominance in the cortex. Are there any functional correlates of these somewhat divergent points of view?

12.6.1.   The Cortical Recruiting Response

The phenomenon with which the intralaminar nuclei are inevitably associated is the cortical "recruiting response." This particular phenomenon was first brought to light in 1942 when Morison and Dempsey discovered that low-frequency electrical stimulation in the vicinity of the anterior pole of the centre médian nucleus of the cat led to long-latency, repetitive, high-voltage, negative electrical waves over much of the cerebral cortex (Fig. 12.13). In the ensuing 20 years the recruiting response and the diffuse or nonspecific thalamocortical projection system that Morison and Dempsey thought it implied, came to dominate a large part of neurophysiology. As Magoun (1950) put it, "This Cinderella-like diffuse projection system has blossomed" and "appears to compete seriously with all the rest of the thalamus together in its functional significance." Some measure of the interest that the recruiting response generated can be obtained from reading the accounts of the discussions that accompanied many of the papers presented at the 1954 Symposium on Brain Mechanisms and Consciousness (Delafresnaye, 1954). In 1960, Jasper could write ". . . the unspecific system . . . seems to be the most important regulator of the spontaneous electrical rhythms of the entire cortex. . . ." Since 1960 when Jasper wrote his definitive review of the subject, interest in the recruiting response has declined but it remains as one of the largely unexplained phenomena in the electrophysiology of the cortex and thalamus.

As originally described, the recruiting response develops in the cortex (often bilaterally) 30–40 msec after stimulating the intralaminar and certain other thalamic nuclei close to the midline with bipolar electrodes at rates between 6 and 12 per second. The threshold for stimulation is low (2–4 volts for 1-msec pulses). As recorded with surface electrodes, the cortical response to each stimulus is a negative wage which progressively increases with each stimulus to a maximum after the first three to five stimuli; responses remain maximal to each stimulus but after 3–5 sec they deline, only to recur, increase, and wane again, and so on (Fig. 12.13). During this time spontaneous cortical rhythms cease.

As studied later with microelectrodes, the wave seems to be generated by fibers lying just beneath the surface of the cortex (Jasper, 1960), leading some workers to relate the afferent fibers shown by Lorente de Nó (1949) in Golgi preparations to end in layer I, with the nonspecific system. According to Jasper (1960) the recruiting wave probably represents summing of dendritic potentials rather than firing of individual neurons because recruiting responses can still be generated when cortical unit firing is depressed as the result of anesthesia. Moreover, although the excitability of cortical neurons in response to a conditioning electrical volley in a peripheral nerve can be enhanced during a recruiting response, the neurons do not discharge as the direct result of the recruiting response (Li *et al.*, 1955a,b).

Morison and Dempsey (1942), early on, noted the similarity between the recruiting response to repetitive stimulation of the intralaminar system, and the spontaneous spindle bursts that occur in the cortex in animals under barbiturate anesthesia. Jasper (1949) noted that cortical spindles resembling barbiturate spindles could be tripped by single 1-msec shocks to the intralaminar nuclei (Fig. 12.13). These workers regarded the nonspecific afferent system that they thought underlay the recruiting phenomenon as exerting a controlling influence over the rhythmic activity of the whole cortex.

As the years went by, it became clear that recruiting responses could be elicited in the cortex by low-frequency stimulation of all the intralaminar nuclei as well as of the medioventral, ventral anterior, interanteromedial, submedial, suprageniculate, and ventral medial nuclei (Jasper and Droogleever-Fortuyn, 1948; Jasper, 1949, 1960).

The recruiting response was obviously quite unlike the surface potentials elicited by stimulation of the principal nuclei of the thalamus. The latter occurred at much shorter latency and were clearly localized to small focal regions in the relevant target cortical area. Apart from their long latency and other temporal properties, recruiting responses could be demonstrated over wide areas of the cortex, from stimulation at a single focus in the intralaminar–midline system. This led Morison and Dempsey to refer to the thalamocortical system supposedly responsible as "nonspecific" and "diffuse." The recruiting responses nevertheless are more readily elicited and have larger amplitudes over motor cortex and over "association areas" of the lateral and suprasylvian gyri of the cat, than over other areas (Starzl and Whitlock, 1952; Fig. 12.13B). This suggests the presence of focal zones of projection for the nonspecific system, though later work was able to record recruiting responses under certain conditions in virtually all areas of the cortex, including the piriform lobule and hippocampus as well as the primary sensory areas (Hanberry and Jasper, 1953; Jasper, 1949, 1960; Jasper *et al.*, 1955).

In their efforts to explain the thalamocortical pathway responsible for the recruiting response, Morison and Dempsey (1942) considered three potential routes: (1) a direct projection from the intralaminar and related nuclei to the cortex; (2) a projection with a synaptic relay in adjacent principal nuclei; (3) a projection with a synaptic relay in other subcortical structures. They did not favor transfer from area to area in the cortex because cortical ablations did not abolish the responses.

Jasper (Fig. 12.14) and his colleagues at the Montreal Neurological Institute, over several years made a concerted effort to track the flow of excitation from the intralaminar nuclei through the thalamus and to identify the route taken to the cortex (Jasper, 1949, 1960; Hanberry and Jasper, 1953; Hanberry et al., 1954) (Fig. 12.15). Placing stimulating electrodes systematically throughout the thalamus of cats and at each point seeking recruiting responses in the cortex, they were able to detect two main outflow streams. The largest commenced in the centre médian nucleus and spread forwards through the central medial and ventral medial nuclei to the ventral part of the ventral anterior nucleus and anterior pole of the reticular nucleus. Stimulation anywhere along this route led to recruiting responses in widespread areas of the cortex, often bilaterally. A smaller system commenced in the midline in medioventral, rhomboid, ventral medial, interanteromedial, and submedial nuclei, spread through the central lateral and paracentral nuclei to the dorsolateral part of the ventral anterior nucleus and posterior part of the reticular nucleus. Stimulation in medial parts of the system led to recruiting responses in frontal and medial cortical areas and in the hippocampal formation; stimulation more laterally led to responses in parietal, posterior, and lateral cortex. It is interesting to note that the regions of projection determined in this way are very comparable to the regions to which the several nuclei mentioned have since been shown to project anatomically (Fig. 12.12 and Chapters 7, 13, and 14). The topography in the projection established by Hanberry and Jasper (1953) was quite detailed and with threshold stimulation with fine (0.1 mm) electrodes they were able to show that a single intralaminar nucleus could project to quite a localized area of cortex. The "diffuse projection system" was, thus, not totally diffuse.

Hanberry and Jasper (1953) ruled out virtually all the principal thalamic nuclei except the ventral anterior as mediators of the thalamocortical recruiting response by selective ablations of the ventral lateral, ventral posterior, mediodorsal, pulvinar, and lateral posterior nuclei and the geniculate bodies. They were able to show that under conditions in which short-latency responses to peripheral stimulation were abolished from a cortical area, recruiting responses to low-frequency stimulation of the intralaminar or ventral anterior nuclei could still be obtained. This made it unlikely that the pathway responsible for the recruiting response relayed in the principal nuclei, a finding confirmed anatomically in much more recent times (Chapter 3).

Prior to Hanberry and Jasper's work, Starzl and Magoun (1951) had reported that recruiting responses identical to those in the cortex could be obtained in the ventral anterior nucleus of the cat by low-frequency stimulation in the vicinity of the centre médian nucleus. Hanberry and Jasper were able to show that the ventral anterior nucleus played a particularly important role in the transmission of the recruiting effect. Low-frequency electrical stimulation of the ventral anterior nucleus led to recruiting responses in very widespread areas of the cortex at latencies 20–30 msec shorter than stimulation elsewhere in the thalamus. A year later, it was shown by Hanberry *et al.* (1954) that recruiting responses elicited in posterolateral regions of the cat cortex by stimulation of the centre médian nucleus, could be abolished by small lesions of the ventral anterior nucleus. Jasper and his group, thus, came to favor the idea that axons ascending from the intralaminar nuclei and other nuclei of the diffuse thalamocortical projection system were interrupted by synapses in the ventral anterior nucleus. However, the idea of a synaptic relay in the ventral anterior nucleus was dismissed by Powell and Cowan (1956) who pointed out that efferent fibers from many thalamic nuclei, including those of the intralaminar complex, simply traverse the ventral anterior nucleus en route to the cortex and striatum. Indeed, even now there is really no definitive evidence for synaptic connections between any other nuclei and the ventral anterior nucleus (Chapters 3 and 7). The role of the ventral anterior nucleus in the thalamocortical response, thus, still remains something of a mystery.

The Montreal group was also obliged to take the reticular nucleus of the thalamus into consideration, because in 1952 it appeared from the work of Rose (1952)

FIGURE 12.14.   Herbert H. Jasper (1906–   ), Professor of Experimental Neurology, Montreal Neurological Institute, who contributed much to the elucidation of the recruiting response and the role of the intralaminar nuclei. Photograph kindly provided by Dr. Jasper.

and Chow (1952) that this nucleus might project in an orderly way on the cortex. Recruiting responses could be elicited in the cortex by stimulation of the reticular nucleus but these appeared more localized than after stimulation of the ventral anterior or intralaminar nuclei (Jasper, 1949; Hanberry and Jasper, 1953; Hanberry *et al.*, 1954). Stimulation of the anterior part of the reticular nucleus, for example, led only to unilateral cortical recruitment and stimulation of the dorsolateral part to recruitment in a very small temporoparietal region. The evidence now

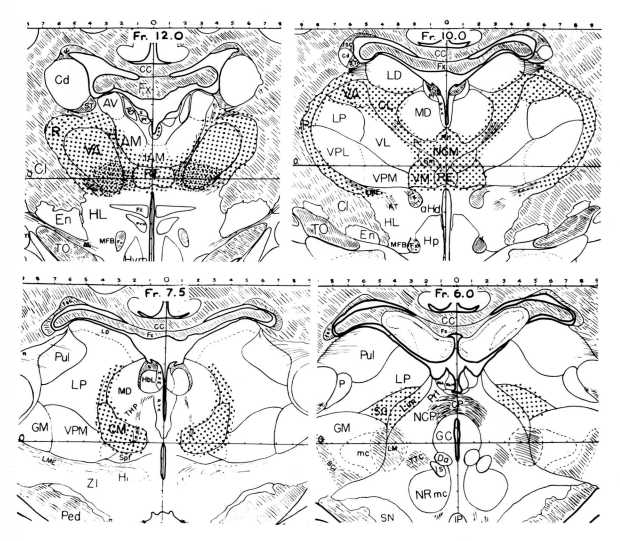

FIGURE 12.15. Stippling indicates thalamic nuclei of the cat which when electrically stimulated at low frequency lead to recruiting responses in the cerebral cortex. Double stippling in VA nucleus indicates region in which anteriorly conducting pathways are most dense. From Jasper (1960).

is strongly against reticular nucleus projections to the cortex (Carman *et al.*, 1964; Jones, 1975c) and it seems to me that Jasper and his colleagues never convincingly integrated the reticular nucleus into their scheme. In his review of 1960, Jasper, though mentioning a possible role for the reticular nucleus, clearly leaned more toward the ventral anterior nucleus as the final common path to cortex for the thalamocortical recruiting system.

12.6.2. The Diffuse Thalamocortical Projection

The weight of anatomical opinion throughout the 1960s was clearly against the concept of a diffuse thalamocortical projection arising in the intralaminar nuclei. McLardy in 1951 had criticized the concept mainly on the grounds that retrograde degeneration could not be demonstrated in the intralaminar nuclei after cortical lesions. He thought that all of the physiological observations could be explained in terms of an intrathalamic diffusion of activity so that the principal nuclei were successively recruited, thus permitting widespread activation of the cortex through their known projections. As we have seen, the absence of significant intrathalamic connections makes this an unlikely proposition. The experiments in which Hanberry and Jasper (1953) ablated many principal nuclei and still obtained recruiting responses also tended to rule it out.

With the growing acceptance of the fact of thalamostriatal projections, at least from the centre médian nucleus, some workers began to consider the possibility that a path from thalamus through striatum to cortex might be involved in the genesis of the recruiting response. We now know, of course, that this was an unlikely possibility, given the absence of projections from striatum to cortex. However, recruiting responses had been obtained in the caudate nucleus following thalamic stimulation (Starzl and Magoun, 1951) and stimulation of the caudate nucleus can elicit recruiting responses in the cortex (Stoupel and Terzuolo, 1954; Jasper, 1954). Against the primary involvement of the striatum in the genesis of the cortical response, it was generally argued that the recruiting responses in the caudate nucleus are of longer latency than in the cortex (Jasper, 1960) and extensive destruction of the striatum does not abolish cortical recruiting responses (Hanberry *et al.*, 1954).

The era of the Nauta technique contributed virtually nothing to the elucidation of a nonspecific thalamocortical projection. In retrospect, Nauta and Whitlock (1954) very clearly demonstrated a widespread, partly layer I projection from the medioventral nucleus of the cat, that virtually mimics the results of Herkenham made 25 years later with autoradiography. However, Nauta and Whitlock could not demonstrate a cortical projection from the centre médian nucleus and felt that even the cortical degeneration ensuing from a lesion of the medioventral nucleus could be contaminated by destruction of fibers of passage from principal nuclei such as the mediodorsal. There, anatomical study of the nonspecific projection essentially ceased. Informed opinion seemed to regard the internal medullary lamina as too difficult a structure to unravel with lesion methods. Though axonal degeneration was commonly demonstrated in layer I as well as in layer IV of the cortex following lesions of the thalamus (e.g., Leonard, 1969; Domesick, 1969; Jones and Powell, 1969a), there was no way of proving convincingly that this arose specifically in the intralaminar or other nuclei of the putative diffuse projection system. Only in the opossum (Killackey and Ebner, 1973) was there some suggestion of a layer I projection from the central lateral nucleus and middle layer projections from the principal nuclei.

Evidence for a cortical projection from the intralaminar and other nuclei of the diffuse projection system came almost immediately after the introduction of horseradish peroxidase as a retrograde anatomical tracer (Jones and Leavitt, 1974; Kuypers *et al.*, 1974; Nauta *et al.*, 1974). It was evident from the study of Jones and Leavitt that the topography of the projections from these nuclei was very similar to that established by Jasper and his colleagues for the system responsible for the cortical recruiting response. The extent of what again started to be called the nonspecific projection, quickly became established from both anterograde and retrograde labeling studies and most of the nuclei implicated by Jasper and his colleagues in the genesis of the recruiting response were brought into the system (Herkenham, 1979, 1980) except for the ventral anterior and the retricular. The cortical target of the ventral anterior nucleus is still not clear but it is obviously not diffuse and the reticular nucleus clearly does not project to the cortex (see Jones, 1975c, and Chapters 3 and 15). It is likely, therefore, that recruiting responses generated by stimulation of these nuclei come from stimulation of fibers passing through them or from spread of current to adjacent nuclei such as the medioventral. The fact that there is huge variation in the conduction velocities of intralaminar fibers passing to different areas of the cat cortex (Steriade and Glenn, 1982) might account for the shorter latency of the recruiting response elicited from the ventral anterior nucleus.

The anatomical studies have indicated a dual character to the nuclei that give rise to the nonspecific projection, for the intralaminar nuclei project to the striatum as well whereas those such as the medioventral, ventromedial, and interanteromedial clearly do not (Jones and Leavitt, 1974; Jones, 1975a; K. Kalil, 1978; Herkenham and Pert, 1981). Initially, it was thought from autoradiographic studies that the intralaminar projection on the cortex was to layer I (Jones, 1975a), thus supporting the idea that this could be the projection responsible for generating the recruiting response. In the rat, however, Herkenham's (1980) brief report suggests that the intralaminar projection is to layer VI while other nuclei of the nonspecific system such as the medioventral and ventromedial provide the appropriately diffuse layer I terminations [see also Glenn *et al.* (1982) in the cat]. It is interesting to note that most neurons of the monkey somatic sensory cortex whose responses altered when the awake animal attended to a vibratory stimulus applied to their receptive fields lay in layer VI and at the border of layers I and II (Hyvärinen *et al.*, 1980). Perhaps this indicates that the layer I and layer VI terminations of the diffuse projection system are indeed active during changes of behavioral state. We are a long way, however, from understanding the relative roles of the intralaminar and other members of the diffuse projecting system in the genesis of cortical activity and there is no information that would lead us to predict whether the intralaminar projection to the striatum is also relevant in this regard.

12.6.3. Intralaminar Nuclei and the Sleep–Waking Cycle

The fact that stimulation of the intralaminar and related nuclei can lead to wave phenomena in the cortex characteristic of sleep or arousal has already been discussed (Chapter 4). Akimoto *et al.* (1956b) using unanesthetized cats with implanted electrodes were the first to recognize the association of low-frequency stimulation in the midline of the thalamus with EEG synchronization and sleep, and the association of high-frequency stimulation with desynchronization and arousal. As pointed out in Chapter 4, these are phenomena that have subsequently been repeated many times in unanesthetized and cerveau isolé animals. The observation that high-frequency stimulation of the upper midbrain reticular formation could also induce desynchronization and arousal (Moruzzi and Magoun, 1949) led naturally to the suggestion that there was a strong interaction between the reticular formation and the intralaminar and other diffuse projecting thalamic nuclei in the regulation of cortical activity (Jasper, 1960) and maintenance of the waking state (Moruzzi, 1972; Steriade, 1981). It is not without significance that the regions of the cat cortex in which activation patterns are most readily elicited following midbrain stimulation are the same as those in which recruiting responses are most readily recorded following thalamic stimulation, the same as those in which maximal barbiturate spindling occurs (Thompson, 1967), and the same as the major zones of regional projection of the intralaminar nuclei (Fig. 12.13B) (Jones and Leavitt, 1974; Macchi *et al.*, 1975, 1977). Can it be then that the intralaminar nuclei and other thalamic nuclei with diffuse cortical projections represent a rostral continuation of the reticular activating system of the brain stem?

The paramedian reticular formation of the pons has been strongly implicated in the sleep–waking cycle (Hobson *et al.*, 1974; Hobson, 1983; Chapter 4). However, its ascending projections to the thalamus are not particularly well defined and its influence over forebrain activities is still largely conjectural (Chapters 4 and 15). By contrast, the electroencephalographic phenomena that accompany desynchronized sleep and the transition from sleep to wakefulness can be mimicked by stimulation of the rostral part of the midbrain reticular formation (Steriade, 1981). Such changes are not dependent on the integrity of the principal thalamic nuclei for they continue to appear over the somatic sensory cortex even after destruction of the ventral posterior nucleus (Steriade and Morin, 1981). The most effective locus for stimulation is the central region of the midbrain, often referred to as the nucleus cuneiformis [(Taber, 1961), a much larger area than the nucleus cuneiformis of Berman (1968), and taking in most of Berman's central tegmental field]. The inputs to this area are difficult to define for the understandable reason that it is traversed by large numbers of fibers of passage. Probably the paramedian pontine reticular formation is one source of input (Graybiel, 1977) and many others have been suggested (Steriade, 1981). In this region in unanesthetized cats with chronically implanted electrodes, many neurons increase their tonic firing rates by more than two times several seconds before the transition from synchronized to desynchronized sleep or to wakefulness and maintain this discharge during the ensuing epoch of desynchronized sleep or wakefulness (Steriade *et al.,* 1980) (Figs. 12.16, 12.17, and 4.35). The discharge rates during these periods are 10–20 times those of actively discharging cells in the paramedian pontine reticular formation. According to Steriade (1981), the greatly enhanced firing of pontine cells during desynchronized sleep (Hobson *et al.,* 1974) is entirely related to rapid eye movements. The mesencephalic cells show a marked diminution or cessation of firing several seconds before the first EEG spindle that heralds the onset of synchronized sleep (Steriade, 1981). These findings suggest that changes in tonic levels of activity of mesencephalic cells may initiate or at least herald the change from one state of alertness to the next.

The outflow of many of the mesencephalic cells that show changes in levels of tonic activity during the sleep–waking cycle, is to the thalamic intralaminar nuclei (Edwards and deOlmos, 1976; Steriade and Glenn, 1982; Glenn and Steriade, 1982). As recorded in cats with chronically implanted electrodes, neurons in the ipsilateral centre médian, parafascicular, and central lateral nuclei can be synaptically driven at short latency by stimulation of the part of the mesencephalic retic-

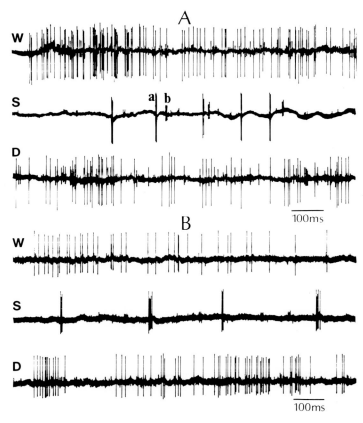

FIGURE 12.16. Discharge patterns of cortically projecting neurons (A, B) in the central lateral or paracentral nucleus of cats during waking (W), synchronized sleep (S), and desynchronized sleep (D). More than one unit seen in each trace. High-frequency spike clusters in synchronized sleep change to sustained discharges in waking and desynchronized sleep (cf. Fig. 4.35). Large spiking cell in (B) was driven synaptically from mesencephalic reticular formation. From Glenn and Steriade (1982).

ular formation containing neurons with state-related tonic discharges* and many of these can be antidromically activated by stimulation of the cerebral cortex or striatum (Chapter 4). The antidromic responsiveness (i.e., the excitability) of the cortically projecting neurons is enhanced following conditioning volleys of the mesencephalic region and also during natural transitions from synchronized sleep to desynchronized sleep or wakefulness (Ropert and Steriade, 1981; Glenn and Steriade, 1982; Steriade and Glenn, 1982) (Fig. 12.17).

These findings present a compelling case for the induction of the changes in cortical rhythmic activity that accompany sleep and wakefulness by the mesencephalic reticular formation, its projection to the thalamic intralaminar nuclei and their projection, in turn, upon the cerebral cortex (see Chapter 4). Steriade (1981) in presenting this case, remarks on the hypersomnolence that occurred in a patient with a medial thalamic lesion involving the intralaminar nuclei (Façon *et al.*, 1958). Whether the system may also be involved in the initiation of changes in the levels of excitability of principal thalamic nuclei under the influence of stimulation of the brain-stem reticular formation remains an open question.

* Stimulation of the many potential fibers of passage was obviated by prior lesioning of the lower brain stem. As a further control, massive depolarization of the mesencephalic cells was induced by injection of kainic acid (which should not affect fibers) and this led to prolonged EEG desynchronization.

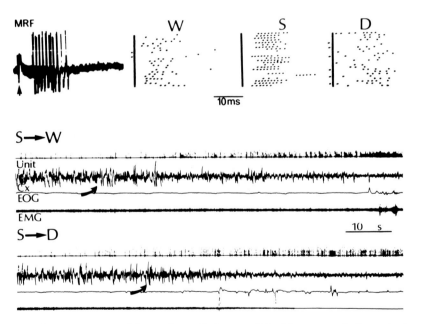

FIGURE 12.17. Upper part of figure shows discharges elicited from a cortically projecting cell in the central lateral nucleus of a cat by stimulation of the mesencephalic reticular formation (MRF). In the rasters, each representing 25 sweeps, note stereotyped burst in synchronized sleep (S) in comparison with waking (W) and desynchronized sleep (D). During S, neuron was less readily activated antidromically by stimulation of cortex. Lower part of figure indicates discharges of same unit (Unit) in relation to cortical (Cx), ocular (EOG), and muscular (EMG) correlates of sleep and wakefulness. Curved arrows indicate periods when increased antidromic responsiveness occurred. From Steriade (1981).

Though the nucleus cuneiformis appears to exert a profound influence over the intralaminar nuclei, it is not clear that all the intralaminar nuclei receive fibers from it. Nor do other nuclei such as the medioventral, ventral medial, and interanteromedial which, from both the physiological and the anatomical point of view, should be intimately involved in the diffuse cortical projection. There are also the several other inputs to the intralaminar system which quantitatively* seem to exceed that from the nucleus cuneiformis. Many of these other inputs are obviously motor related: the deep cerebellar nuclei, the substantia nigra, the globus pallidus, the deep layers of the superior colliculus. Others are sensory: overtly like the spinal cord or less clearly like the pretectum and periaqueductal gray matter. Then there are other inputs from the lower brain-stem reticular formation: some of these centers such as the paramedian pontine area have been implicated in the control of eye movements, others such as the parabrachial nucleus have been implicated in sleep mechanisms (Chapter 4).

There is nothing obvious that these various inputs seem to have in common. What is also true is that many of these inputs terminate in single intralaminar nuclei or among cells that form a subgroup of one nucleus (see Section 12.5). Should one, therefore, be looking for individual functions among the intralaminar nuclei?

12.6.4. Motor Aspects of the Intralaminar Nuclei

Repetitive electrical stimulation of the internal medullary lamina in the general region of the central lateral nucleus in the cat elicits contraversive head turning and saccadic, conjugate, and contraversive eye movements with latencies of about 30 msec (Schlag and Schlag-Rey, 1971, 1984; Hunsperger and Roman, 1976). The effect on eye movements is not abolished by chronic decortication combined with destruction of the striatum but could be abolished by acute lesions in the posterior part of the thalamus. This experiment seems to effectively rule out intralaminar projections to the cortex and striatum being involved in the eye movement effect. However, because there are no descending connections from this part of the thalamus, it is hard to see how the thalamic lesion blocked the effect unless the stimulus antidromically activated an ascending pathway to the intralaminar nuclei from a lower eye-movement center. Visual evoked potentials are enhanced in the cat visual cortex during intralaminar induced eye turning, suggesting activation of the nonspecific projection system (Hunsperger and Roman, 1976).

Other findings also suggest that the intralaminar nuclei are in some way concerned with the control of gaze. Lesions of the internal medullary lamina in the vicinity of the central lateral nucleus on one side lead to contralateral visual neglect and changes in optokinetic nystagmus (Orem *et al.*, 1973). Many neurons in the general vicinity of the central lateral nucleus have visual receptive fields, and are sensitive even to dim visual stimuli (Schlag-Rey and Schlag, 1977; Schlag and Schlag-Rey, 1984) while others fire during and before saccadic eye movements and can be direction selective (Schlag *et al.*, 1974; Schlag and Schlag-Rey,

---

* That is, as judged from the density of anterograde labeling reported in them.

1984). The tonic activity of some cells depends upon eye position and Schlag *et al.* (1980) have determined that the responsiveness of many of these cells depends upon the location of a visual stimulus with respect to the head–body axis, i.e., although a stimulus falls on a part of the retina appropriate for the cell, if the direction of gaze is inappropriate, the cell might not fire. This implied to the investigators that the neurons code visual spatial information in a nonretinal frame of reference, and that the cells could be concerned with the initiation of visually guided movements.

There is little or no information regarding the possible involvement of the intralaminar nuclei in the control of other kinds of movements. Lesions of the intralaminar nuclei in monkeys are far less effective than those of the ventral lateral nuclei in alleviating the dyskinesias induced by prior lesions of the cerebellum or subthalamic nucleus (Carpenter and Brittlin, 1958; Carpenter, 1961; Carpenter *et al.*, 1965). Any influence that the intralaminar nuclei might exert over the striatum is not evident. The results of physiological studies are limited to the observations that the thalamostriatal projection is excitatory (Purpura and Malliani, 1967; Buchwald *et al.*, 1973) and that the thalamic terminations can converge on the same cells as corticostriatal and nigrostriatal terminations (Kitai *et al.*, 1976). The significance of the fine patchwork of thalamostriatal terminations (K. Kalil, 1978; Royce, 1978a; Herkenham and Pert, 1981; Fig. 3.13) remains undetermined.

12.6.5. Sensory Aspects of the Intralaminar Nuclei

As interest in the intralaminar nuclei as the origin of the cortical recruiting response declined, a new interest arose in their putative functions as sensory relay nuclei, mediating nonspecific, modality-convergent, and, particularly, nociceptive information to the telencephalon. Just as Jasper's review of 1960 marked the end of the era of intensive study of the recruiting response, so the review of Albe-Fessard and Besson in 1973 marked the end of another era that had commenced in the middle 1950s with descriptions of heterotopic somatic and polysensory convergence on single neurons in parietal and frontal cortical areas of chloralose-anesthetized cats (Amassian, 1954; Albe-Fessard and Rougeul, 1955). Four areas of visual, somatic sensory, and auditory convergence became well established in the chloralose-anesthetized cat: one overlapping the motor cortex on the precruciate gyrus, one at the anterior end of the lateral gyrus, and one at each end of the middle suprasylvian gyrus [Fig. 12.13B (A); Albe-Fessard and Rougeul, 1955; Thompson and Sindberg, 1960; Thompson *et al.*, 1963a,b; Dubner and Rutledge, 1964, 1965; Bignall *et al.*, 1966; Buser and Bignall, 1967; Dow and Dubner, 1969]. There is evidence for the existence of similar areas in chloralose-anesthetized monkeys (Bignall and Imbert, 1969). None of the polymodal regions can be equated with known architectonic or functional cortical fields in the cat, but it is interesting to note that they correspond rather well with the focal regions of maximal recruiting responses, barbiturate spindles, and responses to stimulation of the midbrain reticular formation, as well as with the zones of regional dominance of inputs from the intralaminar nuclei (Figs. 12.12 and 12.13).

Multimodal responses in these "polysensory areas" survive disconnection of the areas from the primary sensory areas (Thompson *et al.*, 1963a,b; Bignall and Imbert, 1969) and they also survive destruction of the principal sensory relay nuclei of the thalamus (Bignall *et al.*, 1966).

In seeking the route through the thalamus for the polysensory projection on the cortex, Albe-Fessard and her associates observed that short-latency evoked potentials generated from electrical stimulation of all four limbs of chloralose-anesthetized cats and monkeys could be seen to converge in the centre médian, parafascicular, central lateral, and paracentral nuclei (Albe-Fessard and Rougeul, 1958; Kruger and Albe-Fessard, 1960; Albe-Fessard and Bowsher, 1965). Latencies of response were only slightly longer than those of responses occurring in the ventral posterior nucleus (11 msec vs. 7.5 msec for stimulation of the forepaws in cats). Similar effects were later obtained in unanesthetized, decorticated, or chronically implanted animals (Albe-Fessard *et al.*, 1961; Meulders *et al.*, 1963) and it was shown that the amplitude of the responses declined during desynchronized sleep and wakefulness.

Single-unit studies (Albe-Fessard and Kruger, 1962) in chloralose-anesthetized cats revealed that the best type of natural somesthetic stimulus that would excite neurons in the vicinity of the centre médian nucleus was a pinprick or a sharp tap. No units typical of the ventral posterior nucleus were recorded. Suggestions of convergence of visual and auditory inputs into these intralaminar regions were also made (Albe-Fessard and Mallart, 1960; Albe-Fessard and Fessard, 1963; Thompson, 1967). Cells responding to auditory and visual stimulation, however, have only been the subject of detailed study in more recent years (Schlag-Rey and Schlag, 1977; Schlag and Schlag-Rey, 1984; Irvine, 1980).

The demonstration that many caudal intralaminar neurons had large receptive fields, showed spatial and modality convergence, and responded to pain-threatening stimuli, inevitably led to the belief that these nuclei, the centre médian in particular, should play some role in nociception (Albe-Fessard and Besson, 1973). This belief was reinforced by the knowledge that spinothalamic fibers terminated in the posterior intralaminar region (Section 12.5) and by the observation that intraarterially injected bradykinin, a potent algesic, resulted in enhanced neuronal discharge in the centre médian nucleus (Lim *et al.*, 1969). From this belief arose neurosurgical justification for stereotaxic thalamotomy aimed at the centre médian nucleus in efforts to alleviate intractable pain (Mark *et al.*, 1963).

More recent physiological work clearly shows that the spinothalamic tract fibers with properties appropriate for mediating painful sensations terminate in the region of the intralaminar nuclei in monkeys (Applebaum *et al.*, 1979; Giesler *et al.*, 1981). The target of such axons is presumed to be the posterior large-celled component of the central lateral nucleus though this was not specifically determined. Spinothalamic tract axons projecting to the intralaminar nuclei alone arise mostly from high-threshold or wide-dynamic-range neurons (Chapter 7). Those with branches to both the intralaminar nuclei and the ventral posterior nucleus arise mostly from wide-dynamic-range spinal neurons but 27% arise from high-threshold neurons. Receptive fields were small and ipsilateral for cells with axons projecting to both nuclei but large and often bilateral for those projecting to the intralaminar nuclei alone. Both cell types received both A and

C fiber inputs. The small excitatory receptive fields, surrounding larger inhibitory areas and the discharge characteristics of the high-threshold cells projecting to both sites make it likely that they could mediate discriminative aspects of noxious cutaneous stimulation, such as intensity, duration, and localization. The cells with branched axons respond to sural nerve stimulation in the Aα to Aγ range with a short-latency (0–5 msec) burst of discharges lasting approximately 15 msec. Additional stimulation of smaller A and C fibers increased the duration of the burst to 25 msec or so and tonic discharges followed maintained nociceptive stimulation of the skin. Spinal cells projecting to the intralaminar nuclei alone are unlikely to subserve discriminative nociceptive functions because they have large receptive fields with no inhibitory regions and because they respond weakly and at long latency to noxious stimulation even when very intense. Discharges, however, were sustained long after a stimulus ceased. Willis's group believes that such receptive fields and discharge properties could reflect convergent influences descending upon these cells from cells in the bulbar reticular formation (Giesler *et al.*, 1981).

It is relatively easy to conceive of a discriminative nociceptive system in which pain messages eventually rising to consciousness pass to the somatic sensory cortex via the wide-dynamic-range and high-threshold inputs to the ventral posterior nucleus. Given the targets and apparent nature of the outflow from the intralaminar nuclei (Sections 12.5.1 and 12.6.2), it is less easy to conceive of how or where pain messages traversing them would or could rise to consciousness. It seems possible, therefore, that the role of the intralaminar nuclei in the pain process is concerned with affective states engendered by a painful stimulus rather than with the direct appreciation of the stimulus.

Branched and unbranced axons to the intralaminar nuclei also included a few arising from "deep" (i.e., proprioceptive) neurons (Giesler *et al.*, 1981), perhaps in keeping with the idea that, apart from nociception, the intralaminar nuclei may also be involved in movement-related activities. It is interesting to note that the concentration of opiate receptors is particularly high throughout the intralaminar nuclei, not just in the region of spinothalamic tract terminations (Pert *et al.*, 1976; Atweh and Kuhar, 1977; Chapter 5). Could this be in some way related to the fact that morphine and other opiates influence motor behavior in addition to inducing analgesia?

The progressive relative enlargement of the frontal lobes in the phylogenetic scale is by no means so conspicuous as that of the parietal or the temporal lobes. This fact though it does not disprove the preoccupation of the frontal lobes with emotional and personality integration, casts doubt on the common assumption that these lobes are man's greatest developmental achievement and therefore likely to be particularly concerned in his highest mental activities.

Le Gros Clark (1948)

13

The Medial Nuclei

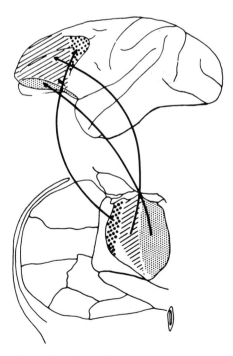

Topographic ordering in the thalamocortical connections of the mediodorsal nucleus in the monkey. From Akert (1964).

13.1.   Definition

For reasons that are rather arbitrary, though based on historical precedent, I consider the medial nuclear complex to consist of the mediodorsal nucleus, the parataenial nucleus, and the medioventral (reuniens) nucleus (Figs. 13.1 and 13.3). J. E. Rose (1942a,b) originally made this definition on the grounds that the three nuclei develop from a common pronuclear mass that is subsequently split by the growth of the intralaminar nuclei (his central commissural system) (see Fig. 6.10). Prior to Rose's work, "medial group" was ill-defined and often included parts of the rostral and caudal intralaminar nuclei, anteromedial nucleus, submedial nucleus, and principal ventral medial nucleus (von Monakow, 1895; Vogt, 1909; Friedemann, 1911; Nissl, 1913; M. Rose, 1935; Walker, 1937, 1938a). Perhaps as J. E. Rose (1942b) intimated, the submedial nucleus should be regarded as an extension of the medioventral nucleus and thus included in the medial group because, like the mediodorsal nucleus, it receives olfactory inputs (Price and Slotnik, 1983). On the other hand, it also receives spinothalamic inputs (Craig and Burton, 1981; Burton and Craig, 1983; Mantyh, 1983) and because of this, I have chosen to include it in the ventral medial group of nuclei (Chapter 7).

13.2.   Mediodorsal Nucleus

13.2.1.   Description

13.2.1.1.   General

The mediodorsal nucleus is usually large and occupies up to two-thirds of the length of the thalamus. In many mammals it has a rather homogeneous cytoarchitecture, consisting for the most part of small to medium, relatively pale-staining cells which are loosely but evenly distributed. Its anterior pole is usually difficult to define for it blends with the parataenial, and other adjacent nuclei. The posterior pole is also often unclear as it is traversed by the habenulopeduncular tract and encroached upon by irregular islands of cells belonging to the parafascicular and central lateral nuclei (see Figs. 12.9 and 12.10).

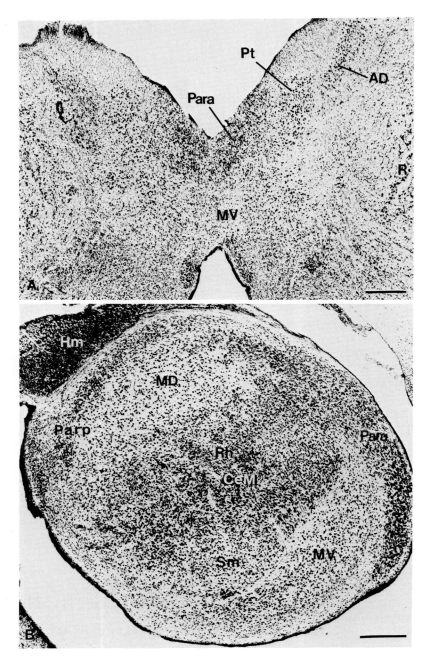

FIGURE 13.1.   Frontal (A) and parasagittal (B) sections through the parataenial (Pt), mediodorsal (MD), and medioventral (MV) or reuniens nuclei in the cat. Thionin stain. Bars = 500 μm.

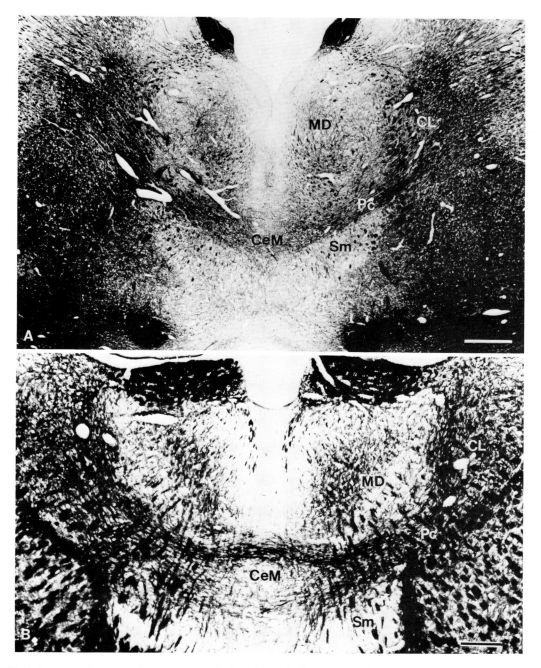

FIGURE 13.2.   Frontal sections through the mediodorsal nuclei of a cat (A) and a rat (B) showing the central fiber-rich division and medial and lateral fiber-poor divisions. Hematoxylin stain. Bars = 1 mm (A), 250 μm (B).

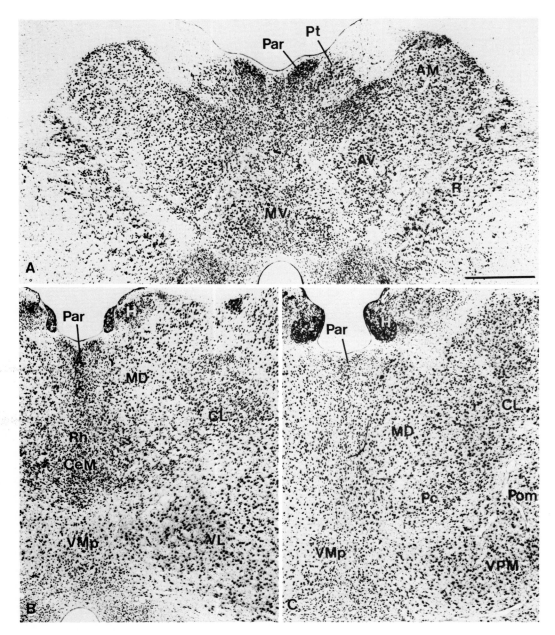

FIGURE 13.3.   Frontal sections through the parataenial (Pt), medioventral (MV), and mediodorsal (MD) nuclei in a marsupial phalanger showing (B, C) the anterior separation of the left and right MD nuclei (B) and their posterior fusion (C). Thionin stain. Bar = 1 mm.

Medially there is usually a cellular condensation where the two mediodorsal nuclei meet in the midline; some workers have referred to this as an "intermediodorsal nucleus (Gurdjian, 1927; Rioch, 1929a), though it seems to me to consist mainly of posterior parts of the rhomboid nucleus (Fig. 12.6). Over a limited posterior region in small mammals, however, there is usually a true union of the two mediodorsal nuclei with no intervening other elements (Fig. 13.3).

Ventrally and laterally, the nucleus abuts upon the internal medullary lamina containing the central medial, paracentral, and central lateral nuclei but the line of junction between them usually becomes less clear posteriorly. Here, large cells of the central lateral nucleus and/or parafascicular nucleus (Figs. 13.5, 13.6, 12.3, 12.4, and 12.8) tend to invade the ventrolateral aspect of the mediodorsal nucleus.

The mediodorsal nucleus of many nonprimates, unlike that of primates, cannot be divided into clearly defined cytoarchitectonic subdivisions. In fiber preparations, however, it is usually possible to see even in the rat that the central part of the nucleus stains much denser than the medial and lateral parts (Krettek

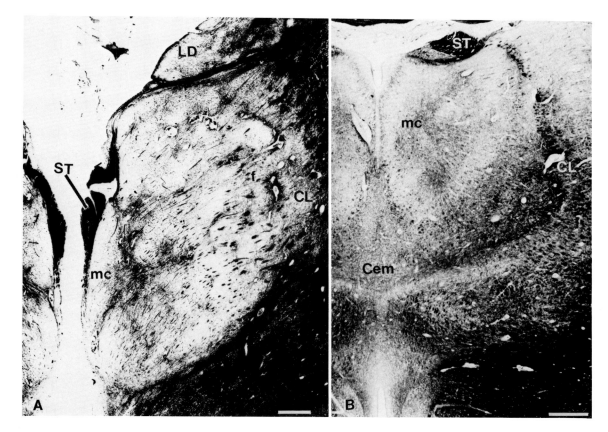

FIGURE 13.4. Frontal sections showing the medial, fiber-rich magnocellular division (mc) of the mediodorsal nucleus in man (A) and the bush baby (B). f in (A) indicates lateral fascicular or multiform division. Hematoxylin stain. Bars = 1 mm (A), 500 μm (B).

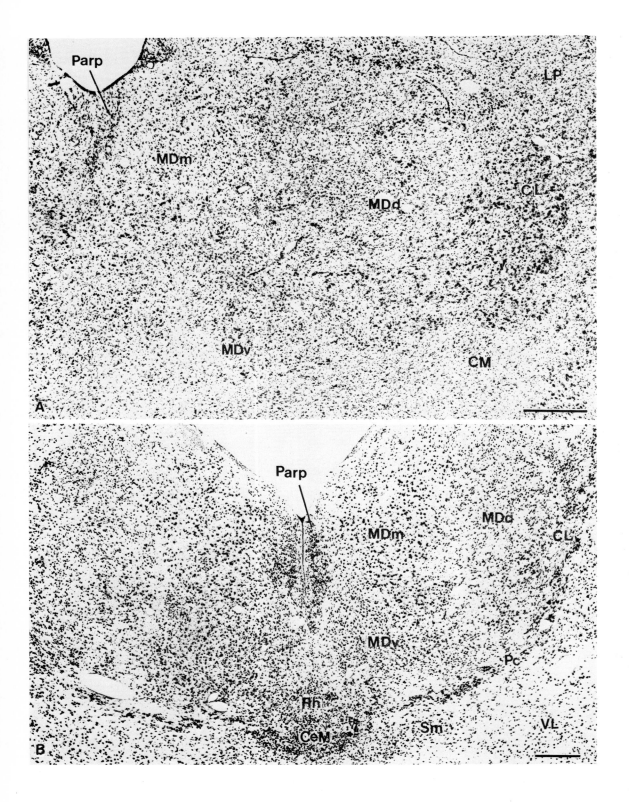

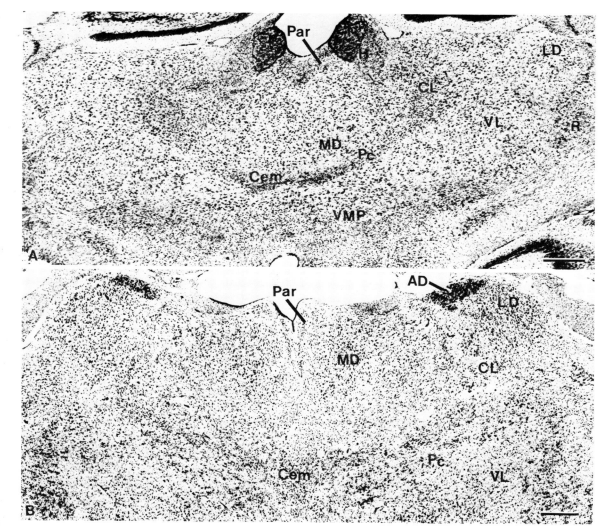

FIGURE 13.6. Mediodorsal nuclei of a moustache bat (A) and a bush baby (B). The nucleus in (A) has a homogeneous cell population but in (B) the typical primate groups of cells, particularly the magnocellular group, are evident. Thionin stain. Bars = 250 μm (A), 500 μm (B).

FIGURE 13.5. Thionin-stained frontal sections from a squirrel monkey (A) and a raccoon (B), showing the magnocellular (MDm), multiform (MDd), and ventral (MDv) divisions of the mediodorsal nucleus. Bars = 250 μm.

and Price, 1977b) (Figs. 13.2 and 13.4). The three parts of the nucleus so delimited have connectional differences (Section 13.3) and seem to represent forerunners of a tripartite division that can be detected cytoarchitectonically as well as myeloarchitectonically and connectionally in primates (Section 13.2.1.2). Some similar cytoarchitectonic distinctions can sometimes be detected in larger non-primates (Fig. 13.5B).

13.2.1.2. The Primate Mediodorsal Nucleus

The mediodorsal nucleus of most primates consists of at least two cytoarchitectonically distinct parts. The medial one-third to one-half is magnocellular in Nissl preparations (Figs. 13.5A and 13.6B) and has a densely fibrous neuropil in myelin preparations (Fig. 13.4). The lateral half to two-thirds has smaller, variably sized, and loosely packed cells; its neuropil is less fibrous though it is traversed by heavy fiber bundles passing to and from the magnocellular, medial division. The lateral division usually extends ventral and posterior to the magnocellular division and forms the posterior pole of the nucleus. These two cyto- and myeloarchitectonic subdivisions have been recognized in a lemur (Pines, 1927), *Tarsius* (Le Gros Clark, 1930a), the squirrel monkey (Emmers and Akert, 1963), Old World monkeys (Friedemann, 1911; Crouch, 1934; Walker, 1938a; Olszewski, 1952; Akert, 1964), and man (Namba, 1958; Hassler, 1959; Gihr, 1964). The lateral division is commonly divided further in the higher primates: Olszewski (1952) recognizes a dorsomedial pars parvocellularis and a ventrolateral pars multiformis, while Hassler (1959) identifies a lateral pars fasciculosus and a posterior pars caudalis with two further subdivisions.

The lateral edge of the lateral division of the mediodorsal nucleus, especially toward its posterior end, is invaded by large, deeply staining cells of the central lateral nucleus (Fig. 12.8). These cells extend around the posterior pole of the mediodorsal nucleus to meet the parafascicular and limitans nuclei (Fig. 12.4). For reasons given in Sections 13.3 and 13.4 and in Chapter 12, I regard these cells as part of the central lateral nucleus, though since the time of Walker (1938a), it has been more customary to regard them as a pars paralamellaris (Walker, 1938a; Hassler, 1959) or pars densocellularis (Olszewski, 1952) of the mediodorsal nucleus.

13.2.2. Terminology

Early workers such as Nissl (1889) and von Monakow (1895), studying non-primates, recognized but a single, homogeneous equivalent of our mediodorsal nucleus. Their other subdivisions of the medial group represent the anteromedial nucleus, parts of the intralaminar system, and the parafascicular nucleus. The term *nucleus medialis dorsalis* appears to have been introduced by Gurdjian (1927) in his study of the rat thalamus and has been translated as both "dorsomedial" (Le Gros Clark, 1932a) and "mediodorsal" (Rose, 1942b) nucleus. The latter translation seems to have become the more widely popular since the work of Rose and Woolsey (1948b). It is probable that the large cells of the central lateral nucleus that invade the posterior part of the mediodorsal nucleus were often included in the mediodorsal nucleus in nonprimates (e.g., Rinvik, 1968a) until Mehler (1966a) clearly pointed out their derivation from the intralaminar system.

The subdivisions of the primate mediodorsal nucleus were detected in some of the earliest studies on the primate thalamus. Vogt (1909) first recognized the densely fibrous character of the medial part of the mediodorsal nucleus in a cercopithecine monkey, calling it *Noyau médiale fibreux*. Then, using Nissl-stained preparations, seemingly from the same animals, Friedemann (1911) correlated this fibrous part with a medial, magnocellular subnucleus. Vogt and Friedemann divided the rest of the monkey mediodorsal nucleus (which they called simply *medial nucleus*) into anterior and posterior parts and also remarked on a further paralamellar part named by Vogt as *nucleus medialis, paralamellaire*, though named by Friedemann, *pars paralamellaris des Kerns der lamella interna*.

Over the ensuing years, the medial magnocellular or fibrous and the lateral mixed cell or less fibrous region came to be generally accepted in primates but the status of the paralamellar region remained somewhat equivocal. Grünthal (1934) included it with the central lateral nucleus in a common "nucleus circularis." Others disagreed: Walker (1938a,c, 1940b) for example, stated that because the paralamellar region underwent retrograde degeneration following destruction of the cerebral cortex, it should be regarded as part of the mediodorsal nucleus rather than an intralaminar nucleus. His position was based upon the belief that intralaminar nuclei survive ablation of the cortex. As we have seen (Chapters 3 and 12), this is not a point of view that can always be satisfactorily defended. Hassler (1959) follows Walker. Olszewski (1952), though naming the paralamellar region *pars densocellularis* of the mediodorsal nucleus, stated that its inclusion in the mediodorsal nucleus was "temporary" because he recognized its similarity to the central lateral nucleus. The observation that the paralamellar region projects upon a focal region of cortex in the vicinity of area 8 (Scollo-Lavizzari and Akert, 1963) in no way rules out classification of the paralamellar cells as part of the central lateral nucleus, for though these nuclei tend to project more diffusely than the principal nuclei, they often still have zones of focal dominance in the cortex (Chapter 12). I prefer to include the paralamellar region in the central lateral nucleus on the grounds that it receives afferent fibers from the spinal cord (Burton and Craig, 1983; Mantyh, 1983). It, thus, appears equivalent to the large cells of the central lateral nucleus that invade the mediodorsal nucleus in nonprimates and receive spinal inputs (Chapter 12; Mehler, 1966a,b; Jones and Burton, 1974; Craig and Burton, 1981).

Some suggestions of finer subdivisions in the mediodorsal nucleus of nonprimates were made by M. Rose (1935) in the rabbit and by J. E. Rose (1942b) in the sheep, pig, rabbit, and cat. However, attention did not become focused on these until Leonard (1969) indicated differences in thalamocortical connections between the medial and lateral parts of the rat nucleus (Section 13.4). Thereafter, Krettek and Price (1977b), examining silver-stained preparations of the rat made by M. Shipley, were able to detect a central fiber-rich zone and medial and lateral fiber-poor zones (as in Fig. 13.2). As indicated in the following section, these zones have different input and output connections.

13.2.3. Connections

13.2.3.1. Background

von Monakow (1895) in the cat and Minkowski (1924) in the monkey early pointed out, from retrograde degeneration studies, the association between the mediodorsal nucleus and the frontal lobe. This was confirmed by Le Gros Clark and Boggon (1933a) using the Marchi technique in the rat and by Walker (1936, 1938a, 1940b, 1949) who, using the retrograde degeneration method, related the nucleus to frontal granular cortex. Other retrograde degeneration studies in monkeys (Mettler, 1947; Pribram *et al.*, 1953) and man (Meyer *et al.*, 1947; Freeman and Watts, 1947) suggested a dissociation in the thalamocortical projection, with the medial, magnocellular part of the nucleus projecting to the lateral aspect of the frontal lobe and the lateral, mixed cell part projecting to the orbital surface. Rose and Woolsey (1948b), working with the retrograde degeneration method in sheep, rabbits, and cats, adopted a rather restricted view of the cortical target of the mediodorsal nucleus, limiting it to a small area at the frontal pole of the rabbit and at the ventral aspect of the gyrus proreus in the sheep and cat (see also Warren *et al.*, 1962, and Narkiewicz and Brutkowski, 1967). Rose and Woolsey named the cortical target of the mediodorsal nucleus, *orbitofrontal cortex,* and described its structure in the three species. Since that time, newer connectional studies have revealed a more extensive projection target for the nucleus and the name *orbitofrontal cortex* has tended to drop out of use.

Afferent connections to the mediodorsal nucleus were described in early studies of normal fiber preparations, to come from the posterior hypothalamus (Rioch, 1931; Le Gros Clark and Boggon, 1933b) but this has not been confirmed. Early axon degeneration studies identified corticothalamic afferents (Meyer, 1949; Auer, 1956; Guillery, 1959; Rinvik, 1968b) and afferents from the septal nuclei and midbrain (Guillery, 1959), but both septal and midbrain afferents were discounted on technical grounds by Raisman (1966). It is only in recent years that the afferent connections of the nucleus have been thoroughly elucidated.

13.2.3.2. Recent Studies

Two studies using the axon degeneration method in the rat are the starting points of our current understanding of the connectivity of the mediodorsal nucleus. First, was the study of Powell *et al.* (1965), showing that fibers from the prepiriform and adjacent areas of paleocortex terminated in the mediodorsal nucleus, thus indicating a route through the thalamus whereby olfactory impulses could reach neocortex. Second, was the study of Leonard (1969, 1972) who showed that central parts of the nucleus projected to frontal cortex in the interior of the rhinal sulcus ("sulcal cortex" of Leonard) and that lateral parts projected to cortex at the border where medial and lateral surfaces of the frontal cortex meet ("shoulder cortex" of Leonard; Fig. 13.7). This established the division of the nonprimate nucleus into parts with different cortical targets. Only the lateral of these ("sulcal cortex") became identified as obviously olfactory following the report of Leonard and Scott (1971) and Heimer (1972) that efferents from the prepiriform cortex terminated only in the central part of the mediodorsal nucleus.

The *central fiber-rich part* of the rat mediodorsal nucleus has since been confirmed both anatomically and physiologically as the target of efferents from olfactory areas of the piriform lobule (Fig. 13.7). Short-latency responses to electrical stimulation of the olfactory bulb occur only in the central part (Jackson and Benjamin, 1974; Price and Slotnik, 1983). Injections of anterograde tracers in the olfactory cortex or underlying endopiriform nucleus label only the central part (Krettek and Price, 1977b) and retrograde labeling from the mediodorsal nucleus itself shows that cells in the polymorph layer of the olfactory tubercle, in the endopiriform part of the claustrum deep to the prepiriform cortex, and in deep layers of periamygdaloid and entorhinal cortex, all contribute (Siegel *et al.*, 1977; Price and Slotnik, 1983).

Olfactory cortex projections have now been demonstrated to the central part of the mediodorsal nucleus of the opossum (Jackson *et al.*, 1977; Benjamin

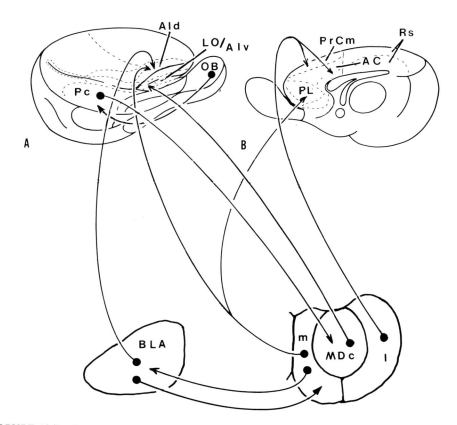

FIGURE 13.7.   Interconnections of the central (MDc), lateral (l), and medial (m) divisions of the rat mediodorsal nucleus, with the basolateral amygdaloid nucleus (BLA) and the cerebral cortex. AC: anterior cingulate area; AId: dorsal agranular insular area; LO/AIv: lateral orbital and ventral agranular insular areas; OB: olfactory bulb; Pc: piriform cortex; PL: prelimbic area; PrCm: medial precentral area; Rs: retrosplenial areas. After Krettek and Price (1977a,b).

et al., 1982) and to the medial half of the nucleus in the rabbit (Jackson and Benjamin, 1974). In monkeys, single- and multiunit recording in response to stimulation of the olfactory bulb very clearly delineates the medial, magnocellular part of the nucleus as the terminus of short-latency olfactory inputs (Benjamin and Jackson, 1974; Tanabe et al., 1975b; Yarita et al., 1980). This implies that the fibrous, magnocellular part of the primate mediodorsal nucleus should be equated with the fibrous central part of the mediodorsal nucleus in nonprimates and that it forms the thalamic, olfactory relay to neocortex.

In the rat the central, olfactory part of the mediodorsal nucleus projects to two neocortical areas in the rhinal sulcus: the lateral orbital cortex and the ventral agranular insular area (Krettek and Price, 1977b; Fig. 13.7). The second of these also receives direct projections from the olfactory cortex (Krettek and Price, 1977a), thus closing a loop between olfactory cortex, thalamus, and neocortex. Similar details have been established for the rabbit (Benjamin et al., 1978) (see also Tobias and Ebner, 1973, in the opossum). In monkeys there are two areas (Fig. 13.8) on the ventral surface of the frontal lobe in which short-latency unitary responses to electrical stimulation of the olfactory bulb or olfactory cortex can

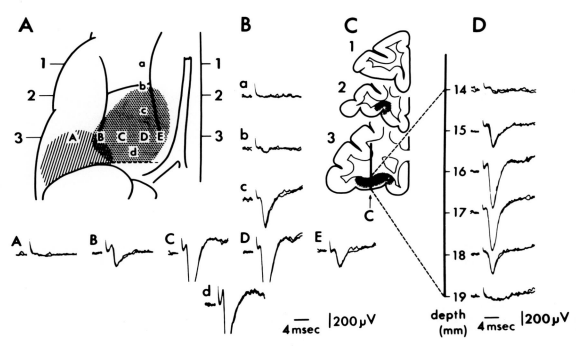

FIGURE 13.8. (A) Diagrammatic representation of the positions of the lateral posterior olfactory field (LPOF, hatching left) and central posterior olfactory field (CPOF, stippling left) on the ventral surface of the frontal lobe of a monkey. CPOF is activated by electrical stimulation of the magnocellular part of the mediodorsal nucleus. (B) Evoked potentials in MDmc from stimulation of sites B–E and c–d in (A). (C) Frontal sections at levels indicated (1–3) in (A), showing position of CPOF. Stimulating electrode in section 3 elicited evoked potentials (D) in MDmc at depths indicated. From Yarita et al. (1980).

be elicited (Tanabe *et al.*, 1974, 1975b; Takagi, 1979; Yarita *et al.*, 1980). One of these, the "lateroposterior portion of the orbitofrontal area" (LPOF), corresponds to parts of Walker's (1940b) areas 12 and 13 and the other, the "centroposterior portion of the orbitofrontal cortex" (CPOF), corresponds to Walker's area 13. Antidromic activation of cells in the magnocellular division of the mediodorsal nucleus can only be elicited from stimulation of area CPOF (Yarita *et al.*, 1980). More direct pathways were proposed for olfactory cortex inputs to area LPOF. It is uncertain, therefore, whether both fields or only CPOF should be regarded as equivalent to the two olfactory-related fields in the rat.

The *medial part* of the mediodorsal nucleus in the rat receives fibers from the basolateral nucleus of the amygdala (Krettek and Price, 1974, 1977a) (Figs. 13.7 and 13.9). It projects to the prelimbic area of cortex medially and to the dorsal agranular insular area in the rhinal sulcus (Fig. 13.7; Krettek and Price, 1977a). The basolateral nucleus of the amygdala completes another loop by also projecting to the dorsal agranular insular area (Krettek and Price, 1977a). In the opossum the basolateral amygdaloid nucleus projects, as in the rat, to a medial part of the mediodorsal nucleus (Porrino *et al.*, 1981). In monkeys the basolateral mass of the amygdala projects to the mediodorsal nucleus (Nauta, 1961) and the location of the terminations in the mediodorsal nucleus has recently been shown to be also in the medial magnocellular division (Aggleton and Mishkin, 1984).

The *lateral part* of the rat mediodorsal nucleus has not yet been characterized in terms of its afferent inputs and there is no comparable work that one could draw on in other species. In the rat the lateral part projects to two medially placed cortical areas: the anterior cingulate area and the medial precentral area (Krettek and Price, 1977b). There is a similarly restricted projection of the lateral part of the rabbit mediodorsal nucleus (Benjamin *et al.*, 1978). Terminations of fibers from all parts of the mediodorsal nucleus are in layers I and III. In the anterior cingulate area, other thalamocortical fibers from the anteromedial nucleus terminate in layers I and VI (Domesick, 1969, 1972; Krettek and Price, 1977b; Chapter 14).

In primates the cortical targets of parts of the mediodorsal nucleus other than the magnocellular, olfactory part, have not been worked out with new techniques. These parts (Figs. 13.4 and 13.5) are somewhat larger than the magnocellular fibrous part and, from the earlier retrograde degeneration studies, one would predict that they would project primarily upon the large areas of cortex on the lateral surface of the frontal lobe (Walker, 1940b; Roberts and Akert, 1963; Akert, 1964; Tobias, 1975).

Corticothalamic afferents to the mediodorsal nucleus apparently arise in areas to which the nucleus projects, as might be expected, but there have been only a few studies since the work of Leonard (1969) and Domesick (1969)—Künzle (1978), Beckstead (1979), and Goldman (1979). In both monkeys and cats there are indications that the corticothalamic projection to parts of the mediodorsal nucleus is bilateral (Rinvik, 1968b; Künzle, 1978; Goldman, 1979). Other afferents to the nucleus have been reported in anatomical studies to come from various parts of the brain stem (Graybiel, 1977; Velayos and Reinoso-Suarez, 1982).

13.2.4. Functions of the Mediodorsal Nucleus

From the connectional data outlined in the preceding section, the medio-dorsal nucleus would seem to be involved in some aspects of olfaction and there is good evidence for this. Using unanesthetized monkeys, Yarita *et al.* (1980) have shown that neurons in the magnocellular part of the mediodorsal nucleus

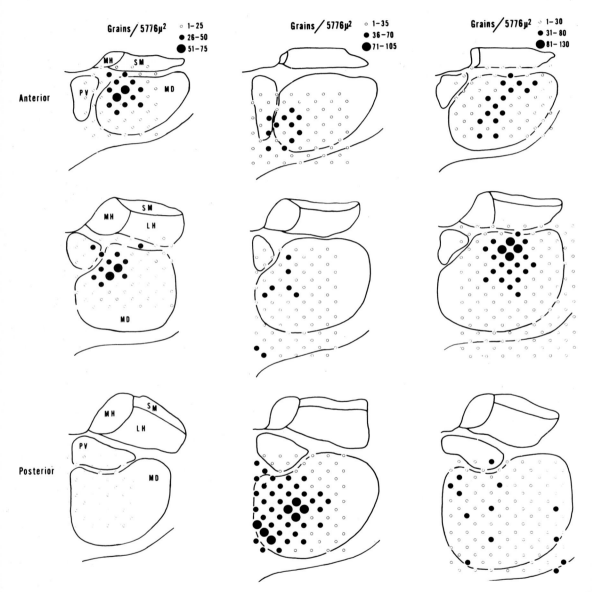

FIGURE 13.9.   Autoradiographic demonstration of the differential termination of fibers from the basolateral amygdaloid nucleus (left), basolateral and central amygdaloid nuclei (middle), and endopiriform nucleus and prepiriform cortex (right) in the mediodorsal nucleus of the rat. Size of dot is proportional to grain density. From Krettek and Price (1977b).

activated by stimulation of the olfactory bulb respond to odoriferous stimuli. Few, however, respond preferentially to a single odor or even to a small group of odors. Most respond to five or seven out of eight. This is also a feature of area CPOF to which the magnocellular part projects (Fig. 13.8). By contrast, in the other cortical olfactory area (LPOF) from which the mediodorsal nucleus could not be activated antidromically (Tanabe *et al.*, 1975b), the number of odors to which 50% of the cells would respond was limited to one. The differences in cellular responsiveness have suggested that the more direct route from pre-piriform cortex to area LPOF is concerned with fine olfactory discriminations, whereas that through the mediodorsal nucleus to area CPOF is involved in some more complex aspect of olfactory behavior (Yarita *et al.*, 1980). There is evidence from rodents to support this idea. In rats, after destruction of the mediodorsal nucleus, simple olfactory discriminations are still possible but more complex olfactory behaviors such as olfactory-guided, male sexual behavior, preference for certain odors, and odor reversal learning are impaired (Sapolsky and Eichenbaum, 1980; Eichenbaum *et al.*, 1980; Slotnik and Kaneko, 1981). Defects of reversal learning related to other modalities are characteristic of the behavior of primates with lesions of the lateral surface of the frontal lobe (Milner, 1974; Gross and Weiskrantz, 1964; Mishkin, 1966). Hence, one may speculate that the defects in olfactory-guided behavior exhibited by rodents, in which the olfactory part of the mediodorsal nucleus exceeds in size that of the other parts, is but one manifestation of a more general behavioral function carried out by the mediodorsal nucleus and its cortical targets.

In man, lesions of the medial part of the thalamus have for some time been associated with defects of memory, von Gudden (1896) perhaps being one of the first to recognize this. Victor *et al.* (1971) state that in Korsakoff's syndrome (among the symptoms of which are confabulation and amnesia for recent events) the only consistent lesions are in the mediodorsal nucleus and in adjacent parts of the anteroventral nucleus and pulvinar, plus the mamillary bodies. The lesions of the mamillary bodies are not regarded by recent authorities as critical for memory dysfunction (Victor *et al.*, 1971; McEntee *et al.*, 1976; Squire and Moore, 1979). Surgical, vascular, or traumatic lesions of the mediodorsal nucleus and its environs and tumors of the third ventricle, on the other hand, are regularly reported to be associated with memory disturbances (Williams and Pennybacker, 1954; Spiegel *et al.*, 1955; Mills and Swanson, 1978; Squire and Moore, 1979; Mair *et al.*, 1979; Schott *et al.*, 1980). A subject with a traumatic lesion, confirmed by CAT scanning to be primarily in the left mediodorsal nucleus (Squire and Moore, 1979), exhibited a chronic amnesia for day-to-day events; informal memory testing showed a deficiency in learning verbal, rather than nonverbal material. The verbal nature of the deficiency with a left-sided lesion is of particular interest in view of Ojemann's (1977, 1981) demonstration of functional asymmetry in the human thalamus. Ojemann has found in studies involving intra-thalamic electrical stimulation, that the left thalamus is specialized for verbal skills and the right for nonverbal. It is unlikely, however, that this functional asymmetry is entirely attributable to the mediodorsal nucleus.

Studies of cats and monkeys with lesions of the mediodorsal nucleus are also suggestive of an involvement of the nucleus in learning and memory (Pectel *et al.*, 1955; Schulmann, 1964; Isseroff *et al.*, 1982). The defect in monkeys following bilateral lesions is one affecting the ability to carry out a delayed response, very similar to that occurring after destruction of the lateral frontal cortex. There is a marked similarity between this, the olfactory behavioral deficit alluded to above, and the character of the memory defect in man. Aggleton and Mishkin (1983a,b) have shown that bilateral lesions primarily of the magnocellular parts of the mediodorsal nuclei lead especially to defects of object reward associative memory. Such deficits are also characteristic of monkeys with lesions of the amygdala (Mishkin, 1982) which (Section 13.2.3.2) provides the input to the magnocellular part. Aggleton and Mishkin consider, however, that damage of the anterior nuclei and secondary degeneration of the mamillary nuclei may be necessary to produce the most severe deficits of memory function.

13.3. Parataenial Nucleus

13.3.1. Description

The parataenial nucleus is at its largest on the anterior surface of the thalamus where it lies posterior to the stria medullaris and between the anterior paraventricular nucleus, the stria terminalis, and the column of the fornix (Figs. 13.1 and 13.10). From there, as it is followed posteriorly along the dorsal aspect of the thalamus, it rapidly narrows to a slender tail intercalated between the anterior paraventricular, anterodorsal, and anteromedial nuclei, and in most places, separated from them by extensions of the internal medullary lamina.

The nucleus terminates by fusing with the mediodorsal nucleus, usually before reaching the anterior pole of the habenular nuclei.

In small mammals, such as marsupials, rodents, bats, and insectivores (Figs. 13.3 and 13.6A), the parataenial nucleus may approach in size the anterior nuclei and it may be hard to divide it from the mediodorsal nucleus because of similar cytoarchitecture. In lagomorphs, carnivores, tree shrews, and lower primates, it becomes relatively much smaller and in higher primates, including man, it is reduced to a thin strip of cells on the dorsal aspect of the central lateral and mediodorsal nuclei (Fig. 13.10) (see, e.g., Le Gros Clark, 1929a,b, 1930a; Pines, 1927; Gurdjian, 1927; Goldby, 1941; Bodian, 1942; Rose, 1942b; Olszewski, 1952; Oswaldo-Cruz and Rocha-Miranda, 1967; Roberts and Akert, 1963; Berman and Jones, 1982).

13.3.2. Terminology

The term *parataenial nucleus* was introduced by Vogt (1909) and Friedemann (1911) in their studies of the guenon brain. Prior to their work, it was probably included in the "anterodorsal medial nucleus" by Nissl (1889, 1913). D'Hollander (1913) first applied the term *parataenial nucleus* to the rabbit and, apart from M. Rose's (1935) appellation, *nucleus supramedialis*, the name seems to have become fixed for all species. All early workers from Vogt on, included the parataenial nucleus in the medial group of nuclei and this position was confirmed by J. E.

Rose (1942a) when he concluded that it has closer developmental affinities with the mediodorsal nucleus than with other nuclei (Chapter 6).

13.3.3. Connections

The connectivity of the parataenial nucleus has not been fully elucidated. Stoffels (1939a,b), Lashley (1941), and Droogleever-Fortuyn (1950) considered from retrograde degeneration experiments in the rat and rabbit that it projected to cortex low on the medial side of the frontal lobe, close to the taenia tecta. This view is supported by certain retrograde (Jones and Leavitt, 1974) and anterograde (Krettek and Price, 1977b) labeling experiments. On the other hand, Powell and Cowan (1954) in their studies with the retrograde degeneration method related the parataenial nucleus of the rat to the nucleus accumbens of the striatum which underlies the taenia tecta. This view is supported by the anterograde and retrograde labeling studies of Swanson and Cowan (1975). The disagreement is more than academic for an association with the striatum should lead to the parataenial nucleus being classified with the intralaminar nuclei rather than in the medial nucleus complex. One problem with any connectional studies of the parataenial nucleus is that it is often difficult to separate it from some of

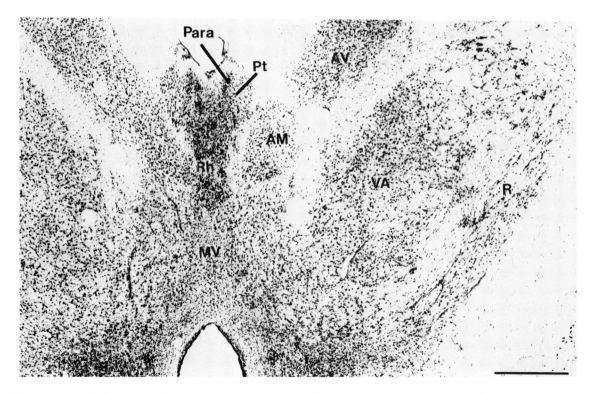

FIGURE 13.10. Thionin-stained frontal section through the small medioventral (MV) and parataenial (Pt) nuclei of a cyno-molgus monkey. Bar = 1 mm.

the more dorsally placed components of the intralaminar complex; hence, a distinction between what is intralaminar and striatal-projecting, and what is not, may be difficult.

There are suggestions, from retrograde labeling studies, that the parataenial nucleus also projects to the hippocampal formation (Wyss *et al.*, 1979; Amaral and Cowan, 1980). Corticothalamic connections have not yet been shown to originate from a particular cortical field and no clear-cut source of subcortical afferents has been described.

13.4. Medioventral (Reuniens) Nucleus

13.4.1. Description

The medioventral nucleus is present over most of the anterior one-third of the thalamus and in frontal sections assumes a bilobed form by fusing with its fellow of the opposite side (Figs. 13.1 and 13.10). The nucleus consists of medium-sized, loosely aggregated cells and its open neuropil makes it stand out in sagittal sections close to the median sagittal plane (Figs. 13.1 and 16.1). The anterior surface of the nucleus protrudes between the reticular nuclei and extends to the anterior end of the thalamus where it invaginates the anterior paraventricular nuclei from behind.

The fused medioventral nuclei are close to the floor of the third ventricle, separated from it only by the ependyma and a few fiber bundles derived from the internal medullary lamina. Dorsally, they abut on the interanteromedial parts of the anteromedial nuclei, often partially separated by lateral winglike extensions of the rhomboid nucleus and/or by parts of the central medial nucleus (Figs. 13.1, 13.3, and 13.10). Posteriorly, the principal ventromedial nuclei fuse across the midline behind them.

The medioventral nucleus is well developed in rodents, ungulates, and lagomorphs (Rose, 1942b; Fig. 13.3 and Chapter 14) but is relatively small in carnivores (Berman and Jones, 1982; Fig. 13.2); it is particularly small though still recognizable in primates (Fig. 13.10) (Walker, 1938a; Olszewski, 1952; Hassler, 1959; Emmers and Akert, 1963). The relative reduction in the size of the nucleus is not only found in the primates, however, for it is much smaller relative to other nuclei in the opossum than in the rat (Gurdjian, 1927; Bodian, 1939), and in some shrews (Le Gros Clark, 1929b, 1932a) and bats may virtually disappear (Fig. 14.8). Even among genera of the same order there may be considerable variation. The nucleus is said, for example, to be much larger in the dog than in the cat (Rioch, 1929a).

13.4.2. Terminology

The fused medioventral nuclei, together, constitute the *nucleus reuniens* of many other authors. The medioventral nucleus seems to have been included by Nissl first (1889) in a large midline nucleus (*MittellinieKern*) and then (1913) in a medial anteroventral nucleus. von Monakow (1895) did not separately identify it. Münzer and Wiener (1902) may have been the first to clearly identify it in the rabbit, calling it *ventraler Schenkel der Nucleus arcuatus* (their dorsal shoulder seems to be the mediodorsal and/or parafascicular nucleus). In her study of the monkey thalamus, Vogt (1909) included the region of the medioventral nucleus in her submedial nucleus (which she abbreviated *MV*) and Friedemann (1911) may have included it in his nucleus ventralis commissurae mediae (*MVV*). The term *nucleus reuniens* is often attributed to Malone (1910) who used it in his study of the human thalamus. However, I have found it applied earlier by Röthig (1909) to the appropriate nucleus of the opossum thalamus. The term, therefore, may have its origins in the comparative anatomical studies of Edinger because Röthig was at that time in Edinger's institute at Frankfurt.

Nucleus reuniens was adopted by Gurdjian (1927) in the rat, Rioch (1929a) and Ingram *et al.* (1932) in the cat, Papez (1932) in the armadillo, Aronson and Papez (1934) in the monkey, and Bodian (1939) in the opossum. Many modern authors look to these descriptions for their provenance.

Medioventral nucleus (nucleus medialis ventralis) was first used by M. Rose (1935) for the left and right nuclei that fuse to form the nucleus reuniens. This term was adopted by J. E. Rose (1942b) on account of the derivation of the medioventral with the mediodorsal nucleus from a common pronucleus (Rose, 1942a) (Chapter 6). Other modern authors, including myself, look to this description for their provenance.

Nucleus medialis ventralis and the translation, *medioventral nucleus*, have, unfortunately, often been used for the principal ventral medial nucleus (Chapter 7) and its interventral fusion across the midline (Le Gros Clark, 1930a; Crouch, 1934; Walker, 1938a,c). At times, probably leading on from Vogt (1909) and Friedemann (1911), *medialis ventralis* has been held to be synonymous with *submedial nucleus* (Rioch, 1929a; Le Gros Clark, 1930a, 1932a). These nomenclatural vagaries have often led to considerable confusion especially in descriptions of primates in which the medioventral (reuniens) nuclei are small and perhaps even undetectable. One can avoid this confusion by adopting J. E. Rose's policy of ensuring that the prefix of a name translated from the Latin specifies nuclear group and the suffix the position of the nucleus within that group.

13.4.3. Connections

In many early accounts, the medioventral (or reuniens) nucleus was regarded as one of the nuclei of the midline and, as such, to be unaffected by destruction of the cerebral cortex (Walker, 1937, 1938a; Waller and Barris, 1937). Later evidence, however, indicated that the nucleus does not survive removal of the cerebral cortex (Rose and Woolsey, 1943, 1949a; Droogleever-Fortuyn, 1950; Cowan and Powell, 1955; Akimoto *et al.*, 1956a) and suggested that it was probably connected with the infralimbic areas close to the projection fields of the mediodorsal and anterior nuclei.

The most thoroughly documented study in recent times of the connections of the medioventral nucleus is that done in the rat by Herkenham (1978). Anterograde tracing indicated that the efferent fibers of the nucleus followed a

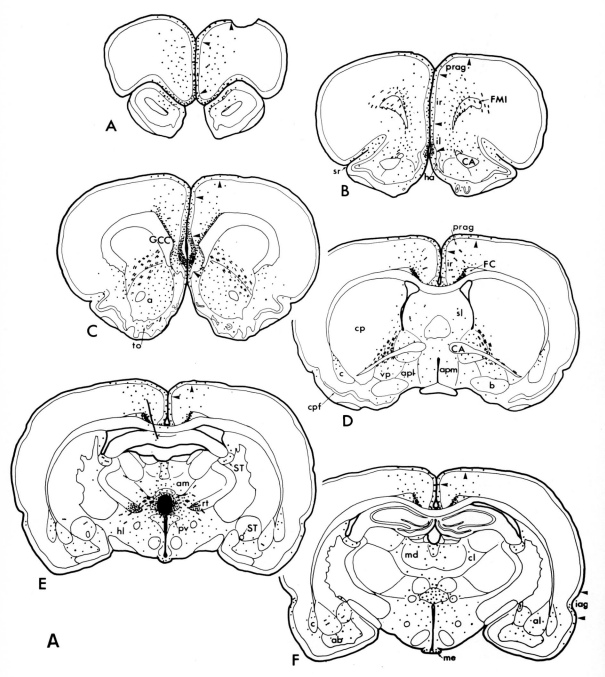

FIGURE 13.11.   Chart showing on frontal sections the distribution of efferent fibers labeled as the result of an injection of tritiated amino acids in the medioventral nuclei of a rat. From Herkenham (1978).

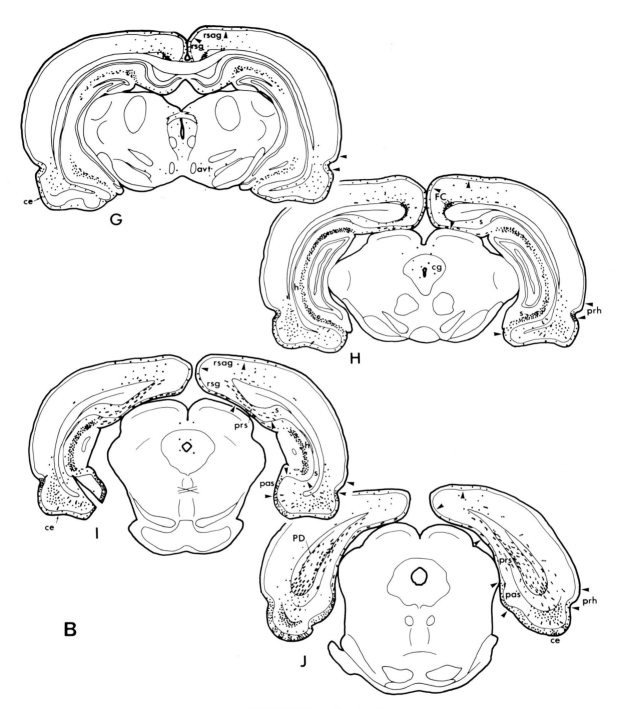

FIGURE 13.11. *(continued)*

course rather similar to that adopted by many of those leaving the anterior nuclei (Chapter 14). That is, instead of joining the internal capsule, they ascend through the postcommissural fornix and around the genu of the corpus callosum to join the cingulum (Fig. 13.11). En route, as would be expected, there is a small projection to the thalamic reticular nucleus. From the cingulum, the efferent fibers distribute terminal ramifications to layer I of most areas of cortex on the medial surface of the hemisphere and continue backwards and downwards to innervate layer I of the pre- and parasubiculum. Fibers continuing on past the hippocampal formation reach layer I of the lateral and possibly medial entorhinal areas and the perirhinal cortex. In the entorhinal cortex there are additional heavy terminations in layer III. Other fibers enter the subiculum and the CA1 field of the hippocampus and terminate in the stratum lacunosum et moleculare. Fields CA2–3 and the dentate gyrus do not receive fibers.

Afferents to the medioventral nucleus had previously been described in the fornix by Nauta (1956). This was confirmed by Herkenham, using retrograde labeling. He showed their origins in the medial cortex and the subiculum. This has subsequently been confirmed by Wyss *et al.* (1979) in the rat and by Amaral and Cowan (1980) in the monkey. Aggleton and Mishkin (1984) in the monkey suggest a small input from the amygdala as well.

There are also suggestions that the medioventral nucleus receives a few afferents from such diverse structures as the septal nuclei, amygdala, raphe nuclei, mesencephalic central gray matter, parabrachial region, and hypothalamus (Herkenham, 1978). Some of these are difficult to substantiate and may result from involvement of adjacent nuclei or of fibers passing through and around the medioventral nucleus in an injection of retrograde tracer. Others may represent true projections to the medioventral nucleus. Similarly, hints of sparse and diffuse projections from the nucleus to an even wider range of basal forebrain centers may also result from spread of anterograde tracers into nuclei adjacent to the medioventral nucleus, but could equally be bona fide. This is an area that needs more work.

The evidence that the medioventral nucleus projects upon virtually all areas of the limbic cortex, including the hippocampus itself, is particularly interesting, especially in view of the seemingly diverse and widespread origins of its inputs. Stimulation of the "midline nuclei" has been known for a long time to elicit changes in EEG activity over widespread areas of cortex, including the limbic regions (Jasper, 1949, 1960; Chapter 12), and it seems likely that the medioventral nuclei would be involved here. It is interesting to note that the laminar terminations of medioventral fibers, even in the hippocampus, are at levels at which they could synapse with the peripheral parts of the apical dendritic systems of pyramidal cells in all layers, an appropriate terminal site for an afferent system involved in maintaining levels of cortical excitability. It is also noteworthy that medioventral inputs to the medial cortex come into relation with the terminations of afferent fibers from the mediodorsal, anterior, ventromedial, and intralaminar nuclei of the thalamus (Chapters 7, 12, and 14), as well as of fibers from the amygdala (Krettek and Price, 1977a) and of serotonin, dopamine, noradrenaline, and possibly cholinergic fibers from the brain stem (Shute and Lewis, 1967; Fuxe *et al.*, 1968; Swanson and Hartman, 1975; Beckstead, 1976; Beaudet and Descarries, 1976). Once again this points up the variety of inputs that converge upon the limbic cortex, in keeping with the complicated functions that are usually attributed to it.

The medioventral nucleus seems to be yet another thalamic nucleus with a dual type of cortical projection: to layer I of several areas and to layers I and III of a more limited number of areas [perhaps only the lateral entorhinal area, as seen in the results of Herkenham (1978)]. The demonstration that the nucleus provides an input to the hippocampal formation has attracted a good deal of attention but this seems to be part of a more widespread and diffuse layer I projection to many areas. It is the projection to the middle layers of the lateral entorhinal area that is the focused, nucleus-to-field projection comparable to that possessed by other principal thalamic nuclei.

They were the homes of the four Ministries between which the entire
apparatus of government was divided. . . . Their names . . . : Minitrue,
Minipax, Miniluv, and Miniplenty.

George Orwell, *Nineteen Eighty-Four*

The Anterior Nuclei and Lateral Dorsal Nucleus

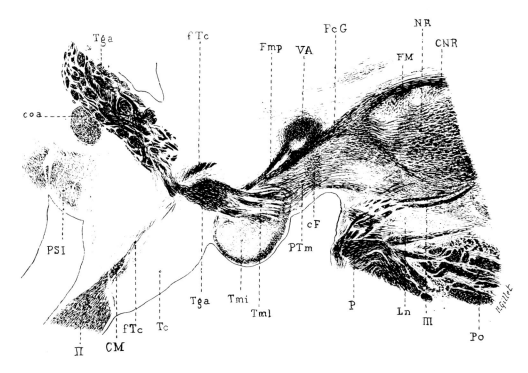

Drawing of a Weigert-stained sagittal section through the human diencephalon and midbrain, showing bifurcation
of the efferent bundles of the mamillary bodies to form the mamillothalamic and mamillotegmental tracts.
From Dejerine (1901).

14.1. Description

Three anterior nuclei are typical of most mammals: anterodorsal, anteroventral, and anteromedial (Figs. 14.1–14.3). They are separated from the rest of the dorsal thalamus by the anterior portion of the internal medullary lamina. A lateral splitting of the lamina further separates off the anterior pole of the lateral dorsal nucleus. Because of this topographic association with the anterior nuclei and because it is connected with many of the same parts of the brain as the anterior nuclei, the lateral dorsal nucleus is probably more appropriately considered part of the anterior nuclear complex. Other nearby nuclei, the parataenial and medioventral, are commonly separated from the anterior nuclei by the internal medullary lamina and are regarded as components of the medial group of thalamic nuclei (Chapter 13). An argument could probably be made on topographic and connectional grounds (Chapter 13) for inclusion of the parataenial nucleus in the anterior group, but I have followed historical precedent and not done so.

The three anterior nuclei can be clearly distinguished in the more common laboratory animals such as rabbits, rats, cats, and most monkeys (Figs. 14.1–14.4). The anterodorsal nucleus, though usually the smallest, is perhaps the most distinctive on account of its large, very deeply staining cells. It lies along the lateral edge of the stria medullaris and is characteristically wedge-shaped in transverse section, with the apex of the wedge extending tenuously toward the midline between the parataenial nucleus (dorsomedially) and the other two anterior nuclei (ventrolaterally). A few cells may even reach the midline, leading some authors to identify a separate "interanterodorsal nucleus" or "commissure" (e.g., Gurdjian, 1927; Rioch, 1929a). The anteroventral nucleus, despite its name, is dorsolaterally situated. Its cells are medium-sized, pale, and moderately densely packed. Anteriorly, it usually bulges into the lateral ventricle beside the anterodorsal nucleus. Posteriorly, it is pushed deeper into the thalamus by the lateral dorsal nucleus. The anteromedial nucleus lies ventral and medial to the other two and in many animals clearly meets its fellow of the opposite side in the midline (between the medioventral and paraventricular nuclei). Its cells are little different from those of the anteroventral nucleus. A thin fiber lamina separates the two nuclei in the rabbit and incompletely in the cat (Figs. 14.1 and 14.2). In the rat and to a lesser extent in monkeys, the cells of the anteromedial nucleus are less densely packed where the nucleus meets the anteroventral nucleus. Near the midline, where the left and right anteromedial nuclei come together, cell packing density usually increases considerably. This region is called an inter-anteromedial nucleus by some workers (e.g., Gurdjian, 1927; Rioch, 1929a; Crouch, 1934). The lateral dorsal nucleus has a cell population similar to the populations of the anteroventral and anteromedial nuclei. It may be recognized by a lower cell packing density but probably it is best distinguished at its anterior pole by its enclosure within the splitting of the internal medullary lamina. Posteriorly, it fades, without a clear line of demarcation, into the dorsal part of the lateral posterior nucleus (Figs. 14.3 and 14.4).

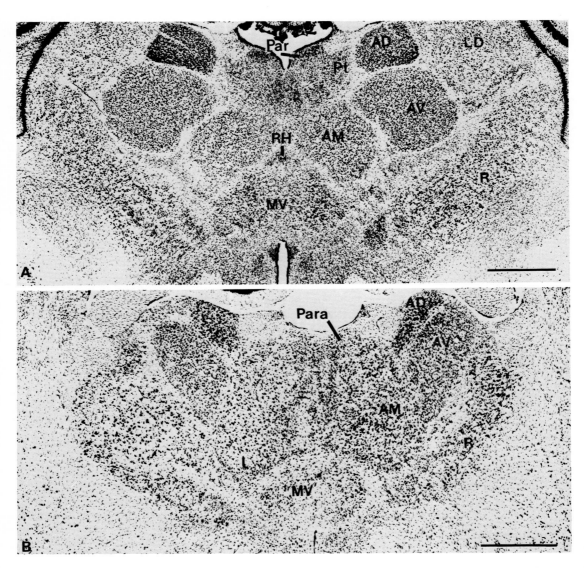

FIGURE 14.1. A gold chloride-stained frontal section of the rabbit thalamus, prepared by Dr. B. B. Stanfield (A), and a thionin-stained section of the rat thalamus (B) showing the anterior and lateral dorsal and adjacent nuclei. Bars = 1 mm.

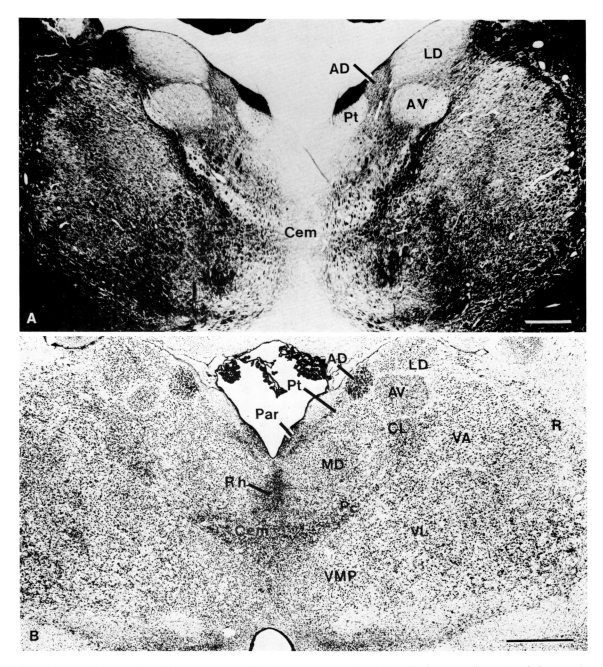

FIGURE 14.2. Hematoxylin- (A) and thionin- (B) stained frontal sections through the posterior parts of the anterior nuclei of the cat thalamus. Bars = 1 mm.

14.2. Terminology

The four nuclei mentioned here can be detected fairly readily in many mammals and there is little disagreement about their names which, with the exception of the lateral dorsal nucleus, date back to Nissl (1889). Other older nomenclatures for the three anterior nuclei, such as those of von Monakow

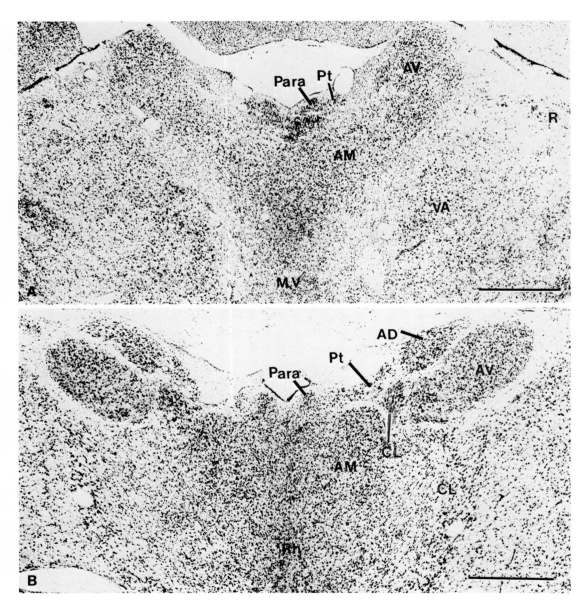

FIGURE 14.3.   Thionin-stained frontal sections through anterior (A) and posterior (B) parts of the anterior nuclei of a cynomolgus monkey. Bars = 1 mm.

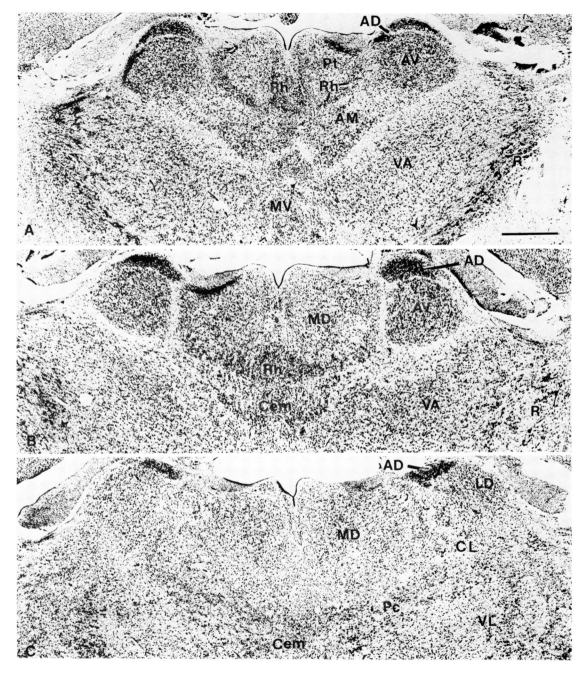

FIGURE 14.4.   Thionin-stained frontal sections in anterior (A) to posterior (C) sequence showing the anterior and lateral dorsal nuclei in a bush baby. Bar = 1 mm.

(1896), Vogt (1909), Friedemann (1911), Ramón y Cajal (1911), and M. Rose (1935), have not survived. Nissl called the lateral dorsal nucleus the *lateral anterior nucleus,* a name that remained in common use for many years and is still sometimes seen (Gurdjian, 1927; Rioch, 1929a; Le Gros Clark, 1930a; Aronson and Papez, 1934; M. Rose, 1935; Krieg, 1948). *Lateral dorsal nucleus* seems to have been introduced by Vogt (1909) and Friedemann (1911). An alternative name, used by Hassler (1959) in man, is *nucleus dorsalis superficialis.*

14.2.1. Species Variations

The three anterior nuclei, though usually present (Berman and Jones, 1982), from species to species show some remarkable variations in their relative sizes and degree of differentiation into subnuclei. This was first recognized by Le Gros Clark (1929b, 1930b) (Fig. 14.5) in the insectivores where he related the size of the anterodorsal nucleus to the degree of development of the retrosplenial cortex. The association of nuclear differentiation with cytoarchitectonic differ-

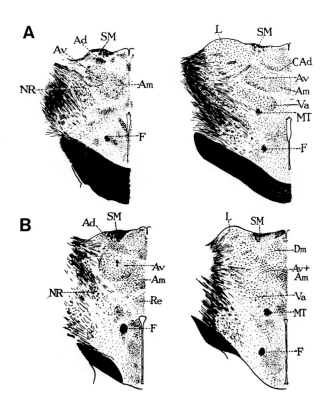

FIGURE 14.5.   Drawings of frontal sections through the anterior nuclei of various insectivores.

entiation of the limbic cortex was further developed by Rose and Woolsey (1948a) in their study of the rabbit and cat.

14.2.1.1. Anteroventral and Anteromedial Nuclei

In many nonprimate mammals, the anteromedial and anteroventral nuclei are distinguishable from one another and are of approximately equal size (Figs. 14.1, 14.6, and 14.7). In the rabbit and cat (Figs. 14.1A and 14.2), differences of cell packing density lead to the clear distinction not only of medial (i.e., interanteromedial) and lateral parts of the anteromedial nucleus but also of dorsal and ventral parts of the anteroventral nucleus (Rose, 1942a; Rose and Woolsey, 1948b; Berman and Jones, 1982). In rodents, ungulates, cetaceans, tree shrews, and marsupials, though the two nuclei can be distinguished and are of approximately equal size, the line of junction between them is less clear (Le Gros Clark, 1930b; Goldby, 1941; Rose, 1942b; Kruger, 1959; Figs. 14.6 and 14.7). In the midline in these animals, the interanteromedial part of the anteromedial nucleus becomes mixed up in the rhomboid and central medial nuclei so that the relative contributions of the three nuclei to this common midline mass of cells sometimes cannot be clearly discerned (Fig. 14.9).

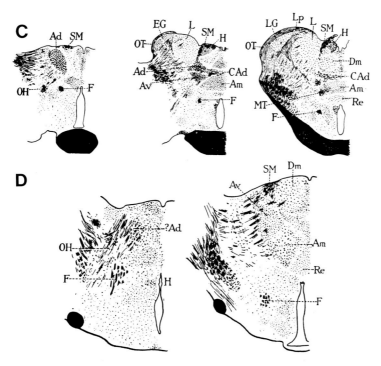

(A) *Tupaia minor;* (B) *Ptilocercus;* (C) *Macroscelides;* (D) *Erinaceus europaeus.* From Le Gros Clark (1929b).

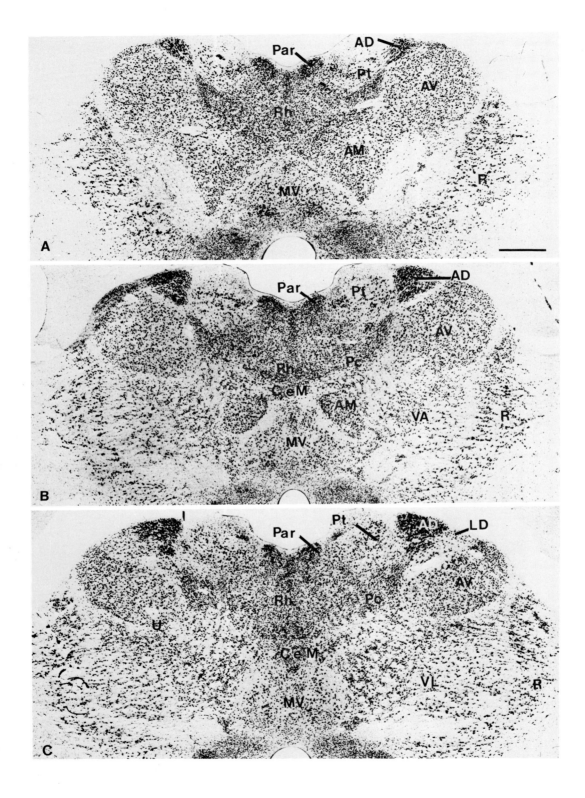

In primates, but also in certain animals with less highly developed brains, the distinction between the anteromedial and the anteroventral nuclei becomes less clear. A small anteromedial nucleus, distinguished by some increased packing density and with an interanteromedial component, can be detected (Fig. 14.3) in man (Powell *et al.*, 1957; Hassler, 1959), chimpanzees (Walker, 1938c), macaque monkeys (Olszewski, 1952), and squirrel monkeys (Emmers and Akert, 1963). However, the anterior nuclear complex is dominated by a large anteroventral nucleus, sometimes, therefore, called *nucleus anterior principalis* (Hassler, 1959) or simply *anterior nucleus* (Grünthal, 1934). In *Tarsius* (Le Gros Clark, 1930a), marmosets (Mann, 1905), and lemurs (Pines, 1927), an anteromedial nucleus cannot be separated from the large anteroventral nucleus (Figs. 14.4 and 14.6). The reduction in size and virtual disappearance of the anteromedial nucleus is not, however, a uniquely primate characteristic. The nucleus is much reduced in some insectivores, such as the tree shrew *Ptilocercus,* and virtually disappears in other insectivores such as the elephant shrew *Macroscelides* (Le Gros Clark, 1929b, 1930b; Fig. 14.6) and in the armadillo (Papez, 1932). In certain small bats (Fig. 14.8), the appearance is very similar to that in the elephant shrew, a large anteroventral nucleus apparently meeting its fellow of the opposite side in the midline and with only a few local concentrations of cells seemingly representing the anteromedial nucleus. So different is the appearance of the anteroventral nucleus in these species that it would be easy to mistake the fused nuclei for an enlarged medioventral (reuniens) nucleus.

14.2.1.2. Anterodorsal Nucleus

In lagomorphs, some rodents, and some insectivores, the three anterior nuclei are of approximately equal size (Figs. 14.1 and 14.6) and the anterodorsal nucleus is divided into a dorsal, smaller-celled part and a ventral, larger-celled part (Le Gros Clark, 1929b; M. Rose, 1935; Rose and Woolsey, 1948b). In most rodents, many marsupials, ungulates, carnivores, primates, and cetaceans, the anterodorsal nucleus has become smaller in size relative to the anteromedial and anteroventral nuclei (Figs. 14.1–14.8; Gurdjian, 1927; Rioch, 1929a; Walker, 1938a; Bodian, 1939; Goldby, 1941; Rose, 1942b; Olszewski, 1952; Holmes, 1953; Hassler, 1959; Kruger, 1959; Emmers and Akert, 1963; Oswaldo-Cruz and Rocha-Miranda, 1967). In certain insectivores such as the hedgehog, the anterodorsal nucleus is said to be virtually absent (Le Gros Clark, 1929b; Powell *et al.*, 1957). In some other insectivores and in small bats (Figs. 14.6–14.8), the anterodorsal nucleus becomes separated from the dorsal surface of the thalamus and is buried deeply in the anterior nuclear complex in association with the enlargement of the lateral dorsal nucleus and the anterior displacement of the dorsal lateral geniculate nucleus.

FIGURE 14.6. Thionin-stained frontal sections in anterior (A) to posterior (C) sequence through the anterior nuclei of a marsupial phalanger (*Trichosurus vulpecula*). Bar = 1 mm.

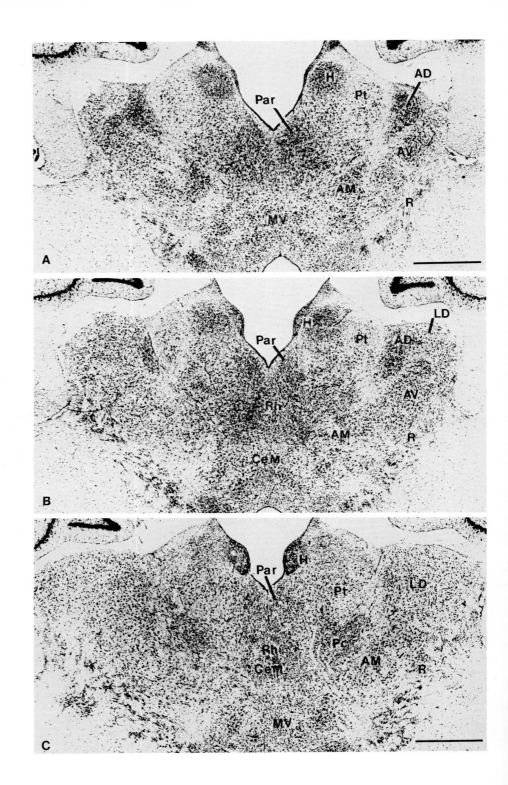

14.2.1.3. Lateral Dorsal Nucleus

685

THE ANTERIOR
NUCLEI AND
LATERAL DORSAL
NUCLEUS

The relative size of the lateral dorsal nucleus and the ease with which it can be distinguished do not change much across species (Figs. 14.4 and 14.5). In some, however, e.g., *Tarsius,* some insectivores, and some bats (Fig. 14.8), it does show an unusual enlargement relative to the anterior nuclei and gives the impression of pushing the anterodorsal and anteroventral nuclei deeper into the thalamus.

14.2.2. Relationships to Limbic Cortex

Rose and Woolsey (1948a) related the degree of differentiation of the anteroventral and anteromedial nuclei and the relative size of the anterodorsal nucleus in different species to the differential development of the three fields of the limbic cortex. This development, they considered to be different for macrosmatic and microsmatic mammals. They assessed the cytoarchitecture of the limbic cortex of the cat and rat in terms of M. Rose's accounts (1929), which had somewhat modified the earlier descriptions of Brodmann (1909) (Fig. 14.10). They then made lesions in the cortical fields situated there and studied the distribution of retrograde degeneration in the anterior nuclei. They found an anterior limbic area lying anterodorsal to the corpus callosum which was agranular and related specifically to the anteromedial nucleus. A posterior limbic area was mostly granular and consisted of two fields: the cingulate area to which the anteroventral nucleus was specifically related and the retrosplenial area which was probably related exclusively to the anterodorsal nucleus. In the opinion of Rose and Woolsey, the granular retrosplenial area which they felt was connected with the anterodorsal nucleus, is very highly developed in rodents but undergoes a regression in primates, hence the reduction in size of the anterodorsal nucleus in primates. By contrast, the pre- and postcingulate fields which are connected with the anteroventral and anteromedial nuclei undergo considerable, though variable, differentiation in primates, accounting for the differences in size and variable cytoarchitecture of these two nuclei.

There has been little attempt to follow up these speculations of Rose and Woolsey. More quantitative studies on a wider range of animals are needed in order to confirm variations in size of the limbic areas; and to relate them to variations in size and degree of differentiation of the anterior nuclei. A better functional variable than simply "macrosmatic" or "microsmatic" is obviously also needed if the structural variations shown are to have behavioral significance. The anterior nuclei and limbic cortex, however, may offer a unique opportunity for assessing thalamocortical relationships in terms of behavioral specializations.

FIGURE 14.7. Thionin-stained frontal sections in anterior (A) to posterior (C) sequence through the anterior nuclei of an opossum. Bar = 1 mm.

14.3. Connections

14.3.1. Afferent Connections

14.3.1.1. Hippocampal Formation, Mamillary Nuclei, and Pretectum

The afferent connections of the anterior nuclei and lateral dorsal nucleus, apart from corticothalamic connections from the limbic cortex (Section 14.3.1.2), are predominately from the hippocampal formation and hypothalamus, as noted by von Gudden (1881b), Ramón y Cajal (1911), and Le Gros Clark and Boggon (1933a) many years ago (Figs. 14.11 and 14.12). The older work and that carried out in the heyday of the Nauta techniques established that the hippocampal region projects via the fornix to the anteroventral and anteromedial nuclei of both sides (Guillery, 1956; Nauta, 1956) and to the lateral dorsal nucleus of the ipsilateral side (Valenstein and Nauta, 1959); different parts of the mamillary complex of the hypothalamus project to the anteroventral, anteromedial, and anterodorsal nuclei, the projection to the anterodorsal nucleus being bilateral

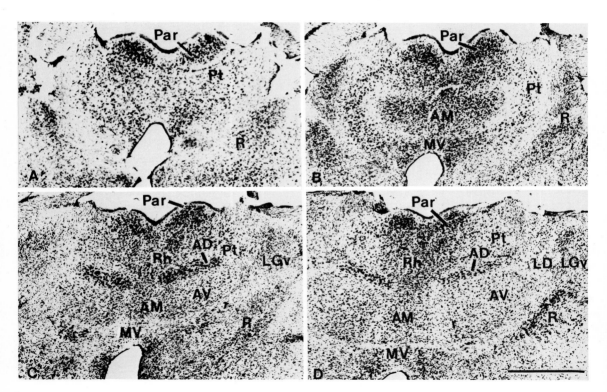

FIGURE 14.8. Relative enlargement of the lateral dorsal nucleus (LD) with internalization of the anterodorsal nucleus (AD) and enlargement of the anteroventral nucleus (AV) in a moustache bat. Frontal sections; thionin stain. Bars = 500 μm.

(Powell and Cowan, 1954; Guillery, 1961; Fry *et al.*, 1963; Fry and Cowan, 1972). Powell (1958) pointed out (Fig. 14.13) that because of the differential distribution of the fornix and mamillary peduncle of the midbrain to the mamillary nuclear complex, the three anterior thalamic nuclei have different relationships with the hippocampal formation and, via the mamillary complex, with the midbrain: the fornix provided input to separate subnuclei of the mamillary complex which projected independently to the anterodorsal and anteroventral nuclei; the mamillary peduncle carrying fibers from the deep and dorsal tegmental nuclei of the midbrain, projected to a third subnucleus which projected in turn to the anterodorsal nuclei and back to the tegmental nuclei. Details of the relationships between hippocampus, midbrain, and anterior nuclei as established at that time

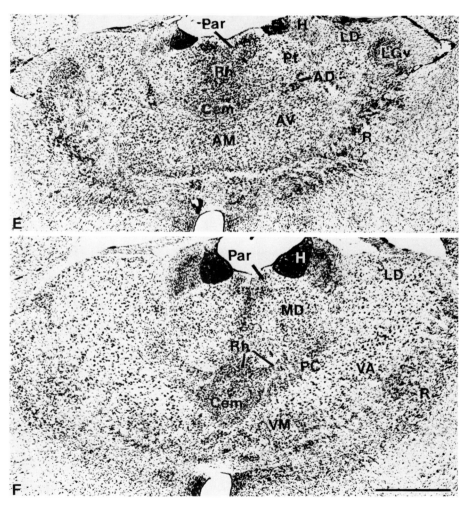

FIGURE 14.8. (*continued*)

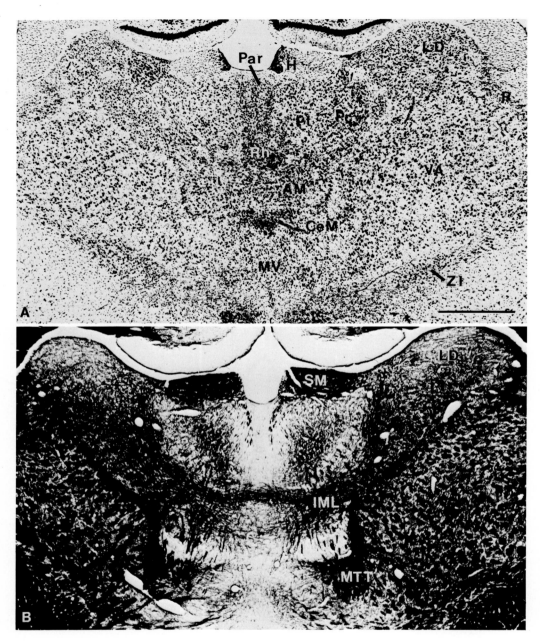

FIGURE 14.9. Fusion of the rhomboid nucleus, central medial nucleus, and interanteromedial parts of the anteromedial nuclei in the internal medullary lamina (IML) of a rat. MTT: mamillothalamic tract; SM: stria medullaris. Frontal sections; (A) thionin stain; (B) hematoxylin stain. Bar = 500 μm.

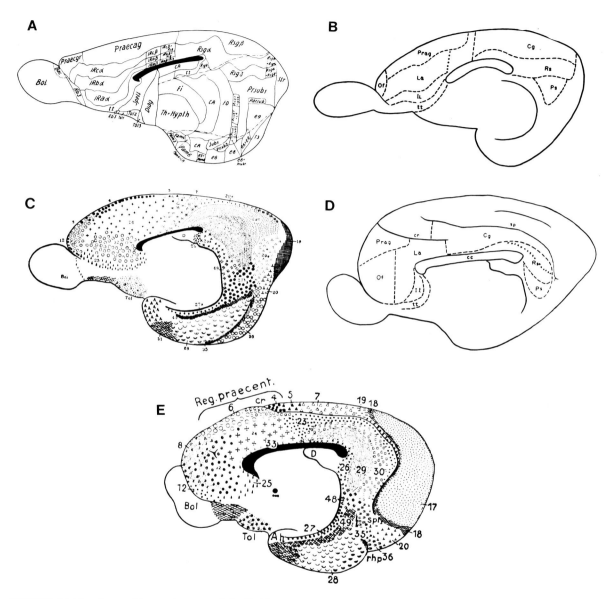

FIGURE 14.10. Cytoarchitectonic maps of the limbic cortex in the rabbit: (A) from M. Rose (1927); (B) from Rose and Woolsey (1948a); (C) from Brodmann (1909); (D) map of the cat from Rose and Woolsey (1948a); (E) Brodmann's (1909) map of a procyonid brain. The retrosplenial (Rs) area in (D) has been subdivided into dorsal and ventral parts, following Robertson and Kaitz (1981). La: anterior limbic area; Cg: cingulate area; Ps: presubiculum.

are to be found in the works of: Guillery (1955, 1957), Nauta (1956, 1958), Powell *et al.* (1957), Cragg and Hamlyn (1960), Morest (1961), Cowan *et al.* (1964), and Raisman *et al.* (1966).

More recent studies have made use of anterograde and retrograde tracing techniques. These have, in general, confirmed the connectional relationships mentioned above but have added a few more details, and certain undiscovered sources of input, especially to the lateral dorsal nucleus, have been established (LaVail, 1975; Cruce, 1975, 1979; Swanson and Cowan, 1977; Meibach and Siegel, 1975, 1977a,b; Krayniak *et al.*, 1979; Kaitz and Robertson, 1981; Veazey *et al.*, 1982; Donovan and Wyss, 1983). The subiculum, the pre- and the para-

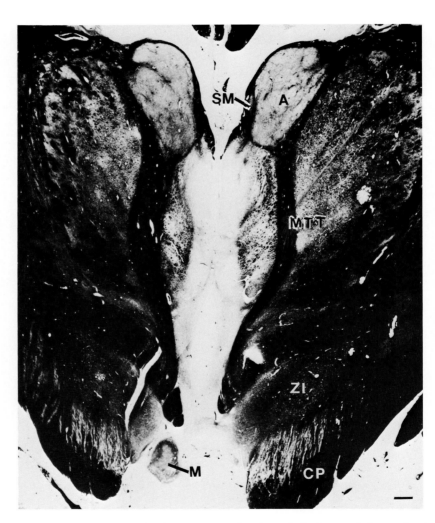

FIGURE 14.11.   Hematoxylin-stained transverse section through the human thalamus showing the mamillothalamic tracts (MTT) and anterior nuclei (A). CP: cerebral peduncle; M: mamillary body. Bar = 1 mm.

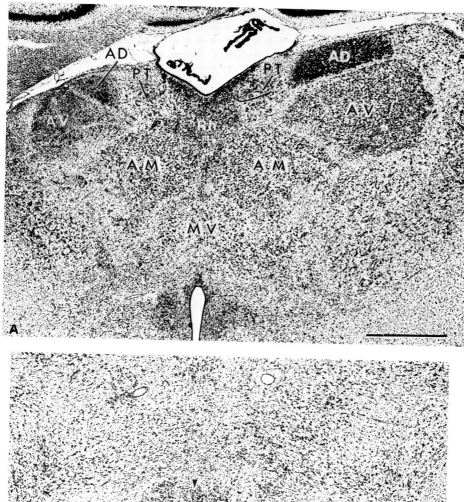

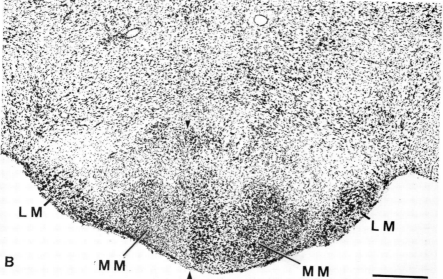

FIGURE 14.12. (A) Retrograde degeneration of the left anterodorsal and anteroventral nuclei of a rabbit from which the cingulate region of the cerebral cortex was removed some weeks previously. (B) Retrograde transneuronal degeneration of the left medial mamillary nucleus (MM) of the hypothalamus in the same rabbit as (A), secondary to the degeneration of the thalamic targets of the medial mamillary cells' axons. LM: lateral mamillary nuclei. Arrowheads indicate midline. Bars = 250 μm (A), 100 μm (B). Preparation by Dr. W. M. Cowan, from Jones and Cowan (1982).

subiculum, rather than the hippocampus proper, have been shown to provide the fornical input to the anterior nuclei. These, in a sense, therefore, represent corticothalamic connections and, like corticothalamic connections to other thalamic nuclei, the fibers arise in the deeper layers of the cortex (Donovan and Wyss, 1983). The heaviest input is to the anteroventral nucleus and this is bilateral. There is also a particularly dense projection from the pre- and parasubiculum and adjacent regions to the lateral dorsal nucleus but, instead of passing by way of the fornix, the fibers forming this projection traverse the internal capsule.

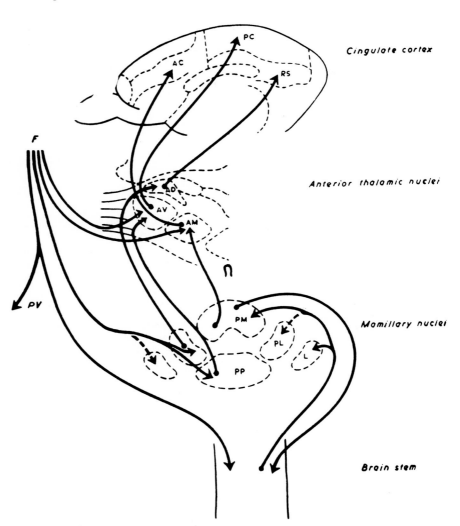

FIGURE 14.13.   Interrelationships of hippocampal formation, anterior thalamic nuclei, mamillary nuclei, and tegmental nuclei of midbrain in rodents. AC: anterior cingulate area; PC: posterior cingulate area; RS: retrosplenial area; F: fornix; PV: precommissural fornix; PM, PL, PP, L: divisions of mamillary nuclei. From Powell (1958).

Fibers of the postcommissural part of the fornix continue past the thalamus to terminate in the mamillary complex of the hypothalamus. It was determined by Guillery (1955) and Powell *et al.* (1957) in several species, including man, that as many as half to two-thirds of the fibers of the postcommissural fornix fail to reach the mamillary nuclei and probably, therefore, end in the anterior nuclei of the thalamus. Ramón y Cajal (1911) had earlier observed collaterals of fornical fibers entering the anterior thalamic nuclei. The extent of collateral projection of single parahippocampal cells to the anterior thalamic and mamillary nuclei has not yet been established by modern double labeling techniques.

Fornical fibers to the mamillary bodies arise in the dorsal subiculum as well as in the pre- and parasubiculum and end bilaterally in the medial mamillary nucleus and ipsilaterally in the lateral mamillary nuclei (Meibach and Siegel, 1977a,b; Swanson and Cowan, 1977; Donovan and Wyss, 1983). The two mamillary nuclei form the subcortical input to the anterior thalamic nuclei. They project to the thalamus by way of the mamillothalamic tract which has been found to contain approximately as many fibers as those innervating the anterior nuclei from the fornix (Powell *et al.*, 1957). Autoradiographic studies by Cruce (1975) in the rat and by Veazey *et al.* (1982) in the monkey show, contrary to the earlier reports, that the medial mamillary nucleus projects only upon the anteromedial and anteroventral thalamic nuclei ipsilaterally and confirm that the lateral mamillary nucleus projects to the anterodorsal nuclei bilaterally (Fig. 14.12). Cruce (1979) and Veazey *et al.* (1982) also confirm earlier evidence of differential projections of the mamillary nuclei on the midbrain; the medial mamillary nucleus projects to the ipsilateral deep tegmental nucleus and the lateral mamillary nucleus to the ipsilateral dorsal tegmental nucleus. It has been suggested from the different degrees of retrograde reaction in the cells of the lateral mamillary nucleus following lesions at different sites in the output pathways, that individual lateral mamillary cells send axon branches into both mamillothalamic and mamillotegmental tracts and that those in the mamillothalamic tract may further branch to innervate both anterodorsal thalamic nuclei (Fry and Cowan, 1972). This has been partially confirmed with modern double labeling techniques (Van der Kooy *et al.*, 1978; Kuypers *et al.*, 1980).

There is no available evidence to suggest that the mamillary nuclei project upon the lateral dorsal nucleus. Its subcortical afferents appear to arise in the pretectum. This particular connection was early mentioned by Benevento and Ebner (1970) in the opossum but was denied or felt to be insignificant by many subsequent workers in several species (Graybiel, 1974b; Weber and Harting, 1980; Berman, 1977; Benevento *et al.*, 1977; Itoh, 1977). Now, others (Arango and Scalia, 1978; Ryska and Heger, 1979; Robertson *et al.*, 1980; Robertson, 1983; Robertson *et al.*, 1983) have presented compelling evidence for such a projection in the cat and rat. All nuclei of the pretectum, both retinal-recipient and non-retinal-recipient, contribute axons to the lateral dorsal nucleus. Double labeling experiments (Robertson *et al.*, 1983; Robertson, 1983) indicate that the cells projecting to the lateral dorsal nucleus are fewer than those projecting to the pulvinar nucleus. The two populations of cells are mingled but none have branched axons to both thalamic nuclei. Because of the projection of visual, somatic sensory, and other sensory pathways upon the pretectum, its projection upon the lateral dorsal nucleus might represent a route whereby exteroceptive information could reach the limbic cortex and hippocampal formation.

14.3.1.2. Corticothalamic Connections

The cortex on the medial surface of the cerebral hemisphere and lying in close relation to the corpus callosum and hippocampal formation can be divided into a number of cytoarchitectonic fields (Figs. 14.10 and 14.14; Brodmann, 1909; M. Rose, 1929; Rose and Woolsey, 1948a; Domesick, 1969, 1972; Leonard, 1972; Krettek and Price, 1977b; Vogt *et al.*, 1979; Robertson and Kaitz, 1981). Those related to the anterior and lateral dorsal thalamic nuclei include the anterior limbic (or anterior cingulate) field, the cingulate field proper, the granular (or ventral) retrosplenial area, the agranular (or dorsal) retrosplenial area, and, as mentioned in Section 14.3.1.1, the pre- and parasubiculum. Many of these fields also project upon the medioventral nucleus (Chapter 13) and, from the published work, it is hard to relate any of them to a single anterior thalamic nucleus on the basis of corticothalamic connections. The anteromedial nucleus, in addition to receiving fibers from the pre- and parasubiculum, also receives fibers from all the other areas (Domesick, 1969, 1972; Kaitz and Robertson, 1981), probably reflecting its diffuse projection to all fields (Domesick, 1969, 1972; Krettek and Price, 1977b). In the cat the cingulate area is the major area projecting to the anteroventral and lateral dorsal nuclei but the retrosplenial areas and pre- and parasubiculum also contribute fibers (Kaitz and Robertson, 1981). In the rat the granular retrosplenial area is reported as providing the major corticothalamic pathway to the anteroventral nucleus and the agranular retrosplenial area, apart from the pre- and parasubiculum, appears to provide the only corticothalamic input to the anterodorsal nucleus (Kaitz and Robertson, 1981).

The lateral dorsal nucleus receives additional corticothalamic fibers from

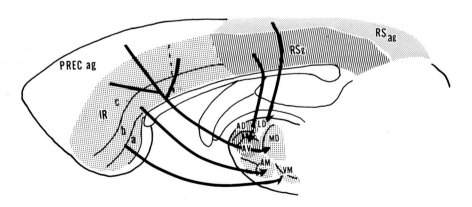

FIGURE 14.14. Connectional relationships of the anterior thalamic nuclei with limbic cortex in the rat. IR: infraradiate cortex; PRECag: precentral agranular cortex; RS$_{ag}$, RS$_{g}$: retrosplenial agranular and granular areas. From Domesick (1969).

parts of the cortex far removed from the limbic areas. In the cat, areas of the

695

THE ANTERIOR
NUCLEI AND
LATERAL DORSAL
NUCLEUS

anterior suprasylvian gyrus have been shown to provide such fibers (Heath and Jones, 1971b; Robertson and Rinvik, 1973; Updyke, 1977, 1981a). Whether this implies a separation of the lateral dorsal nucleus into two parts, one related to limbic and the other to parietal cortex, has not been established.

14.3.2. Thalamocortical Connections

Rose and Woolsey (1948a), it will be recalled (Section 14.2.2), related the anteromedial nucleus to anterior limbic cortex, the anteroventral nucleus to cingulate cortex, and the anterodorsal nucleus provisionally to retrosplenial cortex. In broad terms, this tripartite arrangement has been confirmed by more recent studies though many new details have been added and a finer parcellation of the limbic fields has been made (Fig. 14.14). Domesick (1969, 1972) in studies made with the Nauta technique and Jones and Leavitt (1974) in an early retrograde transport study, largely confirmed Rose and Woolsey's observations in the rat but Domesick and later Shipley and Sørenson (1975) also reported projections from the anteroventral and anterodorsal nuclei to the presubiculum. Leonard (1972), Krettek and Price (1977b), and Beckstead (1976) also presented evidence for rather widespread projections from the anteromedial nucleus which overlapped anteriorly into the projection field of the mediodorsal nucleus.

In cats, Robertson and Kaitz (1981), using both anterograde and retrograde labeling methods, have found that the three anterior nuclei have one area of major (dense) projection but that they may also project less densely to other fields. The principal target of the anteromedial nucleus is the granular retrosplenial area but it has additional projections to anterior limbic, cingulate, and agranular retrosplenial areas and to the pre- and parasubiculum; the principal target of the anterodorsal nucleus also appears to be the granular retrosplenial area (Figs. 14.10, 14.14, and 14.15) with additional projections to the presubiculum; the principal target of the anteroventral nucleus appears to be the agranular retrosplenial area but with additional widespread projections to the granular retrosplenial area, cingulate area, anterior limbic area, and presubiculum. The lateral dorsal nucleus has dense projections to the cingulate area, agranular retrosplenial area, and pre- and parasubiculum but additional projections to anterior limbic and granular retrosplenial areas. [See also Niimi (1978) and Niimi et al. (1978).]

None of the nuclei, therefore, appears to have a single focus of projection in the cortex and the evidence points to a very diffuse system of connections with variable degrees of convergence of two or more nuclei upon the same field or fields. Where two nuclei project upon the same field, there are hints of a laminar segregation in the distribution of their fibers, other than in the pre- and parasubiculum. Fibers from all four nuclei terminate in layer I (Fig. 14.15) and, thus, the inputs converge there, but in deeper layers there is segregation: fibers from the anteroventral and anterodorsal nuclei end at the junction of layers II and III, those from the anteromedial nucleus throughout layer III, and those

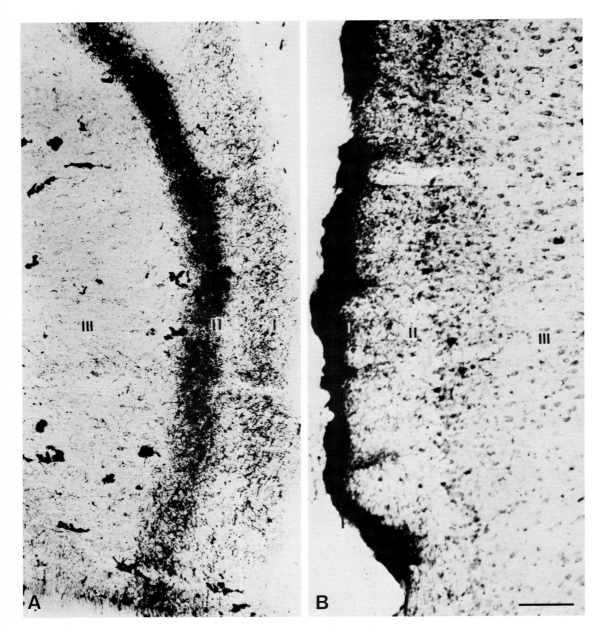

FIGURE 14.15. (Left) Distribution of thalamocortical fibers from the anterior nuclei of the thalamus in the retrosplenial cortex of the rat. Anterograde labeling with horseradish peroxidase. Note relationship to deep part of cholinesterase-stained lamina seen in Fig. 5.12. (Right) Distribution of corticocortical fibers in a pattern largely complementary to (A). Anterograde labeling with horseradish peroxidase and thionin counterstain. Bar = 50 μm.

from the lateral dorsal nucleus terminate at the junction of layers III and IV (Robertson and Kaitz, 1981). The possibility exists, therefore, for differential effects to be exerted by each nucleus upon the various fields of the limbic cortex.

14.3.3. A Cholinergic Projection?

The anterodorsal and anteroventral thalamic nuclei display some of the densest concentrations of acetylcholinesterase staining in the thalamus (Shute and Lewis, 1967; Olivier *et al.*, 1970; Jacobowitz and Palkovits, 1974; Parent and Butcher, 1976; Graybiel and Berson, 1980b; Figs. 5.1, 5.2, and 5.12). Some of this staining is of cell bodies as well as of neuropil. The anteroventral nucleus also has one of the highest thalamic concentrations of muscarinic cholinergic receptors (Rotter *et al.*, 1979; Fig. 5.4). Shute and Lewis (1967) and Lewis and Shute (1967) felt that the anterior thalamic nuclei formed part of a continuous cholinergic system ascending from the midbrain reticular formation, through the thalamus and around the genu of the corpus callosum into the cingulate cortex. Certainly, there is a continuous band of acetylcholinesterase staining from the anterior nuclei around the callosal genu which is continuous with a heavy band of acetylcholinesterase staining in layers I and III of the cingulate and retrosplenial cortex. Many thalamocortical fibers from the anterior nuclei, instead of reaching the cortex via the internal capsule, take a route that follows that of the cholinesterase staining, namely through the postcommissural fornix and the genu of the corpus callosum (Herkenham, 1978; Robertson and Kaitz, 1981). Moreover, the cholinesterase staining in layer III of the cingulate cortex disappears after destruction of the anterodorsal nucleus (Fig. 5.12), suggesting that it is related to the termination of thalamic afferents. While none of these observations necessarily prove that some part of the cortical projection of the anterior nuclei is cholinergic, they are particularly suggestive and await concerted study.

14.3.4. Noradrenaline and the Anterior Nuclei

The dorsal noradrenergic bundle, ascending through the midbrain tegmentum, also has one of its major zones of thalamic ramifications in the anteroventral nucleus (Swanson and Hartman, 1975; Fig. 5.9). Noradrenergic fibers passing by the thalamus continue into the septum and around the genu of the corpus callosum and have some of their heaviest cortical ramifications throughout the target areas of the anterior nuclei (Swanson and Hartman, 1975; Jones and Moore, 1977). Therefore, the anterior nuclei and their cortical projection fields appear to be strongly influenced by this catecholamine-containing ascending system as well as, putatively, by acetylcholine. The anterior nuclei may well represent a route whereby the ascending cholinergic and catecholamine systems can be brought together with those of the hypothalamus before being funneled toward the limbic cortex and hippocampal formation.

... What if the Sun
Be Center to the World, and other Starrs
By his attractive vertue and thir own
Incited, dance about him various rounds?

John Milton, *Paradise Lost* VIII, 122–125

The Ventral Thalamus

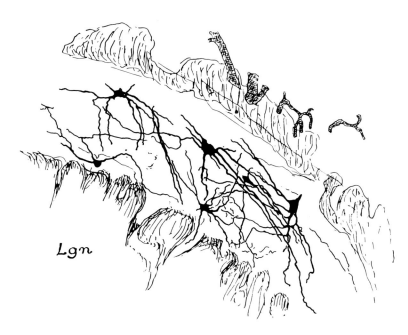

Drawing of a Golgi preparation of the pregeniculate nucleus of an infant rhesus monkey. From Polyak (1957).

15.1. Definition

The ventral thalamus, as usually defined in the adult brain, consists of the reticular nucleus, the zona incerta, and the ventral lateral geniculate nucleus (Fig. 15.1). To this, many authors add the nucleus of the fields of Forel. The subthalamic nucleus, though sometimes included in the ventral thalamus by older authors and by some current describers of the human diencephalon, is excluded on account of its different developmental origin (Rose, 1942a; Chapter 6).

Many aspects of the anatomy and physiology of the reticular nucleus have been covered in other chapters of this book (Chapters 3, 4, and 12) on account of its inextricable functional links with the dorsal thalamus. The close association of the ventral lateral geniculate nucleus with the pathways ending in the dorsal lateral geniculate nucleus has also been mentioned (Chapter 9). The reader may, therefore, wish to read the following account in conjunction with these other chapters.

15.2. The Reticular Nucleus

15.2.1. Description

The reticular nucleus covers most of the anterior, lateral, and ventral surfaces of the dorsal thalamus. The manner in which it enfolds the thalamus can be well appreciated in horizontal and sagittal sections (Figs. 15.2 and 15.3). In larger mammals, it is a thin sheet of large, deeply staining, fusiform cells intervening between the external medullary lamina and the internal capsule and broken up by the many bundles of fibers of the thalamic radiations as they pass through it. The long axes of the cells tend to lie parallel to the surface of the dorsal thalamus. In smaller mammals, the nucleus tends to be thicker relative to the nuclei of the dorsal thalamus, though the same general topographic relatonships are preserved.

The anterior part of the nucleus is mainly in the frontal plane and intervenes between the anterior surface of the dorsal thalamus and the prethalamic nuclei, particularly the bed nucleus of the stria terminalis. Usually it is on the anterior

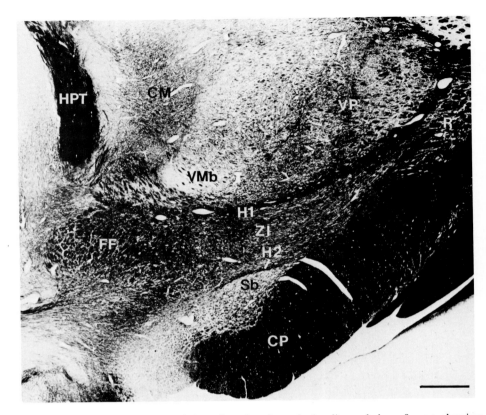

FIGURE 15.1.   Hematoxylin-stained frontal section through the diencephalon of a cat, showing the reticular nucleus (R), zona incerta (ZI), and the fields of Forel (FF, H1, H2). Sb: subthalamic nucleus; CP: cerebral peduncle. Bar = 1 mm.

surface of the ventral nuclear complex of the dorsal thalamus, the anterior nuclear complex lying behind the stria medullaris medial to the reticular nucleus (Fig. 15.2B).

The remainder of the reticular complex curves around the lateral and ventral aspects of the dorsal thalamus and becomes attenuated as it is followed posteriorly, though in many mammals such as the cat (Fig. 15.2A) the ventral part becomes thicker again and its cells larger toward the posterior pole of the ventral posterior nucleus, and where it undershoots the anterior pole of the medial geniculate complex. This thickened part of the reticular nucleus in nonprimates merges with the ventral lateral geniculate nucleus and the two are often difficult to distinguish from one another (Fig. 15.4A).

FIGURE 15.2.   (A) Thionin-stained frontal section through the same part of the cat ventral thalamus as in Fig. 15.1. (B) Thionin-stained horizontal section through the thalamus of a cat showing the reticular nucleus (R) fusing with the ventral lateral geniculate nucleus (LGv). Bars = 1 mm.

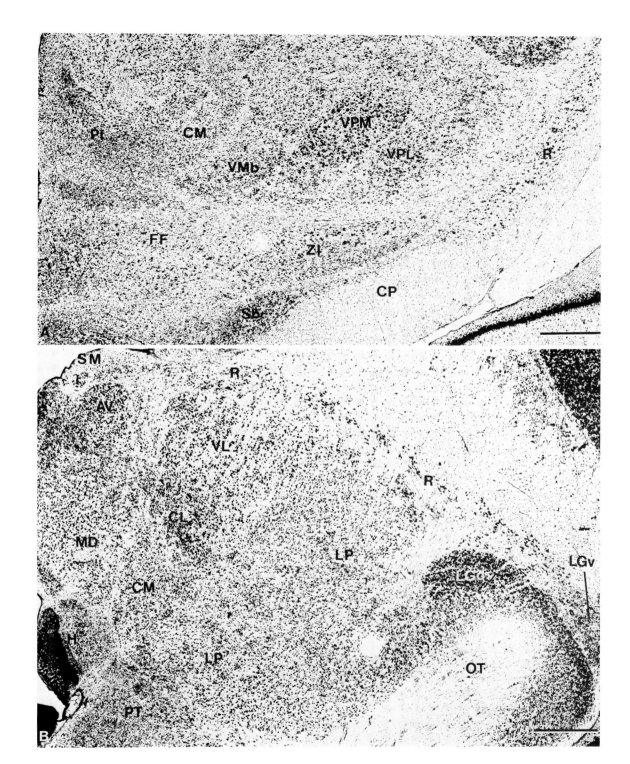

From about the middle of the dorsal thalamus, the ventral part of the reticular nucleus is displaced laterally by the small cells of the zona incerta (Figs. 15.1 and 15.12). The middle part of the reticular complex is also displaced laterally by the dorsal lateral geniculate nucleus, particularly in carnivores and primates in which the lateral geniculate nucleus has bulged laterally and rotated ventrally (Figs. 15.1 and 15.9). In some carnivores, such as the cat, it is possible to see evidence for continuity between the reticular nucleus in this region and the perigeniculate nucleus that surrounds the dorsal lateral geniculate nucleus. In other carnivores such as the raccoon, seal (Fig. 15.5), and mink (Sanderson, 1974), the perigeniculate nucleus is particularly large and dispersed, like the

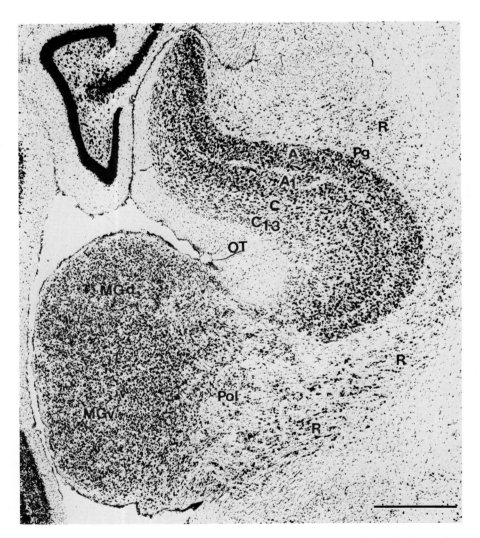

FIGURE 15.3. Thionin-stained sagittal section through the lateral part of the reticular nucleus of a cat, showing its proximity to the dorsal lateral geniculate (LGd) and ventral medial geniculate (MGv) nuclei and tenuous continuity with the perigeniculate nucleus (Pg). Bar = 1 mm.

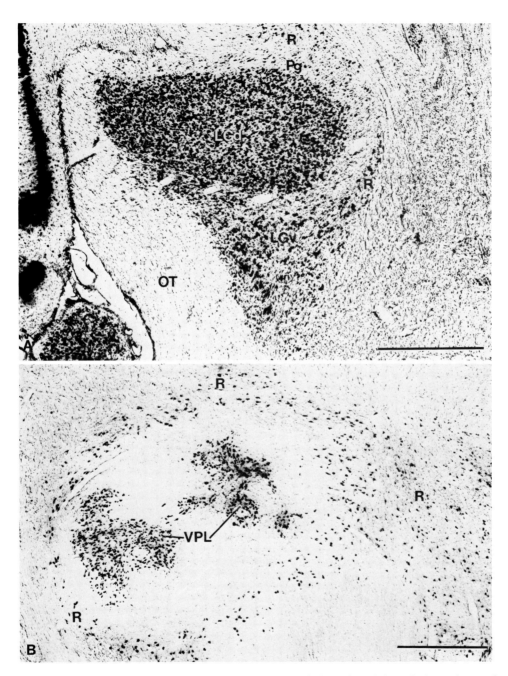

FIGURE 15.4. Thionin-stained sagittal sections (A) through the union of the reticular and ventral lateral geniculate nuclei in a cat and (B) through part of the reticular nucleus in a raccoon. Bars = 1 mm.

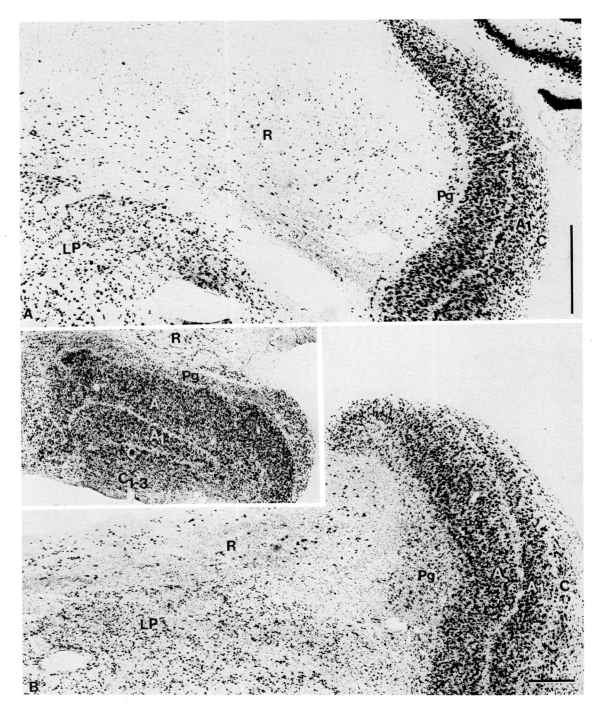

FIGURE 15.5. Thionin-stained sagittal sections showing continuity of the perigeniculate (Pg) nucleus with the reticular nucleus (R) in a raccoon (A) and in a harbor seal (*Phoca vitulina*) (B). Inset shows perigeniculate and reticular nuclei in a frontal section from a seal. Bars = 1 mm.

reticular nucleus itself. The immunocytochemical properties of the perigeniculate nucleus are similar to those of the reticular nucleus (Figs. 15.6 and 15.7) and, physiologically, its cells seem to behave like those of the reticular nucleus (Lindström, 1982; Chapter 4). Therefore, it may be part of the reticular nucleus. Certainly, the tendency has grown up to regard it as such, particularly as a perigeniculate nucleus cannot be identified in noncarnivores.

15.2.2. Terminology

The reticular nucleus was first identified in the human thalamus by Arnold (1838) but until the work of Nissl (1889), it was referred to by Arnold's terms, *reticular stratum* or *reticular zone* (Dejerine, 1901; Marburg, 1904; Vogt, 1909).

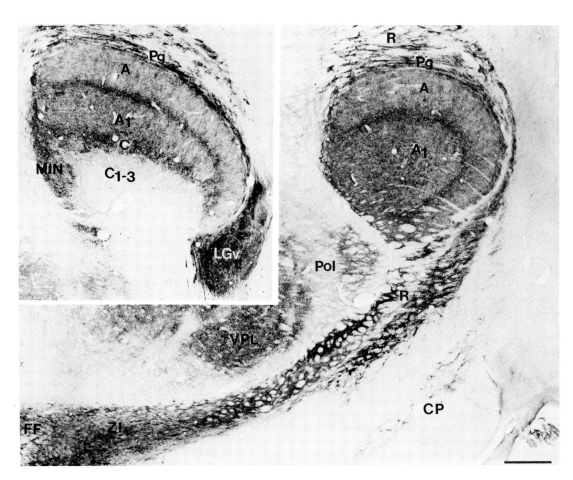

FIGURE 15.6. Frontal sections through the thalamus of a cat showing immunocytochemical staining of perigeniculate (Pg), ventral lateral geniculate (LGv), and reticular nuclei (R), plus zona incerta (ZI), field of Forel (FF), and certain nuclei of dorsal thalamus by monoclonal antibody CAT 301 made against cat spinal cord (see Hockfield *et al.*, 1983, and Hendry *et al.*, 1984). Bar = 100 μm.

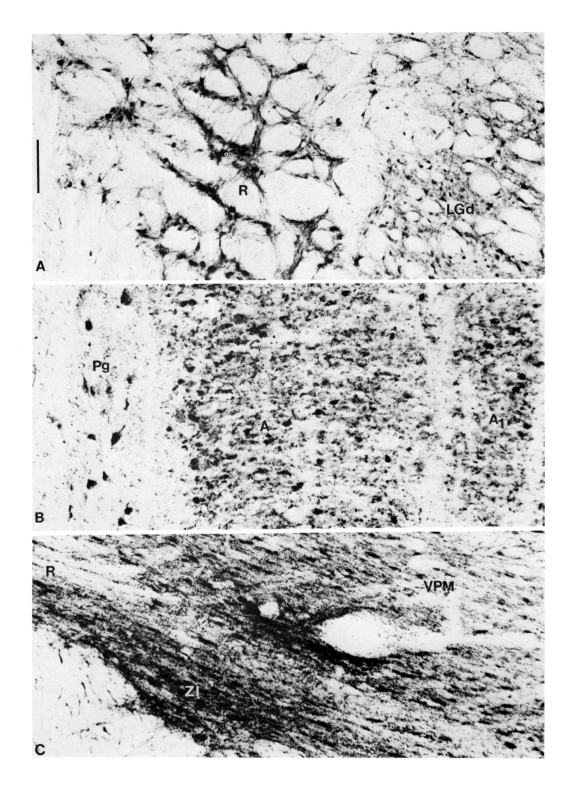

Nissl called it *Gitterschicht* which he divided into a dorsal, a lateral, and a ventral Gitterkern. Of these, however, the dorsal seems to have consisted mainly of the lateral dorsal nucleus. This was recognized by von Monakow (1895) in the cat and rabbit. Von Monakow, noticing the separation of the nucleus by the bulging dorsal lateral geniculate nucleus, referred to dorsal and ventral Gitterkerne. These, together, are the same as the reticular nucleus of modern terminology. The translated term *reticular nucleus* seems to have been used first by Münzer and Wiener (1902) in the rabbit, Mann (1905) in the marmoset, and Friedemann (1911) in the monkey. Ramón y Cajal (1911) translated *Gitterkern* as *noyau rayé* or *noyau grillagé* and the second term is still sometimes seen in the French literature. The nucleus has been recognized in all subsequent accounts of the thalamus from the monotremes (Hines, 1929) (Fig. 17.19) up to man (Hassler, 1959). Some authors, impressed by the variations in cell size and more particularly by alterations in the thickness of the nucleus at various levels, have sought to divide the nucleus into several parts (e.g., M. Rose, 1935) but this seems hardly warranted in view of the connectional similarities of all parts.

Perigeniculate nucleus is derived from Rioch's (1929a) term *substantia grisea perigeniculata*, used in his account of the cat and dog thalamus. Very little attention was paid to it until Szentágothai (1963) pointed out its similarity to the reticular nucleus and Sanderson (1971) reported units in it responding to visual stimulation in the cat. Since then, it has been identified in all species of carnivore that have been studied (Sanderson, 1974; Sanderson *et al.*, 1974; Figs. 15.1–15.5) and, as mentioned in Section 15.2.1, it is particularly large in some members of this order. It has no equivalent in other known mammals other than the reticular nucleus with which it shares many connectional and functional similarities (Ahlsén *et al.*, 1978; Friedlander *et al.*, 1979; Ahlsén and Lindström, 1982a,b; Chapter 4). Some authors have, as a consequence, even started referring to the seemingly equivalent part of the reticular nucleus in noncarnivores as perigeniculate nucleus [e.g., Sumitomo *et al.* (1976) in the rat].

15.2.3. Internal Structure

It is probable that the cellular morphology and fine structure of the reticular nucleus are different from those of the nuclei of the dorsal thalamus. However, I have not been able to find a single electron microscopic description of the reticular nucleus, other than of the perigeniculate nucleus.

Ramón y Cajal (1911) described a single class of cells in the reticular nucleus. They were large, fusiform, or triangular, with long dendrites mainly disposed parallel to the surface of the underlying dorsal thalamus and at right angles to the fibers entering and leaving the dorsal thalamus. The cells had axons which,

FIGURE 15.7. Demonstration of GABAergic neurons in reticular nucleus (R), perigeniculate nucleus (Pg), zona incerta (ZI), dorsal lateral geniculate nucleus (LGd), and adjoining dorsal thalamic nuclei, of a cat by immunocytochemical staining for GAD. Bar = 100 μm. From unpublished work of S. H. C. Hendry, C. Brandon, and the author.

after emitting two or more intranuclear collaterals, entered the dorsal thalamus. There were no local interneurons in the reticular nucleus.

Ramón y Cajal's observations have been reaffirmed by Scheibel and Scheibel (1966a) who showed, possibly for the first time, the wide extent of the ramifications of the axons of reticular nucleus cells within the dorsal thalamus. Cell form and the widespread axonal distribution have also been confirmed by intracellular injection (Yen and Jones, 1983; Fig. 4.14). The terminals of the reticular nucleus axons have also been demonstrated ultrastructurally in the dorsal thalamus and have been shown to contain flattened or pleomorphic synaptic vesicles and to make symmetric synapses, primarily on the dendrites of relay neurons (Ohara *et al.*, 1980; Montero and Scott, 1981) (Fig. 4.16).

The reticular nucleus neurons are GABAergic, as conclusively shown by the immunocytochemical localization of the enzyme GAD (Houser *et al.*, 1980; Oertel *et al.*, 1983; Chapter 5; Fig. 15.7). Somatostatin is reported to be colocalized in some reticular nucleus cells in the cat (Graybiel and Elde, 1983; Oertel *et al.*, 1983) and somatostatin-containing nerve fibers of uncertain origin are found in the reticular nucleus of other species (Fig. 5.14). It has not yet been demonstrated whether the feline perigeniculate nucleus also contains GAD and somatostatin.

Light microscopy indicates that thalamocortical, corticothalamic, and possibly thalamostriatal and pallidothalamic fibers have terminations in the reticular nucleus (Figs. 15.8 and 15.9; Chapters 3 and 4; Ramón y Cajal, 1911; Scheibel and Scheibel, 1966a; Jones, 1975c; Montero *et al.*, 1977). Corticothalamic terminals have been demonstrated electron microscopically (Ohara and Lieberman, 1981). It is commonly assumed that these terminations arise from collateral branches of fibers that traverse the reticular nucleus. Direct proof of this has come from intraaxonal labeling of thalamocortical axons leaving the ventral posterior nucleus (Yen and Jones, 1983; Fig. 4.14). Similarly, labeling of individual thalamocortical fibers leaving the dorsal lateral geniculate nucleus shows that they have collaterals ending in the perigeniculate nucleus (Ferster and LeVay, 1978; Ahlsén *et al.*, 1978; Friedlander *et al.*, 1979, 1981; Stanford *et al.*, 1983). Direct visualization of collaterals of the other three fiber systems mentioned has not yet been reported.

Ide (1982) has examined the perigeniculate nucleus ultrastructurally and finds three kinds of synaptic terminals. One, ending in asymmetric synapses on dendrites, on perikarya, and on other flat vesicle-containing profiles, she thinks are derived from collaterals of geniculocortical axons; the flat vesicle-containing profile has some similarities to the presynaptic dendrites of the dorsal thalamus, though none of the classical thalamic interneurons are found. Other axon terminals ending axodendritically are similar to the terminals of corticothalamic fibers in the dorsal thalamus. A fourth type of axon terminal found in the perigeniculate nucleus is not represented in the dorsal thalamus. It is large and contains flattened synaptic vesicles. Ide, following Ahlsén and Lo (1982), thinks that these terminals may be those of axons arising in the midbrain reticular formation. To my mind, they resemble the terminals of reticular nucleus axons in the dorsal thalamus and could be the terminals of the intranuclear collaterals of these axons. If Ide's observations can be regarded as representative of the fine structure of the reticular nucleus as a whole, then two questions clearly need to be resolved: the origins of the presynaptic dendrites in the absence of a set

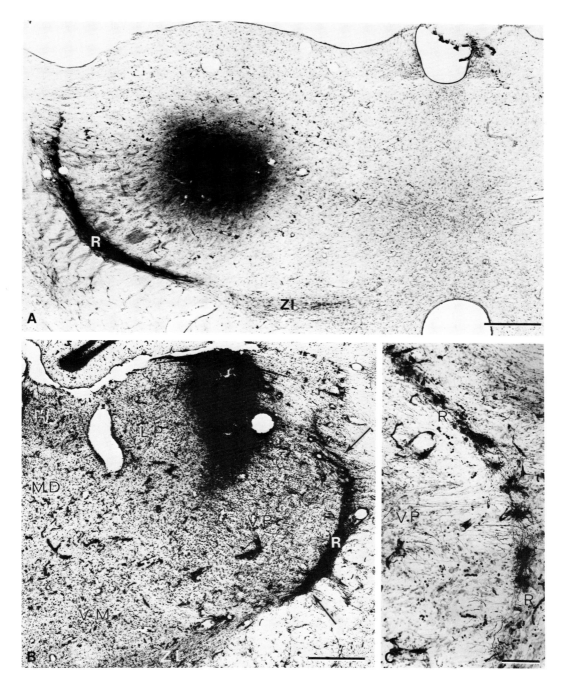

FIGURE 15.8.   (Upper) Anterograde labeling of axons and retrograde labeling of cells in the reticular nucleus (R) and zona incerta (ZI) of a rat that received a large injection of horseradish peroxidase in the dorsal thalamus and pretec-tum. No counterstain. (Lower) Similar preparation from another rat, showing labeling in reticular nucleus only. (C) is 200 μm anterior to (B). Bars = 500 μm (A, B), 100 μm (C). (B,C) from Jones (1975c).

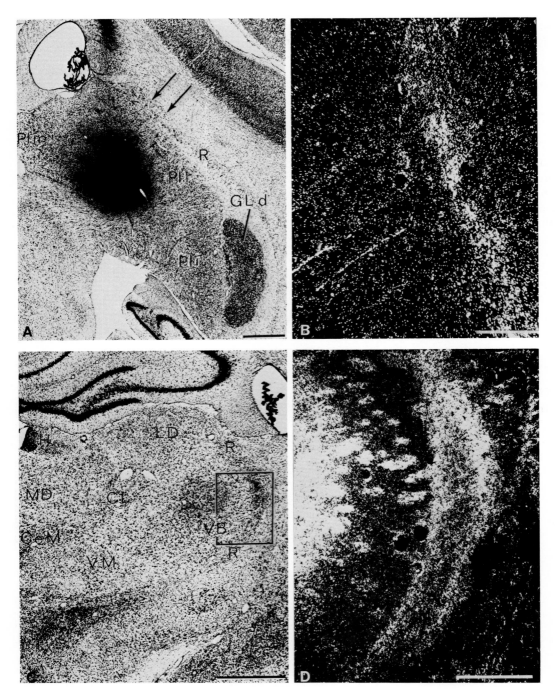

FIGURE 15.9. (A, B) Anterograde labeling of collateral ramifications of thalamocortical fibers in part of reticular nucleus indicated by arrows in (A), following injection of tritiated amino acids in pulvinar nuclei of a squirrel monkey. (C, D) Anterograde labeling of collateral ramifications of corticothalamic fibers in part of reticular nucleus indicated by box in (C), following injection of tritiated amino acids in somatic sensory cortex of a rat. Bars = 1 mm (A, C), 200 μm (B), 250 μm (C). From Jones (1975c).

of intrinsic neurons, and the nature and sites of termination of the intranuclear collateral branches of the reticular nucleus axons.

15.2.4. Connections

15.2.4.1. Cortical Connections

The connections of the reticular nucleus were for a long time difficult to elucidate experimentally. Nissl (1913) had noted that cellular degeneration in the nucleus of the rabbit following destruction of the telencephalon was far less severe than in the nuclei of the dorsal thalamus, few or no cells actually disappearing and large lesions being necessary to cause an effect. Rose and Woolsey (1943, 1949a) confirmed this. Later, Rose (1952) in the cat and rabbit and Chow (1952) in the monkey studied the distribution of cellular degeneration in the reticular nucleus following ablations of different regions of the cerebral cortex. They found that the nucleus is related to the cortex in an organized way: regions of cortex connected with a particular dorsal thalamic nucleus, when ablated, usually lead to cellular changes in the part of the reticular nucleus adjacent to or overlying the relevant affected dorsal thalamic nucleus or nuclei. A similar topographical organization was demonstrated by Carman *et al.* (1964) who studied the distribution of corticofugal fibers within the reticular nucleus with anterograde axonal degeneration techniques. This type of organization has since been confirmed by autoradiography (Jones, 1975c; Figs. 15.9 and 15.10). Rose (1952), though not ruling out a projection to the cortex, had favored the view that the cells of the reticular nucleus underwent a secondary degeneration after cortical ablation as the result of degeneration of the dorsal thalamic nucleus to which they sent axons. Carman *et al.* (1964) argued that the cellular changes which occur in the reticular nucleus after cortical ablations were primarily due to transneuronal degeneration as the result of these cells being deprived of their cortical input. The first modern study, though not resolving the issue of the cause of the cellular reaction in the reticular nucleus, conclusively showed that the reticular nucleus receives fibers from the cortex but does not project to it (Jones, 1975c). In the ensuing years in which retrograde tracers of every kind have been injected into virtually every cortical area of every imaginable mammal, no labeling of reticular nucleus cells has been reported. The perigeniculate nucleus of the cat is like the reticular nucleus in receiving fibers from the visual cortex (Updyke, 1975, 1977) but not projecting to it.

15.2.4.2. Connections with the Dorsal Thalamus

The Golgi work of Ramón y Cajal (1911) and of Scheibel and Scheibel (1966a) suggested that both corticothalamic and thalamocortical fibers passing through the reticular nucleus gave off collaterals as they traversed it. These workers also found that the cells of the reticular nucleus itself sent abundant axons into the underlying dorsal thalamus and Scheibel and Scheibel showed that single axons could ramify widely through several dorsal thalamic nuclei, both principal and intralaminar, and might even descend into the midbrain.

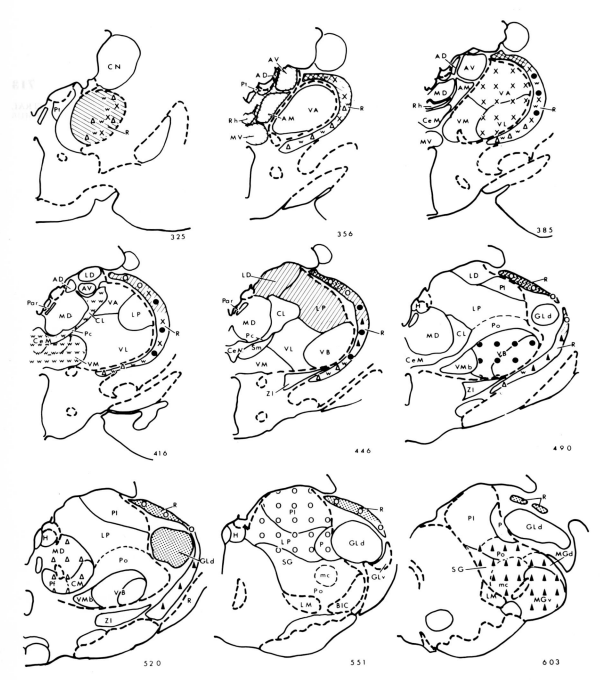

FIGURE 15.10. Scheme of interrelationships between sectors of reticular nucleus of the cat and particular groups of nuclei in underlying dorsal thalamus. Nuclei indicated by particular symbols are related to a similarly indicated sector of the reticular nucleus by collaterals of thalamocortical and of corticothalamic fibers; most of the intrathalamic ramifications of axons leaving that sector of reticular nucleus terminate in the same nuclei. For clarity, symbols appear only once on the various dorsal thalamic nuclei. Small numbers indicate position of each section in a series. From Jones (1975c).

Minderhoud (1971) using anterograde and retrograde degeneration techniques reported a comparable diffuse intrathalamic projection from isolated portions of the cat reticular nucleus.

In the first comprehensive study, involving injection of anterograde and retrograde tracers in the cerebral cortex, in each of the dorsal thalamic nuclei, and in the reticular nucleus itself of cats, rats, and monkeys, Jones (1975c) established the following principles of organization (Fig. 15.10). The reticular nucleus can be thought to consist of a series of overlapping sectors each related to a particular dorsal thalamic nucleus (or nuclei). Each sector is traversed by corticothalamic and thalamocortical fibers passing to and from the dorsal thalamic nuclei related to that sector; the traversing fibers distribute terminals in their sector. The cells of this sector, themselves, send their axons into the dorsal thalamus to terminate primarily in the nuclei to which they are related by traversing corticothalamic and thalamocortical fibers. The consequence of this pattern of organization is that a particular sector of the reticular nucleus can be dominated by a particular sensory (or other) system.

In the years since 1975, anatomical studies on individual dorsal thalamic nuclei have confirmed that each is associated with a constant part of the reticular nucleus (e.g., Montero et al., 1977). Moreover, single-unit recording in the reticular nucleus has established that parts related to a sensory relay nucleus such as the lateral geniculate (Sumitomo et al., 1976; Hale et al., 1982), medial geniculate (Shosaku and Sumitomo, 1983), or ventral posterior (Sugitani, 1979; Pollin and Rokyta, 1982; Yen and Jones, 1983) will contain neurons with receptive fields appropriate for that system. Where sectors such as the somatic sensory and auditory overlap, neurons show mixed receptive fields (Shosaku and Sumitomo, 1983), probably reflecting the length of the dendrites of the reticular nucleus cells and their contact by traversing fibers to both underlying dorsal thalamic nuclei.

Many perigeniculate cells have "on–off"-center receptive fields (Cleland et al., 1971), i.e., they show transient discharges when a visual stimulus applied to the center of their receptive fields is turned on or off. This implies convergent input from on-center and off-center dorsal lateral geniculate cells (Chapter 4) because no adult geniculate cells have on–off-center receptive fields. There is also evidence to suggest that single perigeniculate cells can receive inputs from both X-type and Y-type geniculate relay cells (Dubin and Cleland, 1977; So and Shapley, 1981) in cats. In rats, Hale et al. (1982) found on–off-center receptive fields in the comparable part of the reticular nucleus and suggest that inputs from fast retinal ganglion cells (equivalent to Y cells) may predominate.

Activation of cells in a sensory sector of the reticular nucleus by peripheral stimulation usually occurs at a latency sufficiently longer than that of cells in the underlying relay nucleus to suggest disynaptic activation by way of the relay nucleus. In the case of the perigeniculate nucleus of the cat, some have argued in favor of monosynaptic activation (Schmielau, 1979). Not all agree with this, however, seeing definitely longer latencies in the perigeniculate than in the dorsal lateral geniculate nucleus (e.g., Ahlsén and Lindström, 1982b; Ahlsén et al., 1982b). Ahlsén et al. (1982b) have reported that the part of the reticular nucleus above the perigeniculate nucleus of the cat is actually unresponsive to visual stimuli or to stimulation of the optic tract. They have used these observations to press their case for the perigeniculate nucleus being the visual part of the reticular nucleus. I have sympathy with this point of view but have still seen axon terminal labeling in the dorsal part of the reticular nucleus proper after injections of isotope in the dorsal lateral geniculate nucleus (Fig. 7 of Jones, 1975c) and there is no queston that fibers from the visual cortex terminate there also (e.g., Updyke, 1977). Moreover, stimulation of the visual cortex does lead to EPSPs in this part of the reticular nucleus though at longer latency than in the perigeniculate nucleus (Ahlsén and Lindström, 1982a).

15.2.4.3. Connections with the Midbrain

Although Scheibel and Scheibel (1966a) reported axons of reticular nucleus origin descending through the brachium of the inferior colliculus to the midbrain, this connection could not be confirmed experimentally (Jones, 1975c). On the other hand, there appears to be a significant input to the reticular nucleus and/or perigeniculate nucleus from the midbrain. Undoubtedly, some reports of retrograde labeling in midbrain sites such as the dorsal raphe nucleus, or even in situations such as the cerebellum and spinal cord, after injections of tracer aimed at the reticular or perigeniculate nucleus, stem from involvement of fibers passing through or around those nuclei. However, anterograde labeling from injections in midbrain sites clearly demonstrates a moderately dense projection to the reticular nucleus (Fig. 15.11). This was originally reported as emanating from certain of the deeper-seated, nonvisually connected nuclei of the pretectum (Berman, 1977) but the part of the midbrain tegmentum called the nucleus cuneiformis now seems the more likely site (Edwards and De Olmos, 1976).

The projection from the nucleus cuneiformis may well be cholinergic as both it and the reticular nucleus form part of the ascending, cholinesterase-positive system of Shute and Lewis (1967). Furthermore, the reticular nucleus shows quite dense concentrations of cholinergic receptors (Chapter 5) and spontaneous neuronal discharges of its cells are suppressed by iontophoretic application of acetylcholine in a manner that mimics the effect of electrical stimulation of the nucleus cuneiformis region (Dingledine and Kelly, 1977; Chapter 5).

15.2.5. Functions of the Reticular Nucleus

Aspects of the physiology of the reticular nucleus have been reviewed in Chapters 3, 4, and 5. We can now bring all of the various points together in brief format.

The anatomy of the system (Section 15.2.4) and receptive field mapping in the reticular nucleus indicate that different parts of the nucleus are to a considerable extent concerned with one or other sensory relay nucleus (Sumitomo *et al.*, 1976; Sugitani, 1979; Ahlsén *et al.*, 1982b; Hale *et al.*, 1982; Pollin and Rokyta, 1982; Shosaku and Sumitomo, 1983; Yen and Jones, 1983) though with some overlap. The anatomy and some of the work involving electrical stimulation tells us that whatever a particular sector of the reticular nucleus does in relation to a sensory relay nucleus, other sectors probably carry out the same role in relation to each of the other dorsal thalamic nuclei.

Work on the somatic sensory and auditory parts of the reticular nucleus and on the perigeniculate nucleus which is probably the visual part in cats, indicates that reticular nucleus neurons are excited by peripheral stimulation at latencies sufficiently longer than those in the relay cells of the corresponding sensory nucleus, to imply one synaptic delay in comparison with the sensory nucleus (Sugitani, 1979; Pollin and Rokyta, 1982; Ahlsén and Lindström, 1982a; Hale *et al.*, 1982; Yen and Jones, 1983; Yen *et al.*, 1985). The best evidence that under these circumstances, activation of reticular nucleus cells comes about via collaterals of thalamocortical relay cell axons is in the work on the perigeniculate nucleus. There, Ahlsén and Lindström (1982a) have been able to show collision of excitatory potentials with disynaptic latencies due to optic tract stimulation, with those with monosynaptic latencies due to antidromic stimulation of thala-

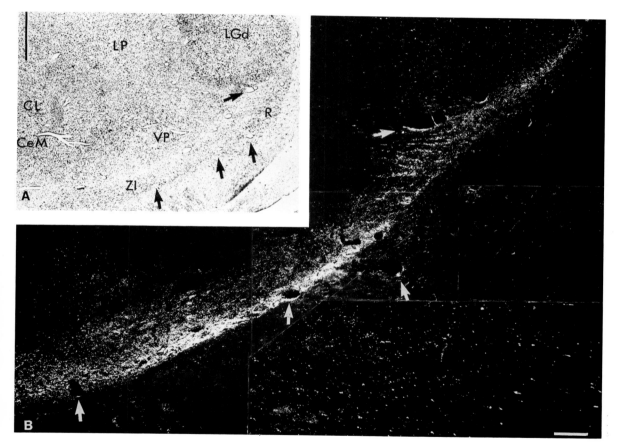

FIGURE 15.11. Main figure shows autoradiographically labeled axon terminations in part of reticular nucleus following injection of tritiated amino acids in nucleus cuneiformis of midbrain reticular formation. Arrows indicate same blood vessels. Bars = 1 mm, 250 μm. After Berman (1977).

mocortical axons. [Orthodromic activation of perigeniculate neurons due to cortical stimulation is unlikely here as visual corticothalamic axons are much slower conducting than thalamocortical (Gilbert, 1977; Ahlsén *et al.*, 1982b).]

Though the close temporal correlation between discharges of reticular nucleus neurons and inhibitory potentials in underlying dorsal thalamic neurons (Purpura and Schofer, 1963; Frigyesi and Schwartz, 1972) has been questioned (Steriade and Wyzinski, 1972), the work on the perigeniculate nucleus at least suggests that some sort of relationship undoubtedly exists. Ahlsén and Lindström (1982a) and Lindström (1982) noted that there was a 1.6- to 2-msec difference in latency between the antidromic invasion of lateral geniculate relay cells by cortical stimulation and the onset of IPSPs in the same cells. Because relay cells generally do not have axon collaterals in the dorsal lateral geniculate nucleus and because any geniculate interneurons do not project to the cortex, these IPSPs can only have been brought about by activation of inhibitory interneurons elsewhere, presumably in the perigeniculate nucleus. These IPSPs in relay neurons occur 0.7–0.8 msec after EPSPs induced in the perigeniculate neurons by the antidromic discharge of thalamocortical axon collaterals passing through the perigeniculate nucleus. Moreover, the development of the IPSP with several ripples seems to correspond to the repetitive firing of the perigeniculate cells. Finally, electrical stimulation of the optic tract within the time of postexcitatory suppression of geniculocortical relay cells due to a previous stimulus, fails to excite these cells or cells in the reticular nucleus (Hale *et al.*, 1982).

There is sufficient work, through by no means so compellingly presented, on other parts of the reticular nucleus to imply that they, like the perigeniculate nucleus, also provide a recurrent inhibition upon thalamocortical relay cells in their related dorsal thalamic nuclei. Certainly the presence of markers for GABA synthesis in probably every reticular nucleus cell (Houser *et al.*, 1980; Fig. 15.10) and the evidence that GABAergic terminals derived from the reticular nucleus terminate on relay cells in the dorsal thalamus (Montero and Scott, 1981; Ohara *et al.*, 1983), support this idea. The clear evidence for mutual inhibition between perigeniculate neurons (Ahlsén and Lindström, 1982b) is in keeping with the GABAergic nature of reticular nucleus cells and the presence of intranuclear collaterals derived from their axons (Yen and Jones, 1983; Fig. 4.14).

All of the results reviewed to this point suggest that the reticular nucleus acts as a source of short-latency recurrent inhibition upon relay neurons in the dorsal thalamus following activation of the latter from the periphery and their discharge of a thalamocortical volley. A direct effect of the reticular nucleus on thalamocortical transmission is indicated by the more rapid recovery of geniculocortical relay cells from postexcitatory suppression after destruction of part of the reticular nucleus (Sumitomo *et al.*, 1976), and by the fact that single shock stimulation of the reticular nucleus can suppress spontaneous and evoked activity in the dorsal lateral and medial geniculate nuclei (Sumitomo *et al.*, 1976; Shosaku and Sumitomo, 1983), with a concomitant reduction of the cortical evoked potential (Yingling and Skinner, 1976). There is little or no evidence from these studies to indicate how the reticular nucleus deals with its collateral corticothalamic inputs and none at all to indicate how it handles collateral inputs from the thalamostriatal and pallidothalamic projections.

The association of the reticular nucleus with the midbrain reticular formation is less easy to assess, although the part of the reticular formation (the nucleus cuneiformis) from which the reticular nucleus receives fibers is that which when stimulated at high frequency can lead at relatively long latency to a facilitation of transmission in many dorsal thalamic nuclei (Purpura *et al.*, 1966; Symmes and Anderson, 1967; Wilson *et al.*, 1973; Singer, 1973a, 1977; Sasaki *et al.*, 1976; Burke and Cole, 1978) (Chapter 4). Under these circumstances intranuclear inhibition is said to be suppressed (Purpura and Schofer, 1963; Singer, 1973a; Singer and Schmielau, 1976; Fukuda and Stone, 1976), though Steriade (1983) notes a sharpening of early inhibition followed by more rapid recovery from inhibition rather than an overall suppression.

It is possible that the reticular nucleus plays a significant role in mediating the effect of stimulation of the reticular formation. While the firing of neurons in the ventral posterior nucleus is being increased by high-frequency stimulation of the midbrain, the spontaneous activity of neurons in adjacent parts of the reticular nucleus is being concomitantly reduced (Dingledine and Kelly, 1977), though at rather long latency. The same neurons of the reticular nucleus can also have their spontaneous activity reduced (though not their bursting behavior which is, in fact, enhanced) by iontophoretic application of small amounts of acetylcholine (Ben-Ari *et al.*, 1976; Dingledine and Kelly, 1977), suggesting that the reticular effect may be mediated by this transmitter. Certainly the reticular nucleus contains acetylcholine receptors (Chapter 5) and an acetylcholinesterase-positive pathway ascends to it from the vicinity of the nucleus cuneiformis (Lewis and Shute, 1967; Chapter 3). However, the effect of stimulating the reticular formation was not readily blocked by atropine (Dingledine and Kelly, 1977) so there are still some grounds for caution in accepting this theory.

Many workers consider that the intermittent (greater than 1 per second), high-frequency burst (200–400 Hz) discharges that are so typical of the behavior of reticular nucleus neurons under barbiturate anesthesia and in desynchronized sleep and wakefulness (Chapter 4), are correlated with similar rhythmic behavior in the underlying dorsal thalamic nuclei (Chapter 4; Schlag and Waszak, 1971; Frigyesi and Schwartz, 1972). Others, however, disagree (Mukhametov *et al.*, 1970a,b; Steriade and Wyzinski, 1972; Steriade, 1978, 1983; Domich *et al.*, 1983). The interrelatedness of the reticular nucleus with the dorsal thalamus and its input from the midbrain reticular formation, as reviewed in the preceding paragraphs, would certainly fit it for a role in controlling transmission in the dorsal thalamus according to states of arousal. On the other hand, it has to be admitted that it is the spontaneous activity of the reticular nucleus that is controlled by the midbrain, not its predilection for discharging bursts of neuronal activity (Dingledine and Kelly, 1977).

Steriade and his associates (see Steriade *et al.*, 1972; Steriade, 1978, 1983) have, over the years, argued that there is no clear temporal correlation or reciprocity between reticular nucleus and dorsal thalamic activity. They have recently discovered (Domich *et al.*, 1983) that reticular nucleus cells behave exactly like relay cells when an animal changes from slow-wave sleep to wakefulness, i.e., both sets of cells show increased excitability and increased spontaneous discharges rather than reciprocal effects. For many years this group seemed to regard the midbrain projections to the intralaminar and ventral medial nuclei, via their diffuse projections to the cortex, to be primary regulators of state-dependent activity. On the other hand, they have recently found (Deschênes *et al.*, 1983; Steriade *et al.*, 1983) that dorsal thalamic neurons separated from the reticular nucleus by knife cuts fail to undergo the low-frequency (7–14 Hz) rhythmic hyperpolarizations that underlie their repetitive spike bursts ("spindles") during anesthesia or slow-wave sleep and, thus, the reticular nucleus may possess some kind of pacemaker activity relevant to them. Because hyperpolarizations still occur following discharges of relay neurons, even in the absence of a reticular nucleus input, Steriade *et al.* (1983) also conclude that intranuclear inhibitory neurons are involved and feel that the reticular nucleus may control the temporal aspects of relay cell discharge, possibly by controlling the intranuclear interneurons themselves. Presumably, on account of the GABAergic nature of the reticular nucleus neurons (Chapter 5 and Section 15.2), this is an inhibitory effect.

We may conclude, I think, that the reticular nucleus is capable of modulating dorsal thalamic neuronal activity, particularly in relation to the passage of afferent volleys through the thalamus. It may also be involved in the modulation of activity during sleep and arousal and this may reflect its own underlying control by the midbrain reticular formation. One is forced to admit, however, that the temporal sequence of events between midbrain, cortex, reticular nucleus, and dorsal thalamus in the sleep–waking cycle is still far from clear and the way in which the reticular nucleus exerts its effects and what this means for the animal are highly conjectural.

15.3. The Zona Incerta

15.3.1. Description

The nucleus of the *zona incerta* consists of small cells which lie between the ventral part of the external medullary lamina of the thalamus and the cerebral peduncle close to the posterior half of the dorsal thalamus, though its exact position can vary from species to species. The cells tend to be arranged in rows lying parallel to the external medullary lamina. The nucleus is tenuously connected with the reticular nucleus anteriorly and islands of the larger reticular cells can invade it (Figs. 15.1, 15.6, and 15.7). The continuity of the two nuclei is reflected in the continuity of acetylcholinesterase staining (Figs. 5.1 and 5.3) and of immunoreactivity for GAD (Fig. 5.7), somatostatin, and certain other substances (Fig. 15.6). Posteriorly, the nucleus of the zona incerta usually becomes separated from the reticular nucleus by a thin fibrous lamina which continues ventral to the nucleus of the zona incerta and separates it then from the subthalamic nucleus.

The medial part of the nucleus of the zona incerta becomes continuous with a rather large, essentially fibrous region lying between the dorsal thalamus, the periventricular hypothalamic area, and the interpeduncular fossa, and extending from the lateral hypothalamic area anteriorly, virtually to the red nucleus posteriorly (Figs. 15.1 and 15.12). This cellular and fibrous region is essentially field H of Forel and I call its cellular elements the nucleus of the fields of Forel (FF).

15.3.2. Terminology

The zona incerta and the three associated fiber fields (H, H_1, and H_2) were originally described by Forel (1877) and confirmed by von Kölliker (1896) in unstained and myelin-stained preparations of human and animal brains (Figs. 15.1 and 15.12). They have been easily recognized ever since then (Dejerine, 1901; Sano, 1910; Foix and Nicolesco, 1925; Kodama, 1928, 1928–1929; Nicolesco and Nicolesco, 1929; Roussy and Mosinger, 1935; Ranson *et al.*, 1941; Brockhaus, 1942; Papez, 1942; Glees and Wall, 1946; Woodburne *et al.*, 1946; Hassler, 1949a,b; Whittier and Mettler, 1949; Laursen, 1955; Emmers and Akert, 1963), though the names strictly refer to the appearances in fiber preparations and, more particularly, to the primate. However, the various subdivisions

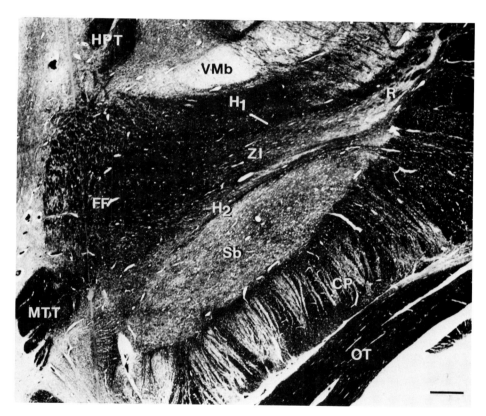

FIGURE 15.12. Hematoxylin-stained transverse section through the ventral part of the human diencephalon showing the fields, FF, H_1, and H_2 of Forel, the zona incerta (ZI), and the subthalamic nucleus (Sb). F: fornix; MTT: mamillothalamic tract. Bar = 1 mm. Compare with Fig. 15.1.

can usually be identified in cellular preparations as well and some hints of the primate pattern can be seen even in the smallest mammals.

Field H₁ is the external medullary lamina intervening between the ventral aspect of the dorsal thalamus and the zona incerta. It contains the *thalmic fasciculus* of von Monakow (1895, 1909–1910) and Foix and Nicolesco (1925), which consists mainly of fibers arising in the internal segment of the globus pallidus, and fibers of the brachium conjunctivum as these approach the thalamus (Mettler, 1945; Vogt and Vogt, 1941; Nauta and Mehler, 1966).

Field H₂ is the capsule of the subthalamic nucleus that intervenes between this nucleus and the nucleus of the zona incerta. It is the *fasciculus lenticularis* of von Monakow (1895, 1909–1910) and Foix and Nicolesco (1925) and contains fibers running from the internal segment of the globus pallidus to the dorsal thalamus and midbrain tegmentum (Nauta and Mehler, 1966).

Field H is the cellular region medial to the zona incerta and is pierced by fibers of the brachium conjunctivum and habenulopeduncular tract.

"Zona incerta" simply refers to a zone of low myelin density (von Monakow, 1895) between fields H₁ and H₂ though it has often been taken to include the H field of Forel (e.g., Jasper and Ajmone-Marsan, 1954). "Nucleus of the zona incerta" was introduced by Berman and Jones (1982) to emphasize the cellular content.

15.3.3. Connections

The nucleus of the zona incerta receives fibers from the sensory motor cortex (Fig. 15.13) (Petras, 1965; Jones and Powell, 1968; Wise and Jones, 1977a), ventral lateral geniculate nucleus (Graybiel, 1974a; Swanson *et al.*, 1974), deep cerebellar nuclei (Mehler, 1966a; Asanuma *et al.*, 1983b; Berkley, 1983), trigeminal complex (Smith, 1973; Erzurumlu and Killackey, 1980), and spinal cord (Lund and Webster, 1967b; Boivie, 1971b; Feldman and Kruger, 1980; Berkley, 1980) but probably not from the globus pallidus (Wilson, 1914; Nauta and Mehler, 1966). The large ascending noradrenergic bundle runs through the zona incerta en route to the rest of the forebrain (Fig. 5.10).

The efferent connections of the zona incerta are but little known. Unlike the reticular nucleus, it does not appear to project into the dorsal thalamus, though it can often be labeled by retrograde tracers injected into the pretectal region (Fig. 15.9). One ill-defined population of cells in it sends axons downwards into the spinal cord (Kuypers and Maisky, 1975) but their sites of termination have not been identified. Other cells, seemingly magnocellular neurosecretory cells related to those in the paraventricular and supraoptic nuclei of the hypothalamus, project to the posterior lobe of the pituitary gland (Kelly and Swanson, 1980).

The nucleus of the fields of Forel receives fibers from the internal segment of the globus pallidus (Vogt and Vogt, 1941; Nauta and Mehler, 1966; Mehler, 1966b), from the spinal cord, and from brain-stem reticular nuclei (Nauta and Kuypers, 1958; Denavit and Kosinski, 1968). Its outputs have not been determined but many spinally projecting cells, loosely described by Kuypers and Maisky (1975) as in the "dorsal hypothalamus," would seem to lie in it.

15.4. The Ventral Lateral Geniculate Nucleus

15.4.1. Description

The ventral lateral geniculate nucleus is obvious in all mammals except monkeys, apes, and man in which it has become the relatively much less overt pregeniculate nucleus (Figs. 15.16 and 9.8) (Vogt, 1909; Friedemann, 1911; Minkowski, 1920; Woollard and Beattie, 1927). In some mammals it may approach in size the dorsal lateral geniculate nucleus. This is particularly true of *Tupaia* and highly visual rodents such as squirrels. As noted by Le Gros Clark (1932b) the size of the ventral lateral geniculate nucleus can usually be correlated with the size of the superior collliculus with which the nucleus is connected. In some bats and insectivores (Fig. 15.14) the ventral lateral geniculate nucleus is anterodorsally placed and comes to lie in association with the anterior nuclei. This position seems to reflect its early developmental position (Chapter 6). In monotremes (Fig. 17.19) this is the position of the nucleus called LGNa by Campbell and Hayhow (1971, 1972). In large carnivores (e.g., seals) the nucleus can be ventrally placed but still well anterior to the dorsal lateral geniculate nucleus.

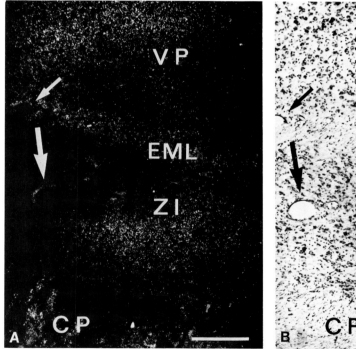
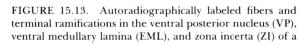
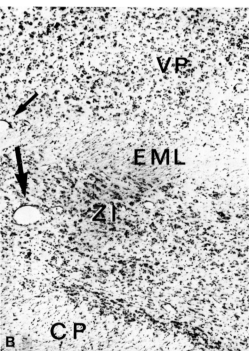

FIGURE 15.13. Autoradiographically labeled fibers and terminal ramifications in the ventral posterior nucleus (VP), ventral medullary lamina (EML), and zona incerta (ZI) of a rat, following injection of tritiated amino acids in the ipsilateral somatic sensory cortex. Bar = 250 μm. From Wise and Jones (1977a).

In most species the ventral lateral geniculate nucleus is more or less continuous with the reticular nucleus and zona incerta (Figs. 15.2B, 15.3, and 15.4A). As a component of the ventral thalamus, it shares a common embryological origin with these structures (Rose, 1942a; Chapter 6). It is traversed by fibers of the optic tract and in animals in which the tract is large the nucleus may lie partially buried in the tract. Large, deeply staining cells and more numerous, small, pale-staining cells are characteristic of the nucleus in all mammals (Jordan and Holländer, 1972; Fig. 15.15). In some highly visual forms such as ungulates and tree shrews the nucleus is distinctly laminated with large and small cells tending to be segregated. A tendency toward lamination can be detected in other forms such as the squirrel and even in the rabbit and rat (Gurdjian, 1927; Le Gros Clark, 1929a; Niimi et al., 1963; Abplanalp, 1970; Tigges, 1970). This lamination to some extent reflects the differential distributions of the retinal and midbrain afferents to the nucleus.

15.4.2. Terminology

The ventral lateral geniculate nucleus was identified in virtually all of the early cytoarchitectonic studies on the thalamus of nonprimates and was named by Nissl (1889) in the first of these. Since Nissl's account there have been no

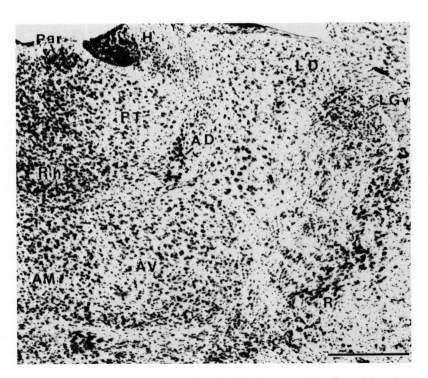

FIGURE 15.14. Thionin-stained frontal section showing the anterodorsally positioned ventral lateral geniculate nucleus of a moustache bat (*Pteronotus parnelli*). Bar = 100 μm.

alternative names put forward. In primates, the pregeniculate nucleus (Fig. 15.16) was named *substance grisé prégéniculeé* by Vogt (1909) in her myeloarchitectonic study of a species of Old World monkey. This may have been the first recognition of the nucleus as an entity warranting a name, but it is clearly visible in several of Dejerine's (1901) drawings of the human thalamus and he seems to have regarded it as part of the reticular nucleus. Friedemann (1911) also called the nucleus *pregeniculate gray substance* in his cytoarchitectonic study of the same monkeys as Vogt. Since then, it has been identified under this name or as *pregeniculate nucleus* in all the higher primates (Grünthal, 1934; Olszewski, 1952; Emmers and Akert, 1963; Niimi *et al.,* 1963) including man (Hassler, 1959). Balado and Franke (1937) may have been the first to discern its bilaminar form (Section 15.4.3) (Fig. 15.16).

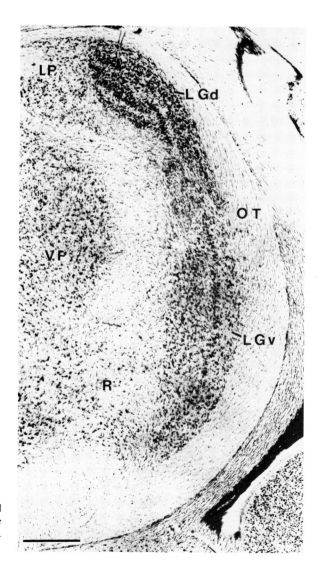

FIGURE 15.15.   Laminated ventral lateral geniculate nucleus of a tree shrew. From a preparation of Dr. M. Conley. Bar = 500 μm.

Recognition of the homology between the ventral lateral geniculate and the pregeniculate nuclei occurred early, though recent writers have tended to regard this as a new discovery. Pines (1927) in his cyto- and myeloarchitectonic study of the lemur thalamus used the term *pregeniculate substance* but clearly demonstrated its ventral rather than dorsal position relative to the dorsal lateral geniculate nucleus in this prosimian. Le Gros Clark (1930a, 1932a,b) in his early studies of tarsioid and lemuroid brains referred to the nucleus as *ventral lateral geniculate nucleus* and explicitly stated that in all mammals it was equivalent to the *corpus pregeniculatum* of higher primates. In this he was influenced not only by his extensive comparative anatomical observations but also by the early developmental position of the nucleus in man, ventral to the anlage of the dorsal lateral geniculate nucleus (Fig. 9.1).

15.4.3. Connections and Functional Properties

15.4.3.1. Retinal Afferents

The ventral lateral geniculate nucleus receives axons from both retinae though the number arising in the contralateral eye is substantially greater than from the ipsilateral eye (Minkowski, 1920; Laties and Sprague, 1966; Garey and

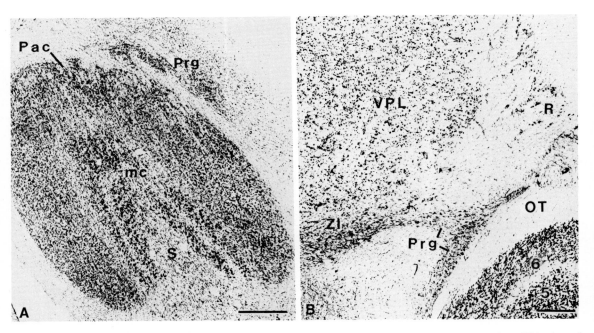

FIGURE 15.16. (A) Sagittal section showing the pregeniculate (Prg; ventral lateral geniculate) nucleus in a squirrel monkey. Anterior to right. (B) Frontal section showing the pregeniculate nucleus in a rhesus monkey. Thionin stain. Bars = 500 μm (A), 250 μm (B).

Powell, 1968; Takahasi *et al.*, 1977; other references in Holländer and Sanides, 1976, and in Hickey and Spear, 1976).

In all nonprimates the terminations of contralateral retinal fibers tend to be concentrated laterally and dorsally, embracing the smaller termination zone of ipsilateral fibers (Fig. 15.17). The region of retinal terminations shows a high concentration of GABAergic interneurons (Ohara *et al.*, 1983) (Fig. 5.15B) and a few neuropeptide Y-immunoreactive cells (Fig. 5.15A).

The pregeniculate nucleus of monkeys and man is bilaminar (Balado and Franke, 1937; Hassler, 1959; Hendrickson, 1973; Babb, 1980). A ventral lamina of small cells adjacent to the dorsal lateral geniculate nucleus receives a significant contralateral and a much weaker ipsilateral retinal input. A dorsal lamina of mixed large and small neurons is closely related to the reticular nucleus and is devoid of retinal inputs (Hendrickson *et al.*, 1970). Possibly it is analogous to the perigeniculate nucleus of cats (Chapters 3 and 9).

In the cat, all retinal inputs to the ventral lateral geniculate nucleus come via slowly conducting, W-type axons (Spear *et al.*, 1977) though in the rat, fast, intermediate, and slow inputs have been described (Hale and Sefton, 1978; Sumitomo *et al.*, 1979). There is a retinotopic organization present, at least in regard to the contralateral eye (Montero *et al.*, 1968; Mathers and Mascitti, 1975; Spear *et al.*, 1977). In the cat the lower visual field is represented anteriorly and the upper visual field posteriorly. Receptive fields in the cat are large, mostly monocular, and cell responses usually "on-tonic." Relatively few show the center–surround organization of the dorsal nucleus (Spear *et al.*, 1977; Hughes and Ater, 1977; Mathers and Mascitti, 1975; Hale and Sefton, 1978; Sumitomo *et al.*, 1979). The cells are subjected to relatively little postexcitatory inhibition, unlike relay cells in the dorsal thalamus.

15.4.3.2. Pretectal and Tectal Afferents

Cells in both the ventral lateral geniculate nucleus of cats and the pregeniculate nucleus of monkeys discharge in relation to saccadic eye movements (Büttner and Fuchs, 1973; Magnin and Fuchs, 1977) and this relationship to eye movements may reflect the strong input that the nucleus receives from the pretectum and superior colliculus (Fig. 15.18). Fibers arising in the ipsilateral superior colliculus come from cells in the superficial and intermediate layers and terminate primarily around the large cells of the nonprimate nucleus and in the dorsal, large-celled layer of the monkey pregeniculate nucleus (Altman and Carpenter, 1961; Tarlov and Moore, 1966; Abplanalp, 1970; Rockel *et al.*, 1972; Harting *et al.*, 1973; Graybiel, 1974b; Graham, 1977; Hughes and Chi, 1981; Weber *et al.*, 1983). Fibers from the ipsilateral pretectum appear to arise mainly in the parts innervated directly by the retina (i.e., the olivary nucleus and the nucleus of the optic tract). In the cat they terminate in relation to both large and small cells of the ventral lateral geniculate nucleus (Berman, 1977; Graybiel, 1977; Graybiel and Berson, 1980a). In the monkey the termination of pretectal fibers in relation to the two cell laminae does not appear to have been described in detail.

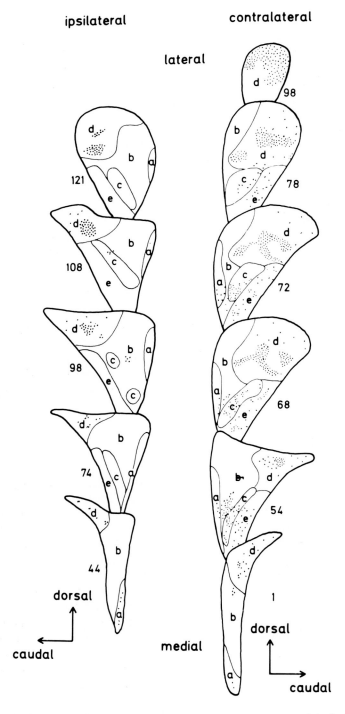

FIGURE 15.17.   Distribution of optic tract axons (dots) on sagittal sections of the ipsi- and contra-lateral ventral lateral geniculate nuclei of the cat. From Holländer and Sanides (1976).

15.4.3.3. Other Subcortical Afferents

Other subcortical afferents to the ventral lateral geniculate nucleus have been described. These include fibers from the cerebellum and subthalamus (Graybiel, 1974a), and the ascending noradrenergic fibers of the locus coeruleus which form a rather dense plexus about the nucleus (Swanson and Hartman, 1975). It is difficult to be sure, however, to what extent these fibers are simply passing around the nucleus en route to other sites.

15.4.3.4. Cortical Afferents

Though the nucleus, as part of the ventral thalamus, does not project to the cerebral cortex, it receives a strong cortical input arising in certan visually-related areas. The ventral lateral geniculate nucleus of rodents and of species with a similarly constituted lateral geniculate complex, receives corticofugal fibers from area 17 and from the adjacent peristriate belt (Nauta and Bucher, 1954; Abplanalp, 1970; Giolli and Guthrie, 1969; Giolli *et al.*, 1978; Holländer *et al.*, 1979).

The ventral lateral geniculate nucleus of the cat receives corticofugal fibers from all parts of areas 18 and 19 of the visual cortex but apparently only from posterior parts of area 17 (Updyke, 1977). In the nucleus, these fibers appear to terminate only partially overlapping the lateral zone in which retinal fibers terminate. Projections from other posterolaterally situated cortical areas, such as the lateral suprasylvian areas (Chapter 10), have sometimes been described but not always confirmed (Kawamura *et al.*, 1974; Updyke, 1981a,b).

Cortical inputs to the pregeniculate nucleus of primates have not been comprehensively described. Area 17 and certain extrastriate areas distribute fibers to the external, non-retinal-recipient part (Spatz *et al.*, 1970; Spatz and Tigges, 1973; Ogren and Henrickson, 1976).

15.4.3.5. Efferent Connections

The outputs of the ventral lateral geniculate nucleus are to a variety of subcortical sites (Fig. 15.19) (Hendrickson, 1973; Graybiel, 1974a,b; Edwards *et al.*, 1974; Swanson *et al.*, 1974; Ribak and Peters, 1975). It is not completely clear whether efferents to a particular site arise only from the large or small cells of the nucleus. Some of the sites of projection are undoubtedly visually related and may reflect a function for the nucleus in eye movement control. The targets include the olivary nucleus and the nucleus of the optic tract in the ipsilateral and contralateral pretectum, and the intermediate layers of the ipsilateral superior colliculus. The projection upon the latter seems to match the retinotopic organization present in the superior colliculus. There is also a projection to the ipsilateral pontine nuclei and to the contralateral ventral lateral geniculate nucleus. Fibers passing to the contralateral ventral lateral geniculate nucleus do so by way of the posterior commissure, continuing the trajectory of fibers destined for the two pretectal regions and ipsilateral superior colliculus.

Other efferent connections of the ventral lateral geniculate nucleus appear to reflect a close relationship with centers involved in the entrainment of circadian

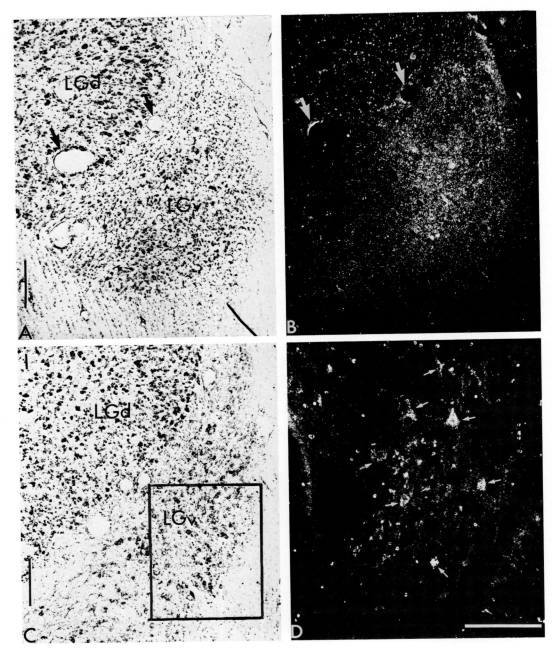

FIGURE 15.18. Pairs of brightfield and darkfield auto-radiographs showing (A, B) distribution of autoradiographically labeled fibers in the ventral lateral geniculate nucleus of a cat following injection of tritiated amino acids in the ipsilateral nucleus of the optic tract and (C, D) distribution of retrogradely labeled cells following injection of horse-

rhythms to cycles of darkness and light (Moore, 1978; Pickard and Turek, 1983). The nucleus projects bilaterally to the lateral terminal nuclei of the accessory optic tracts and to the suprachiasmatic nuclei of the hypothalamus. In both sites the axons terminate in regions that overlap the terminal regions of afferent fibers from the retina though there may not be complete superimposition. The projection to the suprachiasmatic nuclei arises from a small portion of the nucleus among the cells on which optic fibers terminate. The cells in the region and their axons which terminate in the suprachiasmatic nuclei are described as immunoreactive for the peptide, avian pancreatic polypeptide (Card and Moore, 1982), though recent evidence suggests that neuropeptide Y was being localized because of cross-reactivity of the antiserum being employed (Tatemoto *et al.*, 1982; Allen *et al.*, 1983) (Fig. 5.15). Fibers to the suprachiasmatic nuclei traverse the ipsilateral zona incerta in which terminations also occur.

All of the experimental work on efferent connections of the ventral lateral geniculate nucleus appears to have been done on nonprimates. It is assumed, however, that the connections demonstrable there are also representative of those of the pregeniculate nucleus of primates.

There is some possibility that the ventral lateral geniculate nucleus may be involved in the pupillary light reflex (Polyak, 1957) for there is evidence that retinal ganglion cells responding to luminance changes may preferentially project there (Spear *et al.*, 1977). Thence, via its projections to the pretectum, the

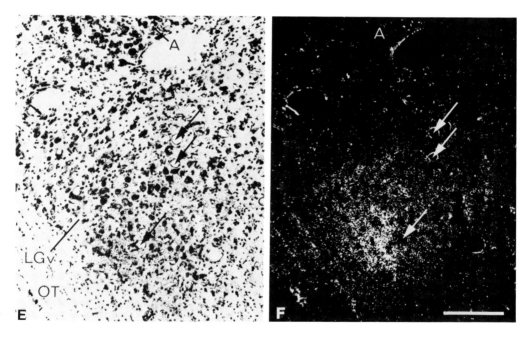

radish peroxidase in the same pretectal nucleus. From Berman (1977). (E, F) Distribution of autoradiographically labeled fibers from the superior colliculus in the ventral lateral geniculate nucleus of a cat. From Graham (1977). Bars = 100 μm.

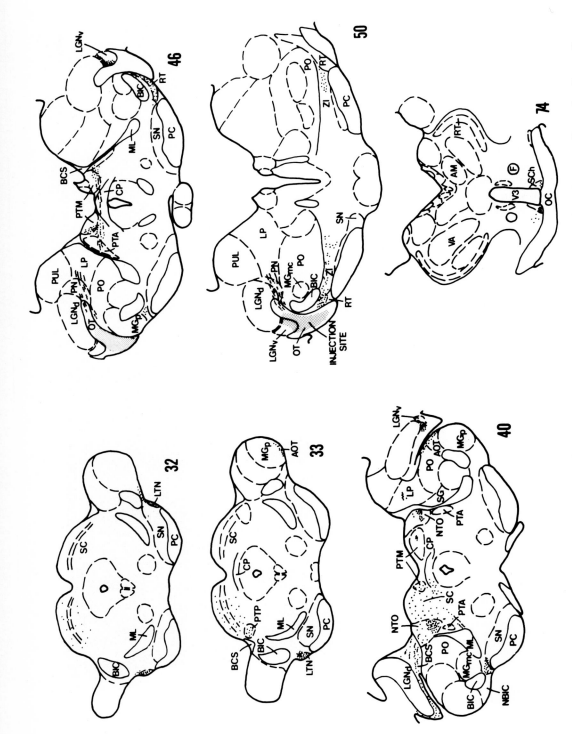

FIGURE 15.19. Distribution of efferent fibers from the ventral lateral geniculate nucleus of the cat as shown by autoradiography. Descending projections to pontine reticular formation not indicated. From Swanson *et al.* (1974).

ventral lateral geniculate nucleus may interface with the pupilloconstrictor centers of the midbrain.

The fact that units in the ventral lateral geniculate (or pregeniculate) nucleus discharge in relation to movements of the head and saccadic movements of the eyes (Büttner and Fuchs, 1973; Putkonen *et al.*, 1973) also implies an involvement with the vestibular and oculomotor systems. This involvement is perhaps reflected in its connections with the superior colliculus and with precerebellar nuclei of the pons (Graybiel, 1974a).

16

The Epithalamus

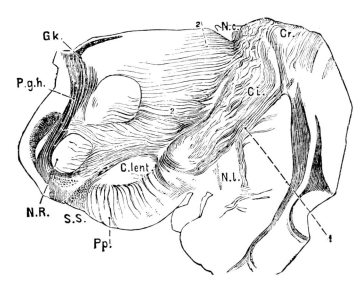

Meynert's drawing of the fasciculus retroflexus or habenulopeduncular tract (P.g.h.). From Meynert (1885).

16.1. Definition

The epithalamus in the adult mammalian brain consists of the anterior and posterior paraventricular nuclei, the medial and lateral habenular nuclei, the stria medullaris thalami, and the pineal body. They develop from a common pronuclear mass from which, early on, a large part splits off to form the nuclei of the pretectum (Rose, 1942b; Chapter 6). Although it may be appropriate to speak of the pretectum as part of the epithalamus during development, it is not customary to do so in the adult brain and it will be omitted from further consideration here. A detailed account of the whole pretectal complex can be found in Berman and Jones (1982). The pineal body is also not considered here because of its tenuous connection with the brain. Reviews of its history, its structure, and its function can be found in Berman (1968), Kappers and Pévet (1979), Reiter (1980), and Matthews and Seamark (1981).

16.2. The Paraventricular Nuclei

16.2.1. Description

The paraventricular nuclei are composed of small, densely packed, and deeply staining cells (Figs. 16.1–16.4). Two nuclei are visible in most mammals: an anterior and a posterior and each lies very close to its neighbor of the opposite side. Throughout most of their extent, they lie just beneath the ependyma of the ventricular system, their cells being separated from the ependyma by a thin, fibrous, stratum zonale. The anterior paraventricular nucleus is C shaped, commencing among the cells of the dorsal hypothalamic area and curving dorsally, up the anterior surface of the thalamus in the floor and posterior wall of the interventricular foramen. From there, it extends posteriorly on the dorsal aspect of the thalamus in close association with the parataenial nucleus and stria medullaris, and fused with its fellow of the opposite side. It usually disappears in the anterior end of the periaqueductal gray matter, ventral to the medial habenular nucleus (Figs. 16.1–16.3).

The posterior paraventricular nucleus, despite its name, extends anteriorly for a considerable distance. It, too, is composed of small, densely packed cells, usually lying ventrolateral and parallel to the anterior paraventricular nucleus and at the medial edge of the parataenial nucleus. The medial edges of the posterior paraventricular nuclei approach one another in the midline ventral to the anterior paraventricular nuclei and commonly fuse.

At its posterior end, the posterior paraventricular nucleus turns ventrally deep to the habenular nuclei and along the caudal pole of the mediodorsal nucleus to enter the periaqueductal gray matter of the midbrain (Figs. 16.1–16.3). In many mammals the distinction between anterior and posterior nuclei is not particularly clear and even the separation into left and right nuclei can be indistinct (Fig. 16.5). In some animals, the fusion of the nuclei in the midline is extended ventrally between the mediodorsal nuclei to meet the rhomboid and/or central medial nucleus (Fig. 16.6); in this position the fused nucleus may be mistaken for an "intermediodorsal nucleus" (e.g., Gurdjian, 1927) or for part of the intralaminar system (e.g., Olszewski, 1952).

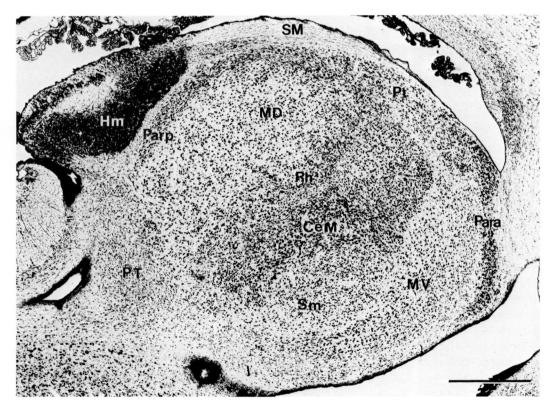

FIGURE 16.1. Thionin-stained sagittal section through the anterior (Para) and posterior (Parp) paraventricular nuclei and the medial (Hm) habenular nuclei of a cat. SM: stria medullaris. Bar = 1 mm.

16.2.2. Terminology

The term *paraventricular nucleus* seems to have been introduced by Gurdjian (1927) in his description of the rat thalamus and was taken up by Papez (1932), Rioch (1929a), and Le Gros Clark (1930a) in their descriptions of the thalamus of the armadillo, of the cat and dog, and of *Tarsius*. One or two pairs of nuclei had been recognized under other names by earlier authors. Nissl (1889) at first included them in the "nuclei of the midline." They correspond to the noyau du raphé used in the rabbit by Ramón y Cajal (1911), to the nucleus dorsalis raphe of Le Gros Clark (1929a) in *Tupaia*, to the medial gray nuclei of Vogt (1909) and Friedemann (1911) in monkeys, to the nucleus paraependymalis of Nissl (1913), nucleus réunissant no. 1 of D'Hollander (1913) and nucleus commissurae mollis dorsalis of Rose (1935) in the rabbit, and to the central gray nuclei of Pines (1927) in a lemur. In man, Hassler (1959) calls them nucleus paramedianus. The nuclei have been recognized under one or other of these names in virtually all mammals, including the higher primates (Crouch, 1934; Walker, 1938a,c;

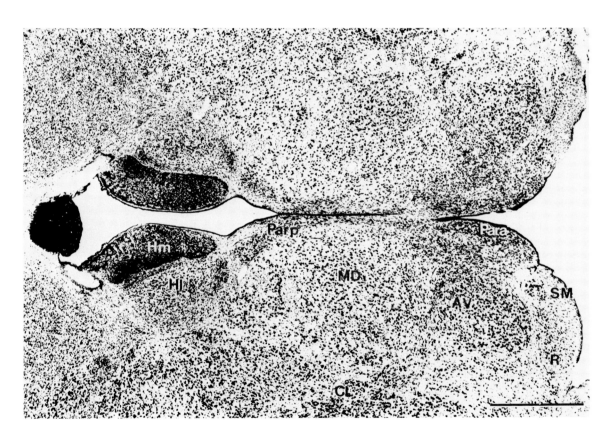

FIGURE 16.2.   Thionin-stained horizontal section through the paraventricular and habenular nuclei of a cat. P: pineal body. Bar = 1 mm.

Sheps, 1945; Krieg, 1948; Olszewski, 1952; Dekaban, 1954; Emmers and Akert, 1963). Fortunately, *paraventricular nuclei* now seems the accepted name. The nuclei are still often loosely regarded as thalamic nuclei of the midline though Nissl (1913) had shown clearly that they do not undergo retrograde degeneration following telencephalic lesions and Rose (1942a,b) showed conclusively that they are developmentally part of the epithalamus and have affinities with the hypothalamus and periaqueductal gray matter. These affinities seem to be reflected in their continuity with both and in the continuous plexus of neuropeptide Y-containing fibers that all three contain (cf. Fig. 16.7 with Figs. 5.16–5.18).

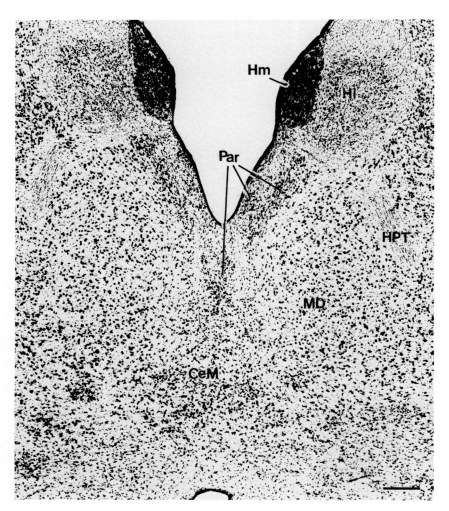

FIGURE 16.3. Thionin-stained frontal section showing the paraventricular nuclei (Par) as they descend toward the periaqueductal gray matter at the posterior margin of the cat thalamus. Bar = 250 μm.

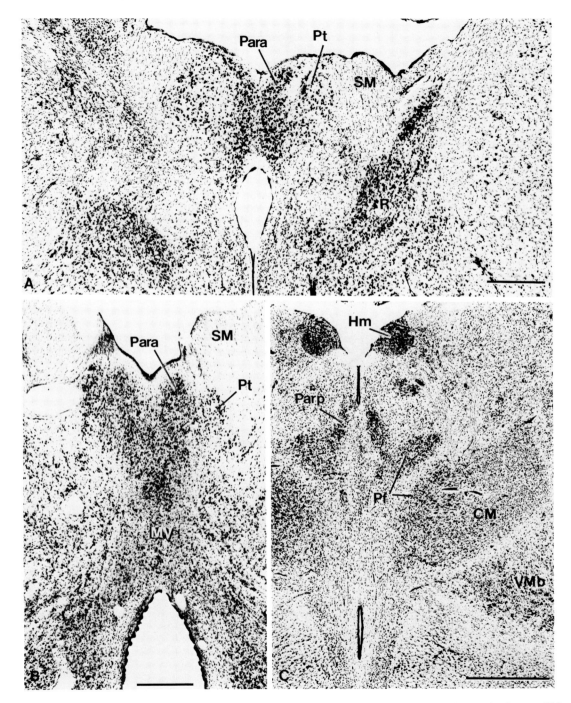

FIGURE 16.4.   Thionin-stained frontal sections through the anterior parts of the paraventricular nuclei of a rat (A) and a cynomolgus monkey (B), and through their posterior parts in a bush baby (*Galago crassicaudatus*) (C). Bars = 500 μm (A, B), 1 mm (C).

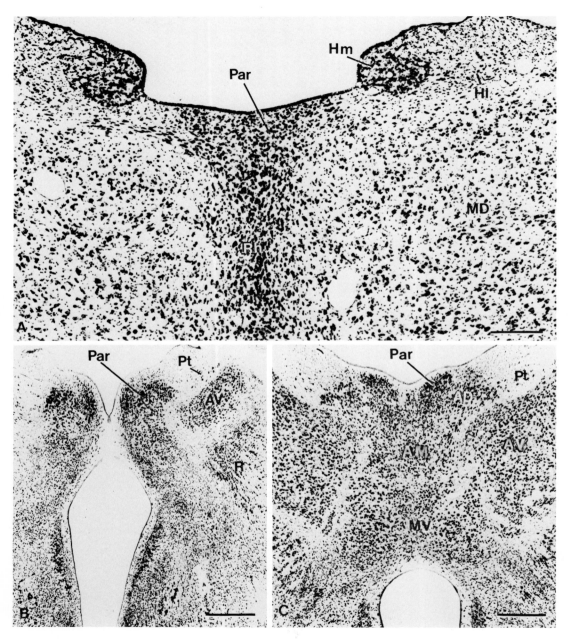

FIGURE 16.5.   Fusion of the paraventricular nuclei in an opossum (A) and a marsupial phalanger (*Trichosurus vulpecula*) (B, C). Frontal sections, thionin stain. Bars = 250 μm (A), 1 mm (B, C).

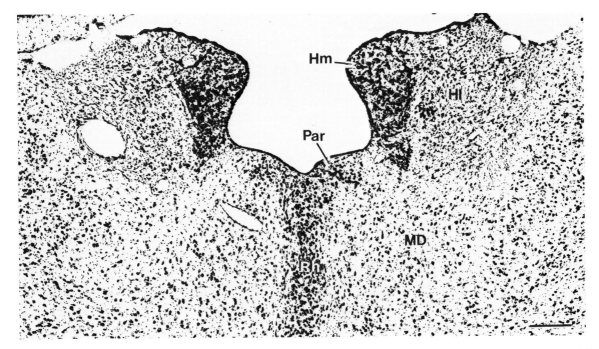

FIGURE 16.6.   Continuity of the paraventricular and rhomboid nuclei in a guinea pig. Frontal section, thionin stain. Bar = 250 μm.

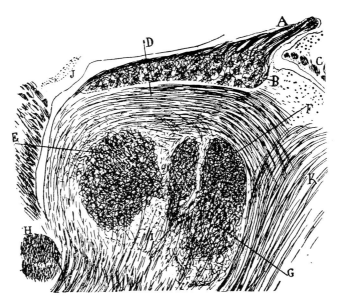

FIGURE 16.7.   Drawing by Ramón y Cajal (1911) of a sagittal section through the epithalamus of a mouse with anterior to the left, showing the medial habenular nucleus (B), habenular commissure (A), and the curving fibers (D) associated with the paraventricular nuclei (cf. Figs. 5.16–5.18). (E, F, G). Fibers associated with midline portions of anterior, rhomboid, and intralaminar nuclei; (H) anterior commissure; (K) periaqueductal fibers.

16.2.3. Connections

According to Walker (1938a) and Le Gros Clark and Boggon (1935), who confirmed that the paraventricular nuclei show no degenerative changes after cortical lesions, these nuclei are connected with certain cell groupings of the hypothalamus. These remarks, however, appear to have been based upon older studies made in normal fiber-stained material (Fig. 16.7). Though there seem to have been no connectional studies directed solely at the nuclei, indications of their connectional relationships have emerged in the course of studies of other structures.

Recent work indicates that both nuclei are connected with the preoptic and anterior or lateral hypothalamic areas (Powell and Cowan, 1954; Cowan and Powell, 1955; Swanson, 1976; Saper *et al.,* 1979a,b) and probably with the hippocampal formation (Swanson and Cowan, 1977; Wyss *et al.,* 1979) and amygdala (Ottersen and Ben-Ari, 1979).

Both nuclei show moderately dense staining for acetylcholinesterase (Chapter 5; see also Graybiel and Berson, 1980b) and receive moderate numbers of noradrenergic (Swanson and Hartman, 1975) and enkephalin-containing fibers (Sar *et al.,* 1978) as well as smaller numbers of β-endorphin-containing fibers of unknown origin (Bloom *et al.,* 1978). There is a rich concentration of neuropeptide Y-containing fibers and a less dense plexus of somatostatin-containing fibers (Figs. 5.16–5.18). The origins of these are not clear but they seem to extend up from the hypothalamus, especially from the vicinity of the paraventricular and suprachiasmatic nuclei and the median eminence (Fig. 5.16).

16.3. The Habenular Nuclei

16.3.1. Description

The medial and lateral habenular nuclei lie just beneath the ependyma of the third ventricle, usually on the posterodorsal aspect of the thalamus, though in small mammals they may extend virtually to the anterior end of the thalamus (Figs. 16.8–16.10). Each nucleus is usually wedge-shaped with a posterior base close to the pretectum, and tapers along the stria medullaris to a narrow apex on the dorsal surface of the mediodorsal and/or parataenial nuclei. The stria medullaris covers the dorsolateral surface of the habenular nuclei and the habenulopeduncular tract emerges posteroventrally.

The *medial habenular nucleus* usually consists of relatively small, deeply staining cells which are closely packed. The nucleus customarily lies ventromedial to the stria medullaris and just beneath the ependyma. The medial habenular nucleus is often a little more extensive, anteroposteriorly, than the lateral nucleus and its posterior part commonly extends along the stria medullaris to the point at which this crosses the midline as the habenular commissure and some of the cells of the nucleus can extend along the commissure close to the pineal body.

The *lateral habenular nucleus* is commonly larger than the medial nucleus, and consists of paler-staining, much more dispersed cells. In some animals islands of cells suggest an incipient architectonic subdivision which becomes clearer in primates (Fig. 16.10). The nucleus lies immediately ventral to the stria medullaris and is separated from the lateral nuclear complex and the mediodorsal nucleus by an encircling fiber lamina formed mainly by fibers entering the habenulopeduncular tract.

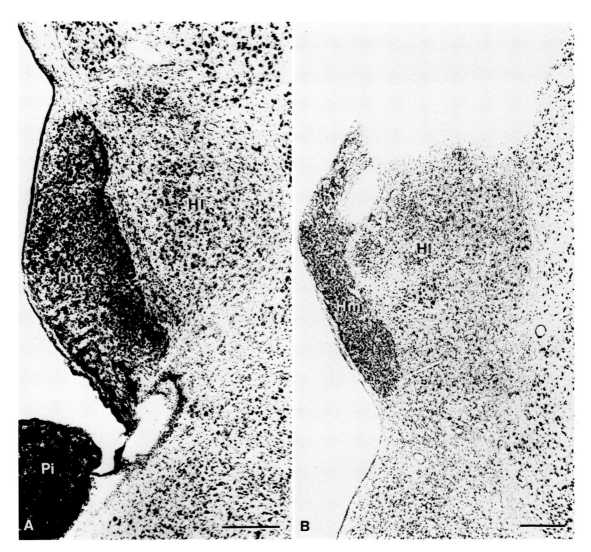

FIGURE 16.8.   Thionin-stained sections showing the habenular nuclei of a cat (A, horizontal plane) and a harbor seal (B, frontal plane). Pi: pineal body. Bars = 250 μm (A), 500 μm (B).

16.3.2. Terminology

The mammalian habenular region is usually said to have been observed for the first time and named by Meynert (1872) who described a small mass of gray matter at the posterior end of the stria medullaris in the human brain. According to Ramón y Cajal (1911), priority should be given to Serres and Stieda who, he said, had seen it some years earlier. Meynert also observed the efferent bundle, the habenulopeduncular, tract or "fasciculus retroflexus," which Forel (1877) later named after him.

The medial and lateral nuclei of the habenula were identified and named by Nissl (1889) in his first cytoarchitectonic description of the rabbit thalamus and their appearances as seen in Golgi preparations from a variety of animals were soon after illustrated by van Gehuchten (1897) and Ramón y Cajal (1911). A few authors for a time referred to the two nuclei simply as *ganglion habenulae* (von Kölliker, 1896; Münzer and Wiener, 1902), sometimes including the para-

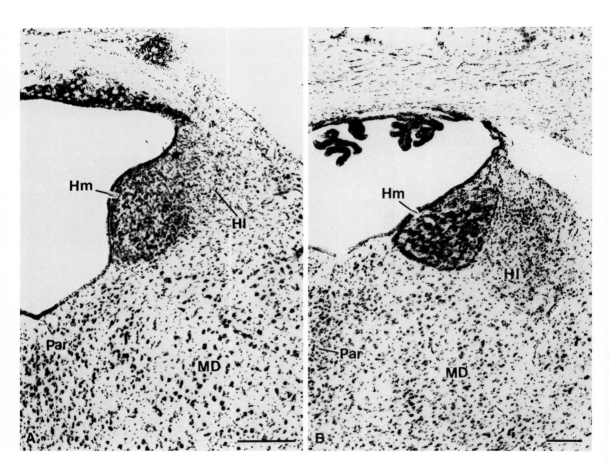

FIGURE 16.9.   Thionin-stained frontal sections showing the habenular nuclei of an opossum (A) and a moustache bat (B). Bars = 250 μm (A), 100 μm (B).

ventricular nuclei in this as well (von Monakow, 1895). However, the recognition of distinct medial and lateral nuclei was quickly established.

Lateral and medial habenular nuclei are not only found in all mammals but are said to be found in virtually all vertebrates (Kappers *et al.*, 1936). I know of no alternative nomenclatures for them. The nuclei are relatively much larger in nonmammalian vertebrates than in mammals, but, nonetheless, undergo some cytological differentiation in primates including man (e.g., Marburg, 1944; Olszewski, 1952; Hassler, 1959) so that in the lateral habenular nucleus, lateral, magnocellular, and medial, parvocellular, subnuclei are recognized. The lateral nucleus has sometimes been further subdivided in nonprimates also [e.g. Rose (1935) in the rabbit]; the grounds for doing this seem rather tenuous, though it is true that some cytoarchitectonic diversity can often be detected and there is some evidence for connectional and neurochemical differences, especially between medial and lateral parts of the lateral nucleus (see Sections 16.3.4 and 16.3.5). The densely packed medial habenular nucleus seems invariant.

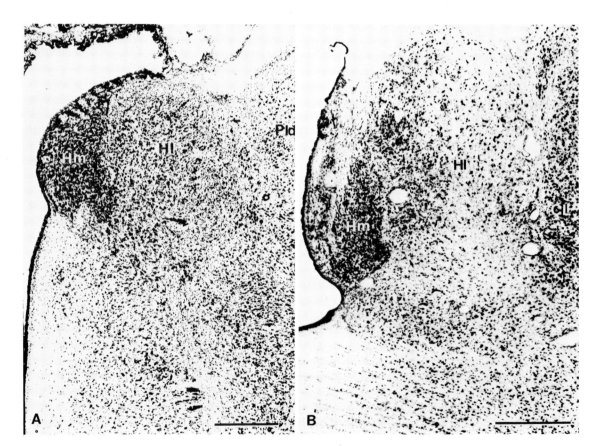

FIGURE 16.10.   Habenular nuclei of a bush baby (A) and a cynomolgus monkey (B). Note cytoarchitectonic divisions in lateral habenular nucleus. Frontal sections; (A, B) thionin stain; (C) hematoxylin stain. Bars = 500 μm (A), 1 mm (B).

16.3.3. Internal Structure

In Golgi preparations Ramón y Cajal (1904, 1911) noted the presence of two populations of nerve cells in the lateral habenular nucleus (Fig. 16.11). A population of large cells sent their axons into the habenulopeduncular tract; a second consisted of small cells that seemed to represent local interneurons (see also Iwahori, 1977). The medial habenular nucleus, on the other hand, appeared to consist of one type of bushy cell only (Fig. 16.11A), all with axons entering the habenulopeduncular tract. Ramón y Cajal was doubtful that any habenular cells sent axons across the midline in the habenular commissure. He regarded the commissure as being derived exclusively from the stria medullaris and this point of view seems to have been borne out by subsequent work. Ramón y Cajal also established that afferent fibers entering the medial habenular nucleus from the stria medullaris end in single, dense, burstlike terminations (Fig. 16.11B). By contrast, in the lateral nucleus, axons of the stria medullaris run anteroposteriorly across the surface of the nucleus, giving off sparsely branched collaterals that at intervals descend into the nucleus (Fig. 16.12). These observations, recently confirmed by Iwahori (1977), suggest a fundamental difference between the organization of afferent inputs to the two nuclei—the one punctate, the other in the form of sagittal slabs.

At the electron microscopic level, the medial habenular nucleus, at least, is

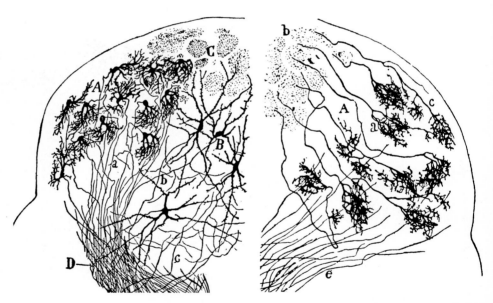

FIGURE 16.11.   (Left) Ramón y Cajal's (1911) drawing of Golgi-impregnated neurons in the medial (A) and lateral (B) habenular nuclei of a dog. C: stria medullaris; D: habenulopeduncular tract. Frontal section. (Right) Ramón y Cajal's drawing of Golgi-impregnated axons entering the medial habenular nucleus (A) of a rabbit from the stria medullaris (b) and terminating in bushy arborizations (a, c); e: habenulopeduncular tract. Transverse section.

unlike the dorsal thalamic nuclei. I can identify no accounts of the fine structure of the lateral habenular nucleus. Tokunaga and Otahi (1978) describe the single neuron type found in the medial nucleus as containing many large, dense-core vesicles with few axosomatic synapses. The neuropil contains axon terminals with clear round vesicles, making asymmetric contacts, others with pleomorphic vesicles making symmetric contacts (both on dendrites), as well as two kinds of large, dense-core vesicle-containing terminals whose synaptic contacts, if any, were not identified. Some of the latter may be neurosecretory, according to the authors. The origins of the parent axons of these four sets of terminals have not so far been identified but from Section 16.3.5, we may suspect that some contain GABA, some acetylcholine, and some substance P.

Despite the close association of the medial habenular nucleus with the third ventricle, I can find no descriptions of structural relationships that would suggest a resemblance to the circumventricular organs such as the area postrema, subcommissural organ, and subfornical organ in which the blood–brain barrier is lacking and over which the ependyma can be denuded.

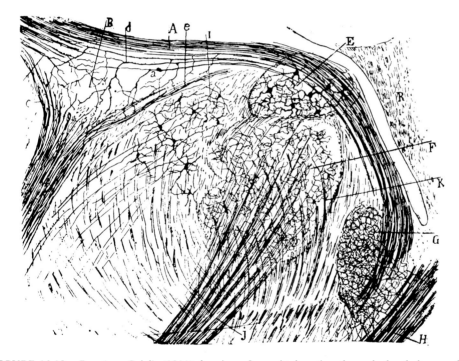

FIGURE 16.12. Ramón y Cajal's (1911) drawing of a sagittal section through the thalamus of a mouse. In the lateral habenular nucleus (B), note the longitudinally running axons (d) of the stria medullaris (A), giving off repeated collaterals which descend into the nucleus. D: habenulopeduncular tract; E: anterodorsal nucleus; F, K: anteromedial nucleus; G: probably part of dorsal hypothalamus or paraventricular nucleus of hypothalamus; H: fornix; I: mediodorsal nucleus; J: mamillothalamic tract.

16.3.4. Connections

16.3.4.1. Afferent Connections

Much of the early work on the afferent connections of habenular nuclei was carried out on reptiles (C. L. Herrick, 1891, 1892; Meyer, 1892; Edinger, 1893, 1896a,b; Vogt, 1895, 1898; C. J. Herrick, 1910, 1917; Crosby, 1917; Huber and Crosby, 1926, 1929). These were later complemented by investigations in mammals (von Kölliker, 1896; Johnston, 1923; Gurdjian, 1925, 1928; Loo, 1931) in which it was established that afferent fibers in the stria medullaris and ending in the habenular nuclei probably arose in olfactorily related regions of the basal forebrain. More recent experimental studies have further defined the sources of afferents. The lateral habenular nucleus has a diverse set of inputs. It receives fibers from the prepiriform cortex of both sides (Guillery, 1959; Nauta, 1961; Powell *et al.*, 1965; Cragg, 1961; Droogleever-Fortuyn *et al.*, 1959) and also bilaterally from the internal segment of the globus pallidus and the basal nucleus of Meynert (Nauta and Mehler, 1966; Nauta, 1974, 1979; Kim *et al.*, 1976; Herkenham and Nauta, 1977; Hendry *et al.*, 1979; Larsen and McBride, 1979; McBride, 1981; Van der Kooy and Carter, 1981). In the monkey, pallidal and basal nucleus fibers terminate respectively in the magnocellular and parvocellular parts of the lateral habenular nucleus; in the cat those from the entopeduncular nucleus (internal segment of globus pallidus) end in the ventrolateral part of the nucleus while those from other sources (see below) end in medial parts. A similar lateral termination of pallidal afferents is found in the rat (Carter and Fibiger, 1978; Nagy *et al.*, 1978). It is probable that the pallidal input, which may arise from branches of fibers directed to the dorsal thalamus (Filion and Harnois, 1978), is GABAergic (Gottesfeld *et al.*, 1977, 1981; Nagy *et al.*, 1978). Other afferents to the lateral habenular nucleus come form the preoptic area (McBride, 1981), lateral hypothalamus (Conrad and Pfaff, 1976a,b; Swanson, 1976; Nauta, 1977, 1979; McBride, 1981), ventral tegmental area of the midbrain (Herkenham and Nauta, 1977; Beckstead *et al.*, 1979), and the lateral dorsal tegmental nucleus or pars compacta of the substantia nigra (Herkenham and Nauta, 1977) and midbrain raphe (McBride, 1981). All appear to terminate in different parts of the nucleus, though primarily medially and with some degree of overlap, both with one another and with the terminations of pallidal afferents.

Herkenham and Nauta (1979; Fig. 16.13) have suggested that the lateral habenular nucleus of the rat can be divided into a medial component dominated by "limbic" inputs, i.e., inputs from the preoptic areas, hypothalamus, and septum, and a lateral component dominated by pallidal inputs. It is in this lateral part that the richest concentration of GABAergic terminals is found (Gottesfeld *et al.*, 1981). Both parts, however, receive substantia nigral inputs. The known afferent fibers to the medial habenular nucleus appear to come only from the septofimbrial nucleus (Nauta, 1956; Raisman, 1966; Herkenham and Nauta, 1977).

Most afferents to the habenular nuclei arrive via the stria medullaris. Those ending bilaterally cross in the habenular commissure. Some fibers of the stria medullaris may traverse the habenular nuclei without terminating and continue toward the midbrain in the habenulopeduncular tract (Herkenham and Nauta,

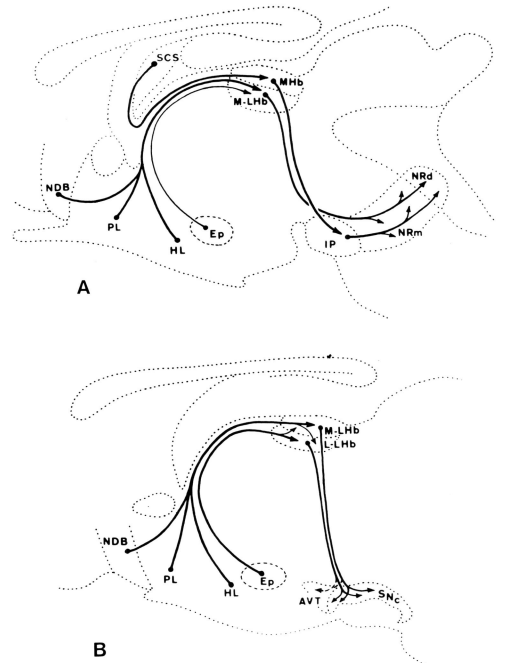

FIGURE 16.13. Schemata of Herkenham and Nauta (1979) showing: (A) relationships of the habenular nuclei with the interpeduncular (IP) and raphe (NRd, NRm) nuclei; (B) relationships of the habenular nuclei with the ventral tegmental area (AVT) and pars compacta of the substantia nigra (SNc). Ep: entopeduncular nucleus; HL: lateral hypothalamus; L-LHb: lateral division of lateral habenular nucleus; MHb: medial habenular nucleus; M-LHb: medial division of lateral habenular nucleus; NDB: nucleus of diagonal band of Broca; PL: lateral preoptic area.

1979). Afferents from the globus pallidus, however, ascend through the ventral medial, intralaminar, and mediodorsal nuclei of the dorsal thalamus (Fig. 16.14).

16.3.4.2. Efferent Connections

The first studies of the efferent connections of the habenular nuclei were made very early. After cutting the habenulopeduncular tract or destroying the interpeduncular nuclei, von Gudden (1881a,b) and van Gehuchten (1894a,b) showed retrograde degeneration in the habenular nuclei and, thus, established the origin of the tract (Fig. 16.15A) from these nuclei.

a. Medial Habenular Nucleus.   Although both habenular nuclei project their axons into the habenulopeduncular tract, recent evidence (Herkenham and Nauta, 1979) suggests that only axons derived from the medial nucleus actually terminate in the interpeduncular nucleus (Fig. 16.13). The fibers of the medial nucleus traverse the center of the tract and end in the interpeduncular nucleus, crossing the midline and recurving through the nucleus in a figure-8 fashion as shown originally by Ramón y Cajal (1911, his Fig. 173). This projection to the interpeduncular nucleus may be the only efferent connection of the medial habenular nucleus. It is possible that it is composed of fibers containing both acetylcholine and substance P (Section 16.3.5). In the anterior part of the interpeduncular nucleus, fibers of the habenulopeduncular tract terminate across

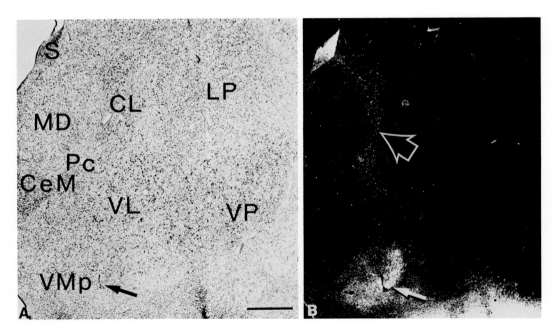

FIGURE 16.14.   Autoradiographically labeled axons (large arrow in B) ascending through the dorsal thalamus, to the stria medullaris (S), en route to the lateral habenular nucleus, following an injection of tritiated amino acids in the ento- peduncular nucleus (lower right) of a cat. Small arrow in (B) indicates axonal terminations in principal ventral medial nucleus (VMp) as a result of spread of injection to substantia nigra. Bar = 1 mm. From Hendry *et al.* (1979).

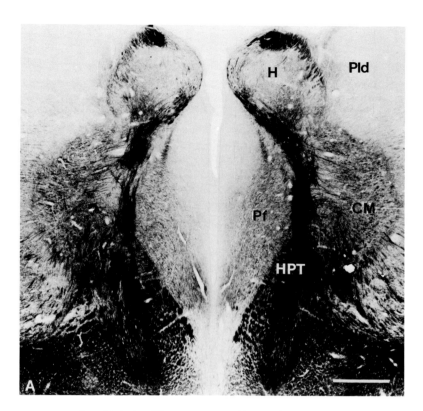

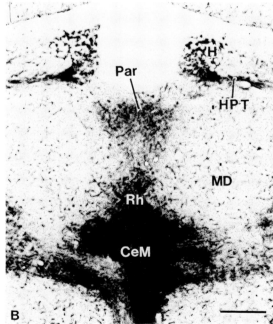

FIGURE 16.15.   (A) Hematoxylin-stained frontal section showing the habenulopeduncular tracts in a bush baby. (B) Acetylthiocholinesterase staining of the paraventricular nuclei and of the commencement of the habenulopeduncular tract in a rat. Bars = 1 mm (A), 250 μm (B).

the full extent of the nucleus but, posteriorly, they end in two zones on either side of the midline (Murray *et al.*, 1979). There, fibers from the tract of each side end on opposing surfaces of crests rising from the surfaces of dendrites (Murray *et al.*, 1979).

b. Lateral Habenular Nucleus. Fibers leaving the lateral habenular nucleus traverse the peripheral part of the habenulopeduncular tract (Herkenham and Nauta, 1979) but leave it in the ventral tegmental area before the tract enters the interpeduncular nucleus (Fig. 16.13). The largest group of fibers passes dorsally and posteriorly and terminates bilaterally in paramedian midbrain regions, especially in the median and dorsal raphe nuclei and in the adjacent tegmental reticular nuclei. Most of these nuclei also receive afferents from the interpeduncular nucleus (Fig. 16.13) (Nauta, 1958) which the efferents of the lateral habenular nuclei bypass. In the raphe nuclei, habenular afferents are known to modulate the activity of the serotonin-producing cells (Aghajanian and Wang, 1977; Wang and Aghajanian, 1977).

A second group of lateral habenular efferents leaving the habenulopeduncular tract, projects bilaterally to the pars compacta of the substantia nigra and adjacent ventral tegmental area, presumably interacting there with dopamine-producing cells; a third passes anteriorly in the medial forebrain bundle and through the adjacent ventral medial nuclei of the dorsal thalamus to terminate bilaterally in lateral, posterior, and dorsomedial hypothalamic regions, and more anteriorly in the lateral preoptic area, in adjacent parts of the septum, and in the substantia innominata.

Herkenham and Nauta (1979; Fig. 16.13) have discovered that lateral and medial parts of the lateral habenular nucleus are relatively independent in terms of some of their efferent as well as their afferent (Secton 16.3.4.1) connections. Most of the efferents to the raphe nuclei arise in the medial part of the nucleus (where hypothalamic and preoptic afferents mainly end) while those to the tegmental reticular nuclei arise in the lateral parts (where pallidal afferents mainly end).

16.3.5. Transmitters and Related Compounds

16.3.5.1. Acetylcholine

The habenulopeduncular projection had been thought for many years to be cholinergic because of its strong staining for acetylcholinesterase (Lewis and Shute, 1967; Parent and Butcher, 1976; Fig. 16.15B), and it has become in recent years the site, par excellence, in the central nervous system in which to study cholinergic transmission (Fibiger, 1982). Cholinergic transmission at the habenulopeduncular synapses seems fairly well supported by neuropharmacological data (Kataoka *et al.*, 1973; Kuhar *et al.*, 1975; Léránth *et al.*, 1975; Sastry, 1978), by immunocytochemical localization of choline acetyltransferase (Houser *et al.*, 1983a) (Fig. 16.16), and by the binding of ligands specific for cholinergic synapses. Binding of ligands to muscarinic receptors occurs at the lateral margins of the nucleus (Rotter *et al.,* 1979) while binding of ligands to probable nicotinic receptors occurs mainly ventrally (Arimatsu *et al.,* 1978; Hunt and Schmidt, 1978). The synapses of habenular fibers with which the receptor binding is presumably associated are asymmetric in type (Lenn, 1976; Murray *et al.,* 1979; Section 16.3.4) but muscarinic receptors at least are not found in the region of crest synapses (Rotter *et al.,* 1979). Acetylcholinesterase-positive neurons are found in both medial and lateral habenular nuclei. Despite some early controversy over the immunocytochemical localization of choline acetyltransferase in cell bodies of habenular neurons contributing to the habenulopeduncular tract (Hattori *et al.,* 1977; Kimura *et al.,* 1981), this has now been confirmed by the use of more specific antisera (see Rossier, 1981) which show choline acetyltransferase-positive cells concentrated in the ventral part of the medial habenular nucleus and stained axons of these cells entering the habenulopeduncular tract (Houser *et al.,* 1983; Fig. 16.16). These latter observations seem to counter the view of Vincent *et al.* (1980) that the cholinergic projection in the habenulopeduncular tract does not arise in the habenular nuclei but is formed by fibers of the stria medullaris passing through the nuclei from the basal forebrain.

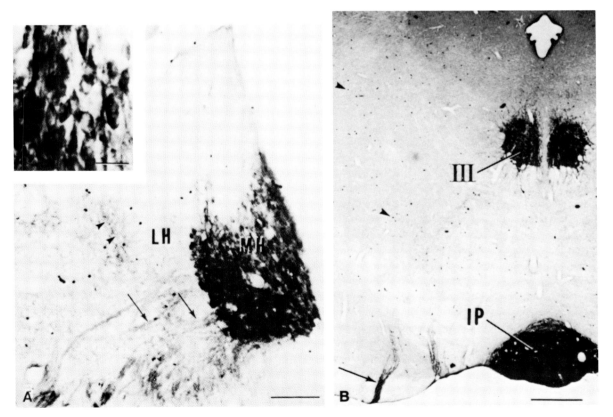

FIGURE 16.16. (A) Immunocytochemical staining of cholinergic nerve cells in the medial habenular nucleus (MH) and of their terminations (B) in the interpeduncular nucleus of a rat. From Houser *et al.* (1983).

16.3.5.2. Substance P

Substance P-containing cell bodies and nerve fibers can be detected immunocytochemically in the medial habenular nucleus (Hökfelt *et al.*, 1975; Cuello *et al.*, 1978); substance P-positive nerve fibers appear in the habenulopeduncular tract and end lateral to the cholinergic fibers in the interpeduncular nucleus (Mroz *et al.*, 1976; Cuello *et al.*, 1978; Hong *et al.*, 1976). There is as yet no evidence for colocalization of substance P and choline acetyltransferase in somata or fibers, and Cuello *et al.* (1978) feel that the two transmitter-related compounds are probably localized in cell populations in different compartments of the medial habenular nucleus.

16.3.5.3. Afferent Transmitters and Receptors

Several transmitters or transmitter-related compounds have been reported in the afferent fibers to the habenula, either by changes in their concentrations in the habenula after lesions affecting various afferent sources, by immunocytochemistry, or by histofluorescence. GABA has been suggested as the transmitter in the pallidal and stria medullaris pathways to the lateral habenular nucleus (Gottesfeld *et al.*, 1977, 1981; Nagy *et al.*, 1978) and dopamine in the projection to the same nucleus from the ventral tegmental area (Kizer *et al.*, 1976). Presumably, that from the pars compacta of the substantia nigra may also be dopaminergic. Acetylcholine may be contained in afferents from the septal areas, substantia innominata, and other basal forebrain sites (Kataoka *et al.*, 1977; Gottesfeld and Jacobowitz, 1979) as well as in the afferents from the lateral dorsal tegmental nucleus which may contain cholinergic cells (Palkovits and Jacobowitz, 1974; Rotter and Jacobowitz, 1981). Serotonin (Kuhar *et al.*, 1972), neurotensin (Uhl *et al.*, 1977), and catecholamine-containing fibers (Lindvall *et al.*, 1974; Swanson and Hartman, 1975) have also been reported. Some hypothalamic afferents are positive for vasopressin and its neurophysin (Buijs, 1978; Sofroniew and Weindl, 1978), and septofimbrial afferents are positive for luteinizing hormone-releasing hormone (Barry, 1978; Silverman and Krey, 1978). Substance P is the only other peptide that seems to have been demonstrated (Section 16.3.5.2).

Among receptors, there is a distinct band of opiate receptors localized at the border between the medial and the lateral habenular nuclei (Pert *et al.*, 1976; Atweh and Kuhar, 1977) and modest accumulations of both muscarinic and nicotinic cholinergic receptors also apper in the lateral habenular nucleus (Rotter *et al.*, 1979; Segal *et al.*, 1978).

16.3.6. Functions

At present there has been insufficient neurophysiological work carried out on single neurons in the habenular nuclei for one to make any meaningful statements about their functions. There is, however, a rather large literature on the effects of lesions of the habenula and its environs. Most studies implicate the habenular nuclei in various kinds of visceral or neuroendocrine functions. These include: control of thyroid (Szentágothai *et al.*, 1968; Ford, 1968) and of gonadal secretion (Motta *et al.*, 1968), regulation of sexual, consummatory, and defense behavior (deGroot, 1965; Donovick *et al.*, 1969; Cooper and Van Hoesen, 1972; Modianos *et al.*, 1974; Reinert, 1964; Rausch and Long, 1974), and of the autonomic responses that accompany them (Kabat, 1936; Cragg, 1961; Lengvari *et al.*, 1970). Presumably all of these are in some way mediated by the various connections that the habenular nuclei maintain with the midbrain and basal forebrain, particularly the hypothalamus and preoptic area. However, no one has yet successfully sorted out the roles of the various connections, especially those with known transmitters, in the different kinds of functions attributed to the nuclei.

One intriguing aspect of the medial habenular nucleus is that, unlike the dorsal thalamus and most other parts of the forebrain, its metabolic activity along with that of its target, the interpeduncular nucleus, does not decline under the influence of anesthetics, provided the stria medullaris is intact (Herkenham, 1981). Herkenham thinks that this may reflect an involvement of the two nuclei in the control of sleep mechanisms via the connections of the interpeduncular nucleus with the midbrain serotonin centers (Section 16.3.4).

I can trace my ancestry back to a
Protoplasmal primordial atomic globule.
Consequently, my family pride is
Something in-conceivable.
　　　　　W. S. Gilbert, *The Mikado*, Act I

Comparative Structure

Bring with you all the nymphes that you can heare
Both of the rivers and the forrests greene,
And of the sea that neighbours to her neare:
All with gay girlands goodly wel beseene.
Edmund Spenser, *Epithalamion*

17

Comparative Anatomy of the Thalamus

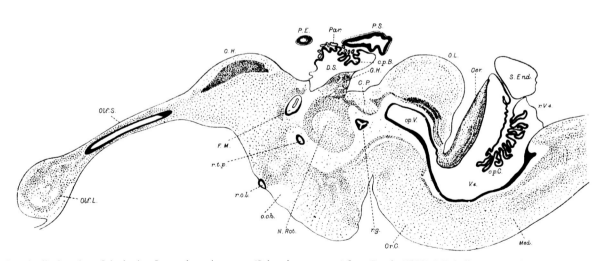

Longitudinal section of the brain of an embryonic tuatara (*Sphenodon punctatum*) from Dendy (1910). P.E. indicates part of the pineal eye, P.S. the pineal sac, G.H. the habenular nuclei, N.Rot. the nucleus rotundus.

17.1. Introduction

No book entitled *The Thalamus* can afford to restrict itself to the mammalian thalamus. One senses, however, that it would be wise to do so, for the field of comparative anatomy of the nonmammalian forebrain is one that has suffered from an absence of adequate experimental material, it has often been characterized by simplistic thinking about evolutionary matters, and it is fraught with controversy. A neuroscientist today can hardly utter the word *homology* without finding a half-dozen other neuroscientists at his throat. Homology in the evolutionary sense is extremely difficult to establish and existing vertebrate forms show such an enormous range of structural and functional adaptations in their forebrains, even within the same order, that many recent comparative neuroanatomists have questioned whether it is relevant to search for common structural relationships. Certainly, no writer in the field today would seriously propose that the brains of existing nonmammalian vertebrates formed a scala naturae leading up to the brains of mammals.

The present chapter, therefore, makes no claims to be dealing with the evolution of the thalamus or any other part of the forebrain. It simply presents perceptions of the general structure of the thalamus in the major orders of nonmammalian vertebrates, gained with the eyes of a novice looking at a limited selection of brains and reading selectively in a rather confusing literature. If one draws parallels between species, orders, or classes and calls things in one species "equivalent to" those in another, he does not intend these to imply homology or a common ancestral origin. But he does not wish to have to qualify statements continually by reference to dubious theories of "homoplasia."

17.2. Background

17.2.1. Telencephalic Development in Vertebrates

Many of the problems associated with the interpretation of patterns of forebrain organization in nonmammals undoubtedly stem from the different developmental history of the telencephalic vesicles (Fig. 17.1). In bony fish the walls of the telencephalon evert and thicken to form large paired masses on either side of a median ventricle (Nieuwenhuys, 1962, 1982; Northcutt and Braford, 1980). In these large masses are situated the apparent equivalents of the cortical and striatal targets of the dorsal thalamus of other vertebrates. In cartilaginous fish and other vertebrates the walls of the telencephalon pouch outwards to form two cerebral hemispheres. In lungfish and amphibians the hemispheres remain relatively unchanged with a pallium (or cortical equivalent) situated dorsomedially and a striatum (or striatal equivalent) situated ventrolaterally. In cartilaginous fish the dorsal part of the pallium on each side thickens considerably at the expense of the lateral ventricles and the thickenings come to form a single, large central nucleus (Figs. 17.1 and 17.5; Northcutt, 1981).

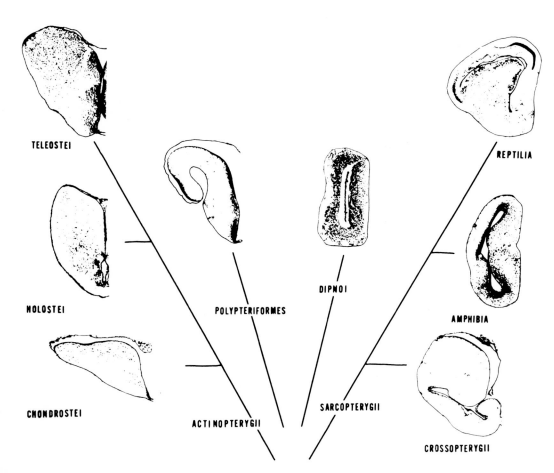

FIGURE 17.1. Variation in the morphological organization of the telencephalon in bony fish, amphibians, and reptiles. Cross-sections of one side shown in each case, medial to right. The ray-finned fish in the left two columns possess everted cerebral hemispheres, seen in its simplest form in the Polypteriformes. There is a single, T-shaped ventricle (cf. Fig. 17.8). The hemispheres of the lungfish (Dipnoi), the single crossopterygian fish, amphibians, and reptiles are paired evaginations, each with a separate ventricle, communicating with a third ventricle in the diencephalon. In the basolateral aspect of the reptilian hemisphere, the dorsal ventricular ridge can be seen protruding into the lateral ventricle. From Northcutt (1981).

In reptiles and birds the lateral wall of the hemisphere enlarges and forms a prominent dorsal ventricular ridge overlying the striatum and protruding into the lateral ventricle (Fig. 17.2). Developmentally, the dorsal ventricular ridge is probably of pallial rather than striatal origin. In birds the dorsal ventricular ridge is enlarged enormously as the neostriatum and as the overlying hyperstriatum which fuses with or replaces most of the pallium and contains several subdivisions. The pallium shows some differentiation into medial, lateral, and dorsal cortical areas in reptiles. In birds it is largely undeveloped except for a thin medial so-called hippocampal and parahippocampal region (Kuhlenbeck, 1938) and an enlarged dorsomedial region, the Wulst (Fig. 17.2) which overlies part of the hyperstriatum.

Mammalian telencephalic development is fundamentally different from that of reptiles and birds: the roof of the telencephalic vesicle forms the cerebral cortex and the floor forms most of the basal ganglia with the probable exception of the globus pallidus, a diencephalic derivative (Rose, 1942a).

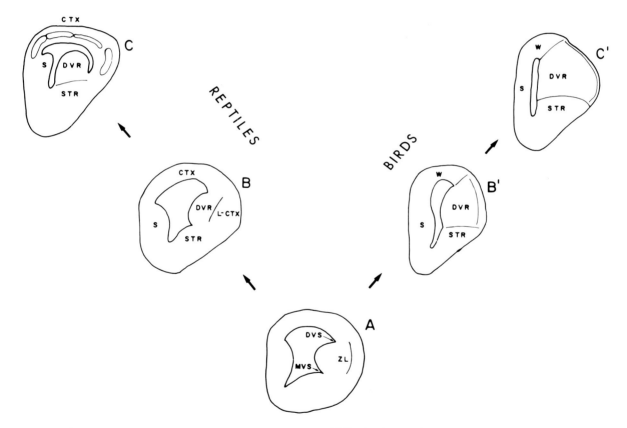

FIGURE 17.2.   Development of the dorsal ventricular ridge (DVR) in reptiles and birds. The single cerebral hemispheres shown are from the side opposite those illustrated in Fig. 17.1; medial to left. Two waves of young neurons migrate from the neuroepithelium between the dorsal (DVS) and the middle (MVS) ventricular sulci. Those migrating lateral to a cell-sparse zona limitans (ZL) contribute to the lateral cortex (L-CTX) of reptiles and the Wulst (W) of birds. Those internal to the zona limitans form the dorsal ventricular ridge. STR: striatum. From Ulinski (1983).

Many similarities are present in the anatomical organization of the dorsal ventricular ridge of reptiles and birds on the one hand and of the neocortex of mammals on the other, as will emerge in the ensuing sections. This has led some workers to postulate a true homology (i.e., a common evolutionary origin) between dorsal ventricular ridge and neocortex (e.g., Karten, 1969; Nauta and Karten, 1970). After thoroughly reviewing all the extant evidence, however, Ulinski (1983) has concluded that although the two display many comparable "design features," it is likely that these design features "have evolved independently in response to common functional demands." This is not a point of view that would have endeared him to past generations of comparative anatomists who sought homology between the brains of mammals and of all nonmammalian vertebrates in an effort to understand the evolutonary history of the mammalian brain. However, it represents a fresh approach that, in cutting through some of the misconceptions of the past, offers the promise of a better understanding of the non-mammalian brain as a functional entity in its own right and not merely a step in some dubious ladder of nature leading to the brain of man.

17.2.2. Some Misconceptions of the Past

It is an old belief, popularized initially by Herrick (1925, 1948) and Kappers *et al.* (1936), that the diencephalon and telencephalon of nonmammals are dominated by secondary and tertiary olfactory connections and that the dorsal thalamus and telencephalon have evolved under the influence of colonization by other afferent systems, particularly the visual. These were thought first to enter the diencephalon in a diffuse or convergent manner so that the early dorsal thalamus was a multimodal one which later in its evolution differentiated into modality-specific nuclei. Efferents from this multimodal thalamus were considered to reach the striatum first in amphibians and then the pallium or rudimentary cortex first in reptiles. In amphibians the whole pallium and in fish virtually the whole telencephalon were thought to receive diffuse olfactory inputs only. The avian telencephalon was held to have evolved with massive further differentiation of the striatum, at the expense of the pallium, whereas the evolution of the mammalian telencephalon was held to involve massive elaboration of the pallium and a more conservative expansion of the striatum (Kappers *et al.*, 1936). There is little reason to question the descriptive basis for claiming that the bird telencephalon structurally resembles something that is likely to have evolved from that of a reptile nor that the telencephalon of a mammal is so fundamentally different that it implies a different evolutionary history. What is questionable, however, is the belief in early olfactory domination of the forebrain and late evolutionary colonization of the telencephalon by nonolfactory, thalamic afferents.

Herrick's case was based upon virtually a lifetime of work on the brain of the tiger salamander, a urodele amphibian (Herrick, 1917, 1948). He had come to the conclusion that the whole dorsal thalamus that he recognized in that creature represented a "nucleus sensitivus" in which virtually all afferent pathways converged. He considered that he could trace into it, in silver-stained normal material, fibers from the optic tract, spinal cord, caudal medulla, tectum, habenula, and hypothalamus (Herrick, 1925, 1933, 1934), though he recognized that the optic system predominated. From the nucleus sensitivus, he felt he could trace efferent axons forwards in the lateral forebrain bundle to the basal parts of the telencephalon but not to the overlying pallium.

Reading the old literature, it is difficult to detect a solid data base for the putative lack of thalamotelencephalic connections in fish nor much evidence for widespread olfactory connections in the forebrain. Similarly, the suggestions of progressive colonization of the telencephalon by fibers of diencephalic origin in reptiles and birds seem to have been based mainly on the assumption that the progressive enlargement and differentiation of the dorsal thalamus and striatum implied this. Perhaps to be fair, it has to be added that in the larger brains of reptiles and birds it was easier to visualize thalamotelencephalic axons in the lateral forebrain bundle directly with the Golgi, reduced silver, and other fiber techniques (e.g., Ramón, 1896; Edinger, 1899, 1908; Hines, 1923; Cairney, 1926; Edinger and Wallenberg, 1899; Craigie, 1928, 1930; Huber and Crosby, 1929). Significant experimental work with the Marchi method or with the retrograde degeneration method came later but it was confined to reptiles (e.g., Goldby, 1937; Powell and Kruger, 1960) and birds (e.g., Powell and Cowan, 1961).

17.2.3. The Modern Era

The advent of the era of the Nauta technique led to a resurgence of interest in the comparative anatomy of the forebrain, for now a technique had become available whereby fiber connections could be more adequately delineated. The earliest studies on reptiles, amphibians, and fish were primarily concerned with visual inputs to the diencephalon (reviewed in Ebbesson, 1972, and Ebbesson *et al.*, 1972). They indicated the presence of significant retinal inputs to regions variously interpreted as ventral or dorsal thalamus even in elasmobranchs and suggested inputs from the spinal cord, tectum, and cerebellum as well. Even in elasmobranchs, a substantial projection was demonstrated from the thalamus to the central nucleus of the telencephalon (Schroeder and Ebbesson, 1974) which is apparently formed as an expanded deep layer of the pallium (Section 17.2.1; Northcutt, 1981). Coincident with this latter was the evidence, mainly from electrophysiological studies, that similar parts of the shark telencephalon could be activated by visual, auditory, and lateral line stimulation (Cohen *et al.*, 1973; Platt *et al.*, 1974; see also Bullock and Corwin, 1979). A thalamotelencephalic connection has now also been confirmed in bony fish (Braford and Northcutt, 1978; Ito and Kishida, 1978; Finger, 1979) but not yet in cyclostomes.

Perhaps the next most significant contribution was that of Scalia and Gregory (1970) and Scalia and Coleman (1975) who showed a dissociation of retinal and tectal afferent terminations in the frog thalamus (Fig. 17.12) and also indicated that the terminal region of retinal fibers projected to the pallium (Scalia, 1976). Subsequent work on the amphibian thalamus has not only confirmed this (Ronan and Northcutt, 1979; Kicliter, 1979) but has indicate more or less segregated terminations of spinal, bulbar, and auditory afferents in the dorsal thalamus (Neary and Wilczynski, 1977a,b; Mudry and Capranica, 1978) and separate thalamic nuclei projecting to pallium and striatum (Northcutt, 1981). These observations indicate the presence of thalamotelencephalic and even thalamo-pallial connections at a much earlier stage of evolution than postulated by Herrick and the presence of a far greater degree of connectional parcellation in the thalamus of amphibians than he had assumed. Furthermore, evidence is mounting to indicate that secondary olfactory connections to the pallium of amphibians and even of teleost fish are not widespread but restricted to certain parts (Scalia *et al.*, 1968; Northcutt and Royce, 1975; Finger, 1975; Northcutt, 1981). Work in many other species, including cyclostomes, has indicated a comparable restriction of certain afferent connections in the thalamus, though evidence for thalamotelencephalic connections has not been forthcoming in every case (Section 17.3).

Many connectional studies were carried out in reptiles in the heyday of the Nauta technique and, like the studies in fish and amphibians, have continued with more recent anatomical techniques. The initial studies dealt primarily with retinal, tectal, and spinal inputs to the thalamus (Ebbesson, 1969; Hall and Ebner, 1970a,b; Ebbesson *et al.*, 1972) and these have been confirmed (Pritz, 1974a,b, 1975; Butler and Northcutt, 1978; Foster and Hall, 1978; Northcutt, 1981); others such as those from the dorsal column nuclei have also been defined (Northcutt and Pritz, 1978; Balaban and Ulinski, 1983). Other early studies demonstrated projections from different thalamic nuclei to specific parts of the reptilian cortex (Hall and Ebner, 1970a,b). This has also been confirmed, with the additional observation of a highly organized set of connections passing to the dorsal ventricular ridge (Johnston, 1915), a structure of pallial origin (Section 17.2.1) that crowns the striatum (Wang and Halpern, 1977; Lohman and van Woerden-Verkley, 1978; Bruce and Butler, 1979; Brauth and Kitt, 1980; Balaban and Ulinski, 1981a,b). It has become evident that part of the reptilian dorsal ventricular ridge is divided into functional zones in many ways comparable to the cortex of mammals and that each receives differential thalamic connections (Reiner *et al.*, 1980; Northcutt, 1981; Ulinski, 1983). Whether there are thalamic connections passing to the underlying striatum as well, however, seems controversial. Northcutt (1981) appears to accept published material as evidence of substantial thalamostriatal connections; Ulinski (1983) is more hesitant.

Some of the best comparative anatomical work carried out initially with the Nauta technique and continuing to this day is that done by Karten and his collaborators on the brain of the pigeon. They have established the differential distribution of tectal, auditory, and somatosensory afferents in the thalamus or in other parts of the basal forebrain (Karten, 1963, 1969; Karten and Revzin, 1966; Benowitz and Karten, 1976; Reiner and Karten, 1978), leading to a thorough understanding of nuclear parcellation in the dorsal thalamus (e.g., Karten and Hodos, 1967), and they have established the differential projections of many of these dorsal thalamic nuclei upon the telencephalon (Karten, 1969; Karten and Hodos, 1970). Their observations have been confirmed and other details added, particularly in regard to the visual and auditory pathways through the thalamus, by Repérant (1973), Repérant *et al.* (1974), Hunt and Künzle (1976), Clarke and Whitteridge (1976), Nottebohm *et al.* (1976, 1982), Kelley and Nottebohm (1979), Bonke *et al.* (1979a), and Pettigrew (1979). Other studies have also established the presence of significant brain-stem pathways to the telencephalon that bypass the thalamus and are probably comparable to the monoamine-containing systems of the mammalian brain (Bertler *et al.*, 1964; Karten and Dubbeldam, 1973; Tohyama *et al.*, 1974; Kitt and Brauth, 1980; Miceli *et al.*, 1980). These pathways are also present in reptiles (see Ulinski, 1981, 1983).

Because there is more comprehensive anatomical information available for the avian forebrain than for that of other nonmammals, it is not surprising that there have been a large number of physiological studies devoted to some aspect of sensory function in the thalamus and telencephalon. These are reviewed in depth by Ulinski (1983).

17.3. Descriptive Anatomy of the Thalamus in Selected Nonmammalian Vertebrates

17.3.1. Introduction

Looking at the diencephalon of a series of fish, amphibians, reptiles, and birds, a basic pattern can be discerned. There is usually a fairly distinct preoptic area with a thick periventricular cell layer, and with more dispersed cells laterally and anteriorly penetrated by medial and lateral forebrain bundles, as these pass toward the telencephalon (Figs. 17.5, 17.8, 17.9, 17.11–17.15). The periventricular layer is usually thickened over the optic chiasm to form a distinct suprachiasmatic nucleus. At about this level the dorsal, middle, and ventral diencephalic sulci of Herrick (Chapter 6) should be visible in many species other than birds, and the habenular nuclei are seen dorsally. Throughout the length of the diencephalon in all except birds and a few reptiles, dorsal thalamus, ventral thalamus, and hypothalamus are represented by a thick periventricular layer of cells, interrupted at the sulci of Herrick and laminated dorsoventrally to varying degrees. In cyclostomes and many urodele amphibians, virtually all neurons are in the periventricular layer (Figs. 17.3 and 17.11). In other groups of nonmammals there are varying degrees of migration of cells into the surrounding neuropil. In most cartilaginous fish, there is relatively little nuclear differentiation in the cells outside the periventricular layer. In teleosts and anuran amphibians, a few distinct nuclei are seen among the looser packed cells; more nuclei become visible in most reptiles (Figs. 17.9, 17.12, and 17.13), while in some reptiles and all birds the periventricular layer has virtually disappeared (Fig. 17.14), the surrounding neuropil has become tightly packed with neurons, and many nuclei, some with further cytoarchitectonic divisions, are distinguishable among them.

There is an obvious parallel between this apparent pattern of progressive migration from the periventricular layer and differentiation in the surrounding neuropil, and the ontogenetic development of the mammalian thalamus (Chapter 6). One has to recognize, however, that no adult mammalian thalamus resembles the thalamus of a bird, reptile, amphibian, or fish. It is true that the dorsal thalamus of monotremes (Fig. 17.19) shows very little nuclear differentiation and the dorsal thalami of many insectivores and some marsupials show much less differentiation than the thalami of other mammals. But even the least differentiated mammalian thalamus bears so little resemblance to that of any living nonmammal that it is virtually impossible to detect among the nonmammals features that would suggest the evolutionary history of the mammalian thalamus.

17.3.2. Cyclostomes

The diencephalon of cyclostomes is a relatively simple tubular structure with a moderately thick periventricular cell layer in which only left and right habenular nuclei are clearly distinct but with incipient division of a dorsal and a ventral thalamus and hypothalamus by middle and ventral diencephalic sulci and corresponding cell-sparse zones (Jansen, 1930; Saito, 1930; Heier, 1948; Kuhlenbeck, 1956; Schober, 1964; Fig. 17.3). The habenular nuclei are fused in the midline in the hagfish and the right is larger than the left in lampreys (Jansen, 1930; Kappers *et al.*, 1936; Kemali and Miralto, 1979).

There is a relatively undifferentiated preoptic area, pierced by the medial and lateral forebrain bundles (Fig. 17.3). Posterior to this comes the diencephalon proper with its periventricular layer and a surrounding thick neuropil layer containing some nerve cells that appear to have migrated from the periventricular layer. Through the neuropil layer runs the efferent tract of the habenular nuclei and a small optic tract (Holmgren, 1919; Kuhlenbeck, 1927).

No cellular masses have been given separate names in the thalamus although Schober (1964) refers to the neuropil lateral to the dorsal thalamus as *lateral geniculate nucleus*. In that region, fibers from the contralateral retina terminate in larval lampreys and fibers from both retinae terminate there in adult lampreys (Kennedy and Rubinson, 1977). In normal material, Heier (1948) felt he could

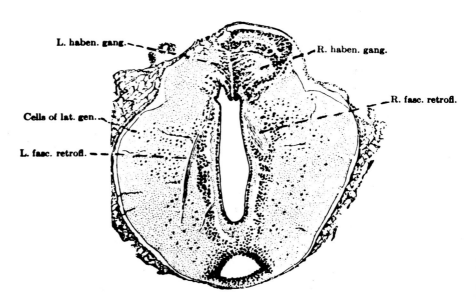

FIGURE 17.3. Cross-section of the diencephalon of a cyclostome, the lamprey (*Petromyzon marinus*), showing the periventricular region and migrated cells. Right habenular nucleus is larger than left. From Kappers *et al.* (1936).

trace fibers forwards from the dorsal thalamus into the small telencephalic vesicles, suggesting the presence of thalamotelencephalic connections, although these have not yet been confirmed experimentally.

17.3.3. Cartilaginous Fish

The diencephalon of sharks and rays is not tubular like that of cyclostomes or urodele amphibians but a rather complicated structure overlain by the optic tectum and containing several well-defined nuclei (Fig. 17.4) (Bergquist, 1932; Gerlach, 1947; Northcutt, 1979).

Anterior to the diencephalon is an extremely elongated preoptic area forming a stalk to the telencephalon which overlies it (Fig. 17.4). The stalk has a monotonous architecture of small, diffusely scattered cells (Fig. 17.5). It is penetrated by the lateral and medial forebrain bundles and down each of its sides runs a thick fiber tract from the lateral telencephalic wall. This pallial tract serves to bury the optic tracts within the diencephalon at levels posterior to the optic chiasm (Fig. 17.5). The descending pallial tracts themselves become buried in the diencephalon caudally and thus tend to divide the expanded ventrolateral part of the hypothalamus (the lobus inferior) off from the rest of the diencephalon (Fig. 17.5).

The rest of the diencephalon shows the commencement of the saccus vasculosus ventrally and large habenular nuclei dorsally. There is a single habenular nucleus on each side, consisting of convoluted sheets of cells and showing various degrees of fusion across the midline and various degrees of asymmetry. There is a moderately thick periventricular layer of cells ventral to the habenula. This is thicker in squalomorph sharks (Northcutt, 1979). Outside the periventricular layer, the diencephalon is dominated by the thick fiber tracts that run through it and by diffusely scattered, mostly small cells. A clear distinction between dorsal and ventral thalamus is not evident either in the periventricular layer or in the outer cells but is recognized by some authors (e.g., Smeets, 1981a,b). Beneath the anterior end of the optic tectum, a number of more or less clearly demarcated nuclear aggregations are seen and extend anteriorly along the medial and lateral surfaces of the optic tract for much of the length of the diencephalon. The largest of these is round in cross-section (Fig. 17.6) and has been called lateral geniculate nucleus by several authors (e.g., Smeets, 1981a,b).

Ebbesson et al. (1972) first mapped the distribution of retinal, tectal, spinal, and cerebellar fibers in the diencephalon of sharks and reported that virtually all the nuclei medial to the optic tract received retinal inputs, that retinal and tectal inputs converged on the large so-called lateral geniculate nucleus (Fig. 17.6), and that spinal and cerebellar fibers ended in small numbers more medially, mostly in the periventricular region. Smeets (1981a) and Northcutt (1979) have reinvestigated the retinal inputs and have identified separate foci of retinal terminations in regions they call dorsal and ventral thalamus as well as in the large, so-called lateral geniculate nucleus. Smeets (1981b) has confirmed the overlap of retinal and tectal terminations in this and has shown tectal projections to an anteromedial part of the periventricular region. Because of the separate foci of retinal terminations in what may be dorsal and ventral thalamus and the presence of a focus of tectal terminations in what may be a part of dorsal thalamus, one wonders whether the so-called lateral geniculate nucleus might not be best regarded as a highly differentiated part of the pretectum.

There is now good evidence to indicate that some or all of the thalamic regions mentioned send fibers to the telencephalon (Schroeder and Ebbesson, 1974; Northcutt and Wathey, 1979). Such fibers end in the large central nucleus of the telencephalon which is now thought to be an expanded part of the pallium (Section 17.2.1) and, thus, comparable to a cerebral cortex. In the central nucleus, electrophysiological studies have provided evidence of substantial visual, auditory, and lateral line projections that are presumably mediated by the thalamotelencephalic connections (Cohen *et al.*, 1973; Platt *et al.*, 1974; Bullock and Corwin, 1979).

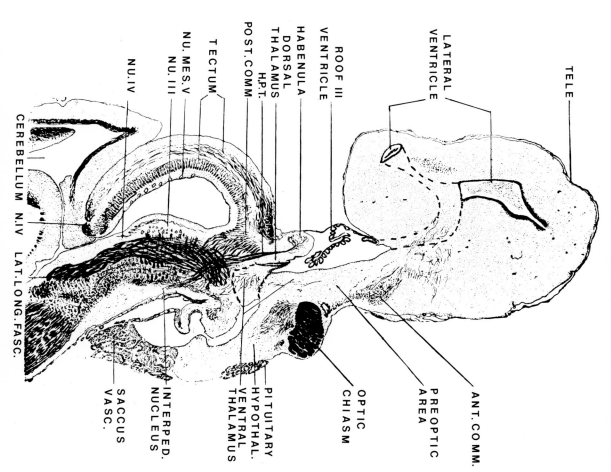

FIGURE 17.4.   Longitudinal section of the forebrain and midbrain of a dogfish (*Scyllium canicula*). Forebrain stalk indicates preoptic region. Dorsal thalamus, ventral thalamus, and habenular region are indicated. From Kappers *et al.* (1936) with labeling altered.

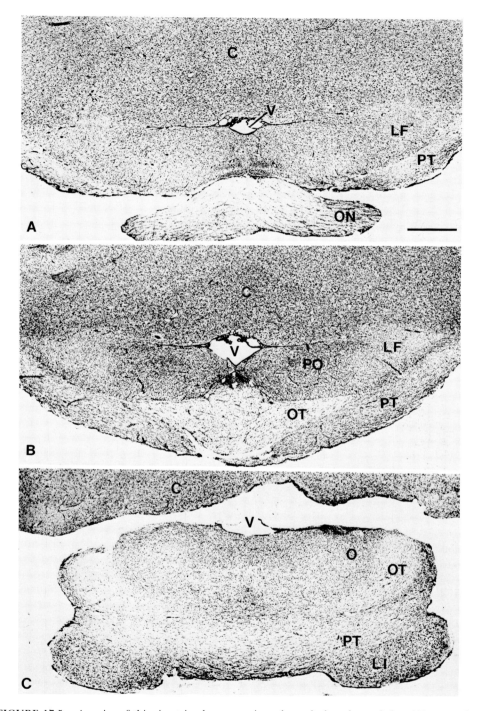

FIGURE 17.5.   A series of thionin-stained cross-sections through the telencephalon (A), preoptic area (B), and diencephalon (C–F) of an Atlantic stingray (*Dasyatus sativa*). From a brain kindly provided by Dr. R. B. Leonard. C: central nucleus of telencephalon; H: habenular nuclei; LF: lateral forebrain bundle; LI: lobus inferius; O: optic nuclei; ON: optic nerve; OT: optic tract; PO: preoptic area; PT: pallial tract; SV: saccus vasculosus; Te: tectum; V: ventricle. Bars = 1 mm.

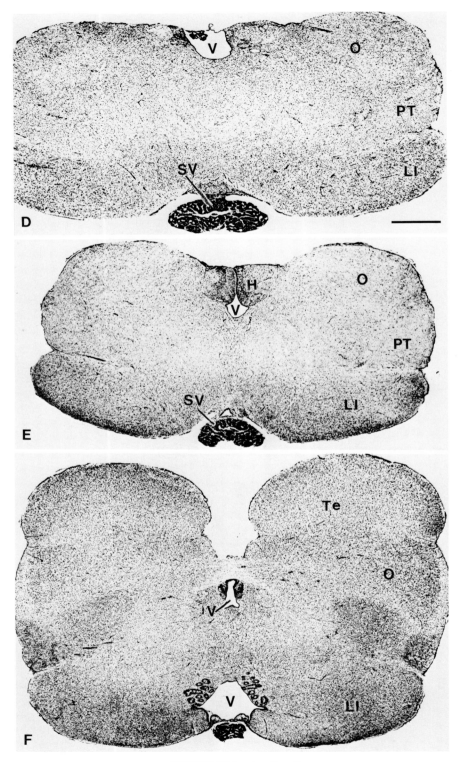

FIGURE 17.5. (*continued*)

17.3.4. Bony Fish

Bony fish show considerable diversity of structure in their diencephalons but, on the whole, the diencephalon is tubular, and resembles that of urodele amphibians more than it does that of cartilaginous fish.

The lungfish (Northcutt, 1977) perhaps have the simplest and most tubular diencephalon, for it is not overlain by the telencephalon or by a large optic tectum. Most neurons are contained within a thick periventricular layer and the

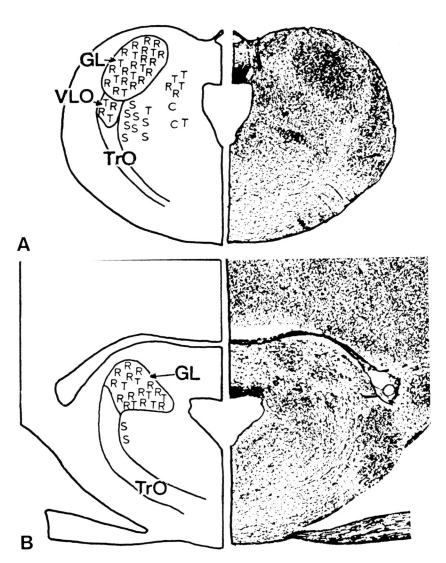

FIGURE 17.6. Distribution of retinal (R), tectal (T), spinal (S), and cerebellar (C) fibers in the diencephalon of a nurse shark. GL: "lateral geniculate" nucleus; TrO: optic tract; LO: ventral optic nucleus. From Ebbesson *et al.* (1972). VLO and GL are represented by O in Fig. 17.5.

surrounding neuropil contains only scattered neurons. In the periventricular layer, only the epithalamus seems to show any degree of cytological differentiation. It consists of left and right habenular nuclei of which one may be larger than the other. On the ventricular wall, dorsal, middle, and ventral diencephalic sulci occur which separate off the epithalamus and, incompletely, dorsal thalamus, ventral thalamus, and hypothalamus. Foci of retinal fiber terminations occur in the neuropil opposite parts of both dorsal and ventral thalamus contralaterally (Northcutt, 1977; Fig. 17.7). This pattern is remarkably similar to that observed in urodeles (Section 17.3.5) and may be similar to the two foci of terminations described in cartilaginous fish (Secton 17.3.3).

Most teleost fish and the Holostei also have a relatively simple diencephalon dominated by large habenular nuclei and a periventricular cell layer, but with varying degrees of migration of neurons into the surrounding neuropil (Miller, 1940; Schnitzlein, 1962; Peter and Gill, 1975). The periventricular layer can be remarkably thin in some species such as the goldfish (Fig. 17.8). In others, small aggregations of neurons can be detected in the neuropil, particularly in relation to the optic tract and habenulopeduncular tract (Fig. 17.9) but in other regions as well. It is difficult, however, to decide whether these nuclei should be regarded as parts of the dorsal thalamus or as parts of the pretectum. They have been given different names by their various describers. The largest group of cells divorced from the periventricular layer is in the inferior lobe, which is part of

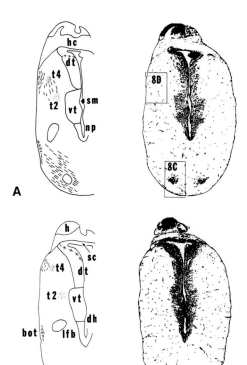

FIGURE 17.7. Distribution of retinal fibers (dashes and dots) in the diencephalon of a lungfish. dt: Dorsal thalamus; vt: ventral thalamus; hc: habenular commissure; np: preoptic nucleus; sm: middle diencephalic sulcus; sc: subcommissural organ; dh: dorsal hypothalamus; lfb: lateral forebrain bundle; bot: basal optic tract; oc: optic chiasma; t2, t4: two of four foci of terminations. From Northcutt (1977).

the hypothalamus. Retinal and tectal fibers have been found to terminate contralaterally and to a lesser extent ipsilaterally among many of the dorsally placed nuclei and in the neuropil opposite the dorsal and ventral thalamic components of the periventricular region (Campbell and Ebbesson, 1969; Ebbesson, 1968; Ebbesson *et al.*, 1972; Sharma, 1972; Vanegas and Ebbesson, 1976; Anders and Hibbard, 1974; Northcutt and Butler, 1976; Repérant *et al.*, 1976; Vanegas and Ito, 1983; Fernald, 1982) (Fig. 17.10). Some of the same or adjacent regions appear to receive tectal, lateral line, auditory, and spinal fibers (Ebbesson, 1970; Ebbesson and Vanegas, 1976; Vanegas and Ebbesson, 1976; Ito and Kishida, 1978; Finger, 1979; Grove and Sharma, 1981; Luiten, 1981; Northcutt, 1981) and the region as a whole, or parts of it, project to dorsal and lateral areas of the telencephalon (Pritz and Northcutt, 1977; Braford and Northcutt, 1978; Ebbesson, 1980; Ito *et al.*, 1980). Some of the diencephalic regions also project to the tectum (Grove and Sharma, 1981; Luiten, 1981; Ito and Kishida, 1977) and cerebellum (Finger, 1978). Perhaps these should be considered ventral thalamic equivalents, by analogy with mammals. For an exhaustive review of the visual associations of the teleost brain, see Vanegas and Ito (1983).

17.3.5. Amphibians

17.3.5.1 Description

In the diencephalon of most amphibians (both anuran and urodele) an epithalamus can be easily recognized and can be seen to consist of dorsal and ventral habenular nuclei (Figs. 17.10–17.12). To me, the dorsal nucleus, because of its fairly dense population of small deeply staining cells, appears comparable to the medial habenular nucleus of other vertebrates. There are reports that the left dorsal habenular nucleus is commonly larger than the right (Frontera, 1952; Kemali and Braitenberg, 1969). A clear habenulopeduncular tract is present (Herrick, 1917; Kemali *et al.*, 1980).

Ventral and posterior to the habenular nuclei, the rest of the diencephalon appears to be composed mostly of a thick periventricular layer of cells often with dorsoventral sublamination (Fig. 17.11). Beyond the lateral margin of this, however, particularly in anuran amphibians, nerve cells have invaded the surrounding white matter and nuclear aggregations can be detected with greater or lesser certainty (Fig. 17.12). Gaupp (1899) originally introduced the concept of periventricular and "migrated" nuclei and this idea, which has much to commend it, was followed by Herrick (1917), Röthig (1923, 1924), and most later authors (e.g., Frontera, 1952; Scalia, 1976; Neary and Northcutt, 1983). Röthig further pointed out that anterior, middle and posterior subdivisions could be detected in both periventricular and migrated nuclei. There has been no consistency in the naming of the nuclei, however, and some unlikely homologies with mammals and reptiles continue to be drawn.

The identification of components of the dorsal and ventral thalamus has been particularly inconsistent. Herrick's longitudinal sulci (Chapter 6) that separate dorsal thalamus from ventral thalamus and ventral thalamus from hypothalamus are usually clearly seen on the ventricular wall of the diencephalon

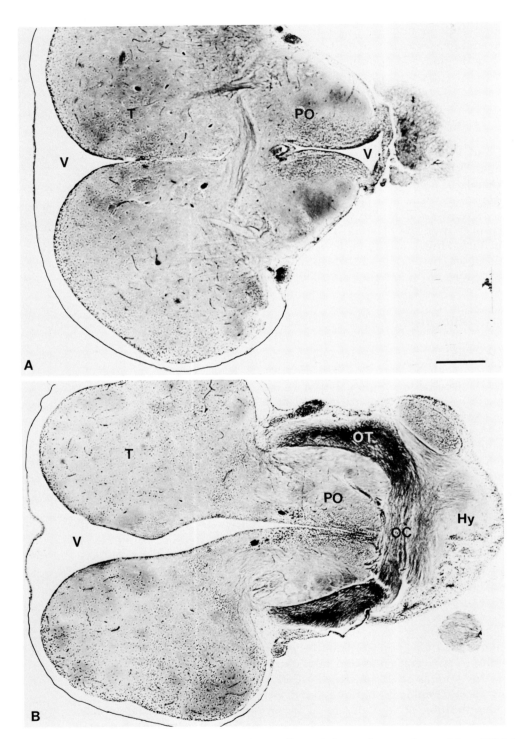

FIGURE 17.8. Cross-sections of the telencephalon, preoptic area, diencephalon, and pretectal area of the goldfish (*Carassius auratus*). (A, B) Bodian stain; from preparations of Dr. David Bodian; (C, D) Luxol fast blue stain; from sections provided by Dr.

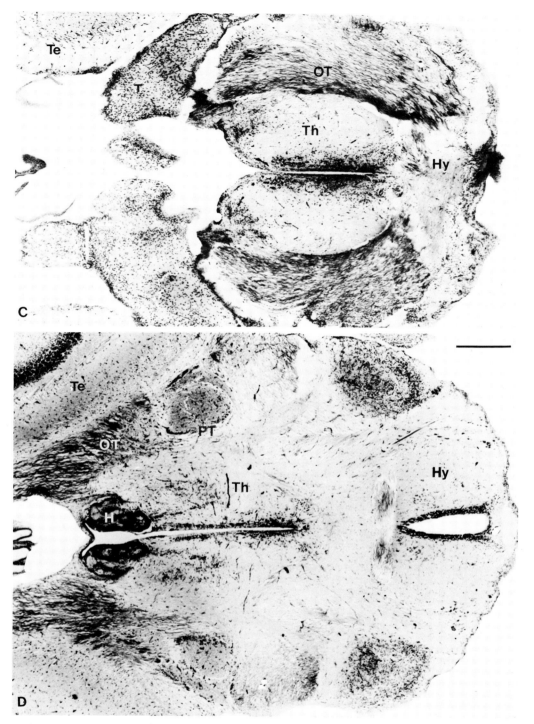

R. B. Leonard. H: habenular nuclei; Hy: hypothalamus; OC: optic chiasm; OT: optic tract; PO: preoptic area; PT: pretectum; T: telencephalon; Te: tectum; Th: thalamus; V: ventricle. Bars = 250 μm.

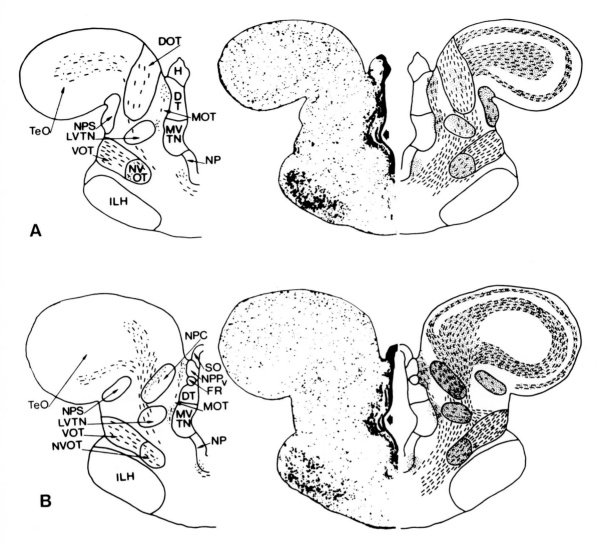

FIGURE 17.9.   Terminations of retinal fibers in the diencephalon of a bony fish, the gar (*Lepisosteus osseus*). (A) is
rostral to (B); side contralateral to removed or labeled eye
is on right. DT: dorsal thalamus; MVTN: medioventral thalamic nucleus; MOT: medial optic tract; LVTN: lateral ventral thalamic nucleus; H: habenula; DOT: dorsal optic tract;
NP: preoptic area; NVOT: nucleus of ventral optic tract;
ILH: inferior lobe of hypothalamus; NPS: part of pretectum; TeO: tectum. From Northcutt and Butler (1976).

FIGURE 17.10.   (A) Transverse section of the diencephalon of a urodele amphibian, the tiger salamander (*Abys-
toma punctatum*). hab.: Elements of habenula; f.retr.: fasciculus retroflexus; p.d.th.: dorsal thalamus; p.v.th.: ventral
thalamus; p.v.hyth.: hypothalamus; tr.op.: optic tract; B:
"crossed, dorsal thalamotegmental tract." From Herrick
(1948). (B) Thionin-stained transverse section of the diencephalon of a urodele amphibian, the mudpuppy (*Necturus
maculosus*). From an animal kindly provided by Dr. R. Miller.
H: habenula; N: neuropil; OT: optic tract; PV: periventricular layer; V: ventricle. Bar = 500 μm.

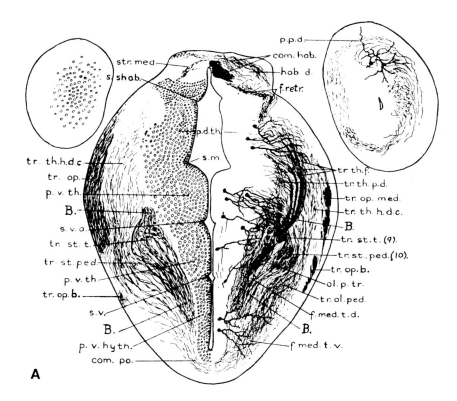

A

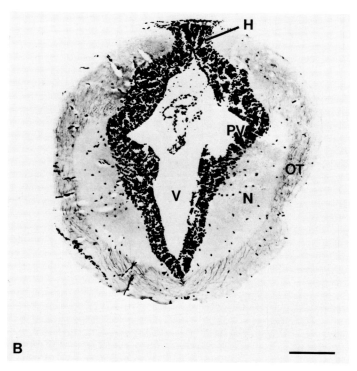

B

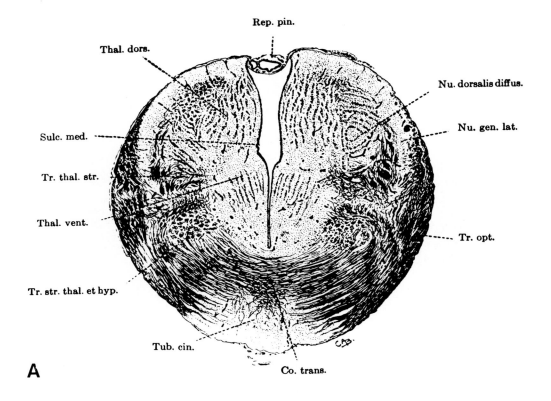

Rep. pin.

Thal. dors.

Nu. dorsalis diffus.

Nu. gen. lat.

Sulc. med.

Tr. thal. str.

Thal. vent.

Tr. opt.

Tr. str. thal. et hyp.

Tub. cin.

Co. trans.

A

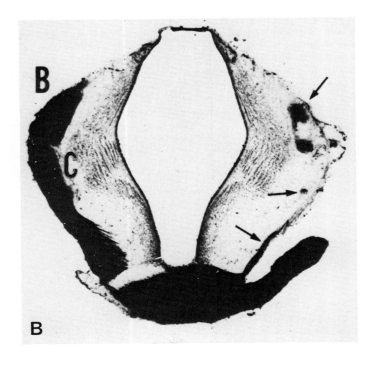

B

C

B

and are matched by relatively cell-free breaks in the periventricular layer. However, lateral to the periventricular layer among the "migrated" nuclei, these breaks are not maintained and until recently there have been no connectional guides to establish the relationships of many of the migrated nuclei to the dorsal or ventral thalamus.

17.3.5.2. Connections

The first modern attempt at systematizing the nuclei of the amphibian diencephalon on the basis of connections was made by Scalia (1976) who in the frog identified nuclei receiving fibers from the retina and those receiving fibers from the optic tectum (Fig. 17.12). He found that an anterior component of the periventricular nuclei was a major recipient of retinal fibers and, moreover, projected to the medial aspect of the pallium. He, therefore, compared this to the retinogeniculocortical system of mammals. Tectal afferents, on the other hand, mainly ended laterally and he compared this system to the extrageniculostriate system of mammals.

Since Scalia's study the amount of connectional data relevant to the amphibian diencephalon has grown considerably, particularly in regard to certain visual connections. Retinal inputs have been shown to be mainly contralateral, with a small ipsilateral component in urodeles and some anurans (Currie and Cowan, 1974; Guillery and Updyke, 1976; Scalia, 1976; Caldwell and Berman, 1977); the fibers end in the neuropil opposite both dorsal and ventral thalamus in a pattern very similar to that in fish (Figs. 17.11 and 17.12) (Guillery and Updyke, 1976; Scalia, 1976; Caldwell and Berman, 1977). Principles of organization are beginning to emerge as a consequence. The most up-to-date review of the extant data is probably that of Neary and Northcutt (1983). Working from their own studies and those of associates, they argue that dorsal and ventral thalami can be readily distinguished on the basis that the nuclei of the dorsal thalamus project to some part of the telencephalon while those of the ventral thalamus do not. This seems to have been the first clear statement of such a distinction in nonmammals. They, therefore, recognize as dorsal thalamic nuclei the anterior, middle, and posterior components of the periventricular region dorsal to the middle diencephalic sulcus, plus anterior, postereoventral, and posterodorsal divisions of the adjacent "migrated" nuclei. Of these, only the anterior periventricular nucleus projects to the pallium while the others, with the possible exceptions of the posterior periventricular and posterodorsal lateral nuclei, project to the striatum. Other cell aggregations not projecting to the

FIGURE 17.11. (A) Transverse section of the diencephalon of an anuran amphibian (*Rana mulgiens*) showing periventricular and migrated nuclei. Nu. gen. lat.: "lateral geniculate nucleus"; Sulc. med: middle sulcus; Thal. dors.: dorsal thalamus; Thal. vent.: ventral thalamus; Tr. opt: optic tract. From Kappers *et al.* (1936). (B) Distribution of retinal fibers in the diencephalon of a frog (*Rana pipiens*) showing heavy contralateral and small, localized ipsilateral (arrows) retinothalamic terminations in neuropil regions referred to by Scalia as nucleus of Bellonci (B) and corpus geniculatum thalamicum (C). Right optic nerve labeled anterogradely by direct application of horseradish peroxidase to cut end. From Scalia (1976).

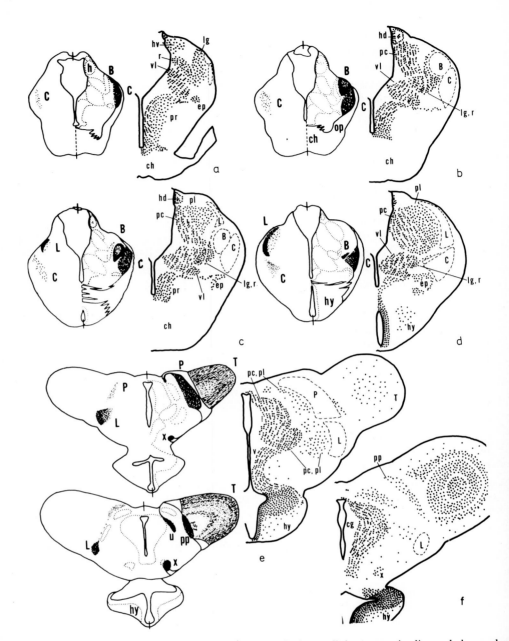

FIGURE 17.12. Right half sections of each pair (a–f) show cellular masses in diencephalon and midbrain of the frog, *Rana pipiens*. Left full sections show distribution of tectodiencephalic (left side) and retinodiencephalic (right side) fibers. Major zones of termination of tectal fibers are in neuropil regions referred to as geniculate body (C), lateral neuropil (L), and posterior thalamic nucleus (P). Major zones of termination of retinal fibers are in neuropil regions referred to as nucleus of Bellonci (B), geniculate body (C), pretectal nucleus (PP), basal optic nucleus (X), and in optic tectum (T). Other abbreviations: ch: optic chiasm; cg: central gray; ep: entopeduncular nucleus; h, hv, hd: elements of habenula; lg,r: lateral geniculate nucleus and nucleus rotundus; pc: posterocentral nucleus; pl: posterolateral nucleus; pr: preoptic area; v: posterior ventral thalamus; vl: ventrolateral nucleus; hy: hypothalamus. From Scalia (1976).

telencephalon are, thus, included in the ventral thalamus or in a "posterior tubercle" of uncertain relationships. It is possible that the posterior periventricular and posterodorsal lateral nuclei may be comparable to the pretectum, for the first of these projects to the anterior periventricular nucleus, to the ventral thalamus, and to the brain stem while both project to the tectum.

In the cortically projecting anterior periventricular nucleus of the dorsal thalamus and the striatally projecting other nuclei, there is obvious similarity to the principal and intralaminar nuclei of the mammalian dorsal thalamus. The similarity can be extended further, for both the anterior periventricular nucleus and the striatally projecting nuclei receive various combinations of afferent fibers, including those from the hypothalamus, pallium, retina, optic tectum, torus semicircularis (inferior colliculi), spinal cord, and the apparent equivalents of the dorsal column nuclei and globus pallidus. There may even be a case for regarding the anterior periventricular nucleus as containing two sets of cortically projecting cells for it appears to project heavily to the medial pallium and diffusely to the dorsal pallium (Scalia and Coleman, 1975; Kicliter, 1979). Unlike most mammalian thalamocortical projections, however, this projection is bilateral.

Though some of these analogies seem superficially quite compelling, it would be incorrect, in the present state of knowledge, to regard them necessarily as representing a fundamental Bauplan upon which the higher vertebrate dorsal thalamus is built. Apart from the bilaterality in the thalamopallial projection mentioned above, there are several features that are not found in mammals. These include efferent connections form the anterior periventricular nucleus to the hypothalamus, to the optic tectum, and to other nuclei such as the posterior periventricular and posterodorsal lateral nucleus (Neary and Northcutt, 1983). There is also the presence of dendrites on the cells of the anterior periventricular nucleus that extend deeply into the lateral, migrated, nuclei, so that they may share the afferent inputs to the latter (Scalia and Gregory, 1970).

The ventral thalamus of Neary and Northcutt (1983) in the frog includes certain nuclei sometimes regarded as belonging to the dorsal thalamus and given names such as *lateral geniculate nucleus* (a mammalian term) and *nucleus of Bellonci* (a sauropsid term). Neither seems appropriate. The ventral thalamus appears to be dominated by visual inputs from the retina (Scalia and Gregory, 1970; Guillery and Updyke, 1976; Caldwell and Berman, 1977; Levine, 1980) and to project primarily to the optic tectum (Trachtenberg and Ingle, 1974; Neary and Wilczynski, 1980). However, other inputs have been demonstrated from the apparent equivalents of the dorsal column nuclei and from the posterior dorsal thalamic nuclei (Neary and Wilczynski, 1977b; Neary and Northcutt, 1983). Other efferent connections have been detected passing to the spinal cord (ten Donkelaar and De Boer-van Huizen, 1981a).

17.3.6. Reptiles

17.3.6.1. Description

The reptilian diencephalon shows a wide range of variation across species. On the whole, a periventricular layer is not distinct except in the hypothalamic region (Fig. 17.13). Dorsally, there are distinct medial and lateral habenular nuclei with cytoarchitectonic appearances similar to those of mammals. In between, there are several well-defined nuclei each with a distinctive cytoarchitecture and sufficiently constant from species to species that a standardized nomenclature is possible. It is not yet clear, however, which are derivatives of the ventral or of the dorsal thalamus. The nomenclature in widest use today is derived largely from that popularized by Huber and Crosby (1926, 1933) (see Armstrong, 1950, 1951; Cruce, 1974; Balaban and Ulinski, 1981a,b; Bass and Northcutt, 1981), though the modified nomenclatures of Papez (1935) and Kuhlenbeck (1931) are also used.

The preoptic region usually displays a periventricular layer of cells but, apart from those cells, it is largely taken up with the medial and lateral forebrain bundles and the stria medullaris. Behind the preoptic area, as the optic tracts start to encircle the diencephalon, the habenular nuclei appear and three or more nuclei become visible in the thalamus. There are distinct anterior dorsomedial and anterior dorsolateral nuclei on each side and a nucleus reuniens (either fused with or separate from a medial nucleus), lying ventrally and across the midline (Fig. 17.13). These fill the anterior half of the thalamus. Other subsidiary nuclei and loosely arranged cells may be intercalated between the major nuclei.

The posterior half or two-thirds of most reptilian thalami is dominated by a large, medial or central nucleus rotundus; this is round as the name suggests and commonly consists of dispersed large cells and an open neuropil (Rainey, 1979). In some species it is encapsulated by small cells. In others such as snakes (Ulinski, 1977), it may be small, dispersed, or even absent. Laterally, lying internal to the optic tract is a large lateral geniculate complex, usually divided into dorsal and ventral lateral geniculate nuclei and sometimes showing further degrees of cytoarchitectonic specialization such as incipient lamination. The anterior dorsomedial and anterior dorsolateral thalamic nuclei are compressed dorsally by the nucleus rotundus, and other small cell groups of varying distinctness lie beneath the nucleus rotundus or between it and the lateral geniculate complex (Fig. 17.13). Some of these may well be derivatives of the ventral thalamus.

Posterior to the nucleus rotundus and lateral geniculate complex appear a series of pretectal nuclei, overlain by a large optic tectum.

17.3.6.2. Connections

Many studies of afferent connections to the reptilian thalamus have now been made. One of the most important contributions has been the realization that the nucleus rotundus is the target of the optic tectum (bilaterally) while the lateral geniculate complex is the target (ipsi- and contralaterally) of the retina (Kosareva, 1974; Knapp and Kang, 1968a,b; Hall and Ebner, 1970a; Cruce and Cruce, 1975; Butler and Northcutt, 1978; Repérant et al., 1978; Balaban and Ulinski, 1981b; Bass and Northcutt, 1981; Rainey and Ulinski, 1982). The nucleus rotundus projects upon the dorsal ventricular ridge and possibly the underlying striatum. It is uncertain about the lateral geniculate complex. Some have argued that parts of it project upon the dorsal cortex of the turtle (Hall and Ebner, 1970b) while others have argued that most of the dorsal and ventral lateral geniculate nuclei do not project upon the telecephalon at all. Instead they have postulated a projection to the visually active cortex from a small, retinal-recipient nucleus lying dorsal to the dorsal lateral geniculate nucleus (Northcutt, 1981), or shown a projection from the caudal part of the dorsal lateral geniculate nucleus to the dorsal ventricular ridge, not to the cortex (Wang and Halpern, 1977; Lohman and van Woerden-Verkley, 1978). Whatever the resolution of this controversy it is clear that there are two visual pathways through the reptilian thalamus—one from the optic tectum and the other from the retina direct—and that these seem to project to different parts of the telencephalon. In this, some workers have seen parallels to the "two visual systems" (Chapter 10) of mammals.

Of the other thalamic nuclei, seemingly different parts of the medial or reuniens nuclei receive bilateral, presumed auditory, inputs from the torus semi-circularis (Pritz, 1974a,b, 1975; Ulinski, 1977; Foster and Hall, 1978; Balaban and Ulinski, 1980), and from the spinal cord and dorsal column nuclei (Goldby and Robinson, 1962; Northcutt and Pritz, 1978; Cruce, 1979; Ulinski, 1983; Molenaar and Fizaan-Oostven, 1980; Hoogland, 1981; Ebbesson and Goodman, 1981).

The dorsal ventricular ridge, to which these and other thalamic nuclei project, is divided into anterior and posterior parts and the anterior part is further subdivided into three longitudinal bands with different input connections and functional relationships (Ulinski, 1983). The medial band receives fibers from the nucleus reuniens of the thalamus (Pritz, 1975; Foster and Hall, 1978; Lohman and van Woerden-Verkley, 1978) and is auditory in function (Weibach and Schwartzkopf, 1967). The intermediate zone receives fibers from the part of the thalamus innervated by the spinal cord and dorsal column nuclei (Pritz and Northcutt, 1980) and is presumed to be somatosensory in function. The lateral zone is the target of the nucleus rotundus (Belekhova *et al.*, 1978; Balaban and Ulinski, 1981a,b). This projection appears to be topographically organized and units in the appropriate part of the dorsal ventricular ridge have visual receptive fields (Dünser *et al.*, 1981).

In addition to these projections to specific parts of the dorsal ventricular ridge, there are diffuse projections from the dorsomedial thalamic nucleus (Ulinski, 1983) to all three anterior zones and to cortex, septum, and hypothalamus as well.

All thalamic nuclei that project to the dorsal ventricular ridge may also project to the striatum (Goldby, 1937; Powell and Kruger, 1960; Parent, 1976), though this is debated (Ulinski, 1983); some of those not yet shown to have dorsal ventricular ridge connections may also send fibers to the striatum (North-cutt, 1981; Ulinski, 1983) or to the nucleus accumbens (Distel and Ebbesson, 1975). Moreover, parts of the striatum possibly equivalent to the mammalian globus pallidus project back upon the nucleus rotundus (Reiner *et al.*, 1980).

Thus, the reptilian thalamus displays a highly structured pattern of input–output connections and there is a corresponding cytoarchitectonic patterning. The cytoarchitecture is so different from that of a mammal that it would be unwise to make any suggestions as to homology. However, many of the

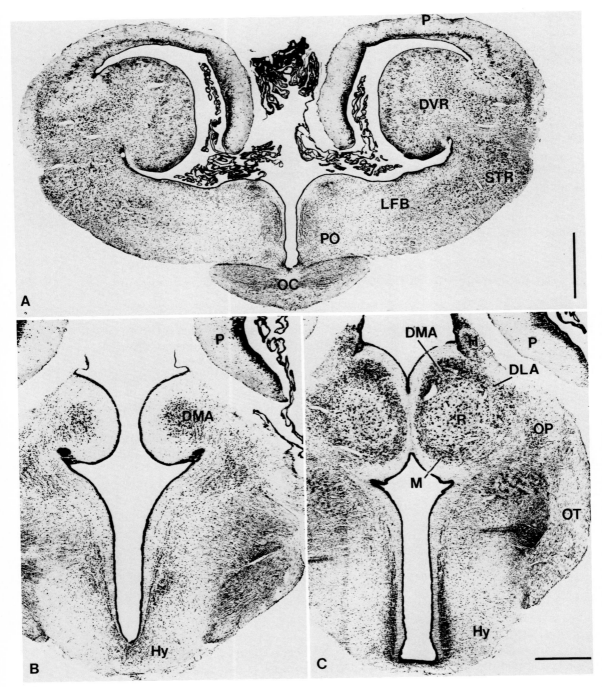

FIGURE 17.13. Thionin-stained transverse sections through the preoptic area (A), diencephalon (B–E), and pretectal area (F) of a turtle (*Pseudemys scripta*). DMA, DLA: dorsomedial anterior and dorsolateral anterior nuclei; DVR: dorsal ventricular ridge; H: habenular nuclei; Hy: hypothalamus; LFB: lateral forebrain bundle; OC: optic chiasm; Ot: optic tract; OP: optic nuclei; M: medial nucleus; P: pallium; PO: preoptic area; PT: pretectum; R: nucleus rotundus; STR: striatum; V: ventricle. Preparations made in the author's laboratory by Dr. S. P. Wise. Bars = 1 mm (A, F), 500 μm (B–E).

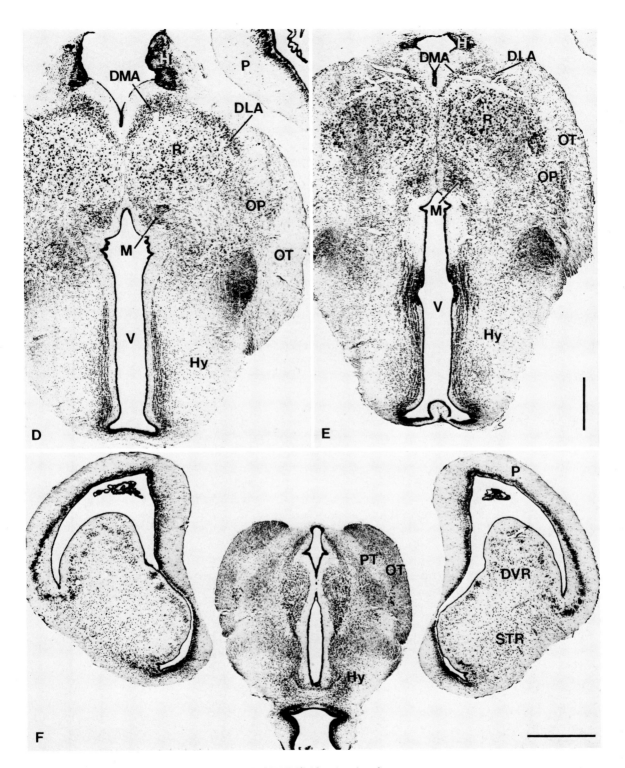

FIGURE 17.13.   (*continued*)

patterns of connection with the dorsal ventricular ridge and striatum have similarities with those between the mammalian thalamus and the cerebral cortex and striatum. They also bear comparison with the telencephalic connections of the dorsal thalamus in amphibians (Section 17.3.5).

What is lacking in reptiles at the present time is some distinction between dorsal and ventral thalamic nuclei and more finite information on the nature of thalamic connections with the cerebral cortex in which architectonic and functional areas of various kinds are thought to exist (Rose, 1923; Goldby and Gamble, 1957; Kruger and Berkowitz, 1960; Ulinski and Rainey, 1980; Ulinski, 1979). With further connectional studies, this information will undoubtedly come to hand.

17.3.7. Birds

17.3.7.1. Description

The thalamus of birds, though usually larger and with clearer nuclear differentiation, does not differ in fundamental organization from that of reptiles (Fig. 17.14). A modern, comprehensive atlas of the pigeon brain (Karten and Hodos, 1967) clearly demonstrates these nuclei and older accounts of other species on which this atlas draws for terminology can be found in Edinger and Wallenberg (1899), Craigie (1928, 1930, 1932), Huber and Crosby (1929), Papez (1935), and Kappers *et al.* (1936).

The habenular nuclei of birds are not disproportionately large as they are in reptiles, amphibians, and fish. Distinct medial and lateral habenular nuclei with cytoarchitectural features comparable to the nuclei in mammals are found. The preoptic area and hypothalamus have thick periventricular layers but with some migrated nuclei as well. The thalamus is composed of several nuclei of varying size with clear-cut borders and there are relatively few zones of diffuse cells and neuropil or of cytoarchitectonic transition. The thalamus of most birds is, therefore, as clearly parcellated as the thalamus of most mammals. The arrangement of nuclei is, however, totally different from that of the mammalian thalamus. Instead, it appears like an elaborated reptilian thalamus, and reptiles and birds are the only classes where one can feel relatively comfortable about homologies in the thalamus.

There are anterior dorsolateral and anterior dorsomedial groups of nuclei probably corresponding to the nuclei of the same name in reptiles (Section 17.3.6) but showing further parcellation into subnuclei, including posterior extensions that warrant the names *posterior dorsolateral* and *posterior dorsomedial nuclei* (Karten and Hodos, 1967). On the dorsolateral surface of the diencephalon superficial to these nuclei are a series of "superficial" nuclei. The dorsal and the superficial nuclei occupy the dorsal one-third or so of the thalamus. The remainder, especially in the posterior two-thirds, is occupied by a nucleus rotundus and a lateral geniculate or optic nuclear complex which show varying degrees of relative size, and a number of smaller though no less distinct nuclei.

In some birds (e.g., pigeons) the nucleus rotundus is particularly large and occupies most of the lateral half of the cross-sectional area of the thalamus (Fig. 17.15). In doing so, it compresses, as it were, the principal optic nucleus ("dorsal lateral geniculate nucleus" of some workers) into a narrow peripheral shell adjacent to the encircling optic tract. A small part of the principal optic nucleus, placed anterolateral to the nucleus rotundus (Fig. 17.14), is called *nucleus lateralis anterior* by Karten and Hodos (1967). The rest of the geniculate complex is a bilaminar strip of cells lying on the optic tract ventrolateral to the nucleus rotundus (Fig. 17.14). Traditionally, it has been called *lateral geniculate nucleus*, though nowadays it is more common to refer to it as *ventral lateral geniculate nucleus*.

In other birds, such as the owls (Karten *et al.*, 1973), the principal optic nucleus is particularly large and contains several subnuclear divisions. If one can think of it as an enlarged nucleus lateralis anterior, then its enlargement appears to have displaced the relatively smaller nucleus rotundus posteriorly. A ventral lateral geniculate nucleus is still present.

Medial to the nucleus rotundus is a small, densely cellular nucleus ovoidalis from which a second dense-celled nucleus extends ventrolaterally beneath the nucleus rotundus as the nucleus subrotundus (Fig. 17.14). In this region other less distinct, posterior nuclei have also been described.

Dorsal to the nucleus rotundus is often a small nucleus triangularis and posterior to the nucleus rotundus a small, dense-celled nucleus spiriformis which may be part of the pretectal complex of nuclei, though Karten and Finger (1976) include it in the thalamus.

Apart from the ventral lateral geniculate nucleus and some diffuse cells that extend from it dorsomedially across the hypothalamus, there is no distinct ventral thalamus (Fig. 17.14).

17.3.7.2. Visual Connections

The several afferent pathways converging on the thalamus of the bird terminate to a large extent in different nuclei. As in other nonmammals, the visual system has been the most thoroughly studied. Different populations of retinal ganglion cells appear to project upon the optic tectum and upon the thalamus (Bravo and Pettigrew, 1981).

The central gray layer of the optic tectum projects bilaterally to the nucleus rotundus (Karten and Revzin, 1966; Karten and Hodos, 1970; Benowitz and Karten, 1976; Hunt and Künzle, 1976; Reiner and Karten, 1978) (Fig. 17.16). This projection appears to be diffuse, i.e., not retinotopically organized, though different parts of the nucleus rotundus receive fibers from different depths in the central gray layer of the tectum (Benowitz and Karten, 1976). In keeping with the lack of retinotopy, rotundal units recorded extracellularly, have large receptive fields (Crossland, 1972; Revzin, 1979). Many, however, are sensitive to moving stimuli and can be direction selective (Crossland, 1972; Revzin, 1979). Inhibitory effects can also be identified (De Britto *et al.*, 1975) and some units are color sensitive, sometimes showing opponent-color responses comparable to those demonstrable in the mammalian visual system (Chapter 9) (Yazulla and

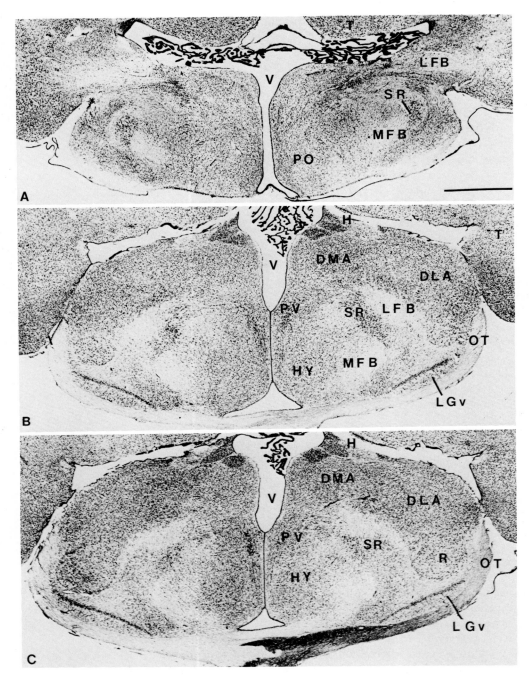

FIGURE 17.14. Thionin-stained transverse sections through the preoptic area (A, B) and diencephalon (C–F) of a domestic chicken (*Gallus domesticus*). The sections are autoradiographs and labeling of the optic tract contralateral to an injection of tritiated amino acids in an eye can be detected in (C–F). Abbreviations as in Fig. 17.13 plus: DMP, DLP: dorsomedial posterior and dorsolateral posterior nuclei; SPc: superficial nucleus; SRT: nucleus subrotundus; T: nucleus triangularis; Te: tectum; SR: superior reticular nucleus; MFB: medial forebrain bundle; Ov: nucleus ovoidalis; PV: periventricular nuclei; LGv: ventral lateral geniculate nucleus. Preparations made in the author's laboratory by Dr. S. P. Wise. Bars = 1 mm.

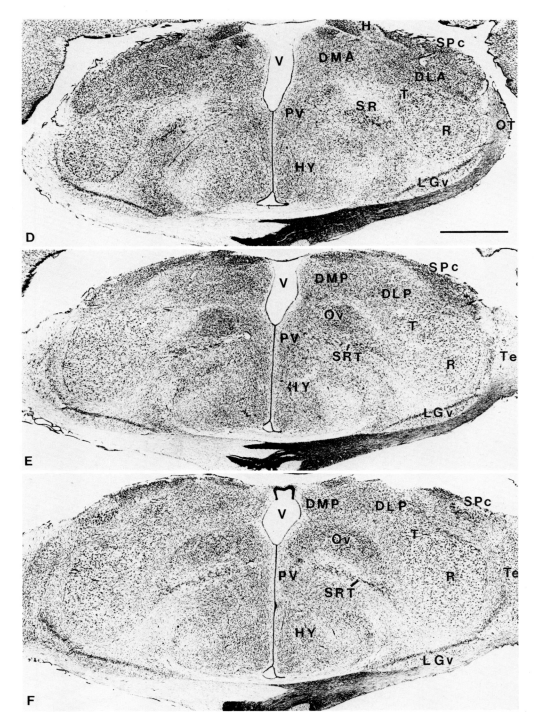

FIGURE 17.14. *(continued)*

Granda, 1973; Maxwell and Granda, 1979). Convincing evidence for a telencephalic projecton of the nucleus rotundus was provided by Powell and Cowan (1961) who used the retrograde degeneration technique (Fig. 17.15). Since then, it has been established that the target of the nucleus rotundus is the ipsilateral ectostriatum (Fig. 17.16) (Karten and Hodos, 1970; Facciolli and Minelli, 1975; De Brito *et al.*, 1975; Benowitz and Karten, 1976), a component of the avian "striatum" which can be regarded as the avian dorsal ventricular ridge (Ulinski, 1983). The ectostriatum shows enhanced [^{14}C]-2-deoxy-D-glucose uptake following visual stimulation (Streit *et al.*, 1980) and contains neurons with response properties similar to those of the nucleus rotundus (Kimberly *et al.*, 1971).

Gamlin and Cohen (1982) have recently shown that a small posterior dorsolateral nucleus of the pigeon is responsive to visual stimuli and also receives fibers from the central gray stratum of the tectum. It projects to a small part of

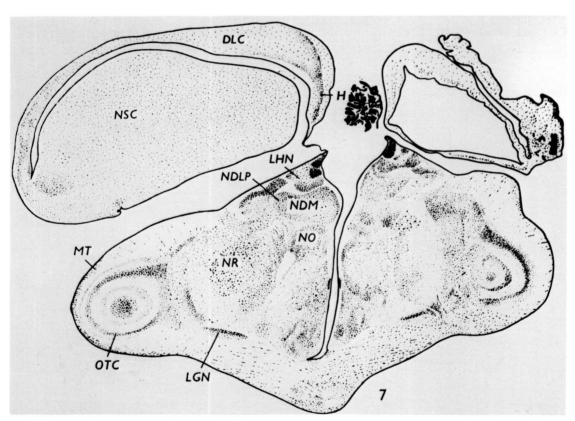

FIGURE 17.15. Retrograde degeneration (on right) of nucleus rotundus (NR), nucleus ovoidalis (NO), and anterior dorsomedial nucleus (NDM) in a pigeon from which most of the ipsilateral telencephalon was removed 44 days previously. Severe degeneration also occurred in the anterior dorsolateral nucleus (not shown). LHN: lateral habenular nucleus; NDLP: posterior dorsolateral nucleus; MT: marginal tract; OTC: optic tectum; LGN: (ventral) lateral geniculate nucleus; DLC: dorsolateral cortex; NSC: caudal neostriatum; H: hippocampus. From Powell and Cowan (1961).

the dorsal ventricular ridge lying medial and posterior to the ectostriatum. This appears to represent a second tectothalamodorsal ventricular ridge visual pathway.

The retina itself projects to most of the subnuclei of the lateral geniculate (or principal optic) complex and to other parts of the anterior dorsolateral nucleus (Fig. 17.16). It is mainly a contralateral projection, though with an ipsilateral component of variable size that might be absent in some species (Cowan *et al.*, 1961; Karten and Revzin, 1966; Karten and Nauta, 1968; Repérant, 1973;

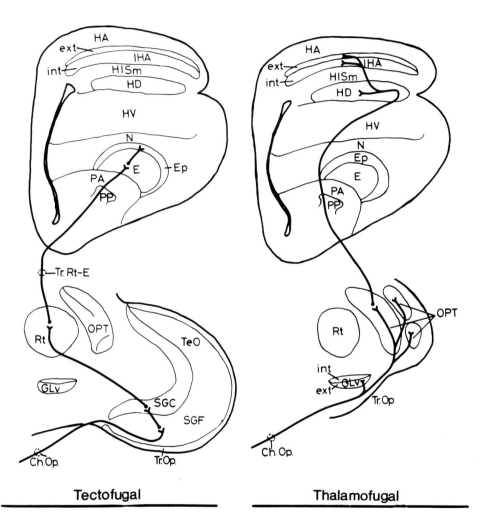

Tectofugal   **Thalamofugal**

FIGURE 17.16.   Schemes showing the retinothalamotelencephalic pathways (right) and the retinotectorotundal (left) pathways in birds. HA, IHA, HISm, HD: parts of Wulst. HV: ventral hyperstriatum; E: ectostriatum; Ep: peripheral zone; PA, PP: parts of paleostriatum; N: neostriatum; TeO: optic tectum; GLv: ventral lateral geniculate nucleus; OPT: principal optic complex (parts of DLA nucleus in Fig. 17.14). From Karten *et al.* (1973).

Mihailovic *et al.*, 1974; Bravo and Pettigrew, 1981). The projection to the principal optic nucleus is retinotopically organized (Pettigrew, 1979). The principal optic nucleus projects bilaterally to the visual part of the Wulst (Section 17.2.1) where its axons end retinotopically in the prominent bilaminated granule cell stratum called *nucleus intercalatus hyperstriatum accessorium* (Figs. 17.16 and 17.17) (Hunt and Webster, 1972; Karten *et al.*, 1973; Miceli *et al.*, 1975, 1979; Cooper and Pettigrew, 1979). This layered structure has much in common both anatomically and physiologically with the mammalian visual cortex (Pettigrew and Konishi, 1976). For example (Pettigrew, 1979), it is retinotopically organized, units have receptive fields that can be characterized as monocular, concentric, and nonoriented; as binocular, simple, complex, and hypercomplex. Binocular neurons can show orientation selectivity, movement selectivity, and selectivity for binocular disparities. Ocular dominance columns and orientation columns have been reported.

The visual part of the Wulst gives rise to efferent connections that return to a part of the principal optic nucleus of the thalamus, as well as to the internal lamina of the ventral lateral geniculate nucleus, to the pretectum, and to the

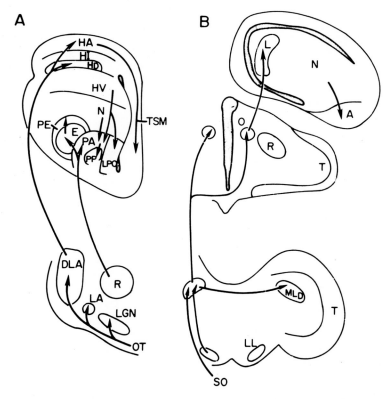

FIGURE 17.17.   Scheme of the central auditory pathways (B) in the pigeon and connections of the visual Wulst (HA) (A). L: field L of neostriatum; Ov: nucleus ovoidalis of thalamus; MLD: nucleus mesencephali lateralis dorsalis. From Nauta and Karten (1970).

Pettigrew, 1981). In many ways, therefore, the visual part of the Wulst and its connectivity parallel the mammalian visual cortex, the retinogeniculocortical pathway of mammals, and the cortical–subcortical connections of the mammalian visual cortex.

17.3.7.3. Auditory Connections

Auditory afferents to the thalamus in birds (Fig. 17.17) arise in the midbrain nucleus mesencephalicus lateralis, pars dorsalis, a region comparable to the torus semicircularis of reptiles and possibly equivalent, in turn, to the mammalian inferior colliculus (Boord, 1969; Knudsen and Konishi, 1978). This nucleus in birds projects bilaterally to the nucleus ovoidalis of the thalamus in pigeons (Karten, 1967). The nucleus ovoidalis in pigeons, canaries, and guinea fowl has been shown to project ipsilaterally to a particular posteromedially situated part of the "neostriatum" (i.e., to a part of the dorsal ventricular ridge, Fig. 17.17) (Karten, 1969; Nottebohm *et al.*, 1976; Kelley and Nottebohm, 1979; Bonke *et al.*, 1979b). In this region, click-evoked potentials can be recorded (Erulkar, 1955), and exposure to white noise or pure tones leads to enhanced $[^{14}C]$-2-deoxy-D glucose uptake (Lippe and Masterton, 1980; Scheich and Maier, 1981). Single-unit studies indicate a tonotopic organization with isofrequency bands across the area and complex patterns of unit reponse similar to those reported in the mammalian auditory cortex. The major area is a cytoarchitectonic region described as field L by Rose (1914) (Zaretsky and Konishi, 1976; Zaretsky, 1978; Leppelsack, 1978; Scheich *et al.*, 1979). In field L, units also respond to species-specific calls or to other complex sound patterns (Leppelsack and Vogt, 1976; Bonke *et al.*, 1979b, 1981; Leppelsack, 1981; Langner *et al.*, 1981a,b; Margoliash, 1982). Two other auditory resposive regions in adjacent areas have been described but their thalamic inputs, if any, are not yet known (Delius *et al.*, 1979; Kirsch *et al.*, 1980).

Emanating from the auditory responsive areas of the telencephalon are multisynaptic pathways through other parts of the dorsal ventricular ridge that eventually gain access to the hypoglossal nucleus and, thus, to the control of vocalization (Nottebohm *et al.*, 1976; Bonke *et al.*, 1979a,b).

17.3.7.4. Somatic Sensory Connections

Somatic sensory pathways ending in the avian thalamus emanate from the spinal cord and from the dorsal column nuclei (Karten, 1963). There appear to be both direct connections and indirect connections via brain-stem relays. The terminal regions of the somatic sensory pathways in the bird thalamus have not been defined anatomically but evoked potentials have been recorded in posterior parts of the thalamus in response to small movements of the feathers (Delius and Bennetto, 1972). It is possible that this thalamic region projects to a somatic sensory responsive area that has been demonstrted electrophysiologically in posterior parts of the neostriatum, close to the auditory projection field (Erulkar, 1955; Delius and Bennetto, 1972). Another somatic sensory responsive area has been briefly reported in part of the Wulst of owls and it is said to receive input from thalamic nuclei that are innervated by the dorsal column nuclei (Karten *et al.*, 1978). Neurons in this part of the Wulst respond only to stimulation of the toes, leading to ideas about eye–toe coordination in the Wulst. The toe part of the Wulst projects back to thalamus and brain stem and thus resembles the somatic sensory cortex of mammals.

The large somatic sensory pathway ascending from the trigeminal nuclear complex (Wallenberg, 1906) appears to bypass the thalamus and end directly in the nucleus basalis, a deeply seated part of the dorsal ventricular ridge complex (Zeigler and Karten, 1973; Cohen and Karten, 1974; Dubbeldam *et al.*, 1981). The projection is bilateral and the nucleus basalis contains a detailed representation of the beak and interior of the mouth (Berkhoudt *et al.*, 1981).

Other afferents to the avian thalamus come from parts of the brain stem near the muscles innervating the larynx (Nottebohm *et al.*, 1982). They end in a small thalamic nucleus ("uvaeformis") that appears to be connected with vocal control regions of the hyperstriatum (Nottebohm *et al.*, 1976).

17.3.7.5. Other Afferent Connections

The other inputs of significance to the avian thalamus come from the septal nuclei (Krayniak and Siegel, 1978) and from deeply seated parts of the striatum (Karten and Dubbeldam, 1973; Brauth *et al.*, 1978). The "striatal" afferents arise in the paleostriatum primitivum and nucleus interpeduncularis (Fig. 17.18).

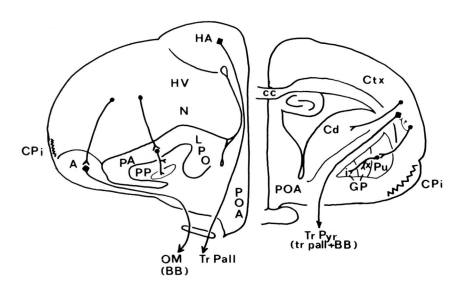

FIGURE 17.18. Comparative scheme of connections between components of the striatum and the thalamus in the pigeon (left) and a mammal (right). HA: hyperstriatum accessorium; HV: hyperstriatum ventrale; N: neostriatum; CPi: piriform cortex; PA: paleostriatum augmentatum; PP: paleostriatum primitivum; LPO: lobus parolfactorius; POA: preoptic area; OM: tractus occipitomesencephalicus; TrPall: pallidal efferent tract. From Karten and Dubbeldam (1973).

These have afferent and efferent connections that suggest their equivalence to the external and internal divisions, respectively, of the mammalian globus pallidus (Karten and Dubbeldam, 1973). Their efferent fibers reach the thalamus via an obvious ansa lenticularis and end in two nuclei (dorsalis ,intermedius posterior and spiriformis lateralis) that Karten and Dubbeldam, therefore, suggest are comparable to the mammalian ventral lateral and centre médian nuclei. The inputs to the two putative pallidal homologs come from the paleostriatum augmentatum which for this and other connectional reasons (Fig. 17.18) the authors suggest may represent the equivalent of the mammalian striatum. The only thing lacking in this plausible, mammallike circuit is a projecton from the thalamus to the paleostriatum augmentatum. Parent (1976) has described such a connection but Revzin and Karten (1966) have denied it and Ulinski (1983) is also skeptical of its existence. Indeed, Ulinski feels that striatal–thalamic associations were only effectively established during the evolutionary transition from therapsid reptiles to mammals.

17.3.7.6. Conclusions

The avian thalamus presents a picture of nuclear differentiation and connectional specificity as clearly defined as that of any mammal. In its patterns of input–output connections can be seen many parallels with the mammal. For example, the tectal–rotundal–hyperstriatal connection can bear comparison with the "second" mammalian visual pathway passing from the superior colliculus through the lateral posterior or inferior pulvinar nucleus to the extrastriate cortex. Similarly, the retinogeniculo-Wulst connection can be compared to the retinogeniculostriate pathway and there are comparable patterns of organization in the somatic sensory, auditory, and possibly the basal ganglionic systems.

However, these are but parallel patterns of organization and they do not necessarily imply homology. The bird thalamus is sufficiently similar to that of reptiles to suggest the common evolutionary origin of the two classes. But the sauropsid thalamus and its telencephalic targets are so fundamentally different from those of mammals that one would need to be very brave to assert that parallel chains of connections indicated a common origin.

17.3.8. Summing Up

This brief survey of the nonmammalian thalamus indicates the diversity of structure exhibited by the diencephalon in other vertebrates and shows that many thalami can be as different from one another as they are from those of mammals. Some basic patterns do seem to emerge, however. A periventricular thalamus with afferent fibers terminating in a lateral neuropil is common to several classes and it is not difficult to envisage this evolving into a thalamus dominated by nuclei that have "migrated" into the neuropil. This, in fact, seems to have occurred in different representatives of the same class such as anuran amphibians in comparison with urodeles. The common origin of the reptilian and avian thalamus also seems clear. Indeed, the differences between reptiles with large, highly differentiated thalami and birds with small, poorly differentiated thalami are not much greater than those between certain mammals.

Certain connectional patterns also appear to be fundamental and present in virtually all vertebrate thalami. These include: afferent connections to the thalamus from all sensory systems, each with circumscribed and largely or wholly nonoverlapping regions of termination; thalamic projections to circumscribed parts of the pallium or its overtly nonmammal-like derivatives such as the central telencephalic nucleus of sharks, the dorsal ventricular ridge of reptiles, and the hyperstriatum of birds. Thalamostriate connections may also be present in all species but the complete evidence is not yet in for birds or reptiles; telencephalic projections to the thalamus also seem to be present in a majority of the animals so far studied. Whether these are fundamentally homologous aspects or examples of parallel evolution can only be guessed at, but they show far greater parallelism between the mammalian and the nonmammalian thalamus than had been anticipated from the studies of Herrick and his contemporaries.

17.4. Mammals

17.4.1. Introduction

The mammalian thalamus has dominated this book and because of the fundamentally similar pattern of nuclear organization and connectivity, it has usually been possible to describe it in general terms rather than by repeated reference to different species. So obvious is the basic structural pattern that anyone familiar with the rodent, cat, and monkey thalamus and perhaps with some understanding of that of the prosimians can usually find his or her way around the thalamus of an unfamiliar mammalian species without difficulty. The only mammalian thalamus that might present difficulties is that of the monotremes for, in them, the cytoarchitecture is somewhat monotonous and the nuclear divisions of marsupials and eutherian mammals are far less clear. It seems appropriate, therefore, to devote a brief description to the thalamus of the two monotremes, the platypus and echidna. These have formerly been described in more or less detail by Ziehen (1908), Hines (1929), Abbie (1934), Lende (1964), Campbell and Hayhow (1971, 1972), Welker and Lende (1980), and Ulinski (1984).

17.4.2. The Monotreme Thalamus

Two habenular nuclei typical of those in other mammals are present on each side (Fig. 17.19) and a region of small compacted cells runs along the floor of the dorsal recess of the third ventricle in a manner very similar to the paraventricular nuclei but without a clear distinction into anterior and posterior or left and right nuclei (Fig. 17.19).

A distinct reticular nucleus and zona incerta with cytoarchitectures comparable to those in other mammals can be seen but they are anteriorly placed and the reticular nucleus does not encompass the whole dorsal thalamus. The retinal-recipient nuclei have been identified by the tracing of connections from the eye (Campbell and Hayhow, 1971, 1972). One of these nuclei called LGNa by Campbell and Hayhow is the larger and is situated posteriorly beside the cerebral peduncle; the other called LGNb is dorsally placed in the floor of the lateral ventricle.* The posteriorly situated LGNa contains large and small cells which can even be partially laminated, resembling the ventral lateral geniculate of other mammals (Campbell and Hayhow, 1971). The LGNb undergoes retrograde degeneration after ablation of the visual cortex whereas the LGNa does not (Welker and Lende, 1980). Therefore, the LGNb is probably equivalent to the dorsal and the LGNa to the ventral lateral geniculate nuclei. Th caudal pole of the thalamus dorsal to the LGNa has been called the *medial geniculate body* (Campbell and Hayhow, 1972) and there are some islands of large cells in that region (Fig. 17.19) that suggest an incipient architectonic division, as well as an additional posterior nucleus (Figs. 17.19H,I).

In the dorsal thalamus proper there is no distinct intralaminar group of nuclei so that it is hard to make a distinction between medial, lateral, and anterior nuclear goups, though Campbell and Hayhow (1971, 1972) draw outlines of what they call *anterior, medial,* and *lateral nuclear regions.* The architecture, overall, is rather monotonous with few distinctive nuclei. There is a medial concentration of smaller, more densely staining cells that are not particularly separated from the paraventricular nuclei. Ulinski (1984) calls these *ventral medial nucleus,* Lende (1964) *mediodorsal nucleus,* and Campbell and Hayhow (1971) *nucleus reuniens* but it would be hard to rule out it being simply an enlarged paraventricular nucleus. The anterior thalamic region is a homogeneous mass of cells, without the distinctive large-celled anterodorsal nucleus characteristic of all other mammals. In the lateral posterior regions of the thalamus the only truly distinctive dorsal thalamic nucleus is found. It is composed of medium-sized, relatively deeply staining cells that appear to coalesce out of the ventral nuclear region and it has a prominent, medial, toelike edge (Fig. 17.19 and Chapter 7). This resembles an enlarged basal ventral medial nucleus though it might well represent a ventral posteromedial nucleus becuse both the echidna and the platypus have very large trigeminal nerves and the trigeminal representation in the cerebral cortex is large. The rest of the lateral nuclear mass is very large but with a homogeneous architecture. Parts of it are known to project to a somatosensory responsive region in the cerebral cortex (Lende, 1964). This cortical region appears to consist of two architectonic fields but only a single representation of the body surface has so far been mapped in it (Lende, 1964; Bohringer and Rowe, 1977). The anterior of the two fields receives inputs from dorsal parts of the lateral nuclear mass and the posterior from ventral parts (Ulinski, 1984). No field equivalent to the second somatic sensory area (SII), however, has been identified. Anterior to the somatosensory area is a motor area and visual and auditory

---

* Campbell and Hayhow (1972) describe LGNa as posterodorsally placed in the platypus and posteroventrally placed in the echidna.

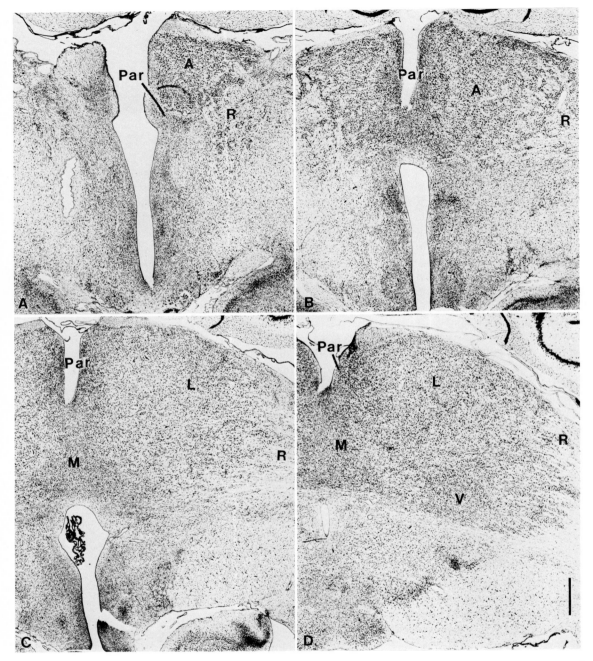

FIGURE 17.19. Thionin-stained frontal sections through the thalamus of a monotreme, the echidna (*Tachyglossus aculeatus*). Photographs kindly provided by Dr. W. I. Welker. Tentative parcellations: A: anterior nuclei; H: habenular nuclei; L: lateral nuclei; LGb: dorsal lateral geniculate nucleus; LGa: ventral lateral geniculate nucleus; M: medial nuclei; MG: medial geniculate complex; Po: posterior complex; M: medial nucleus; R: reticular nucleus; V: ventral nuclei; PT: pretectum; Par: paraventricular nuclei. Bars = 1 mm.

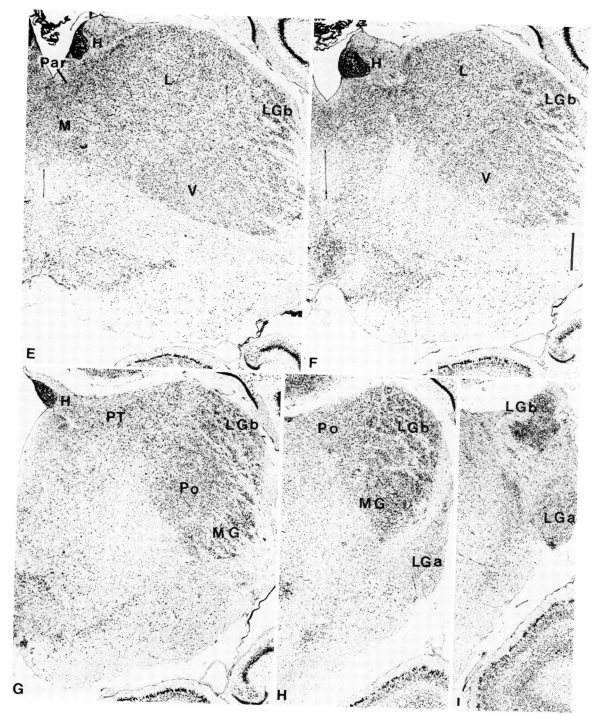

FIGURE 17.19. (continued)

cortical fields have also been defined (Lende, 1964; Allison and Goff, 1972; Bohringer and Rowe, 1977). The connectivity of these areas, however, has not been demonstrated.

The sensory areas are situated near the posterior pole of the cortex in the echidna, which therefore possesses a relatively enormous frontal cortex. Lende (1964) has suggested, as a consequence, that the greater part of the medial half of the thalamus is a large mediodorsal nucleus. But in the absence of a clearly delimiting internal medullary lamina, this is hard to determine.

There can be little doubt that the monotreme thalamus is fundamentally mammalian in its organization. Though it lacks nuclear definition, it seems probable from the little connectional work that has been done, that patterns of thalamocortical connectivity similar to those in marsupials and eutherian mammals are present.

... there are many things to be learnt in natural philosophy which abundantly reward the pains of the curious with delight and advantage. But these, I think, are rather to be found amongst such writers as have employed themselves in making rational experiments and observations, than writing speculative systems.

John Locke, Letter to Edward Clarke (1686)

Conclusions

Even a fool can farm
When he lights on fertile ground.
Rich crops have no need
Of merit in the sower.
          Vísākhadatta, *Rakshasa's Ring*,
          translated by Michael Coulson

18

Concluding Remarks

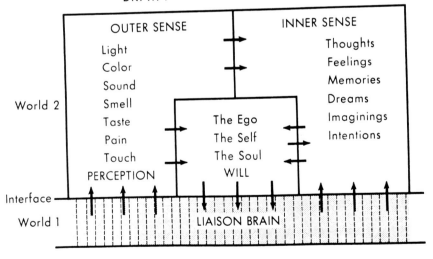

Eccles's (1977b) schematic view of mind–brain interactions. Presumably, the thalamus is part of the liaison brain.

18.1. Introduction

Walker in 1938 was able to end his book *The Primate Thalamus* with a lengthy chapter entitled "The Anatomical, Physiological and Clinical Significance of the Thalamus." In this he concluded that the thalamus was composed of three fundamentally different groups of nuclei: midline and intralaminar nuclei projecting to other diencephalic structures; three relay nuclei (ventral posterior and the geniculate bodies); and a phylogenetically recent group including mediodorsal, lateral posterior, and pulvinar nuclei not receiving fibers from the principal afferent pathways. It was his viewpoint that these latter nuclei and the thalamus in general was an integrative center for all incoming stimuli, elaborating them before presentation "to the highest hierarchy of the central nervous system, the cerebral cortex, as complex and at least partially synthesized impulses." His belief in the thalamus as a sensory integrative center seems to have derived from the clinical observation that lesions at upper levels of the neuraxis rarely lead to a disturbance of single sensory modalities. He recognized nevertheless that spatial relationships would be preserved through the thalamus because of the topographic ordering of inputs and of thalamocortical projections.

To Walker, different nuclei played different roles in the integrative functions that he envisaged for the thalamus. The midline nuclei he felt were concerned with vasomotor control through connections with the hypothalamus, but might also be concerned with pain mechanisms. The anterior nuclei were predominately olfactory in character. He could make little of the intralaminar nuclei. The mediodorsal, lateral posterior, and pulvinar nuclei were probably involved in integrating afferent sensory and visceral impulses as "sensory concepts" rather than as "modalities." These nuclei were thought to receive their inputs primarily from nuclei of the ventral group. The ventral group of nuclei also relayed "crude modalities of sensation" to the motor and sensory areas of the cerebral cortex.

We have advanced a long way since Walker's time. When he made his anatomical studies, the only data of a functional kind upon which he could build a case for the significance of the thalamus, were clinicopathological. In recent times we have not only seen a huge increase in the resolving power of anatomical techniques but also the introduction and advance of physiological methods which offer greater prospects for understanding what the thalamus does.

18.2. Types of Thalamic Nuclei

Although we have had to discard Walker's classification of the thalamic nuclei, mainly on account of the evidence that each receives some kind of subcortical input and projects to the cerebral cortex, it is clear that thalamic nuclei have different relationships with the telencephalon. The work reviewed in Chapter 3 forces us to think in terms of nuclei that project only to cortex and those (the intralaminar nuclei) that project to both cortex and striatum. But this is an inadequate characterization, for those projecting to cortex do so in different ways. It seems likely that the classical, topographically organized projection focused on layer IV of a single cortical area (or at most a pair of areas) is the least common. Because this type of projection is found in the sensory relay nuclei, the lateral geniculate, medial geniculate, and ventral posterior, it has tended to dominate our thoughts. But many other thalamic nuclei project upon several cortical areas and often do so, it would appear, in a rather diffuse manner. Such diffuse projections can terminate in one or more of several layers other than layer IV. Of these, layers I and VI may be the commonest to receive diffuse inputs but no layer is necessarily exempt from a thalamic input: it depends on the area of cortex in which the layer is situated and on the thalamic nucleus or nuclei projecting to it. Macchi's (1983) characterization of the thalamic nuclei according to their type of cortical projection (Fig. 3.3), therefore, has much to commend it.

It was pointed out in Chapter 3 that the division of thalamic nuclei into those with focused (or "specific") and those with diffuse (or "nonspecific") cortical projections is not simply a division into principal and intralaminar nuclei. Although most available evidence indicates that no nuclei other than the intralaminar project upon the striatum, many nuclei other than the intralaminar project diffusely upon the cortex. And it is at this point that our attempts at classifying thalamic nuclei start to fail, for it seems clear that certain nuclei, such as the dorsal lateral geniculate, the lateral posterior or pulvinar, the ventral posterior, and certain of the anterior nuclei, contain two or more populations of cells, one projecting in a focused manner upon a single cortical area and others projecting diffusely upon several areas. What are the relative contributions that each of these make to cortical activity? Let us try to explore this question in the following sections.

18.3. Selective Channels through the Relay Nuclei

Needless to say, the focused or "specific" thalamocortical projections of the three principal sensory relay nuclei have been more thoroughly studied than the diffuse or "nonspecific" projections from these or other nuclei. It has become very clear from work in the visual system that relatively independent channels of communication exist in the dorsal lateral geniculate nucleus whereby different functional classes of retinal ganglion cells can gain access to the visual cortex. This relates not only to the W, X, Y groupings of ganglion cells but also to those with on-center and off-center receptive fields among them (Chapters 4 and 9). Even at the level of the cortex itself, these channels may remain distinct on account of the axons of the geniculate cells ending in different layers and even on different cell types. There is less information available for other relay nuclei but the submodalities of somatic sensation appear to be projected to the cortex through separate channels in the ventral posterior nucleus (Chapters 3 and 7). In one case at least (that of the Meissner corpuscle) an individual receptor class appears to be projected independently upon the somatic sensory cortex. It seems likely that individual channels comparable to those mentioned may be found in other relay nuclei. The segregation of terminations of axons from the different deep cerebellar nuclei in the ventral lateral complex (Chapter 7) comes to mind here. What these seemingly independent channels of input mean for the motor cortex now becomes a question worth exploring.

18.4. Common Principles of Topographic Organization

The precision with which the axons of thalamic cells seek out appropriate cortical cells on which to terminate is clearly not simply a matter of topographic organization. The evidence of competition for synaptic space during the development of the geniculostriate projection (Chapter 6) speaks against that. Furthermore, if it should hold up that the axons of different functional classes of thalamic cells end on different classes of cortical neurons, as intimated from work in the visual cortex (Gilbert and Wiesel, 1979, 1983; McConnell and LeVay, 1983), then this, too, speaks against topographic ordering being the sole determinant of the precision of thalamocortical projections at the single-cell level. Accepting this, however, and the active cell–cell recognition process that it implies developmentally, it is still evident that there are some common principles of topographic ordering in all the thalamic relay nuclei. At the crudest level, there is a lamella-to-band pattern of thalamocortical projection in which a lamellar arrangement of relay cells representing a body part in the ventral posterior nucleus or a narrow band of frequencies in the ventral medial geniculate nucleus projects upon a comparable band in the somatic sensory or auditory cortex. In the visual system, this becomes an arc-to-arc pattern in which an arclike arrangement of cells representing a few degrees of visual field eccentricity projects upon a comparable arc of visual cortex (Chapter 3). It seems likely that there are similar patterns of projection in other nuclei such as the ventral lateral and lateral posterior or pulvinar. This organizational pattern is of the crudest but it can be dissected further into what I have called a rod-to-column pattern of projection (Chapter 3). In this, an elongated group of cells along a projection line spanning the laminae of the dorsal lateral geniculate nucleus or along an anteroposterior trajectory in a lamella of the ventral posterior or ventral medial geniculate nuclei, projects to a narrow column in the cortex (to a pair of ocular dominance columns in the visual cortex).

In the ventral posterior nucleus it is evident that different rods of cells receive inputs from different constellations of peripheral receptors, thus maintaining the segregation of submodalities in the nucleus and the projection of these to independent columns in the somatic sensory cortex. In the ventral medial geniculate nucleus there is a comparable segregation of EE and EI cells in different thalamic rods and, thus, in independent cortical columns (Chapter 8). In the visual systems there may be a further fractionation in view of the new evidence that on-center and off-center geniculate relay cells may be segregated in different laminae or sublaminae of the dorsal lateral geniculate nucleus of some animals and that on-center and off-center cells project to different columns in the visual cortex. That is, parts of a rod of cells along a projection column in the dorsal lateral geniculate nucleus, though projecting within the same pair of ocular dominance columns in the cortex, seek out different parts of the ocular dominance columns according to whether they are off-center or on-center in type. One assumes that within each on-center or off-center column, X-type and Y-type axon terminations may be further segregated. It is at this level that we must switch our thinking from the topographic ordering of groups of thalamocortical projection neurons to a pattern based upon the specificity of the connections made by individual neurons.

18.5.   State-Dependent Activity

Circuitry still tends to dominate our thoughts on the thalamus and cortex and the conclusion of the preceding paragraph in a sense represents the reductionist approach to this. It is very difficult to do otherwise as most of the data we have speak eloquently for the isolation of afferent channels through the relay nuclei. Should we regard the integrative actions of the thalamus as being solely confined to what Poggio and Mountcastle (Chapter 7) called the dynamic properties of ventral posterior neurons? That is, properties that reside in the temporal rather than the spatial domain? Or is there some other property that can be identified that the thalamus imposes on the afferent streams traversing it?

One obvious possibility in this regard is to consider those activities of thalamic neurons that are state dependent, i.e., those properties that covary with the level of consciousness of the individual. From the work reviewed in Chapters 4, 9, and 12, it is very clear that a thalamic relay neuron is a more efficient transmitter of sensory impulses to the cortex in an awake than in a drowsy animal. Unfortunately, this is an area where circuit analysis lets us down. We think that the brain-stem reticular formation controls the levels of excitability of thalamic relay neurons but if we take a hard, critical look at the anatomical data presented in support of this claim, they are altogether rather tenuous (Chapters 4 and 9). As Brodal (1981) has cogently put it: " 'the activating system' is a functional concept, the 'reticular formation' a morphological one, and it has been obvious for many years that these do not correspond."

It is conceivable that we know more about how the thalamus controls the level of excitability of the cortex than about how that of the thalamus is controlled. Interest in the effects exerted on the cortex by the nonspecific thalamocortical projection systems has been rekindled lately by the evidence presented by Steriade and his co-workers (Chapters 4 and 12), of a close correlation between the activities of cells in the intralaminar and ventral medial nuclei and those of the midbrain reticular formation in which a change of activity seems to lead the change in behavioral state. It is still rather difficult to assess the manner in which the nonspecific systems exert their effects on the cortex and the evidence that there are multiple nonspecific or diffuse thalamocortical projection systems targeted at different cortical regions and different cortical layers makes them doubly difficult to analyze.

18.6. Intrinsic Properties of Thalamic Neurons

We tend to think automatically of the thalamus as a center that takes an afferent input and relays it more or less faithfully to the cerebral cortex. We admit, however, that the afferent drive may be more or less effective depending upon the state of arousal of the possessor of that thalamus. But, here too, our tendency is to consider this state-dependent effect being itself dependent upon inputs from other brain centers. These may be the reticular formation, the cerebral cortex, or some other system depending on one's point of view. The thalamus, it seems, is not permitted to have a will of its own. Yet there is growing evidence that thalamic relay neurons do possess innate properties that may determine some aspects of their behavior, independent of afferent driving.

The issue of state-dependent properties of dorsal thalamic neurons is bound up with the question of what determines the rhythmic, oscillatory discharges that characterize their spontaneous behavior. Spontaneous rhythmic discharges with a frequency of 7–10 per second, occurring in bursts or "spindles," were noted in some of the earliest extracellular recordings made in the thalamus of anesthetized animals (Morison et al., 1943a,b; Dempsey and Morison, 1943; Galambos et al., 1952; Verzeano and Calma, 1954) and were thought even then to underlie spontaneous cortical waves with a similar frequency (Bremer, 1958). These cortical spindles were shown to be recorded best in the same areas in which recruiting responses were obtained from low-frequency stimulation of the intralaminar nuclei (Chapter 12).

There have been many attempts to define the underlying mechanism of thalamic spindling. The work of Purpura and Cohen (1962), of Andersen et al. (1964a), and of Andersen and Sears (1964) showed that regular sequences of EPSPs and IPSPs could be induced in thalamic relay cells by low-frequency stimulation in the midline of the thalamus or by a single afferent volley in a peripheral nerve (Chapters 4 and 12). In analyzing their intracellular records of spontaneous spindles and spindles evoked by stimulation of peripheral nerves or of the subcortical white matter, Andersen and Sears (1964) determined that the underlying basis of thalamic spindling was a sequence of large, augmenting and summing IPSPs, each lasting about 120 msec. During each IPSP the cell was inhibited from firing but spikes occurred regularly on the crests between successive IPSPs. They concluded that the IPSPs were induced by a recurrent inhibitory pathway in which the inhibitory interneurons were activated in the thalamus by recurrent collaterals of thalamic relay cell axons (Chapter 4). Through this means, they thought, several thalamic cells could be induced to discharge spindles more or less in synchrony and that this might be the origin of the α rhythm seen in the electrocorticogram (Andersen and Andersson, 1968).

Many of the early workers showed that thalamic spindle activity does not depend on the cerebral cortex because spontaneous and evoked spindles could still be recorded after removal of the cortex. Spontaneous spindling also survives separation of the thalamus from the brain stem (Dempsey and Morison, 1943) and decortication plus separation of the VPL from the intralaminar nuclei (Andersen and Sears, 1964). To Andersen and Sears, therefore, thalamic spindling was the result of recurrent inhibition and did not depend upon a pacemaker nucleus.

Like with the recruiting response (Chapter 12), interest in the mechanism of thalamic spindling declined considerably in the 1970s. Only Steriade and his co-workers seem to have maintained an active experimental approach. These workers, while accepting that oscillatory behavior was an intrinsic characteristic of thalamic relay neurons, have nevertheless felt that they could demonstrate that corticothalamic connections have the capacity to trigger spindle bursts in the thalamus, probably by an effect on inhibitory interneurons. They, therefore, conclude that the cortex may reinforce rhythmic oscillations of the thalamus especially in the act of falling asleep, with the concomitant induction of spindle waves in the cortex itself (Steriade *et al.*, 1972; Steriade, 1983).

The most recent contributions to this topic come from the studies of Llinás and his colleagues (Jahnsen and Llinás, 1982, 1984a,b) on *in vitro* slices of the guinea pig thalamus. These studies clearly indicate that rhythmic oscillatory discharges are an intrinsic property of thalamic neurons and may not depend in the first instance upon other (inhibitory) interneurons at all. Their basic observation is that all thalamic neurons have the capacity to switch their firing behavior between two states depending on their membrane potential. At membrane potentials negative to minus 60 mV, the cells exhibit bursting behavior whether stimulated directly, antidromically, or by means of afferent synapses. The underlying mechanism of this burst behavior appears to be a change in calcium conductance presumably located in the soma membrane and, therefore, not obviously dependent on IPSPs. After activation, there is a refractory period of 150–180 msec which resembles the postactivation refractoriness well known from *in vivo* studies and usually attributed to IPSPs. At membrane potentials positive to minus 55 mV, the cells, when stimulated, fire the repetitive fast spikes typical of thalamocortical relay cells and with a marked hyperpolarization following each spike. It was thought that the afterhyperpolarization depended upon a calcium-dependent, potassium conductance mechanism residing in the dendritic membranes of the cell, and that this second behavior could put the cell in an oscillatory state that would produce low-frequency waves in the cerebral cortex. These findings indicate that thalamic cells have the capacity to switch from one integrated state to another simply with changes in their membrane potential. It is not necessary, therefore, to invoke particular kinds of circuitry in attempting to explain the basis of thalamic spindling and through it the generation of the α rhythm and of recruiting responses in the cerebral cortex. Indirectly, circuitry must be involved in driving the thalamic cells' membrane potentials toward one or the other functional state but it seems likely that the exact nature of the circuitry involved may vary from time to time. Recent experiments (Deschênes *et al.*, 1984; Roy *et al.*, 1984) indicate that mechanisms similar to those elucidated by Jahnsen and Llinás may be detectable *in vivo*.

Because of the indications that recruiting responses and barbiturate spindles are generated in regions of cortex which are the foci of major inputs from the intralaminar nuclei (Chapter 12), the tendency has grown up to consider the intralaminar projection as the principal determinant of cortical slow-wave activity and thus of states of sleep and wakefulness. The recognition that other thalamic nuclei have diffuse projections upon the cortex makes it necessary to consider that these, too, may be involved in maintaining levels of cortical activity. But the indications that spindling behavior is typical of all thalamic neurons, no matter what nucleus they are situated in, seem to imply that cortical slow-wave activity might be induced by any thalamic nucleus, not by a single defined set of pacemaker nuclei. The only thing we can be reasonably confident of is that the thalamic nuclei do talk to the cortex. The lack of internuclear connections, with the exception of the reticular nucleus, ensures that they cannot talk directly to one another.

18.7. Similarity among Thalamic Nuclei

The work of Jahnsen and Llinás reviewed in the preceding section indicates that, from the point of view of their electrophysiological behavior, all thalamic cells are similar, irrespective of the nucleus in which they are located. Studies of the fine structural organization of the thalamic nuclei indicate that, anatomically, too, all the nuclei of the dorsal thalamus are fundamentally similar. It is evident that each possesses a complement of relay neurons and a population of GABAergic and, therefore, inhibitory interneurons. The only detectable differences between nuclei and between mammalian species seem to relate to the proportions of each population. The synaptic circuitry, whether studied anatomically or physiologically, between these two groups of cells and between them and the afferent fibers entering a nucleus from subcortical sites and from the cerebral cortex is also identical for all dorsal thalamic nuclei. For each nucleus, a second set of GABAergic inhibitory neurons is added by virtue of the sector of the reticular nucleus related to the nucleus (Chapters 3 and 15). The intrinsic interneurons appear to operate in a feedforward manner, though their effect, apparently being mediated by presynaptic dendrites, makes this somewhat atypical. The reticular nucleus interneurons appear to operate in a typical feedback manner as the result of their apparent activation by collateral branches of thalamocortical relay cell axons. Looked at from the point of view of corticothalamic fibers, however, they can be thought of as acting in a feedforward manner.

GABA is the only conventional transmitter that has been conclusively identified in the dorsal and ventral thalamus. GABA is a major transmitter, too, for there are no excitatory interneurons in the thalamus and there is an almost complete lack of recurrent collaterals from thalamocortical fibers, except in the reticular nucleus. The other major thalamic transmitter would presumably be that contained in afferent synapses but there are no indications as to what this transmitter might be. Among other afferent pathways, it seems likely that acetylcholine may be the transmitter in the projection of the nucleus cuneiformis to the reticular nucleus though this is still somewhat tentative. The apparent noradrenergic and serotoninergic inputs to nuclei such as the anterior and the ventral lateral geniculate have not been effectively analyzed anatomically or physiologically. It is likely (Chapter 5) that aspartate or glutamate may be the transmitter used by corticothalamic fibers but there are still no hints as to the nature of the transmitter in the major sets of afferent fibers converging on each nucleus, nor about the nature of the thalamocortical transmitter.

The dorsal thalamus appears singularly deficient in any of the neuropeptides that have been identified to date. Somatostatin and neuropeptide Y appear in cells and/or fibers in the ventral thalamus and in the epithalamus, and substance P and β-endorphin are also found in the latter. The dorsal thalamus, however, always allowing for the fact that some hitherto unidentified neuropeptide may be found there, seems to be a privileged site with a relatively small complement of neurotransmitters or neuromodulators. It is all the more exasperating, therefore, that the identities of the transmitters concerned with the flow of information from periphery to cortex continue to elude us.

18.8. Morphological Diversity and Parallelism among Thalamic Nuclei?

Work on the dorsal lateral geniculate nucleus has led to the morphological identification of different functional classes of afferent fibers and of thalamocortical relay neurons (Chapters 4 and 9). The exciting findings in this system have naturally raised hopes that afferent fibers and relay neurons in other thalamic nuclei such as the ventral posterior might also show morphological differences related to their functional status, whether it be in terms of the submodality served or some other characteristic such as tonic or phasic discharge properties. At present, this hope has not been realized. Although there are some indications of morphological diversity among ventral posterior neurons (Chapter 4), these are not as striking as in the case of the dorsal lateral geniculate nucleus and it has not been possible to relate the differences to unique physiological properties. Once again, the worker on the lateral geniculate nucleus has the advantage.

Where the visual system presents diversity in the afferent channels through the dorsal lateral geniculate nucleus, it also displays diversity in the routes that it uses to transfer information through the thalamus to the cortex. The early anatomists had some inkling that the lateral posterior–pulvinar complex was in some way concerned with visual functions on account of its association with the elaboration of the primate temporal lobe. But until quite recent times it could still be argued that this large mass of thalamic neurons was divorced from the peripheral visual apparatus and under the influence of other thalamic nuclei through some nebulous set of intrathalamic connections. In recent years it has become evident that, with the exception of the reticular nucleus, no thalamic nucleus makes connections with its neighbors. Moreover, every thalamic nucleus, including the lateral posterior–pulvinar complex, receives a subcortical input. We can drop, therefore, the whole idea of some thalamic nuclei being "association nuclei." All are, in a sense, relay nuclei.

For the visual system, the lateral posterior–pulvinar complex provides several additonal routes through the thalamus to the cerebral cortex. These additional channels, commencing in the superior colliculus or pretectum, are no less highly organized than the retinogeniculostriate system itself (Chaptes 4, 9, and 10). Although a rigid separation of functions between geniculostriate and extrageniculostriate systems, along the lines originally suggested by Schneider (1969) in his concept of the two visual systems, is now unacceptable, there is sufficient behavioral evidence to suggest that these different visual pathways do have different functional attributes.

These multiple visual pathways may have their equivalents in the organization of the auditory system in which there seem to be at least two and possibly more parallel pathways through the subdivisions of the medial geniculate complex (Chapter 8). As yet, however, there is virtually no information available about the functional capacities of nuclei other than the ventral medial geniculate nucleus.

In the somatic sensory system there are certainly segregated channels through the ventral posterior nucleus for the major submodalities (Chapters 4 and 7). The evidence for somatic sensory pathways through other nuclei, comparable to the visual routes through the lateral posterior–pulvinar complex, is less evident. Possibly the relay of spinothalamic, cerebellar, and vestibular pathways through the posterior ventral lateral nucleus to the motor cortex can be regarded as a parallel route. There is also the anterior pulvinar nucleus and part of the lateral posterior nucleus which project upon cortical fields adjacent to and connected with the sensory motor fields. Unfortunately, we know too little about their afferent inputs to regard them as somatic sensory in function.

What seems evident in this brief summary of data presented in full elsewhere in this book, is that virtually no dorsal thalamic nucleus stands divorced from a subcortical input and in the case of many nuclei, this input is little removed from the peripheral sensory and motor apparatus. The contribution that the extrageniculostriate visual pathways make to cortical function is starting to be analyzed and it is to be hoped that those pathways which hold a similar parallel relationship to the other main sensory nuclei will soon start to yield up their functional secrets also.

18.9.   The Thalamus and Pain

A large part of Walker's concluding chapter was devoted to the question of the thalamic pain syndrome (Chapter 1) and its causation. Though we are no nearer an explanation of why necrosis of the posterolateral part of the thalamus due to interference with the thalamogeniculate vessels, leads to intractable pain, we do seem to have come nearer an understanding of the routes taken by the classical pain pathways through the thalamus (Chapters 4, 7, and 11).

The view that painful sensations might rise to consciousness in the thalamus has no adherents nowadays. Instead, investigators are looking toward the cerebral cortex. It has become evident that significant numbers of neurons in the ventral posterior nucleus are innervated by the spinothalamic tract and, through it, they are connected specifically with high-threshold (nociceptive) receptors at the periphery. There is little reason to doubt, therefore, that impulses leading ultimately to the awareness and localization of a painful stimulus are relayed to the somatic sensory cortex. Because of this, it is somewhat surprising that the ventral posterior nucleus is singularly lacking in innervation by substance P- or enkephalin-containing fibers and that it lacks opiate receptors, all features that characterize regions associated with the pain pathways and their modulation in the brain stem and spinal cord.

Another major region of the thalamus to which fibers arising from spinothalamic tract cells with high-threshold inputs project, is the intralaminar nuclei—not the centre médian nucleus as many have postulated, but the greater part of the rostral group of intralaminar nuclei (Chapter 12). These are characterized by the presence of significant numbers of opiate receptors (Chapter 5) but their heavy striatal and modest and diffuse cortical projection seems to make them unsatisfactory candidates for transmitting information with any precision upon the cortex.

Of the other thalamic nuclei that receive spinothalamic inputs, neither the posterior ventral lateral nucleus (projecting to motor cortex) nor the medial nucleus of the posterior complex (projecting to opercular cortex) seems to contain neurons with properties now thought to be appropriate for the mediation of pain (Chapter 11). And it is difficult to conceive of the small submedial nucleus (projecting to frontal cortex) being a significant pain relay.

Looked at simply from the point of view of the targets of the spinothalamic tract, we have a constellation of several thalamic nuclei and perhaps it is the variety in these that contributes to that complex sensory phenomenon with its host of affective overtones that we call pain. It is clear that Henry Head's (1920) hypothesis of the cause of the thalamic pain syndrome has little validity and it was being attacked even at the time that Walker (1938a) was writing. In Head's view corticothalamic fibers suppressed the lateral thalamus which, in turn, suppressed the medial thalamus in which some phylogenetically old, protopathic component of the spinothalamic system ended. Destruction of the lateral thalamus, thus, unleashed protopathic sensations in the medial thalamus. One still reads about a medial, paleospinothalamic system and a lateral, neospinothalamic system. However, it seems to me that these make very little sense phylogenetically or functionally. They certainly do not help us to understand central mechanisms of pain. The fact that intralaminar cells with inputs from high-threshold nociceptors (Chapter 12) do not have properties suitable for mediating intensity or localization of a stimulus, makes theories of a phylogenetically old, protopathic, medially projecting spinothalamic system rather unlikely. Indeed, if one had to predict what the intralaminar nuclei might do in relation to pain, one would be likely to opt for their having a role in the affective aspect; this would seem to represent a quality more highly represented in man than in creatures lower on the rather dubious phylogenetic scale of living animals.

18.10. Corticothalamic Reciprocity

At least in regard to the three principal sensory relay nuclei we can probably feel that we have a fairly good idea of the nature and quality of much of the information that they transmit to the cerebral cortex. When we come to consider the inverse relationship, i.e., the nature of the information relayed from the cortex back to the thalamus, we have to confess to almost total ignorance. Though the principle of corticothalamic reciprocity (Chapters 3 and 4) is well established anatomically and there is plenty of morphological evidence that the number of corticothalamic synapses made in a thalamic nucleus is quite considerable, so little effect seems to be exerted by these under experimental conditions that they can be virtually ignored, even when the cortex is stimulated to drive thalamocortical neurons antidromically. Can it be that such a highly organized and numerically dense projection has virtually no functional significance? One doubts it. The very anatomical precision speaks against it. Every dorsal thalamic nucleus receives fibers back from all the cortical areas to which it projects; the nature of the thalamocortical projection to a particular area can be predicted from where the somata of the corticothalamic neurons reside: in layer VI for a specific thalamocortical projection and in layer V for a nonspecific; in the monkey visual system, different sublayers of layer VI even project upon different laminae of the dorsal lateral geniculate nucleus (Chapters 3 and 9). Experimental studies have been singularly unsuccessful in revealing the functional role of this ubiquitous projection. As to theories, we do have Singer's interesting idea on the role of corticothalamic fibers in setting up mechanisms of stereopsis in the lateral geniculate nucleus (Chapter 4). This, and the ideas of Steriade on the control of thalamic spindling by corticothalamic fibers, depend upon a rather selective activation of thalamic interneurons by the corticothalamic fibers. At the moment we have no morphological confirmation of this. What we do know is that corticothalamic fibers end in large numbers of synapses on dendrites of relay neurons distal to the concentrated region of terminals made by the principal afferent pathway to the nucleus and by the presynaptic dendrites of interneurons. Presumably, therefore, their effect (which from the morphology of their synapses and their possible use of aspartate or glutamate as a transmitter is likely to be an excitatory one) will be less powerful than that of the subcortical afferents. But this begs the question of their overall functional effect and takes no cognizance of effects that they might mediate through other terminations on inhibitory interneurons and on inhibitory neurons in the reticular nucleus.

It seems paradoxical that the thalamus with its protected channels of information flow, and its clearly defined circuitry consisting of a limited number of elements, is less well understood from the point of view of synaptology than a much more complicated structure like the cerebral cortex. The difficulty of uncovering the effect of corticothalamic fibers is but one of a number of difficulties. The way in which the presynaptic dendrites of the interneurons operate is another problem that has scarcely been touched. What function the reticular nucleus really effects is also largely a matter of conjecture. The thalamic interneuron, at least, with its presynaptic dendrites and seemingly a conventional axon as well, is unique among mammalian neurons. It presents a challenge that surely cannot be ignored much longer.

Walker (1938a) concluded his book with the words: "The thalamus . . . holds the secret of much that goes on within the cerebral cortex." At the moment we might argue that those secrets that the cortex has yielded up to the neuroscientist have told us relatively little about what the thalamus really does. It would almost be possible to study the cortex without considering the thalamus as anything more than a direct relay that neither added to nor subtracted from the information flowing through it. To adopt this attitude, however, indicates a failure to rise to the challenge of defining the critical questions to ask of the thalamus. We have had strong indications for many years that thalamic excitability is state dependent. In the work of Gottschaldt *et al.* (1983a,b) (Chapters 4 and 7), we have indications that the output of a thalamic relay neuron need not necessarily be a faithful replication of its input and that it may be determined by how and where in the peripheral receptive field a stimulus is applied. Then, in the important new work of Jahnsen and Llinás (1984a,b) (Section 6), we have indications that thalamic neurons possess unique biophysical properties, the operations of which may lead to effects which mimic synaptic driving by other cells when, in fact, this is absent. It is in these areas that the future challenge to the investigator of the thalamus lies.

There is a dead medical literature and there is a live one. The dead is not all ancient, the live is not all modern.
O. W. Holmes, Sr., *The Autocrat of the Breakfast Table*

References

Abbie, A. A., 1934, The brainstem and cerebellum of echidna, *Philos. Trans. R. Soc. London Ser. B* **224:**1–74.

Abe, Y., 1952, Zur Cytoarchitecktonik des Thalamus beim Elephanten, *Folia Psychiatr. Neurol. Jpn.* **5:**213–239.

Ables, M. F., and Benjamin, R. M., 1960, Thalamic relay for taste in albino rat, *J. Neurophysiol.* **23:**376–382.

Abplanalp, P., 1970, Some subcortical connections of the visual system in three shrews and squirrels, *Brain Behav. Evol.* **3:**155–168.

Abplanalp, P., 1971, The neuroanatomical organization of the visual system in the tree shrew, *Folia Primatol.* **16:**1–34.

Adrian, E. D., 1940, Double representation of the feet in the sensory cortex of the cat, *J. Physiol. (London)* **98:**16.

Adrian, H. O., Lifschitz, W. M., Tavitas, R. J., and Galli, F. P., 1966, Activity of neural units in medical geniculate body of cat and rabbit, *J. Neurophysiol.* **29:**1046–1060.

Aggleton, J. P., and Mishkin, M., 1983a, Memory impairments following restricted medial thalamic lesions in monkeys, *Exp. Brain Res.* **52:**199–209.

Aggleton, J. P., and Mishkin, M., 1983b, Visual recognition impairment following medial thalamic lesions in monkeys, *Neuropsychologia* **21:**189–197.

Aggleton, J. P., and Mishkin, M., 1984, Projections of the amygdala to the thalamus in the cynomolgus monkey, *J. Comp. Neurol.* **222:**56–68.

Aghajanian, G. K., and Wang, R. Y., 1977, Habenular and other midbrain raphe afferents demonstrated by a modified retrograde tracing technique, *Brain Res.* **122:**229–242.

Ahlsén, G., and Lindström, S., 1982a, Excitation of perigeniculate neurones via axon collaterals of principal cells, *Brain Res.* **236:**477–481.

Ahlsén, G., and Lindström, S., 1982b, Mutual inhibition between perigeniculate neurones, *Brain Res.* **236:**482–486.

Ahlsén, G., and Lo, F.-S., 1982, Projections of brain stem neurons to the perigeniculate nucleus and the lateral geniculate nucleus in the cat, *Brain Res.* **238:**433–438.

Ahlsén, G., Lindström, S., and Sybirska, E., 1978, Subcortical axon collaterals of principal cells in the lateral geniculate body of the cat, *Brain Res.* **156:**106–109.

Ahlsén, G., Grant, K., and Lindström, S., 1982a, Monosynaptic excitation of principal cells in the lateral geniculate nucleus by corticofugal fibers, *Brain Res.* **234:**454–458.

Ahlsén, G., Lindström, S., Lo, F.-S., 1982b, Functional distinction of perigeniculate and thalamic reticular neurons in the cat, *Exp. Brain Res.* **46:**118–126.

Aitkin, L. M., 1973, Medial geniculate body of the cat: Responses to tonal stimuli of neurons in medial division, *J. Neurophysiol.* **36:**275–283.

Aitkin, L. M., and Dunlop, C. W., 1969, Inhibition in the medial geniculate body of the cat, *Exp. Brain Res.* **7:**68–83.

Aitkin, L. M., and Gates, G. R., 1983, Connections of the auditory cortex of the brush-tailed possum, *Trichosurus vulpecula, Brain Behav. Evol.* **22:**75–88.

Aitkin, L. M., and Moore, D. R., 1975, Inferior colliculus. II. Development of tuning characteristics and tonotopic organization in central nucleus of the neonatal cat, *J. Neurophysiol.* **38:**1208–1216.

Aitkin, L. M., and Prain, S. M., 1974, Medial geniculate body: Unit responses in the awake cat, *J. Neurophysiol.* **37:**512–521.

Aitkin, L. M., and Webster, W. R., 1972, Medical geniculate body of the cat: Organization and responses to tonal stimuli of neurons in ventral division, *J. Neurophysiol.* **35:**365–380.

Aitkin, L. M., Webster, W. R., Veale, J. L., and Crosby, D. C., 1975, Inferior colliculus. I. Comparison of response properties of neurons in central, pericentral, and external nuclei of adult cat, *J. Neurophysiol.* **38:**1196–1207.

Aitkin, L. M., Dickhaus, H., Schult, W., and Zimmerman, M., 1978, External nucleus of inferior colliculus: Auditory and spinal somatosensory afferents and their interactions, *J. Neurophysiol.* **41:**837–847.

Akers, R. M., and Killackey, H. P., 1979, Segregation of cortical and trigeminal afferents to the ventrobasal complex of the neonatal rat, *Brain Res.* **161:**527–532.

Akert, K., 1964, Comparative anatomy of the frontal cortex and thalamofrontal connections, in: *The Frontal Granular Cortex and Behavior* (J. M. Warren and K. Akert, eds.), McGraw–Hill, New York, pp. 372–394.

Akimoto, H., Negishi, K., and Yamada, K., 1956a, Studies on thalamocortical connection in cat by means of retrograde degeneration method, *Folia Psychiatr. Neurol. Jpn.* **10:**39–83.

Akimoto, H., Yamaguchi, N., Okabe, K., Nakagawa, I., Abe, K., Torii, H., and Masahashi, K., 1956b, On sleep induced through electrical stimulation of dog thalamus, *Folia Psychiatr. Neurol. Jpn.* **10:**117–153.

Albano, J. E., Norton, T. T., and Hall, W. C., 1979, Laminar origin of projections from the superficial layers of the superior colliculus in the tree shrew, *Tupaia glis, Brain Res.* **173:**1–11.

Albe-Fessard, D., and Besson, J. M., 1973, Convergent thalamic and cortical projections—The non-specific system, in: *Handbook of Sensory Physiology,* Volume II: *Somatosensory System* (A. Iggo, ed.), Springer, Berlin, pp. 490–560.

Albe-Fessard, D., and Bowsher, B., 1965, Responses of monkey thalamus to somatic stimuli under chloralose anaesthesia, *Electroencephalogr. Clin. Neurophysiol.* **19:**1–15.

Albe-Fessard, D., and Fessard, A., 1963, Thalamic integrations and their consequences at the telencephalic level, *Prog. Brain Res.* **1:**115–148.

Albe-Fessard, D., and Kruger, L., 1962, Duality of unit discharges from cat centrum medianum in response to natural and electrical stimulation, *J. Neurophysiol.* **25:**3–20.

Albe-Fessard, D., and Liebeskind, J., 1966, Origines des messages somato-sensitifs activant les cellules du cortex motor chez le singe, *Exp. Brain Res.* **1:**127–146.

Albe-Fessard, D., and Mallart, A., 1960, Dualité des responses du centre médian a la stimulation visuelle, *C.R. Acad. Sci.* **251:**1191–1193.

Albe-Fessard, D., and Rougeul, A., 1955, Activités bilatérales tardives évoquées sur le cortex du chat sours chloralose par stimulation d'une voie somesthésique, *J. Physiol. (Paris)* **47:**69–72.

Albe-Fessard, D., and Rougeul, A., 1958, Activités d'origine, somesthésique évoquées sur le cortex non-spécifique du chat anesthésié au chloralose: Role du centre médian du thalamus, *Electroencephalogr. Clin. Neurophysiol.* **10:**131–151.

Albe-Fessard, D., Mallart, A., and Aleonard, P., 1961, Réduction au cours du comportement attentif de l'amplitude des réponses évoquées dans le centre médian du thalamus chez le chat éveillé libré, porteur de l'électrodes à demeure, *C.R. Acad. Sci.* **252:**187–189.

Albe-Fessard, D., Stutinsky, F., and Libouban, S., 1966, *Atlas stereotaxique du Rat Blanc,* CNRS, Paris.

Allen, G. I., and Tsukahara, N., 1974, Cerebrocerebellar communication systems, *Physiol. Rev.* **54**:957–1006.

Allen, W. F., 1924, Distribution of fibers originating from the different basal cerebellar nuclei, *J. Comp. Neurol.* **36**:399–439.

Allen, Y. S., Adrian, T. E., Allen, J. M., Tatemoto, K., Crow, T. J., Bloom, S. R., and Polak, J. M., 1983, Neuropeptide Y distribution in the rat brain, *Science* **221**:877–879.

Allison, T., and Goff, W. R., 1972, Electrophysiological studies of the echidna, *Tachyglossus aculeatus.* III. Sensory and interhemispheric evoked responses, *Arch. Ital. Biol.* **110**:185–194.

Allman, J. M., and Kaas, J. H., 1971, Representation of the visual field in striate and adjoining cortex of the owl monkey *(Aotus trivirgatus)*, *Brain Res.* **35**:89–106.

Allman, J. M., and Kaas, J. H., 1974a, The organization of the second visual area (VII) in the owl monkey: A second order transformation of the visual hemifield, *Brain Res.* **76**:247–265.

Allman, J. M., and Kaas, J. H., 1974b, A crescent-shaped cortical visual area surrounding the middle temporal area (MT) in the owl monkey *(Aotus trivirgatus)*, *Brain Res.* **81**:199–213.

Allman, J. M., and Kaas, J. H., 1976, Representation of the visual field on the medial wall of occipital–parietal cortex in the owl monkey, *Science* **191**:572–575.

Allman, J. M., Kaas, J. H., Lane, R. H., and Miezin, F. M., 1972, A representation of the visual field in the inferior nucleus of the pulvinar in the owl monkey *(Aotus trivirgatus)*, *Brain Res.* **40**:291–302.

Allman, J. M., Kaas, J. H., and Lane, R. H., 1973, The middle temporal visual area (MT) in the bushbaby, *Galago senegalensis, Brain Res.* **57**:197–202.

Allon, N., Yeshurun, Y., and Wollberg, Z., 1981, Responses of single cells in the medial geniculate body of awake squirrel monkeys, *Exp. Brain Res.* **41**:222–232.

Altman, J., 1962, Some fiber projections to the superior colliculus in the cat, *J. Comp. Neurol.* **119**:77–96.

Altman, J., and Bayer, S. A., 1979a, Development of the diencephalon in the rat. IV. Quantitative study of the time of origin of neurons and the intranuclear chronological gradients in the thalamus, *J. Comp. Neurol.* **188**:455–472.

Altman, J., and Bayer, S. A., 1979b, Development of the diencephalon in the rat. V. Thymidine-radiographic observations on internuclear and intranuclear gradients in the thalamus, *J. Comp. Neurol.* **188**:473–500.

Altman, J., and Carpenter, M. B., 1961, Fiber projections of the superior colliculus in the cat, *J. Comp. Neurol.* **116**:157–177.

Amaral, D. G., and Cowan, W. M., 1980, Subcortical afferents to the hippocampal formation in the monkey, *J. Comp. Neurol.* **189**:573–591.

Amassian, V. E., 1951a, Cortical representation of visceral afferents, *J. Neurophysiol.* **14**:433–444.

Amassian, V. E., 1951b, Fiber groups and pathways of cortically represented visceral afferents, *J. Neurophysiol.* **14**:445–462.

Amassian, V. E., 1954, Studies on organisation of a somesthetic association area, including a single unit analysis, *J. Neurophysiol.* **17**:39–58.

Anders, J. J., and Hibbard, E., 1974, The optic system of the teleost *Cichlasoma biocellatum, J. Comp. Neurol.* **158**:145–154.

Andersen, P., and Andersson, S. A., 1968, *Physiological Basis of the Alpha Rhythm*, Appleton–Century–Crofts, New York.

Andersen, P., and Curtis, D. R., 1964, The excitation of thalamic neurones by acetylcholine, *Acta Physiol. Scand.* **61**:85–99.

Andersen, P., and Sears, T. A., 1964, The role of inhibition in the phasing of spontaneous thalamocortical discharge, *J. Physiol. (London)* **173**:459–480.

Andersen, P., Brooks, C. M., Eccles, J. C., and Sears, T. A., 1964a, The ventro-basal nucleus of the thalamus: Potential fields, synaptic transmission and excitability of both presynaptic and postsynaptic components, *J. Physiol. (London)* **174**:348–369.

Andersen, P., Eccles, J. C., and Sears, T. A., 1964b, The ventro-basal complex of the thalamus: Types of cells, their responses and their functional organization, *J. Physiol. (London)* **174**:370–399.

Andersen, P., Andersson, S. A., and Landgren, S., 1966, Some properties of the thalamic relay cells in the spino-cervico-lemniscal path, *Acta Physiol. Scand.* **68**:72–83.

Andersen, P., Junge, K., and Sveen, O., 1967, Cortico-thalamic facilitation of somatosensory impulses, *Nature (London)* **214**:1011–1012.

Andersen, R. A., Knight, P. L., and Merzenich, M. M., 1980a, The thalamocortical and cortico-thalamic connections of AI, AII, and the anterior auditory field (AFF) in the cat: Evidence for two largely segregated systems of connections, *J. Comp. Neurol.* **194:**663–701.

Andersen, R. A., Roth, G. L., Aitkin, L. M., and Merzenich, M. M., 1980b, The efferent projections of the central nucleus and the pericentral nucleus of the inferior colliculus in the cat, *J. Comp. Neurol.* **194:**649–662.

Anderson, M. E., and Yoshida, M., 1980, Axonal branching patterns and location of nigrothalamic and nigrocollicular neurons in the cat, *J. Neurophysiol.* **43:**883–895.

Andersson, S. A., 1962, Projection of different spinal pathways to the second somatic sensory area of the cat, *Acta Physiol. Scand.* **56**(Suppl. 194):1–74.

Andersson, S. A., and Norrsell, U., 1973, Unit activation in the post central gyrus of the monkey via ventral spinal pathways, *Acta Physiol. Scand.* **87:**47–48.

Andersson, S. A., Landgren, S., and Wolsk, D., 1966, The thalamic relay and cortical projection of group I muscle afferents from the forelimb of the cat, *J. Physiol. (London)* **183:**576–591.

Andersson, S. A., Norrsell, K., and Norrsell, U., 1972, Spinal pathways projecting to the cerebral first somatosensory area in the monkey, *J. Physiol. (London)* **225:**589–597.

Andô, S., 1937, Zur Zytoarchitektonik des Thalamus beim Kaninchen, *Okajimas Folia Anat. Jpn.* **15:**361–410.

Andô, S., 1938, Zur Cytoarchitektonik der Thalamuskerne der Maus, *Z. Mikrosk. Anat. Forsch.* **43:**245–265.

Andres, K. H., 1966, Über die Feinstruktur der Rezeptoren an Sinushaaren, *Z. Zellforsch. Mikrosk. Anat.* **75:**339–365.

André-Thomas, *Cerebellar Functions* (translated by W. Conyers Herring), Journal of Nervous and Mental Disease Publishing Co., New York.

Andrew, B. L., and Dodt, E., 1953, The development of sensory nerve endings at the knee joint of the cat, *Acta Physiol. Scand.* **28:**287–296.

Angaut, P., 1970, The ascending projections of the nucleus interpositus posterior of the cat cerebellum: An experimental anatomical study using silver impregnation methods, *Brain Res.* **24:**377–394.

Angel, A., Magni, F., and Strata, P., 1965, Excitability of intrageniculate optic tract fibers after reticular stimulation in the mid-pontine pretrigeminal cat, *Arch. Ital. Biol.* **103:**668–693.

Angevine, J. B., Jr., 1970, Time of neuron origin in the diencephalon of the mouse: An autoradiographic study, *J. Comp. Neurol.* **139:**129–188.

Anker, R. L., 1977, The prenatal development of some of the visual pathways in the cat, *J. Comp. Neurol.* **173:**185–204.

Anker, R. L., and Cragg, B. G., 1974, Development of the extrinsic connections of the visual cortex in the cat, *J. Comp. Neurol.* **154:**29–42.

Applebaum, A. E., Beall, J. E., Foreman, R. D., and Willis, W. D., 1975, Organization and receptive fields of primate spinothalamic tract neurons, *J. Neurophysiol.* **38:**572–586.

Applebaum, A. E., Leonard, R. B., Kenshalo, D. R., Jr., Martin, R. F., and Willis, W. D., 1979, Nuclei in which functionally identified spinothalamic tract neurons terminate, *J. Comp. Neurol.* **188:**575–585.

Arai, H., 1939, Zur Cytoarchitektonik des Thalamus der Ziege, *Z. Mikrosk. Anat. Forsch.* **45:**563–630.

Arango, V., and Scalia, F., 1978, Differential projections of the nuclei of the pretectal complex of the rat, *Anat. Rec.* **190:**327–328.

Archer, S. M., Dubin, M. W., and Strick, L. A., 1982, Abnormal development of kitten retinogeniculate connectivity in the absence of action potentials, *Science* **217:**743–745.

Arimatsu, Y., Seto, A., and Amano, T., 1978, Localization of alpha-bungarotoxin binding sites in mouse brain by light and electron microscopic autoradiography, *Brain Res.* **147:**165–169.

Armstrong, J. A., 1950, An experimental study of the visual pathways in a reptile (*Lacerta vivipara*), *J. Anat.* **84:**147–167.

Armstrong, J. A., 1951, An experimental study of the visual pathways in a snake (*Natrix natrix*), *J. Anat.* **85:**275–288.

Armstrong-James, M., 1975, The functional status and columnar organization of single cells responding to cutaneous stimulation in neonatal rat somatosensory cortex S1, *J. Physiol. (London)* **246:**501–538.

Arnold, F., 1838, *Bemerkungen über den Bau des Hirns und Rückenmarks nebst Beiträgen zur Physiologie des zehnten und elften Hirnnerven, mehren kritischen Mittheilungen sowei verschiedenen pathologischen und anatomischen Beobachtungen,* Höhr, Zurich.

Aronson, L. R., and Papez, J. W., 1934, Thalamic nuclei of *Pithecus (Macacus) rhesus.* II. Dorsal thalamus, *Arch. Neurol. Psychiatry (Chicago)* **32:**27–44.

Asanuma, C., Thach, W. T., and Jones, E. G., 1983a, Cytoarchitectonic delineation of the ventral lateral thalamic region in monkeys, *Brain Res. Rev.* **5:**219–235.

Asanuma, C., Thach, W. T., and Jones, E. G., 1983b, Anatomical evidence for segregated focal groupings of efferent cells and their terminal ramifications in the cerebellothalamic pathway of the monkey, *Brain Res. Rev.* **5:**267–297.

Asanuma, C., Thach, W. T., and Jones, E. G., 1983c, Distribution of cerebellar terminations and their relation to other afferent terminations in the thalamic ventral lateral region of the monkey, *Brain Res. Rev.* **5:**237–265.

Asanuma, H., Larsen, K. D., and Yumiya, H., 1979a, Receptive fields of thalamic neurons projecting to the motor cortex in the cat, *Brain Res.* **172:**217–228.

Asanuma, H., Larsen, K., and Zarzecki, P., 1979b, Peripheral input pathways projecting to the motor cortex in the cat, *Brain Res.* **172:**197–208.

Asanuma, H., Larsen, K., and Yumiya, H.,1980, Peripheral input pathways to the monkey motor cortex, *Exp. Brain Res.* **38:**349–355.

Atlas, O., and Ingram, W. R., 1937, Topography of the brain stem of the rhesus monkey with special reference to the diencephalon, *J. Comp. Neurol.* **66:**263–289.

Atweh, S., and Kuhar, M., 1977, Autoradiographic localization of opiate receptors in rat brain. II. The brain stem, *Brain Res.* **129:**1–12.

Auen, E. L., Poulos, D. A., Hirata, H., and Molt, J. T., 1980, Location and organization of thalamic thermosensitive neurons responding to cooling the cat oral-facial regions, *Brain Res.* **191:**260–264.

Auer, J., 1956, Terminal degeneration in the diencephalon after ablation of frontal cortex in the cat, *J. Anat.* **90:**30–41.

Avanzini, G., Franceschitti, S., and Spreafico, R., 1983, Lemniscal input to the pulvinar–lateralis posterior complex of the cat: An intracellular study, in: *Somatosensory Integration in the Thalamus* (G. Macchi, A. Rustioni, and R. Spreafico, eds.), Elsevier, Amsterdam, pp. 165–182.

Axelrad, H. E., Farkas, A., Triller, A., and Verley, R., 1977, Maturation fonctionnelle du neocortex somato-sensoriel S1 chez le rat: Étude electrophysiologique et histo-enzymologique, *Rev. Electroencephalogr. Neurophysiol. Clin.* **3:**346–350.

Babb, R. S., 1980, The pregeniculate nucleus of the monkey (*Macaca mulatta*). 1. A study at the light microscopy level, *J. Comp. Neurol.* **190:**651–672.

Bachmann, R., 1950, Zwischenhirnstudien. VIII. Zur Karyoarchitektonik des Zwischenhirns der weissen Maus, *Dtsch. Z. Nervenheilkd.* **163:**529–537.

Bacq, Z. M., 1975, *Chemical Transmission of Nerve Impulses: A Historical Sketch,* Pergamon Press, Elmsford, N.Y.

Baizer, J. S., Robinson, D. L., and Dow, B. M., 1977, Visual responses of area 18 neurons in awake, behaving monkey, *J. Neurophysiol.* **40:**1024–1037.

Baker, F. H., and Malpeli, J. G., 1977, Effects of cryogenic blockade of visual cortex on the responses of lateral geniculate neurons in the monkey, *Exp. Brain Res.* **29:**433–444.

Baker, J. F., Petersen, S. E., Newsome, W. T., and Allman, J. M., 1981, Visual response properties of neurons in four extrastriate visual areas of the owl monkey (*Aotus trivirgatus*): A quantitative comparison of medial, dorsomedial, dorsolateral, and middle temporal areas, *J. Neurophysiol.* **45:**397–416.

Baker, M. A., 1971, Spontaneous and evoked activity of neurones in the somatosensory thalamus of the waking cat, *J. Physiol. (London)* **217:**359–380.

Balaban, C. D., and Ulinski, P. S., 1981a, Organization of thalamic afferents to anterior dorsal ventricular ridge in turtles. I. Projections of thalamic nuclei, *J. Comp. Neurol.* **200:**95–130.

Balaban, C. D., and Ulinski, P. S., 1981b, Organization of thalamic afferents to anterior dorsal ventricular ridge in turtles. II. Properties of the rotundo-dorsal map, *J. Comp. Neurol.* **200:**131–150.

Balado, M., and Franke, E., 1937, *Das Corpus Geniculatum Externum,* Springer, Berlin.

Baldissera, F., and Margnelli, M., 1979, Disynaptic IPSPs of lemniscal origin in relay-cells of thalamic ventrobasal nuclear complex of the cat, *Brain Res.* **163:**171–175.

Baleydier, C., and Maugière, F., 1977, Pulvinar–lateroposterior afferents to cortical area 7 in monkeys demonstrated by horseradish peroxidase tracing technique, *Exp. Brain Res.* **27:**501–507.

Bantli, H., and Bloedel, J. R., 1976, Characteristics of the output from the dentate nucleus to spinal neurons via pathways which do not involve the primary sensorimotor cortex, *Exp. Brain Res.* **25:**199–220.

Barbas, H., and Mesulam, M. M., 1981, Organization of afferent input to subdivisions of area 8 in the rhesus monkey, *J. Comp. Neurol.* **200:**407–431.

Barlow, H. B., 1981, Critical limiting factors in the design of the eye and visual cortex, *Proc. R. Soc. London Ser. B* **212:**1–34.

Barris, R. W., 1935, Disposition of fibers of retinal origin in the lateral geniculate body: Course and termination of fibers of the optic system in the brain of the cat, *Arch. Ophthalmol. (N.Y.) [Arch. Ophthalmol. (Chicago)]* **14:**61–70.

Barry, J., 1978, Septo-epithalamo-habenular LRF-reactive neurons in monkeys, *Brain Res.* **151:**183–187.

Bartlett, J. R., Doty, R. W., Pecci-Saavedra, J., and Wilson, P. D., 1973, Mesencephalic control of lateral geniculate nucleus in primates. III. Modifications with state of alertness, *Exp. Brain Res.* **18:**214–224.

Bass, A. H., and Northcutt, R. G., 1981, Retinal recipient nuclei in the painted turtle, *Chrysemys picta:* An autoradiographic and HRP study, *J. Comp. Neurol.* **199:**97–112.

Batton, R. R., III, Jayaraman, A., Ruggiero, D., and Carpenter, M. B., 1977, Fastigial efferent projections in the monkey: An autoradiographic study, *J. Comp. Neurol.* **174:**281–306.

Bauchot, R., 1959, Étude des structures cytoarchitectonics du diencéphale de *Talpa europaea* (Insectivora, Talpidae), *Acta Anat.* **39:**90–140.

Baughman, R. W., and Gilbert, C. D., 1981, Aspartate and glutamate as possible neurotransmitters in the visual cortex, *J. Neurosci.* **4:**427–439.

Beaudet, A., and Descarries, L., 1976, Quantitative data on serotonin nerve terminals in adult rat neocortex, *Brain Res.* **111:**301–309.

Bechterew, W., 1895, Über die Schleifenschicht, *Arch. Anat. Physiol. (Anat. Abt.)* **1895:**379–395.

Beckstead, R. M., 1976, Convergent thalamic and mesencephalic projections to the anterior medial cortex in the rat, *J. Comp. Neurol.* **166:**403–416.

Beckstead, R. M., 1979, An autoradiographic examination of cortico-cortical and subcortical projections of the mediodorsal projection (prefrontal) cortex in the rat, *J. Comp. Neurol.* **184:**43–62.

Beckstead, R. M., Domesick, V. B., and Nauta, W. J. H., 1979, Efferent connections of the substantia nigra and ventral tegmental area in the rat, *Brain Res.* **175:**191–218.

Beckstead, R. M., Morse, J. R., and Norgren, R., 1980, The nucleus of the solitary tract in the monkey: Projections to the thalamus and brain stem nuclei, *J. Comp. Neurol.* **190:**259–282.

Belekhova, M. G., Kosareva, A. A., Veselkin, N. P., and Ermakova, T. V., 1978, Telencephalic afferent connections in the turtle *Emys orbicularis:* A peroxidase study, *J. Evol. Biochem. Physiol. (USSR)* **15:**97–103.

Belford, G. R., and Killackey, H. P., 1979, Vibrissae representation in subcortical trigeminal centers of the neonatal rat, *J. Comp. Neurol.* **183:**305–322.

Bell, C., 1811, Idea of a New Anatomy of the Brain; Submitted for the Observation of his Friends, privately printed.

Ben-Ari, Y., Dingledine, R., Kanazawa, I., and Kelly, J. S., 1976, Inhibitory effects of acetylcholine on neurones in the feline nucleus reticularis thalami, *J. Physiol. (London)* **261:**647–671.

Bender, D. B., 1981, Retinotopic organization of macaque pulvinar, *J. Neurophysiol.* **46:**672–693.

Bender, D. B., 1982, Receptive-field properties of neurons in the macaque inferior pulvinar, *J. Neurophysiol.* **48:**1–17.

Benevento, L. A., and Davis, B., 1977, Topographical projection of the prestriate cortex to the pulvinar nuclei in the macaque monkey: An autoradiographic study, *Exp. Brain Res.* **30:**405–424.

Benevento, L. A., and Ebner, F. F., 1970, Pretectal, tectal, retinal and cortical projections to thalamic nuclei of the Virginia opossum in stereotaxic coordinates, *Brain Res.* **18:**171–175.

Benevento, L. A., and Ebner, F. F., 1971, The contribution of the dorsal lateral geniculate nucleus to the total pattern of thalamic terminations in striate cortex of the Virginia opossum, *J. Comp. Neurol.* **143:**243–260.

Benevento, L. A., and Fallon, J. H., 1975a, The ascending projections of the superior colliculus in the rhesus monkey (*Macaca mulatta*), *J. Comp. Neurol.* **160**:339–361.

Benevento, L. A., and Fallon, J. H., 1975b, The projection of occipital cortex to the dorsal lateral geniculate nucleus in the rhesus monkey (*Macaca mulatta*), *Exp. Neurol.* **46**:409–417.

Benevento, L. A., and Rezak, M., 1975, Extrageniculate projections to layers VI and I of the striate cortex (area 17) in the rhesus monkey (*Macaca mulatta*), *Brain Res.* **96**:51–55.

Benevento, L. A., and Rezak, M., 1976, The cortical projections of the inferior pulvinar and adjacent lateral pulvinar in the rhesus monkey (*Macaca Mulatta*): An autoradiographic study, *Brain Res.* **108**:1–24.

Benevento, L. A., and Standage, G. P., 1983, The organization of projections of the retinorecipient and nonretinorecipient nuclei of the pretectal complex and layers of the superior colliculus to the lateral pulvinar and medial pulvinar in the macaque monkey, *J. Comp. Neurol.* **217**:307–336.

Benevento, L. A., and Yoshida, K., 1981, The afferent and efferent organization of the lateral geniculo-prestriate pathways in the macaque monkey, *J. Comp. Neurol.* **203**:455–474.

Benevento, L. A., Rezak, M., and Santos-Anderson, R., 1977, An autoradiographic study of the projections of the pretectum in the rhesus monkey (*Macaca mulatta*): Evidence for sensorimotor links to the thalamus and oculomotor nuclei, *Brain Res.* **127**:197–218.

Benjamin, R. M., 1963, Some thalamic and cortical mechanisms of taste, in: *Olfaction and Taste* (Y. Zotterman, ed.), Volume 1, Pergamon Press, Elmsford, N.Y., pp. 309–329.

Benjamin, R. M., and Akert, K., 1959, Cortical and thalamic areas involved in taste discrimination in the albino rat, *J. Comp. Neurol.* **111**:231–260.

Benjamin, R. M., and Burton, H., 1968, Projection of taste nerve afferents to anterior opercular–insular cortex in squirrel monkey (*Saimiri sciureus*), *Brain Res.* **7**:221–231.

Benjamin, R. M., and Jackson, J. C., 1974, Unit discharges in the mediodorsal nucleus of the squirrel monkey evoked by electrical stimulation of the olfactory bulb, *Brain Res.* **75**:181–191.

Benjamin, R. M., Emmers, R., and Blomquist, A. J., 1968, Projection of tongue nerve afferents to somatic sensory area I in squirrel monkey (*Saimiri sciureus*), *Brain Res.* **7**:208–220.

Benjamin, R. M., Jackson, J. C., and Golden, G. T., 1978, Cortical projections of the thalamic mediodorsal nucleus in the rabbit, *Brain Res.* **141**:251–265.

Benjamin, R. M., Jackson, J. C., Golden, G. T., and West, C. H. K., 1982, Sources of olfactory inputs to opossum mediodorsal nucleus identified by horseradish peroxidase and autoradiographic methods, *J. Comp. Neurol.* **207**:358–368.

Benowitz, L. I., and Karten, H. J., 1976, The tractus infundibuli and other afferents to the parahippocampal region of the pigeon, *Brain Res.* **102**:174–180.

Bentivoglio, M., Van der Kooy, D., and Kuypers, H. G. J. M., 1979, The organization of the efferent projections of the substantia nigra in the rat: A retrograde fluorescent double labeling study, *Brain Res.* **174**:1–17.

Bentivoglio, M., Macchi, G., and Albanese, A., 1981, The cortical projections of the thalamic intralaminar nuclei, as studied in cat and rat with multiple fluorescent retrograde tracing technique, *Neurosci. Lett.* **26**:5–10.

Bentivoglio, M., Molinari, M., Minchiacchi, D., and Macchi, G., 1983, Organization of the cortical projections of the posterior complex and intralaminar nuclei of the thalamus as studied by means of retrograde tracers, in: *Somatosensory Integration in the Thalamus* (G. Macchi, A. Rustioni, and R. Spreafico, eds.), Elsevier, Amsterdam, pp. 337–364.

Beresford, W. A., 1961, Fibre degeneration following lesions of the visual cortex of the cat, in: *Neurophysiologie und Psychophysik des visuallen Systems* (R. Jung and H. Kornhuber, eds.), Springer-Verlag, Berlin, pp. 247–255.

Bergquist, H., 1932, Zur Morphologie des Zwischenhirn bei niederen Wirbeiltieren, *Acta Zool. (Stockholm)* **13**:57–303.

Bergquist, H., and Källén, B., 1954, Notes on the early histogenesis and morphogenesis of the central nervous system in vertebrates, *J. Comp. Neurol.* **100**:627–659.

Berkhoudt, H., Dubbeldam, J. L., and Zeilstra, S., 1981, Studies on the somatotopy of the trigeminal system in the mallard, *Anas platyrhynchos*, IV. Tactile representation in the nucleus basalis, *J. Comp. Neurol.* **196**:407–420.

Berkley, K. J., 1973, Response properties of cells in ventrobasal and posterior group nuclei of the cat, *J. Neurophysiol.* **36**:940–952.

Berkley, K. J., 1975, Different targets of different neurons in nucleus gracilis of the cat, *J. Comp. Neurol.* **163**:285–303.

Berkley, K. J., 1980, Spatial relationships between the terminations of somatic sensory and motor pathways in the rostral brainstem of cats and monkeys. I. Ascending somatic sensory inputs to lateral diencephalon, *J. Comp. Neurol.* **193**:283–317.

Berkley, K. J., 1983, Spatial relationships between the terminations of sensory motor pathways in the rostral brainstem of cats and monkeys. II. Cerebellar projections compared with those of the ascending somatic sensory pathways in lateral diencephalon, *J. Comp. Neurol.* **220**:229–251.

Berkley, K. J., and Hand, P. J., 1978, Efferent projections of the gracile nucleus in the cat, *Brain Res.* **153**:263–283.

Berkley, K. J., and Parmer, R., 1974, Somatosensory cortical involvement in responses to noxious stimulation in the cat, *Exp. Brain Res.* **20**:363–374.

Berlucchi, G., Sprague, J. M., Levy, J., and DiBerardino, A. C., 1972, Pretectum and superior colliculus in visually guided behavior and in flux and form discrimination in the cat, *J. Comp. Physiol. Psychol. Mono. Monogr.* **78**:123–172.

Berman, A. L., 1968, *The Brain Stem of the Cat: A Cytoarchitectonic Atlas with Stereotaxic Coordinates,* University of Wisconsin Press, Madison.

Berman, A. L., and Jones, E. G., 1982, *The Thalamus and Basal Telencephalon of the Cat: A Cytoarchitectonic Atlas with Stereotaxic Coordinates,* University of Wisconsin Press, Madison.

Berman, N., 1977, Connections of the pretectum in the cat, *J. Comp. Neurol.* **174**:227–254.

Berman, N., and Jones, E. G., 1977, A retino-pulvinar projection in the cat, *Brain Res.* **134**:237–248.

Berman, N., and Payne, B. R., 1982, Contralateral corticofugal projection from the lateral, suprasylvian and ectosylvian gyri in the cat, *Exp. Brain Res.* **47**:234–238.

Berson, D. M., and Graybiel, A. M., 1978, Parallel thalamic zones in the LP–pulvinar complex of the cat identified by their afferent and efferent connections, *Brain Res.* **147**:139–148.

Bertler, Å, Falck, B., Gottfries, C.-G., Ljungren, L., and Rosengren E., 1964, Some observations on adrenergic connections between the mesencephalon and the cerebral hemispheres, *Acta Pharmacol. Toxicol.* **21**:283–298.

Bianchi, V., 1909, Anatomische Untersuchungen über die Entwicklungsgeschichte der Kerne des Thalamus opticus des Kaninchens, *Monatsschr. Psychiatr. Neurol.* **25**(Suppl.):425–471.

Bielschowsky, M., 1895, Obere Schleife und Hirnrinde, *Neurol. Centralbl.* **14**:205–207.

Biemond, A., 1930, Experimentell-anatomische Untersuchungen über die corticofugalen optischen Verbindungen bei Kaninchen und Affen, *Z. Gesamte Neurol. Psychiatr.* **129**:65–127.

Biemond, A., 1956, The conduction of pain above the level of the thalamus opticus, *Arch. Neurol. Psychiatry* **75**:231–244.

Bignall, K. E., and Imbert, M., 1969, Polysensory and cortico-cortical projections to frontal lobe of squirrel and rhesus monkey, *Electroencephalogr. Clin. Neurophysiol.* **26**:206–215.

Bignall, K. E., Imbert, M., and Buser, P., 1966, Optic projections to nonvisual cortex of the cat, *J. Neurophysiol.* **29**:396–409.

Bishop, P. O., 1953, Synaptic transmission: An analysis of the electrical activity of the lateral geniculate nucleus in the cat after optic nerve stimulation, *Proc. R. Soc. London Ser. B* **141**:362–392.

Bishop, P. O., 1983, Processing of visual information within the retinostriate system, in: *Handbook of Physiology,* Section I, Volume III, Part 1 (J. M. Brookhart, V. B. Mountcastle, and I. Darian-Smith, eds.), American Physiological Society, Washington, D. C., pp. 341–424.

Bishop, P. O., Burke, W., and Davis, R., 1959a, Synapse discharge by single fibre in mammalian visual system, *Nature (London)* **182**:728–730.

Bishop, P. O., Burke, W., and Davis, R., 1959b, Activation of single lateral geniculate cells by stimulation of either optic nerve, *Science* **130**:506–507.

Bishop, P. O., Kozak, W., Levick, W. R., and Vakkur, G. J., 1962, The determination of the projection of the visual field on to the lateral geniculate nucleus in the cat, *J. Physiol. (London)* **163**:503–539.

Biziere, K., and Coyle, J. T., 1978, Influence of cortico-striatal afferents on striatal kainic acid neurotoxicity, *Neurosci. Lett.* **8**:303–310.

Bizzi, E., 1965, Changes in the orthodromic and antidromic response of the optic tract during eye movements of sleep, *Physiologist* **8**:113.

Bizzi, E., 1966, Discharge patterns of single geniculate neurons during the rapid eye movements of sleep, *J. Neurophysiol.* **29**:1087–1095.

Bizzi, E., and Brooks, D. C., 1963, Pontine reticular formation: Relation to lateral geniculate nucleus during deep sleep, *Science* **141:**270–272.

Blakemore, C., and van Sluyters, R. C., 1974, Reversal of the physiological effects of monocular deprivation in kittens: Further evidence for a sensitive period, *J. Physiol. (London)* **237:**195–216.

Blakemore, C., and van Sluyters, R. C., 1975, Innate and environmental factors in the development of the kitten's visual cortex, *J. Physiol. (London)* **248:**663–716.

Bliss, T. V. P., and Gardner-Medwin, A. R., 1973, Long-lasting potentiation of synaptic transmission in the dentate area of the unanaesthetized rabbit following stimulation of the perforant path, *J. Physiol. (London)* **232:**357–374.

Bliss, T. V. P., and Lømø, T., 1973, Long-lasting potentiation of synaptic transmission in the dentate area of the anaesthetized rabbit following stimulation of the perforant path, *J. Physiol. (London)* **232:**331–356.

Block, C. H., and Schwartzbaum, J. S., 1983, Ascending efferent projections of the gustatory parabrachial nuclei in the rabbit, *Brain Res.* **259:**1–10.

Blomquist, A. J., Benjamin, R. M., and Emmers, R., 1962, Thalamic localization of afferents from the tongue in squirrel monkey (*Saimiri sciureus*), *J. Comp. Neurol.* **118:**77–87.

Blomquist, A., Flink, R., Bowsher, D., Griph, S., and Westman, S., 1978, Tectal and thalamic projections of dorsal column and lateral cervical nuclei: A quantitative study in the cat, *Brain Res.* **141:**325–341.

Bloom, F. E., Rossier, J., Battenburg, E. L. F., Bayon, A., French, E., Henriksen, S. J., Siggins, G. R., Segal, D., Browne, R., Ling, N., and Guillemin, R., 1978, Beta-endorphin: Cellular localization, electrophysiological and behavior effects, in: *The Endorphins: Advanced Biochemistry and Psychopharmacology* (E. Costa and M. Trabucchi, eds.), Volume 1, Raven Press, New York, pp. 89–110.

Blum, P. S., Abraham, L. D., and Gilman, S., 1979, Vestibular, auditory and somatic input to the posterior thalamus of the cat, *Exp. Brain Res.* **34:**1–9.

Bobillier, P., Seguin, S., Petitjean, F., Salvert, D., Touret, M., and Jouvet, M., 1976, The raphe nuclei of the cat brain stem: A topographical atlas of their efferent projections as revealed by auto-radiography, *Brain Res.* **113:**449–486.

Bodian, D., 1937, An experimental study of the optic tracts and retinal projection of the Virginia opossum, *J. Comp. Neurol.* **66:**113–144.

Bodian, D., 1939, Studies on the diencephalon of the Virginia opossum. Part I. The nuclear pattern in the adult, *J. Comp. Neurol.* **71:**259–324.

Bodian, D., 1942, Studies on the diencephalon of the Virginia opossum. Part III. The thalamo-cortical projection, *J. Comp. Neurol.* **77:**525–575.

Bohringer, R. C., and Rowe, M. J., 1977, The organization of the sensory and motor areas of the cerebral cortex in the platypus (*Ornithorhynchus anatinus*), *J. Comp. Neurol.* **174:**1–14.

Boivie, J., 1970, The termination of the cervicothalamic tract in the cat: An experimental study with silver impregnation methods, *Brain Res.* **19:**333–360.

Boivie, J., 1971a, The termination in the thalamus and zona incerta of fibres from the dorsal column nuclei (DCN) in the cat: An experimental study with silver impregnation methods, *Brain Res.* **28:**459–490.

Boivie, J., 1971b, The termination of the spinothalamic tract in the cat: An experimental study with silver impregnation methods, *Exp. Brain Res.* **112:**331–353.

Boivie, J., 1978, Anatomical observations on the dorsal column nuclei, their thalamic projection and the cytoarchitecture of some somatosensory thalamic nuclei in the monkey, *J. Comp. Neurol.* **178:**17–48.

Boivie, J., 1979, An anatomical reinvestigation of the termination of the spinothalamic tract in the monkey, *J. Comp. Neurol.* **186:**343–370.

Boivie, J., 1980, Thalamic projections from lateral cervical nucleus in monkey: a degeneration study, *Brain Res.* **198:**13–26.

Boivie, J., and Boman, K., 1981, Termination of a separate (proprioceptive?) cuneothalamic tract from external cuneate nucleus in monkey, *Brain Res.* **224:**235–246.

Bombardieri, R. A., Jr., Johnson, J. I., Jr., and Campos, G. B., 1975, Species differences in mechano-sensory projections from the mouth to the ventrobasal thalamus, *J. Comp. Neurol.* **163:**41–64.

Bonke, B. A., Bonke, D., and Scheich, H., 1979a, Connectivity of the auditory forebrain nuclei in the guinea fowl (*Numida meleagris*), *Cell Tissue Res.* **200:**101–121.

Bonke, D., Scheich, H., and Langner, G., 1979b, Responsiveness of units in the auditory neostriatum of the guinea fowl (*Numida meleagris*) to species-specific calls and synthetic stimuli. I. Tonotopy and functional zones of field L, *J. Comp. Physiol.* **132A**:243–255.

Bonke, D., Bonke, B. A., Langner, G., and Scheich, H., 1981, Some aspects of functional organization of the auditory neostriatum (field L) in the guinea fowl, in: *Neuronal Mechanisms of Hearing* (J. Syka and L. Aitkin, eds.), Plenum Press, New York, pp. 323–327.

Boord, R. L., 1969, Ascending projections of the primary cochlear nuclei and nucleus laminaris in the pigeon, *J. Comp. Neurol.* **133**:523–541.

Bos, J., and Benevento, L. A., 1975, Projections of the medial pulvinar to orbital cortex and frontal eye fields in the rhesus monkey (*Macaca mulatta*), *Exp. Neurol.* **49**:487–496.

Bowling, D. B., and Michael, C. R., 1980, Projection patterns of single physiologically characterized optic tract fibers in cat, *Nature (London)* **286**:899–902.

Bowling, D. B., and Michael, C. R., 1984, Terminal patterns of single, physiologically characterized optic tract fibers in the cat's lateral geniculate nucleus, *J. Neurosci.* **4**:198–216.

Boycott, B. B., and Wässle, H., 1974, The morphological types of ganglion cells of the domestic cat's retina, *J. Physiol. (London)* **240**:397–419.

Boyd, I. A., and Roberts, T. D. M., 1953, Proprioceptive discharges from stretch receptors in the knee joint of the cat, *J. Physiol. (London)* **122**:38–58.

Braford, M. R., Jr., and Northcutt, R. G., 1978, Correlation of telencephalic afferents and SDH distribution in the bony fish *Polypterus, Brain Res.* **152**:157–160.

Brauth, S. E., and Kitt, C. A., 1980, The paleostriatal system of *Caiman crocodilus, J. Comp. Neurol.* **189**:437–466.

Brauth, S. E., Ferguson, J. L., and Kitt, C. A., 1978, Prosencephalic pathways related to the paleostriatum of the pigeon (*Columba livia*), *Brain Res.* **147**:205–221.

Bravo, H., and Pettigrew, J. D., 1981, The distribution of neurons projecting from the retina and visual cortex to the thalamus and tectum opticum of the barn owl, *Tyto alba,* and the burrowing owl, *Speotyto cunicularia, J. Comp. Neurol.* **199**:419–441.

Bremer, F., 1958, Cerebral and cerebellar potentials, *Physiol. Rev.* **38**:357–388.

Bright, R., 1837, Cases and observations illustrative of diagnosis where tumors are situated at the basis of the brain; or where other parts of the brain and spinal cord suffer lesion from disease, *Guy's Hosp. Rep.* **2**:279–310.

Brinkman, J., Bush, B. M., and Porter, R., 1978, Deficient influences of peripheral stimuli on precentral neurones in monkeys with dorsal column lesions, *J. Physiol. (London)* **276**:27–48.

Brockhaus, H., 1942, Beitrag zur normalen Anatomie des Hypothalamus und der Zona incerta beim Menschen, *J. Psychol. Neurol.* **51**:96–196.

Brodal, A., 1981, *Neurological Anatomy in Relation to Clinical Medicine,* 3rd ed., Oxford University Press, London, pp. 394–447.

Brodal, A., and Pompeiano, O., 1957, The vestibular nuclei in the cat, *J. Anat.* **91**:438–454.

Brodal, A., and Rexed, B., 1953, Spinal afferents to the lateral cervicular nucleus in the cat: An experimental study, *J. Comp. Neurol.* **98**:179–212.

Brodmann, K., 1906, Beiträge zur histologischen Lokalisation der Grosshirnrinde. Fünfte Mitteilung: Über den allgemeinen Bauplan des Cortex pallii bei den Mammaliern und Zwei homologe Rindenfelder in besondern, *J. Psychol. Neurol.* **6**:275–400.

Brodmann, K., 1909, *Vergleichende Lokalisationslehre der Grosshirnrinde in ihren Prinzipien dargestellt auf Grund des Zellenbaues,* Barth, Leipzig, p. 324.

Brooks, D. C., 1967, Effect of bilateral optic nerve section on visual system monophasic wave activity in the cat, *Electroencephalogr. Clin. Neurophysiol.* **23**:134–141.

Brouwer, B., 1923, Experimentell-anatomische Untersuchungen über die corticofugalen Verbindungen bei Kaninchen und Affen, *Z. Gesamte Neurol. Psychiatr.* **129**:65–127.

Brouwer, B., 1939, Ueber die Projektion der Makula auf die Area striata des Menschen, *J. Psychol. Neurol.* **40**:147–159.

Brouwer, B., and Zeeman, W. P. C., 1926, The projection of the retina in the primary optic neuron in monkeys, *Brain* **49**:1–35.

Brown, A. G., House, C. R., Rose, P. K., and Snow, P. J., 1976, The morphology of spinocervical tract neurones in the cat, *J. Physiol. (London)* **260**:719–739.

Brown, D. L., and Salinger, W. L., 1975, Loss of X-cells in lateral geniculate nucleus with monocular paralysis: Neural plasticity in the adult cat, *Science* **189**:1011–1012.

Brown, T. S., and Marco, L. A., 1967, Effects of stimulation of the superior colliculus and lateral thalamus on visual evoked potentials, *Electroencephalogr. Clin. Neurophysiol.* **22**:150–158.

Browne, J. C., 1875, The functions of the thalami optici, *West Riding Lunatic Asylum Med. Rep.* **5**:227–256.

Bruce, C., Desimone, R., and Gross, C. G., 1981, Visual properties of neurons in a polysensory area in superior temporal sulcus of the macaque, *J. Neurophysiol.* **46**:369–384.

Bruce, L. L., and Butler, A. B., 1979, Afferent projections to the anterior dorsal ventricular ridge in the lizard *Iguana iguana*, *Neurosci. Abstr.* **5**:140.

Brückner, G., Mareš, V., and Biesold, D., 1976, Neurogenesis in the visual system of the rat: An autoradiographic investigation, *J. Comp. Neurol.* **166**:245–256.

Bruesch, S. R., and Arey, L. B., 1942, The number of myelinated and unmyelinated fibers in the optic nerve of vertebrates, *J. Comp. Neurol.* **77**:631–665.

Brugge, J. F., Dubrovsky, N. A., Aitkin, L. M., and Anderson, D. J., 1969, Sensitivity of single neurons in auditory cortex of cat to binaural tonal stimulation: Effects of varying interaural time and intensity, *J. Neurophysiol.* **32**:1005–1024.

Brugge, J. F., Anderson, D. J., and Aitkin, L. M., 1970, Responses of neurons in the dorsal nucleus of the lateral lemniscus of cat to binaural tonal stimulation, *J. Neurophysiol.* **33**:441–458.

Brugge, J. F., Javel, E., and Kitzes, L. M., 1978, Signs of functional maturation of peripheral auditory system in discharge patterns of neurons in anteroventral cochlear nucleus of kitten, *J. Neurophysiol.* **41**:1557–1579.

Brunner, H., and Spiegel, E. A., 1918, Vergleichend-anatomische Studien am Hapalidengehirn, *Folia Neurobiol.* **11**:171–203.

Brunso-Bechtold, J. K., and Casagrande, V. A., 1981, Effect of bilateral enucleation on the development of layers in the dorsal lateral geniculate nucleus, *Neuroscience* **6**:2579–2586.

Brunso-Bechtold, J. K., and Casagrande, V. A., 1982, Early postnatal development of laminar characteristics in the dorsal lateral geniculate nucleus of the tree shrew, *J. Neurosci.* **2**:589–597.

Brunso-Bechtold, J. K., Florence, S. L., and Casagrande, V. A., 1983, The role of retinogeniculate afferents in the development of connections between visual cortex and the dorsal lateral geniculate nucleus, *Dev. Brain Res.* **10**:33–40.

Bryan, R. N., Trevino, D. L., Coulter, J. D., and Willis, W. D., 1973, Location and somatotopic organization of the cells of origin of the spino-cervical tract, *Exp. Brain Res.* **17**:177–189.

Bryan, R. N., Coulter, J. D., and Willis, W. D., 1974, Cells of origin of the spinocervical tract in the monkey, *Exp. Neurol.* **42**:547–586.

Buchwald, N. A., Price, D. D., Vernon, L., and Hull, C. D., 1973, Caudate intracellular response to thalamic and cortical inputs, *Exp. Neurol.* **38**:311–323.

Buijs, R. M., 1978, Intra- and extrahypothalamic vasopressin and oxytocin pathways in the cat, *Cell Tissue Res.* **192**:432–435.

Buisseret, J., and Imbert, M., 1976, Visual cortical cells: Their developmental properties in normal and dark-reared kittens, *J. Physiol. (London)* **255**:511–525.

Bullier, J., and Henry, G. H., 1979, Laminar distribution of first-order neurons and afferent terminals in cat striate cortex, *J. Neurophysiol.* **42**:1271–1281.

Bullier, J., and Norton, T. T., 1979, X and Y relay cells in cat lateral geniculate nucleus: Quantitative analysis of receptive-field properties and classification, *J. Neurophysiol.* **42**:244–273.

Bullock, T. H., and Corwin, J. T., 1979, Acoustic evoked activity in the brain in sharks, *J. Comp. Physiol.* **129**:223–234.

Bunt, A. H., Lund, R. D., and Lund, J. S., 1974, Retrograde axonal transport of horseradish peroxidase by ganglion cells of the albino rat retina, *Brain Res.* **73**:215–228.

Bunt, A. H., Hendrickson, A. E., Lund, J. S., Lund, R. D., and Fuchs, A. F., 1975, Monkey retinal ganglion cells: Morphometric analysis and tracing of axonal projections, with a consideration of the peroxidase technique, *J. Comp. Neurol.* **164**:265–286.

Bunt, A. H., Minckler, D. S., and Johanson, G. W., 1977, Demonstration of bilateral projection of the central retina of the monkey with horseradish peroxidase neuronography, *J. Comp. Neurol.* **171**:619–630.

Bunt, S., Lund, R. D., and Land, P. W., 1981, The prenatal development of the primary optic projection in hooded and albino rats, *Invest. Ophthalmol. Vis. Sci.* **20**(Suppl.):174.

Burchfiel, J. L., and Duffy, F. H., 1972, Muscle afferent input to single cells in primate somatosensory cortex, *Brain Res.* **45**:241–249.

Burchfiel, J. L., and Duffy, F. H., 1974, Corticofugal influence upon cat thalamic ventrobasal complex, *Brain Res.* **70**:395–411.

Burdach, K. F., 1819–1826, *Vom Baue und Leben des Gehirns*, 3 vols. (Vol. 2, 1822). Dyk'schen Buchhandlung, Leipzig.

Burgess, P. R., and Clark, F. J., 1969, Characteristics of knee joint receptors in the cat, *J. Physiol. (London)* **203**:317–335.

Burgess, P. R., and Perl, E. R., 1973, Cutaneous mechanoreceptors and nociceptors, in: *Handbook of Sensory Physiology*, Volume II: *Somatosensory System* (A. Iggo, ed.), Springer, Berlin, pp. 29–78.

Burke, W., and Cole, A. M., 1978, Extraretinal influences on the lateral geniculate nucleus, *Rev. Physiol. Biochem, Pharmacol.* **80**:106–166.

Burke, W., and Sefton, A. J., 1966, Discharge patterns of principal cells and interneurones in lateral geniculate nucleus of rat, *J. Physiol. (London)* **187**:201–212.

Burrows, G. R., and Hayhow, W. R., 1971, The organization of the thalamo-cortical visual pathways in the cat: An experimental degeneration study, *Brain Behav. Evol.* **4**:220–272.

Burton, H., and Benjamin, R. M., 1971, Central projections of the gustatory system, in: *Handbook of Sensory Physiology*, Volume IV, Part 2: *Taste* (L. M. Beidler, ed.), Springer-Verlag, Berlin, pp. 148–164.

Burton, H., and Craig, A. D., Jr., 1979, Distribution of trigeminothalamic projection cells in cat and monkey, *Brain Res.* **161**:515–521.

Burton, H., and Craig, A. D., 1983, Spinothalamic projections in cat, raccoon and monkey: A study based on anterograde transport of horseradish peroxidase, in: *Somatosensory Integration in the Thalamus* (G. Macchi, A. Rustioni, and R. Spreafico, eds.), Elsevier: Amsterdam, pp. 17–42.

Burton, H., and Earls, F., 1969, Cortical representation of the ipsilateral chorda tympani nerve in the cat, *Brain Res.* **16**:520–523.

Burton, H., and Jones, E. G., 1976, The posterior thalamic region and its cortical projection in New World and Old World monkeys, *J. Comp. Neurol.* **168**:249–301.

Burton, H., and Kopf, E. M., 1984, Connections between the thalamus and the somatosensory areas of the anterior ectosylvian gyrus in the cat, *J. Comp. Neurol.* **224**:173–205.

Burton, H., Forbes, D. J., and Benjamin, R. M., 1970, Thalamic neurons responsive to temperature changes of glabrous hand and foot skin in squirrel monkey, *Brain Res.* **24**:179–180.

Burton, H., Craig, A. D., Jr., Poulos, D. A., and Molt, J. T., 1979, Efferent projections from temperature sensitive recording loci within the marginal zone of the nucleus caudalis of the spinal trigeminal complex in the cat, *J. Comp. Neurol.* **183**:753–778.

Burton, H., Mitchell, G., and Brent, D., 1982, Second somatic area in the cerebral cortex of cats: Somatotopic organization and cytoarchitecture, *J. Comp. Neurol.* **210**:109–135.

Buser, P., and Bignall, K. E., 1967, Nonprimary sensory projections on cat neocortex, *Int. Rev. Neurobiol.* **10**:111–165.

Bushnell, M. C., Goldby, M. E., and Robinson, D. L., 1981, Behavioral enhancement of visual responses in monkey cerebral cortex. I. Modulation in posterior parietal cortex related to selective visual attention, *J. Neurophysiol.* **46**:755–772.

Butler, A. B., and Northcutt, R. G., 1978, New thalamic visual nuclei in lizards, *Brain Res.* **149**:469–476.

Büttner, U., and Fuchs, A. F., 1973, Influence of saccadic eye movements on unit activity in simian lateral geniculate and pregeniculate nuclei, *J. Neurophysiol.* **36**:127–141.

Büttner, U., and Henn, V., 1976, Thalamic unit activity in the alert monkey during natural vestibular stimulation, *Brain Res.* **103**:127–132.

Büttner, U., Henn, V., and Oswald, H. P., 1977, Vestibular related neuronal activity in the thalamus of the alert monkey during sinusoidal rotation in the dark, *Exp. Brain Res.* **30**:435–444.

Cabral, R. J., and Johnson, J. I., 1971, The organization of mechanoreceptive projections in the ventrobasal thalamus of the sheep, *J. Comp. Neurol.* **141**:17–36.

Cairney, J., 1926, A general survey of the forebrain of *Sphenodon punctatum*, *J. Comp. Neurol.* **42**:255–348.

Caldwell, J. H., and Berman, N., 1977, The central projections of the retina in *Necturus maculosus*, *J. Comp. Neurol.* **171**:455–464.

Caldwell, R. B., and Mize, R. R., 1981, Superior colliculus neurons which project to the cat lateral posterior nucleus have varying morphologies, *J. Comp. Neurol.* **203**:53–66.

Calford, M. B., 1983, The parcellation of the medial geniculate body of the cat defined by the auditory response properties of single units, *J. Neurosci.* **3**:2350–2364.

Calford, M. B., and Aitkin, L. M., 1983, Ascending projections to the medial geniculate body of the cat: Evidence for multiple, parallel auditory pathways through thalamus, *J. Neurosci.* **3:**2365–2380.

Calford, M. B., and Webster, W. R., 1981, Auditory representation within principal division of cat medical geniculate body: An electrophysiological study, *J. Neurophysiol.* **45:**1013–1028.

Calma, N., 1965, The activity of the posterior group of thalamic nuclei in the cat, *J. Physiol. (London)* **180:**350–370.

Caminiti, R., Innocenti, G. M., and Manzoni, T., 1979, The anatomical substrate of callosal messages from SI and SII in the cat, *Exp. Brain Res.* **35:**295–314.

Campbell, A. W., 1905, *Histological Studies on the Localisation of Cerebral Function,* Cambridge University Press, London.

Campbell, B., and Ryzen, M., 1953, The nuclear anatomy of the diencephalon of *Sorex cinerus, J. Comp. Neurol.* **99:**1–22.

Campbell, C. B. G., 1969, The visual system of insectivores and primates, *Ann. N.Y. Acad. Sci.* **167:**388–403.

Campbell, C. B. G., and Ebbesson, S. O. E., 1969, The optic system of a teleost: *Holocentrus* reexamined, *Brain Behav. Evol.* **2:**415–430.

Campbell, C. B. G., and Hayhow, W. R., 1971, Primary optic pathways in the echidna, *Tachyglossus aculeatus:* An experimental degeneration study, *J. Comp. Neurol.* **143:**119–136.

Campbell, C. B. G., and Hayhow, W. R., 1972, Primary optic pathways in the duckbill platypus, *Ornithorhynchus anatinus:* An experimental degeneration study, *J. Comp. Neurol.* **145:**195–208.

Campbell, C. B. G., Jane, J. A., and Yashon, D., 1967, The retinal projections of the tree shrew and hedgehog, *Brain Res.* **5:**406–418.

Campos-Ortega, J. A., 1970, The distribution of retinal fibres in the brain of the pig, *Brain Res.* **19:**306–312.

Campos-Ortega, J. A., and Clüver, P. F. de V., 1968, The distribution of optic fibers in *Galago crassicaudatus, Brain Res.* **7:**487–489.

Campos-Ortega, J. A., and Hayhow, W. R., 1971, A note on the connexions and possible significance of Minkowski's "intermediäre Zellgruppe" in the lateral geniculate body of cercopithecid primates, *Brain Res.* **26:**177–183.

Campos-Ortega, J. A., Glees, P., and Neuhoff, V., 1968, Ultrastructural analysis of individual layers in the lateral geniculate body of the monkey, *Z. Zellforsch. Mikrosk. Anat.* **87:**82–100.

Campos-Ortega, J. A., Hayhow, W. R., and Clüver, P. F. de V., 1972, A note on the problem of retinal projections to the inferior pulvinar of primates, *Brain Res.* **22:**126–130.

Card, J. P., and Moore, R. Y., 1982, Ventral lateral geniculate nucleus efferents to the rat suprachiasmatic nucleus exhibit avian pancreatic polypeptide-like immunoreactivity, *J. Comp. Neurol.* **206:**397–416.

Carey, R. G., Fitzpatrick, D., and Diamond, I. T., 1979, Layer I of striate cortex of *Tupaia glis* and *Galago senegalensis:* Projections from thalamus and claustrum revealed by retrograde transport of horseradish peroxidase, *J. Comp. Neurol.* **186:**393–438.

Carman, J. B., Cowan, W. M., and Powell, T. P. S., 1964, Cortical connexions of the thalamic reticular nucleus, *J. Anat.* **98:**587–598.

Carpenter, M. B., 1961, Brain stem and infratentorial neuraxis in experimental dyskinesia, *Arch. Neurol.* **5:**504–524.

Carpenter, M. B., 1981, Anatomy of the corpus striatum and brain stem integrating system, in: *Handbook of Physiology—The Nervous System II* (J. M. Brookhart and V. B. Mountcastle, eds.), American Physiological Society, Washington, D. C.

Carpenter, M. B., and Brittlin, G. M., 1958, Subthalamic hyperkinesia in the rhesus monkey: Effects of secondary lesions in the red nucleus and brachium conjunctivum, *J. Neurophysiol.* **21:**400–413.

Carpenter, M. B., and Peter, P., 1972, Nigrostriatal and nigrothalamic fibers in the rhesus monkey, *J. Comp. Neurol.* **144:**933–1116.

Carpenter, M. B., Strominger, N. L., and Weiss, A. M., 1965, Effects of lesions of the intralaminar nuclei upon subthalamic dyskinesia: A study in the rhesus monkey, *Arch. Neurol.* **13:**113–125.

Carpenter, M. B., Nakano, K., and Kim, R., 1976, Nigrothalamic projections in the monkey demonstrated by autoradiographic technics, *J. Comp. Neurol.* **165:**401–415.

Carreras, M., and Andersson, S. A., 1963, Functional properties of neurons of the anterior ectosylvian gyrus of the cat, *J. Neurophysiol.* **26:**100–126.

Carstens, E., and Trevino, D. L., 1978, Laminar origins of spinothalamic projections in the cat as determined by the retrograde transport of horseradish peroxidase, *J. Comp. Neurol.* **182:** 151–165.

Carter, D. A., and Fibiger, H. C., 1978, The projections of the entopeduncular nucleus and globus pallidus in rat as demonstrated by autoradiography and horseradish peroxidase histochemistry, *J. Comp. Neurol.* **177:**113–124.

Casagrande, V. A., 1974, The laminar organization and connections of the lateral geniculate nucleus in tree shrew (*Tupaia glis*), *Anat. Rec.* **178:**323.

Casagrande, V. A., and Diamond, I. T., 1974, Ablation study of the superior colliculus in the tree shrew (*Tupaia glis*), *J. Comp. Neurol.* **156:**207–237.

Casagrande, V. A., and Harting, J. K., 1975, Transneuronal transport of tritiated fucose and proline in the visual pathways of tree shrew *Tupaia glis, Brain Res.* **96:**367–372.

Casagrande, V. A., and Joseph, R., 1980, Morphological effects of monocular deprivation and recovery on the dorsal lateral geniculate nucleus in *Galago, J. Comp. Neurol.* **194:**413–426.

Casagrande, V. A., Harting, J. K., Hall, W. C., Diamond, I. T., and Martin, G. F., 1972, Superior colliculus of the tree shrew (*Tupaia glis*): Evidence for a structural and functional subdivision into superficial and deep layers, *Science* **177:**444–447.

Casagrande, V. A., Guillery, R. W., and Harting, J. K., 1978, Differential effects of monocular deprivation seen in different layers of the lateral geniculate nucleus, *J. Comp. Neurol.* **179:** 469–486.

Casey, K. L., 1966, Unit analysis of nociceptive mechanisms in the thalamus of the awake squirrel monkey, *J. Neurophysiol.* **29:**727–750.

Casey, K. L., and Morrow, T. J., 1983, Ventral posterior thalamic neurons differentially responsive to noxious stimulation of the awake monkey, *Science* **221:**675–677.

Casseday, J. H., Diamond, I. T., and Harting, J. K., 1976, Auditory pathways to the cortex in *Tupaia glis, J. Comp. Neurol.* **166:**303–340.

Casson, L., 1971, *Ships and Seamanship in the Ancient World*, Princeton University Press, Princeton, N. J., pp. 28–441.

Catsman-Berrevoets, C. E., and Kuypers, H. G. J. M., 1978, Differential laminar distribution of corticothalamic neurons projecting to the VL and the center median: An HRP study in the cynomolgus monkey, *Brain Res.* **154:**359–365.

Catsman-Berrevoets, C. E., and Kuypers, H. G. J. M., 1981, A search for corticospinal collaterals to thalamus and mesencephalon by means of multiple retrograde fluorescent tracers in cat and rat, *Brain Res.* **218:**15–34.

Cavalcante, L. A., and Rocha-Miranda, C. E., 1978, Postnatal development or retinogeniculate, retinopretectal and retinotectal projections in the opossum, *Brain Res.* **146:**231–248.

Caviness, V. S., Jr., and Frost, D. O., 1980, Tangential organization of thalamic projections to the neocortex in the mouse, *J. Comp. Neurol.* **194:**335–368.

Cesaro, P., Nguyen, J., Berger, B., Alvarez, C., Albe-Fessard, D., 1979, Double labelling of branched neurons in the central nervous system of the rat by retrograde axonal transport of horseradish peroxidase and iron dextran complex, *Neurosci. Lett.* **15:**1–7.

Chacko, L. W., 1948, The laminar pattern of the lateral geniculate body in the primates, *J. Neurol. Neurosurg. Psychiatry* **11:**211–219.

Chacko, L. W., 1954a, The lateral geniculate body in Lemuroidea, *J. Anat. Soc. India* **3:**24–35.

Chacko, L. W., 1954b, The lateral geniculate body in the New World monkeys, *J. Anat. Soc. India* **3:**62–74.

Chacko, L. W., 1954c, The lateral geniculate body of *Tarsius spectrum, J. Anat. Soc. India* **3:**75–77.

Chacko, L. W., 1955a, The lateral geniculate body of the chimpanzee, *J. Anat. Soc. India* **4:**10–13.

Chacko, L. W., 1955b, The lateral geniculate body of the gibbon *Hylobates hoolock, J. Anat. Soc. India* **4:**69–81.

Chalupa, L. M., Coyle, R. S., and Lindsley, D. B., 1976, Effect of pulvinar lesions on visual pattern discrimination in monkeys, *J. Neurophysiol.* **39:**354–369.

Chalupa, L. M., Williams, R. W., and Hughes, M. J., 1983, Visual response properties in the tecto-recipient zone of the cat's lateral posterior–pulvinar complex: A comparison with the superior colliculus, *J. Neurosci.* **3:**2587–2596.

Changeux, J.-P., and Danchin, A., 1976, Selective stabilization of developing synapses as a mechanism for the specification of neuronal networks, *Nature (London)* **264:**705–712.

Chan-Palay, V., 1977, *Cerebellar Dentate Nucleus: Organization, Cytology and Transmitters*, Springer, Berlin.

Cheek, M. D., Rustioni, A., and Trevino, D. L., 1975, Dorsal column nuclei projections to the cerebellar cortex in cats as revealed by the use of retrograde transport of horseradish peroxidase, *J. Comp. Neurol.* **164**:31–46.

Chevalier, G., and Deniau, J. M., 1982, Inhibitory nigral influence on cerebellar evoked responses in the rat ventromedial thalamic nucleus, *Exp. Brain Res.* **48**:369–376.

Chiray, M., Foix, C., and Nicolesco, I., 1923, Hémitremblement du type de la sclérose en plaques par lésion rubro-thalamo-sous-thalamique, *Ann. Med. (Paris)* **14**:173–191.

Choudhury, B. P., and Whitteridge, D., 1965, Visual field projection on the dorsal nucleus of the lateral geniculate body in the rabbit, *Q. J. Exp. Physiol.* **50**:104–112.

Chow, K. L., 1952, Regional degeneration of the thalamic reticular nucleus following cortical ablations in monkeys, *J. Comp. Neurol.* **97**:36–60.

Chow, K. L., 1954, Lack of behavioral effects following destruction of some thalamic association nuclei in monkey, *Arch. Neurol. Psychol. (Chicago)* **71**:762–771.

Chow, K. L., 1955, Failure to demonstrate changes in the visual system of monkeys kept in darkness or in colored lights, *J. Comp. Neurol.* **102**:597–606.

Chow, K. L., and Dewson, J. H., III, 1966, Numerical estimates of neurons and glia in lateral geniculate body during retrograde degeneration, *J. Comp. Neurol.* **128**:63–74.

Chow, K. L., and Stewart, D. L., 1972, Reversal of structural and functional effects of long-term visual deprivation in cats, *Exp. Neurol.* **34**:409–433.

Chow, K. L., Blum, J. S., and Blum, R. A., 1950, Cell ratios in the thalamo-cortical visual system of *Macaca mulatta*, *J. Comp. Neurol.* **92**:227–240.

Christensen, B. N., and Perl, E. R., 1970, Spinal neurons specifically excited by noxious or thermal stimuli: Marginal zone of the dorsal horn, *J. Neurophysiol.* **33**:293–307.

Chu, H. N., 1932, The cell masses of the diencephalon of the opossum *Didelphys virginiana*, Monogr. Natl. Res. Inst. Psychol., Peiping, No. 2.

Chung, J. M., Kenshalo, D. R., Jr., Gerhart, K. D., and Willis, W. D., 1979, Excitation of primate spinothalamic neurons by cutaneous C fiber volleys, *J. Neurophysiol.* **42**:1354–1369.

Ciemens, V., 1970, Localized thalamic hemorrhage a cause of aphasia, *Neurology* **20**:776–782.

Clare, M. H., and Bishop, G. H., 1954, Responses from an association area secondarily activated from optic cortex, *J. Neurophysiol.* **17**:271–277.

Clark, F. J., and Burgess, P. R., 1975, Slowly adapting receptors in cat knee joint: Can they signal joint angle?, *J. Neurophysiol.* **38**:1448–1463.

Clark, W. E. Le Gros: *see* Le Gros Clark, W. E.

Clarke, P. G. H., and Whitteridge, D., 1976, The projection of the retina, including the "red area," in the pigeon, *Q. J. Exp. Physiol.* **61**:351–358.

Cleland, B. G., and Levick, W. R., 1974, Brisk and sluggish concentrically organised ganglion cells in the cat's retina, *J. Physiol. (London)* **240**:421–456.

Cleland, B. G., Dubin, M. W., and Levick, W. R., 1971, Sustained and transient neurones in the cat's retina and lateral geniculate nucleus, *J. Physiol. (London)* **217**:473–496.

Cleland, B. G., Levick, W. R., and Sanderson, K. J., 1973, Properties of sustained and transient ganglion cells in the cat retina, *J. Physiol. (London)* **228**:649–680.

Cleland, B. G., Levick, W. R., and Wässle, H., 1975a, Physiological identification of a morphological class of cat retinal ganglion cells, *J. Physiol. (London)* **248**:151–171.

Cleland, B. G., Morstyn, R., Wagner, H. G., and Levick, W. R., 1975b, Long-latency retinal input to lateral geniculate neurones of the cat, *Brain Res.* **91**:306–310.

Cleland, B. G., Levick, W. R., Morstyn, R., and Wagner, H. G., 1976, Lateral geniculate relay of slowly conducting retinal afferents to the cat visual cortex, *J. Physiol. (London)* **225**:299–320.

Coenen, A. M. L., Gerrits, H. J. M., and Vendrik, A. J. H., 1972, Analysis of the response characteristics of optic tract and geniculate units and their mutual relationship, *Exp. Brain Res.* **15**:452–471.

Coggeshall, R. E., 1964, A study of diencephalic development in the albino rat, *J. Comp. Neurol.* **122**:241–269.

Cohen, B., and Feldman, M., 1968, Relationship of electrical activity in pontine reticular formation and lateral geniculate body to rapid eye movements, *J. Neurophysiol.* **31**:806–817.

Cohen, B., Housepian, E. M., and Purpura, D. P., 1962, Intrathalamic regulation of activity in a cerebellocortical projection pathway, *Exp. Neurol.* **6**:492–506.

Cohen, D. H., and Karten, H. J., 1974, The structural organization of avian brain: An overview, in: *Birds: Brain and Behavior* (I. J. Goodman and M. W. Schein, eds.), Academic Press, New York, pp. 29–73.

Cohen, D. H., Duff, T. A., and Ebbesson, S. O. E., 1973, Electrophysiological identification of a visual area in shark telencephalon, *Science* **182**:492–494.

Cohen, M. J. S., Landgren, S., Strom, L., and Zotterman, Y., 1957, Cortical reception of touch and taste in the cat: A study of single cortical cells, *Acta Physiol. Scand.* **40**(Suppl. 135):1–50.

Cohn, R., 1956, Laminar electrical responses in lateral geniculate body of cat, *J. Neurophysiol.* **19**:317–324.

Cole, A. M., Griffin, A. E., and Burke, W., 1980, All-or-nothing and graded slow waves in the lateral geniculate nucleus to stimulation of the midbrain reticular formation, *Exp. Neurol.* **69**:528–542.

Cole, F. J., 1949, *A History of Comparative Anatomy*, Macmillan & Co., London.

Coleman, J., and Clerici, W. J., 1980, Extrastriate projections from thalamus to posterior occipital–temporal cortex in rat, *Brain Res.* **194**:205–209.

Coleman, J., and Clerici, W. J., 1981, Organization of thalamic projections to visual cortex in opossum, *Brain Behav. Evol.* **18**:41–59.

Coleman, J., Diamond, I. T., and Winer, J. A., 1977, The visual cortex of the opossum: The retrograde transport of horseradish peroxidase to the lateral geniculate and lateral posterior nuclei, *Brain Res.* **137**:233–252.

Condé, F., and Condé, H., 1978, Thalamic projections of the vestibular nuclei in the cat as revealed by retrogade transport of horseradish peroxidase, *Neurosci. Lett.* **9**:141–146.

Conley, M., Fitzpatrick, D., and Diamond, I. T., 1984, The laminar organization of the lateral geniculate body and the striate cortex in the tree shrew (*Tupaia glis*), *J. Neurosci.* **4**:171–197.

Conrad, B., Meyer-Lohmann, J., Matsunami, K., and Brooks, V. B., 1975, Precentral unit activity following torque pulse injections into elbow movements, *Brain Res.* **94**:219–236.

Conrad, L. C. A., and Pfaff, D. W., 1976a, Efferents from medial basal forebrain and hypothalamus in the rat. I. An autoradiographic study of the medial preoptic area, *J. Comp. Neurol.* **169**:185–220.

Conrad, L. C. A., and Pfaff, D. W., 1976b, Efferents from medial basal forebrain and hypothalamus in the rat. II. An autoradiographic study of the anterior hypothalamus, *J. Comp. Neurol.* **169**:221–262.

Conrad, L. C., Leonard, C. M., and Pfaff, D. W., 1974, Connections of the median and dorsal raphe nuclei in the rat: An autoradiographic and degeneration study, *J. Comp. Neurol.* **156**:179–205.

Constantine-Paton, M., 1979, Axonal navigation, *Bioscience* **29**:526–541.

Conway, J., and Schiller, P. H., 1983, Laminar organization of tree shrew dorsal lateral geniculate nucleus, *J. Neurophysiol.* **50**:1330–1342.

Conway, J., Schiller, P. H., and Misler, L., 1980, Functional organization of the tree shrew lateral geniculate nucleus, *Neurosci. Abstr.* **6**:583.

Cook, W. H., Walker, J. H., and Barr, M. L., 1951, A cytological study of transneuronal atrophy in the cat and rabbit, *J. Comp. Neurol.* **94**:267–292.

Coombs, C. M., 1949, Fiber and cell degeneration in the albino rat brain after hemicortication, *J. Comp. Neurol.* **90**:373–401.

Coombs, C. M., 1951, The distribution and temporal course of fiber degeneration after experimental lesions in the rat brain, *J. Comp. Neurol.* **94**:123–175.

Cooper, E. R. A., 1945, The development of the human lateral geniculate body, *Brain* **68**:222–237.

Cooper, I. S., Amin, I., Chandra, R., and Waltz, J. M., 1973, A surgical investigation of the clinical physiology of the LP–pulvinar complex in man, *J. Neurol. Sci.* **18**:89–110.

Cooper, M. L., and Pettigrew, J. D., 1979, A neurophysiological determination of the vertical horopter in cat and owl, *J. Comp. Neurol.* **184**:1–25.

Cooper, S., and Sherrington, C. S., 1940, Gower's tract and spinal border cells, *Brain* **63**:123–134.

Cooper, W. E., and Van Hoesen, G. W., 1972, Stria medullaris–habenular lesions and gnawing behavior in rats, *J. Comp. Physiol. Psychol.* **79**:151–155.

Cotter, J. R., and Pierson Pentney, R. J., 1979, Retinofugal projections of nonecholocating (*Pteropus giganteus*) and echolocating (*Myotis lucifugus*) bats, *J. Comp. Neurol.* **184**:381–400.

Cowan, W. M., 1973, Neuronal death as a regulative mechanism in the control of cell number in the nervous system, in: *Development and Aging in the Nervous System* (M. Rockstein, ed.), Academic Press, New York, pp. 19–41.

Cowan, W. M., and Powell, T. P. S., 1954, An experimental study of the relation between the medial mamillary nucleus and the cingulate cortex, *Proc. R. Soc. London Ser. B* **143**:114–125.

Cowan, W. M., and Powell, T. P. S., 1955, The projection of the midline and intralaminar nuclei of the thalamus of the rabbit, *J. Neurol. Neurosurg. Psychiatry* **18**:266–279.

Cowan, W. M., Adamson, L., and Powell, T. P. S., 1961, An experimental study of the avian visual system, *J. Anat.* **95**:545–563.

Cowan, W. M., Guillery, R. W., and Powell, T. P. S., 1964, The origin of the mamillary peduncle and other hypothalamic connexions from the midbrain, *J. Anat.* **98**:345–363.

Cowey, A., and Perry, V. H., 1979, The projection of the temporal retina in rats, studied by retrograde transport of horseradish peroxidase, *Exp. Brain Res.* **35**:457–464.

Cragg, B. G., 1961, The connections of the habenula in the rabbit, *Exp. Neurol.* **3**:388–409.

Cragg, B. G., 1975, The development of synapses in the visual system of the cat, *J. Comp. Neurol.* **160**:147–166.

Cragg, B. G., and Hamlyn, L. H., 1960, Histological connections and visceral actions of components of the fimbria in the rabbit, *Exp. Neurol.* **2**:581–597.

Craig, A. D., Jr., 1978, Spinal and medullary input to the lateral cervical nucleus, *J. Comp. Neurol.* **181**:729–744.

Craig, A. D., Jr., and Burton, H., 1979, The lateral cervical nucleus in the cat; anatomic organization of cervicothalamic neurons, *J. Comp. Neurol.* **185**:329–346.

Craig, A. D., Jr., and Burton, H., 1981, Spinal and medullary lamina I projection to nucleus submedius in medial thalamus: A possible pain center, *J. Neurophysiol.* **45**:443–466.

Craig, A. D., Jr., and Tapper, D. N., 1978, Lateral cervical nucleus in the cat: Functional organization and characteristics, *J. Neurophysiol.* **41**:1511–1534.

Craig, A. D., Jr., Wiegand, S. J., and Price, J. L., 1982, The thalamo-cortical projection of the nucleus submedius in the cat, *J. Comp. Neurol.* **206**:28–48.

Craigie, E. H., 1928, Observations on the brain of the humming bird (*Chrysolampis mosquitus* Linn, and *Chlorostilbon caribaeus* Lawr), *J. Comp. Neurol.* **45**:377–481.

Craigie, E. H., 1930, Studies on the brain of the kiwi (*Apteryx australis*), *J. Comp. Neurol.* **49**:223–357.

Craigie, E. H., 1932, The cell structure of the cerebral hemisphere in the humming bird, *J. Comp. Neurol.* **56**:135–168.

Crain, B. J., and Hall, W. C., 1980, The organization of the lateral posterior nucleus of the golden hamster after neonatal superior colliculus lesions, *J. Comp. Neurol.* **193**:383–401.

Crawford, J. M., 1970, The sensitivity of cortical neurones to acidic amino acids and acetylcholine, *Brain Res.* **17**:287–296.

Creel, D. J., and Giolli, R. A., 1972, Retino-geniculostriate projections in guinea pigs: Albino and pigmented strains compared, *Exp. Neurol.* **36**:420–425.

Creel, D., Hendrickson, A. E., and Leventhal, A. G., 1982, Retinal projections in tyrosinase-negative albino cats, *J. Neurosci.* **2**:907–911.

Crighel, E., Cooper, I. S., Amin, I., Waltz, J. M., and Orzuchowski, J., 1974, Preliminary electrophysiological investigations of pulvinar and LP nucleus in man, in: *The Pulvinar–LP Complex* (I. S. Cooper, M. Riklan, and P. Rakic, eds.), Thomas, Springfield, Ill., pp. 254–270.

Cropper, E. C., Eisenman, J. S., and Azmitia, E. C., 1984, An immunocytochemical study of the serotoninergic innervation of the thalamus of the rat, *J. Comp. Neurol.* **224**:38–50.

Crosby, E. C., 1917, The forebrain of *Alligator mississipiensis*, *J. Comp. Neurol.* **27**:325–402.

Crossland, W. J., 1972, Receptive field characteristics of some thalamic visual nuclei of the pigeon (*Columba livia*), Doctoral dissertation, University of Illinois, Urbana.

Crouch, R. L., 1934, The nuclear configuration of the thalamus of *Macacus rhesus*, *J. Comp. Neurol.* **59**:451–485.

Cruce, J. A. F., 1974, A cytoarchitectonic study of the diencephalon of the Tegu lizard, *Tupinambis nigro-punctatus*, *J. Comp. Neurol.* **153**:215–238.

Cruce, J. A. F., 1975, An autoradiographic study of the projections of the mammillothalamic tract in the rat, *Brain Res.* **85**:211–220.

Cruce, W. L. R., 1979, Spinal cord in lizards: in: *Biology of the Reptilia*, Volume 10 (C. Gans, R. G. Northcutt, and P. S. Ulinski, eds.), Academic Press, New York, pp. 111–131.

Cruce, W. L. R., and Cruce, J. A. F., 1975, Projections from the retina to the lateral geniculate nucleus and mesencephalic tectum in a reptile (*Tupinambis nigropunctatus*): A comparison of anterograde transport and anterograde degeneration, *Brain Res.* **85**:221–228.

Cuello, A. C., Emson, P. E., Paxinos, G., and Jessell, T., 1978, Substance P containing and cholinergic projections from the habenula, *Brain Res.* **149**:413–429.

Cullen, M. J., and Kaiserman-Abramof, I. R., 1976, Cytological organization of the dorsal lateral geniculate nuclei in mutant anophthalamic and postnatally enucleated mice, *J. Neurocytol.* **5:**407–424.

Cummings, J. F., and de Lahunta, A., 1969, An experimental study of the retinal projections in the horse and sheep, *Ann. N. Y. Acad. Sci.* **167:**293–318.

Cunningham, T. J., and Lund, R. D., 1971, Laminar patterns in the dorsal division of the lateral geniculate nucleus of the rat, *Brain Res.* **34:**394–398.

Cunningham, T. J., Huddelston, C., and Murray, M., 1979, Modification of neuron numbers in the visual system of the rat, *J. Comp. Neurol.* **184:**423–434.

Curcio, C. A., and Harting, J. K., 1978, Organization of pulvinar afferents to area 18 in the squirrel monkey: Evidence for stripes, *Brain Res.* **143:**155–161.

Currie, J., and Cowan, W. M., 1974, Evidence for the late development of the uncrossed retinothalamic projections in the frog, *Rana pipiens, Brain Res.* **71:**133–139.

Curry, M. J., 1972a, The exteroceptive properties of neurones in the somatic part of the posterior group (PO), *Brain Res.* **44:**439–462.

Curry, M. J., 1972b, The effects of stimulating the somatic sensory cortex on single neurones in the posterior group (PO) of the cat, *Brain Res.* **44:**463–481.

Curry, M. J., and Gordon, G., 1972, The spinal input to the posterior group in the cat: An electrophysiological investigation, *Brain Res.* **44:**417–437.

Curtis, D. R., and Johnston, G. A. R., 1974, Amino acid transmitters in the mammalian central nervous system, *Ergeb. Physiol.* **69:**97–188.

Curtis, D. R., and Tebēcis, 1972, Bicuculline and thalamic inhibition, *Exp. Brain Res.* **16:**210–218.

Curtis, D. R., Duggan, A. W., Felix, D., and Johnston, G. A. R., 1970, GABA bicuculline and thalamic inhibition, *Nature (London)* **226:**1222–1224.

Da Fano, C., 1909, Studien über Veränderungen im Thalamus opticus bei Defektpsychosen, *Monatsschr. Psychiatr. Neurol.* **26:**4–36.

Dahlstrom, A., and Fuxe, K., 1964, Evidence for the existence of monoamine-containing neurons in the central nervous system. I. Demonstration of monoamines in the cell bodies of brainstem neurons, *Acta Physiol. Scand.* **62**(Suppl. 232):1–55.

Daniel, P. M., and Whitteridge, D., 1961, The representation of the visual field on the cerebral cortex in monkeys, *J. Physiol. (London)* **159:**203–221.

Daniels, J. D., Pettigrew, J. D., and Norman, J. L., 1978, Development of single-neuron responses in kitten's lateral geniculate nucleus, *J. Neurophysiol.* **41:**1373–1393.

Darian-Smith, I., Isbister, J., Mok, H., and Yokota, T., 1966, Somatic sensory cortical projection areas excited by tactile stimulation of the cat: A triple representation, *J. Physiol. (London)* **182:**671–689.

Davies, P., and Verth, A. H., 1978, Regional distribution of muscarinic acetylcholine receptor in normal and Alzheimer's-type dementia brains, *Brain Res.* **138:**385–392.

Daw, N. W., 1972, Color-coded cells in goldfish, cat and rhesus monkey, *Invest. Ophthalmol.* **11:**411–416.

Daw, N. W., and Pearlman, A. L., 1970, Cat colour vision: Evidence for more than one cone process, *J. Physiol. (London)* **211:**125–137.

De Britto, L. R. G., Brunelli, M., Francesconi, W., and Magni, F., 1975, Visual response pattern of thalamic neurons in the pigeon, *Brain Res.* **97:**337–343.

Deecke, L., Schwarz, D. W. F., and Frederickson, J. M., 1975, Nucleus ventroposterior inferior (VPI) as the vestibular thalamic relay in the rhesus monkey. I. Field potential investigation, *Exp. Brain Res.* **20:**88–100.

deGroot, J., 1965, The influence of limbic structures on pituitary functions related to reproduction, in: *Sex and Behavior* (F. A. Beach, ed.), Wiley, New York, pp. 495–511.

De Haene, A., 1936, Recherches anatomo-expérimentales sur les connexions thalamo-corticales, *Cellule* **44:**315–348.

Dejerine, J. (with collaboration of Mme Dejerine-Klumpke), 1901, *Anatomie des Centres Nerveux,* Volume 2, Fasc. 1, Rueff, Paris, pp. 344–411.

Dejerine, J., and Dejerine-Klumpke, Mme J., 1895, Sur les connexions du ruban de Reil avec la corticalité cérébrale, *C. R. Soc. Biol.* **2**(Ser. 10):285–291.

Dejerine, J., and Egger, M., 1903, Contribution à l'étude de la physiologie pathologique de l'incoordination motrice, *Rev. Neurol.* **11:**397–405.

Dejerine, J., and Roussy, G., 1906, Le Syndrome thalamique, *Rev. Neurol.* **14**:521–532.

Dekaban, A., 1953, Human thalamus: An anatomical, developmental and pathological study, *J. Comp. Neurol.* **99**:639–683.

Dekaban, A., 1954, Human thalamus. An anatomical, developmental and pathological study: Development of the human thalamic nuclei, *J. Comp. Neurol.* **100**:63–97.

Dekker, J. J., and Kuypers, H. G. J. M., 1976, Quantitative EM study of projection terminals in the rat's AV thalamic nucleus: Autoradiographic and degeneration techniques compared, *Brain Res.* **117**:399–422.

Delafresnaye, J. F. (ed.), 1954, *Brain Mechanisms and Consciousness*, Blackwell, Oxford.

DeLange, S. J., 1913, Das Zwischenhirn und das Mittelhirn der Reptilien, *Folia Neurobiol.* **7**:67–93.

Delius, J. D., and Bennetto, K., 1972, Cutaneous sensory projections to the avian forebrain, *Brain Res.* **37**:205–222.

Delius, J. D., Runge, T. E., and Oechinghaus, H., 1979, Short-latency auditory projection to the frontal telencephalon of the pigeon, *Exp. Neurol.* **63**:594–609.

Dell, P., 1952, Corrélations entre le système végétatif et le système de la vie de relation: Mésencéphale, diencéphale et cortex cérébral, *J. Physiol. (Paris)* **44**:471–557.

DeMonasterio, F. M., 1978, Properties of concentrically organized X and Y ganglion cells of macaque retina, *J. Neurophysiol.* **41**:1394–1417.

Dempsey, E. W., and Morison, R. S., 1943, The electrical activity of a thalamo-cortical relay system, *Am. J. Physiol.* **138**:283–296.

Denavit, M., and Kosinski, E., 1968, Somatic afferents to the cat subthalamus, *Arch. Ital. Biol.* **106**:391–411.

Dendy, A., 1910, On the structure, development and morphological interpretation of the pineal organs and adjacent parts of the brain in the tuatara (*Sphenodon punctatus*), *Philos. Trans. R. Soc. London Ser. B* **201**:227–331.

Deniau, J. M., Lackner, D., and Feger, J., 1978, Effect of substantia nigra stimulation on identified neurons in the VL–VA thalamic complex: Comparison between intact and chronically decorticated cats, *Brain Res.* **145**:27–36.

De Renzi, E., 1982, *Disorders of Space Exploration and Cognition*, Wiley, New York.

Descartes, R., 1664, *L'Homme*, in: *Ouevres de Descartes* (C. E. Adam and P. Tannery, eds.), Volume 11, pp. 119–215, Cerf, Paris (1909).

Deschênes, M., Landry, P., and Clercq, M., 1982, A reanalysis of the ventrolateral input in slow and fast pyramidal tract neurons of the cat motor cortex, *Neuroscience* **7**:2149–2157.

Deschênes, M., Roy, J.-P., and Steriade, M., 1983, Nature of 7–14 Hz rhythmic hyperpolarizations in thalamic relay neurons, *Neurosci. Abstr.* **9**:678.

Deschênes, M., Paradis, M., Roy, J. P., and Steriade, M., 1984, Electrophysiology of neurons of lateral thalamic nuclei in cat: resting properties and burst discharges, *J. Neurophysiology* **51**:1196–1219.

De Valois, R. L., 1972, Processing of intensity and wavelength information by the visual system, *Invest. Ophthalmol.* **11**:417–427.

De Valois, R. L., Smith, C. J., Kitai, S. T., and Karoly, A. J., 1958, Responses of single cells in monkey lateral geniculate nucleus to monochromatic light, *Science* **127**:238–239.

De Vito, J. L., 1967, Thalamic projection of the anterior ectosylvian gyrus (somatic area II) in the cat, *J. Comp. Neurol.* **131**:67–78.

De Vito, J. L., 1969, Contralateral projections from the cerebral cortex to intralaminar nuclei in monkey, *J. Comp. Neurol.* **136**:193–202.

De Vito, J. L., 1978, A horseradish peroxidase-autoradiographic study of parietopulvinar connections in *Saimiri sciureus*, *Exp. Brain Res.* **32**:581–590.

De Vito, J. L., and Anderson, M. E., 1982, An autoradiographic study of efferent connections of the globus pallidus in *Macaca mulatta*, *Exp. Brain Res.* **46**:107–117.

Dews, P. D., and Wiesel, T. N., 1970, Consequences of monocular deprivation on visual behaviour in kittens, *J. Physiol. (London)* **206**:437–455.

D'Hollander, F., 1913, Recherches anatomiques sur les couches optiques: La topographic des noyaux thalamiques, *Névraxe* **14–15**:469–519.

D'Hollander, F., 1922, Recherches anatomiques sur les couches optiques: Les voies corticothalamiques et les voies corticotectales, *Arch. Biol.* **32**:249–344.

D'Hollander, F., 1928, Sur les fonctions des couches optiques, *Bruxelles Méd.* **8:**1563–1569.

D'Hollander, F., and Gerebtzoff, M. A., 1939, Les couches optiques et leurs connexions: Synthèse de nos recherches anatomo-expérimentales, *Bull. Acad. R. Méd. Belg. Ser. 6* **4:**305–314.

Diamond, I. T., 1979, The subdivisions of neocortex: A proposal to revise the traditional view of sensory, motor and association areas, *Prog. Psychobiol. Physiol. Psychol.* **8:**1–43.

Diamond, I. T., and Hall, W. C., 1969, Evolution of neocortex, *Science* **164:**251–262.

Diamond, I. T., and Utley, J. D., 1963, Thalamic retrograde degeneration study of sensory cortex in opossum, *J. Comp. Neurol.* **120:**129–160.

Diamond, I. T., Chow, K. L., and Neff, W. D., 1958, Degeneration of caudal medial geniculate body following cortical lesion ventral to auditory area II in cat, *J. Comp. Neurol.* **109:**349–362.

Diamond, I. T., Jones, E. G., and Powell, T. P. S., 1969, The projection of the auditory cortex upon the diencephalon and brain stem in the cat, *Brain Res.* **15:**205–340.

Diamond, I. T., Snyder, M., Killackey, H., Jane, J., and Hall, W. C., 1970, Thalamo-cortical projections in the tree shrew (*Tupaia glis*), *J. Comp. Neurol.* **139:**273–306.

DiChiara, G., Proceddu, M. L., Morelli, M., Mulas, M. L., and Gessa, G. L., 1979, Evidence for a GABAergic projection from the substantia nigra to the ventromedial thalamus and to the superior colliculus of the rat, *Brain Res.* **176:**273–284.

Dilly, P. N., Wall, P. D., and Webster, K. E., 1968, Cells of origin of the spinothalamic tract in the cat and rat, *Exp. Neurol.* **21:**550–562.

Dingledine, R., and Kelly, J. S., 1977, Brain stem stimulation and the acetylcholine-evoked inhibition of neurones in the feline nucleus reticularis thalami, *J. Physiol. (London)* **271:**135–154.

Distel, H., and Ebbesson, S. O. E., 1975, Connections of the thalamus in the monitor lizard, *Neurosci. Abstr.* **1:**559.

Divac, I., 1975, Magnocellular nuclei of the basal forebrain project to neocortex, brainstem and olfactory bulb: Review of some functional correlates, *Brain Res.* **93:**385–398.

Divac, I., Fonnum, F., and Storm-Mathisen, J., 1977a, High affinity uptake of glutamate in terminals of corticostriate axons, *Nature (London)* **266:**377–378.

Divac, I., LaVail, J. H., Rakic, P., and Winston, K. R., 1977b, Heterogeneous afferents to the inferior parietal lobule of the rhesus monkey revealed by the retrograde transport method, *Brain Res.* **123:**197–207.

Domesick, V. B., 1969, Projections from the cingulate cortex in the rat, *Brain Res.* **12:**296–320.

Domesick, V. B., 1970, The fasciculus cinguli in the rat, *Brain Res.* **20:**19–32.

Domesick, V. B., 1972, Thalamic relationships of the medial cortex in the rat, *Brain Behav. Evol.* **6:**457–483.

Domich, L., Steriade, M., Oakson, G., and Hada, J., 1983, Reticularis thalami neurons exhibit tonically increased rates of spontaneous firing and enhanced synaptic excitability during EEG-desynchronized behavioral states, *Neurosci. Abstr.* **9:**1214.

Donaldson, L., Hand, P. J., and Morrison, A. R., 1975, Corticothalamic relationships in the rat, *Exp. Neurol.* **47:**448–458.

Donoghue, J. P., and Ebner, F. F., 1981a, The organization of thalamic projections to the parietal cortex of the Virginia opossum, *J. Comp. Neurol.* **198:**365–388.

Donoghue, J. P., and Ebner, F. F., 1981b, The laminar distribution and ultrastructure of fibers projecting from three thalamic nuclei to the somatic sensory-motor cortex of the opossum, *J. Comp. Neurol.* **198:**389–420.

Donoghue, J. P., Kerman, K. L., and Ebner, F. F., 1979, Evidence for two organizational plans within the somatic sensory-motor cortex of the rat, *J. Comp. Neurol.* **183:**647–664.

Donovan, M. K., and Wyss, J. M., 1983, Evidence for some collateralization between cortical and diencephalic efferent axons of the rat subicular cortex, *Brain Res.* **259:**181–192.

Donovick, P. J., Burright, R. G., Kaplan, J., and Rosenstreich, N., 1969, Habenular lesions, water consumption, and palatability of fluids, in the rat, *Physiol. Behav.* **4:**45–47.

Doty, R. W., 1971, Survival of pattern vision after removal of striate cortex in the adult cat, *J. Comp. Neurol.* **143:**341–369.

Doty, R. W., Glickstein, M., and Calvin, W. H., 1966, Lamination of the lateral geniculate nucleus in the squirrel monkey, *Saimiri sciureus*, *J. Comp. Neurol.* **127:**335–340.

Doty, R. W., Wilson, P. D., Bartlet, J. R., and Pecci-Saavedra, J., 1973, Mesencephalic control of lateral geniculate nucleus in primates. I. Electrophysiology, *Exp. Brain Res.* **18:**189–203.

Douglas, R. M., and Goddard, G. V., 1975, Long-term potentiation of the perforant path–granule cell synapse in the rat hippocampus, *Brain Res.* **86:**205–215.

Dow, B. M., and Dubner, R., 1969, Visual receptive fields and responses to movement in an association area of cat cerebral cortex, *J. Neurophysiol.* **32:**773–784.

Downman, C. B. B., and Evans, M. H., 1957, The distribution of splanchnic afferents in the spinal cord of the cat, *J. Physiol. (London)* **137:**66–79.

Dräger, U. C., 1974, Autoradiography of tritiated proline and fucose transported transneuronally from the eye to the visual cortex in pigmented and albino mice. *Brain Res.* **82:**284–292.

Dräger, U. C., and Olsen, J. F., 1980, Origins of crossed and uncrossed retinal projections in pigmented and albino mice, *J. Comp. Neurol.* **191:**383–412.

Dreher, B., and Sefton, A. J., 1979, Properties of neurons in cat's dorsal lateral geniculate nucleus: A comparison between medial interlaminar and laminated parts of the nucleus, *J. Comp. Neurol.* **183:**47–64.

Dreher, B., Fukada, Y., and Rodieck, R. W., 1976, Identification, classification and anatomical segregation of cells with X-like and Y-like properties in the lateral geniculate nucleus of Old World primates, *J. Physiol. (London)* **258:**433–452.

Dreher, B., Leventhal, A. G., and Hale, P. T., 1980, Geniculate input to cat visual cortex: A comparison of area 19 with areas 17 and 18, *J. Neurophysiol.* **44:**804–826.

Dreyer, D. A., Schneider, R. J., Metz, C. B., and Whitsel, B. L., 1974, Differential contributions of spinal pathways to body representation in postcentral gyrus of *Macaca mulatta, J. Neurophysiol.* **37:**119–145.

Droogleever-Fortuyn, A. B., 1912, Die Ontogenie der Kerne des Zwischenhirns beim Kaninchen, *Arch. Anat. Physiol. (Anat. Abt.)* **1912:**303–352.

Droogleever-Fortuyn, J., 1950, On the configuration and the connections of the medioventral area and the midline-cells in the thalamus of the rabbit, *Folia Psychiatr. Neurol. Neurochir. Neerl.* **53:**213–254.

Droogleever-Fortuyn, J., 1953, Anatomical basis of cortico-subcortical relationships, *Electroencephalogr. Clin. Neurophysiol.* **4:**149–162.

Droogleever-Fortuyn, J., and Stefens, R., 1951, On the anatomical relations of the intralaminar and midline cells of the thalamus, *Electroencephalogr. Clin. Neurophysiol.* **3:**393–400.

Droogleever-Fortuyn, J., Hiddema, F., and Sanders-Woudstra, J. A. R., 1959, A note on rhinencephalic components of the dorsal thalamus: The parataenial and dorso-medial nuclei, in: *Recent Neurological Research* (A. Biemond, ed.), Elsevier, Amsterdam, pp. 46–53.

Dubbeldam, J. L., Brauch, C. S. M., and Don, A. 1981, Studies on the somatotopy of the trigeminal system in the mallard, *Anas platyrhynchos,* L. III. Afferents and organization of the nucleus basalis, *J. Comp. Neurol.* **196:**391–405.

Dubin, M. W., and Cleland, B. G., 1977, Organization of visual inputs to interneurons of lateral geniculate nucleus of the cat, *J. Neurophysiol.* **40:**410–427.

Dubner, R., and Rutledge, L. T., 1964, Recording and analysis of converging input upon neurons in cat association cortex, *J. Neurophysiol.* **27:**620–634.

Dubner, R., and Rutledge, L. T., 1965, Intracellular recording of the convergence of input upon neurons in cat association cortex, *Exp. Neurol.* **12:**349–369.

Dubner, R., and Zeki, S. M., 1971, Response properties and receptive fields of cells in an anatomically defined region of the superior temporal sulcus in the monkey, *Brain Res.* **35:**528–532.

Duffy, F. H., and Burchfiel, J. L., 1975, Eye movement-related inhibition of primate visual neurons, *Brain Res.* **89:**121–132.

Duggan, A. W., and Hall, J. G., 1975, Inhibition of thalamic neurons by acetylcholine, *Brain Res.* **100:**445–449.

Dunlop, C. W., Itzkowic, D. J., and Aitkin, L. M., 1969, Tone-burst response patterns of single units in the cat medial geniculate body, *Brain Res.* **16:**149–164.

Dünser, K. R., Granda, A. M., Maxwell, J. H., and Fulbrook, J. E., 1981, Visual properties of cells in anterior dorsal ventricular ridge of turtle, *Neurosci. Lett.* **25:**281–285.

Dürsteler, M. R., Blakemore, C., and Garey, L. J., 1979, Projections to the visual cortex in the golden hamster, *J. Comp. Neurol.* **183:** 185–204.

Duval, M. M., 1879, De l'emploi du collodion humide pour la pratique des coupes microscopiques, *J. Anat. (Paris)* **15:**185–188.

Dykes, R. W., Dudar, J. D., Tanji, D. G. and Publicover, N. G., 1977, Somatotopic projectons of mystacial vibrissae on cerebral cortex of cats, *J. Neurophysiol.* **40**:997–1014.

Dykes, R. W., Rasmusson, D. D., and Hoeltzell, P. B., 1980, Organization of primary somatosensory cortex in the cat, *J. Neurophysiol.* **43**:1527–1546.

Dykes, R. W., Sur, M., Merzenich, M. M., Kaas, J. H., and Nelson, R. J., 1981, Regional segregation of neurons responding to quickly adapting, slowly adapting, deep and Pacinian receptors within thalamic ventroposterior lateral and ventroposterior inferior nucei in the squirrel monkey (*Saimiri sciureus*), *Neuroscience* **6**:1687–1692.

Ebbesson, S. O. E., 1968, Retinal projections in two teleost fishes (*Opsanus tau* and *Gymnothorax funebris*): An experimental study with silver impregnation methods, *Brain Behav. Evol.* **1**:134–154.

Ebbesson, S. O. E., 1969, Brain stem afferents from the spinal cord in a sample of reptilian and amphibian species, *Ann. N.Y. Acad. Sci.* **167**:80–101.

Ebbesson, S. O. E., 1970, On the organization of central visual pathways in vertebrates, *Brain Behav. Evol.* **3**:178–194.

Ebbesson, S. O. E., 1972, A proposal for a common nomenclature for some optic nuclei in vertebrates and the evidence for a common origin of two such cell groups, *Brain Behav. Evol.* **6**:75–91.

Ebbesson, S. O. E., 1980, On the organization of the telencephalon in elasmobranchs, in: *Comparative Neurology of the Telencephalon* (S. O. E. Ebbesson, ed.), Plenum Press, New York, pp. 1–16.

Ebbesson, S. O. E., and Goodman, D. C., 1981, Organization of ascending spinal projections in *Caiman crocodilus, Cell Tissue Res.* **215**:383–396.

Ebbesson, S. O. E., Jane, J. A., and Schroeder, D. M., 1972, A general overview of major interspecific variations in thalamic organization, *Brain Behav. Evol.* **6**:92–130.

Ebner, F. F., 1967, Afferent connections to neocortex in the opossum (*Didelphis virginiana*), *J. Comp. Neurol.* **129**:241–267.

Ebner, F. F., Donoghue, J. P., Foster, R. E. and Christensen, B. N., 1976, The organization of opossum somatic sensory-motor cortex, *Neurosci. Abstr.* **2**:135.

Eccles, J. C., 1977a, Under the spell of the synapse, in: *The Neurosciences: Paths of Discovery* (F. G. Worden, J. P. Swazey, and G. Adelman, eds.), MIT Press, Cambridge, Mass., pp. 159–180.

Eccles, J. C., 1977b, *The Understanding of the Brain*, 2nd ed., McGraw–Hill, New York.

Eccles, J. C., Eccles, R. M., and Lundberg, A., 1960, Types of neurons in and around the intermediate nucleus of the lumbosacral cord, *J. Physiol. (London)* **154**:89–114.

Eccles, J. C., Rantucci, T., Rosén, I., Scheid, P., and Táboříková, H., 1974a, Somatotopic studies on cerebellar interpositus neurons, *J. Neurophysiol.* **37**:1449–1559.

Eccles, J. C., Sabah, N. H. and Táboříková, H., 1974b, The pathways responsible for excitation and inhibition of fastigial neurons, *Exp. Brain Res.* **19**:78–99.

Edinger, L., 1885, *Vorlesungen über den Bau der nervösen Zentralorgone des Menschen und der Tiere*, Vogel, Leipzig.

Edinger, L., 1890, Einiges vom Verlauf der Gefuhlsbahnen im centralen Nervensystem, *Dtsch. Med. Wochenschr.* **16**:421–426.

Edinger, L., 1891, Giebt es zentral entstehende Krankheiten?, *Dtsch. Z. Nervenheilkd.* **1**:262–282.

Edinger, L., 1893, Vergleichend-entwicklungsgeschichtliche und anatomische Studien im Bereiche der Hirnanatomie. 3. Reichapparat und Ammonshorn, *Anat. Anz.* **8**:305–321.

Edinger, L., 1896a, Untersuchungen über die vergleichende Anatomie des Gehirns. 3. Neue Studien über das Vorderhirn der Reptilian, *Abh. Sencken. Naturforsch. Ges.* **19**:313–386.

Edinger, L., 1896b, *Vorlesungen über den Bau der nervösen Zentralorgane des Menschen und der Thiere*, 5th ed., Vogel, Leipzig.

Edinger, L., 1899, Untersuchungen über die vergleichende Anatomie des Gehirns. 4. Studien über das Zwischenhirn der Reptilian, *Abh. Sencken. Naturforsch. Ges.* **20**:161–197.

Edinger, L., 1900, *Vorlesungen über den Bau der nervösen Zentralorgane des Menschen und der Tiere*, 6th ed., Vogel, Leipzig.

Edinger, L., 1908, *Vorlesungen über den Bau der nervösen Zentralorgane des Menschen und der Tiere*. Volume 2, *Vergleichende Anatomie des Gehirns*, 7th ed., Vogel, Leipzig.

Edinger, L., and Wallenberg, A., 1899, Untersuchungen über das Gehirn der Tauben, *Anat. Anz.* **15**:245–426.

Edwards, S. B., 1972, The ascending and descending projection of the red nucleus in the cat: An experimental study using an autoradiographic tracing method, *Brain Res.* **48**:45–63.

Edwards, S. B., and De Olmos, J. S., 1976, Autoradiographic studies of the projections of the midbrain reticular formation: Ascending projections of nudleus cuneiformis, *J. Comp. Neurol.* **165:**417–431.

Edwards, S. B., Rosenquist, A. C., and Palmer, L. A., 1974, An autoradiographic study of the ventral lateral geniculate projections in the cat, *Brain Res.* **72:**282–287.

Eichenbaum, H., Shedlack, K. J., and Eckmann, K. W., 1980, Thalamocortical mechanisms in odor-guided behavior. I. Effects of lesions of the mediodorsal thalamic nucleus and frontal cortex on olfactory discrimination in the rat, *Brain Behav. Evol.* **17:**255–275.

Elde, R., Hökfelt, T., Johansson, O., and Terenius, L., 1976, Immunohistochemical studies using antibodies to leucine-enkephalin: Initial observatons on the nervous system of the rat, *Neuroscience* **1:**349–351.

Emmers, R., 1964, Localization of thalamic projection of afferents from the tongue in the cat, *Anat. Rec.* **148:**67–74.

Emmers, R., 1965, Organization of the first and second somesthetic regions (SI and SII) in the rat thalamus, *J. Comp. Neurol.* **124:**215–227.

Emmers, R., 1966, Separate relays of tactile, pressure, thermal and gustatory modalities in the cat thalamus, *Proc. Soc. Exp. Biol. Med.* **121:**527–531.

Emmers, R., and Akert, K., *A Stereotaxic Atlas of the Brain of the Squirrel Monkey (Saimiri sciureus),* University of Wisconsin Press, Madison.

Emmers, R., Benjamin, R. M., and Blomquist, A. J., 1962, Thalamic localization of afferents from the tongue in albino rat, *J. Comp. Neurol.* **118:**43–48.

Emson, P. C., 1979, Peptides as neurotransmitter candidates in the mammalian CNS, *Prog. Neurobiol.* **13:**61–116.

Enroth-Cugell, C., and Robson, J. G., 1966, The contrast sensitivity of retinal ganglion cells of the cat, *J. Physiol. (London)* **187:**517–552.

Erickson, R. P., Hall, W. C., Jane, J. A., Snyder, M., and Diamond, I. T., 1967, Organization of the posterior dorsal thalamus of the hedgehog, *J. Comp. Neurol.* **131:**103–130.

Erulkar, D. S., 1955, Tactile and auditory areas of the brain of the pigeon: An experimental study by means of evoked potentials, *J. Comp. Neurol.* **103:**421–458.

Erulkar, D. S., and Fillenz, M., 1960, Single-unit activity in the lateral geniculate body of the cat, *J. Physiol. (London)* **154:**206–218.

Erzurumlu, R. S., and Killackey, H. P., 1980, Diencephalic projections of the subnucleus interpolaris of the brainstem trigeminal complex in the rat, *Neuroscience* **5:**1891–1901.

Evarts, E. V., and Fromm, C., 1977, Sensory responses in motor cortex neurons during precise motor control, *Neurosci. Lett.* **5:**267–272.

Evarts, E. V., and Thach, W. T., 1969, Motor mechanisms of the CNS: Cerebrocerebellar interrelations, *Annu. Rev. Physiol.* **31:**451–498.

Eysel, U. T., 1976, Quantitative studies of intracellular postsynaptic potentials in the lateral geniculate nucleus of the cat with respect to optic tract stimulus response latencies, *Exp. Brain Res.* **25:**469–486.

Eysel, U. T., and Wolfhard, U., 1983, Morphological fine tuning of retinotopy within the cat lateral geniculate nucleus, *Neurosci. Lett.* **39:**15–20.

Eysel, U. T., Grüsser, O., and Hoffmann, K.-P., 1979, Monocular deprivation and the signal transmission by X- and Y-neurons of the cat lateral geniculate nucleus, *Exp. Brain Res.* **34:**521–540.

Faccioli, G., and Minelli, G., 1975, Diencephalic degenerations following ablation of a telencephalic hemisphere in *Coturnix coturnix japonica, Boll. Zool.* **42:**1–7.

Façon, E., Steriade, M., and Wertheim, N., 1958, Hypersomnie prolongée engendrée par des lésions bilatérales du système activateur médial: Le syndrome thrombotique de la bifurcation du tronc basilaire, *Rev. Neurol.* **98:**117–133.

Famiglietti, E. V., Jr., and Peters, A., 1972, The synaptic glomerulus and the intrinsic neuron in the dorsal lateral geniculate nucleus of the cat, *J. Comp. Neurol.* **144:**285–334.

Faull, R. L. M., and Carman, J. B., 1968, Ascending projections of the substantia nigra in the rat, *J. Comp. Neurol.* **132:**73–92.

Faull, R. L. M., and Carman, J. B., 1978, The cerebellofugal projections in the brachium conjunctivum of the rat. I. The contralateral ascending pathway, *J. Comp. Neurol.* **178:**495–518.

Faull, R. L. M., and Mehler, W. R., 1978, The cells of origin of nigrotectal, nigrothalamic and nigrostriatal projections in the rat, *Neuroscience* **3:**989–1002.

Feldman, S. G., and Kruger, L., 1980, An axonal transport study of the ascending projection of medial lemniscal neurons in the rat, *J. Comp. Neurol.* **192:**427–454.

Felleman, D. J., Wall, J. T., Cusick, C. G., and Kaas, J. H., 1983, The representation of the body surface in S-I of cats, *J. Neurosci.* **3:**1648–1669.

Feremutsch, K., 1963, Thalamus, in: *Primatologia* (H. Hofer, A. H. Schultz, and D. Starck, eds.), Volume 2, Part 2, No. 6, Karger, Basel, pp. 1–226.

Feremutsch, K., and Simma, K., 1953, Strukturanalysen des menschlichen Thalamus. I. Das ventrikelnahe Grau (griseum periventriculare thalami), *Monatsschr. Psychiatr. Neurol.* **126:**209–229.

Feremutsch, K., and Simma, K., 1954a, Strukturanalysen des menschlichen Thalamus. II. Die intralaminaren Kerne (intralaminar nuclei), *Monatsschr. Psychiatr. Neurol.* **127:**88–102.

Feremutsch, K., and Simma, K., 1954b, Strukturanalysen des menschlichen Thalamus. III. Der Lateralkernkomplex (nucleus lateralis thalami), *Monatsschr. Psychiatr. Neurol.* **128:**365–396.

Feremutsch, K., and Simma, K., 1955, Struckturanalysen des menschlichen Thalamus. IV. Nucleus anterior, nucleus mediodorsalis, nucleus pulvinaris thalami, *Monatsschr. Psychiatr. Neurol.* **130:**347–359.

Feremutsch, K., and Simma, K., 1958, Strukturanalysen des menschlichen Thalamus. V. Epithalamus und Regio subthalamica, *Psychiatr. Neurol.* **135:**94–106.

Feremutsch, K., and Simma, K., 1959, Strukturanalysen des menschlichen Thalamus. VI. Die morphologischen Elemente der thalamocortikalen Systeme, *Psychiatr. Neurol.* **137:**103–127.

Fernald, R. D., 1982, Retinal projections in the African cichlid fish *Haplochromis burtoni, J. Comp. Neurol.* **206:**379–389.

Fernández, V., 1969, An autoradiographic study of the development of the anterior thalamic group and limbic cortex in the rabbit, *J. Comp. Neurol.* **136:**423–452.

Fernández, V., and Bravo, H., 1973, Autoradiographic study of the development of the medial thalamic nuclei in the rabbit, *Brain Behav. Evol.* **7:**453–465.

Ferraro, A., and Barrera, S. E., 1935, Summary of clinical and anatomical findings following lesions in the dorsal column system of Macacus rhesus monkeys: Sensation: Its mechanisms and disturbances, *Res. Publ. Assoc. Nerv. Ment. Dis.* **15:**371–395.

Ferrier, D., 1874, The localisation of function in the brain, *Proc. R. Soc. London* **22:**229–232.

Ferrier, D., 1876, *The Functions of the Brain,* Smith Elder, London.

Ferrier, D., and Turner, W. A., 1894, A record of experiments illustrative of the symptomatology and degenerations following lesions of the cerebellum and its peduncles and related structures in monkeys, *Philos. Trans. R. Soc. London. Ser. B* **185:**719–778.

Ferrington, D. G., and Rowe, M., 1980, Differential contributions to coding of cutaneous vibratory information by cortical somatosensory areas I and II, *J. Neurophysiol.* **43:**310–331.

Ferster, D., and LeVay, S., 1978, The axonal arborizations of lateral geniculate neurons in the striate cortex of the cat, *J. Comp. Neurol.* **182:**923–944.

Fibiger, H. C., 1982, The organization and some projections of cholinergic neurons of the mammalian forebrain, *Brain Res. Rev.* **4:**327–388.

Filion, M., and Harnois, C., 1978, A comparison of projections of entopeduncular neurons to the thalamus, the midbrain and the habenula in the cat, *J. Comp. Neurol.* **181:**763–780.

Finger, T. E., 1975, The distribution of the olfactory tracts in the bullhead catfish, *Ictalurus nebulosus, J. Comp. Neurol.* **161:**125–142.

Finger, T. E., 1978, Cerebellar afferents in teleost catfish (Ictaluridae), *J. Comp. Neurol.* **181:**173–182.

Finger, T. E., 1979, A thalamic relay nucleus for the lateral line system in teleost fish, *Neurosci. Abstr.* **5:**141.

Fisher, G. R., Freeman, B., and Rowe, M.J., 1983, Organization of parallel projections from Pacinian afferent fibers to somatosensory cortical areas I and II in the cat, *J. Neurophysiol.* **49:**75–97.

Fitzpatrick, D., and Diamond, I. T., 1980, Distribution of acetylcholinesterase in the geniculo-striate system of *Galago senegalensis* and *Aotus trivirgatus:* Evidence for the origin of the reaction product in the lateral geniculate body, *J. Comp. Neurol.* **194:**703–720.

Fitzpatrick, D., Carey, R. G., and Diamond, I. T., 1980, The projection of the superior colliculus upon the lateral geniculate body in *Tupaia glis* and *Galago senegalensis, Brain Res.* **194:**494–499.

Fitzpatrick, D., Itoh, K., and Diamond, I. T., 1983, The laminar organization of the lateral geniculate body and the striate cortex in the squirrel monkey (*Saimiri sciureus*), *J. Neurosci.* **3:**673–702.

Fitzpatrick, K. A., and Imig, T. J., 1978, Projections of auditory cortex upon the thalamus and midbrain in the owl monkey, *J. Comp. Neurol.* **177:**537–556.

Flechsig, P. E., 1886, Zur Lehr vom centralen Verlauf der Sinnesnerven, *Neurol. Zentralbl.* **5:**545–551.

Flechsig, P. E., and Hösel, O., 1890, Die Centralwindungen im Centralorgan der Hinterstränge, *Neurol. Zentralbl.* **9:**417–419.

Florence, S. L., and Casagrande, V. A., 1978, A note on the evolution of ocular dominance columns in primates, *Invest. Ophthalmol.* **17:**291–292.

Flourens, M. J. P., 1824, *Recherches Expérimenteles sur les Propriétés et les Fonctions due Système Nerveux*, 2nd ed., Paris, Crevot.

Flumerfelt, B. A., Otabe, S., and Courville, J., 1973, Distinct projections to the red nucleus from the dentate and interposed nuclei in the monkey, *Brain Res.* **50:**408–414.

Foerster, O., and Gagel, O., 1932, Die Vorderseitenstrangdurschreidung beim Menschen: Eine Klinische-pathologisch-pathophysiologisch-anatomische Studie, *Z. Gesamte Neurol. Psychiatr.* **138:**1-92.

Foix, C., and Bariéty, M., 1926, Hemichorée d'origine thalamique, *Rev. Neurol.* **2:**598.

Foix, C., and Nicolesco, J., 1925, *Anatomie Cérébrale: Les Noyaux Gris Centraux et la Région Mésencéphalo-Sous-optique suivi d'un Appendice sur l'Anatomie Pathologique de la Maladie de Parkinson*, Masson, Paris.

Foix, C., Chavanay, J.-A., and Hillemand, P., 1926, Syndrome cérébello-thalamique supériere, *Rev. Neurol.* **2:**598–604.

Foley, V., and Soedel, N., 1981, Ancient oared warships, *Sci. Am.* **244:**148–163.

Foote, W. E., Maciewicz, R. J., and Mordes, J. P., 1974, Effect of midbrain raphe and lateral mesencephalic stimulation on spontaneous and evoked activity in the lateral geniculate of the cat, *Exp. Brain Res.* **19:**124–130.

Ford, D. H., 1968, Central nervous system–thyroid interrelationships, *Brain Res.* **7:**329–349.

Forel, A., 1872, Beiträge zur Kenntniss des Thalamus opticus und der ihn umgebenden Gebilde bei den Säugethieren, *Sber. Akad. Wiss. Wien* **66**(Part 3):25–58 (Reprinted in *Gesammelte hirnanatomische Abhandlungen*, 1907, pp. 17–43, Reinhardt, Munich).

Forel, A., 1877, Untersuchungen über die Haubenregion und ihre oberen Verknüpfungen im Gehirne des Menschen und einiger Säugethiere, mit Beiträgen zu den Methoden der Gehirnuntersuchung, *Arch. Psychiatr. Nervenkr.* **7:**393–495.

Foster, R. E., and Hall, W. C., 1978, The organization of the central auditory pathways in a reptile, *Iguana iguana, J. Comp Neurol.* **178:**783–831.

Fournié, E., 1873, *Recherches Expérimentales sur le Fonctionnement du Cerveau*, Delahaye, Paris.

Frazer, J. E., 1931, *Manual of Embryology: The Development of the Human Body*, London: Baillière, Tindall, and Cox.

Fredrickson, J. M., Figge, U., Scheid, P., and Kornhuber, H. H., 1966, Vestibular nerve projection to the cerebral cortex of the rhesus monkey, *Exp. Brain Res.* **2:**318–327.

Freeman, W., and Watts, J. W., 1947, Retrograde degeneration of the thalamus following prefrontal lobotomy, *J. Comp. Neurol.* **86:**65–93.

Friedemann, M., 1911, Die Cytoarchitektonik des Zwischenhirns der Cercopitheken mit besonderer Berücksichtigung des Thalamus opticus, *J. Psychol. Neurol.* **18:**309–378.

Friedlander, M. J., 1982, Structure of physiologically classified neurons in the kitten dorsal lateral geniculate nucleus, *Nature (London)* **300:**180–183.

Friedlander, M. J., Lin, C.-S, and Sherman, S. M., 1979, Structure of physiologically identified X and Y cells in the cat's lateral geniculate nucleus, *Science* **204:**1114–1117.

Friedlander, M. J., Lin, C.-S., Stanford, L. R., and Sherman, S. M., 1981, Morphology of functionally identified neurons in lateral geniculate nucleus of the cat, *J. Neurophysiol.* **46:**80–129.

Friedlander, M. J., Stanford, L. R., and Sherman, S. M., 1982, Effects of monocular deprivation on the structure–function relationship of individual neurons in the cat's lateral geniculate nucleus, *J. Neurosci.* **2:**321–330.

Friedman, D. P., and Jones, E. G., 1980, Focal projection of electrophysiologically defined groupings of thalamic cells on monkey somatic sensory cortex, *Brain Res.* **191:**249–252.

Friedman, D. P., and Jones, E. G., 1981, Thalamic input to areas 3a and 2 in monkeys, *J. Neurophysiol.* **45:**59–85.

Friedman, D. P., Jones, E. G., and Burton, H., 1980, Representation pattern in the second somatic sensory area of the monkey cerebral cortex, *J. Comp. Neurol.* **192:**21–41.

Friedman, D. P., Murray, E. A., and O'Neill, J. B., 1983, Thalamic connectivity of the somatosensory cortical fields of the lateral sulcus of the monkey, *Neurosci. Abstr.* **9:**921.

Frigyesi, T. L., and Schwartz, R., 1972, Cortical control of sensorimotor relay activities in the cat and the squirrel monkey, in: *Thalamo-cortical Projections and Sensorimotor Activities* (T. L. Frigyesi, E. Rinvik, and M. D. Yahr, eds.), Raven Press, New York, pp. 161–195.

Frontera, J. G., 1952, A study of the anuran diencephalon, *J. Comp. Neurol.* **96:**51–70.

Frost, D. O., 1981, Orderly anomalous retinal projections to the medial geniculate, ventrobasal, and lateral posterior nuclei of the hamster, *J. Comp. Neurol.* **203:**227–256.

Frost, D. O., 1982, Anomalous visual connections to somatosensory and auditory systems following brain lesions in early life, *Dev. Brain Res.* **3:**627–636.

Frost, D. O., and Caviness, V. S., Jr., 1980, Radial organization of thalamic projections to the neocortex in the mouse, *J. Comp. Neurol.* **194:**369–394.

Frost, D. O., So, K.-F., and Schneider, G. E., 1979, Postnatal development of retinal projections in Syrian hamsters: A study using auoradiographic and anterograde degeneration techniques, *Neuroscience* **4:**1649–1678.

Fry, F. J., and Cowan, W. M., 1972, A study of retrograde degeneration in the lateral mammillary nucleus of the cat, with special reference to the role of axonal branching in the preservation of the cell, *J. Comp. Neurol.* **144:**1–24.

Fry, W. J., Krummins, R., Fry, F. J., Thomas, G., Borbely, S., and Ades, H. 1963, Origins and distribution of some efferent pathways from the mammillary nuclei of the cat, *J. Comp. Neurol.* **120:**195–258.

Fukada, Y., 1971, Receptive field organization of cat optic nerve fibers with special reference to conduction velocity, *Vision Res.* **11:**209–226.

Fukuda, Y., 1973, Differentiation of principal cells of the rat lateral geniculate body into two groups; fast and slow cells, *Exp. Brain Res.* **17:**242–260.

Fukuda, Y., 1977, A three-group classification of rat retinal ganglion cells; histological and physiological studies, *Brain Res.* **119:**327–344.

Fukuda, Y., and Iwama, K., 1971, Reticular inhibition of internuncial cells in the rat lateral geniculate body, *Brain Res.* **35:**107–118.

Fukuda, Y., and Saito, H., 1972, Phasic and tonic cells in the cat's lateral geniculate nucleus, *Tohoku J. Exp. Med.* **106:**209–210.

Fukuda, Y., and Stone, J., 1974, Retinal distribution and central projections of Y-, X- and W-cells of the cat's retina, *J. Neurophysiol.* **37:**749–772.

Fukuda, Y., and Stone, J., 1976, Evidence of differential inhibitory influences of X- and Y-type relay cells in the cat's lateral geniculate nucleus, *Brain Res.* **113:**188–196.

Fukushima, T., and Kerr, F. W. L., 1979, Organization of trigeminothalamic tracts and other thalamic afferent systems of the brainstem in the rat: Presence of gelatinosa neurons with thalamic connections, *J. Comp. Neurol.* **183:**169–184.

Fulton, J. F., 1949, *Physiology of the Nervous System*, 3rd ed., Oxford University Press, London.

Fuster, J. M., Creutzfeldt, O. D., and Straschill, M., 1965a, Intracellular recording of neuronal activity in the visual system, *Z. Vgl. Physiol.* **49:**605–622.

Fuster, J. M., Herz, A., and Creutzfeldt, O. D., 1965b, Interval analysis of cell discharge in spontaneous and optically modulated activity in the visual system, *Arch. Ital. Biol.* **103:**159–177.

Fuxe, K., Hamberger, B., and Hökfelt, T., 1968, Distribution of noradrenaline nerve terminals in cortical areas of the rat, *Brain Res.* **8:**125–131.

Fuxe, K., Ganten, D., Hökfelt, T., and Bohne, P., 1976, Immunohistochemical evidence for the existence of angiotensin II-containing nerve terminals in the brain and spinal cord of the rat, *Neurosci. Lett.* **2:**229–234.

Galambos, R., Rose, J. E., Bromiley, R. B., and Hughes, J. R. 1952, Microelectrode studies on medial geniculate body of cat. II. Response to clicks, *J. Neurophysiol.* **15:**359–380.

Galen, 1906, *Sieben Bucher Anatomie des Galen*, translated from Greek and Arabic texts, Simon, Leipzig.

Galen, 1956, *On Anatomical Procedures (De Anatomicis Administrationibus*, Books 1–9, Part 1), translated by C. Singer, Wellcome Historical Medical Museum and Oxford University Press London.

Galen, 1962, *On Anatomical Procedures*, Later Books, translated by W. L. Duckworth (M. C. Lyons and B. Towers, eds.), Cambridge University Press, London.

Gall, F. J., and Spurzheim, G., 1809, *Recherches sur le Système Nerveux en Général*, Schoell & Nicholle, Paris.

Gall, F. J. and Spurzheim, G., 1810, *Anatomie et Physiologie du Système Nerveux en Général et du Cerveau en Particulier*, 4 vols., Schoell, Paris.

Gamlin, P. D. R., and Cohen, D. H., 1982, A possible second ascending avian tectofugal pathway, *Neurosci. Abstr.* **8:**206.

Ganchrow, D., 1978, Intratrigeminal and thalamic projections of nucleus caudalis in the squirrel monkey *(Saimiri sciureus):* A degeneration and autoradiographic study, *J. Comp. Neurol.* **178:**281–312.

Ganchrow, D., and Erickson, R. P., 1972, Thalamocortical relations in gustation, *Brain Res.* **36:**289–305.

Ganser, S., 1882, Vergleichend-anatomische Studien über das Gehirn des Maulwulfs, *Morphol Jahrb.* **7:**591–725.

Ganz, L., Fitch, M., and Satterberg, J. A., 1968, The selective effect of visual deprivation and receptive field shape determined neurophysiologically, *Exp. Neurol.* **22:**614–637.

Gardner, E. P., and Costanzo, R. M., 1981, Properties of kinesthetic neurons in somatosensory cortex of awake monkeys, *Brain Res.* **214:**301–320.

Garey, L. J., and Blakemore, C., 1977, The effects of monocular deprivation on different neuronal classes in the lateral geniculate nucleus of the cat, *Exp. Brain Res.* **28:**259–278.

Garey, L. J., and Powell, T. P. S., 1967, The projection of the lateral geniculate nucleus upon the cortex in the cat, *Proc. R. Soc. London Ser. B* **169:**107–126.

Garey, L. J., and Powell, T. P. S., 1968, The projection of the retina in the cat, *J. Anat.* **102:**189–222.

Garey, L. J., and Powell, T. P. S., 1971, An experimental study of the termination of the lateral geniculo-cortical pathway in the cat and monkey, *Proc. R. Soc. London Ser. B* **179:**41–63.

Garey, L. J., and Saini, K. D., 1981, Golgi studies of the normal development of neurons in the lateral geniculate nucleus of the monkey, *Exp. Brain Res.* **44:**117–128.

Garey, L. J., Jones, E. G., and Powell, T. P. S., 1968, Interrelationships of striate and extrastriate cortex with the primary relay sites of the visual pathway, *J. Neurol. Neurosurg. Psychiatry* **31:**135–157.

Garey, L. J., Fisken, R. A., and Powell, T. P. S., 1973, Effects of experimental deafferentation on cells in the lateral geniculate nucleus of the cat, *Brain Res.* **52:**363–369.

Gattass, R., Oswaldo-Cruz, E., and Sousa, A. P. B., 1978a, Visuotopic organization of the cebus pulvinar: A double representation of the contralateral hemifield, *Brain Res.* **152:**1–16.

Gattass, R., Oswaldo-Cruz, E., and Sousa, A. P. B., 1978b, Single unit response type in the pulvinar of the cebus monkey to multisensory stimulation, *Brain Res.* **158:**75–88.

Gattass, R., Oswaldo-Cruz, E., and Sousa, A. P. B., 1979, Visual receptive fields of units in the pulvinar of cebus monkey, *Brain Res.* **160:**413–430.

Gaupp, E., 1899, *Anatomie des Frosches,* Vieweg, Braunschweig.

Gaze, R. M., and Gordon, G., 1954, The representation of cutaneous sense in the thalamus of the cat and monkey, *Q. J. Exp. Physiol.* **39:**279–304.

Geisert, E. E., Jr., 1980, Cortical projections of the lateral geniculate nucleus in the cat, *J. Comp. Neurol.* **190:**793–812.

Geisert, E. E., Langsetmo, A., and Spear, P. D., 1981, Influence of the cortico-geniculate pathway on response properties of cat lateral geniculate neurons, *Brain Res.* **208:**409–415.

Geisert, E. E., Spear, P. D., Zetlan, S. R., and Langsetmo, A., 1982, Recovery of Y-cells in the lateral geniculate nucleus of monocularly deprived cats, *J. Neurosci.* **2:**577–588.

Gerebtzoff, M. A., 1936, Le pédoncules cérébelleux supérieur et les terminaisons réelles de la voie cérébello-thalamique, *Mém. Acad. R. Méd. Belg.* **25:**1–58.

Gerebtzoff, M. A., 1937, Systèmatisation des connexions thalamo-corticales, *Cellule* **46:**5–54.

Gerebtzoff, M. A., 1939, Les voies centrales de la sensibilité et du gout et leurs terminaisons thalamiques, *Cellule* **48:**89–146.

Gerebtzoff, M. A., 1940a, Recherches sur l'écorce cérébrale et le thalamus de cobaye. I. Étude architectonique, *Cellule* **48:**335–352.

Gerebtzoff, M. A., 1940b, Les connexions thalamo-striées: Le noyaux parafasciculaire et le centre médian, *J. Belge Neurol. Psychiatr.* **40:**407–416.

Gerhardt, K. D., Yezierski, R. P, Giesler, G. J., Jr., and Willis, W. D., 1981, Inhibitory receptive fields of primate spinothalamic tract cells, *J. Neurophysiol.* **46:**1309–1332.

Gerlach, J., 1947, Beiträge zur vergleichenden Morphologie des Selachierhirnes, *Anat. Anz.* **96:**79–165.

Gerren, R. A., and Weinberger, N. M., 1983, Long term potentiation in the magnocellular medial geniculate nucleus of the anesthetized cat, *Brain Res.* **265:**138–142.

Getz, B., 1952, The termination of spinothalamic fibres in the cat as studied by the method of terminal degeneration, *Acta Anat.* **16:**271–290.

Giesler, G. J., Menétrey, D., and Basbaum, A. I., 1979, Differential origins of spinothalamic tract projections to medial and lateral thalamus in the rat, *J. Comp. Neurol.* **184:**107–126.

Giesler, G. J., Jr., Spiel, H. R., and Willis, W. D., 1981, Organization of spinothalamic tract axons within the rat spinal cord, *J. Comp. Neurol.* **195:**243–252.

Gihr, M., 1964, Die Zellformen des Nucleus medialis dorsalis thalami des Menschen, *Prog. Brain Res.* **5**:74–87.

Gilbert, C. D., 1977, Laminar differences in receptive field properties of cells in cat primary visual cortex, *J. Physiol. (London)* **268**:391–421.

Gilbert, C. D., and Kelly, J. P., 1975, The projections of cells in different layers of the cat's visual cortex, *J. Comp. Neurol.* **163**:81–105.

Gilbert, C. D., and Wiesel, T. N., 1979, Morphology and intracortical projections of functionally characterised neurones in the cat visual cortex, *Nature (London)* **280**:120–125.

Gilbert, C. D., and Wiesel, T. N., 1983, Clustered intrinsic connections in cat visual cortex, *J. Neurosci.* **3**:1116–1133.

Gilbert, M. S., 1934, The early development of the human diencephalon, *J. Comp. Neurol.* **62**:81–116.

Gillingham, F. J., Walsh, E. G., and Zogada, L. F., 1977, Stereotaxic lesions of the pulvinar for hypertonus and dyskinesias, *Acta Neurochir.* **24**(Suppl.):15–20.

Giolli, R. A., and Creel, D. J., 1973, The primary optic projections in pigmented and albino guinea pigs; An experimental degeneration study, *Brain Res.* **55**:25–39.

Giolli, R. A., and Guthrie, M. D., 1969, The primary optic projections in the rabbit: An experimental degeneration study, *J. Comp. Neurol.* **136**:99–126.

Giolli, R. A., and Tigges, J., 1970, The primary optic pathways and nuclei in primates, *Adv. Primatol.* **1**:29–54.

Giolli, R. A., Towns, L. C., Takahashi, T. T., Karamanlidis, A. N., and Williams, D. D., 1978, An autoradiographic study of the projections of visual cortical area 1 to the thalamus, pretectum and superior colliculus of the rabbit, *J. Comp. Neurol.* **180**:743–752.

Glees, P., 1945, The interrelation of the strio-pallidum and the thalamus in the macaque monkey, *Brain* **68**:331–346.

Glees, P., and Le Gros Clark, W. E., 1941, The termination of optic fibres in the lateral geniculate body of the monkey, *J. Anat.* **75**:295–303.

Glees, P., and Wall, P., 1946, Fibre connexions of the subthalamic region, *J. Anat.* **80**:240–241.

Glendenning, K. K., Hall, J. A., Diamond, I. T., and Hall, W. C., 1975, The pulvinar nucleus of *Galago senegalensis*, *J. Comp. Neurol.* **161**:419–458.

Glendenning, K. K., Kofron, E. A., and Diamond, I. T., 1976, Laminar organization of projections of the lateral geniculate nucleus to the striate cortex in *Galago, Brain Res.* **105**:538–546.

Glenn, L. L., and Steriade, M., 1982, Discharge rate and excitability of cortically projecting intralaminar thalamic neurons during waking and sleep states, *J. Neurosci.* **10**:1387–1404.

Glenn, L. L., Hada, J., Roy, J. P., Deschênes, M., and Steriade, M., 1982, Anterograde tracer and field potential analysis of the neocortical layer I projection from nucleus ventralis medialis of the thalamus in cat, *Neuroscience* **7**:1861–1878.

Glickstein, M., 1967, Laminar structure of the dorsal lateral geniculate nucleus in the tree shrew (*Tupaia glis*), *J. Comp. Neurol.* **131**:93–102.

Glickstein, M., King, R. A., Miller, J., and Berkley, M., 1967, Cortical projections from the dorsal lateral geniculate nucleus of cat, *J. Comp. Neurol.* **130**:55–76.

Glorieux, P., 1929, Anatomie et connexions thalamiques chez le chien, *J. Neurol. Psychiatr. Brux.* **29**:525–554.

Godement, P., Saillour, P., and Imbert, M., 1979, Thalamic afferents to the visual cortex in congenitally anophthalamic mice, *Neurosci. Lett.* **13**:271–278.

Godement, P., Saillour, P., and Imbert, M., 1980, The ipsilateral optic pathway to the dorsal lateral geniculate nucleus and superior colliculus in mice with prenatal or postnatal loss of one eye, *J. Comp. Neurol.* **190**:611–626.

Godfraind, J. M., 1975, Micro-electrophoretic studies in the cat pulvinar region: Effect of acetylcholine, *Exp. Brain Res.* **22**:243–254.

Godfraind, J. M., 1978, Acetylcholine and somatically evoked inhibition on perigeniculate neurones in the cat, *Br. J. Pharmacol.* **63**:295–302.

Godfraind, J. M., Meulders, M., and Veraart, C., 1972, Visual properties of neurons in pulvinar, nucleus lateralis posterior and nucleus suprageniculatus thalami in the cat. I. Qualitative investigation, *Brain Res.* **44**:503–526.

Goldberg, J. M., and Brown, P. B., 1968, Functional organization of the dog superior olivary complex: An anatomical and electrophysiological study, *J. Neurophysiol.* **31**:639–656.

Goldberg, J. M., and Brown, P. B., 1969, Response of binaural neurons of dog superior olivary complex to dichotic tonal stimuli: Some physiological mechanisms of sound localization, *J. Neurophysiol.* **32:**613–636.

Goldberg, J. M., and Moore, R. Y., 1967, Ascending projections of the lateral lemniscus in the cat and monkey, *J. Comp. Neurol.* **129:**143–156.

Goldberg, J. M., and Neff, W. D., 1961, Frequency discrimination after bilateral section of the brachium of the inferior colliculus, *J. Comp. Neurol.* **116:**265–289.

Goldberger, M. E., and Crowdon, J. H., 1973, Patterns of recovery following cerebellar deep nuclear lesions in monkeys, *Exp. Nuerol.* **39:**307–322.

Goldby, F., 1937, An experimental investigation of the cerebral hemisphere of *Lacerta viridis*, *J. Anat.* **71:**332–355.

Goldby, F., 1941, The normal histology of the thalamus in the phalanger, *Trichosurus vulpecula*, *J. Anat.* **75:**197–224.

Goldby, F., 1957, A note on transneuronal atrophy in the human lateral geniculate body, *J. Neurol. Neurosurg. Psychiatry* **20:**202–207.

Goldby, F., and Gamble, H. J., 1957, The reptilian cerebral hemispheres, *Biol. Rev.* **32:**383–420.

Goldby, F., and Robinson, L. R., 1962, The central connections of dorsal spinal nerve roots and the ascending tracts in the spinal cord of *Lacerta viridis*, *J. Anat.* **96:**153–170.

Goldman, P. S., 1979, Contralateral projections to the dorsal thalamus from frontal association cortex in the rhesus monkey, *Brain Res.* **166:**166–171.

Goldman, P. S., and Nauta, W. J. H., 1977, An intricately patterned prefronto-caudate projection in the rhesus monkey, *J. Comp. Neurol.* **171:**369–386.

Golovchinsky, V., Kruger, L., Saporta, S. A., Stein, B. E., and Young,, D. W., 1981, Properties of velocity-mechanosensitive neurons of the cat ventrobasal thalamic nucleus with special reference to the concept of convergence, *Brain Res.* **209:**355–374.

Gordon, G., and Jukes, M. G. M., 1962, Correlation of different excitatory and inhibitory influences on cells in the nucleus gracilis of the cat, *Nature (London)* **196:**1183–1185.

Gordon, G., and Jukes, M. G. M., 1964, Dual organization of the exteroceptive components of the cat's gracile nucleus, *J. Physiol. (London)* **173:**263–290.

Gordon, G., and Manson, J. R., 1967, Cutaneous receptive fields of single nerve cells in the thalamus of the cat, *Nature (London)* **215:**597–599.

Gordon, G., and Seed, W. A., 1961, An investigation of nucleus gracilis of the cat by antidromic stimulation, *J. Physiol. (London)* **155:**589–601.

Goto, A., Kosaka, K., Kubota, K., Nakamura, R., and Narabayashi, H., 1968, Thalamic potentials from muscle afferents in the human, *Arch. Neurol. (Chicago)* **19:**302–309.

Gottesfeld, Z., and Jacobowitz, D. M., 1979, Cholinergic projections from the septal-diagonal band area to the habenular nuclei, *Brain Res.* **176:**391–394.

Gottesfeld, Z., Massari, V. J., Muth, E. A., and Jacobowitz, D. M., 1977, Stria medullaris: A possible pathway containing GABAergic afferents to the lateral habenula, *Brain Res.* **130:** 184–189.

Gottesfeld, Z., Brandon, C., and Wu, J.-Y., 1981, Immunocytochemistry of glutamate decarboxylase in the deafferented habenula, *Brain Res.* **208:**181–186.

Gottschaldt, K.-M., Vahle-Hinz, C., and Hicks, T. P., 1983a, Electrophysiological and micropharmacological studies on mechanisms of input–output transformations in single neurones of the somato-sensory thalamus, in: *Somatosensory Integration in the Thalamus* (G. Macchi, A. Rustioni, and R. Spreafico, eds.), Elsevier, Amsterdam, pp. 199–216.

Gottschaldt, K.-M., Vahle-Hinz, C., and Young, D. W., 1983b, The extraction of specific stimulus information out of complex afferent signals in subcortical relay stations of the cat's sinus hair system, *Forschr. Zool.* **28:**113–128.

Gould, H. J., III, Hall, W. C., and Ebner, F. F., 1978, Connections of visual cortex in the hedgehog (*Paraechinus hypomelas*). I. Thalamocortical projections, *J. Comp. Neurol.* **177:**445–472.

Gould, J. E., Chalupa, L. M., and Lindsley, D. B., 1974, Modification of pulvinar and geniculocortical evoked potentials during visual discrimination learning in monkeys, *Electroencephalogr. Clin. Neurophysiol.* **36:**639–649.

Gouras, P., 1969, Antidromic responses of orthodromically identified ganglion cells in monkey retina, *J. Physiol. (London)* **204:**407–419.

Gowers, W. R., 1886, On the antero-lateral ascending tract of the spinal cord, *Lancet* **1:**1153–1154.

Graef, W., and Volker, H., 1968, Zytologie der Zwischen-Mittelhirnregion beim Meerschweinchen *(Cavia porcellus* L.), *J. Hirnforsch.* **10:**17–37.

Graham, J., 1977, An autoradiographic study of the efferent connections of the superior colliculus in the cat, *J. Comp. Neurol.* **173:**629–654.

Graham, J., 1982, Some topographical connections of the striate cortex with subcortical structures in *Macaca fascicularis, Exp. Brain Res.* **47:**1–14.

Graham, J., and Berman, N., 1981, Origins of the projections of the superior colliculus to the dorsal lateral geniculate nucleus and the pulvinar in the rabbit, *Neurosci. Lett.* **26:**101–106.

Graham, J., and Casagrande, V. A., 1980, A light microscopic and electron microscopic study of the superficial layers of the superior colliculus of the tree shrew *(Tupaia glis), J. Comp. Neurol.* **191:**133–151.

Graham, J., Lin, C.-S., and Kaas, J. H. 1979, Subcortical projections of six visual cortical areas in the owl monkey, *Aotus trivirgatus, J. Comp. Neurol.* **187:**557–580.

Grant, G., Boivie, J., and Brodal, A., 1968, The question of a cerebellar projection from the lateral cervical nucleus re-examined, *Brain Res.* **9:**95–102.

Grant, G., Boivie, J., and Silfvenius, H., 1973, Course and termination of fibres from the nucleus Z of the medulla oblongata: An experimental light microscopical study in the cat, *Brain Res.* **55:**55–70.

Gratiolet, L. P., 1857, *Anatomie Comparée du Système Nerveux, Considéré dans ses Rapport avec l'Intelligence* (F. Leuret and L. P. Gratiolet, eds.), Vol. 2 (Vol. 1, 1839), Baillière, Paris.

Graybiel, A. M., 1970, Some thalamocortical projections of the pulvinar–posterior system of the thalamus in the cat, *Brain Res.* **22:**131–136.

Graybiel, A. M., 1972, Some extrageniculate visual pathways in the cat, *Invest. Ophthalmol.* **11:**322–332.

Graybiel, A. M., 1974a, Visuo-cerebellar and cerebello-visual connections involving the ventral lateral geniculate nucleus, *Exp. Brain Res.* **20:**303–306.

Graybiel, A. M., 1974b, Studies on the anatomical organization of posterior association cortex, in: *The Neurosciences: Third Study Program* (F. O. Schmitt and F. G. Worden, eds.), MIT Press, Cambridge, Mass., pp. 205–214.

Graybiel, A. M., 1975, Anatomical organization of retinotectal afferents in the cat: An autoradiographic study, *Brain Res.* **96:**1–23.

Graybiel, A. M., 1976, Evidence for banding of the cat's ipsilateral retinotectal connection, *Brain Res.* **114:**318–327.

Graybiel, A. M., 1977, Direct and indirect preoculomotor pathways of the brainstem: An autoradiographic study of the pontine reticular formation in the cat, *J. Comp. Neurol.* **175:**37–78.

Graybiel, A. M., and Berson, D. M., 1980a, Autoradiographic evidence for a projection from the pretectal nucleus of the optic tract to the dorsal lateral geniculate complex in the cat, *Brain Res.* **195:**1–12.

Graybiel, A. M., and Berson, D. M., 1980b, Histochemical identification and afferent connections of subdivisions in the lateralis posterior–pulvinar complex and related thalamic nuclei in the cat, *Neuroscience* **5:**1175–1238.

Graybiel, A. M., and Elde, R. P., 1983, Somatostatin-like immunoreactivity characterizes neurons of the nucleus reticularis thalami in the cat and monkey, *J. Neurosci.* **3:**1308–1321.

Graybiel, A. M., and Nauta, W. J. H., 1971, Some projections of superior colliculus and visual cortex upon the posterior thalamus in the cat, *Anat. Rec.* **169:**328.

Graybiel, A. M., and Ragsdale, C. W., Jr., 1978, Histochemically distinct compartments in the striatum of human, monkey and cat demonstrated by acetylthiocholinesterase staining, *Proc. Natl. Acad. Sci. USA* **75:**5723–5726.

Graybiel, A. M., and Ragsdale, C. W., Jr., 1982, Pseudocholinesterase staining in the primary visual pathway of the macaque monkey, *Nature (London)* **299:**439–442.

Graybiel, A. M., Ragsdale, C. W., Jr. and Moon-Edley, S., 1979, Compartments in the striatum of the cat observed by retrograde cell labeling, *Exp. Brain Res.* **34:**189–195.

Graybiel, A. M., Pickel, V. M., Joh, T. H., Reis, D. J., and Ragsdale, C. W., Jr., 1981a, Direct demonstration of a correspondence between the dopamine islands and acetylcholinesterase patches in the developing striatum *Proc. Natl. Acad. Sci., USA* **78:**5871–5875.

Graybiel, A. M., Ragsdale, C. W., Jr., Yoneoka, E. S., and Elde, R. P., 1981b, An immunohistochemical study of enkephalins and other neuropeptides in the striatum of the cat with evidence that the opiate peptides are arranged to form mosaic patterns in register with the striosomal compartments visible by acetylcholinesterase staining, *Neuroscience* **6:**377–398.

Groenewegen, H. J., Boesten, A. J. P., and Voogd, J., 1975, The dorsal column nuclear projections to the nucleus ventralis posterior lateralis thalami and the inferior olive in the cat: An autoradiographic study, *J. Comp. Neurol.* **162:**505–517.

Groenewegen, H. J., Becker, N. E. H. M., and Lohman, A. H. M., 1980, Subcortical afferents of the nucleus accumbens septi in the cat, studied with retrograde axonal transport of horseradish peroxidase and bisbenzimid, *Neuroscience* **5:**1903–1916.

Gröschel, G., 1930, Über die Cytoarchitektonik und Histologie der Zwischenhirnbasis beim Hund, *Dtsch. Z. Nervenheilkd.* **112:**108–123.

Gross, C., and Weiskrantz, L., 1964, Some changes in behavior produced by lateral frontal lesions in the macaque, in: *The Frontal Granular Cortex and Behavior* (J. M. Warren and K. Akert, eds.), McGraw–Hill, New York, pp. 74–101.

Gross, N. B., Lifschitz, W. S., and Anderson D. J., 1974, Tonotopic organization of the auditory thalamus of the squirrel monkey *(Saimiri sciureus)*, *Brain Res.* **65:**323–332.

Grove, B. G., and Sharma, S. C., 1979, Tectal projections in the goldfish *(Carassius auratus):* A degeneration study, *J. Comp. Neurol.* **184:**435–454.

Grünthal, E., 1934, Der Zellbau im Thalamus der Säuger und des Menschen, *J. Psychol. Neurol.* **46:**41–112.

Grünthal, E., 1945, Die Rindenprojektion der Thalamuskerne bei der Maus: Eine experimentell-anatomische Untersuchung, *Monatsschr. Psychiatr. Neurol.* **110:**245–268.

Guilbaud, G., Peschanski, M., Gautron, M., and Binder, D., 1980, Neurones responding to noxious stimulation in VB complex and caudal adjacent regions in the thalamus of the rat, *Pain* **8:**303–318.

Guillery, R. W., 1955, A quantitative study of the mamillary bodies and their connexions, *J. Anat.* **89:**19–32.

Guillery, R. W., 1956, Degeneration in the post-commissural fornix and the mamillary peduncle of the rat, *J. Anat.* **90:**350–370.

Guillery, R. W., 1957, Degeneration in the hypothalamic connexions of the albino rat, *J. Anat.* **91:**91–115.

Guillery, R. W., 1959, Afferent fibres to the dorso-medial thalamic nucleus in the cat, *J. Anat.* **93:**403–419.

Guillery, R. W., 1961, Fibre degeneration in the efferent mamillary tracts of the cat, in: *Cytology of Nervous Tissue*, Taylor & Francis, London, pp. 64–67.

Guillery, R. W., 1966, A study of Golgi preparations from the dorsal lateral geniculate nucleus of the adult cat, *J. Comp. Neurol.* **128:**21–49.

Guillery, R. W., 1967, Patterns of fiber degeneration in the dorsal lateral geniculate nucleus of the cat following lesions in the visual cortex, *J. Comp. Neurol.* **130:**197–222.

Guillery, R. W., 1969a, A quantitative study of synaptic interconnections in the dorsal lateral geniculate nucleus of the cat, *Z. Zellforsch. Mikrosk. Anat.* **96:**39–48.

Guillery, R. W., 1969b, The organization of synaptic interconnections in the laminae of the dorsal lateral geniculate nucleus of the cat, *Z. Zellforsch. Mikrosk. Anat.* **96:**1–38.

Guillery, R. W., 1969c, An abnormal retinogeniculate projection in Siamese cats, *Brain Res.* **14:**739–741.

Guillery, R. W., 1970, The laminar distribution of retinal fibers in the dorsal lateral geniculate nucleus of the cat: A new interpretation, *J. Comp. Neurol.* **138:**339–367.

Guillery, R. W., 1971a, An abnormal retinogeniculate projection in the albino ferret *(Mustela furo)*, *Brain Res.* **22:**482–485.

Guillery, R. W., 1971b, Patterns of synaptic interconnections in the dorsal lateral geniculate nucleus of cat and monkey: A brief review, *Vision Res. Suppl.* **3:**211–227.

Guillery, R. W., 1971c, Survival of large cells in the dorsal lateral geniculate laminae after interruption of retinogeniculate afferents, *Brain Res.* **28:**541–544.

Guillery, R. W., 1972a, Experiments to determine whether retino-geniculate axons can form translaminar collateral sprouts in the lateral geniculate nucleus of the cat, *J. Comp. Neurol.* **146:**407–420.

Guillery, R. W., 1972b, Binocular competition in the control of geniculate cell growth, *J. Comp. Neurol.* **148:**417–429.

Guillery, R. W., 1973, Quantitative studies of transneuronal atrophy in the dorsal lateral geniculate nucleus of cats and kittens, *J. Comp. Neurol.* **149:**423–438.

Guillery, R. W., 1979, A speculative essay on geniculate lamination and its development, *Prog. Brain Res.* **51:**403–418.

Guillery, R. W., and Casagrande, V. A., 1977, Studies of the modifiability of the visual pathways in Midwestern Siamese cats, *J. Comp. Neurol.* **174:**15–46.

Guillery, R. W., and Colonnier, M., 1970, Synaptic patterns in the dorsal lateral geniculate nucleus of the monkey, *Z. Zellforsch. Mikrosk. Anat.* **103:**90–108.

Guillery, R. W., and Kaas, J. H., 1971, A study of normal and congenitally abnormal retinogeniculate projections in cats, *J. Comp. Neurol.* **143:**73–99.

Guillery, R. W., and Kaas, J. H., 1973, Genetic abnormality of the visual pathways in a "white" tiger, *Science* **180:**1287–1289.

Guillery, R. W., and Kaas, J. H., 1974, The effects of monocular lid suture upon the development of the lateral geniculate nucleus in squirrels (*Sciureus carolinensis*), *J. Comp. Neurol.* **154:**433–441.

Guillery, R. W., and Oberdorfer, M. D., 1977, A study of fine and coarse retino-fugal axons terminating in the geniculate C laminae and in the medial interlaminar nucleus of the mink, *J. Comp. Neurol.* **176:**515–525.

Guillery, R. W., and Scott, G. L., 1971, Observations on synaptic patterns in the dorsal lateral geniculate nucleus of the cat: The C laminae and the perikaryal synapses, *Exp. Brain Res.* **12:**184–203.

Guillery, R. W., and Stelzner, D. J., 1970, The differential effects of unilateral lid closure upon the monocular and binocular segments of the dorsal lateral geniculate nucleus in the cat, *J. Comp. Neurol.* **139:**413–422.

Guillery, R. W., and Updyke, B. V., 1976, Retinofugal pathways in normal and albino axolotls, *Brain Res.* **109:**235–244.

Guillery, R. W., Adrian, H. O., Woolsey, C. N., and Rose, J. E., 1966, Activation of somatosensory areas I and II of cat's cerebral cortex by focal stimulation of the ventrobasal complex, in: *The Thalamus* (D. P. Purpura and M. D. Yahr, eds.), Columbia University Press, New York, pp. 197–206.

Guillery, R. W., Sitthi-Amorn, C., and Eighmy, B. B., 1971, Mutants with abnormal visual pathways: An explanation of anomalous geniculate laminae, *Science* **174:**831–832.

Guillery, R. W., Casagrande, V. A., and Oberdorfer, M. D., 1974, Congenitally abnormal vision in Siamese cats, *Nature (London)* **252:**195–199.

Guillery, R. W., Okoro, A. N., and Witkop, C. J., Jr., 1975, Abnormal visual pathways in the brain of a human albino, *Brain Res.* **96:**373–377.

Guillery, R. W., Geisert, E. E., Jr., Polley, E. H., and Mason, C. A., 1980, An analysis of the retinal afferents to the cat's medial interlaminar nucleus and its rostral thalamic extension, the "geniculate wing," *J. Comp. Neurol.* **194:**117–142.

Gurdjian, E. S., 1925, Olfactory connections of the albino rat with special reference to stria medullaris and anterior commissure, *J. Comp. Neurol.* **38:**127–163.

Gurdjian, E. S., 1927, The diencephalon of the albino rat: Studies on the brain of the rat, No. 2, *J. Comp. Neurol.* **43:**1–114.

Gurdjian, E. S., 1928, The corpus striatum of the rat: Studies on the brain of the rat, No. 3, *J. Comp. Neurol.* **45:**249–281.

Haberly, L. B., and Price, J. L., 1978, Association and commissural fiber systems of the olfactory cortex of the rat. I. Systems originating in the piriform cortex and adjacent areas, *J. Comp. Neurol.* **178:**711–740.

Haight, J. R., 1972, The general organization of somatotopic projections to SII cerebral neocortex in the cat, *Brain Res.* **44:**483–502.

Haight, J. R., and Neylon, L., 1978, An atlas of the dorsal thalamus of the marsupial brush-tailed possum, *Trichosurus vulpecula*, *J. Anat.* **126:**225–246.

Haight, J. R., Sanderson, K. J., Neylon, L., and Patten G. S., 1980, Relationships of the visual cortex in the marsupial brush-tailed possum, *Trichosurus vulpecula*, a horseradish peroxidase and autoradiographic study, *J. Anat.* **131:**387–413.

Hajdu, F., Somogyi, G., and Tömböl, T., 1974, Neuronal and synaptic arrangements in the lateralis posterior–pulvinar complex of the thalamus in the cat, *Brain Res.* **73:**89–104.

Haldeman, S., and McLennan, H. 1973, The action of two inhibitors of glutamic acid uptake upon amino acid-induced and synaptic excitations of thalamic neurones, *Brain Res.* **63:**123–130.

Hale, P. T., and Sefton, A. J., 1978, A comparison of the visual and electrical response properties of cells in the dorsal and ventral lateral geniculate nuclei, *Brain Res.* **153:**591–595.

Hale, P. T., Sefton, A. J., Baur, L. A., and Cottee, L. J., 1982, Interrelations of the rat's thalamic reticular and dorsal lateral geniculate nuclei, *Exp. Brain Res.* **45:**217–229.

Hall, R. D., and Lindholm, E. P., 1974, Organization of motor and somatosensory neocortex in the albino rat, *Brain Res.* **66:**23–38.

Hall, W. C., and Diamond, I. T., 1968, Organization and function of the visual cortex in hedge hog. 1. Cortical cytoarchitecture and thalamic retrograde degeneration, *Brain Behav. Evol.* **1:** 181–214.

Hall, W. C., and Ebner, F. F., 1970a, Parallels in the visual afferent projections of the thalamus in the hedgehog (*Paraechinus hypomelas*) and the turtle (*Pseudemys scripta*), *Brain Behav. Evol.* **3:** 135–154.

Hall, W. C., and Ebner, F. F., 1970b, Thalamo-telencephalic projections in the turtle *Pseudemys scripta, J. Comp. Neurol.* **140:**101–122.

Haller, B., 1900, Vom Bau des Wirbelthiergehirns. Theil III. Mus nebst Bemerkungen über das Hirn von Echidna, *Morphol. Jahrb.* **28:**347–477.

Hamburger, V., 1975, Cell death in the development of the lateral motor column of the chick embryo, *J. Comp. Neurol.* **160:**535–546.

Hamilton, B. L., 1973, Projections of the nuclei of the periaqueductal gray matter in the cat, *J. Comp. Neurol.* **152:**45–58.

Hámori, J., Pasik, T., Pasik, P., and Szentágothai, J., 1974, Triadic synaptic arrangements and their possible significance in the lateral geniculate nucleus of the monkey, *Brain Res.* **80:**379–393.

Hanberry, J., and Jasper, H. H., 1953, Independence of diffuse thalamo-cortical projection system shown by specific nuclear destructions, *J. Neurophysiol.* **16:**252–271.

Hanberry, J., Ajmone-Marsan, C., and Dilworth, M., 1954, Pathways of non-specific thalamo-cortical projection system, *Electroencephalogr. Clin. Neurophysiol.* **6:**103–118.

Hancock, M. B., Foreman, R. D., and Willis, W. D., 1975, Convergence of visceral and cutaneous input onto spinothalamic tract cells in the thoracic spinal cord of the cat, *Exp. Neurol.* **47:**240–248.

Hand, P. J., and Morrison, A. R., 1970, Thalamocortical projections from the ventrobasal complex to somatic sensory areas I and II, *Exp. Neurol.* **26:**291–308.

Hand, P. J., and Morrison, A. R., 1972, Thalamocortical relationships in the somatic sensory system as revealed by silver impregnation techniques, *Brain Behav. Evol.* **5:**273–302.

Hand, P. J., and Van Winkle, T., 1977, The efferent connections of the feline nucleus cuneatus, *J. Comp. Neurol.* **171:**83–109.

Harding, B. N., 1973a, An ultrastructural study of the centre median and ventrolateral thalamic nuclei of the monkey, *Brain Res.* **54:**335–340.

Harding, B. N., 1973b, An ultrastructural study of the termination of afferent fibres within the ventrolateral and centre median nuclei of the monkey thalamus, *Brain Res.* **54:**341–346.

Harding, B. N., and Powell, T. P. S., 1977, An electron microscopic study of the centre-median and ventrolateral nuclei of the thalamus in the monkey. *Philos. Trans. R. Soc. London Ser. B* **279:**357–412.

Harris, F. A., 1970, Population analysis of somatosensory thalamus in the cat, *Nature (London)* **225:**559–562.

Harris, F. A., 1978a, Functional subsets of neurons in somatosensory thalamus of the cat, *Exp. Neurol.* **58:**149–170.

Harris, F. A., 1978b, Regional variations of somatosensory input convergence in nucleus VPL of the cat thalamus, *Exp. Neurol.* **58:**170–189.

Harting, J. K., and Guillery, R. W., 1976, Organization of retinocollicular pathways in the cat, *J. Comp. Neurol.* **166:**133–144.

Harting, J. K., and Huerta, M. F., 1983, The geniculostriate projection in the grey squirrel: Preliminary autoradiographic data, *Brain Res.* **272:**341–349.

Harting, J. K., and Noback, C. R., 1971, Subcortical projections from the visual cortex in tree shrew (*Tupaia glis*), *Brain Res.* **25:**21–33.

Harting, J. K., Hall, W. C., and Diamond, I. T., 1972, Evolution of the pulvinar, *Brain Behav. Evol.* **6:**424–452.

Harting, J. K., Hall, W. C., Diamond, I. T., and Martin, G. F., 1973, Anterograde degeneration study of the superior colliculus in *Tupaia glis:* Evidence for a subdivision between superficial and deep layers, *J. Comp. Neurol.* **148:**361–386.

Harting, J. K., Casagrande, V. A., and Weber, J. T., 1978, The projection of the primate superior colliculus upon the dorsal lateral geniculate nucleus: Autoradiographic demonstraton of interlaminar distribution of tectogeniculate axons, *Brain Res.* **150:**593–599.

Harting, J. K., Huerta, M. F., Frankfurter, A. J., Strominger, N. L. and Royce, G. J., 1980, Ascending pathways from the monkey superior colliculus: An autoradiographic analysis, *J. Comp. Neurol.* **192:**853–882.

Harvey, A. R., 1978, Characteristics of corticothalamic neurons in area 17 of the cat, *Neurosci. Lett.* **7:**177–181.

Harvey, R. J., Porter, R., and Rawson, J. A., 1979, Discharges of intracerebellar nuclear cells in monkeys, *J. Physiol. (London)* **297:**559–580.

Hassler, R., 1948, Forels Haubenfaszikel als vestibuläre Empfindungsbahn mit Bemerkungen über einige andere sekundäre Bahnen des Vestibularis und Trigeminus, *Arch. Psychiatr. Nervenkr.* **180:**23–53.

Hassler, R., 1949a, Über die Rinden-und Stammhirnanteile des menschlichen Thalamus, *Psychiatr. Neurol. Med. Psychol.* **1:**181–187.

Hassler, R., 1949b, Über die afferenten Bahnen und Thalamuskerne des motorischen Systems des Grosshirns. I. Mitteilung. Bindearm und Fasciculus Thalamicus, *Arch. Psychiatr. Nervenkr.* **182:**759–785.

Hassler, R., 1949c, Über die afferenten Bahnen und Thalamuskerne des motorischen Systems des Grosshirns. II. Mitteilung. Weitere Bahnen aus Pallidum, Ruber, vestibulären System zum Thalamus; Übersicht und Besprechung der Ergebnisse, *Arch. Psychiatr. Nervenkr.* **182:** 786–818.

Hassler, R., 1950, Die Anatomie des Thalamus, *Arch. Psychiatr. Nervenkr.* **184:**249–256.

Hassler, R., 1959, Anatomy of the thalamus, in: *Introduction to Stereotaxis with an Atlas of the Human Brain* (G. Schaltenbrand and P. Bailey, eds.), Volume 1, Thieme, Stuttgart, pp. 230–290.

Hassler, R., 1960, Die zentralen Systeme des Schmerzes, *Acta Neurochir.* **8:**353–423.

Hassler, R., 1962, New aspects of brain functions revealed by brain diseases, in: *Frontiers in Brain Research,* Columbia University Press, New York, pp. 242–285.

Hassler, R., 1964, Spezifische und unspezifische Systeme des menschlichen Zwischenhirns, *Prog. Brain Res.* **5:**1–32.

Hassler, R., 1966, Comparative anatomy of the central visual systems in day- and night-active primates, in: *Evolution of the Forebrain* (R. Hassler and H. Stephan, eds.), Thieme, Stuttgart, pp. 419–434.

Hassler, R., and Muhs-Clement, K., 1964, Architektonischer Aufbau des sensomotorischen und parietalen Cortex der Katze, *J. Hirnforsch.* **6:**377–420.

Hattori, T., McGeer, E. G., Singh, V. K. and McGeer, P. L., 1977, Cholinergic synapse of the interpeduncular nucleus, *Exp. Neurol.* **55:**666–679.

Hayhow, W. R., 1958, The cytoarchitecture of the lateral geniculate body in the cat in relation to the distribution of crossed and uncrossed optic fibers, *J. Comp. Neurol.* **110:**1–64.

Hayhow, W. R., 1967, The lateral geniculate nucleus of the marsupial phalanger, *Trichosurus vulpecula:* An experimental study of cytoarchitecture in relation to the intranuclear optic nerve projection fields, *J. Comp. Neurol.* **131:**571–603.

Hayhow, W. R., Sefton, A., and Webb, C., 1962, Primary optic centers of the rat in relation to the terminal distribution of the crossed and uncrossed optic nerve fibers, *J. Comp. Neurol.* **118:**295–321.

Hayward, J. N., 1975, Response of ventrobasal thalamic cells to hair displacement on the face of the waking monkey, *J. Physiol. (London)* **250:**385–407.

Hazlett, J. C., Dutta, C. R., and Fox, C. A., 1976, The neurons in the centromedian–parafascicular complex of the monkey *(Macaca mulatta):* A Golgi study, *J. Comp. Neurol.* **168:**41–73.

Head, H., 1920, *Studies in Neurology,* 2 vols., Henry Frowde, Oxford University Press, Hodder and Stoughton, London.

Head, H., and Holmes, G., 1911, Sensory disturbances from cerebral lesions, *Brain* **34:**102–254, 255–271.

Headon, M. P., and Powell, T. P. S., 1973, Cellular changes in the lateral geniculate nucleus of infant monkeys after suture of the eyelids, *J. Anat.* **116:**135–145.

Headon, M. P., and Powell, T. P. S., 1978, The effect of bilateral eye closure upon the lateral geniculate nucleus in infant monkeys, *Brain Res.* **143:**147–154.

Headon, M. P., Sloper, J. J., Hiorns, R. W., and Powell, T. P. S., 1979, Cell size changes in undeprived laminae of monkey lateral geniculate nucleus after monocular closure, *Nature (London)* **281:**572–574.

Headon, M. P., Sloper, J. J., Hiorns, R. W. and Powell, T. P. S., 1981a, Cell sizes in the lateral geniculate nucleus of normal infant and adult rhesus monkeys, *Brain Res.* **229:**183–186.

Headon, M. P., Sloper, J. J., Hiorns, R. W., and Powell, T. P. S., 1981b, Shrinkage of cells in undeprived laminae of the monkey lateral geniculate nucleus following late closure of one eye, *Brain Res.* **229:**187–192.

Headon, M. P., Sloper, J. J., and Powell, T. P. S., 1982, Initial hypertrophy of cells in undeprived laminae of the lateral geniculate nucleus of the monkey following early monocular visual deprivation, *Brain Res.* **238:**439–444.

Heath, C. J., 1970, Distribution of axonal degeneration following lesions of the posterior group of thalamic nuclei in the cat, *Brain Res.* **21:**435–438.

Heath, C. J., and Jones, E. G., 1971a, An experimental study of ascending connections from the posterior group of thalamic nuclei in the cat, *J. Comp. Neurol.* **141:**397–426.

Heath, C. J., and Jones, E. G., 1971b, Anatomical organization of the suprasylvian gyrus of the cat, *Ergeb. Anat. Entwicklungsgesch.* **45:**1–64.

Heath, C. J., Hore, J., and Phillips, C. G., 1976, Inputs from low threshold muscle and cutaneous afferents of hand and forearm to areas 3a and 3b of baboon's cerebral cortex, *J. Physiol. (London)* **257:**199–227.

Hedreen, J. C., 1970, Diencephalic projection of the retina in the cat, *Anat. Rec.* **166:**317.

Heffner, H., 1978, Effect of auditory cortex ablation on localization and discrimination of brief sounds, *J. Neurophysiol.* **41:**963–976.

Heier, P., 1948, Fundamental principles in the structure of the brain: A study of the brain of *Petromyzon fluviatilis, Acta Anat.* **5:**1–213.

Heimer, L. 1972, The olfactory connections of the diencephalon in the rat: An experimental light and electron microscopic study with special emphasis on the problem of terminal degeneration, *Brain Behav. Evol.* **6:**484–523.

Heiner, J. R., 1960, A reconstruction of the diencephalic nuclei of the chimpanzee, *J. Comp. Neurol.* **114:**217–238.

Hellon, R. F., and Misra, N. K., 1973, Neurones in the ventrobasal complex of the rat thalamus responding to scrotal skin temperature changes, *J. Physiol. (London)* **232:**389–399.

Hellon, R. F., and Mitchell, D., 1975, Convergence in a thermal afferent pathway in the rat, *J. Physiol. (London)* **248:**359–376.

Hellweg, F.-C., Schultz, W., and Creutzfeldt, O. D., 1977, Extracellular and intracellular recordings from cat's cortical whisker projection area: Thalamocortical response transformation, *J. Neurophysiol.* **40:**463–479.

Hendrickson, A. E., 1969, Electron microscopic radioautography: Identification of origin of synaptic terminals in normal nervous tissue, *Science* **165:**194–196.

Hendrickson, A. E., 1973, The pregeniculate nucleus of the monkey, *Anat. Rec.* **175:**341.

Hendrickson, A. E., and Rakic, P., 1977, Histogenesis and synaptogenesis in the dorsal lateral geniculate nucleus (LGd) of the fetal monkey brain, *Anat. Rec.* **187:**602.

Hendrickson, A. E., Wilson, M. E., and Toyne, M. J., 1970, The distribution of optic nerve fibers in *Macaca mulatta, Brain Res.* **23:**425–427.

Hendrickson, A. E., Wilson, J. R., and Ogren, M. P., 1978, The neuroanatomical organization of pathways between the dorsal lateral geniculate nucleus and visual cortex in Old World and New World primates, *J. Comp. Neurol.* **182:**123–135.

Hendrickson, A. E., Hunt, S. P., and Wu, J.-Y., 1981, Immunocytochemical localization of glutamic acid decarboxylase in monkey striate cortex, *Nature (London)* **292:**605–606.

Hendry, S. H. C., and Jones, E. G., 1981, Sizes and distributions of intrinsic neurons incorporating tritiated GABA in monkey sensory-motor cortex, *J. Neurosci.* **1:**390–408.

Hendry, S. H. C., and Jones, E. G., 1983a, The organization of pyramidal and non-pyramidal cell dendrites in relation to thalamic afferent terminations in the monkey somatic sensory cortex, *J. Neurocytol.* **12:**277–298.

Hendry, S. H. C., and Jones, E. G., 1983b, Thalamic inputs to identified commissural neurons in the monkey somatic sensory cortex, *J. Neurocytol.* **12:**299–316.

Hendry, S. H. C., Jones, E. G., and Graham, J., 1979, Thalamic relay nuclei for cerebellar and certain related fiber systems in the cat, *J. Comp. Neurol.* **185:**679–714.

Hendry, S. H. C., Jones, E. G., Hockfield, S., and McKay, R., 1984, Monoclonal antibody that identified subsets of neurones in the central visual system of monkey and cat, *Nature (London)* **307:**267–269.

Henle, G. J., 1871, *Handbuch der Systematischen Anatomie des Menschen,* Volume 3, Part 2, Vieweg, Braunschweig.

Herbert, J., 1963, Nuclear structure of the thalamus of the ferret. *J. Comp. Neurol.* **120:**105–127.

Herkenham, M., 1978, The connections of the nucleus reuniens thalami: Evidence for a direct thalamo-hippocampal pathway in the rat, *J. Comp. Neurol.* **177:**589–610.

Herkenham, M., 1979, The afferent and efferent connections of the ventromedial thalamic nucleus in the rat, *J. Comp. Neurol.* **183**:487–518.

Herkenham, M. 1980, Laminar organization of thalamic projections to rat neocortex, *Science* **207**:532–534.

Herkenham, M., 1981, Anesthetics and the habenulo-interpeduncular system: Selective sparing of metabolic activity, *Brain Res.* **210**:461–466.

Herkenham, M., and Nauta, W. J. H., 1977, Afferent connections of the habenular nuclei in the rat: A horseradish peroxidase study, with a note on the fiber-of-passage problem, *J. Comp. Neurol.* **173**:123–146.

Herkenham, M., and Nauta, W. J. H., 1979, Efferent connections of the habenular nuclei in the rat, *J. Comp. Neurol.* **187**:19–48.

Herkenham, M., and Pert, C. B., 1981, Mosaic distribution of opiate receptors, parafascicular projections and acetylcholinesterase in rat striatum, *Nature (London)* **291**:415–418.

Herrick, C. J., 1910, The morphology of the forebrain in Amphibia and Reptilia, *J. Comp. Neurol.* **20**:413–547.

Herrick, C. J., 1917, The internal structure of the midbrain and thalamus of *Necturus, J. Comp. Neurol.* **28**:215–348.

Herrick, C. J., 1918, *An Introduction to Neurology,* 2nd ed., Saunders, Philadellphia.

Herrick, C. J., 1925, The amphibian forebrain. III. The optic tracts and centers of Amblystoma and the frog, *J. Comp. Neurol.* **39**:433–489.

Herrick, C. J., 1926, *Brains of Rats and Men,* University of Chicago Press, Chicago.

Herrick, C. J., 1933, The amphibian forebrain. VI. *Necturus, J. Comp. Neurol.* **58**:1–288.

Herrick, C. J., 1934, The amphibian forebrain. IX. Neuropil and other interstitial nervous tissue, *J. Comp. Neurol.* **59**:93–116.

Herrick, C. J., 1948, *The Brain of the Tiger Salamander,* University of Chicago Press, Chicago.

Herrick, C. L., 1891, Studies on the brains of some American fresh-water fishes, *J. Comp. Neurol.* **1**:228–245, 333–358.

Herrick, C. L., 1892, Contribution to the morphology of the brain of bony fishes, II, *J. Comp. Neurol.* **2**:21–72.

Herron, P., 1978, Somatotopic organization of mechanosensory projections to SII cerebral neocortex in the raccoon *(Procyon lotor), J. Comp. Neurol.* **181**:717–728.

Herron, P., 1982, The connections of cortical somatosensory areas I and II with separate nuclei in the ventroposterior thalamus in the raccoon, *Neuroscience* **8**:243–257.

Hess, A., 1955, The nuclear topography and architectonics of the thalamus of the guinea pig, *J. Comp. Neurol.* **103**:385–419.

Hess, A., 1957, Optic centers and pathways after eye removal in fetal guinea pigs, *J. Comp. Neurol.* **109**:91–116.

Hickey, T. L., 1975, Translaminar growth of axons in the kitten dorsal lateral geniculate nucleus following removal of one eye, *J. Comp. Neurol.* **161**:359–382.

Hickey, T. L., 1977a, Variability in patterns of lamination in the human lateral geniculate nucleus, *Neurosci. Abstr.* **3**:562.

Hickey, T. L., 1977b, Postnatal development of the human lateral geniculate nucleus: Relationship to a critical period for the visual system, *Science* **198**:836–838.

Hickey, T. L., 1980, Development of the dorsal lateral geniculate nucleus in normal and visually deprived cats, *J. Comp. Neurol.* **189**:467–481.

Hickey, T. L., and Cox, N. R., 1979, Cell birth in the lateral geniculate nucleus of the cat: A ^{3}H-thymidine study, *Soc. Neurosci. Abstr.* **5**:788.

Hickey, T. L., and Guillery, R. W., 1974, An autoradiographic study of retinogeniculate pathways in the cat and the fox, *J. Comp. Neurol.* **156**:239–254.

Hickey, T. L., and Guillery, R. W., 1979, Variability of laminar patterns in the human lateral geniculate nucleus, *J. Comp. Neurol.* **183**:221–246.

Hickey, T. L., and Spear, P. D., 1976, Retinogeniculate projections in hooded and albino rats: An autoradiographic study, *Exp. Brain Res.* **24**:523–529.

Hickey, T. L., Spear, P. D., and Kratz, K. E., 1977, Quantitative studies of cell size in the cat's dorsal lateral geniculate nucleus following visual deprivation, *J. Comp. Neurol.* **172**:265–282.

Hicks, T. P., Lee, B. B., and Vidyasagar, T. R., 1983, The responses of cells in macaque lateral geniculate nucleus to sinusoidal gratings, *J. Physiol. (London)* **337**:183–200.

Hiley, C. R., and Burgen, A. S. V., 1974, The distribution of muscarinic receptor sites in the nervous system of the dog, *J. Neurochem.* **22**:159–162.

Hill, R. G., Pepper, C. M., and Mitchell, J. F., 1976, Depression of nociceptive and other neurones in the brain by iontophoretically applied Met-enkephalin, *Nature (London)* **262**:604–606.

Hind, J. E., Goldberg, J. M., Greenwood, D. D., and Rose, J. E., 1963, Some discharge characteristics of single neurons in the inferior colliculus of the cat. II. Timing of the discharges and observations on binaural stimulation, *J. Neurophysiol.* **26**:321–341.

Hines, M., 1923, The development of the telencephalon in *Sphenodon punctatum*, *J. Comp Neurol.* **35**:483–537.

Hines, M., 1929, The brain of *Ornithorhynchus anatinus*, *Philos. Trans. R. Soc. London Ser. B* **217**:155–288.

Hirokawa, S., 1941, Über die Thalamuskerne der Katze, *Jpn. J. Med. Sci. 1* **9**:119–147.

His, W., 1893, Über das frontale Ende des Gehirnrohres, *Z. Anat. Entwicklungsgesch* **157**:171–194.

His, W., 1904, *Die Entwicklung des Menschlichen Gehirns Wahrend der Ersten Monate*, Hirzel, Leipzig.

Hitchcock, P. F., and Hickey, T. L, 1980, Prenatal development of the human lateral geniculate nucleus, *J. Comp. Neurol.* **194**:395–412.

Hitchcock, P. F., and Hickey, T. L., 1983, Morphology of C-laminae neurons in the dorsal lateral geniculate nucleus of the cat: A Golgi impregnation study, *J. Comp. Neurol.* **220**:137–146.

Hobson, J. A., 1984, How does the cortex know when to do what? A neurological theory of state control, in: *Dynamic Aspects of Neocortical Function* (G. Edelman, W. E. Gall, and W. M. Cowan, eds.), Wiley, New York, pp. 219–257.

Hobson, J. A., McCarley, R. W., Pivik, R. T., and Freedman, R., 1974, Selective firing by cat pontine brain-stem neurons in desynchronized sleep, *J. Neurophysiol.* **37**:497–511.

Hockfield, S., and Gobel, S., 1978, Neurons in and near nucleus caudalis with long ascending projection axons demonstrated by retrograde labeling with horseradish peroxidase, *Brain Res.* **139**:333–339.

Hockfield, S., McKay, R. D., Hendry, S. H. C., and Jones, E. G., 1983, A surface antigen that identifies ocular dominance columns in the visual cortex and laminar features of the lateral geniculate nucleus, *Cold Spring Harbor Symp. Quant. Biol.* **48**:877–890.

Hoffmann, K.-P., 1973, Conduction velocity in pathways from retina to superior colliculus in the cat: A correlation with receptive-field properties, *J. Neurophysiol.* **36**:409–424.

Hoffmann, K.-P., and Cynader, M., 1977, Functional aspects of plasticity in the visual system of adult cats after early monocular deprivation, *Philos. Trans. R. Soc. London Ser. B* **278**:411–424.

Hoffmann, K.-P., and Holländer, H., 1978, Physiological and morphological changes in cells of the lateral geniculate nucleus in monocularly-deprived and reverse-sutured cats, *J. Comp. Neurol.* **177**:145–157.

Hoffmann, K.-P., and Stone, J., 1971, Conduction velocity of afferents to cat visual cortex: A correlation with cortical receptive field properties, *Brain Res.* **32**:460–466.

Hoffmann, K.-P., Stone, J., and Sherman, S. M., 1972, Relay of receptive-field properties in dorsal lateral geniculate nucleus of the cat, *J. Neurophysiol.* **35**:518–531.

Hökfelt, T., Fuxe, K., Goldstein, M., and Johansson, O., 1974, Immunohistochemical evidence for the existence of adrenaline neurons in the rat brain, *Brain Res.* **66**:235–251.

Hökfelt, T., Keller, J. O., Nilsson, G., and Pernow, W., 1975, Substance P localization in the central nervous system and in some primary sensory neurons, *Science* **190**: 889–890.

Hökfelt, T., Johansson, O., Ljungdahl, A., Lundberg, J. M., and Schultzberg, M., 1980, Peptidergic neurones, *Nature (London)* **284**:515–521.

Holländer, H., 1974, Projection from the striate cortex to the diencephalon in the squirrel monkey (*Saimiri sciureus*): A light microscopic radioautographic study following intracortical injections of H^3 leucine, *J. Comp. Neurol.* **155**:425–440.

Holländer, H., and Hälbig, W., 1980, Topography of retinal representation in the rabbit cortex: An experimental study using transneuronal and retrograde tracing techniques, *J. Comp. Neurol.* **93**:701–710.

Holländer, H., and Martinez-Millan, L., 1975, Autoradiographic evidence for a topographically organized projection from the striate cortex to the lateral geniculate nucleus in the rhesus monkey (*Macaca mulatta*), *Brain Res.* **100**:407–411.

Holländer, H., and Sanides, D., 1976, The retinal projection to the ventral part of the lateral geniculate nucleus: An experimental study with silver-impregnation methods and axoplasmic protein tracing, *Exp. Brain Res.* **26**:329–342.

Holländer, H., and Vanegas, H., 1977, The projection from the lateral geniculate nucleus onto the visual cortex in the cat: A quantitative study with horseradish-peroxidase, *J. Comp. Neurol.* **173:**519–536.

Holländer, H., Tietze, J., and Distel, H., 1979, An autoradiographic study of the subcortical projections of the rabbit striate cortex in the adult and during postnatal development, *J. Comp. Neurol.* **184:**783–794.

Holmes, R. L., 1953, Nuclear studies on the thalamus of the mouse, *J. Comp. Neurol.* **99:**377–413.

Holmgren, N., 1919, Zur Anatomie des Gehirns von Myxine, *Kungl. Sven. Vetenskapakad. Handl.* **60:**1096.

Honda, C. N., Mense, S., and Perl, E. R., 1983, Neurons in ventrobasal region of cat thalamus selectively responsive to noxious mechanical stimulation, *J. Neurophysiol.* **49:**662–673.

Hong, J. S., Costa, E., and Yang, H.-Y. T., 1976, Effects of habenular lesions on the substance P content of various brain regions, *Brain Res.* **118:**523–525.

Hoogland, P. V., 1981, Spinothalamic projections in a lizard, *Varanus exanthematicus:* An HRP study, *J. Comp. Neurol.* **198:**7–12.

Hoover, D. B., and Baisden, R. H., 1980, Localization of putative cholinergic neurons innervating the anteroventral thalamus, *Brain Res. Bull.* **5:**519–524.

Hoover, D. B., and Jacobowitz, D. M., 1979, Neurochemical and histochemical studies of the effect of a lesion of the nucleus cuneiformis on the cholinergic innervation of discrete areas of the rat brain, *Brain Res.* **170:**113–122.

Hopf, A., 1971, Index of nuclei, in: *Anatomy of the Normal Human Thalamus: Topometry and Standardized Nomenclature* (A. Dewulf, ed.), Elsevier, Amsterdam, pp. 150–158.

Hopkins, D. A., and Lawrence, D. G., 1975, On the absence of a rubrothalamic projection in the monkey with observations on some ascending mesencephalic projections, *J. Comp. Neurol.* **161:**269–293.

Hore, J., Preston, J. B., Durkovic, R. G., and Cheney, P. D. 1976, Response of cortical neurons (areas 3a and 4) to ramp stretch of hindlimb muscles in the baboon, *J. Neurophysiol.* **39:**484–500.

Horne, M. K., and Porter, R., 1980, The discharges during movement of cells in the ventrolateral thalamus of the conscious monkey, *J. Physiol. (London)* **304:**349–372.

Horne, M. K., and Tracey D. J., 1979, The afferents and projections of the ventroposterolateral thalamus in the monkey, *Exp. Brain Res.* **36:**129–141.

Hornet, T., 1933, Vergleichend anatomische Untersuchungen über das Corpus geniculatum mediale (c.g.m.), *Arb. Neurol. Inst. Wien Univ.* **35:**76–92.

Horton, J. C., and Hubel, D. H. 1981, Regular patchy distributions of cytochrome oxidase staining in primary visual cortex of macaque monkey, *Nature (London)* **292:**762–763.

Horvath, F. E., and Buser, P., 1976, Relationships between two types of thalamic unit patterned discharges and cortical spindle waves, *Brain Res.* **103:**560–567.

Houser, C. R., Vaughn, J. E., Barber, R. P., and Roberts, E., 1980, GABA neurons are the major cell type of the nucleus reticularis thalami, *Brain Res.* **200:**341–354.

Houser, C. R., Crawford, G. D., Barber, R. P., Salvaterra, P. M., and Vaughn, J. E., 1983, Organization and morphological characteristics of cholinergic neurons: An immunocytochemical study with a monoclonal antibody to choline acetyltransferase, *Brain Res.* **266:**97–119.

Hu, J. W., Dostrovsky, J. O., and Sessle, B. J., 1981, Functional properties of neurons in cat trigeminal subnucleus caudalis (medullary dorsal horn). I. Responses to oral-facial noxious and non-noxious stimuli and projections to thalamus and subnucleus oralis, *J. Neurophysiol.* **45:**173–192.

Hubel, D. H., 1960, Single unit activity in lateral geniculate body and optic tract of unrestrained cats, *J. Physiol. (London)* **150:**91–104.

Hubel, D. H., 1975, An autoradiographic study of the retino-cortical projections in the tree shrew (*Tupaia glis*), *Brain Res.* **96:**41–50.

Hubel, D. H., and Freeman, D. C., 1977, Projection into the visual field of ocular dominance columns in macaque monkey, *Brain Res.* **122:**336–343.

Hubel, D. H., and Wiesel, T. N., 1959, Receptive fields of single neurones in the cat's striate cortex, *J. Physiol. (London)* **148:**574–591.

Hubel, D. H., and Wiesel, T. N., 1960, Receptive fields of optic nerve fibres in the spider monkey, *J. Physiol. (London)* **154:**572–580.

Hubel, D. H., and Wiesel, T. N., 1961, Integrative action in the cat's lateral geniculate body, *J. Physiol. (London)* **155:**385–398.

Hubel, D. H., and Wiesel, T. N., 1962, Receptive fields, binocular interaction and functional architecture in the cat's visual cortex, *J. Physiol. (London)* **160**:106–154.

Hubel, D. H., and Wiesel, T. N., 1965, Receptive fields and functional architecture in two nonstriate visual areas (18 and 19) of the cat, *J. Neurophysiol.* **28**:229–289.

Hubel, D. H., and Wiesel, T. N., 1968, Receptive fields and functional architecture of monkey striate cortex, *J. Physiol. (London)* **195**:215–243.

Hubel, D. H., and Wiesel, T. N., 1969, Anatomical demonstration of columns in the monkey striate cortex, *Nature (London)* **221**:747–750.

Hubel, D. H., and Wiesel, T. N., 1970, The period of susceptibility to the physiological effects of unilateral eye closure in kittens, *J. Physiol. (London)* **206**:419–436.

Hubel, D. H., and Wiesel, T. N., 1971, Aberrant visual projections in the Siamese cat, *J. Physiol. (London)* **218**:33–62.

Hubel, D. H., and Wiesel, T. N., 1972, Laminar and columnar distribution of geniculo-cortical fibers in the macaque monkey, *J. Comp. Neurol.* **146**:421–443.

Hubel, D. H., and Wiesel, T. N., 1974, Uniformity of monkey striate cortex: A parallel relationship between field size, scatter and magnification factor, *J. Comp. Neurol.* **15**:295–306.

Hubel, D. H., and Wiesel, T. N., 1977, Functional architecture of macaque monkey visual cortex, *Proc. R. Soc. London Ser. B* **198**:1–59.

Hubel, D. H., Wiesel, T. N., and LeVay, S., 1977, Plasticity of ocular dominance columns in monkey striate cortex, *Philos. Trans. R. Soc. London Ser. B.* **278**:377–409.

Huber, G. C., and Crosby, E. C., 1926, On thalamic and tectal nuclei and fiber paths in the brain of the American alligator, *J. Comp. Neurol.* **40**:97–227.

Huber, G. C., and Crosby, E. C., 1929, The nuclei and fiber paths of the avian diencephalon, with consideration of telencephalic and certain mesencephalic centers and connections, *J. Comp. Neurol.* **48**:1–225.

Huber, G. C., and Crosby, E. C., 1933, The reptilian optic tectum, *J. Comp. Neurol.* **57**:57–164.

Huerta, M. F., and Harting, J. K., 1983, Sublamination within the superficial gray layer of the squirrel monkey: An analysis of the tectopulvinar projection using anteograde and retrograde transport methods, *Brain Res.* **261**:119–126.

Hughes, A., 1981, Population magnitudes and distribution of the major modal classes of cat retinal ganglion cells as estimated from HRP filling and a systematic survey of the soma diameter spectra for classical neurons, *J. Comp. Neurol.* **197**:303–339.

Hughes, C. P., and Ater, S. B., 1977, Receptive field properties in the ventral lateral geniculate nucleus of the cat, *Brain Res.* **132**:163–166.

Hughes, C. P., and Chi, D. Y. K., 1981, Afferent projections in the ventral lateral geniculate nucleus in the cat, *Brain Res.* **207**:445–448.

Hughes, H. C., 1977, Anatomical and neurobehavioral investigations concerning the thalamocortical organization of the rat's visual system, *J. Comp. Neurol.* **175**:311–336.

Hughes, H. C., 1980, Efferent organization of the cat pulvinar complex, with a note on bilateral claustrocortical and reticulocortical connections, *J. Comp. Neurol.* **193**:937–963.

Hull, E. M., 1968, Corticofugal influence in the macaque lateral geniculate nucleus, *Vision Res.* **8**:1285–1298.

Humphrey, A. L., Albano, J. E., and Norton, T. T., 1977, Organization of ocular dominance in tree shrew striate cortex, *Brain Res.* **134**:225–236.

Humphrey, N. K., 1974, Vision in a monkey without striate cortex: A case study, *Perception* **3**:241–255.

Hunsperger, R. W., and Roman, D., 1976, The integrative role of the intralaminar system of the thalamus in visual orientation and perception in the cat, *Exp. Brain Res.* **25**:231–246.

Hunt, S. P., and Künzle, H., 1976, Observations on the projection and instrinsic organization of the pigeon optic tectum: An autoradiographic study based on anterograde and retrograde, axonal and dendritic flow, *J. Comp. Neurol.* **170**:153–172.

Hunt, S. P., and Schmidt, J., 1978, Some observations on the binding patterns of α-bungarotoxin in the central nervous system of the rat, *Brain Res.* **157**:213–232.

Hunt, S. P., and Webster, K. E., 1972, Thalamo-hyperstriate interrelations in the pigeon, *Brain Res.* **44**:647–651.

Hunter, J., 1825, Case of fungus haematodes of the brain, *Med. Chir. Trans.* **13**:88–96.

Hyvärinen, J., 1982, *The Parietal Cortex of Monkey and Man*, Springer-Verlag, Berlin.

Hyvärinen, J., Poranen, A., and Jokinen, Y., 1980, Influence of attentive behavior on neuronal responses to vibration in primary somatosensory cortex of the monkey, *J. Neurophysiol.* **43**:870–882.

Hyyppä, M., 1969, Differentiation of hypothalamic nuclei during ontogenetic development in the rat, *Z. Anat. Entwicklungsgesch.* **129**:41–52.

Ibrahim, M., and Shanklin, W. M., 1941, The diencephalon of the coney, *Hyrax syriaca*, *J. Comp. Neurol.* **75**:427–485.

Ide, L. S., 1982, The fine structure of the perigeniculate nucleus in the cat, *J. Comp. Neurol.* **210**:317–334.

Iggo, A., and Ogawa, H., 1977, Correlative physiological and morphological studies of rapidly adapting mechanoreceptors in cat's glabrous skin, *J. Physiol. (London)* **266**:275–296.

Ikeda, H., and Wright, M. J., 1972, Receptive field organisation of "sustained" and "transient" retinal ganglion cells which subserve different functional roles, *J. Physiol. (London)* **227**:769–800.

Illing, R.-B., and Wässle, H., 1981, The retinal projection to the thalamus in the cat: A quantitative investigation and a comparison with the retinotectal pathway, *J. Comp. Neurol.* **202**:265–286.

Imig, T. J., and Adrian, H. O., 1977, Binaural columns in the primary field (AI) of cat auditory cortex, *Brain Res.* **138**:241–257.

Imig, T. J., and Brugge, J. F., 1978, Sources of terminations of callosal axons related to binaural and frequency maps in primary auditory cortex of the cat, *J. Comp. Neurol.* **183**:637–660.

Imig, T. J., and Reale, R. A., 1981, Ipsilateral corticocortical projections related to binaural columns in cat primary auditory cortex, *J. Comp. Neurol.* **203**:1–14.

Imig, T. J., and Weinberger, N. M., 1973, Relationships between rate and pattern of unitary discharges in medial geniculate body of the cat in response to click and amplitude-modulated white-noise stimulation, *J. Neurophysiol.* **36**:385–397.

Inagaki, S., Kubota, Y., Shinoda, K., Kawai, Y., and Tohyama, M., 1983, Neurotensin-containing pathway from the endopiriform nucleus and the adjacent prepiriform cortex to the dorsomedial thalamic nucleus in the rat, *Brain Res.* **260**:143–146.

Ingram, W. R., Hannett, F. I., and Ranson, S. W., 1932, The topography of the nuclei of the diencephalon of the cat, *J. Comp. Neurol.* **55**:333–394.

Ionescu, D. A., and Hassler, R., 1968, Six cell layers in the lateral geniculate body in the night-active prosimian, *Galago crassicaudatus*, *Brain Res.* **10**:281–284.

Irvine, D. R. F., 1980, Acoustic properties of neurons in posteromedial thalamus of cat, *J. Neurophysiol.* **43**:395–408.

Isseroff, H., Rosvold, H. E., Galkin, T. W., and Goldman-Rakic, P. S., 1982, Spatial memory impairments following damage to the mediodorsal nucleus of the thalamus in rhesus monkeys, *Brain Res.* **232**:97–113.

Ito, H., and Kishida, R., 1977, Tectal afferent neurons identified by the retrograde HRP method in the carp telencephalon, *Brain Res.* **130**:142–145.

Ito, H., and Kishida, R., 1978, Telencephalic afferent neurons identified by the retrograde HRP method in the carp diencephalon, *Brain Res.* **149**:211–215.

Ito, H., Morita, Y., Sakamoto, N., and Ueda, S., 1980, Possibility of telencephalic visual projection in teleosts, *Holocentridae*, *Brain Res.* **197**:219–222.

Itoh, K., 1977, Efferent projections of the pretectum in the cat, *Brain Res.* **30**:89–105.

Itoh, K., and Mizuno, N., 1977, Topographical arrangement of thalamocortical neurons in the centrolateral nucleus (CL) of the cat, with special reference to a spino-thalamo-motor cortical path through the CL, *Exp. Brain Res.* **30**:471–480.

Itoh, K., and Mizuno, N., Sugimoto, T., Nomura, S., Nakamura Y., and Konishi, A., 1979, A cerebello-pulvino-cortical and a retino-pulvino-cortical pathway in the cat as revealed by the use of anterograde and retrograde transport of horseradish peroxidase, *J. Comp. Neurol.* **187**:349–358.

Itoh, K., Conley, M., and Diamond, I. T., 1981, Different distributions of large and small retinal ganglion cells in the cat after HRP injections of single layers of the lateral geniculate body and the superior colliculus, *Brain Res.* **207**:147–152.

Ivy, G. O., Akers, R. M., and Killackey, H. P., 1979, Differential distribution of callosal projection neurons in the neonatal and adult rat, *Brain Res.* **173**:532–537.

Iwahori, N., 1977, A Golgi study on the habenular nucleus of the cat, *J. Comp. Neurol.* **171**:319–344.

Iwama, K., Sakakura, H., and Kasamatsu, T., 1965, Presynaptic inhibition in the lateral geniculate body induced by stimulation of the cerebral cortex, *Jpn. J. Physiol.* **15**:310–322.

Iwamura, Y., and Inubushi, S., 1974a, Functional organization of receptive fields in thalamic ventrobasal neurons examined by neuronal response to iterative electrical stimulation of skin, *J. Neurophysiol.* **37:**920–926.

Iwamura, Y., and Inubushi, S., 1974b, Regional diversity in excitatory and inhibitory receptive-field organization of cat thalamic ventrobasal neurons, *J. Neurophysiol.* **37:**910–919.

Jabbur, S. J., Baker, M. A., and Towe, A. L., 1972, Wide-field neurons in thalamic nucleus ventralis posterolateralis of the cat, *Exp. Neurol.* **36:**213–238.

Jackson, J. C., and Benjamin, R. M., 1974, Unit discharges in the mediodorsal nucleus of the rabbit evoked by electrical stimulation of the olfactory bulb, *Brain Res.* **75:**193–201.

Jackson, J. C., Golden, G. T., and Benjamin, R. M., 1977, The distribution of olfactory input in the opossum mediodorsal nucleus, *Brain Res.* **138:**229–240.

Jackson, J. H., 1864, Illustrations of diseases of the nervous system: Clinical lectures and reports by the medical and surgical staff of the London hospital, *London Hosp. Rep.* **1:**337–387.

Jackson, J. H., 1866, Note on the functions of the optic thalamus, *London Hosp. Clin. Lect. Rep.* **3:**373–377.

Jackson, J. H., 1875, Autopsy on a case of hemianopia with hemiplegia and hemianaesthesia, *Lancet* **1:**722.

Jacob, C., 1895, Ein Beitrag zur Lehre vom Schleifenverlauf (obere, Rinden-, Thalamusschleife), *Neurol. Centralbl.* **14:**308–310.

Jacobowitz, D. M., and Palkovits, M., 1974, Topographic atlas of catecholamine- and acetylcholinesterase-containing neurons in the rat brain. I. Forebrain (telencephalon, diecephalon), *J. Comp. Neurol.* **157:**13–28.

Jacobs, B. L., and Jones, B. E., 1978, The role of central monoamine and acetylcholine systems in sleep–wakefulness states: Mediation or modulation?, in: *Cholinergic–Monoaminergic Interactions in the Brain* (L. L. Butcher, ed.), Academic Press, New York, pp. 97–138.

Jacobs, G. H., 1969, Transneuronal changes in the lateral geniculate nucleus of the squirrel monkey, *Saimiri sciureus, J. Comp. Neurol.* **135:**81–84.

Jacobson, S., and Trojanowski, J. Q., 1975, Corticothalamic neurons and thalamocortical terminal fields: An investigation in rat using horseradish peroxidase and autoradiography, *Brain Res.* **85:**385–401.

Jahns, R., 1975, Types of neuronal responses in the rat thalamus to peripheral temperature changes, *Exp. Brain Res.* **23:**157–166.

Jahnsen, H., and Llinás, R., 1982, Electrophysiological properties of guinea pig thalamic neurons studied *in vitro, Neurosci. Abstr.* **8:**413.

Jahnsen, H., and Llinás R., 1984a, Electrophysiological studies of guinea-pig thalamic neurones: An *in vitro* study, *J. Physiol. (London)* **349**205–226.

Jahnsen, H., and Llinás, R., 1984b, Ionic basis for the electroresponsiveness and oscillatory properties of guinea-pig thalamic neurones *in vitro, J. Physiol. (London)* **349:**227–248.

Jakob, C. (with C. Onelli), 1911, *Vom Tierhirn zum Menschenhirn.* Part I. *Tafelwirk nebst Einführung in die Geschichte der Hirnrinde,* Lehmann, Munich.

Jänig, W., Schmidt, R. F., and Zimmerman, M., 1968, Single unit responses and the total afferent outflow from the cat's foot pad upon mechanical stimulation, *Exp. Brain Res.* **6:**100–115.

Jansen, J., 1930, The brain of *Myxine glutinosa, J. Comp. Neurol.* **49:**359–507.

Jasper, H. H., 1949, Diffuse projection systems: The integrative action of the thalamic reticular system, *Electroencephalogr. Clin. Neurophysiol.* **1:**405–420.

Jasper, H. H., 1954, Functional properties of the thalamic reticular system, in: *Brain Mechanisms and Consciousness* (J. F. Delafresnaye, ed.), Blackwell, Oxford, pp. 374–401.

Jasper, H. H., 1960, Unspecific thalamocortical relations, in: *Handbook of Physiology,* Section 1, Volume 2 (J. Field, H. W. Magoun, and V. E. Hall, eds.), American Physiological Society, Washington, D. C., pp. 1307–1321.

Jasper, H. H., and Ajmone-Marsan, C., 1954, *A Stereotaxic Atlas of the Diencephalon of the Cat,* National Research Council, Canada, Ottawa.

Jasper, H. H., and Bertrand, G., 1966, Thalamic units involved in somatic sensation and voluntary and involuntary movements in man, in: *The Thalamus* (D. P. Purpura and M. D. Yahr, eds.), Columbia University Press, New York, pp. 365–390.

Jasper, H. H., and Droogleever-Fortuyn, J., 1948, Thalamocortical system and the electrical activity of the brain, *Fed. Proc.* **7:**61–62.

Jasper, H. H., Naquet, R., and King, E. E., 1955, Thalamocortical recruiting responses in sensory receiving areas in the cat, *Electroencephalogr. Clin. Neurophysiol.* **7:**99–114.

Jeanmonod, D., Rice, F. L., and Van der Loos, H., 1977, Mouse somatosensory cortex: Development of the alterations in the barrel field which are caused by injury to the vibrissal follicles, *Neurosci. Lett.* **6:**151–156.

Jeffery, G., Cowey, A., and Kuypers, H. G. J. M., 1981, Bifurcating retinal ganglion cell axons in the rat demonstrated by retrograde double labelling, *Exp. Brain Res.* **44:**34–40.

Jimenez-Castellanos, J., 1949, Thalamus of the cat in Horsley–Clarke coordinates, *J. Comp. Neurol.* **91:**307–330.

Jinnai, K., and Matsuda, Y., 1981, Thalamocaudate projection neurons with a branching axon to the cerebral motor cortex, *Neurosci. Lett.* **26:**95–99.

Johnson, J. I., and Marsh, M. P., 1969, Laminated lateral geniculate nucleus in the nocturnal marsupial, *Petaurus breviceps* (sugar glider), *Brain Res.* **15:**250–254.

Johnson, J. I., Rubel, E. W., and Horton, G. I., 1974, Mechano-sensory projections to cerebral cortex of sheep, *J. Comp. Neurol.* **158:**81–108.

Johnston, J. B., 1915, The cell masses in the forebrain of the turtle, *Cistudo carolina*, *J. Comp. Neurol.* **25:**393–468.

Johnston, J. B., 1923, Further contributions to the study of the evolution of the forebrain, *J. Comp. Neurol.* **35:**337–481.

Johnston, M. V., McKinney, M., and Coyle, J. T., 1981, Neocortical cholinergic innervation: A description of extrinsic and intrinsic components in the rat, *Exp. Brain Res.* **43:**159–172.

Jones, A. E., 1964, The lateral geniculate nucleus of *Ateles ater*, *J. Comp. Neurol.* **123:**205–210.

Jones, A. E., 1966, The lateral geniculate complex of the owl monkey, *Aotus trivirgatus*, *J. Comp. Neurol.* **126:**171–180.

Jones, B. E., and Moore, R. Y., 1977, Ascending projections of the locus coeruleus in the rat. II. Autoradiographic study, *Brain Res.* **127:**23–53.

Jones, E. G., 1967, Pattern of cortical and thalamic connexions of the somatic sensory cortex, *Nature (London)* **216:**704–705.

Jones, E. G., 1972, The development of the "muscular sense" concept in the nineteenth century and the work of H. Charlton Bastian, *J. Hist. Med.* **27:**298–311.

Jones, E. G., 1974, The anatomy of extrageniculostriate visual mechanisms, in: *The Neurosciences: Third Study Program* (F. O. Schmitt and F. B. Worden, eds.), MIT Press, Cambridge, Mass., pp. 215–227.

Jones, E. G., 1975a, Possible determinants of the degree of retrograde neuronal labeling with horseradish peroxidase, *Brain Res.* **85:**249–253.

Jones, E. G., 1975b, Lamination and differential distribution of thalamic afferents within the sensory-motor cortex of the squirrel monkey, *J. Comp. Neurol.* **160:**167–204.

Jones, E. G., 1975c, Some aspects of the organization of the thalamic reticular complex, *J. Comp. Neurol.* **162:**285–308.

Jones, E. G., 1975d, Varieties and distribution of non-pyramidal cells in the somatic sensory cortex of the squirrel monkey, *J. Comp. Neurol.* **160:**205–267.

Jones, E. G., 1981a, Functional subdivision and synaptic organization of the mammalian thalamus, in: *International Review of Physiology: Neurophysiology IV* (R. Porter, ed.), University Park Press, Baltimore, pp. 173–245.

Jones, E. G., 1981b, Development of connectivity in the cerebral cortex, in: *Studies in Developmental Neurobiology in Honor of Victor Hamburger* (W. M. Cowan, ed.), Oxford University Press, London, pp. 354–394.

Jones, E. G., 1981c, Anatomy of cerebral cortex: Columnar input–output organization, in: *The Organization of the Cerebral Cortex* (F. O. Schmitt, F. G. Worden, G. Adelman, and S. G. Dennis, eds.), MIT Press, Cambridge, Mass., pp. 199–236.

Jones, E. G., 1983a, Organization of the thalamocortical complex and its relation to sensory processes, in: *Handbook of Physiology: Neurophysiology/Sensory Processes* (I. Darian-Smith, ed.), American Physiological Society, Washington, D.C., pp. 149–212.

Jones, E. G., 1983b, Distribution patterns of individual medial lemniscal axons in the ventrobasal complex of the monkey thalamus, *J. Comp. Neurol.* **215:**1–16.

Jones, E. G., 1983c, History of cortical cytology, in: *Cerebral Cortex,* Volume 1 (A. Peters and E. G. Jones, eds.), Plenum Press, New York, pp. 1–32.

Jones, E. G., 1983d, Summing up, in: *Somatosensory Integration in the Thalamus* (G. Macchi, A. Rustioni, and R. Spreafico, eds.), Elsevier, Amsterdam, pp. 385–392.

Jones, E. G., 1983e, Identification and classification of intrinsic circuit elements in the neocortex, in: *Dynamic Aspects of Neocortical Organization* (G. Edelman, W. E. Gall, and W. M. Cowan, eds.), Wiley, New York, pp. 1–40.

Jones, E. G., 1983f, Laminar distribution of cortical efferent cells, in: *Cerebral Cortex,* Volume 1 (A. Peters and E. G. Jones, eds.), Plenum Press, New York, pp. 521–553.

Jones, E. G., 1983g, Lack of collateral thalamocortical projection to fields of the first somatic sensory cortex in monkeys, *Exp. Brain Res.* **52:**375–384.

Jones, E. G., 1983h, The thalamus, in: *Chemical Neuroanatomy* (P. C. Emson, ed.), Raven Press, New York, pp. 257–293.

Jones, E. G., and Burton, H., 1974, Cytoarchitecture and somatic sensory connectivity of thalamic nuclei other than the ventrobasal complex in the cat, *J. Comp. Neurol.* **154:**395–432.

Jones, E. G., and Burton, H., 1976a, A projection from the medial pulvinar to the amygdala in primates, *Brain Res.* **104:**142–147.

Jones, E. G., and Burton, H., 1976b, Areal differences in the laminar distribution of thalamic afferents in cortical fields of the insular, parietal and temporal regions of primates, *J. Comp. Neurol.* **168:**197–247.

Jones, E. G., and Cowan, W. M., 1982, Nervous tissue, in: *Histology,* Fifth Edition (L. Weiss, ed.), Elsevier, New York, pp. 282–369.

Jones, E. G., and Friedman, D. P., 1982, Projection pattern of functional components of thalamic ventrobasal complex on monkey somatosensory cortex, *J. Neurophysiol.* **48:**521–544.

Jones, E. G., and Hendry, S. H. C., 1984, Cytochrome oxidase staining reveals functional organization of monkey ventral posterior thalamic nucleus, *Neurosci. Abstr.* **10:**945.

Jones, E. G., and Leavitt, R. Y., 1973, Demonstration of thalamo-cortical connectivity in the cat somato-sensory system by retrograde axonal transport of horseradish peroxidase, *Brain Res.* **63:**414–418.

Jones, E. G., and Leavitt, R. Y., 1974, Retrograde axonal transport and the demonstration of non-specific projections to the cerebral cortex and striatum from thalamic intralaminar nuclei in the rat, cat and monkey, *J. Comp. Neurol.* **154:**349–378.

Jones, E. G., and Porter, R., 1980, What is area 3a?, *Brain Res. Rev.* **2:**1–43.

Jones, E. G., and Powell, T. P. S., 1968, The projection of the somatic sensory cortex upon the thalamus in the cat, *Brain Res.* **10:**369–391.

Jones, E. G., and Powell, T. P. S., 1969a, The cortical projection of the ventroposterior nucleus of the thalamus in the cat, *Brain Res.* **13:**298–318.

Jones, E. G., and Powell, T. P. S., 1969b, An electron microscopic study of the mode of termination of cortico-thalamic fibres within the thalamic relay nuclei of the cat, *Proc. R. Soc. London Ser. B* **172:**173–185.

Jones, E. G., and Powell, T. P. S., 1969c, Electron microscopy of synaptic glomeruli in the thalamic relay nuclei of the cat, *Proc. R. Soc. London Ser. B* **172:**153–171.

Jones, E. G. and Powell, T. P. S., 1970a, Connexions of the somatic sensory cortex of the rhesus monkey, III. Thalamic connexions, *Brain* **93:**37–56.

Jones, E. G., and Powell, T. P. S., 1970b, Electron microscopy of the somatic sensory cortex of the cat. I. Cell types and synaptic organization, *Philos. Trans. R. Soc. London Ser. B* **257:**1–11.

Jones, E. G., and Rockel, A. J., 1971, The synaptic organization in the medial geniculate body of afferent fibres ascending from the inferior colliculus, *Z. Zellforsch. Mikrosk. Anat.* **113:**44–66.

Jones, E. G., and Wise, S. P., 1977, Size, laminar and columnar distribution of efferent cells in the sensory-motor cortex of monkeys, *J. Comp. Neurol.* **175:**391–438.

Jones, E. G., Burton, H., Saper, C. B., and Swanson, L. W., 1976, Midbrain, diencephalic and cortical relationships of the basal nucleus of Meynert and associated structures in primates, *J. Comp. Neurol.* **167:**385–420.

Jones, E. G., Coulter, J. D., Burton, H., and Porter, R., 1977, Cells of origin and terminal distribution of corticostriatal fibers arising in the sensory-motor cortex of monkeys, *J. Comp. Neurol.* **173:**53–80.

Jones, E. G., Wise, S. P., and Coulter, J. D., 1979, Differential thalamic relationships of sensory-motor and parietal cortical fields in monkeys, *J. Comp. Neurol.* **183:**833–882.

Jones, E. G., Schreyer, D. J., and Wise, S. P. 1982a, Growth and maturation of the rat corticospinal tract, *Prog. Brain Res.* **57:**361–379.

Jones, E. G., Valentino, K. L., and Fleshman, J. W., Jr., 1982b, Adjustment of connectivity in rat neocortex after prenatal destruction of precursor cells of layers II–IV, *Dev. Brain Res.* **2:**425–431.

Jones, E. G., Friedman, D. P., and Hendry, S. H. C., 1982c, Thalamic basis of place- and modality-specific columns in monkey somatosensory cortex: A correlative anatomical and physiological study, *J. Neurophysiol.* **48:**545–568.

Jordan, H., 1973, The structure of the medial geniculate nucleus (MGN): A cyto- and myeloarchitectonic study in the squirrel monkey, *J. Comp. Neurol.* **148:**469–480.

Jordan, H., and Holländer, H., 1971, The structure of the ventral part of the lateral geniculate nucleus: A cyto- and myeloarchitectonic study in the cat, *J. Comp. Neurol.* **145:**259–272.

Jouvet, M., 1962, Recherches sur les structures nerveuses et les mécanismes responsible des differentes phases du somneil physiologique, *Arch. Ital. Biol.* **100:**125–206.

Jouvet, M., 1972, The role of monoamines and acetylcholine-containing neurons in the regulation of the sleep–waking cycle, *Ergeb. Physiol.* **64:**166–307.

Juraska, J. M., and Fifková, E., 1979, An electron microscopic study of the early postnatal development of the visual cortex of the hooded rat, *J. Comp. Neurol.* **183:**257–268.

Kaas, J. H., and Guillery, R. W., 1973, The transfer of abnormal visual field representations from the dorsal lateral geniculate nucleus to the visual cortex in Siamese cats, *Brain Res.* **59:**61–96.

Kaas, J., Hall, W. C., and Diamond, I. T., 1970, Cortical visual areas I and II in the hedgehog: Relation between evoked potential maps and architectonic subdivisions, *J. Neurophysiol.* **33:**595–613.

Kaas, J. H., Hall, W. C., and Diamond, I. T., 1972a, Visual cortex of the grey squirrel (*Sciurus carolinensis*): Architectonic subdivisions and connections from the visual thalamus, *J. Comp. Neurol.* **145:**273–306.

Kaas, J. H., Hall, W. C., Killackey, H., and Diamond, I. T., 1972b, Visual cortex of the tree shrew (*Tupaia glis*): Architectonic subdivisions and representations of the visual field, *Brain Res.* **42:**491–496.

Kaas, J. H., Guillery, R. W., and Allman, J. M., 1972c, Some principles of organization in the dorsal lateral geniculate nucleus, *Brain Behav. Evol.* **6:**253–299.

Kaas, J. H., Guillery, R. W., and Allman, J. M., 1973, Discontinuities in the dorsal lateral geniculate nucleus corresponding to the optic disc: A comparative study, *J. Comp. Neurol.* **147:**163–179.

Kaas, J. H., Lin, C.-S., and Casagrande, V. A., 1976, The relay of ipsilateral and contralateral retinal input from the lateral geniculate nucleus to striate cortex in the owl monkey: A transneuronal transport study, *Brain Res.* **106:**371–378.

Kaas, J. H., Huerta, M. F., Weber, J. T., and Harting, J. K., 1978, Patterns of retinal terminations and laminar organization of the lateral geniculate nucleus of primates, *J. Comp. Neurol.* **182:**517–554.

Kaas, J. H., Nelson, R. J., Sur, M., Lin, C.-S., and Merzenich, M. M., 1979, Multiple representations of the body within the primary somatosensory cortex of primates, *Science* **204:**521–523.

Kabat, H., 1936, Electrical stimulation of points in the forebrain and midbrain: The resultant alterations in respiration, *J. Comp. Neurol.* **64:**187–208.

Kaelber, W. W., 1966, Nuclear configuration of the diencephalon of *Tamandua tetradactyla* and *Myrmecophaga jubata*, *J. Comp. Neurol.* **128:**133–170.

Kahle, W., 1956, Zur Entwicklung des menschlichen Zwischenhirns, *Dtsch. Z. Nervenheilkd* **62:**3–34.

Kaiserman-Abramoff, I. R., Graybiel, A. M., and Nauta, W. J. H., 1980, The thalamic projection of cortical area 17 in a congenitally anophthalmic mouse strain, *Neuroscience* **5:**41–52.

Kaitz, S. S. and Robertson, R. T., 1981, Thalamic connections with limbic cortex. II. Corticothalamic projections, *J. Comp. Neurol* **195:**527–545.

Kalil, K., 1978, Patch-like termination of thalamic fibers in the putamen of the rhesus monkey: An autoradiographic study, *Brain Res.* **140:**333–339.

Kalil, K., 1981, Projections of the cerebellar and dorsal column nuclei upon the thalamus of the rhesus monkey, *J. Comp. Neurol.* **195:**25–50.

Kalil, R., 1978, Development of the dorsal lateral geniculate nucleus in the cat, *J. Comp. Neurol.* **182:**265–292.

Kalil, R., 1980, A quantitative study of the effects of monocular enucleation and deprivation on cell growth in the dorsal lateral geniculate nucleus of the cat, *J. Comp. Neurol.* **189:**438–524.

Kalil, R. E., 1972, Formation of new retino-geniculate projections in kittens after removal of one eye, *Anat. Rec.* **172**:339–340.

Kalil, R. E., and Chase, R., 1970, Corticofugal influence on acitivity of lateral geniculate neurons in the cat, *J. Neurophysiol.* **33**:459–474.

Kalil, R. E., and Schneider, G. E., 1975, Abnormal synaptic connections of the optic tract in the thalamus after midbrain lesions in newborn hamsters, *Brain Res.* **100**:690–698.

Kanagasuntheram, R., and Krishnamurti, A., 1970, The termination of optic fibers in the lateral geniculate nucleus of some primates, *Brain Res.* **17**:129–132.

Kanagasuntheram, R., and Wong, W. C., 1968, Nuclei of the diencephalon of *Hylobatidae, J. Comp. Neurol.* **134**:265–286.

Kanagasuntheram, R., Wong, W. C., and Krishnamurti, A., 1968, Nuclear configuration of the diencephalon in some lorisoids, *J. Comp. Neurol.* **133**:241–268.

Kanagasuntheram, R., Krishnamurti, A., and Wong, W. C., 1969, Observations on the lamination of the lateral geniculate body in some primates, *Brain Res.* **14**:623–631.

Kaplan, E., and Shapley, R. M., 1982, X and Y cells in the lateral geniculate nucleus of macaque monkeys, *J. Physiol. (London)* **330**:125–143.

Kappers, C. U. A., 1908, Weitere Mitteilungen über die Phylogenese des Corpus striatum und Thalamus, *Anat. Anz.* **33**:321–336.

Kappers, C. U. A., Huber, G. C., and Crosby, E. C., 1936, *The Comparative Anatomy of the Nervous System of Vertebrates, Including Man,* 2 vols., Macmillan Co., New York.

Kappers, J. A., and Pévet, P. (eds.), 1979, *The Pineal Gland of Vertebrates Including Man, Prog. Brain Res.* **52**, Elsevier, Amsterdam.

Karamanlidis, A. N., and Giolli, R. A., 1977, Thalamic inputs to the rabbit visual cortex: Identification and organization using horseradish peroxidase (HRP), *Exp. Brain Res.* **29**:191–199.

Karamanlidis, A. N., and Magras, J., 1972, Retinal projections in domestic ungulates. I. The retinal projections in the sheep and the pig, *Brain Res.* **44**:127–145.

Karamanlidis, A. N., and Magras, J., 1974, Retinal projections in domestic ungulates. II. The retinal projections in the horse and ox, *Brain Res.* **66**:209–225.

Karamanlidis, A. N., and Voogd, J., 1970, Trigemino-thalamic fibre connections in the goat: An experimental anatomical study, *Acta Anat.* **75**:596–622.

Karamanlidis, A. N., Michaloudi, H., Mangana, O., and Saigal, R. P., 1978, Trigeminal ascending projections in the rabbit, studied with horseradish peroxidase, *Brain Res.* **156**:110–116.

Karamanlidis, A. N., Saigal, R. P., Giolli, R. A., Mangana, O., and Michaloudi, H., 1979, Visual thalamocortical connections in sheep studied by means of the retrograde transport of horseradish-peroxidase, *J. Comp. Neurol.* **187**:245–260.

Karlsson, U., 1967, Observations on the postnatal development of neuronal structures in the lateral geniculate nucleus of the rat by electron microscopy, *J. Ultrastruct. Res.* **17**:158–175.

Karten, H., 1963, Ascending pathways from the spinal cord in the pigeon, *Columba livia, Proc. 16th Int. Congr. Zool.* **2**:23.

Karten, H., 1967, The organization of the ascending auditory pathway in the pigeon (*Columba livia*). I. Diencephalic projections of the inferior colliculus (nucleus mesencephalic lateralis, pars dorsalis), *Brain Res.* **6**:409–427.

Karten, H. J., 1969, The organization of the avian telencephalon and some speculations on the phylogeny of the amniote telencephalon, *Ann. N.Y. Acad. Sci.* **167**:164–179.

Karten, H. J., 1971, Efferent projections of the wulst of the owl, *Anat. Rec.* **169**:353.

Karten, H. J., and Dubbeldam, J. L., 1973, The organization and projections of the paleostriatal complex in the pigeon (*Columba livia*), *J. Comp. Neurol.* **148**:61–90.

Karten, H. J., and Finger, T. E., 1976, A direct thalamo-cerebellar pathway in pigeon and catfish, *Brain Res.* **102**:335–338.

Karten, H. J. and Hodos, W., 1967, *A Stereotaxic Atlas of the Brain of the Pigeon (Columba livia),* Johns Hopkins University Press, Baltimore.

Karten, H. J., and Hodos, W., 1970, Telencephalic projections of the nucleus rotundus in the pigeon (*Columba livia*), *J. Comp. Neurol.* **140**:35–52.

Karten, H. J., and Nauta, W. J. H., 1968, Organization of retinothalamic projections in the pigeon and owl, *Anat. Rec.* **160**:373.

Karten, H. J., and Revzin, A. M., 1966, The afferent connections of the nucleus rotundus in the pigeon, *Brain Res.* **2**:368–377.

Karten, H. J., Hodos, W., Nauta, W. J. H., and Revzin, A. M., 1973, Neural connections of the "visual Wulst" of the avian telencephalon: Experimental studies in the pigeon (*Columba livia*) and owl (*Speotyto cunicularia*), *J. Comp. Neurol.* **150:**253–278.

Karten, H. J., Konishi, M., and Pettigrew, J., 1978, Somatosensory representation in the anterior Wulst of the owl (*Speotyto cunicularia*), *Neurosci. Abstr.* **4:**554.

Kasdon, D. L., and Jacobson, S., 1978, The thalamic afferents to the inferior parietal lobule of the rhesus monkey, *J. Comp. Neurol.* **177:**685–706.

Kataoka, K., Nakamura, Y., and Hassler, R., 1973, Habenulo-interpeduncular tract: A possible cholinergic neuron in rat brain, *Brain Res.* **62:**264–267.

Kataoka, K., Sorimachi, M., Okuno, S., and Mizuno, N., 1977, Cholinergic and GABAergic fibers in the stria medullaris of the rabbit, *Brain Res. Bull.* **2:**461–464.

Katz, M. J., and Lasek, R. J., 1979, Substrate pathways which guide growing axons in *Xenopus* embryos, *J. Comp. Neurol.* **183:**817–832.

Kawamura, S., Sprague, J. M., and Niimi, K., 1974, Corticofugal projections from the visual cortices to the thalamus, pretectum and superior colliculus in the cat, *J. Comp. Neurol.* **158:**339–362.

Kawamura, S., Fukushima, N., Hattori, S., and Kudo, M., 1980, Laminar segregation of cells of origin of ascending projections from the superficial layers of the superior colliculus in the cat, *Brain Res.* **184:**486–490.

Kelley, D. B., and Nottebohm, F., 1979, Projections of a telencephalic auditory nucleus—field L— in the canary, *J. Comp. Neurol.* **183:**455–470.

Kelly, J., and Swanson, L. W., 1980, Additional forebrain regions projecting to the posterior pituitary: Preoptic region, bed nucleus of the stria terminalis and zona incerta, *Brain Res.* **197:**1–10.

Kelly, J. P., and Gilbert, C. D., 1975, The projections of different morphological types of ganglion cells in the cat retina, *J. Comp. Neurol.* **163:**65–80.

Kelly, J. S., Godfraind, J.-M., and Maruyama, S., 1979, The presence and nature of inhibition in small slices of the dorsal lateral geniculate nucleus of rat and cat incubated *in vitro, Brain Res.* **168:**388–392.

Kemali, M., and Braitenberg, V., 1969, *Atlas of the Frog Brain,* Springer, Berlin.

Kemali, M., and Miralto, A., 1979, The habenular nuclei of the elasmobranch "Scyllium stellare": Myelinated perikarya, *Am. J. Anat.* **155:**147–152.

Kemali, M., Guglielmotti, V., and Gioffré, D., 1980, Neuroanatomical identification of the frog habenular connections using peroxidase (HRP), *Exp. Brain Res.* **38:**341–347.

Kemp, J. M., and Powell, T. P. S., 1971, The connexions of the striatum and globus pallidus: Synthesis and speculation, *Philos. Trans. R. Soc. London Ser. B.* **262:**441–457.

Kennedy, C., Des Rosiers, M. H., Jehle, J. W., Reivich, M., Sharp, F., and Sokoloff, L., 1976, Mapping of functional neural pathways by autoradiographic survey of local metabolic rate with [^{14}C] deoxyglucose, *Science* **187:**850–853.

Kennedy, H., and Baleydier, C., 1977, Direct projections from thalamic intralaminar nuclei to extra-striate visual cortex in the cat traced with horseradish peroxidase, *Exp. Brain Res.* **28:**133–139.

Kennedy, M. C., and Rubinson, K., 1977, Retinal projections in larval, transforming and adult sea lamprey, *Petromyzon marinus, J. Comp. Neurol.* **171:**465–480.

Kenshalo, D. R., Jr., Leonard, R. B., Chung, J. M., and Willis, W. D., 1979, Response of primate spinothalamic neurons to graded and to repeated noxious heat stimuli, *J. Neurophysiol.* **42:**1370–1389.

Kenshalo, D. R., Jr., Giesler, G. J., Jr., Leonard, R. B., and Willis, W. D., 1980, Responses of neurons in primate ventral posterior lateral nucleus to noxious stimuli, *J. Neurophysiol.* **43:**1594–1614.

Kerr, F. W. L., 1975, The ventral spinothalamic tract and other ascending systems of the ventral funiculus of the spinal cord, *J. Comp. Neurol.* **159:**335–356.

Kerr, F. W. L., and Lippman, H. H., 1974, The primate spinothalamic tract as demonstrated by anterolateral cordotomy and commissural myelotomy, *Adv. Neurol.* **4:**147–156.

Keyser, A., 1972, The development of the diencephalon of the Chinese hamster, *Acta Anat. Suppl.* (59) **83:**1–181.

Khachaturian, H., Lewis, M. E., and Watson, S. J., 1983, Enkephalin systems in diencephalon and brainstem of the rat, *J. Comp. Neurol.* **220:**310–320.

Kicliter, E., 1979, Some telencephalic connections in the frog, *Rana pipiens, J. Comp. Neurol.* **185:**75–86.

Kievit, J., and Kuypers, H. G. J. M., 1975, Subcortical afferents to the frontal lobe in the rhesus monkey studied by means of retrograde horseradish peroxidase transport, *Brain Res.* **85:**261–266.

Killackey, H. P., and Belford, G. R., 1979, The formation of afferent patterns in the somatosensory cortex of the neonatal rat, *J. Comp. Neurol.* **183:**285–304.

Killackey, H., and Diamond, I. T., 1971, Visual attention in the tree shrew: An ablation study of the striate and extrastriate visual cortex, *Science* **171:**696–699.

Killackey, H., and Ebner, F., 1973, Convergent projection of three separate thalamic nuclei on to a single cortical area, *Science* **179:**283–285.

Killackey, H., and Shinder, A., 1981, Central correlates of peripheral pattern alterations in the trigeminal system of the rat. II. The effect of nerve section, *Dev. Brain Res.* **1:**121–126.

Killackey, H., Wilson, M., and Diamond, I. T., 1972, Further studies of the striate and extrastriate visual cortex in the tree shrew, *J. Comp. Physiol. Psychol.* **81:**45–63.

Killackey, H., Ivy, G. O., and Cunningham, T. J., 1978, Anomalous organization of SMI somatotopic map consequent to vibrissae removal in the newborn rat, *Brain Res.* **155:**136–140.

Kim, R., Nakano, K., Jayaraman, A., and Carpenter, M. B., 1976, Projections of the globus pallidus and adjacent structures: An autoradiographic study in the monkey, *J. Comp. Neurol.* **169:**263–290.

Kimberly, R. P., Holden, A. L., and Bambrough, P., 1971, Response characteristics of pigeon forebrain cells to visual stimulation, *Vision Res.* **11:**475–478.

Kimura, H., McGeer, P. L., Peng, J. H., and McGeer, E. G., 1981, The central cholinergic system studied by choline acetyltransferase immunohistochemistry in the cat, *J. Comp. Neurol.* **200:** 151–201.

Kingsbury, B. F., 1920, The extent of the floor-plate of His and its significance, *J. Comp. Neurol.* **32:**113–135.

Kinston, W. J., Vadas, M. A., and Bishop, P. O., 1969, Multiple projection of the visual field to the medial portion of the dorsal lateral geniculate nucleus and the adjacent nuclei of the thalamus of the cat, *J. Comp. Neurol.* **136:**295–316.

Kirk, D. L., Cleland, B. G., and Levick, W. R., 1975, Axonal conduction latencies of cat retinal ganglion cells, *J. Neurophysiol.* **38:**1395–1402.

Kirk, D. L., Levick, W. R., and Cleland, B. G., 1976a, The crossed or uncrossed destination of axons of sluggish-concentric and non-concentric cat retinal ganglion cells, with an overall synthesis of the visual field representation, *Vision Res.* **16:**233–236.

Kirk, D. L., Levick, W. R., Cleland, B. G., and Wässle, H., 1976b, Crossed and uncrossed representation of the visual field by brisk-sustained and brisk-transient cat retinal ganglion cells, *Vision Res.* **16:**225–232.

Kirkwood, P. A., and Shatz, C. J., 1982, Development of binocular inputs to dorsal lateral geniculate neurones in the fetal cat, *J. Physiol. (London)* **336:**27–28P.

Kirsch, M., Coles, R. B., and Leppelsack, H.-J., 1980, Unit recordings from a new auditory area in the frontal neostriatum of the awake starling (*Sturnus vulgaris*), *Exp. Brain Res.* **38:**375–380.

Kitai, S. T., Kocsis, J. D., Preston, R. J., and Sugimori, M., 1976, Monosynaptic input to caudate neurons identified by intracellular injection of horseradish peroxidase, *Brain Res.* **109:**601–606.

Kitt, C. A., and Brauth, S. E., 1980, Telencephalic projections from catecholamine cell groups in the pigeon, *Neurosci. Abstr.* **6:**630.

Kizer, J. S., Palkovits, M., and Brownstein, M. J., 1976, The projections of the A8, A9 and A10 dopaminergic cell bodies: Evidence for a nigral-hypothalamic-median eminence dopaminergic pathway, *Brain Res.* **108:**363–370.

Knapp, H., and Kang, D. S., 1968a, The visual pathways of the snapping turtle (*Chelydra serpentina*), *Brain Behav. Evol.* **1:**19–42.

Knapp, H., and Kang, D. S., 1968b, The retinal projections of the side-necked turtle (*Podocnemis unifilis*) with some notes on the possible origin of the pars dorsalis of the lateral geniculate body, *Brain Behav. Evol.* **1:**369–404.

Kniffki, K.-D., and Mizumura, K., 1983, Responses of neurons in VPL and VPL–VL region of the cat to algesic stimulation of muscle and tendon, *J. Neurophysiol.* **49:**649–661.

Knight, P. A., 1977, Representation of the cochlea within the anterior auditory field (AAF) of the cat, *Brain Res.* **130:**447–467.

Knighton, R. S., 1950, Thalamic relay nucleus for the second somatic sensory receiving area in the cerebral cortex of the cat, *J. Comp. Neurol.* **92:**183–191.

Knox, R., 1832, *Engravings of the Nerves*, 3rd ed., Maclachlan & Stewart, Edinburgh.

Knudsen, E. I., and Konishi, M., 1978, Space and frequency are represented separately in auditory midbrain of the owl, *J. Neurophysiol.* **41:**870–884.

Kobayashi, R. M., Palkovits, M., Hruska, R. E., Rothschild, R., and Yamamura, H. I, 1978, Regional distribution of muscarinic cholinergic receptors in rat brain, *Brain Res.* **154**:13–23.

Kodama, S., 1928, Beiträge zur normalen Anatomie des Corpus Luysi beim Menschen, *Arb. Anat. Inst. Sendai (Tohoku Daigaku)* **13**:221–254.

Kodama, S., 1928–1929, Über die sogenannten Basalganglien (morphogenetische und pathologisch-anatomische Untersuchungen), *Schweiz. Arch. Neurol. Psychiatr.* **23**:38–100, 179–265.

Kolle, K., 1956, 1959, 1963, *Grosse Nervenärzte*, 3 vols., Thieme, Stuttgart.

König, J. F. R., and Klippel, R. M., 1963, *The Rat Brain,* Williams & Wilkins, Baltimore.

Kosar, E., and Hand, P. J., 1981, First somatosensory cortical columns and associated neuronal clusters of nucleus ventralis posterolateralis of the cat: An anatomical demonstration, *J. Comp. Neurol.* **198**:515–539.

Kosareva, A. A., 1974, Afferent and efferent connections of the nucleus rotundus in the tortoise *Emys orbicularis, Zh. Evol. Biokhim. Fiziol.* **10**:395–399.

Kotchabhakdi, N., Rinvik, E., Yingchareon, K., and Walberg, F., 1980, Afferent projections to the thalamus from the perihypoglossal nuclei, *Brain Res.* **187**:457–461.

Kozak, W., Rodieck, R. W., and Bishop, P. O., 1965, Responses of single units in lateral geniculate nucleus of cat to moving visual patterns, *J. Neurophysiol.* **28**:19–47.

Kratz, K. E., Spear, P. D., and Smith, D. C., 1976, Postcritical-period reversal of effects of monocular deprivation on striate cortex cells in the cat, *J. Neurophysiol.* **39**:501–511.

Kratz, K. E.., Webb, S. V., and Sherman, S. M., 1978a, Studies of the cat's medial interlaminar nucleus: A subdivision of the dorsal lateral geniculate nucleus, *J. Comp. Neurol.* **181**:601–614.

Kratz, K. E., Webb, S. V., and Sherman, S. M., 1978b, Effects of early monocular lid suture upon neurons in the cat's medial interlaminar nucleus, *J. Comp. Neurol.* **181**:615–626.

Kratz, K. E., Webb, S. V., and Sherman, S. M., 1978c, Electrophysiological classification of X- and Y-cells in the cat's lateral geniculate nucleus, *Vision Res.* **18**:489–492.

Kratz, K. E., Mangel, S. C., Lehmkuhle, S., and Sherman, S. M., 1979, Retinal X- and Y-cells in monocular lid-sutured cats: Normality of spatial and temporal properties, *Brain Res.* **172**:545–551.

Krayniak, P. F., and Siegel, A., 1978, Efferent connections of the septal area in the pigeon, *Brain Behav. Evol.* **15**:389–404.

Krayniak, P. F., Siegel, A., Meibach, R. C., Fruchtman, D., and Scrimenti, M., 1979, Origin of the fornix system in the squirrel monkey, *Brain Res.* **160**:401–411.

Krettek, J. E., and Price, J. L., 1974, A direct input from the amygdala to the thalamus and the cerebral cortex, *Brain Res.* **67**:169–174.

Krettek, J. E., and Price, J. L, 1977a, Projections from the amygdaloid complex to the cerebral cortex and thalamus in the rat and cat, *J. Comp. Neurol.* **172**:687–722.

Krettek, J. E., and Price, J. L., 1977b, The cortical projections of the mediodorsal nucleus and adjacent thalamic nuclei in the rat, *J. Comp. Neurol.* **171**:157–191.

Krieg, W. J. S., 1944, The medial region of the thalamus of the albino rat, *J. Comp. Neurol.* **80**:381–415.

Krieg, W. J. S., 1947, Connections of the cerebral cortex. I. The albino rat. C. Extrinsic connections, *J. Comp. Neurol.* **86**:267–394.

Krieg, W. J. S., 1948, A reconstruction of the diencephalic nuclei of *Macacus rhesus, J. Comp. Neurol.* **88**:1–51.

Krishnamurti, A., Kanagasuntheram, R., and Wong, W. C., 1972, Functional significance of the fibrous laminae in the ventrobasal complex of the thalamus of slow loris, *J. Comp. Neurol.* **145**:515–524.

Kristt, D. A., 1978, Neuronal differentiation in somatosensory cortex of the rat. I. Relationship to synaptogenesis in the first postnatal week, *Brain Res.* **150**:467–486.

Kristt, D. A., 1979, Development of neocortical circuitry: Histochemical localization of acetylcholinesterase in relation to the cell layers of rat somatosensory cortex, *J. Comp. Neurol.* **186**:1–16.

Krnjević, K. 1974, Chemical nature of synaptic transmission in vertebrates, *Physiol. Rev.* **54**:418–540.

Krnjević, K., and Phillis, J. W., 1963, Acetyl choline-sensitive cells in the cerebral cortex, *J. Physiol. (London)* **166**:296–327.

Krnjević, K., and Puil, E., 1975, Electrophysiological studies on actions of taurine, in: *Taurine* (R. Huxtable and A. Barbeau, eds.), Raven Press, New York, pp. 179–189.

Kromer, L. F., and Moore, R. Y., 1980, A study of the organization of the locus coeruleus projections to the lateral geniculate nuclei in the albino rat, *Neuroscience* **5**:255–271.

Kruger, L., 1959, The thalamus of the dolphin *(Tursiops truncatus)* and comparison with other mammals, *J. Comp. Neurol.* **111**:133–194.

Kruger, L, and Albe-Fessard, D., 1960, Distribution of responses to somatic afferent stimuli in the diencephalon of the cat under chloralose anesthesia, *Exp. Neurol.* **2**:442–467.

Kruger, L, and Berkowitz, E. C., 1960, The main afferent connections of the reptilian telencephalon as determined by degeneration and electrophysiological methods, *J. Comp. Neurol.* **115**:125–141.

Kruger, L., Saporta, S., and Feldman, S. G., 1977, Axonal transport studies of the sensory trigeminal complex, in: *Pain in the Trigeminal Region* (D. J. Anderson and B. M. Matthews, eds.), Elsevier, Amsterdam, pp. 191–202.

Kudo, M, and Niimi, K., 1978, Ascending projections of the inferior colliculus onto the medial geniculate body in the cat studied by anterograde and retrograde tracing techniques, *Brain Res.* **155**:113–117.

Kudo, M., and Niimi, K., 1980, Ascending projection of the inferior colliculus in the cat: An autoradiographic study, *J. Comp. Neurol.* **191**:545–556.

Kuffler, S., 1953, Discharge patterns and functional organization of mammalian retina, *J. Neurophysiol.* **16**:37–68.

Kuhar, M. J., Aghajanian, G. K., and Roth, R. H., 1972, Tryptophan hydroxylase activity and synaptosomal uptake of serotonin in discrete brain regions: Correlations with serotonin levels and histochemical fluorescence, *Brain Res.* **44**:165–176.

Kuhar, M. J., Dehaven, R. N., Yamamura, H. I., Rommelspacher, H., and Simon, J. R., 1975, Further evidence for cholinergic habenulo-interpeduncular neurons: Pharmacologic and functional characteristics, *Brain Res.* **97**:265–275.

Kuhlenbeck, H., 1927, *Vorlesungen über das Zentralnervensystem der Wirbeltiere. Eine Einführung in die Gehirnanatomie auf vergleichender Grundlage,* Fischer, Jena.

Kuhlenbeck, H., 1931, Über die Grundbestandteile des Zwischenhirnbauplans bei Reptilien, *Morphol. Jahrb.* **66**:244–317.

Kuhlenbeck, H., 1938, The ontogenetic development and phylogenetic significance of the cortex telencephali in the chick, *J. Comp. Neurol.* **69**:273–301.

Kuhlenbeck, H., 1948, The derivatives of the thalamus ventralis in the human brain and their relation to the so-called subthalamus, *Mil. Surg.* **102**:433–447.

Kuhlenbeck, H., 1954, The human diencephalon: A summary of development, structure, function, and pathology, *Confin. Neurol.* **14**(Suppl.):1–230.

Kuhlenbeck, H., 1956, Die Formbeständteile der Regio praetectalis des Anamnier-Gehirns und ihre Beziehungen zum Hirnbauplan, *Okajima's Folia Anat. Jpn.* **28**:23–44.

Kumazawa, T., and Perl, E. R., 1978, Excitation of marginal and substantia gelatinosa neurons in the primate spinal cord: Indications of their place in dorsal horn functional organization, *J. Comp. Neurol.* **177**:417–434.

Kunze, W., McKenzie, J. S., and Bendrups, A. P., 1979, An electrophysiological study of thalamo-caudate neurones in the cat, *Exp. Brain Res.* **36**:233–244.

Künzle, H., 1975, Bilateral projections from precentral motor cortex to the putamen and other parts of the basal ganglia: An autoradiographic study in *Macaca fascicularis, Brain Res.* **88**:195–210.

Künzle, H., 1976, Thalamic projections from the precentral motor cortex in *Macaca fascicularis, Brain Res.* **105**:253–267.

Künzle, H., 1978, An autoradiographic analysis of the efferent connections from premotor and adjacent prefrontal regions (areas 6 and 9) in *Macaca fascicularis, Brain Behav. Evol.* **15**:185–234.

Künzle, H., and Akert, K., 1977, Efferent connections of cortical area 8 (frontal eye field) in *Macaca fascicularis:* A reinvestigation using the autoradiographic technique, *J. Comp. Neurol.* **173**:147–164.

Kuo, J. S., and Carpenter, M. B., 1973, Organization of pallidothalamic projections in the rhesus monkey, *J. Comp. Neurol.* **151**:201–236.

Kupfer, C., 1965, The distribution of cell size in the lateral geniculate nucleus of man following transneuronal cell atrophy, *J. Neuropathol. Exp. Neurol.* **24**:653–661.

Kupfer, C., and Palmer, P., 1964, Lateral geniculate nucleus histological and cytochemical changes following afferent denervation and visual deprivation, *Exp. Neurol.* **9**:400–409.

Kurepina, M., 1966, Cytoarchitektonik des Sehhügels (Thalamus dorsalis) der Chiroptera, in: *Evolution of the Forebrain* (R. Hassler and H. Stephan, eds.), Thieme, Stuttgart, pp. 356–364.

Kuru, M., 1949, *Sensory Paths in the Spinal Cord and Brain Stem of Man,* Sogensya, Tokyo.

Kuypers, H. G. J. M., 1966, Discussion, in: *The Thalamus* (D. P. Purpura and M. D. Yahr, eds.), Columbia University, New York, pp. 122–127.

Kuypers, H. G. J. M., and Maisky, V. A., 1975, Retrograde axonal transport of horseradish peroxidase from spinal cord to brain stem cell groups in the cat, *Neurosci. Lett.* **1:**9–14.

Kuypers, H. G. J. M., and Tuerk, J. D., 1964, The distribution of the cortical fibers within the nuclei cuneatus and gracilis of the cat, *J. Anat.* **98:**143–162.

Kuypers, H. G. J. M., Kievet, M. J., and Groen-Klevant, A. C., 1974, Retrograde axonal transport of horseradish peroxidase in rat's forebrain, *Brain Res.* **67:**211–218.

Kuypers, H. G. J. M., Bentivoglio, M., Catsman-Berrevoets, C. E., and Bharos, A. T., 1980, Double retrograde labeling through divergent axon collaterals, using two fluorescent tracers with the same excitation wavelength which label different features of the cell, *Exp. Brain Res.* **40:**383–392.

Laemle, L. K., 1968, Retinal projections of *Tupaia glis, Brain Behav. Evol.* **1:**473–499.

Laemle, L. K., and Noback, C. R., 1970, The visual pathways of the lorisid lemurs *(Nycticebus coucang* and *Galago crassicaudatus), J. Comp. Neurol.* **138:**49–62.

Lamarre, Y., Filion, M., and Cordeau, J. P., 1971, Neuronal discharges of the ventrolateral nucleus of the thalamus during sleep and wakefulness in the cat. I. Spontaneous activity, *Exp. Brain Res.* **12:**480–498.

LaMotte, C. C., Snowman, A., Pert, C. B., and Snyder, S. N., 1978, Opiate receptor binding in rhesus monkey brain: Association with limbic structures, *Brain Res.* **155:**374–379.

Land, P. W., Polley, E. H., and Kernis, M. M., 1976, Patterns of retinal projections to the lateral geniculate nucleus and superior colliculus of rats with induced unilateral congenital eye defects, *Brain Res.* **103:**394–399.

Landgren, S., 1957, Convergence of tactile, thermal, and gustatory impulses on single cortical cells, *Acta Physiol. Scand.* **40:**210–221.

Landgren, S., 1960, Thalamic neurons responding to cooling of the cat's tongue, *Acta Physiol. Scand.* **48:**255–267.

Landgren, S., and Silfvenius, H., 1970, The projection of group I muscle afferents from the hindlimb to the contralateral thalamus of the cat, *Acta Physiol. Scand.* **80:**10A.

Landgren, S., and Silfvenius, H., 1971, Nucleus Z, the medullary relay in the projection path to the cerebral cortex of group I muscle afferents from the cat's hindlimb, *J. Physiol. (London)* **218:**551–571.

Landgren, S., Nordwall, A., and Wengström, C., 1965, The location of the thalamic relay in the spino-cervico-lemniscal path, *Acta Physiol. Scand.* **65:**164–175.

Landgren, S., Silfvenius, H., and Wolsk, D., 1967, Vestibular, cochlear and trigeminal projections to the cortex in the anterior suprasylvian sulcus of the cat, *J. Physiol. (London)* **191:**561–573.

Lange, W., Büttner-Ennever, J. A., and Büttner, U., 1979, Vestibular projections to the monkey thalamus: An autoradiographic study, *Brain Res.* **177:**3–18.

Langner, G., Bonke, D., and Scheich, H., 1981a, Neuronal discrimination of natural and synthetic vowels in field L of trained mynah birds, *Exp. Brain Res.* **42:**11–24.

Langner, G., Bonke, D., and Scheich, H., 1981b, Selectivity of auditory neurons for vowels and consonants in the forebrain of the mynah bird, in: *Neuronal Mechanisms of Hearing* (J. Syka and L. Aitkin, eds.), Plenum Press, New York, pp. 317–321.

Larsen, K. D., and Asanuma, H., 1979, Thalamic projections to the feline motor cortex studied with horseradish peroxidase, *Brain Res.* **172:**209–215.

Larsen, K. D., and McBride, R. L., 1979, The organization of feline entopeduncular nucleus projections: Anatomical studies, *J. Comp. Neurol.* **184:**293–308.

Lashley, K. S., 1934a, The mechanism of vision. III. The projection of the retina upon the primary optic centers of the rat, *J. Comp. Neurol.* **59:**341–374.

Lashley, K. S., 1934b, The mechanism of vision. XVI. The functioning of small remnants of the visual cortex, *J. Comp. Neurol.* **70:**45–67.

Lashley, K. S., 1941, Thalamo-cortical connections of the rat's brain, *J. Comp. Neurol.* **75:**67–121.

Lasurski, 1897, Ueber die Scheifenbahnen, *Neurol. Centralbl.* **16:**526.

Laties, A. M., and Sprague, J. M., 1966, The projection of optic fibers to the visual centers in the cat, *J. Comp. Neurol.* **127:**35–70.

Laurent, J.-P., Guerrero, F. A., and Jouvet, M., 1974, Reversible suppression of the geniculate PGO waves and of the concomitant increase of excitability of the intrageniculate optic nerve terminals in cats, *Brain Res.* **81:**558–563.

Laursen, A. M., 1955, An experimental study of the pathways from the basal ganglia, *J. Comp. Neurol.* **102**:1–25.

LaVail, J. H., 1975, Retrograde cell degeneration and retrograde transport techniques, in: *The Use of Axonal Transport for Studies of Neuronal Connectivity* (W. M. Cowan and M. Cuénod, eds.), Elsevier, Amsterdam, pp. 217–248.

Leger, L., Sakai, K., Salvert, D., Touret, M., and Jouvet, M., 1975, Delineation of dorsal lateral geniculate afferents from the cat brain stem as visualised by the horseradish peroxidase technique, *Brain Res.* **93**:490–496.

Le Gros Clark, W. E., 1928, On the brain of the *Macroscelidae*, *J. Anat.* **62**:245–274.

Le Gros Clark, W. E., 1929a, The thalamus of *Tupaia minor*, *J. Anat.* **63**:177–206.

Le Gros Clark, W. E., 1929b, Studies of the optic thalamus of the *Insectivora:* The anterior nuclei, *Brain* **52**:334–358.

Le Gros Clark, W. E., 1930a, The thalamus of *Tarsius*, *J. Anat.* **64**:371–414.

Le Gros Clark, W. E., 1930b, The anterior nuclei of the thalamus in insectivores, *J. Anat.* **64**:125.

Le Gros Clark, W. E., 1931, The brain of *Microcebus murinus*, *Proc. Zool. Soc. London* **1**:463–486.

Le Gros Clark, W. E., 1932a, The structure and connections of the thalamus, *Brain* **55**:406–470.

Le Gros Clark, W. E., 1932b, A morphological study of the lateral geniculate body, *Br. J. Ophthalmol.* **16**:264–284.

Le Gros Clark, W. E., 1933, The medial geniculate body and the nucleus isthmi, *J. Anat.* **67**:536–548.

Le Gros Clark, W. E., 1936a, The thalamic connections of the temporal lobe of the brain in the monkey, *J. Anat.* **70**:447–464.

Le Gros Clark, W. E., 1936b, The termination of ascending tracts in the thalamus of the macaque monkey, *J. Anat.* **71**:7–40.

Le Gros Clark, W. E., 1940, Anatomical basis of colour vision, *Nature (London)* **146**:558–559.

Le Gros Clark, W. E., 1941a, The laminar organization and cell content of the lateral geniculate body in the monkey, *J. Anat.* **75**:419–433.

Le Gros Clark, W. E., 1941b, The lateral geniculate body in the platyrrhine monkeys, *J. Anat.* **76**:131–140.

Le Gros Clark, W. E., 1942, The visual centres of the brain and their connexions, *Physiol. Rev.* **22**:205–232.

Le Gros Clark, W. E., 1943, The anatomy of cortical vision, *Trans. Ophthalmol. Soc. U.K.* **62**:229–245.

Le Gros Clark, W. E., 1948, The connexions of the frontal lobes of the brain, *Lancet* **1**:353–356.

Le Gros Clark, W. E., 1949, The laminar pattern of the lateral geniculate nucleus considered in relation to colour vision, *Doc. Ophthalmol.* **3**:57–64.

Le Gros Clark, W. E., 1952, A note on cortical cyto-architectonics, *Brain* **75**:96–104.

Le Gros Clark, W. E., 1962, *The Antecedents of Man: An Introduction to the Evolution of the Primates*, 2nd ed., Edinburgh University Press, Edinburgh.

Le Gros Clark, W. E., 1968, *Chant of Pleasant Exploration*, Livingston, Edinburgh.

Le Gros Clark, W. E., and Boggon, R. H., 1933a, On the connections of the anterior nucleus of the thalamus, *J. Anat.* **67**:215–226.

Le Gros Clark, W. E., and Boggon, R. H., 1933b, On the connections of the medial cell groups of the thalamus, *Brain* **56**:83–98.

Le Gros Clark, W. E., and Boggon, R. H., 1935, The thalamic connections of the parietal and frontal lobes of the brain in the monkey, *Philos. Trans. R. Soc. London Ser. B* **224**:313–359.

Le Gros Clark, W. E., and Northfield, D. W. C., 1937, The cortical projection of the pulvinar in the macaque monkey, *Brain* **60**:126–142.

Le Gros Clark, W. E., and Penman, G. G., 1934, The projection of the retina in the lateral geniculate body, *Proc. R. Soc. London Ser. B* **114**:291–313.

Le Gros Clark, W. E., and Powell, T. P. S., 1953, On the thalamo-cortical connexions of the general sensory cortex of *Macaca*, *Proc. R. Soc. London Ser. B* **141**:467–487.

Le Gros Clark, W. E., and Russell, D. S., 1940, Atrophy of the thalamus in a case of acquired hemiplegia associated with diffuse porencephaly and sclerosis of the left cerebral hemisphere, *J. Neurol. Psychiatry* **3**:123–140.

Le Gros Clark, W. E., and Russell, W. R., 1939, Observations of the efferent connexions of the centre median nucleus, *J. Anat.* **73**:255–262.

Lehmkuhle, S., Kratz, K. E., Mangel, S. C., and Sherman, S. M., 1980, Spatial and temporal sensitivity of X- and Y-cells in dorsal lateral geniculate nucleus of the cat, *J. Neurophysiol.* **43**:520–541.

Lemon, R. N., and Porter, R., 1976, Afferent input to movement-related precentral neurones in conscious monkeys, *Proc. R. Soc. London Ser. B* **194**:313–339.

Lemon, R. N., and van der Burg, J., 1979, Short-latency peripheral inputs to thalamic neurones projecting to the motor cortex in the monkey, *Exp. Brain Res.* **36**:445–462.

Lende, R. A., 1963, Sensory representation in the cerebral cortex of the opossum *(Didelphys virginiana)*, *J. Comp. Neurol.* **121**:395–404.

Lende, R. A., 1964, Representation in the cerebral cortex of a primitive mammal: Sensorimotor, visual and auditory fields in the echidna *(Tachyglossus aculeatus)*, *J. Neurophysiol* **27**:37–48.

Lende, R. A., Kirsch, W. M., and Druckman, R., 1971, Relief of facial pain after combined removal of precentral and postcentral cortex, *J. Neurosurg.* **34**:537–543.

Lengvari, I., Koves, K., and Halász, B., 1970, The medial habenular nucleus and the control of salt and water balance, *Acta Biol. Acad. Sci. Hung.* **21**:75–83.

Lenn, N. J., 1976, Synapses in the interpeduncular nucleus: Electron microscopy of normal and habenula lesioned rats, *J. Comp. Neurol.* **166**:73–100.

Lennie, P., 1980a, Parallel visual pathways: A review, *Vision Res.* **20**:561–594.

Lennie, P., 1980b, Perceptual signs of parallel pathways, *Philos. Trans. R. Soc. London Ser. B* **290**:23–37.

Lent, R., Cavalcante, L. A., and Rocha-Miranda, C. E., 1976, Retinofugal projections in the opossum: An anterograde degeneration and radioautographic study, *Brain Res.* **107**:9–26.

Leonard, C. M., 1969, The prefrontal cortex of the rat. I. Cortical projection of the mediodorsal nucleus. II. Efferent connections, *Brain Res.* **12**:321–343.

Leonard, C. M., 1972, The connections of the dorsomedial nuclei, *Brain Behav. Evol.* **6**:524–541.

Leonard, C. M., and Scott, J. W. , 1971, Origin and distribution of the amygdalofugal pathways in the rat: An experimental neuroanatomical study, *J. Comp. Neurol.* **141**:313–330.

Leontovich, T. A., and Zhukova, G. P., 1963, The specificity of the neuronal structure and topography of the reticular formation in the brain and spinal cord of Carnivora, *J. Comp. Neurol.* **121**:347–380.

Leppelsack, H.-J., 1978, Unit responses to species-specific sounds in the auditory forebrain center of birds, *Fed. Proc. Am. Soc. Exp. Biol.* **37**:2336–2341.

Leppelsack, H.-J., and Vogt, M., 1976, Responses of auditory neurons in the forebrain of a songbird to stimulation with species-specific sounds, *J. Comp. Physiol.* **107**:263–274.

Léranth, C., Brownstein, M. J., Záborsky, L., Járányi, Z. S., and Palkovits, M., 1975, Morphological and biochemical changes in the rat interpeduncular nucleus following the transection of the habenulo-interpeduncular tract, *Brain Res.* **99**:124–128.

LeVay, S., and Ferster, D., 1977, Relay cell classes in the lateral geniculate nucleus of the cat and the effects of visual deprivation, *J. Comp. Neurol.* **172**:563–584.

LeVay, S., and Ferster, D., 1979, Proportion of interneurons in the cat's lateral geniculate nucleus, *Brain Res.* **164**:304–308.

LeVay, S., and Gilbert, C. D., 1976, Laminar patterns of geniculocortical projection in the cat, *Brain Res.* **113**:1–20.

LeVay, S., and McConnell, S. K., 1982, ON and OFF layers in the lateral geniculate nucleus of the mink, *Nature (London)* **300**:350–351.

LeVay, S., and Sherk, H., 1981, The visual claustrum of the cat. I. Structure and connections, *J. Neurosci.* **1**:956–980.

LeVay, S., Hubel, D. H., and Wiesel, T. N., 1975, The pattern of ocular dominance columns in macaque visual cortex revealed by a reduced silver stain, *J. Comp. Neurol.* **159**:559–576.

LeVay, S., Stryker, M. P., and Shatz, C. J., 1978, Ocular dominance columns and their development in layer IV of the cat's visual cortex: A quantitative study, *J. Comp. Neurol.* **179**:223–244.

LeVay, S., Wiesel, T. N., and Hubel, D. H., 1980, The development of ocular dominance columns in normal and visually deprived monkeys, *J. Comp. Neurol.* **191**:1–51.

Leventhal, A. G., 1979, Evidence that the different classes of relay cells of the cat's lateral geniculate nucleus terminate in different layers of the striate cortex, *Exp. Brain Res.* **37**:349–372.

Leventhal, A. G., 1982, Morphology and distribution of retinal ganglion cells projecting to different layers of the dorsal lateral geniculate nucleus in normal and Siamese cats, *J. Neurosci.* **2**:1024–1042.

Leventhal, A. G., and Hirsch, H. V. B., 1983a, Effects of visual deprivation upon the morphology of retinal ganglion cells projecting to the dorsal lateral geniculate nucleus of the cat, *J. Neurosci.* **3**:332–344.

Leventhal, A. G., and Hirsch, H. V. B., 1983b, Effects of visual deprivation upon the geniculocortical W-cell pathway in the cat: Area 19 and its afferent input, *J. Comp. Neurol.* **214**:59–71.

Leventhal, A. G., Keens, J., and Tork, I., 1979, Evidence for an extrageniculate thalamic relay to cortical areas 19 and Clare–Bishop in the cat, *Neurosci. Abstr.* **5**:793.

Leventhal, A. G., Keens, J., and Tork, I., 1980a, The afferent ganglion cells and cortical projections of the retinal recipient zone (RRZ) of the cat's "pulvinar complex," *J. Comp. Neurol.* **194**:535–554.

Leventhal, A. G., Rodieck, R. W., and Dreher, B., 1980b, Morphology and central projections of different types of retinal ganglion cells in cats and Old World monkey (*M. fascicularis*), *Neurosci. Abstr.* **6**:582.

Leventhal, A., Rodieck, R. W., and Dreher, B., 1981, Retinal ganglion cell classes in the Old World monkey: Morphology and central projections, *Science* **213**:1139–1142.

Levey, N. H., Harris, J., and Jane, J. A., 1973, Effects of visual cortical ablation on pattern discrimination in the ground squirrel (*Citellus tridecemlineatus*), *Exp. Neurol.* **39**:270–276.

Levick, W. R., Cleland, B. G., and Dubin, M. W., 1972, Lateral geniculate neurons of cat: Retinal inputs and physiology, *Invest. Ophthalmol.* **11**:302–310.

Levine, R. L., 1980, An autoradiographic study of the retinal projections in *Xenopus laevis* with comparisons to *Rana*, *J. Comp. Neurol.* **189**:1–29.

Lewandowsky, M., 1904, Untersuchungen über die Leitungsbahnen des Truncus cerebri und ihren Zusammenhang mit denen der Medulla spinalis und des Cortex cerebri, *Denkschr. Med. Naturwiss. Ges. Jena* **10**:63–150.

Lewis, P. R., and Shute, C. C. D., 1967, The cholinergic limbic system: Projection to hippocampal formation, medial cortex, nuclei of the ascending cholinergic reticular system, and the subfornical organ and supra-optic crest, *Brain* **90**:521–540.

Lhermitte, J., 1921, Un cas de syndrome thalamique a evolution regressive; l'ataxie residuelle, *Rev. Neurol.* **37**:1256–1259.

Lhermitte, J., 1925, Les syndromes thalamiques dissociés: Les formes analgique et hemialgique, *Ann. Méd. (Paris)* **17**:488–501.

Lhermitte, J., 1929, Le lobe frontal: Données expérimentales anatomo-cliniques et psychopathologiques, *Encéphale* **24**:87–118.

Lhermitte, J., 1932, El Síndrome de la capa óptica en fisiologia y en clínica, *Med. Argent.* **11**:1361–1370.

Lhermitte, J., 1936, Symptomatologie de l'hémorragie du thalamus, *Rev. Neurol.* **65**:89–93.

Lhermitte, J., and Cornil, L., 1929, La forme hémialgique du syndrome thalamique, *Gaz. Hôp. Paris* **102**:1017–1022.

Li, C.-L., Cullen, C., and Jasper, H. H., 1955a, Laminar microelectrode analysis of cortical unspecific recruiting responses and spontaneous rhythms, *J. Neurophysiol.* **19**:131–143.

Li, C.-L., Cullen, C., and Jasper, H. H., 1955b, Laminar microelectrode studies of specific somatosensory cortical potentials, *J. Neurophysiol.* **19**:111–130.

Lieberman, A. R., and Webster, K. E., 1972, Presynaptic dendrites and a distinctive class of synaptic vesicle in the rat dorsal lateral geniculate nucleus, *Brain Res.* **42**:196–200.

Liedgren, S. R. C., Milne, A. C., Rubin, A. M., Schwarz, D. W. F., and Tomlinson, R. D., 1976, Representation of vestibular afferents in somatosensory thalamic nuclei of the squirrel monkey (*Saimiri sciureus*), *J. Neurophysiol.* **39**:601–612.

Lim, R. K. S., Krauthamer, G., Guzman, F., and Fulp, R. R., 1969, Central nervous system activity associated with the pain evoked by bradykinin and its alternation by morphine and aspirin, *Proc. Natl. Acad. Sci. USA* **63**:705–712.

Lin, C.-S., and Kaas, J. H., 1977, Projections from cortical visual areas 17, 18 and MT onto the dorsal lateral geniculate nucleus in owl monkeys, *J. Comp. Neurol.* **173**:457–474.

Lin, C.-S., and Kaas, J. H., 1979, The inferior pulvinar complex in owl monkeys: Architectonic subdivisions and patterns of input from the superior colliculus and subdivisions of visual cortex, *J. Comp. Neurol.* **187**:655–678.

Lin. C.-S., and Kaas, J. H., 1980, Projections from the medial nucleus of the inferior pulvinar complex to the middle temporal area of the visual cortex, *Neuroscience* **5**:2219–2228.

Lin, C.-S., and Sherman, S. M., 1978, Effects of early monocular eyelid suture upon development of relay cell classes in the cat's lateral geniculate nucleus, *J. Comp. Neurol.* **181**:809–832.

Lin, C.-S., Wagor, E., and Kaas, J. H., 1974, Projections from the pulvinar to the middle temporal visual area (MT) in the owl monkey, *Aotus trivirgatus*, *Brain Res.* **76**:145–149.

Lin, C.-S., Kratz, K. E., and Sherman, S. M., 1978, Percentage of relay cells in the cat's lateral geniculate nucleus, *Brain Res.* **131:**167–173.

Lin, C.-S., Merzenich, M. M., Sur, M., and Kaas, J. H., 1979, Connections of areas 3b and 1 of the parietal somatosensory strip with the ventroposterior nucleus in the owl monkey (*Aotus trivirgatus*), *J. Comp. Neurol.* **185:**355–372.

Linden, D. C., Guillery, R. W., and Cucchiaro, J., 1981, The dorsal lateral geniculate nucleus of the normal ferret and its postnatal development, *J. Comp. Neurol.* **203:**189–211.

Lindström, S., 1982, Synaptic organization of inhibitory pathways to principal cells in the lateral geniculate nucleus of the cat, *Brain Res.* **234:**447–453.

Lindvall, O., and Björklund, A., 1974, The organization of the ascending catecholamine neuron systems in the rat brain as revealed by the glyoxylic acid fluorescence method, *Acta Physiol. Scand. Suppl.* **412:**1–48.

Lindvall, O., Björklund, A., Moore, R. Y., and Stenevi, U., 1974, Mesencephalic dopamine neurons projecting to neocortex, *Brain Res.* **81:**325–331.

Lippe, W. R., and Masterton, R. B., 1980, The distribution of activity in the avian auditory system to monoaural sound stimulation examined with ^{14}C-2-deoxy-D-glucose autoradiography, *Neurosci. Abstr.* **6:**555.

Lisney, S. J. W., 1978, Some anatomical and electrophysiological properties of tooth-pulp afferents in the cat, *J. Physiol. (London)* **284:**19–36.

Livingstone, M. S., and Hubel, D. H., 1981, Effects of sleep and arousal on the processing of visual information in the cat, *Nature (London)* **291:**554–561.

Livingstone, M. S., and Hubel, D. H., 1982, Thalamic inputs to cytochrome oxidase-rich regions in monkey visual cortex, *Proc. Natl. Acad. Sci. USA* **79:**6098–6101.

Lo, F.-S., 1981, Synaptic organization of the lateral geniculate nucleus of the rabbit: Lack of feed-forward inhibition, *Brain Res.* **221:**387–392.

Lo, F.-S., 1983, Both fast and slow relay cells in lateral geniculate nucleus of rabbits receive recurrent inhibition, *Brain Res.* **271:**335–338.

Loe, P. R., and Benevento, L. A., 1969, Auditory–visual interaction in single units in the orbito-insular cortex of the cat, *Electroencephalogr. Clin. Neurophysiol.* **26:**395–398.

Loe, P. R., Whitsel, B. L., Dreyer, D. A., and Metz, C. B., 1977, Body representation in ventrobasal thalamus of macaque: A single-unit analysis, *J. Neurophysiol.* **40:**1339–1355.

Loewy, A. D., Wallach, J. H., and McKellar, S., 1981, Efferent connections of the ventral medulla oblongata in the rat, *Brain Res. Rev.* **3:**63–80.

Lohman, A. H. M., and van Woerden-Verkley, I., 1978, Ascending connections to the forebrain in the tegu lizard, *J. Comp. Neurol.* **182:**555–594.

Long, E., 1899, *Les Voies Centrales de la Sensibilité Générale*, Steinhill, Paris.

Loo, Y. T., 1931, The forebrain of the opossum, *Didelphis virginiana*. Part II. Histology, *J. Comp. Neurol.* **52:**1–148.

Lorente de Nó, R., 1949, Cerebral cortex: Architectonics, intracortical connections, in: *Physiology of the Nervous System* (J. F. Fulton, ed.), 3rd ed., Oxford University Press, London, pp. 274–301.

Lucier, G. E., Rüegg, D. C., and Weisendanger, M., 1975, Responses of neurones in motor cortex and in area 3A to controlled stretches of forelimb muscles in cebus monkeys, *J. Physiol. (London)* **251:**833–853.

Ludwig, C., 1858, *Lehrbuch der Physiologie des Menschen*, Volume I, Winter'sche Verlagshandlung, Leipzig.

Luiten, P. G. M., 1981, Afferent and efferent connections of the optic tectum in the carp (*Cyprinus carpio* L.), *Brain Res.* **220:**51–65.

Lund, J. S., Lund, R. D., Hendrickson, A. E., Bunt, A. H., and Fuchs, A. F., 1975, The origin of efferent pathways from the primary visual cortex, area 17, of the macaque monkey as shown by retrograde transport of horseradish peroxidase, *J. Comp. Neurol.* **164:**287–303.

Lund, J. S., Hendrickson, A. E., Ogren, M. P., and Tobin, E. A., 1981, Anatomical organization of primate visual cortex area VII, *J. Comp. Neurol.* **202:**19–45.

Lund, R. D., 1965, Uncrossed visual pathways of hooded and albino rats, *Science* **149:**1506–1507.

Lund, R. D., and Bunt, A. H., 1976, Prenatal development of central optic pathways in albino rats, *J. Comp. Neurol.* **165:**247–264.

Lund, R. D., and Lund, J. S., 1965, The visual system of the mole, *Talpa europaea*, *Exp. Neurol.* **13:**302–316.

Lund, R. D., and Mustari, M. J., 1977, Development of the geniculocortical pathway in rats, *J. Comp. Neurol.* **173**:289–306.

Lund, R. D., and Webster, K. E., 1967a, Thalamic afferents from the dorsal column nuclei: An experimental anatomical study in the rat, *J. Comp. Neurol.* **130**:301–312.

Lund, R. D., and Webster, K. E., 1967b, Thalamic afferents from the spinal cord and trigeminal nuclei: An experimental anatomical study in the rat, *J. Comp. Neurol.* **130**:313–328.

Lund, R. D., Cunningham, T. J., and Lund, J. S., 1973, Modified optic projections after unilateral eye removal in young rats, *Brain Behav. Evol.* **8**:51–72.

Lund, R. D., Lund, J. S., and Wise, R. P., 1974, The organization of the retinal projection to the dorsal lateral geniculate nucleus in pigmented and albino rats, *J. Comp. Neurol.* **158**:383–404.

Lund-Karlsen, R., and Fonnum, F., 1978, Evidence for glutamate as a neurotransmitter in the corticofugal fibres to the dorsal lateral geniculate body and the superior colliculus in rats, *Brain Res.* **151**:457–468.

Luys, J., 1865, *Recherches sur le Système Nerveux Cérébro-spinal: Sa Structure, ses Fonctions et ses Maladies,* 2 vols., Baillière, Paris.

Luys, J., 1876, *Le Cerveau,* Baillière & Fils, Paris.

Luys, J., 1881, *The Brain and its Functions,* Kegan Paul, Trench, London.

Lynch, J. C., 1980, The functional organization of posterior parietal association cortex, *Behav. Brain Sci.* **3**:485–534.

Lynch, J. C., Mountcastle, V. B., Talbot, W. H., and Yin, T. C. T., 1977, Parietal lobe mechanisms for directed attention, *J. Neurophysiol.* **40**:362–389.

McAllister, J. P., II, and Das, G. D., 1977, Neurogenesis in the epithalamus, dorsal thalamus and ventral thalamus of the rat: An autoradiographic and cytological study, *J. Comp. Neurol.* **172**:647–686.

McBride, R. L., 1981, Organization of afferent connections of the feline lateral habenular nucleus, *J. Comp. Nuerol.* **198**:89–100.

McCall, W. D., Jr., Farias, M. A. C., Williams, W. J., and DeMent, S. L., 1974, Static and dynamic responses of slowly adapting joint receptors, *Brain Res.* **70**:221–243.

McCance, I., Phillis, J. W., and Westerman, R. A., 1968, The pharmacology of acetylcholine excitation of thalamic neurones, *Br. J. Pharmacol. Chemother.* **32**:652–662.

Macchi, G., 1971, Index of nuclei, in: *Anatomy of the Normal Human Thalamus: Topometry and Standardized Nomenclature* (A. Dewulf, ed.), Elsevier, Amsterdam, pp. 169–178.

Macchi, G., 1983, Old and new anatomo-functional criteria in the subdivision of the thalamic nuclei, in: *Somatosensory Integration in the Thalamus* (G. Macchi, A. Rustioni, and R. Spreafico, eds.), Elsevier, Amsterdam, pp. 3–16.

Macchi, G., Angeleri, F., and Guazzi, G., 1959, Thalamo-cortical connections of the first and second somatic sensory areas in the cat, *J. Comp. Neurol.* **111**:387–405.

Macchi, G., Quattrini, A., Chinzari, P., Marchesi, G., and Capocchi, G., 1975, Quantitative data on cell loss and cellular atrophy of intralaminar nuclei following cortical and subcortical lesions, *Brain Res.* **89**:43–60.

Macchi, G., Bentivoglio, M., D'Atena, C., Rossini, P., and Tempesta, E., 1977, The cortical projection of the thalamic intralaminar nuclei restudied by means of the HRP retrograde axonal transport, *Neurosci. Lett.* **4**:121–126.

Macchi, G., Bentivoglio, M., Molinari, M., and Minciacchi, D., 1984, The thalamo-caudate versus thalamo-cortical projections as studied in the cat with fluorescent retrograde double labeling, *Exp. Brain Res.* **54**:225–239.

McConnell, S. K., and LeVay, S., 1983, Geniculocortical afferents in the mink: Evidence for on/off and ocular dominance patches, *Neurosci. Abstr.* **9**:617.

McEntee, W. J., Biber, M. P., Perl, D. P., and Benson, D. F., 1976, Diencephalic amnesia: A reappraisal, *J. Neurol. Neurosurg. Psychiatry* **39**:436–441.

McGeer, P. L., McGeer, E. G., Scherer, U., and Singh, K., 1977, A glutamatergic corticostriatal path?, *Brain Res.* **128**:369–373.

Maciewicz, R., 1975, Thalamic afferents to areas 17, 18, and 19 of cat cortex traced with horseradish peroxidase, *Brain Res.* **84**:308–312.

Maciewicz, R., Phipps, B. S., Bry, J., and Highstein, S. M., 1982, The vestibulothalamic pathway: Contribution of the ascending tract of Deiters, *Brain Res.* **252**:1–12.

McIlwain, J. T., 1972, Nonretinal influences on the lateral geniculate nucleus, *Invest. Opthalmol.* **11**:311–322.

McIlwain, J. T., and Creutzfeldt, O. D., 1967, Microelectrode study of synaptic excitation and inhibition in the lateral geniculate nucleus of the cat, *J. Neurophysiol.* **30**:1–21.

McIlwain, J. T., and Lufkin, R. B., 1976, Distribution of direct Y-cell inputs to the cat's superior colliculus: Are there spatial gradients?, *Brain Res.* **103**:133–138.

McIntyre, A. K., 1962, Cortical projection of impulses in the interosseous nerve of the cat's hindlimb, *J. Physiol. (London)* **163**:46–60.

Mackay-Sim, A., Sefton, A. J., and Martin, P. R., 1983, Subcortical projections to lateral geniculate and thalamic reticular nuclei in the hooded rat, *J. Comp. Neurol.* **213**:24–35.

McKellar, S., and Loewy, A. D., 1982, Efferent connections of the A1 catecholamine cell group in the rat: An autoradiographic study, *Brain Res.* **241**:11–30.

McKenna, T. M., Whitsel, B. L., and Dreyer, D. A., 1982, Anterior parietal cortical topographic organization in the macaque monkey: A reevaluaton, *J. Neurophysiol.* **48**:289–317.

McLardy, T., 1948, Projection of the centromedian nucleus of the human thalamus, *Brain* **71**:290–303.

McLardy, T., 1951, Diffuse thalamic projection to cortex: An anatomical critique, *Electroencephalogr. Clin. Neurophysiol.* **3**:183–188.

McLardy, T., 1963, Thalamic microneurons, *Nature (London)* **199**:820–821.

McLeod, J. G., 1958, The representation of the splanchnic afferent pathways in the thalamus of the cat, *J. Physiol. (London)* **140**:462–478.

Macpherson, J. M., Rasmusson, D. D., and Murphy, J. T., 1980, Activities of neurons in "motor" thalamus during control of limb movement in the primate, *J. Neurophysiol.* **44**:11–28.

Madarász, M., Tömböl, T., Hajdu, F., and Somogyi, G., 1981, Some comparative quantitative data on the different (relay and associative) thalamic nuclei in the cat, *Anat. Embryol.* **162**:363–378.

Madarász, M., Somogyi, J., Silakov, V. L., and Hámori, J., 1983, Residual neurons in the lateral geniculate nucleus of adult cats following chronic disconnection from the cortex, *Exp. Brain Res.* **52**:363–374.

Maekawa, K., and Purpura, D. P., 1967a, Properties of spontaneous and evoked synaptic activities of thalamic ventrobasal neurons, *J. Neurophysiol.* **30**:360–381.

Maekawa, K., and Purpura, D. P., 1967b, Intracellular study of lemniscal and non-specific synaptic interactions in thalamic ventrobasal neurons, *Brain Res.* **4**:308–323.

Maendly, R., Rüegg, D. G., Wiesendanger, M., Wiesendanger, R., Lagowska, J., and Hess, B., 1981, Thalamic relay for group I muscle afferents of forelimb nerves in the monkey, *J. Neurophysiol.* **46**:901–917.

Maffei, L., Moruzzi, G., and Rizzolatti, G., 1965, Influence of sleep and wakefulness on the response of lateral geniculate units to sinewave photic stimulation, *Arch. Ital. Biol.* **103**:596–608.

Magendie, F., 1823, On the functions of the corpora striata and corpora quadrigemina, *Lancet* **1**:343–345.

Magendie, F., 1841, *Leçons sur les Fonctions et les Maladies due Système Nerveux*, 2 vols., Lecaplain, Paris.

Magnin, M., and Fuchs, A. F., 1977, Discharge properties of neurons in the monkey thalamus tested with angular acceleration, eye movement and visual stimuli, *Exp. Brain Res.* **28**:293–299.

Magoun, H. W., 1950, Caudal and cephalic influences of the brainstem reticular formation, *Physiol. Rev.* **30**:459–474.

Mahaim, A., 1983, Ein Fall von Sekundärer Erkrankung des Thalamus opticus und der Regio subthalamica, *Arch. Psychiatr. Nervenkr.* **25**:343–376.

Mair, W. G. P., Warrington, E. K., and Weiskrantz, L., 1979, Memory disorder in Korsakoff's psychosis: A neuropathological and neuropsychological investigation of two cases, *Brain* **102**:749–783.

Majorossy, K., and Réthelyi, M., 1968, Synaptic architecture in the medial geniculate body (ventral division), *Exp. Brain Res.* **6**:306–323.

Majorossy, K., Réthelyi, M., and Szentágothai, J., 1965, The large glomerular synapse of the pulvinar, *J. Hirnforsch.* **7**:415–432.

Malcolm, L. J., Bruce, I. S. C., and Burke, W., 1970, Excitability of the lateral geniculate nucleus in the alert, non-alert and sleeping cat, *Exp. Brain Res.* **10**:283–297.

Malis, L. I., Pribram, K. H., and Kruger, L., 1953, Action potentials in "motor" cortex evoked by peripheral nerve stiumlation, *J. Neurophysiol.* **16**:161–167.

Mallart, A., 1968, Thalamic projection of muscle nerve afferents in the cat, *J. Physiol. (London)* **194**:337–353.

Malone, E., 1910, Über die Kerne des menschlichen Diencephalon, Verhandl. Königl. Akad. Wiss., Berlin.

Malpeli, J. G., and Baker, F. H., 1975, The representation of the visual field in the lateral geniculate nucleus of *Macaca mulatta, J. Comp. Neurol.* **161**:569–594.

Mancia, M., Broggi, G., and Margnelli, M., 1971, Brain stem reticular effects on intralaminar thalamic neurons in the cat, *Brain Res.* **25**:638–641.

Mann, G., 1905, On the thalamus, *Br. Med. J.* **1**:289–291.

Manson, J., 1969, The somatosensory cortical projections of single nerve cells in the thalamus of the cat, *Brain Res.* **12**:489–492.

Mantyh, P. W., 1983, The spinothalamic tract in the primate: A re-examination using WGA-HRP, *Neuroscience* **9**:847–862.

Mantyh, P., and Hunt, S. P., 1983, Neuropeptides are present in projection neurons at all levels in visceral and taste pathways from periphery to sensory cortex, *Brain Res.* **299**:297–312.

Mantyh, P., and Kemp, J. W., 1983, The distribution of putative neurotransmitters in the lateral geniculate nucleus of the rat, *Brain Res.* **288**:344–348.

Marburg, O., 1904, *Mikroskopisch-topographischer Atlas des menschlichen Zentralnervensystems*, 2 vols., Deuticke, Leipzig.

Marburg, O., 1944, The structure and fiber connections of the human habenula, *J. Comp. Neurol.* **80**:211–234.

Marchi, V., 1891, Sull Origine e Decorso dei Peduncoli Cerebellari e sui Loro Rapporti Cogli altri Centri Nerevosi, Florence (Quoted from a review by E. Belmondo, 1891, *Neurol. Zentralbl.* **10**:237–238).

Marchi, V., and Algeri, G., 1886, Sulle degenerazioni discendenti consecutive a lesioni sperimentali in diverse zone della corteccia cerebrale, *Riv. Sper. Freniatr.* **12**:208–252.

Marcotte, R. R., and Updyke, B. V., 1981, Thalamic projections onto the visual areas of the middle suprasylvian sulcus in the cat, *Anat. Rec* **199**:160A.

Margoliash, D., 1982, Specificity and selectivity of neuronal responses to song in a vocal control nucleus of white-crowned sparrows *(Zonotrichia leucophrys), Neurosci. Abstr.* **8**:1022.

Mark, V. H., Ervin, F. R., and Yakovlev, P. I., 1963, Stereotactic thalamotomy. III. The verification of anatomical lesion sites in the human thalamus, *Arch. Neurol. (Chicago)* **8**:528–538.

Marrocco, R. T., and Brown, J. B., 1975, Correlation of receptive field properties of monkey LGN cells with the conduction velocity of retinal afferent input, *Brain Res.* **92**:137–144.

Marrocco, R. T., McClurkin, J. W., and Young, R. A., 1981, Spatial properties of superior colliculus cells projecting to the inferior pulvinar and parabigeminal nucleus of the monkey, *Brain Res.* **222**:150–154.

Marshall, K. C., and McLennan, H., 1972, The synaptic activation of neurones of the feline ventrolateral thalamic nucleus: Possible cholinergic mechanisms, *Exp. Brain Res.* **15**:472–483.

Maruyama, N., Kawasaki, T., Abe, J., Katoh, I., and Yamazaki, H., 1966, Unitary response to tone stimuli recorded from the medial geniculate body of cats, *Intern. Audiol.* **5**:184–188.

Mason, C. A., and Robson, J. A., 1979, Morphology of retino-geniculate axons in the cat, *Neuroscience* **4**:79–97.

Mason, R., 1975, Cell properties in the medial interlaminar nucleus of the cat's lateral geniculate nucleus in relation to the transient/sustained classification, *Exp. Brain Res.* **22**:327–329.

Mason, R., 1976, Functional organization in the cat's dorsal lateral geniculate complex, *J. Physiol. (London)* **258**:66–67P.

Mason, R., 1978, Functional organization in the cat's pulvinar complex, *Exp. Brain Res.* **31**:51–66.

Mason, R., 1981, Differential responsiveness of cells in the visual zones of the cat's LP–pulvinar complex to visual stimuli, *Exp. Brain Res.* **43**:25–33.

Massion, J., and Rispal-Padel, L., 1972, Spatial organization of the cerebellothalamocortical pathway, *Brain Res.* **40**:61–65.

Mathers, L. H., 1972a, The synaptic organization of the cortical projection to the pulvinar of the squirrel monkey, *J. Comp. Neurol.* **146**:43–60.

Mathers, L. H., 1972b, Ultrastructure of the pulvinar of the squirrel monkey, *J. Comp. Neurol.* **146**:15–42.

Mathers, L. H., and Mascitti, G. G., 1975, Electrophysiological and morphological properties of neurons in the ventral lateral geniculate nucleus of the rabbit, *Exp. Neurol.* **46**:506–520.

Mathers, L. H., and Rapisardi, S. C., 1973, Visual and somatosensory receptive fields of neurons in the squirrel monkey pulvinar, *Brain Res.* **64**:65–83.

Matthews, C. D., and Seamark, R. F. (eds.), 1981, *Pineal Function*, Elsevier, Amsterdam.

Matthews, M. R., 1964, Further observations on transneuronal degeneration in the lateral geniculate nucleus of the macaque monkey, *J. Anat.* **98:**255–263.

Matthews, M. R., Cowan, W. M., and Powell, T. P. S., 1960, Transneuronal cell degeneration in the lateral geniculate nucleus of the macaque monkey, *J. Anat.* **94:**145–169.

Maunsell, J. H. R., and Van Essen, D. C., 1983a, Functional properties of neurons in middle temporal visual area of the macaque monkey. I.Selectivity for stimulus direction, speed and orientation, *J. Neurophysiol.* **49:**1127–1147.

Maunsell, J. H. R., and Van Essen, D. C., 1983b, Functional properties of neurons in middle temporal visual area of the macaque monkey. II. Binocular interactions and sensitivity to binocular disparity, *J. Neurophysiol.* **49:**1148–1167.

Maxwell, J. H., and Granda, A. M., 1979, Receptive fields of movement-sensitive cells in the pigeon thalamus, in: *Neural Mechanisms of Behavior in the Pigeon* (A. M. Granda and J. H. Maxwell, eds.), Plenum Press, New York, pp. 177–198.

May, M. T., 1968, *Galen, On the Usefulness of the Parts of the Body*, Volume 1, Cornell University Press, Ithaca, N.Y., pp. 400–401.

Mayeda, Y., 1941, Zur vergleichenden Cytoarchitektonik des Thalamus des Affen, *Macacus rhesus,* *Z. Mikrosk. Anat. Forsch.* **49:**549–588.

Mehler, W. R., 1966a, Further notes on the center median nucleus of Luys, in *The Thalamus* (D. P. Purpura and M. D. Yahr, eds.), Columbia University Press, New York, pp. 109–122.

Mehler, W. R., 1966b, Some observations on secondary ascending afferent systems in the central nervous system, in: *Pain* (R. S. Knighton and P. R. Dumke, eds.), Little, Brown, Boston, pp. 11–32.

Mehler, W. R., 1966c, The posterior thalamic region in man, *Confin. Neurol.* **27:**18–29.

Mehler, W. R., 1969, Some neurological species differences—a posteriori, *Ann. N.Y. Acad. Sci.* **167:**424–468.

Mehler, W. R., 1971, Idea of a new anatomy of the thalamus, *J. Psychiatr. Res.* **8:**203–217.

Mehler, W. R., and Nauta, W. J. H., 1974, Connections of the basal ganglia and of the cerebellum, *Confin. Neurol.* **36:**205–222.

Mehler, W. R., Feferman, M. E., and Nauta, W. J. H., 1960, Ascending axon degeneration following anterolateral cordotomy: An experimental study in the monkey, *Brain* **83:**718–750.

Meibach, R. C., and Siegel, A., 1975, The origin of fornix fibers which project to the mammillary bodies in the rat: A horseradish peroxidase study, *Brain Res.* **88:**508–512.

Meibach, R. C., and Siegel, A., 1977a, Efferent connections of the hippocampal formation in the rat, *Brain Res.* **124:**197–224.

Meibach, R. C., and Siegel, A., 1977b, Thalamic projections of the hippocampal formation: Evidence for an alternative pathway involving the internal capsule, *Brain Res.* **134:**1–12.

Menétrey, D., Giesler, G. J., and Besson, J. M., 1977, An analysis of response properties of spinal cord dorsal horn neurons to non-noxious and noxious stimuli in the spinal rat, *Exp. Brain Res.* **27:**15–33.

Merzenich, M. M., and Brugge, J. F., 1973, Representation of the cochlear partition on the superior temporal plane of the macaque monkey, *Brain Res.* **50:**275–296.

Merzenich, M. M., Knight, P. L., and Roth, G. L., 1975, Representation of cochlea within primary auditory cortex in the cat, *J. Neurophysiol.* **38:**231–249.

Merzenich, M. M., Colwell, S. A., and Andersen, R. A., 1982, Thalamocortical and corticothalamic connections in the auditory system of the cat, in: *Cortical Sensory Organization*, Volume III (C. N. Woolsey, ed.), Humana Press, Clifton, N.J., pp. 43–57.

Mesulam, M. M., and Pandya, D. N., 1973, The projections of the medial geniculate complex within the sylvian fissure of the rhesus monkey, *Brain Res.* **60:**315–333.

Mesulam, M. M., and Van Hoesen, G. W., 1976, Acetylcholinesterase-rich projections from the basal forebrain of the rhesus monkey to neocortex, *Brain Res.* **109:**152–157.

Mettler, F. A., 1935a–d, Corticofugal fiber connections of the cortex of *Macaca mulatta, J. Comp. Neurol.* **61:**221–256, 509–542; **62:**263–291; **63:**25–47.

Mettler, F. A., 1945, Fiber connections of the corpus striatum of the monkey and baboon, *J. Comp. Neurol.* **82:**L169–204.

Mettler, F. A., 1947, Extracortical connections of the primate frontal cerebral cortex. I. Thalamo-cortical connections, *J. Comp. Neurol.* **86:**95–117.

Meulders, M., and Godfraind, J. M., 1969, Influence du reveil d'origine réticulaire sur l'étendue des champs visuels des neurons de la région genouille chez le chat avec cerveau intact ou avec cerveau isolé, *Exp. Brain Res.* **9**:201–220.

Meulders, M., Massion, J., Colle, J., and Albe-Fessard, D., 1963, Effet d'ablations telencéphaliques sur l'amplitude des potentiels évoques dans le centre médian par stimulation somatique, *Electroencephalogr. Clin. Neurophysiol* **15**:29–38.

Meyer, A(dolf), 1892, Über das Vorderhirn einiger Reptilien, *Z. Wiss. Zool. Abt. A* **55**:63–133.

Meyer, A(lfred), 1971, *Historical Aspects of Cerebral Anatomy*, Oxford University Press, London.

Meyer, A., Beck, E., and McLardy, T., 1974, Prefrontal leucotomy: A neuroanatomical report, *Brain* **70**:18–49.

Meyer, D. R., and Woolsey, C. N., 1952, Effects of localized cortical destruction on auditory discriminative conditioning in cat, *J. Neurophysiol.* **15**:149–162.

Meyer, G., and Albus, K., 1981, Topography and cortical projections of morphologically identified neurons in the visual thalamus of the cat, *J. Comp. Neurol.* **201**:353–374.

Meyer, M., 1949, A study of efferent connections of the frontal lobe in the human brain after leucotomy, *Brain* **72**:265–296.

Meyer, R. L., and Sperry, R. W., 1976, Retinotectal specificity: Chemoaffinity theory, in: *Studies on the Development of Behavior and the Nervous System* (G. Gottlieb, ed.), Academic Press, New York, pp. 111–149.

Meynert, T., 1867, Der Bau der Grosshirnrinde und seine ortlichen Verschiedenheiten, *Vjschr. Psychiatr.* **1**:77–93, 126–170, 198–217.

Meynert, T., 1872, Vom Gehirne der Säugethiere, in: *Handbuch der Lehre von den Geweben des Menschen und der Thiere* (S. Stricker, ed.), Volume 2, Engelmann, Leipzig, pp. 694–808.

Meynert, T., 1884, *Psychiatrie*, Klinik der Erkrankungen des Vorderhirns, Vienna.

Meynert, T., 1885, *Psychiatry: A Clinical Treatise on the Diseases of the Fore-Brain Based upon a Study of its Structure, Functions and Nutrition*, Part I (translated by B. Sachs), Putnam, New York.

Miceli, D., Peyrichoux, J., and Repérant, J., 1975, The retino-thalamo-hyperstriatal pathway in the pigeon *(Columba livia), Brain Res.* **100**:125–131.

Miceli, D., Gioanni, H., Repérant, J., and Peyrichoux, J., 1979, The avian visual Wulst. I. An anatomical study of afferent and efferent pathways. II. An electrophysiological study of the functional properties of single neurons, in: *Neural Mechanisms of Behavior in the Pigeon* (A. M. Granda and J. H. Maxwell, eds.), Plenum Press, New York, pp. 223–254.

Miceli, D., Peyrichoux, J., Repérant, J., and Weidner, C., 1980, Étude anatomique des afferences due telencéphale rostral chez le poussin de *Gallus domesticus, J. Hirnforsch.* **21**:627–646.

Michael, C. R., 1968, Receptive fields of single optic nerve fibers in a mammal with an all-cone retina. III. Opponent color units, *J. Neurophysiol.* **31**:268–282.

Michael, C. R., 1971, Dual opponent-color cells in the lateral geniculate of the ground squirrel, *J. Gen. Physiol.* **57**:254.

Michael, C. R., 1973, Opponent-color and opponent-contrast cells in lateral geniculate nucleus of the ground squirrel, *J. Neurophysiol.* **36**:536–550.

Michael, C. R., 1978, Color vision mechanisms in monkey striate cortex: Dual-opponent cells with concentric receptive fields, *J. Neurophysiol.* **41**:572–588.

Michael, C. R., 1981, Columnar organization of color cells in monkey's striate cortex, *J. Neurophysiol.* **46**:587–604.

Michael, C.R., 1983, Functional classes of neurons in the monkey's lateral geniculate nucleus have distinctive morphology, *Neurosci. Abstr.* **9**:1047.

Michael, C. R., and Bowling, D. B., 1982, Projection patterns of single physiologically characterized optic tract fibres in cat, *Nature (London)* **286**:899–902.

Michail, S. and Karamanlidis, A. N., 1970, Trigeminothalamic fibre connexions in the dog and pig: An experimental study with retrograde degeneration method, *J. Anat.* **107**:557–566.

Mickle, W. A., and Ades, H. W., 1952, A composite sensory projection area in the cerebral cortex of the cat, *Am. J. Physiol.* **170**:682–689.

Mickle, W. A., and Ades, H. W., 1954, Rostral projection pathway of the vestibular system, *Am. J. Physiol.* **176**:243–246.

Middlebrooks, J. C., and Zook, J. M., 1983, Intrinsic organization of the cat's medial geniculate body identified by projections to binaural response-specific bands in the primary auditory cortex, *J. Neurosci.* **3**:203–224.

Middlebrooks, J. C., Dykes, R. W., and Merzenich, M. M., 1980, Binaural response-specific bands in primary auditory cortex (AI) of the cat: Topographical organization orthogonal to isofrequency contours, *Brain Res.* **181**:31–48.

Mihailovic, J., Perisic, M., Bergonzi, R., and Meier, R. E., 1974, The dorsolateral thalamus as a relay in the retino-Wulst pathway in pigeon (*Columba livia*): An electrophysiological study, *Exp. Brain Res.* **21**:229–240.

Mihailovic, L. T., Cupic, D., and Dekleva, N., 1971, Changes in the numbers of neurons and glial cells in the lateral geniculate nucleus of the monkey during retrograde cell degeneration, *J. Comp. Neurol.* **142**:223–230.

Millar, J., and Basbaum, A. I., 1975, Topography of the projection of the body surface of the cat to cuneate and gracile nuclei, *Exp. Neurol.* **49**:281–290.

Miller, J. W., Buschmann, M. B. T., and Benevento, L. A., 1980, Extrageniculate thalamic projections to the primary visual cortex, *Brain Res.* **189**:221–227.

Miller, R. N., 1940, The diencephalon cell masses in the teleost, *Corydora paliatus*, *J. Comp. Neurol.* **73**:345–378.

Mills, R. P., and Swanson, P. D., 1978, Vertical oculomotor apraxia and memory loss, *Ann. Neurol.* **4**:149–153.

Milner, B. L., 1974, Hemispheric specialization: Scope and limits, in: *The Neurosciences: Third Study Program* (F. O. Schmitt and F. G. Worden, eds.), MIT Press, Cambridge, Mass., pp. 75–89.

Minderhoud, J. M., 1971, An anatomical study of the efferent connections of the thalamic reticular nucleus, *Exp. Brain Res.* **12**:435–446.

Minkowski, M., 1913, Experimentelle Untersuchungen über die Beziehungen der Grosshirnrinde und der Netzhaut zu den primären optischen Zentren, besonders zum Corpus geniculatum externum, *Arb. Hirnanat. Inst. Zurich* **7**:255–362.

Minkowski, M., 1920, Über den Verlauf, die Endigung und die zentrale Repräsentation von gekreuzten und ungekreuzten Sehnervenfasern bei einigen Säugetieren und beim Menschen, *Schweiz. Arch. Neurol. Psychiatr.* **6**:201–252; **7**:268–303.

Minkowski, M., 1922, Sur les conditions anatomiques de la vision binoculaire dans les voies optiques centrales, *Encéphale* **17**:65–96.

Minkowski, M., 1923–1924, Étude sur les connexions anatomiques des circonvolutions rolandiques, parietales et frontales, *Schweiz. Arch. Neurol. Psychiatr.*, 1923 **12**:71–104, 227–268; 1924 **14**:255–278, **15**:97–132.

Mishkin, M., 1966, Visual mechanisms beyond the striate cortex, in: *Frontiers in Physiological Psychology* (R. W. Russel, ed.), Academic Press, New York, pp. 93–119.

Mishkin, M., 1972, Cortical visual areas and their interactions, in: *Brain and Human Behavior* (H. G. Karczmas and J. C. Eccles, eds.), Springer, Berlin, pp. 187–208.

Mishkin, M., 1982, A memory system in the monkey, *Philos. Trans. R. Soc. London Ser. B* **298**:85–95.

Mitzdorf, V., and Singer, W., 1977, Laminar segregation of afferents to lateral geniculate nucleus of the cat: An analysis of current source density, *J. Neurophysiol.* **40**:1227–1244.

Miura, R., 1933, Über die Differenzierung der Grundbestandteile im Zwischenhirn des Kaninchens, *Anat. Anz.* **77**:1–65.

Mizuno, N., 1970, Projection fibers from the main sensory trigeminal nucleus and the supratrigeminal region, *J. Comp. Neurol.* **139**:457–472.

Mizuno, N., Nakano, K., Imaizumi, M., and Okamoto, M., 1967, The lateral cervical nucleus of the Japanese monkey (*Macaca fuscata*), *J. Comp. Neurol.* **129**:375–384.

Mizuno, N., Konishi, A., Sato, M., Kawaguchi, S., Yamamoto, T., Kawamura, S., and Yamawaki, M., 1975, Thalamic afferents to the rostral portions of the middle suprasylvian gyrus in the cat, *Exp. Neurol.* **48**:79–87.

Modianos, D. T., Hitt, J. C., and Flexman, J., 1974, Habenular lesions produce decrements in feminine, but not masculine, sexual behavior in rats, *Behav. Biol.* **10**:75–87.

Molinari, M., Minciacchi, D., Bentivoglio, M., and Macchi, G., 1985, Efferent fibers from the motor cortex terminate bilaterally in the thalamus of rats and cats, *Exp. Brain Res.* **57**:305–312.

Molennar, G. J., and Fizaan-Oosteen, J. L. F. P., 1980, Ascending projections from the lateral descending and common sensory trigeminal nuclei in python, *J. Comp. Neurol.* **189**:555–572.

Molliver, M. E., Kostovic, I., and Van der Loos, H., 1973, The development of synapses in cerebral cortex of the human fetus, *Brain Res.* **50**:403–407.

Molotchnikoff, S., and Lessard, I., 1980, The conduction velocity and central projections of retinofugal fibers in the rabbit, *Brain Res. Bull.* **5**:687–692.

Molotchnikoff, S., Richard, D., and Lachapelle, P., 1980, Influence of the visual cortex upon receptive field organization of lateral geniculate cells in rabbits, *Brain Res.* **193**:383–399.

Mondino Da Luzzi (fl. 1315), Anathomia overo dissectione del corpo humano, in: *Fasciculo di medicina* (Petrus De Montagnana, ed.), Venice, 1493; English translation by C. Singer in the *Fasciculo di medicina*, Lier, Florence, 1925.

Montero, V. M., and Guillery, R. W., 1968, Degeneration in the dorsal lateral geniculate nucleus of the rat followng interruption of the retinal or cortical connections, *J. Comp. Neurol.* **134**:211–241.

Montero, V. M., and Scott, G. L., 1981, Synaptic terminals in the dorsal lateral geniculate nucleus from neurons of the thalamic reticular nucleus: A light and electron microscopic autoradiographic study, *Neuroscience* **6**:2561–2577.

Montero, V. M., Brugge, J. F., and Beitel, R. E., 1968, Relation of the visual field to the lateral geniculate body of the albino rat, *J. Neurophysiol.* **31**:221–236.

Montero, V. M., Guillery, R. W., and Woolsey, C. N., 1977, Retinotopic organization within the thalamic reticular nucleus demonstrated by a double label autoradiographic technique, *Brain Res.* **138**:407–422.

Mooney, R. D., Dubin, M. W., and Rusoff, A. C., 1979, Interneuron circuits in the lateral geniculate nucleus of monocularly deprived cats, *J. Comp. Neurol.* **187**:533–543.

Moore, C. L., Kalil, R., and Richards, W., 1976, Development of myelination in optic tract of the cat, *J. Comp. Neurol.* **165**:125–136.

Moore, D. T., and Murphy, E. H., 1976, Differential effects of two visual cortical lesions in the rabbit, *Exp. Neurol.* **53**:21–30.

Moore, J. K., Karapas, F., and Moore, R. Y., 1977, Projections of the inferior colliculus in insectivores and primates, *Brain Behav. Evol.* **14**:301–327.

Moore, R. Y., 1978, Central neural control of circadian rhythms, *Front. Neuroendocrinol* **5**:185–206.

Moore, R. Y., and Goldberg, J. M., 1960, Projections of the inferior colliculus in the monkey, *Exp. Neurol.* **14**:429–438.

Moore, R. Y., and Goldberg, J. M., 1963, Ascending projections of the inferior colliculus in the cat, *J. Comp. Neurol.* **121**:109–135.

Morest, D. K., 1961, Connexions of the dorsal tegmental nucleus in the rat and rabbit, *J. Anat.* **95**:229–246.

Morest, D. K., 1964, The neuronal arcchitecture of the medial geniculate body of the cat, *J. Anat.* **98**:611–630.

Morest, D. K., 1965a, The laminar structure of the medial geniculate body of the cat, *J. Anat.* **99**:143–160.

Morest, D. K., 1965b, The lateral tegmental system of the midbrain and the medial geniculate body: Study with Golgi and Nauta methods in cat, *J. Anat.* **99**:611–634.

Morest, D. K., 1969, The growth of dendrites in the mammalian brain, *Z. Anat. Entwicklungsgesch.* **128**:290–317.

Morest, D. K., 1975, Synaptic relationships of Golgi type II cells in the medial geniculate body of the cat, *J. Comp. Neurol.* **162**:157–193.

Morin, F., and Catalano, J. V., 1955, Central connections of a cervical nucleus (nucleus cervicalis lateralis of the cat), *J. Comp. Neurol.* **103**:17–32.

Morin, F., Schwartz, H. G., and O'Leary, J. L., 1951, Experimental study of the spinothalamic and related tracts, *Acta Psychiatr. Neurol.* **26**:371–396.

Morison, R. S., and Dempsey, E. W., 1942, A study of thalamo-cortical relations, *Am. J. Physiol.* **135**:281–292.

Morison, R. S., Finley, K. H., and Lothrop, G. N., 1943a, Spontaneous electrical activity of the thalamus and other forebrain structures, *J. Neurophysiol.* **6**:243–254.

Morison, R. S., Finley, K. H., and Lothrop, G. N., 1943b, Influence of basal forebrain areas on the electrocorticogram, *Am. J. Physiol.* **139**:410–416.

Morrison, A. R., and Pompeiano, O., 1966, Vestibular influences during sleep. IV. Functional relation between vestibular nuclei and lateral geniculate nucleus during desynchronized sleep, *Arch. Ital. Biol.* **104**:425–458.

Morrison, A. R., Hand, P. J., and O'Donoghue, J., 1970, Contrasting projections from the posterior and ventrobasal thalamic nuclear complexes to the anterior ectosylvian gyrus of the cat, *Brain Res.* **21**:115–121.

Morrison, J. S., and Williams, R. T., 1968, *Greek Oared Ships 900–322 B.C.*, Cambridge University Press, London.

Moruzzi, G., 1972, The sleep–waking cycle, *Ergeb. Physiol. Biol. Chem. Exp. Pharmakol.* **64**:1–165.

Moruzzi, G., and Magoun, H. W., 1949, Brain stem reticular formation and activation of the EEG, *Electroencephalogr. Clin. Neurophysiol.* **1**:455–473.

Moskowitz, N., and Noback, C. R., 1962, The human lateral geniculate body of normal development and congenital unilateral anophthalamia, *J. Neuropathol. Exp. Neurol.* **21**:377.

Mott, F. W., 1892, Ascending degenerations resulting from lesions of the spinal cord in monkeys, *Brain* **15**:215–229.

Mott, F. W., 1895, Experimental enquiry upon the afferent tracts of the central nervous system of the monkey, *Brain* **18**:1–20.

Motta, M., Franschini, F., Giuliani, G., and Martini, L., 1968, The central nervous system, estrogen and puberty, *Endocrinology* **83**:1101–1107.

Mountcastle, V. B., 1957, Modality and topographic properties of single neurons of cat's somatic sensory cortex, *J. Neurophysiol.* **20**:408–434.

Mountcastle, V. B., 1982, Neural mechanisms in somesthesia, in: *Medical Physiology* (V. B. Mountcastle, ed.), 14th ed., Volume 1, Mosby, St. Louis, pp. 307–347.

Mountcastle, V. B., and Edelman, G., 1977, *The Mindful Brain: Cortical Organization and the Group-Selective Theory of Higher Brain Function*, MIT Press, Cambridge, Mass.

Mountcastle, V. B., and Henneman, E., 1949, Pattern of tactile representation in thalamus of cat, *J. Neurophysiol* **12**:85–100.

Mountcastle, V. [B.], and Henneman, E., 1952, The representation of tactile sensibility in the thalamus of the monkey, *J. Comp. Neurol.* **97**:409–440.

Mountcastle, V. B., and Powell, T. P. S., 1959, Neural mechanisms subserving cutaneous sensibility, with special reference to the role of afferent inhibition in sensory perception and discrimination, *Bull. Johns Hopkins Hosp.* **105**:201–232.

Mountcastle, V. B., Poggio, G. F., and Werner, G., 1963, The relation of thalamic cell response to peripheral stimuli varied over an intensive continuum, *J. Neurophysiol.* **26**:807–834.

Mountcastle, V. B., Talbot, W. H., Sakata, H., and Hyvärinen, J., 1969, Cortical neuronal mechanisms in flutter-vibration studied in unanesthetized monkeys: Neuronal periodicity and frequency discrimination, *J. Neurophysiol.* **32**:452–484.

Mountcastle, V. B., Lynch, J. C., Georgopoulos, A., Sakata, H., and Acuna, C., 1975, Posterior parietal association cortex of the monkey: Command functions for operations within extrapersonal space, *J. Neurophysiol.* **38**:871–908.

Mountcastle, V. B., Andersen, R. A., and Motter, B. C., 1981, The influence of attentive fixation upon the excitability of the light-sensitive neurons of the posterior parietal cortex, *J. Neurosci.* **1**:1218–1235.

Mountcastle, V. B., Motter, B. C., Steinmetz, M. A., and Duffy, C.J., 1983, Looking and seeing: The visual functions of the parietal lobe, in: *Dynamic Aspects of Neocortical Functions* (G. M. Edelman, W. E. Gall, and W. M. Cowan, eds.), Wiley, New York, pp. 159–193.

Movshon, J. A., and van Sluyters, R. C., 1981, Visual neural development, *Annu. Rev. Psychol.* **32**:477–522.

Mroz, E. A., Brownstein, M. J., and Leeman, S. E., 1976, Evidence for substance P in the habenulo-interpeduncular tract, *Brain Res.* **113**:597–599.

Muakkassa, K. F., and Strick, P. L., 1979, Frontal lobe inputs to primate motor cortex: Evidence for four somatotopically organized "premotor" areas, *Brain Res.* **177**:176–182.

Mücke, L., Norita, M., Benedek, G., and Creutzfeldt, O., 1982, Physiologic and anatomic investigation of a visual cortical area situated in the ventral bank of the anterior ectosylvian sulcus of the cat, *Exp. Brain Res.* **46**:1–11.

Mudry, K. M., and Capranica, R. R., 1978, Electrophysiological evidence for auditory responsive areas in the diencephalon and telencephalon of the bullfrog, *Rana catesbeiana, Neurosci. Abstr.* **4**:101.

Mukhametov, L. M., Rizzolatti, G., and Seitun, A., 1970a, An analysis of the spontaneous activity of lateral geniculate neurons and of optic tract fibres in free moving cats, *Arch. Ital. Biol.* **108**:325–347.

Mukhametov, L. M., Rizzolatti, G., and Tradardi, V., 1970b, Spontaneous activity of neurons of nucleus reticularis thalami in freely behaving cats, *J. Physiol. (London)* **210**:651–667.

Müller, F. W. P., 1921, Die Zellgruppen im Corpus geniculatum mediale des Menschen, *Monatsschr. Psychiatr. Neurol.* **49**:251–271.

Münzer, E., and Wiener, H., 1902, Das Zwischen- und Mittelhirn des Kaninchens und die Beziehungen dieser Teile zum übrigen Centralnervensystem, mit besonderer Berücksichtigung der Pyramidenbahn und Schleife, *Monatsschr. Psychiatr. Neurol.* **12:**241–279.

Murray, M., 1966, Degeneration of some intralaminar thalamic nuclei after cortical removals in the cat, *J. Comp. Neurol.* **127:**341–367.

Murray, M., Zimmer, J., and Raisman, G., 1979, Quantitative electron microscopic evidence for reinnervation in the adult rat interpeduncular nucleus after lesions of the fasciculus retroflexus, *J. Comp. Neurol.* **187:**447–468.

Mussen, A. T., 1923, A cytoarchitectural atlas of the brain stem of the *Macacus rhesus, J. Psychol. Neurol.* **29:**451–518.

Mussen, A. T., 1927, Experimental investigations on the cerebellum, *Brain* **50:**313–349.

Nagy, J. I., Carter, D. A., Lehman, J., and Fibiger, H. C., 1978, Evidence for a GABA-containing projection from the entopeduncular nucleus to the lateral habenula in the rat, *Brain Res.* **145:**360–364.

Nakai, Y., and Takaori, S., 1974, Influence of norepinephrine-containing neurons derived from the locus coeruleus on lateral geniculate neuronal activities of cat, *Brain Res.* **71:**47–60.

Namba, M., 1958, Über die feineren Strukturen des medio-dorsalen Supranucleus und der Lamella medialis des Thalamus beim Menschen, *J. Hirnforsch.* **4:**1–42.

Narkiewicz, O., and Brutkowski, S., 1967, The organization of projections from the thalamic mediodorsal nucleus to the prefrontal cortex of the dog, *J. Comp. Neurol.* **129:**361–374.

Nauta, H. J. W., 1974, Evidence of a pallidohabenular pathway in the cat, *J. Comp. Neurol.* **156:**19–28.

Nauta, H. J. W., 1979, Projections of the pallidal complex: An autoradiographic study in the cat, *Neuroscience* **4:**1853–1874.

Nauta, H. J. W., Pritz, M. B., and Lasek, R. J., 1974, Afferents to the rat caudatoputamen studied with horseradish peroxidase: An evaluation of a retrograde neuroanatomical research method, *Brain Res.* **67:**219–238.

Nauta, W. J. H., 1956, An experimental study of the fornix system in the rat, *J. Comp. Neurol.* **104:**247–272.

Nauta, W. J. H., 1958, Hippocampal projections and related neural pathways to the midbrain in the cat, *Brain* **81:**319–340.

Nauta, W. J. H., 1961, Fibre degeneration following lesions of the amygdaloid complex in the monkey, *J. Anat.* **95:**515–531.

Nauta, W. J. H., and Bucher, V. M., 1954, Efferent connections of the striate cortex in the albino rat, *J. Comp. Neurol.* **100:**257–285.

Nauta, W. J. H., and Karten, H. J., 1970, A general profile of the vertebrate brain with sidelights on the ancestry of cerebral cortex, in: *The Neurosciences: Second Study Program* (F. O. Schmitt, ed.), The Rockefeller University Press, New York, pp. 7–26.

Nauta, W. J. H., and Kuypers, H. G. J. M., 1958, Some ascending pathways in the brain stem reticular formation, in: *Reticular Formation of the Brain* (H. H. Jasper, L. D. Proctor, R. S. Knighton, W. C. Noshay, and R. T. Costello, eds.), Little, Brown, Boston, pp. 3–30.

Nauta, W. J. H., and Mehler, W. R., 1961, Some efferent connections of the lentiform nucleus in monkey and cat, *Anat. Rec.* **139:**260.

Nauta, W. J. H. and Mehler, W. R., 1966, Projections of the lentiform nucleus in the monkey, *Brain Res.* **1:**3–42.

Nauta, W. J. H. and Whitlock, D. G., 1954, An anatomical analysis of the non-specific thalamic projection system, in: *Brain Mechanisms and Consciousness* (J. F. Delafresnaye, ed.), Thomas, Springfield, Ill., pp. 81–116.

Neary, T. J., and Northcutt, R. G., 1983, Nuclear organization of the bullfrog diencephalon, *J. Comp. Neurol.* **213:**262–278.

Neary, T. J., and Wilczynski, W., 1977a, Autoradiographic demonstration of hypothalamic efferents in the bullfrog, *Rana catesbeiana, Anat. Rec.* **187:**665.

Neary, T. J., and Wilczynski, W., 1977b, Ascending thalamic projections from the obex region in ranid frogs, *Brain Res.* **138:**529–533.

Neary, T. J., and Wilczynski, W., 1980, Descending inputs to the optic tectum in ranid frogs, *Neurosci. Abstr.* **6:**629.

Negishi, L., Lu, E. S., and Verzeano, M., 1962, Neuronal activity in the lateral geniculate body and the nucleus reticularis of the thalamus, *Vision Res.* **1**:343–353.

Nelson, P. G., and Erulkar, S. D., 1963, Synaptic mechanisms of excitation and inhibition in the central auditory pathway, *J. Neurophysiol.* **26**:908–923.

Nelson, R. J., Sur, M., and Kaas, J. H., 1979, The organization of the second somatosensory area (SmII) of the grey squirrel, *J. Comp. Neurol.* **184**:473–490.

Nelson, R. J., Sur, M., Felleman, D. J., and Kaas, J. H., 1980, Representations of the body surface in postcentral parietal cortex of *Macaca fascicularis*, *J. Comp. Neurol.* **192**:611–643.

Ng, A. Y. K., and Stone, J., 1982, The optic nerve of the cat: Appearance and loss of axons during normal development, *Dev. Brain Res.* **5**:263–271.

Nichterlein, O. E., and Goldby, F., 1944, An experimental study of optic connections in the sheep, *J. Anat.* **78**:59–67.

Nicolesco, I., and Nicolesco, M., 1929, Quelques données sur les centres végétatifs de la région infundibulo-tubérienne et de la frontière diencéphalo-télencéphalique, *Rev. Neurol.* **2**:289–317.

Nieuwenhuys, R., 1962, Trends in the evolution of the actinopterygian forebrain, *J. Morphol.* **111**:69–88.

Nieuwenhuys, R., 1982, An overview of the organization of the brain of actinopterygian fishes, *Am. Zool.* **22**:287–310.

Niimi, K., 1978, Cortical projections of the anterior thalamic nuclei in the cat, *Exp. Brain Res.* **31**:403–416.

Niimi, K., and Kuwahara, E., 1973, The dorsal thalamus of the cat and comparison with monkey and man, *J. Hirnforsch.* **14**:303–325.

Niimi, K., and Naito, F., 1974, Cortical projections of the medial geniculate body in the cat, *Exp. Brain Res.* **19**:326–342.

Niimi, K., and Sprague, J. M., 1970, Thalamo-cortical organization of the visual system in the cat, *J. Comp. Neurol.* **138**:219–249.

Niimi, K., Katayama, K., Kanaseki, T., and Morimoto, K., 1960, Studies on the derivation of the centre médian nucleus of Luys, *Tokushima J. Exp. Med.* **7**:261–268.

Niimi, I., Harada, I., Kusaka, Y., and Kishi, S., 1961, The ontogenetic development of the diencephalon of the mouse, *Tokushima J. Exp. Med.* **8**:203–238.

Niimi, K., Kanaseki, T., and Takimoto, T., 1963, The comparative anatomy of the ventral nucleus of the lateral geniculate body in mammals, *J. Comp. Neurol.* **121**:313–323.

Niimi, K., Niimi, M., and Okada, Y., 1978, Thalamic afferents to the limbic cortex in the cat studied with the method of retrograde axonal transport of horseradish peroxidase, *Brain Res.* **145**:225–238.

Niimi, K., Matsuoka, H., Aisaka, T., and Okada, Y., 1981a, Thalamic afferents to the prefrontal cortex in the cat traced with horseradish peroxidase, *J. Hirnforsch.* **22**:221–241.

Niimi, K., Matsuoka, H., Yamazaki, Y., and Matsumoto, H., 1981b, Thalamic afferents to the visual cortex in the cat studied by retrograde axonal transport of horseradish peroxidase, *Brain Behav. Evol.* **18**:114–139.

Nissl, F., 1885, Über die Untersuchungsmethoden der Grosshirnrinde, *Ber. Vers. Dtsch. Naturforsch. Arzte (Strasb.)* **58**:506–507.

Nissl, F., 1889, Die Kerne des Thalamus beim Kaninchen, *Neurol. Zentralbl.* **8**:549–550.

Nissl, F., 1892, Über die Veranderungen der Ganglienzellen am Facialiskern des Kaninchens auch Ausreissung der Nerven, *Allg. Z. Psychiatr.* **48**:197–198.

Nissl, F., 1903, *Die Neuronlehre und ihre Anhanger: Ein Beitrag zur Lösung des Problems der Beziehungen zwischen Nervenzelle, Faser und Grau*, Fischer, Jena.

Nissl, F., 1913, Die Grosshirnanteile des Kaninchens, *Arch Psychiatr. Nervenkr.* **52**:867–953.

Norden, J. J., 1979, Some aspects of the organization of the lateral geniculate nucleus in *Galago senegalensis* revealed by using horseradish peroxidase to label relay neurons, *Brain Res.* **174**:193–206.

Norden, J. J., and Kaas, J. H., 1978, The identification of relay neurons in the dorsal lateral geniculate nucleus of monkeys using horseradish peroxidase, *J. Comp. Neurol.* **182**:707–726.

Norgren, R., 1976, Taste pathways to hypothalamus and amygdala, *J. Comp. Neurol.* **166**:17–29.

Norgren, R., and Leonard, C. M., 1971, Taste pathways in rat brainstem, *Science* **173**:1136–1139.

Norgren, R., and Leonard, C. M., 1973, Ascending central gustatory pathways, *J. Comp. Neurol.* **150**:217–237.

Norgren, R., and Wolf, G., 1975, Projections of thalamic gustatory and lingual areas in the rat, *Brain Res.* **92**:123–129.

Norita, M., and Creutzfeldt, O. D., 1982, An HRP and ^{3}H-apo-HRP study of the thalamic projections to visual cortical areas in the cat, *Acta Biol. Acad. Sci. Hung.* **33**:269–275.

Norman, J. L., Pettigrew, J. D. and Daniels, J. D., 1977, Early development of X-cells in kitten LGN, *Science* **198**:202–204.

Norrsell, U., 1966, An evoked potential study of spinal pathways projecting to the cerebral somatosensory areas in the dog, *Exp. Brain Res.* **2**:261–268.

Norrsell, U., and Wolpaw, E. R., 1966, An evoked potential study of different pathways from hindlimb to somatosensory areas in the cat, *Acta Physiol. Scand.* **66**:19–33.

Northcutt, R. G., 1977, Retinofugal projections in the lepidosirenid lungfishes, *J. Comp. Neurol.* **174**:553–574.

Northcutt, R. G., 1979, Retinofugal pathways in fetal and adult spiny dogfish, *Squalus acanthias*, *Brain Res.* **162**:219–230.

Northcutt, R. G., 1981, Evoluton of the telencephalon in nonmammals, *Annu. Rev. Neurosci.* **4**:301–350.

Northcutt, R. G., and Braford, M. R., Jr., 1980, New observations on the organization and evolution of the telencephalon of actinopterygian fishes, in: *Comparative Neurology of the Telencephalon* (S. O. E. Ebbesson, ed.), Plenum Press, New York, pp. 41–98.

Northcutt, R. G., and Butler, A. B., 1976, Retinofugal pathways in the long nose gar *(Lepisosteus osseus Linnaeus)*, *J. Comp. Neurol.* **166**:1–16.

Northcutt, R. G., and Pritz, M. B., 1978, A spinothalamic pathway to the dorsal ventricular ridge in the spectacled caiman, *Caiman crocodilus*, *Anat. Rec.* **190**:618–619.

Northcutt, R. G., and Royce, G. J., 1975, Olfactory bulb projections in the bullfrog, *Rana catesbeiana*, *J. Morphol.* **145**:251–268.

Northcutt, R. G., and Wathey, J. C., 1979, Some connections of the skate dorsal and medial pallia, *Neurosci. Abstr.* **5**:145.

Norton, T. T., and Casagrande, V. A., 1982, Laminar organization of receptive-field properties in lateral geniculate nucleus of bush baby (*Galago crassicaudatus*), *J. Neurophysiol.* **47**:715–741.

Norton, T. T., Casagrande, V. A., and Sherman, S. M., 1977, Loss of Y-cells in the lateral geniculate nucleus of monocularly deprived tree shrews, *Science* **197**:784–786.

Nothnagel, C. W. H., 1873, Experimentelle Untersuchungen über die Funktion des Gehirns, *Virchows Arch. A* **57**:184–214; **58**:420–436.

Nottebohm, F., Stokes, T. M., and Leonard, C. M., 1976, Central control of song in the canary, *Serinus canarius*, *J. Comp. Neurol.* **165**:457–486.

Nottebohm, F., Kelley, D. B., and Paton, J. A., 1982, Connections of vocal control nuclei in the canary telencephalon, *J. Comp. Neurol.* **207**:344–357.

Oakley, B., and Pfaffmann, C., 1962, Electrophysiologically monitored lesions in the gustatory thalamic relay of the albino rat, *J. Comp. Physiol. Psychol.* **55**:155–160.

Oberdorfer, M. D., Guillery, R. W., and Murphy, E. H., 1977, Geniculo-cortical pathways in mink that have abnormal retino-geniculate pathways, *Anat. Rec.* **187**:669–670.

Ödkvist, L. M., Schwartz, D. W. F., Fredrickson, J. M., and Hassler, R., 1974, Projection of the vestibular nerve to the area 3a arm field in the squirrel monkey (*Saimiri sciureus*), *Exp. Brain Res.* **21**:97–105.

Oertel, W. H., Graybiel, A. M., Mugnaini, E., Elde, R. P., Schmechel, D. E., and Kopin, I. J., 1983, Coexistence of glutamic acid decarboxylase- and somatostatin-like immunoreactivity in neurons of the feline nucleus reticularis thalami, *J. Neurosci.* **3**:1322–1332.

Ogren, M. P., and Hendrickson, A. E., 1976, Pathways between striate cortex and subcortical regions in *Macaca mulatta* and *Saimiri sciureus:* Evidence for a reciprocal pulvinar connection, *Exp. Neurol.* **53**:780–800.

Ogren, M. P., and Hendrickson, A. E., 1977, The distribution of pulvinar terminals in visual areas 17 and 18 of the monkey, *Brain Res.* **137**:343–350.

Ogren, M. P., and Hendrickson, A. E., 1979, The structural organization of the inferior and lateral subdivisions of the *Macaca* monkey pulvinar, *J. Comp. Neurol.* **188**:147–177.

Ogren, M. P., and Rakic, P., 1981, The prenatal development of the pulvinar in the monkey: ^{3}H-thymidine autoradiographic and morphometric analyses, *Anat. Embryol.* **162**:1–20.

Ohara, P. T., and Lieberman, A. R., 1981, Thalamic reticular nucleus; Anatomical evidence that cortico-reticular axons establish monosynaptic contact with reticulo-geniculate projection cells, *Brain Res.* **207**:153–156.

Ohara, P. T., Sefton, A. J., and Lieberman, A. R., 1980, Mode of termination of afferents from the thalamic reticular nucleus in the dorsal lateral geniculate nucleus of the rat, *Brain Res.* **197**:503–506.

Ohara, P. T., Lieberman, A. R., Hunt, S. P., and Wu, J.-Y., 1983, Neural elements containing glutamic acid decarboxylase (GAD) in the dorsal lateral geniculate nucleus of the rat: Immunohistochemical studies by light and electron microscopy, *Neuroscience* **8:**189–212.

Ojemann, G. A., 1977, Asymmetric function of the thalamus in man, *Ann. N.Y. Acad. Sci.* **299:**380–396.

Ojemann, G., 1981, Interrelationship in the localization of language, memory and motor mechanisms in human cortex and thalamus, in: *New Perspectives in Cerebral Localization* (R. Thomson, ed.), Raven Press, New York, pp. 157–175.

Ojemann, G. A., 1983, Brain organization for language from the perspective of electrical stimulation mapping, *Behav. Brain Sci.* **6:**189–230.

Ojemann, G., and Fedio, P., 1968, Effect of stimulation of the human thalamus and parietal and temporal white matter on short-term memory, *J. Neurosurg.* **29:**51–59.

Ojemann, G., Fedio, P., and Van Buren, J., 1968, Anomia from pulvinar and subcortical parietal stimulation, *Brain* **91:**99–116.

Olavarria, J., 1979, A horseradish peroxidase study of the projections from the lateroposterior nucleus to three lateral peristriate areas in the rat, *Brain Res.* **173:**137–141.

O'Leary, J. L., 1940, A structural analysis of the lateral geniculate nucleus of the cat, *J. Comp. Neurol.* **73:**405–430.

Oliver, D. L., and Hall, W. C., 1975, Subdivisions of the medial geniculate body in the tree shrew (*Tupaia glis*), *Brain Res.* **86:**217–227.

Oliver, D. L., and Hall, W. C., 1978a, The medial geniculate body of the tree shrew, *Tupaia glis*. I. Cytoarchitecture and midbrain connections, *J. Comp. Neurol.* **182:**423–458.

Oliver, D. L., and Hall, W. C., 1978b, The medial geniculate body of the tree shrew *Tupaia glis*. II. Connections with the neocortex, *J. Comp. Neurol.* **182:**459–494.

Olivier, A., Parent, A., and Poirier, L. J., 1970, Identification of the thalamic nuclei on the basis of their cholinesterase content in the monkey, *J. Anat.* **106:**37–50.

Olson, C. R., and Graybiel, A. M., 1981, A visual area in the anterior ectosylvian sulcus of the cat, *Neurosci. Abstr.* **7:**831.

Olszewski, J., 1952, *The Thalamus of the Macaca mulatta: An Atlas for Use with the Stereotaxic Instrument*, Karger, Basel.

Orban, G., Vandenbussche, E., and Callens, M., 1972, Electrophysiological evidence for the existence of connections between the brain stem oculomotor areas and the visual system in the cat, *Brain Res.* **41:**225–229.

Orem, J., Schlag-Rey, M., and Schlag, J., 1973, Unilateral visual neglect and thalamic intralaminar lesions in the cat, *Exp. Neurol.* **40:**784–797.

Osen, K. K., 1972, Projection of the cochlear nuclei on the inferior colliculus in the cat, *J. Comp. Neurol.* **144:**355–372.

Oswaldo-Cruz, E., and Rocha-Miranda, C. E., 1967, The diencephalon of the opossum in stereotaxic coordinates. I. The epithalamus and dorsal thalamus, *J. Comp. Neurol.* **129:**1–37.

Ottersen, O. P. and Ben-Ari, Y., 1979, Afferent connections to the amygdaloid complex of the rat and cat. I. Projections from the thalamus, *J. Comp. Neurol.* **187:**401–424.

Overbosch, J. F. A., 1927, Experimenteel-anatomische onderzoekingen over de projectie der retine in het centrale zeruwstelsel, Dissertation, H. J. Paris, Amsterdam.

Packer, A. D., 1941, An experimental investigation of the visual system in the phalanger, *Trichosurus vulpecula*, *J. Anat.* **75:**309–329.

Padmos, P., and Norren, D. V., 1975, Cone systems interaction in single neurons of the lateral geniculate nucleus of the macaque, *Vision Res.* **15:**617–619.

Palacios, J. M., Wamsley, J. K., and Kuhar, M. J., 1981, High affinity GABA receptors—autoradiographic localization, *Brain Res.* **222:**285–308.

Palkovits, M., and Jacobowitz, D. M., 1974, Topographic atlas of catecholamine and acetylcholinesterase-containing neurons in the rat brain, *J. Comp. Neurol.* **157:**29–42.

Palmer, L. A., Rosenquist, A. C., and Tusa, R., 1975, Visual receptive fields in the lam LGNd, MIN and PN of the cat, *Neurosci. Abstr.* **1:**54.

Palmer, L. A., Rosenquist, A. C., and Tusa, R. J., 1978, The retinotopic organization of lateral suprasylvian visual areas in the cat, *J. Comp. Neurol.* **177:**237–256.

Papez, J. W., 1932, The thalamic nuclei of the nine-banded armadillo (*Tatusia novemcincta*), *J. Comp. Neurol.* **56:**49–103.

Papez, J. W., 1935, Thalamus of turtles and thalamic evolution, *J. Comp. Neurol.* **61:**433–475.

Papez, J. W., 1938, Thalamic connections in a hemidecorticate dog, *J. Comp. Neurol.* **69:**103–120.

Papez, J. W., 1942, A summary of fiber connections of the basal ganglia with each other and with other portions of the brain, in: *The Diseases of the Basal Ganglia*, Volume 21, Williams & Wilkins, Baltimore, pp. 21–68.

Papez, J. W., and Rundles, W., 1937, The dorsal trigeminal tract and the centre median nucleus of Luys, *J. Nerv. Ment. Dis.* **85:**509–519.

Pappas, C. L., and Strick, P. L., 1981, Anatomical demonstration of multiple representation in the forelimb region of the cat motor cortex, *J. Comp. Neurol.* **200:**491–500.

Parent, A., 1976, Striatal afferent connections in the turtle *(Chrysemys picta)* as revealed by retrograde axonal transport of horseradish peroxidase, *Brain Res.* **108:**25–36.

Parent, A., and Butcher, L. L., 1976, Organization and morphologies of acetylcholinesterase-containing neurons in the thalamus and hypothalamus of the rat, *J. Comp. Neurol.* **170:** 205–226.

Parent, A., and DeBellefeuille, L., 1982, Organization of efferent projections from the internal segment of globus pallidus in primate as revealed by fluorescence retrograde labeling method, *Brain Res.* **245:**201–213.

Parent, A., and DeBellefeuille, L., 1983, The pallidointralaminar and pallidonigral projections in primate as studied by retrograde double-labeling method, *Brain Res.* **278:**11–28.

Parent, A., Mackey, A., Smith, Y., and Boucher, R., 1983, The output organization of the substantia nigra in primate as revealed by a retrograde double labeling method, *Brain Res. Bull.* **10:**529–537.

Parnavelas, J. G., Bradford, R., Mounty, E. J., and Lieberman, A. R., 1977a, Postnatal growth of neuronal perikarya in the dorsal lateral geniculate nucleus of the rat, *Neurosci. Lett.* **5:**33–37.

Parnavelas, J. G., Mounty, E. J., Bradford, R., and Lieberman, A. R., 1977b, The postnatal development of neurons in the dorsal lateral geniculate nucleus of the rat: A Golgi study, *J. Comp. Neurol.* **171:**481–500.

Partlow, G. D., Colonnier, M., and Szabo, J., 1977, Thalamic projections of the superior colliculus in the rhesus monkey, *Macaca mulatta:* A light and electron microscopic study, *J. Comp. Neurol.* **171:**285–318.

Pasantes-Morales, H., Salceda, R., and Lopez-Colombe, A. M., 1975, The role of taurine in retina: Factors affecting its release, in: *Taurine* (R. Huxtable and A. Barbeau, eds.), Raven Press, New York, pp. 191–200.

Pasik, P., Pasik, T., Hámori, J., and Szentágothai, J., 1973, Golgi type II interneurons in the neuronal circuit of the monkey lateral geniculate nucleus, *Exp. Brain Res.* **17:**18–34.

Pasik, P., Pasik, T., and Hámori, J., 1976, Synapses between interneurons in the lateral geniculate nucleus of monkeys, *Exp. Brain Res.* **25:**1–13.

Patton, H. D., and Amassian, V. E., 1951, Thalamic relay of the splanchnic afferent fibers, *Am. J. Physiol.* **167:**815–816.

Paul, R. L., Merzenich, M., and Goodman, 1972, Representation of slowly adapting and rapidly adapting cutaneous mechano-receptors of the hand in Brodmann's areas 3 and 1 of *Macaca mulatta, Brain Res.* **36:**229–249.

Pearson, J. C., and Haines, D. E., 1980, Somatosensory thalamus of a prosimian primate *(Galago senegalensis),* II. An HRP and Golgi study of the ventral postero-lateral nucleus (VPL), *J. Comp. Neurol.* **190:**559–580.

Pearson, L. J., Sanderson, K.J., and Wells, R. T., 1976, Retinal projections in the ring-tailed possum *Pseudocheirus peregrinus, J. Comp. Neurol.* **170:**227–240.

Pearson, R. C. A., Brodal, P., and Powell, T. P. S., 1978, The projection of the thalamus upon the parietal lobe in the monkey, *Brain Res.* **144:**143–148.

Pearson, R. C. A., Gatter, K. C., Brodal, P., and Powell, T. P. S., 1983, The projection of the basal nucleus of Meynert upon the neocortex in the monkey, *Brain Res.* **259:**132–136.

Pecci-Saavedra, J., Vaccarezza, O. L., and Reader, T. A., 1968, Ultrastructure of cells and synapses in the parvocellular portion of the cebus monkey lateral geniculate nucleus, *Z. Zellforsch. Mikrosk. Anat.* **89:**462–477.

Pectel, C., Masserman, J. H., Schreiner, L., and Levitt, M., 1955, Differential effects of lesions in the mediodorsal nuclei of the thalamus on normal and neurotic behavior in the cat, *J. Nerv. Ment. Dis.* **121:**26–33.

Peichel, L., and Wässle, H., 1981, Morphological identification of on- and off-centre brisk transient (Y) cells in the cat retina, *Proc. R. Soc. London Ser. B* **212:**139–156.

Penfield, W., and Rasmussen, T., 1950, *The Cerebral Cortex of Man*, Macmillan Co., New York.

Penny, G. R., Itoh, K., and Diamond, I. T., 1982, Cells of different sizes in the ventral nuclei project to different layers of the somatic cortex in the cat, *Brain Res.* **242**:55–65.

Penny, G. R., Fitzpatrick, D., Schmechel, D., and Diamond, I. T., 1983, Glutamic acid decarboxylase immunoreactive neurons and horseradish peroxidase-labeled projection neurons in the ventral posterior nucleus of the cat and *Galago senegalensis*, *J. Neurosci.* **3**:1868–1887.

Pennyman, K. M., Lindsley, D. F., and Lindsley, D. B., 1980, Pulvinar neuron responses to spontaneous and trained eye movements and to light flashes in squirrel monkeys, *Electroencephalogr. Clin. Neurophysiol.* **49**:152–161.

Percheron, G., 1977, The thalamic territory of cerebellar afferents and lateral region of the thalamus of the macaque in stereotaxic ventricular coordinates, *J. Hirnforsch.* **18**:375–400.

Perl, E. R., and Whitlock, D. G., 1961, Somatic stimuli exciting spinothalamic projections to thalamic neurons in cat and monkey, *Exp. Neurol.* **3**:256–296.

Perry, J. G., and Thach, W. T., 1979, The somatotopy of the deep cerebellar nuclei as revealed by active movement of individual joints, *Neurosci. Abstr.* **5**:1285.

Perry, V. H., 1979, The ganglion cell layer of the retina of the rat: A Golgi study, *Proc. R. Soc. London Ser. B* **204**:363–375.

Pert, C. B., Aposhian, D., and Snyder, S. H., 1974, Phylogenetic distribution of opiate receptor binding, *Brain Res.* **75**:356–361.

Pert, C. B., Kuhar, M. J., and Snyder, S. H., 1976, Autoradiographic localization of opiate receptors in rat brain, *Proc. Natl. Acad. Sci. USA* **73**:3729–3733.

Peter, R. E., and Gill, V. E., 1975, A stereotaxic atlas and technique for forebrain nuclei of the goldfish, *Carassius auratus*, *J. Comp. Neurol.* **159**:69–102.

Peters, A., and Palay, S. L., 1966, The morphology of laminae *A* and *A1* of the dorsal nucleus of the lateral geniculate body of the cat, *J. Anat.* **100**:451–486.

Peters, A., Feldman, M., and Saldanha, J., 1976, The projection of the lateral geniculate nucleus on area 17 of the rat cerebral cortex. II. Terminations upon neuronal perikarya and dendritic shafts, *J. Neurocytol.* **5**:85–107.

Petras, J. M., 1965, Some efferent connections of the precentral and postcentral cortex with the basal ganglia, thalamus and subthalamus, *Trans. Am. Neurol. Assoc.* **90**:274–275.

Petras, J. M., 1969, Some efferent connections of the motor and somatosensory cortex of simian primates and felid, canid and procyonid carnivores, *Ann. N.Y. Acad. Sci.* **167**:469–505.

Pettigrew, J. D., 1972, The importance of early visual experience for neurons of the developing geniculo-striate system, *Invest. Ophthalmol.* **11**:386–393.

Pettigrew, J. D., 1979, Binocular visual processing in the owl's telencephalon, *Proc. R. Soc. London Ser. B* **204**:435–454.

Pettigrew, J. D., and Konishi, M., 1976, Neurons selective for orientation and binocular disparity in the visual Wulst of the barn owl (*Tyto alba*), *Science* **193**:675–678.

Phillis, J. W., 1970, *The Pharmacology of Synapses*, Pergamon Press, Elmsford, New York.

Phillis, J. W., Tebēcis, A. K., and York, D. H., 1967, A study of cholinoceptive cells in the lateral geniculate nucleus, *J. Physiol. (London)* **192**:695–713.

Pickard, G. E., and Turek, F. W., 1983, The suprachiasmatic nuclei: Two circadian clocks?, *Brain Res.* **268**:201–210.

Pickel, V. M., Segal, M., and Bloom, F. E., 1974, A radioautographic study of the efferent pathways of the nucleus locus coeruleus, *J. Comp. Neurol.* **155**:15–41.

Pines, J. L., 1927, Zur Architectonik des Thalamus opticus beim Halbaffen (*Lemur catta*), *J. Psychol. Neurol.* **33**:31–72.

Platt, C. J., Bullock, T. H., Czeh, G., Kovacevic, N., Konjevic, D. and Gojkovic, M., 1974, Comparison of electroreceptor, mechanoreceptor, and optic evoked potentials in the brain of some rays and sharks, *J. Comp. Physiol.* **95**:323–355.

Poggio, G. F., and Fischer, B., 1977, Binocular interaction and depth sensitivity in striate and prestriate cortex of behaving rhesus monkey, *J. Neurophysiol.* **40**:1392–1405.

Poggio, G. F., and Mountcastle, V. B., 1960, A study of the functional contributions of the lemniscal and spinothalamic systems to somatic sensibility: Central nervous mechanisms in pain, *Bull. Johns Hopkins Hosp.* **106**:266–316.

Poggio, G. F., and Mountcastle, V. B., 1963, The functional properties of ventrobasal thalamic neurons studied in unanesthetized monkeys, *J. Neurophysiol.* **26**:775–806.

Polley, E. H., and Guillery, R. W., 1980, An anomalous uncrossed retinal input to lamina A of the cat's dorsal lateral geniculate nucleus, *Neuroscience* **5:**1603–1608.

Pollin, B., and Albe-Fessard, D., 1979, Organization of somatic thalamus in monkeys with and without section of dorsal spinal tracts, *Brain Res.* **173:**431–449.

Pollin, B., and Rokyta, R., 1982, Somatotopic organization of nucleus reticularis thalami in chronic awake cats and monkeys, *Brain Res.* **250:**211–221.

Polyak, S., 1927, An experimental study of the association, callosal, and projection fibers of the cerebral cortex of the cat, *J. Comp. Neurol.* **44:**197–258.

Polyak, S., 1932, *The Main Afferent Fiber Systems of the Cerebral Cortex in Primates,* University of California Press, Berkeley.

Polyak, S., 1933, A contribution to the cerebral representation of the retina, *J. Comp. Neurol.* **57:** 541–617.

Polyak, S., 1957, *The Vertebrate Visual System* (H. Klüver, ed.), University of Chicago Press, Chicago.

Pomeranz, B., Wall, P. D., and Weber, W. V., 1968, Cord cells responding to fine myelinated afferents in viscera, muscle and skin, *J. Physiol. (London)* **199:**511–532.

Pontes, C., Reis, F. F., and Sousa-Pinto, A., 1975, The auditory cortical projections onto the medial geniculate body in the cat: An experimental anatomical study with silver and autoradiographic methods, *Brain Res.* **91:**43–63.

Porrino, L. J., Crane, A. M., and Goldman-Rakic, P. S., 1981, Direct and indirect pathways from the amygdala to the frontal lobe in rhesus monkeys, *J. Comp. Neurol.* **198:**121–136.

Porter, R., and Rack, P. M. H., 1976, Timing of the responses in the motor cortex of monkeys to an unexpected disturbance of finger position, *Brain Res.* **103:**201–213.

Poulos, D. A., and Benjamin, R. M., 1968, Response of thalamic neurons to thermal stimulation of the tongue, *J. Neurophysiol.* **31:**28–43.

Poulos, D. A., Burton, H., Molt, J. T., and Barron, K. D., 1979, Localization of specific thermoreceptors in spinal trigeminal nucleus of the cat, *Brain Res.* **165:**144–148.

Powell, T. P. S., 1952, Residual neurons in the human thalamus following hemidecortication, *Brain* **75:**571–583.

Powell, T. P. S., 1958, The organization and connexions of the hippocampal and intralaminar systems, *Recent Prog. Psychiatry* **3:**54–74.

Powell, T. P. S., and Cowan, W.M., 1954, The connexions of the midline and intralaminar nuclei of the thalamus of the rat, *J. Anat.* **88:**307–319.

Powell, T. P. S., and Cowan, W. M., 1956, A study of thalamo-striate relations in the monkey, *Brain* **79:**364–390.

Powell, T. P. S., and Cowan, W. M., 1961, The thalamic projection upon the telencephalon in the pigeon *(Columba livia), J. Anat.* **95:**78–109.

Powell, T. P. S., and Cowan, W. M., 1967, The interpretation of the degenerative changes in the intralaminar nuclei of the thalamus, *J. Neurol. Neurosurg. Psychiatry* **30:**140–153.

Powell, T. P. S., and Kruger, L., 1960, The thalamic projection upon the telencephalon in *Lacerta viridis, J. Anat.* **94:**528–542.

Powell, T. P. S., and Mountcastle, V. B., 1959, Some aspects of the functional organization of the cortex of the postcentral gyrus of the monkey: A correlation of findings obtained in a single unit analysis with cytoarchitecture, *Bull. Johns Hopkins Hosp.* **105:**133–162.

Powell, T. P. S., Cowan, W. M., and Raisman, G., 1965, The central olfactory connexions, *J. Anat.* **99:**791–813.

Powell, T. P. S., Guillery, R. W., and Cowan, W. M., 1957, A quantitative study of the fornix-mamillo-thalamic system, *J. Anat.* **91:**419–437.

Přecechtěl, A., 1925, Some notes upon the finer anatomy of the brain stem and basal ganglia of *Elephas indicus, Proc. Sect. Sci. K. Ned. Akad. Wet.* **28:**81–93.

Pribram, K. H., Chow, K. L., and Semmes, J., 1953, Limit and organization of the cortical projection from the medial thalamic nucleus in monkey, *J. Comp. Neurol.* **98:**433–448.

Price, D. D., Dubner, R., and Hu, J. W., 1976, Trigeminothalamic neurons in nucleus caudalis response to tactile, thermal, and nociceptive stimulation of monkey's face, *J. Neurophysiol.* **39:**936–953.

Price, D. D., Hayes, R. L., Ruda, M. A., and Dubner, R., 1978, Spatial and temporal transformations of input to spinothalamic tract neurons and their relation to somatic sensations, *J. Neurophysiol.* **41:**933–947.

Price, D. D., Hayashi, H., Dubner, R., and Ruda, M. A., 1979, Functional relationships between neurons of marginal and substantia gelatinosa layers of primate dorsal horn, *J. Neurophysiol.* **42**:1590–1608.

Price, J. L., and Slotnik, B. M., 1983, Dual olfactory representation in the rat thalamus: An anatomical and electrophysiological study, *J. Comp. Neurol.* **215**:63–77.

Pritz, M. B., 1974a, Ascending connections of a midbrain auditory area in a crocodile, *Caiman crocodilus*, *J. Comp. Neurol.* **153**:179–198.

Pritz, M. B., 1974b, Ascending connections of a thalamic auditory area in a crocodile, *Caiman crocodilus*, *J. Comp. Neurol.* **153**:199–214.

Pritz, M. B., 1975, Anatomical identification of a telencephalic visual area in crocodiles: Ascending connections of nucleus rotundus in *Caiman crocodilus*, *J. Comp. Neurol.* **164**:323–338.

Pritz, M. B., and Northcutt, R. G., 1977, Succinate dehydrogenase activity in the telencephalon of crocodiles correlates with the projection areas of sensory thalamic nuclei, *Brain Res.* **124**:357–360.

Pritz, M. B., and Northcutt, R. G., 1980, Anatomical evidence for an ascending somatosensory pathway to the telencephalon in crocodiles, *Caiman crocodilus*, *Exp. Brain Res.* **40**:342–345.

Probst, M., 1898, Experimentelle Untersuchungen über das Zwischenhirn und dessen Verbindungen, besonders die sogenannte Rindenschleife, *Dtsch. Z. Nervenheilkd.* **13**:384–408.

Probst, M., 1900a, Experimentelle Untersuchungen über die Schleifenendigung, die Haubenbahn, das dorsale Längsbündel und die hintere Commissur, *Arch. Psychiatr. Nervenkr.* **33**:1–57.

Probst, M., 1900b, Physiologische, anatomische und pathologisch-anatomische Untersuchungen des Sehhügels, *Arch. Psychiatr. Nervenkr.* **3**:721–817.

Probst, M., 1900c, Experimentelle Untersuchungen über die Anatomie und Physiologie des Sehhügels, *Monatsschr. Psychiatr. Neurol.* **7**:387–404.

Probst, M., 1901a, Zur Kenntnis des Bindearmes, der Haubenstrahlung und der Regio subthalamica, *Monatsschr. Psychiatr. Neurol.* **10**:288–309.

Probst, M., 1901b, Ueber den Bau des vollständig balkenlösen Grosshirnes sowie über Mikrogyrie und Heterotopie der grauen Substanz, *Arch. Psychiatr. Nervenkr.* **34**:709–786.

Probst, M., 1901c, Über den Verlauf und die Endigung der Rindensehhügelfasern des Parietallappens, *Arch. Anat. Physiol. (Anat. Abt.)* **1901**:357–368.

Probst, M., 1903, Über die Leitungsbahnen des Grosshirns mit besonderer Berücksichtigung der Anatomie und Physiologie des Sehhügels, *Jahrb. Psychiatr. Neurol.* **23**:18–106.

Probst, M., 1906, Über die zentralen Sinnesbahnen und die Sinneszentren des menschlichen Gehirns, *S.-B. Akad. Wiss. Wien Math.-Nat. Kl.* **115**:103–176.

Pubols, B. H., Jr., 1968, Retrograde degeneration study of somatic sensory thalamocortical connections in brain of Virginia opossum, *Brain Res.* **7**:232–251.

Pubols, B. H., Jr., 1977, The second somatic sensory area (SmII) of opossum neocortex, *J. Comp. Neurol.* **174**:71–78.

Pubols, B. H., and Pubols, L. M., 1966, Somatic sensory representation in the thalamic ventrobasal complex of the Virginia opossum, *J. Comp. Neurol.* **127**:19–34.

Pubols, L. M., 1968, Somatic sensory representation in the thalamic ventrobasal complex of the spider monkey (*Ateles*), *Brain Behav. Evol.* **1**:305–323.

Pujol, R., 1972, Development of tone-burst responses along the auditory pathway in the cat, *Acta Oto-Laryngol.* **74**:383–391.

Pujol, R., and Hilding, D., 1973, Anatomy and physiology of the onset of auditory function, *Acta Oto-Laryngol.* **76**:1–10.

Pujol, R., and Marty, R., 1968, Structural and physiological relationships of the maturing auditory system, in: *Ontogenesis of the Brain* (L. Jilek and S. Trojan, eds.), Universita Karlova, Prague, pp. 337–385.

Purpura, D. P., 1972, Intracellular studies of synaptic organization in the mammalian brain, in: *Structure and Function of Synapses* (G. D. Pappas and D. P. Purpura, eds.), Raven Press, New York, pp. 257–302.

Purpura, D. P., and Cohen, B., 1962, Intracellular recording from thalamic neurons during recruiting reponses, *J. Neurophysiol.* **25**:621–635.

Purpura, D. P., and Malliani, A., 1967, Synaptic potentials and discharge characteristics of caudate neurons activated by thalamic stimulation, *Brain Res.* **6**:325–340.

Purpura, D. P., and Schofer, R. J., 1963, Intracellular recording from thalamic neurons during reticulocortical activation, *J. Neurophysiol.* **26**:494–505.

Purpura, D. P., McMurtry, J. G., and Maekawa, K., 1966, Synaptic events in ventrolateral thalamic neurons during suppression of recruiting responses by brain stem reticular stimulation, *Brain Res.* **1**:63–76.

Purves, D., and Lichtman, J. W., 1980, Elimination of synapses in the developing nervous system, *Science* **210**:153–157.

Putkonen, P., Magnin, T. S., and Jeannerod, M., 1973, Directional responses to head rotation in neurons from the ventral nucleus of the lateral geniculate body, *Brain Res.* **61**:407–411.

Quensel, F., and Kohnstamm, O., 1907, Präparate mit activen Zelldegenerationen nach Hirnstamm-verletzung beim Kaninchen, *Neurol. Zentralbl.* **26**:1138–1139.

Quinlan, J. T., and Phillips, M. I., 1981, Immunoreactivity for an angiotensin II-like peptide in the human brain, *Brain Res.* **205**:212–218.

Raczkowski, D., and Diamond, I. T., 1978, Connections of the striate cortex in *Galago senegalensis, Brain Res.* **144**:383–388.

Raczkowski, D., and Diamond, I. T., 1981, Projections from the superior colliculus and the neocortex to the pulvinar nucleus in *Galago, J. Comp. Neurol.* **200**:231–254.

Raczkowski, D., and Rosenquist, A. C., 1980, Connections of the parvocellular C laminae of the dorsal lateral geniculate nucleus with the visual cortex in the cat, *Brain Res.* **199**:447–451.

Raczkowski, D., and Rosenquist, A. C., 1981, Retinotopic organization in the cat lateral posterior–pulvinar complex, *Brain Res.* **221**:185–191.

Raczkowski, D., and Rosenquist, A. C., 1983, Connections of the multiple visual cortical areas with the lateral posterior pulvinar complex and adjacent thalamic nuclei in the cat, *J. Neurosci.* **3**:1912–1942.

Rafols, J. A., and Valverde, F., 1973, The structure of the dorsal lateral geniculate nucleus in the mouse: A Golgi and electron microscopic study, *J. Comp. Neurol.* **150**:303–332.

Ragsdale, C. W., Jr., and Graybiel, A. M., 1981, The fronto-striatal projection in the cat and monkey and its relationship to inhomogeneities established by acetylcholinesterase histochemistry, *Brain Res.* **208**:259–266.

Rainey, W. T., 1979, Organization of nucleus rotundus, a tectofugal thalamic nucleus in turtles. I. Nissl and Golgi analyses, *J. Morphol.* **160**:121–142.

Rainey, W. T., and Jones, E. G., 1983, Spatial distribution of individual medial lemniscal axons in the thalamic ventrobasal complex of the cat, *Exp. Brain Res.* **49**:229–246.

Rainey, W. T., and Ulinski, P. S., 1982, Organization of nucleus rotundus, a tectofugal thalamic nucleus in turtles. III. The tectorotundal projection, *J. Comp. Neurol.* **209**:208–223.

Raisman, G., 1966, The connexions of the septum, *Brain* **89**:317–348.

Raisman, G., Cowan, W. M., and Powell, T. P. S., 1966, An experimental analysis of the efferent projection of the hippocampus, *Brain* **89**:83–108.

Rakic, P., 1972, Mode of cell migration to the superficial layers of fetal monkey neocortex, *J. Comp. Neurol.* **145**:61–84.

Rakic, P., 1974, Embryonic development of the pulvinar–LP complex in man, in: *The Pulvinar–LP Complex* (I. S. Cooper, M. Riklan, and P. Rakic, eds.), Thomas, Springfield, Ill., pp. 3–30.

Rakic, P., 1976, Prenatal genesis of connections subserving ocular dominance in the rhesus monkey, *Nature (London)* **261**:467–471.

Rakic, P., 1977a, Differences in the time of origin and in eventual distribution of neurons in areas 17 and 18 of the visual cortex in rhesus monkey, *Exp. Brain Res. Suppl.* **1**:244–248.

Rakic, P., 1977b, Genesis of the dorsal lateral geniculate nucleus in the rhesus monkey: Site and time of origin, kinetics of proliferation, routes of migration and pattern of distribution of neurons, *J. Comp. Neurol.* **176**:23–51.

Rakic, P., 1977c, Prenatal development of the visual system in rhesus monkey, *Philos. Trans. R. Soc. London Ser. B* **278**:245–260.

Rakic, P., 1979, Genesis of visual connections in the rhesus monkey, in: *Developmental Neurobiology of Vision* (R. D. Freeman, ed.), Plenum Press, New York, pp. 249–276.

Rakic, P., 1981, Development of visual centers in the primate brain depends on binocular competition before birth, *Science* **214**:928–931.

Rakic, P., and Riley, K. P., 1983, Overproduction and elimination of retinal axons in the fetal rhesus monkey, *Science* **219**:1441–1444.

Rakic, P., and Sidman, R. L., 1968, Supravital DNA synthesis in the developing human and mouse brain, *J. Neuropathol. Exp. Neurol.* **27**:246–276.

Rakic, P., and Sidman, R. L., 1969, Telencephalic origin of pulvinar neurons in the fetal human brain, *Z. Anat. Entwicklungsgesch.* **129**:53–82.

Ralston, H. J., III, 1969, The synaptic organization of lemniscal projections to the ventrobasal thalamus of the cat, *Brain Res.* **14**:99–115.

Ralston, H. J., III, 1971, Evidence for presynaptic dendrites and a proposal for their mechanism of action, *Nature (London)* **230**:585–587.

Ralston, H. J., III, 1983, The synaptic organization of the ventrobasal thalamus in the rat, cat and monkey, in: *Somatosensory Integration in the Thalamus* (G. Macchi, A. Rustioni, and R. Spreafico, eds.), Elsevier, Amsterdam, pp. 241–250.

Ralston, H. J., III, and Herman, M. M., 1969, The fine structure of neurons and synapses in the ventrobasal thalamus of the cat, *Brain Res.* **14**:77–97.

Ramón, P., 1896, Estructura del encefalo del camaleon, *Rev. Trim. Micrografica* **1**:146–182.

Ramón-Moliner, E., 1975, Specialized and generalized dendritic patterns, in: *Golgi Centennial Symposium: Perspectives in Neurobiology* (M. Santini, ed.), Raven Press, New York, pp. 87–100.

Ramón y Cajal, S., 1899–1904, *Textura del Sistema Nervioso del Hombre y de los Vertebrados*, 2 vols., Moya, Madrid.

Ramón y Cajal, S.,1909–1911, *Histologie du Système Nerveux de l'Homme et des Vertébrés* (translated by L. Azoulay), 2 vols., Maloine, Paris.

Ranson, S. W., 1921, *The Anatomy of the Nervous System*, Saunders, Philadelphia.

Ranson, S. W., and Ingram, W. R., 1932, The diencephalic course and termination of the medial lemniscus and the brachium conjunctivum, *J. Comp. Neurol.* **56**:257–276.

Ranson, S. W., Ranson, S. W., Jr., and Ranson, M., 1941, Fiber connections of the corpus striatum as seen in Marchi preparations, *Arch. Neurol. Psychiatry (Chicago)* **46**:230–249.

Rapaport, D. H., and Wilson, P. D., 1983, Retinal ganglion cells size groups projecting to the superior colliculus and the dorsal lateral geniculate nucleus in the North American opossum, *J. Comp. Neurol.* **213**:74–85.

Rasminsky, M., Mauro, A., and Albe-Fessard, D., 1973, Projections of medial thalamic nuclei to the putamen and cerebral frontal cortex in the cat, *Brain Res.* **61**:69–77.

Rausch, L. J., and Long, C. J., 1974, Habenular lesions and avoidance learning deficits in albino rats, *Physiol. Psychol.* **2**:352–365.

Raymond, J., Demêmes, D., and Marty, R., 1976, Voies et projections vestibulaires ascendantes émanant des noyaux primaires: Étude radioautographique, *Brain Res.*, **111**:1–12.

Reale, R. A., and Imig, T. J., 1980, Tonotopic organization in auditory cortex of the cat, *J. Comp. Neurol.* **192**:265–292.

Reep, R. L., and Winans, S. S., 1982, Efferent connections of dorsal and ventral agranular insular cortex in the hamster, *Mesocricetus auratus*, *Neuroscience* **7**:2609–2635.

Reese, B. E., and Jeffery, G., 1983, Crossed and uncrossed visual topography in dorsal lateral geniculate nucleus of the pigmented rat, *J. Neurophysiol.* **49**:877–884.

Reil, J. C., 1809, Untersuchungen über den Bau des grossen Gehirns im Menschen, *Arch. Anat. Physiol.* **9**:136–524.

Reiner, A., and Karten, H. J., 1978, A bisynaptic retinocerebellar pathway in the turtle, *Brain Res.* **150**:163–169.

Reiner, A., Brauth, S. E., Kitt, C. A., and Karten, H. J., 1980, Basal ganglionic pathways to the tectum: Studies in reptiles, *J. Comp. Neurol.* **193**:565–589.

Reinert, H., 1964, Defence reaction from the habenular nuclei, stria medullaris and fasciculus retroflexus, *J. Physiol. (London)* **170**:28–29P.

Reiter, R. J., 1980, The pineal and its hormones in the control of reproduction in mammals, *Endocrine Rev.* **1**:109–175.

Repérant, J., 1973, Nouvelles données sur les projectons visuelles chez le pigeon (*Columba livia*), *J. Hirnforsch.* **14**:151–187.

Repérant, J., Raffin, J.-P., and Miceli, D., 1974, La voie retino-thalamo-hyperstriatale chez le pouissin (*Gallus domesticus* L.), *C.R. Acad. Sci.* **279**:279–281.

Repérant, J., Lemire, M., Miceli, D., and Peyrichoux, J., 1976, An autoradiographic study of the visual system in fresh water teleosts following intraocular injection of tritiated fucose and proline, *Brain Res.* **118**:123–131.

Repérant, J., Rio, J. P., Miceli, D., and Lemire, M., 1978, A radioautographic study of retinal projections in type I and type II lizards, *Brain Res.* **142**:401–411.

Reubi, J. C., and Cuénod, M., 1979, Glutamate release in vitro from corticostriatal terminals, *Brain Res.* **176**:185–188.

Revzin, A. M., 1979, Functional localization in the nucleus rotundus, in: *Neural Mechanisms of Behavior in the Pigeon* (A. M. Granda and J. H. Maxwell, eds.), Plenum Press, New York, pp. 165–176.

Revzin, A. M., and Karten, H. J., 1966, Rostral projections of the optic tectum and the nucleus rotundus in the pigeon, *Brain Res.* **3**:264–276.

Rexed, B., and Brodal, A., 1951, The nucleus cervicalis lateralis: A spinocerebellar relay nucleus, *J. Neurophysiol.* **14**:399–407.

Rezak, M., and Benevento, L. A., 1979, A comparison of the organization of the projections of the dorsal lateral geniculate nucleus, the inferior pulvinar and adjacent lateral pulvinar to primary visual cortex (area 17) in the macaque monkey, *Brain Res.* **167**:19–40.

Ribak, C. E., and Peters, A., 1975, An autoradiographic study of the projections from the lateral geniculate body of the rat, *Brain Res.* **92**:341–368.

Ricardo, J. A., and Koh, E. T., 1978, Anatomical evidence of direct projections from the nucleus of the solitary tract to the hypothalamus, amygdala and other forebrain structures in the rat, *Brain Res.* **153**:1–26.

Richard, D., Gioanni, Y., Kitsikis, Y., and Buser, P., 1975, A study of geniculate unit activity during cryogenic blockade of the primary visual cortex in the cat, *Exp. Brain Res.* **22**:235–242.

Rinvik, E., 1968a, A re-evaluation of the cytoarchitecture of the ventral nuclear complex of the cat's thalamus on the basis of corticothalamic connections, *Brain Res.* **8**:237–254.

Rinvik, E., 1968b, The corticothalamic projection from the gyrus proreus and the medial wall of the rostral hemisphere in the cat: An experimental study with silver impregnation methods, *Exp. Brain Res.* **5**:129–152.

Rinvik, E., 1968c, The corticothalamic projection from the pericruciate and coronal gyri in the cat: An experimental study with silver-impregnation methods, *Brain Res.* **10**:79–119.

Rinvik, E., 1968d, The corticothalamic projection from the second somatosensory cortical area in the cat: An experimental study with silver impregnation methods, *Exp. Brain Res.* **5**:153–172.

Rinvik, E., 1972, Organization of thalamic connections from motor and somatosensory cortical areas in the cat, in: *Corticothalamic Projections and Sensorimotor Activities* (T. Frigyesi, E. Rinvik, and M. D. Yahr, eds.), Raven Press, New York, pp. 57–90.

Rinvik, E., 1975, Demonstration of nigrothalamic connections in the cat by retrograde axonal transport of horseradish peroxidase, *Brain Res.* **90**:313–318.

Rioch, D. M., 1929a, Studies on the diencephalon of Carnivora. Part I. The nuclear configuration of the thalamus, epithalamus, and hypothalamus of the dog and cat, *J. Comp. Neurol.* **49**:1–119.

Rioch, D. M., 1929b, Studies on the diencephalon of Carnivora. Part II. Certain nuclear configurations and fiber connections of the subthalamus and midbrain of the dog and cat, *J. Comp. Neurol.* **49**:121–153.

Rioch, D. M., 1931, Studies on the diencephalon of Carnivora. Part III. Certain myelinated-fiber connections of the diencephalon of the dog *(Canis familiaris)*, cat *(Felis domestica)* and aevisa *Crossarchus obscurus), J. Comp. Neurol.* **53**:319–388.

Riolan, J. (the younger, 1577–1657), 1610, *Anatomia Corporis Humani* (In *Opera Omnia*, Plantin, Paris).

Roberts, T. S., and Akert, K., 1963, Insular and opercular cortex and its thalamic projection in *Macaca mulatta*, *Schweiz. Arch. Neurol. Neurochir. Psychiatr.* **92**:1–43.

Robertson, R. T., 1977, Thalamic projections to parietal cortex, *Brain Behav. Evol.* **14**:161–184.

Robertson, R. T., 1983, Efferents of the pretectal complex: Separate populations of neurons project to lateral thalamus and to inferior olive, *Brain Res.* **258**:91–95.

Robertson, R. T., and Cunningham, T. J., 1981, Organization of corticothalamic projections from parietal cortex in cat, *J. Comp. Neurol.* **199**:569–586.

Robertson, R. T., and Feiner, A. R., 1982, Diencephalic projections from the pontine reticular formation: Autoradiographic studies in the cat, *Brain Res.* **239**:3–16.

Robertson, R. T., and Kaitz, 1981, Thalamic connections with limbic cortex. I. Thalamocortical projections, *J. Comp. Neurol.* **195**:501–525.

Robertson, R. T., and Rinvik, E., 1973, The corticothalamic projections from parietal regions of the cerebral cortex: Experimental degeneration studies in the cat, *Brain Res.* **51**:61–79.

Robertson, R. T., Kaitz, S. S., and RoBards, M. J., 1980, A subcortical pathway links sensory and limbic systems of the forebrain, *Neurosci. Lett.* **16**:161–165.

Robertson, R. T., Thompson, S. M., and Kaitz, S. S., 1983, Projections from the pretectal complex to the thalamic lateral dorsal nucleus of the cat, *Exp. Brain Res.* **51**:157–171.

Robertson, T. W., Hickey, T. L., and Guillery, R. W., 1980, Development of the dorsal lateral geniculate nucleus in normal and visually deprived Siamese cats, *J. Comp. Neurol.* **191**:573–580.

Robinson, C. J., and Burton, H., 1980a, Somatotopographic organization in the second somatosensory area of *M. fascicularis*, *J. Comp. Neurol.* **192**:43–68.

Robinson, C. J., and Burton, H., 1980b, Organization of somatosensory receptive fields in cortical areas 7b, retroinsular, postauditory and granular insula of *M. fascicularis*, *J. Comp. Neurol.* **192**:69–92.

Robinson, C. J., and Burton, H., 1980c, Somatic submodality distribution within the second somatosensory (SII), 7b, retroinsular, postauditory, and granular insular cortical areas of *M. fascicularis*, *J. Comp. Neurol.* **192**:93–108.

Robinson, D. L., 1973, Electrophysiological analysis of interhemispheric relations in the second somatosensory cortex of the cat, *Exp. Brain Res.* **18**:131–144.

Robson, J. A., 1981, Abnormal axonal growth in the dorsal lateral geniculate nucleus of the cat, *J. Comp. Neurol.* **195**:453–475.

Robson, J. A., and Hall, W. C., 1975, Connections of layer VI in striate cortex of the grey squirrel *(Sciurus carolinensis)*, *Brain Res.* **93**:133–139.

Robson, J. A., and Hall, W. C., 1976, Projections from the superior colliculus to the dorsal lateral geniculate nucleus of the grey squirrel *(Sciurus carolinensis)*, *Brain Res.* **113**:379–385.

Robson, J. A., and Hall, W. C., 1977, The organization of the pulvinar in the grey squirrel *Sciurus carolinensis*). I. Cytoarchitecture and connections, *J. Comp. Neurol.* **173**:355–388.

Robson, J. A., and Mason, C. A., 1979, The synaptic organization of terminals traced from individual labeled retino-geniculate axons in the cat, *Neuroscience* **4**:99–112.

Robson, J. A., Mason, C. A., and Guillery, R. W., 1978, Terminal arbors of axons that have formed abnormal connections, *Science* **201**:635–637.

Rockel, A. J., Heath, C. J., and Jones, E. G., 1972, Afferent connections to the diencephalon in the marsupial phalanger and the question of sensory convergence in the "posterior group" of the thalamus, *J. Comp. Neurol.* **145**:105–130.

Rodieck, R. W., 1979, Visual pathways, *Annu. Rev. Neurosci.* **2**:193–225.

Rodieck, R. W., and Dreher, B., 1979, Visual suppression from non-dominant eye in the lateral geniculate nucleus: A comparison of cat and monkey, *Exp. Brain Res.* **35**:465–477.

Rogawski, M. A., and Aghajanian, G. K., 1980, Modulation of lateral geniculate excitability by noradrenaline microiontophoresis or locus coeruleus stimulation, *Nature (London)* **287**:731–734.

Rogawski, M. A., and Aghajanian, G. K., 1982, Activation of lateral geniculate neurons by locus coeruleus or dorsal noradrenergic bundle stimulation: Selective blockade by the alpha 1-adrenoreceptor antagonist prazosin, *Brain Res.* **250**:31–40.

Rogers, R. C., Novin, D., and Butcher, L. L., 1979, Hepatic sodium and osmoreceptors activate neurons in ventrobasal thalamus, *Brain Res.* **168**:398–403.

Roland, P. E., Larsen, B., Lassen, N. A., and Skinhøj, E., 1980, Supplementary motor area and other cortical areas in organization of voluntary movements in man, *J. Neurophysiol.* **43**:118–136.

Rolls, E. T., Perrett, D., Thorpe, S. J., Puerto, A., Roper-Hall, A., and Maddison, S., 1979, Responses of neurons in area 7 of the parietal cortex to objects of different significance, *Brain Res.* **169**:194–198.

Romand, R., and Marty, R., 1975, Postnatal maturation of the cochlear nuclei in the cat: A neurophysiological study, *Brain Res.* **83**:225–233.

Ronan, M. C., and Northcutt, R. G., 1979, Afferent and efferent connections of the bullfrog medial pallium, *Neurosci. Abstr.* **5**:146.

Ropert, N., and Steriade, M., 1981, Input–output organization of the midbrain reticular core, *J. Neurophysiol.* **46**:17–31.

Rose, J. E., 1942a, The ontogenetic development of the rabbit's diencephalon, *J. Comp. Neurol.* **77**:61–129.

Rose, J. E., 1942b, The thalamus of the sheep: Cellular and fibrous structure and comparison with pig, rabbit and cat, *J. Comp. Neurol.* **77**:469–523.

Rose, J. E., 1949, The cellular structure of the auditory region of the cat, *J. Comp. Neurol.* **91**:409–440.

Rose, J. E., 1952, The cortical connections of the reticular complex of the thalamus, *Res. Publ. Assoc. Res. Nerv. Ment. Dis.* **30**:454–479.

Rose, J. E., 1972, Introductory remarks, *Brain Behav. Evol.* **6**:17–20.

Rose, J. E., and Galambos, R., 1952, Microelectrode studies on medial geniculate body of cat. I. Thalamic region activated by click stimuli, *J. Neurophysiol.* **15**:343–357.

Rose, J. E., and Malis, L. I., 1965, Geniculo-striate connections in the rabbit. II. Cytoarchitectonic structure of the striate region and of the dorsal lateral geniculate body; organization of the geniculo-striate projections, *J. Comp. Neurol.* **125**:121–140.

Rose, J. E., and Mountcastle, V. B., 1952, The thalamic tactile region in rabbit and cat, *J. Comp. Neurol.* **97**:441–489.

Rose, J. E., and Mountcastle, V. B., 1954, Activity of single neurons in the tactile thalamic region of the cat in response to a transient peripheral stimulus, *Bull. Johns Hopkins Hosp.* **94**:238–282.

Rose, J. E., and Mountcastle, V. B., 1959, Touch and kinesthesis, in: *Handbook of Physiology*, Section 1, Volume 1 (J. Field, H. W. Magoun, and V. E. Hall, eds.), American Physiological Society, Washington, D.C., pp. 387–429.

Rose, J. E., and Woolsey, C. N., 1943, A study of thalamocortical relations in the rabbit, *Bull. Johns Hopkins Hosp.* **73**:65–128.

Rose, J. E., and Woolsey, C. N., 1984a, Structure and relations of limbic cortex and anterior thalamic nuclei in rabbit and cat, *J. Comp. Neurol.* **89**:79–347.

Rose, J. E., and Woolsey, C. N., 1948b, The orbitofrontal cortex and its connections with the mediodorsal nucleus in rabbit, sheep and cat, *Res. Publ. Assoc. Res. Nerv. Ment. Dis.* **27**:210–232.

Rose, J. E., and Woolsey, C. N., 1949a, Organization of the mammalian thalamus and its relationships to the cerebral cortex, *Electroencephalogr. Clin. Neurophysiol.* **1**:391–403.

Rose, J. E., and Woolsey, C. N., 1949b, The relations of thalamic connections, cellular structure and evocable electrical activity in the auditory region of the cat, *J. Comp. Neurol.* **91**:441–466.

Rose, J. E., and Woolsey, C. N., 1958, Cortical connections and functional organization of the thalamic auditory system of the cat, in: *Biological and Biochemical Bases of Behavior* (H. F. Harlow and C. N. Woolsey, eds.), University of Wisconsin Press, Madison, pp. 127–150.

Rose, J. E., Gross, N. B., Geisler, C. D., and Hind, J. E., 1966, Some neural mechanisms in the inferior colliculus of the cat which may be relevant to localization of a sound source, *J. Neurophysiol.* **29**:288–314.

Rose, M., 1914, Über die cytoarchitektonische Gliederung des Vorderhirns der Vögel, *J. Psychol. Neurol.* **21**:278–352.

Rose, M., 1923, Histologische Lokalization des Vorderhirns der Reptilien, *J. Psychol. Neurol.* **29**:219–272.

Rose, M., 1927, Gyrus limbicus anterior und Regio retrosplenialis (Cortex holoprotoptychos quinquestratificatus): Vergleichende Architektonik bei Tier und Mensch, *J. Psychol. Neurol.* **35**:65–173.

Rose, M., 1929, Cytoarchitektonischer Atlas der Grosshirnrinde der Maus, *J. Psychol. Neurol.* **40**:1–51.

Rose, M., 1935, Das Zwischenhirn des Kaninchens, *Mem. Acad. Pol. Sci. (Ser. B) No. 8.*

Rosén, I., 1969a, Excitation of group I activated thalamocortical relay neurones in the cat, *J. Physiol. (London)* **205**:237–255.

Rosén, I., 1969b, Localization in caudal brain stem and cervical spinal cord of neurones activated from forelimb group I afferents in the cat, *Brain Res.* **16**:55–71.

Rosén, I., and Asanuma, H., 1972, Peripheral afferent inputs to the forelimb area of the monkey motor cortex: Input–output relations, *Exp. Brain Res.* **14**:257–273.

Rosenquist, A. C., Edwards, S. B., and Palmer, L. A., 1974, An autoradiographic study of the projections of the dorsal lateral geniculate nucleus and the posterior nucleus in the cat, *Brain Res.* **80**:71–93.

Rossier, J., 1981, Serum monospecificity: A prerequisite for reliable immunohistochemical localization of neuronal markers including choline acetyl transferase, *Neuroscience* **6**:989–991.

Roth, G. L., Aitkin, L. M., Andersen, R. A., and Merzenich, M. M., 1978, Some features of the spatial organization of the central nucleus of the inferior colliculus of the cat, *J. Comp. Neurol.* **182**:661–680.

Röthig, P., 1909, Riechbahnen, Septum und Thalamus bei *Didelphys marsupialis*, *Abh. Senckenb. Naturforsch. Ges.* **31**:1–19.

Röthig, P., 1923, Beiträge zum Studium des Zentralnervensystems der Wirbeltiere. VIII. Über das Zwischenhirn der Amphibien, *Arch. Mikrosk. Anat.* **98**:616–645.

Röthig, P., 1924, Beiträge zum Studium des Zentralnervensystems der Wirbeltiere. IX. Über die Faser-züge im Zwischenhirn der Urodelen, *Z. Mikrosk. Anat. Forsch.* **1**:5–40.

Rotter, A., and Jacobowitz, D. M., 1981, Neurochemical identification of cholinergic forebrain projection sites of the nucleus tegmentalis dorsalis lateralis, *Brain Res. Bull.* **6**:525–529.

Rotter, A., Birdsall, N. J. M., Burgen, A. S. V., Field, P. M., Hulme, E. C., and Raisman, G., 1979, Muscarinic receptors in the central nervous system of the rat. I. Technique for autoradiographic localization of the binding of [³H]propylbenzilylcholine mustard and its distribution in the forebrain, *Brain Res. Rev.* **1**:141–166.

Roussy, G., and Mosinger, M., 1935, L'hypothalamus chez l'homme et chez le chien, *Rev. Neurol.* **63**:1–35.

Rowe, M. H., and Dreher, B., 1982, Retinal W-cell projections to the medial intralaminar nucleus in the cat: Implications for ganglion cell classification, *J. Comp. Neurol.* **204**:117–133.

Rowe, M. H., and Stone, J., 1977, Naming of neurones, *Brain Behav. Evol.* **14**:185–216.

Rowe, M. J., and Sessle, B. J., 1968, Somatic afferent input to posterior thalamic neurones and their axon projection to the cerebral cortex in the cat, *J. Physiol. (London)* **196**:19–35.

Roy, J. P., Clerq, M., Steriade, M., and Deschênes, M., 1984, Electrophysiology of neurons of lateral thalamic nuclei in cat: Mechanisms of long-lasting hyperpolarizations, *J. Neurophysiol.* **51**:1220–1235.

Royce, G. J., 1978a, Autoradiographic evidence for a discontinuous projection to the caudate nucleus from the centromedian nucleus in the cat, *Brain Res.* **146**:145–150.

Royce, G. J., 1978b, Cells of origin of subcortical afferents to the caudate nucleus: A horseradish peroxidase study in the cat, *Brain Res.* **153**:465–476.

Royce, G. J., 1983, Cortical neurons with collateral projections to both the caudate nucleus and the centromedian–parafascicular thalamic complex: A fluorescent retrograde double labelling study in cat, *Exp. Brain Res.* **50**:157–165.

Royce, G. J., Ward, J. P., and Harting, J. K., 1976, Retinofugal pathways in two marsupials, *J. Comp. Neurol.* **170**:391–414.

Rubel, E. W., 1971, A comparison of somatotopic organization in sensory neocortex of newborn kittens and adult cats, *J. Comp. Neurol.* **143**:447–480.

Ruda, M., 1976, Autoradiographic study of the efferent projections of the midbrain central gray in the cat, Thesis, University of Pennsylvania.

Ruderman, M. I., Morrison, A. R., and Hand, P. J., 1972, A solution to the problem of cerebral cortical localization of taste in the cat, *Exp. Neurol.* **37**:522–537.

Rushton, W. A. H., 1977, Some memories of visual research in the past 50 years, in: *The Pursuit of Nature: Informal Essays on the History of Physiology*, Cambridge University Press, London, pp. 85–104.

Rusoff, A. C., and Dubin, M. W., 1977, Development of receptive-field properties of retinal ganglion cells in kittens, *J. Neurophysiol.* **40**:1188–1198.

Russell, G. V., 1954, The dorsal trigemino-thalamic tract in the cat reconsidered as a lateral reticulo-thalamic system of connections, *J. Comp. Neurol.* **101**:237–264.

Ryska, A., and Heger, M., 1979, Afferent connections of the laterodorsal thalamic nucleus in the rat, *Neurosci. Lett.* **15**:61–64.

Ryugo, D. K., and Killackey, H. P., 1974, Differential telencephalic projections of the medial and ventral divisions of the medial geniculate body of the rat, *Brain Res.* **82**:173–177.

Ryugo, D. K., and Weinberger, N. W., 1976, Corticofugal modulation of the medial geniculate body, *Exp. Neurol.* **51**:377–391.

Ryugo, D. K., and Weinberger, N. M., 1978, Differential plasticity of morphologically distinct neuron population in the medial geniculate body of the cat during classical conditioning, *Behav. Biol.* **22**:275–301.

Sachs, E., 1909a, Eine vergleichende anatomische Studie des Thalamus opticus der Säugetiere, *Arb. Neurol. Inst. Wien. Univ.* **17**:280–306.

Sachs, E., 1909b, On the structure and functional relations of the optic thalamus, *Brain* **32**:95–186.

Saito, H.-A., 1983, Morphology of physiologically identified X-, Y-, and W-type retinal ganglion cells of the cat, *J. Comp. Neurol.* **221**:279–288.

Saito, T., 1930, Über das Gehirn des japanischen Glussneunauges (*Entosphenus japanicus Martens*), *Folia Anat. Jpn.* **8**:189–263.

Sakai, K., and Jouvet, M., 1980, Brainstem PGO-on cells projecting directly to the cat dorsal lateral geniculate nucleus, *Brain Res.* **194**:500–505.

Sakai, S. T., 1982, The thalamic connectivity of the primary motor cortex (MI) in the raccoon, *J. Comp. Neurol.* **204**:238–252.

Sakakura, H., and Iwama, K., 1965, Presynaptic inhibition and postsynaptic facilitation in lateral geniculate body and so-called deep sleep wave activity, *Tohoku J. Exp. Med.* **87**:40–51.

Sakakura, H., and Iwamura, K., 1967, Effects of bilateral eye enucleation upon single unit activity of the lateral geniculate body in free behaving cats, *Brain Res.* **6**:667–678.

Sakata, H., Shibutani, H., and Kawano, K., 1980, Spatial properties of visual fixation neurons in posterior parietal association cortex of the monkey, *J. Neurophysiol.* **43**:1654–1672.

Salinger, W. L., Schwartz, M. A., and Wilkerson, P. R., 1977, Selective loss of lateral geniculate cells in the adult cat after chronic monocular paralysis, *Brain Res.* **125**:257–263.

Sanderson, J. B., 1874, Note on the excitation of the surface of the cerebral hemisphere by induced currents, *Proc. R. Soc. London* **22**:368–370.

Sanderson, K. J., 1971, The projection of the visual field to the lateral geniculate and medial interlaminar nuclei in the cat, *J. Comp. Neurol.* **143**:101–118.

Sanderson, K. J., 1974, Lamination of the dorsal lateral geniculate nucleus in carnivores of the weasel (*Mustelidae*), raccoon (*Procyonidae*) and fox (*Canidae*) families, *J. Comp. Neurol.* **153**:239–266.

Sanderson, K. J., and Kaas, J. H., 1974, Thalamocortical interconnections of the visual system of the mink, *Brain Res.* **70**:139–143.

Sanderson, K.J., and Pearson, L.J. 1977, Retinal projections in the native cat, *Dasyurus viverrinus*, *J. Comp. Neurol.* **174**:347–358.

Sanderson, K. J., Bishop, P. O., and Darian-Smith, I., 1971, The properties of the binocular receptive fields of lateral geniculate neurons, *Exp. Brain Res.* **13**:178–207.

Sanderson, K. J., Guillery, R. W., and Shackelford, R. M., 1974, Congenitally abnormal visual pathways in mink (*Mustela vison*) with reduced retinal pigment, *J. Comp. Neurol.* **154**:225–248.

Sanderson, K. J., Pearson, L. J., and Haight, J. R., 1979, Retinal projections in the Tasmanian devil, *Sarcophilus harrisii*, *J. Comp. Neurol.* **188**:335–346.

Sanderson, K. J., Haight, J. R., and Pearson, L. J., 1980, Transneuronal transport of tritiated fucose and proline in the visual pathways of the brushtailed possum, *Trichosurus vulpecula*, *Neurosci. Lett.* **20**:243–248.

Sanderson, K. J., Welker, W., and Shambes, G. M., 1984, Reevaluation of motor cortex and of sensorimotor overlap in cerebral cortex of albino rats, *Brain Res.* **292**:251–260.

Sano, T., 1910, Beitrag zur vergleichenden Anatomie der Substantia nigra, des Corpus Luysii und der Zona incerta, *Monatsschr. Psychiatr. Neurol.* **27**:110–127, 274–283, 381–389, 476–488; **28**:26–34, 129–133, 269–278, 367–375.

Sans, A., Raymond, J., and Marty, R., 1970, Responses thalamiques et corticales a la stimulation électrique du nerf vestibulaire chez le Chat, *Exp. Brain Res.* **10**:265–275.

Santorini, G. D., 1724, *Observationes Anatomicae*, J. B. Recurti, Venice.

Santorini, G. D., 1775, *Tabulae Anatomicae* (M. Girardi, ed.), Parma.

Saper, C. B., and Loewy, A. F., 1980, Efferent connections of the parabrachial nucleus in the rat, *Brain Res.* **197**:291–318.

Saper, C. B., Swanson, L. W., and Cowan, W. M., 1979a, An autoradiographic study of the efferent connections of the lateral hypothalamic area in the rat, *J. Comp. Neurol.* **183**:689–706.

Saper, C. B., Swanson, L. W., and Cowan, W. M., 1979b, Some efferent connections of the rostral hypothalamus in the squirrel monkey (*Saimiri sciureus*) and cat, *J. Comp. Neurol.* **184**:205–242.

Sapolsky, R. M., and Eichenbaum, H., 1980, Thalamocortical mechanisms in odor-guided behavior. II. Effects of lesions of the mediodorsal thalamic nucleus and frontal cortex on odor preferences and sexual behavior in the hamster, *Brain Behav. Evol.* **17**:276–290.

Saporta, S., and Kruger, L., 1977, The organization of thalamocortical relay neurons in the rat ventrobasal complex studied by the retrograde transport of horseradish peroxidase, *J. Comp. Neurol.* **174**:187–208.

Saporta, S., and Kruger, L., 1979, The organization of projections to selected points of somatosensory cortex from the cat ventrobasal complex, *Brain Res.* **178**:275–296.

Sar, M., Stumpf, W. E., Miller, R. J., Chang, K.-J., and Cuatrecasas, P., 1978, Immunohistochemical localization of enkephalin in rat brain and spinal cord, *J. Comp. Neurol.* **182**:17–38.

Sasaki, K., Matsuda, Y., Kawaguchi, S., and Mizuno, N., 1972, On the cerebellothalamo-cerebral pathway for the parietal cortex, *Exp. Brain Res.* **16**:89–103.

Sasaki, K., Matsuda, Y., Oka, H., and Mizuno, N., 1975, Thalamocortical projections for recruiting responses and spindling-like responses in the parietal cortex, *Exp. Brain Res.* **22:**87–96.

Sasaki, K., Shimono, T., Oka, H., Yamamoto, T., and Matsuda, Y., 1976, Effects of stimulation of the midbrain reticular formations upon thalamocortical neurones responsible for cortical recruiting responses, *Exp. Brain Res.* **26:**261–273.

Sastry, B. R., 1978, Effects of substance P, acetylcholine and stimulation of habenula on rat interpenduncular neuronal activity, *Brain Res.* **144:**404–410.

Satinsky, D., 1968, Reticular influences on lateral geniculate neurons activity, *Electroencephalogr. Clin. Neurophysiol.* **25:**543–549.

Sato, M., Itoh, K., and Mizuno, N., 1979, Distribution of thalamo-caudate neurons in the cat as demonstrated by horseradish peroxidase, *Exp. Brain Res.* **34:**143–154.

Sauer, M. E., 1959, Radioautographic study of the location of newly synthesized deoxyribonucleic acid in the neural tube of the chick embryo: Evidence for intermitotic migration of nuclei, *Anat. Rec.* **133:**456.

Scalia, F., 1976, The optic pathway of the frog: Nuclear organization and connections, in: *Frog Neurobiology* (R. Llinás and W. Precht, eds.), Springer-Verlag, Berlin, pp. 386–406.

Scalia, F., and Coleman, D. R., 1975, Identification of telencephalic–afferent thalamic nuclei associated with the visual system of the frog, *Neurosci. Abstr.* **1:**46.

Scalia, F., and Gregory, K., 1970, Retinofugal projections in the frog: Location of the postsynaptic neurons, *Brain Behav. Evol.* **3:**16–29.

Scalia, F., Halpern, M., Knapp, H., and Riss, W., 1968, The afferent connexions of the olfactory bulb in the frog: A study of degenerating unmyelinated fibers, *J. Anat.* **103:**245–262.

Scheibel, M. E., and Scheibel, A. B., 1966a, The organization of the nucleus reticularis thalami: A Golgi study, *Brain Res.* **1:**43–62.

Scheibel, M. E., and Scheibel, A. B., 1966b, Patterns of organization in specific and nonspecific thalamic fields, in: *The Thalamus* (D. P. Purpura and M. D. Yahr, eds.), Columbia University Press, New York, pp. 13–46.

Scheich, H., and Maier, V., 1981, ^{14}C-Deoxyglucose labeling of the auditory neostriatum in young and adult guinea fowl, in: *Neuronal Mechanisms of Hearing* (I. Syka and L. Aitkin, eds.), Plenum Press, New York, pp. 329–334.

Scheich, H., Bonke, B. A., Bonke, D., and Langner, G., 1979, Functional organization of some auditory nuclei in the guinea fowl demonstrated by the 2-deoxyglucose technique, *Cell Tissue Res.* **204:**17–27.

Schiller, P. H., 1982, Central connections of the retinal ON and OFF pathways, *Nature (London)* **297:**580.

Schiller, P. H., and Malpeli, J. G., 1978, Functional specificity of lateral geniculate nucleus laminae of the rhesus monkey, *J. Neurophysiol.* **41:**788–797.

Schingnitz, G., 1981, Neuronal responses in the rat's thalamus to scrotal heating, *Exp. Brain Res.* **43:**419–421.

Schlag, J., and Schlag-Rey, M., 1971, Induction of oculomotor responses from thalamic internal medullary lamina in the cat, *Exp. Neurol.* **33:**498–508.

Schlag, J. and Schlag-Rey, M., 1984, Visuomotor functions of central thalamus in monkey. II. Unit activity related to visual events, targeting and fixation, *J. Neurophysiol.* **51:**1175–1195.

Schlag, J., and Waszak, M., 1971, Characteristics of unit response in nucleus reticularis thalami, *Brain Res.* **32:**79–97.

Schlag, J., Lethinen, I., and Schlag-Rey, M., 1974, Neuronal activity before and during eye movements in thalamic internal medullary lamina of the cat, *J. Neurophysiol.* **37:**982–995.

Schlag, J., Schlag-Rey, M., Peck, C. K., and Joseph, J. P., 1980, Visual responses of thalamic neurons depending on the direction of gaze and the position of targets in space, *Exp. Brain Res.* **40:**170–184.

Schlag-Rey, M., and Schlag, J., 1977, Visual and presaccadic neuronal activity in thalamic internal medullary lamina of cat: A study of targeting, *J. Neurophysiol.* **40:**156–173.

Schmielau, F., 1979, Integration of visual and nonvisual information in nucleus reticularis thalami of the cat, in: *Developmental Neurobiology* (R. D. Freeman, ed.), Plenum Press, New York, pp. 205–226.

Schmielau, F., and Singer, W., 1977, The role of visual cortex for binocular interactions in the cat lateral geniculate nucleus, *Brain Res.* **120:**354–361.

Schneider, G. E., 1969, Two visual systems, *Science* **163:**795–802.

Schneider, G. E., 1970, Mechanisms of functional recovery following lesions of visual cortex or superior colliculus in neonate and adult hamsters, *Brain Behav. Evol.* **3**:295–323.

Schneider, G. E., 1973, Early lesions of the superior colliculus: Factors affecting the formation of abnormal retinal projections, *Brain Behav. Evol.* **8**:73–109.

Schnitzlein, H. N., 1962, The habenula and the dorsal thalamus of some teleosts, *J. Comp. Neurol.* **118**:225–267.

Schober, W., 1964, Vergleichend-anatomische Untersuchungen am Gehirn der Larven und adulten Tiere von *Lampetra fluviatilis* (Linne, 1758) und *Lampetra planeri* (Bloch, 1784), *J. Hirnforsch* **7**:107–209.

Schott, B., Maugiere, F., Laurent, B., Serclerat, O., and Fischer, C., 1980, L'amnesie thalamique, *Rev. Neurol.* **136**:117–130.

Schreyer, D. J., and Jones, E. G., 1982, Growth and target finding by axons of the corticospinal tract in prenatal and postnatal rats, *Neuroscience* **7**:1837–1853.

Schroeder, D. M., and Ebbesson, S. O. E., 1974, Nonolfactory telencephalic afferents in the nurse shark *(Ginglymostoma cirratum)*, *Brain Behav. Evol.* **9**:121–155.

Schulmann, S., 1964, Impaired delay response from thalamic lesions: Studies in monkeys, *Arch. Neurol. (Chicago)* **11**:477–499.

Schultz, W., Montgomery, E. B., Jr., and Marini, R., 1976, Stereotyped flexion of forelimb and hindlimb to microstimulation of dentate nucleus in cebus monkeys, *Brain Res.* **107**:151–155.

Schwarz, D. W. F., and Frederickson, J. M., 1971, Rhesus monkey vestibular cortex: A bimodal primary projection field, *Science* **172**:280–281.

Schwarz, D. W. F., Deecke, L., and Frederickson, J. M., 1973, Cortical projection of group I muscle afferents to areas 2, 3a and the vestibular field in the rhesus monkey, *Exp. Brain Res.* **17**:516–526.

Scollo-Lavizzari, G., and Akert, K., 1963, Cortical area 8 and its thalamic projection in *Macaca mulatta*, *J. Comp. Neurol.* **121**:259–269.

Segal, M., and Landis, S., 1974, Afferents to the hippocampus of the rat studied with the method of retrograde transport of horseradish peroxidase, *Brain Res.* **78**:1–15.

Segal, M., Dudai, Y., and Amsterdam, A., 1978, Distribution of α-bungarotoxin-binding cholinergic nicotinic receptor in rat brain, *Brain Res.* **148**:105–120.

Semple, M. N., and Aitkin, L. M., 1979, Representation of sound frequency and laterality by units in central nucleus of cat inferior colliculus, *J. Neurophysiol.* **42**:1626–1639.

Seneviratne, K. N., and Whitteridge, D., 1962, Visual evoked responses in the lateral geniculate nucleus, *Electroencephalogr. Clin. Neurophysiol.* **14**:785.

Sesma, M. A., Irvin, G. E., Kuyk, T. K., Norton, T. T., and Casagrande, V. A., 1984, Effects of monocular deprivation on the lateral geniculate nucleus in a primate, *Proc. Natl. Acad. Sci. USA* **81**:2255–2259.

Sharma, S. C., 1972, The retinal projection in the goldfish: An experimental study, *Brain Res.* **39**:213–223.

Shatz, C., 1977, A comparison of visual pathways in Boston and Midwestern Siamese cats, *J. Comp. Neurol.* **171**:205–228.

Shatz, C. J., 1981, Inside-out pattern of neurogenesis of the cat's lateral geniculate nucleus, *Soc. Neurosci. Abstr.* **7**:140.

Shatz, C. J. 1983, The prenatal development of the cat's retinogeniculate pathway, *J. Neurosci.* **3**:482–499.

Shatz, C. J., and Kliot, M., 1982, Prenatal misrouting of the retinogeniculate pathway in the Siamese cat, *Nature (London)* **300**:525–529.

Shatz, C. J., and Rakic, P., 1981, The genesis of efferent connections from the visual cortex of the fetal rhesus monkey, *J. Comp. Neurol.* **196**:287–308.

Shatz, C. J., and Stryker, M. P., 1978, Ocular dominance in layer IV of the cat's visual cortex and the effects of monocular deprivation, *J. Physiol. (London)* **281**:267–283.

Shatz, C. J., Lindström, S., and Wiesel, T. N., 1977, The distribution of afferents representing the right and left eyes in the cat's visual cortex, *Brain Res.* **131**:103–106.

Sheps, J. G., 1945, The nuclear configuration and cortical connections of the human thalamus, *J. Comp. Neurol.* **83**:1–56.

Sherman, S. M., 1973, Visual field defects in monocularly and binocularly deprived cats, *Brain Res.* **49**:25–45.

Sherman, S. M., 1979, The functional significance of X- and Y-cells in normal and visually deprived cats, *Trends Neurosci.* **2**:192–195.

Sherman, S. M., and Spear, P. D., 1982, Organization of visual pathways in normal and visually deprived cats, *Physiol. Rev.* **62**:738–855.

Sherman, S. M., and Wilson, J. R., 1975, Behavioral and morphological evidence for binocular competition in the postnatal development of the dog's visual system, *J. Comp. Neurol.* **161**:183–196.

Sherman, S. M., and Wilson, J. R., 1981, Further evidence for an early critical period in the development of the cat's dorsal lateral geniculate nucleus, *J. Comp. Neurol.* **196**:459–470.

Sherman, S. M., Hoffmann, K.-P., and Stone, J., 1972, Loss of a specific cell type from dorsal lateral geniculate nucleus in visually deprived cats, *J. Neurophysiol.* **35**:532–541.

Sherman, S. M., Guillery, R. W., Kaas, J. H., and Sanderson, K. J., 1974, Behavioral electrophysiological, and morphological studies of binocular competition in the development of the geniculocortical pathways of cats, *J. Comp. Neurol.* **158**:1–18.

Sherman, S. M., Norton, T. T., and Casagrande, V. A., 1975, X- and Y-cells in the dorsal lateral geniculate nucleus of the tree shrew (*Tupaia glis*), *Brain Res.* **93**:152–157.

Sherman, S. M., Wilson, J. R., Kaas, J. H., and Webb, S. V., 1976, X- and Y-cells in the dorsal lateral geniculate nucleus of the owl monkey (*Aotus trivirgatus*), *Science* **192**:475–477.

Shigenaga, Y., Nakatani, Z., Nishimori, T., Suemune, S., Kuroda, R., and Matano, S., 1983, The cells of origin of cat trigeminothalamic projections: Especially in the caudal medulla, *Brain Res.* **277**:201–222.

Shimazu, H., Yanagisawa, N., and Garoutte, B., 1965, Corticopyramidal influences on thalamic somatosensory transmission in the cat, *Jpn. J. Physiol.* **15**:101–124.

Shinoda, Y., 1964, in: *Neurophysiological Approaches to Higher Brain Functions* (E. V. Evarts, Y. Shinoda, and S. P. Wise, eds.), John Wiley, New York, pp. 105–114.

Shinoda, Y., Yamazaki, M., and Futami, T., 1982, Convergent inputs from the dentate and the interpositus nuclei to pyramidal tract neurons in the motor cortex, *Neurosci. Lett.* **34**:111–115.

Shipley, M. T., and Sørenson, K. E., 1975, On the laminar organization of the anterior thalamus projections to the presubiculum in the guinea pig, *Brain Res.* **86**:473–477.

Sholl, D. A., 1956, *The Organization of the Cerebral Cortex*, Methuen, London, and Wiley, New York.

Shosaku, A., and Sumitomo, I., 1983, Auditory neurons in the rat thalamic reticular nucleus, *Exp. Brain Res.* **49**:432–442.

Shute, C. C. D., and Lewis, P. R., 1967, The ascending cholinergic reticular system: Neocortical, olfactory and subcortical projections, *Brain* **90**:497–520.

Sidman, R. L., and Miale, I., 1959, Histogenesis of mouse cerebellum: Studies by autoradiography with tritiated thymidine, *Anat. Rec.* **133**:429–430.

Sidman, R. L., and Rakic, P., 1973, Neuronal migration with special reference to developing human brain: A review, *Brain Res.* **62**:1–35.

Siegal, A., Fukushima, T., Meibach, R., Burke, L., Edinger, H., and Weiner, S., 1977, The origin of the afferent supply to the mediodorsal thalamic nucleus: Enhancement of HRP transport by selective lesions, *Brain Res.* **135**:11–23.

Sillito, A. M., and Kemp, J. A., 1983, The influence of GABAergic inhibitory processes on the receptive field structure of X and Y cells in cat dorsal lateral geniculate nucleus (dLGN), *Brain Res.* **277**:63–78.

Sillito, A. M., Kemp, J. A., and Barardi, N., 1983, The cholinergic influence on the function of the cat dorsal lateral geniculate nucleus (dLGN), *Brain Res.* **280**:299–308.

Silverman, A. J., and Krey, L. C., 1978, The luteinizing hormone-releasing (LH-RH) neuronal networks of the guinea pig brain. I. Intra- and extra-hypothalamic projections, *Brain Res.* **157**:233–246.

Simantov, R., Snowman, A. M., and Snyder, S. H., 1976, A morphine-like factor "enkephalin" in rat brain: Subcellular localization, *Brain Res.* **107**:650–657.

Simma, K., 1950, Zur Cytoarchitektik des menschlichen Centrum medianum thalami, *Monatsschr. Psychiatr. Neurol.* **120**:119–130.

Simma, K., 1957, Der Thalamus der Menschenaffen: Eine vergleichend-anatomische Untersuchung, *Psychiatr. Neurol.* **134**:145–175.

Simmons, R. M. T., 1980, The morphology of the diencephalon in the Prosimii. II. The Lemuroidea and Lorisoidea. Part I. Thalamus and metathalamus, *J. Hirnforsch.* **21**:449–491.

Singer, C., 1925: *see* Mondino Da Luzzi.

Singer, W., 1970, Inhibitory binocular interaction in the lateral geniculate body of the cat, *Brain Res.* **18:**165–170.

Singer, W., 1973a, The effect of mesencephalic reticular stimulation on intracellular potentials of cat lateral geniculate neurones, *Brain Res.* **61:**35–54.

Singer, W., 1973b, Inhibitory interaction between X and Y units in the cat lateral geniculate nucleus, *Brain Res.* **49:**291–307.

Singer, W., 1977, Control of thalamic transmission by corticofugal and ascending reticular pathways in the visual system, *Physiol. Rev.* **57:**386–420.

Singer, W., 1979, Central core control of visual cortex function, in: *The Neurosciences: Fourth Study Program* (F. O. Schmitt and F. G. Worden, eds.), MIT Press, Cambridge, Mass., pp. 1093–1110.

Singer, W., and Bedworth, N., 1973, Inhibitory interaction between X and Y units in the cat LGN, *Brain Res.* **49:**291–307.

Singer, W., and Bedworth, N., 1974, Correlation between the effects of brain stem stimulation and saccadic eye movements on transmission in the cat lateral geniculate nucleus, *Brain Res.* **72:**185–202.

Singer, W., and Creutzfeldt, O., 1970, Reciprocal lateral inhibition of on- and off-centre neurones in the lateral geniculate body of the cat, *Exp. Brain Res.* **10:**311–330.

Singer, W., and Dräger, U., 1972, Postsynaptic potentials in relay neurons of cat lateral geniculate nucleus after stimulation of the mesencephalic reticular formation, *Brain Res.* **41:**214–220.

Singer, W., and Lux, H. D., 1973, Presynaptic depolarization and extracellular potassium in the cat lateral geniculate nucleus, *Brain Res.* **64:**17–33.

Singer, W., and Phillips, W. A., 1974, Function and interaction of on- and off-transients in vision. II. Neurophysiology, *Exp. Brain Res.* **19:**507–521.

Singer, W., and Schmielau, F., 1976, The effect of reticular stimulation on binocular inhibition in the cat lateral geniculate body, *Exp. Brain Res.* **25:**221–223.

Singer, W., Pöppel, E., and Creutzfeldt, O., 1972, Inhibitory interacting in the cat's lateral geniculate nucleus, *Exp. Brain Res.* **14:**210–226.

Singer, W., Tretter, F., and Cynader, M., 1976, The effect of reticular stimulation on spontaneous and evoked activity in the cat visual cortex, *Brain Res.* **102:**71–90.

Skoglund, S., 1956, Anatomical and physiological studies on knee joint innervation in the cat, *Acta Physiol. Scand.* **36**(Suppl. 124)**1:**101.

Sleeswyk, A. W., 1982, A new reconstruction of the Attic trieres and bireme, *Int. J. Naut. Archeol. Underwater Explor.* **11:**35–46.

Sloane, M. W. M., 1951, The diencephalon of the mink. I. The nuclear pattern of the dorsal thalamus, *J. Comp. Neurol.* **95:**463–519.

Slotnik, B. M., and Kaneko, N., 1981, Role of mediodorsal thalamic nucleus in olfactory discrimination learning in rats, *Science* **214:**91–92.

Smeets, W. J. A. J., 1981a, Retinofugal pathways in two chondrichthyans, the shark *Scyliorhinus canicula* and the ray *Raja clavata*, *J. Comp. Neurol.* **195:**1–12.

Smeets, W. J. A. J., 1981b, Efferent tectal pathways in two chondrichthyans, the shark *Scyliorhinus canicula* and the ray *Raja clavata*, *J. Comp. Neurol.* **195:**13–24.

Smith, D. C., Spear, P. D., and Kratz, K. E., 1978, Role of visual experience in postcritical-period reversal of effects of monocular deprivation in cat striate cortex, *J. Comp. Neurol.* **178:**313–328.

Smith, R. L., 1973, The ascending fiber projections from the principal sensory trigeminal nucleus in the rat, *J. Comp. Neurol.* **148:**423–445.

Smith, R. L., 1975, Axonal projections and connections of the principal sensory trigeminal nucleus in the monkey, *J. Comp. Neurol.* **163:**347–375.

Snider, R. S., and Niemer, W. T., 1961, *A Stereotaxic Atlas of the Cat Brain,* University of Chicago Press, Chicago.

Snider, R. S., and Stowell, A., 1944a, Receiving areas of the tactile, auditory and visual systems in the cerebellum, *J. Neurophysiol.* **7:**331–358.

Snider, R. S., and Stowell, A., 1944b, Electroanatomical studies on a tactile system in the cerebellum of monkey (*Macaca mulatta*), *Anat. Rec.* **88:**457.

Snyder, M., and Diamond, I. T., 1968, The organization and function of the visual cortex in the tree shrew, *Brain Behav. Evol.* **1:**244–288.

Snyder, S. H., 1980, Brain peptides as neurotransmitters, *Science* **209:**976–983.

Snyder, S. H., and Childers, S. R., 1979, Opiate receptors and opioid peptides, *Annu. Rev. Neurosci.* **2:**35–64.

So, K.-F., Schneider, G. E., and Frost, D. O., 1978, Postnatal development of retinal projections to the lateral geniculate body in Syrian hamsters, *Brain Res.* **142:**343–352.

So, K.-F., Schneider, G. E., and Ayres, S., 1981, Lesions of the brachium of the superior colliculus in neonate hamsters: Correlation of anatomy with behavior, *Exp. Neurol.* **72:**379–400.

So, Y. T., and Shapley, R., 1981, Spatial tuning of cells in and around lateral geniculate nucleus of the cat: X and Y relay cells and perigeniculate interneurons, *J. Neurophysiol.* **45:**107–120.

Soemmering, S. T., 1778, *De Basi Encephali et Originibus Nervorum Cranio Egredientium Libri Quinque,* A. Vandenhoeck, Göttingen.

Sofroniew, M. V., and Weindl, A., 1978, Projections from the parvocellular vasopressin- and neurophysin-containing neurons of the suprachiasmatic nucleus, *Am. J. Anat.* **153:**391–430.

Solnitzky, O., 1938, The thalamic nuclei of *Sus scrofa, J. Comp. Neurol.* **69:**121–171.

Solnitzky, O., and Harman, P. J., 1946, A comparative study of the central and peripheral sectors of the visual cortex in primates with observations on the lateral geniculate body, *J. Comp. Neurol.* **85:**313–419.

Somogyi, G., Hajdu, F., and Tömböl, T., 1978, Ultrastructure of the anterior ventral and anterior medial nuclei of the cat thalamus, *Exp. Brain Res.* **31:**417–431.

Søreide, A. J., and Fonnum, F., 1980, High affinity uptake of D-aspartate in the barrel subfield of the mouse somatic sensory cortex, *Brain Res.* **201:**427–430.

Sousa-Pinto, A., 1973, Cortical projections of the medial geniculate body in the cat, *Ergeb. Anat. Entwicklungsgesch.* **48:**1–40.

Špaček, J., and Lieberman, A. R., 1974, Ultrastructure and three-dimensional organization of synaptic glomeruli in rat somatosensory thalamus, *J. Anat.* **117:**487–516.

Spatz, H., 1961, Franz Nissl (1860–1919), in: *50 Jahre Neuropathologie in Deutschland 1885–1935* (W. Scholz, ed.), Thieme, Stuttgart, pp. 43–66.

Spatz, W. B., and Erdmann, G., 1974, Striate cortex projections to the lateral geniculate and other thalamic nuclei: A study using degeneration and autoradiographic tracing methods in the marmoset *(Callithrix jacchus), Brain Res.* **82:**91–108.

Spatz, W. B., and Tigges, J., 1973, Studies on the visual area MT in primates. II. Projection fibers to subcortical structures, *Brain Res.* **61:**374–378.

Spatz, W. B., Tigges, J., and Tigges, M., 1970, Subcortical projections, cortical associations, and some intrinsic interlaminar connections of the striate cortex in the squirrel monkey *(Saimiri), J. Comp. Neurol.* **140:**155–173.

Spear, P. D., and Barbas, H., 1975, Recovery of pattern discrimination ability in rats receiving serial or one-stage visual cortex lesions, *Brain Res.* **94:**337–346.

Spear, P. D., and Baumann, T. P., 1975, Receptive field characteristics of single neurons in lateral suprasylvian visual area of the cat, *J. Neurophysiol.* **38:**1403–1420.

Spear, P. D., and Baumann, T. P., 1979, Effects of visual cortex removal on receptive-field properties of neurons in lateral suprasylvian visual area of the cat, *J. Neurophysiol.* **42:**31–56.

Spear, P. D., and Braun, J. J., 1969, Pattern discrimination following removal of visual cortex in cat, *Exp. Neurol.* **25:**331–348.

Spear, P. D., and Hickey, T. L., 1979, Postcritical-period reversal of effects of monocular deprivation on dorsal lateral geniculate cell size in the cat, *J. Comp. Neurol.* **185:**317–328.

Spear, P. D., Smith, D. C., and Williams, L. L., 1977, Visual receptive-field properties of single neurons in cat's ventral-lateral geniculate nucleus, *J. Neurophysiol.* **40:**390–409.

Spear, P. D., Langsetmo, A., and Smith, D. C., 1980, Age-related changes in effects of monocular deprivation on cat striate cortex neurons, *J. Neurophysiol.* **43:**559–580.

Spiegel, E. A., Wycis, H. T., Orchink, C. W., and Freed, H., 1955, The thalamus and temporal orientation, *Science* **121:**771–772.

Spillane, J. D., 1981, *The Doctrine of the Nerves: Chapters in the History of Neurology,* Oxford University Press, London.

Sprague, J. M., 1972, The superior colliculus and pretectum in visual behavior, *Invest. Ophthalmol.* **11:**473–482.

Sprague, J. M., Berlucchi, G., and DiBerardino, A., 1970, The superior colliculus and pretectum in visually guided behavior and visual discrimination in the cat, *Brain Behav. Evol.* **3:**285–294.

Sprague, J. M., Levy, J., DiBerardino, A., and Berlucchi, G., 1977, Visual cortical areas mediating form discrimination in the cat, *J. Comp. Neurol.* **172:**441–448.

Sprague, J. M., Hughes, H. C., and Berlucchi, G., 1981, Cortical mechanisms in pattern and form perception, in: *Brain Mechanisms of Awareness and Purposeful Behavior* (O. Pompeiano and C. Ajmone-Marsan, eds.), Raven Press, New York, pp. 192–210.

Spreafico, R., Hayes, N. L., and Rustiono, A., 1981, Thalamic projections to the primary and secondary somatosensory cortices in cat: Single and double retrograde tracer studies, *J. Comp. Neurol.* **203**:67–90.

Spreafico, R., Schmechel, D. E., Ellis, L. C., Jr., and Rustioni, A., 1983, Cortical relay neurons and interneurons in the N. ventralis posterolateralis of cats: A horseradish peroxidase, electron-microscopic, Golgi and immunocytochemical study, *Neuroscience* **9**:491–510.

Squire, L. R., and Moore, R. Y., 1979, Dorsal thalamic lesion in a noted case of human memory dysfunction, *Ann. Neurol.* **6**:503–506.

Stretavan, D., and Dykes, R. W., 1983, The organization of two subcutaneous modalities in the forearm region of area 3b of cat somato-sensory cortex, *J. Comp. Neurol.* **213**:381–398.

Stanford, L. R., Friedlander, M. J., and Sherman, S. M., 1981, Morphology of physiologically identified W-cells in the C-laminae of the cat's lateral geniculate nucleus, *J. Neurosci.* **6**:578–584.

Stanford, L. R., Friedlander, M. J., and Sherman, S. M., 1983, Morphological and physiological properties of geniculate W-cells of the cat: A comparison with X- and Y-cells, *J. Neurophysiol.* **50**:582–608.

Stanton, G. B., 1980, Topographical organization of ascending cerebellar projections from the dentate and interposed nuclei in *Macaca mulatta:* An anterograde degeneration study, *J. Comp. Neurol.* **190**:699–733.

Starzl, T. E., and Magoun, H. W., 1951, Organization of the diffuse thalamic projection system, *J. Neurophysiol.* **14**:133–146.

Starzl, T. E., and Whitlock D. G., 1952, Diffuse thalamic projection in monkey, *J. Neurophysiol.* **15**:449–468.

Stefens, R., and Droogleever-Fortuyn, J., 1953, Contribution à l'étude de la structure et de quelques connexions des noyaux intermédiaires du thalamus chez le lapin, *Schweiz. Arch. Neurol. Psychiatr.* **72**:299–318.

Stein, S. A. W., 1834, *De Thalamo et Origine Nervi Optici in Homine et Animalibus Vertebratis,* S. Trier, Copenhagen.

Stensen, N., 1669, *Discours de Monsieur Stenon sur l'Anatomie du Cerveau à Messieurs de l'Assemblie qui se Fait chez Monsier Thévenot,* R. de Ninville, Paris.

Steriade, M., 1970, The cerebello-thalamo-cortical pathway: Ascending (specific and unspecific) and corticofugal controls, *Int. J. Neurol.* **7**:177–200.

Steriade, M., 1978, Cortical long-axoned cells and putative interneurons during the sleep–waking cycle, *Behav. Brain Sci.* **3**:465–514.

Steriade, M., 1981, Mechanisms underlying cortical activation: Neuronal organization and properties of the midbrain reticular core and intralaminar thalamic nuclei, in: *Brain Mechanisms and Perceptual Awareness* (O. Pompeiano and C. Ajmone-Marsan, eds.), Raven Press, New York, pp. 327–377.

Steriade, M., 1984, The excitatory–inhibitory response sequence of thalamic and cortical neurons: State related changes and regulatory systems, in: *Dynamic Aspects of Neocortical Function* (G. Edelman, W. E. Gall, and W. M. Cowan, eds.), Wiley, New York pp. 107–157.

Steriade, M., and Glenn, L. L., 1982, Neocortical and caudate projections of intralaminar thalamic neurons and their synaptic excitation from midbrain reticular core, *J. Neurophysiol.* **48:** 352–371.

Steriade, M., and Morin, D., 1981, Reticular influences on primary and augmenting responses in primary somatosensory cortex, *Brain Res.* **205**:67–80.

Steriade, M., and Wyzinski, P., 1972, Cortically elicited activities in thalamic reticularis neurons, *Brain Res.* **42**:514–520.

Steriade, M., Wyzinski, P., and Apostol, V., 1972, Corticofugal projections governing rhythmic thalamic activity, in: *Corticothalamic Projections and Sensorimotor Activities* (T. Frigyesi, E. Rinvik, and M. D. Yahr, eds.), Raven Press, New York, pp. 221–271.

Steriade, M., Diallo, A., Oakson, G., and White-Guay, B., 1977a, Some synaptic inputs and ascending projections of lateralis posterior thalamic neurons, *Brain Res.* **131**:39–53.

Steriade, M., Oakson, G., and Diallo, A., 1977b, Reticular influence on lateralis posterior thalamic neurons, *Brain Res.* **131**:55–71.

Steriade, M., Ropert, N., Kitsikis, A., and Oakson, G., 1980, Ascending activating neuronal networks in midbrain reticular core and related rostral systems, in: *The Reticular Formation Revisited* (J. A. Hobson and M. A. B. Brazier, eds.), Raven Press, New York, pp. 125–167.

Steriade, M., Deschênes, M., and Domich, L.,1983, Abolition of spindling rhythmicity in thalamo-cortical cells disconnected from the reticularis thalami nucleus, *Neurosci. Abstr.* **9:**1213.

Sterling, P., and Davis, T. L., 1981, Neurons in cat lateral geniculate nucleus that concentrate exogenous [³H]-gamma-aminobutyric acid (GABA), *J. Comp. Neurol.* **192:**737–750.

Stevens, J. K., and Gerstein, G. L., 1976, Interactions between cat lateral geniculate neurons, *J. Neurophysiol.* **39:**239–256.

Stilling, B., 1857–1859, *Neue Untersuchungen über den Bau des Rückenmarks,* Hotop, Kassel.

Stoffels, J., 1939a, La projection des noyaux antérieurs du thalamus sur l'écorce interhémisphérique: Étude anatomo-expérimentale, *Mém. Acad. R. Méd. Belg. Ser. 2* **1**(Fasc. 2):1–59.

Stoffels, J., 1939b, Organisation du thalamus et du cortex cérébral chez le lapin: Synthèse finale, *J. Belge Neurol. Psychiatr.* **39:**557–575.

Stone, J., and Clarke, R., 1980, Correlation between soma size and dendritic morphology in cat retinal ganglion cells: Evidence of further variation in the gamma-cell class, *J. Comp. Neurol.* **192:**211–217.

Stone, J., and Dreher, B., 1973, Projection of X- and Y-cells of the cat's lateral geniculate nucleus to area 17 and 18 of visual cortex, *J. Neurophysiol.* **36:**551–567.

Stone, J., and Fukuda, Y., 1974, Properties of cat retinal ganglion cells: A comparison of W-cells with X- and Y-cells, *J. Neurophysiol.* **37:**722–748.

Stone, J., and Hansen, S. M., 1966, The projection of the cat's retina on the lateral geniculate nucleus, *J. Comp. Neurol.* **126:**601–624.

Stone, J., and Hoffmann, K.-P., 1971, Conduction velocity as a parameter of organization in the cat's lateral geniculate nucleus, *Brain Res.* **32:**454–459.

Stone, J., and Hoffmann, K.-P., 1972, Very slow-conducting ganglion cells in the cat's retina: A major new functional type, *Brain Res.* **43:**610–616.

Stone, J., Leicester, J., and Sherman, S. M., 1973, The naso-temporal division of the monkey's retina, *J. Comp. Neurol.* **150:**333–348.

Stone, J., Dreher, B., and Leventhal, A., 1979, Hierarchical and parallel mechanisms in the organization of visual cortex, *Brain Res. Rev.* **1:**345–394.

Stone, J., Rapaport, D. H., Williams, R. W., and Chalupa, L., 1982, Uniformity of cell distribution in the ganglion cell layer of prenatal cat retina: Implications for mechanisms of retinal development, *Dev. Brain Res.* **2:**231–242.

Stoupel, N., and Terzuolo, C., 1954, Étude electrophysiologique des connexions et de la physiologie du noyau caudé, *Acta Neurol. Psychiatr. Belg.* **54:**239–248.

Streit, P., 1980, Selective retrograde labeling indicating the transmitter of neuronal pathways, *J. Comp. Neurol.* **191:**429–464.

Streit, P., Burkhalter, A., Stella, M., and Cuénod, M., 1980, Patterns of activity in pigeon brain's visual relays as revealed by the [¹⁴C] 2-deoxyglucose method, *Neuroscience* **5:**1053–1066.

Strick, P. L., 1970, Cortical projections of the feline thalamic nucleus ventralis lateralis, *Brain Res.* **20:**130–134.

Strick, P. L., 1973, Light microscopic analysis of the cortical projection of the thalamic ventrolateral nucleus in the cat, *Brain Res.* **55:**1–24.

Strick, P. L., 1975, Multiple sources of thalamic input to the primate motor cortex, *Brain Res.* **88:**372–377.

Strick, P. L., 1976a, Anatomical analysis of ventrolateral thalamic input in primate motor cortex, *J. Neurophysiol.* **39:**1020–1031.

Strick, P. L., 1976b, Activity of ventrolateral thalamic neurons during arm movement, *J. Neurophysiol.* **39:**1032–1044.

Strick, P. L., and Preston, J. B., 1978a, Multiple representations in the primate motor cortex, *Brain Res.* **154:**366–370.

Strick, P. L., and Preston, J. B., 1978b, Sorting of somatosensory afferent information in primate motor cortex, *Brain Res.* **156:**364–368.

Ströer, W. F. H., 1956, Studies on the diencephalon. I. The embryology of the diencephalon of the rat, *J. Comp. Neurol.* **105:**1–24.

Strominger, N. L., Nelson, L. R., and Dougherty, W. J., 1977, Second order auditory pathways in the chimpanzee, *J. Comp. Neurol.* **172:**349–366.

Stryker, M. P., and Zahs, K. R., 1983, ON and OFF sublaminae in the lateral geniculate nucleus of the ferret, *J. Neurosci.* **3:**1943–1951.

Sugitani, M., 1979, Electrophysiological and sensory properties of the thalamic reticular neurones related to somatic sensation in rats, *J. Physiol. (London)* **290:**79–95.

Sumitomo, I., Ide, K., Iwama, K., and Arikuni, T., 1969, Conduction velocity of optic nerve fibres innervating lateral geniculate body and superior colliculus in the rat, *Exp. Neurol.* **25:**378–392.

Sumitomo, I., Nakamura, M., and Iwama, K., 1976, Location and function of the so-called interneurons of rat lateral geniculate body, *Exp. Neurol.* **51:**110–173.

Sumitomo, I., Sugitani, M., Fukuda, Y., and Iwama, K., 1979, Properties of cells responding to visual stimuli in the rat ventral lateral geniculate nucleus, *Exp. Neurol.* **66:**721–736.

Sur, M., and Sherman, S. M., 1982a, Linear and nonlinear W-cells in C-laminae of the cat's lateral geniculate nucleus, *J. Neurophysiol.* **47:**869–884.

Sur, M., and Sherman, S. M., 1982b, Retinogeniculate terminations in cats: Morphological differences between X and Y cell axons, *Science* **218:**389–391.

Sur, M., Wall, J. T., and Kaas, J. H., 1981a, Modular segregation of functional cell classes within the postcentral somatosensory cortex of monkeys, *Science* **212:**1059–1061.

Sur, M., Weller, R. E., and Kaas, J. H., 1981b, The organization of somatosensory area II in tree shrews, *J. Comp. Neurol.* **201:**121–134.

Sur, M., Humphrey, A. L., and Sherman, S. M., 1982, Monocular deprivation affects X- and Y-cell retinogeniculate terminations in cats, *Nature (London)* **300:**183–185.

Suzuki, H., and Kato, E., 1966, Binocular interaction of cat's lateral geniculate body, *J. Neurophysiol.* **29:**909–920.

Suzuki, H., and Taira, N., 1961, Effect of reticular stimulation upon synaptic transmission in cat's lateral geniculate body, *Jpn. J. Physiol.* **11:**641–655.

Suzuki, H., and Takahashi, M., 1970, Organization of lateral geniculate neurons in binocular inhibition, *Brain Res.* **23:**261–264.

Swadlow, H. A., and Weyand, T. G., 1981, Efferent systems of the rabbit visual cortex: Laminar distribution of the cells of origin, axonal conduction velocities, and identification of axonal branches, *J. Comp. Neurol.* **203:**799–822.

Swanson, L. W., 1976, An autoradiographic study of the efferent connections of the preoptic region in the rat, *J. Comp. Neurol.* **167:**227–256.

Swanson, L. W., and Cowan, W. M., 1975, A note on the connections and development of the nucleus accumbens, *Brain Res.* **92:**324–330.

Swanson, L. W., and Cowan, W. M., 1977, An autoradiographic study of the organization of the efferent connections of the hippocampal formation in the rat, *J. Comp. Neurol.* **172:**49–84.

Swanson, L. W., and Hartman, B. K., 1975, The central adrenergic system: An immunofluorescence study of the location of cell bodies and their efferent connections in the rat utilizing dopamine-beta-hydroxylase as a marker, *J. Comp. Neurol.* **163:**467–506.

Swanson, L. W., Cowan, W. M., and Jones, E. G., 1974, An autoradiographic study of the efferent projections of the ventral lateral geniculate nucleus in the albino rat and the cat, *J. Comp. Neurol.* **156:**143–163.

Sychowa, B., 1962, Medial geniculate body of the dog, *J. Comp. Neurol.* **118:**355–371.

Symmes, D., and Anderson, K. V., 1967, Reticular modulation of higher auditory centers in monkey, *Exp. Neurol.* **18:**161–176.

Symonds, L. L., and Kaas, J. H., 1978, Connections of striate cortex in the prosimian, *Galago senegalensis, J. Comp. Neurol.* **181:**477–512.

Symonds, L. L., Rosenquist, A., Edwards, S., and Palmer, L., 1978, Thalamic projections to electrophysiologically defined visual areas in the cat, *Neurosci. Abstr.* **4:**647.

Symonds, L. L., Rosenquist, A. C., Edwards, S. B., and Palmer, L. A., 1981, Projections of the pulvinar–lateral posterior complex to visual cortical areas in the cat, *Neuroscience* **6:**1995–2020.

Szentágothai, J., 1963, The structure of the synapse in the lateral geniculate body, *Acta Anat.* **55:**166–185.

Szentágothai, J., 1973, Neuronal and synaptic architecture of the lateral geniculate body, in: *Handbook of Sensory Physiology,* Volume 7, Part 3B (R. Jung, ed.), Springer-Verlag, Berlin, pp. 141–176.

Szentágothai, J., Hámori, J., and Tömböl, T., 1966, Degeneration and electron microscopic analysis of the synaptic glomeruli in the lateral geniculate body, *Exp. Brain Res.* **2:**283–301.

Szentágothai, J., Flerkő, B., Mess, B., and Halász, B., 1968, *Hypothalamic Control of the Anterior Pituitary,* 2nd ed., Akademiai Kiado, Budapest.

Taber, E., 1961, The cytoarchitecture of the brain stem of the cat. I. Brain stem nuclei of cat, *J. Comp. Neurol.* **116:**27–69.

Takagi, S. F., 1979, Dual systems for sensory olfactory processing in higher primates, *Trends Neurosci.* **2:**313–315.

Takahashi, E. S., Hickey, T. L., and Oyster, C. W., 1977, Retinogeniculate projections in the rabbit: An autoradiographic study, *J. Comp. Neurol.* **175:**1–12.

Talairach, J., Hecaen, H., David, M., Monnier, M., and de Ajuriaguerra, J., 1949, Recherches sur la coagulation therapeutique des structures sous-corticale chez l'homme, *Rev. Neurol.* **81:**4–24.

Talbot, S. A., and Marshall, W. M., 1941, Physiological studies on neural mechanisms of visual localization and discrimination, *Am. J. Ophthalmol.* **24:**1255–1264.

Talbot, W. H., Darian-Smith, I., Kornhuber, H. H., and Mountcastle, V. B., 1968, The sense of flutter-vibration: Comparison of the human capacity with response patterns of mechano-receptive afferents from the monkey hand, *J. Neurophysiol.* **31:**301–334.

Tanabe, T., Iino, M., Ooshima, Y., and Takagi, S. F., 1974, An olfactory area in the prefrontal lobe, *Brain Res.* **80:**127–130.

Tanabe, T., Iino, M., and Takagi, S. F., 1975a, Discrimination of odors in olfactory bulb, pyriform–amygdaloid areas, and orbitofrontal cortex of the monkey, *J. Neurophysiol.* **38:**1284–1296.

Tanabe, T., Yarita, H., Iino, M., Ooshima, Y., and Takagi, S. F., 1975b, An olfactory projection area in orbitofrontal cortex of the monkey, *J. Neurophysiol.* **38:**1269–1283.

Tanji, D. G., Wise, S. P., Dykes, R. W., and Jones, E. G., 1977, Cytoarchitecture and thalamic connectivity of third somatic area of cat cerebral cortex, *J. Neurophysiol.* **41:**268–284.

Tanji, J., and Wise, S. P., 1981, Submodality distribution in sensorimotor cortex of the unanesthetized monkey, *J. Neurophysiol.* **45:**467–481.

Tanji, J., Taniguchi, K., and Saga, T., 1980, Supplementary motor area: Neuronal response to motor instructions, *J. Neurophysiol.* **43:**60–68.

Tarlov, E., 1969, The rostral projections of the primate vestibular nuclei: An experimental study in macaque, baboon and chimpanzee, *J. Comp. Neurol.* **135:**27–56.

Tarlov, E. C., and Moore, R. Y., 1966, The tecto-thalamic connections in the brain of the rabbit, *J. Comp. Neurol.* **126:**403–421.

Tatemoto, K., Carlquist, M., and Mutt, V., 1982, Neuropeptide-Y, a novel brain peptide with structural similarities to peptide YY and pancreatic polypeptide, *Nature (London)* **296:**659–660.

Tebēcis, A. K., 1972, Cholinergic and non-cholinergic transmission in the medial geniculate nucleus of the cat, *J. Physiol. (London)* **226:**153–172.

Tebēcis, A. K., and DiMaria, A., 1972, A re-evaluation of the mode of action of 5-hydroxytryptamine on lateral geniculate neurones: Comparison with catecholamines and LSD, *Exp. Brain Res.* **14:**480–493.

Teitelbaum, H., Sharpless, S. K., and Byck, R., 1968, Role of somatosensory cortex in interhemispheric transfer of tactile habits, *J. Comp. Physiol. Psychol.* **66:**623–632.

Tello, F., 1904, Disposición macroscópica y estructura del cuerpo geniculado externo, *Trabajos Lab. Invest. Biol. Madrid* **3:**39–62.

ten Donkelaar, H. J., and De Boer-van Huizen, R., 1981a, Basal ganglia projections to the brainstem in the lizard *Varanus exanthematicus* as demonstrated by retrograde transport of horseradish peroxidase, *Neuroscience* **6:**1567–1590.

ten Donkelaar, H. J., and De Boer-van Huizen, R., 1981b, Ascending projections of the brainstem reticular formation in a nonmammalian vertebrate (the lizard *Varanus exanthematicus*), with notes on the afferent connections of the forebrain, *J. Comp. Neurol.* **200:**501–528.

Thach, W. T., 1978, Correlation of neural discharge with pattern and force of muscular activity, joint position, and direction of intended next movement in motor cortex and cerebellum, *J. Neurophysiol.* **41:**654–676.

Thach, W. T., and Jones, E. G., 1979, The cerebellar dentatothalamic connection: Terminal field, lamellae, rods and somatotopy, *Brain Res.* **169:**168–172.

Thompson, R., and Myers, R. E., 1971, Brainstem mechanisms underlying visually guided responses in the rhesus monkey, *J. Comp. Physiol. Psychol.* **74:**479–512.

Thompson, R. F., 1967, *Foundations of Physiological Psychology,* Harper & Row, New York, pp. 474–528.

Thompson, R. F., and Sindberg, R. M., 1960, Auditory response fields in association and motor cortex of cat, *J. Neurophysiol.* **23:**87–105.

Thompson, R. F., Johnson, R. H., and Hoopes, J. J., 1963a, Organization of auditory, somatic sensory, and visual projection to association fields of cerebral cortex in the cat, *J. Neurophysiol.* **26:**343–364.

Thompson, R. F., Smith, H. E., and Bliss, D., 1963b, Auditory, somatic sensory, and visual response interactions and interrelations in association and primary cortical fields of the cat, *J. Neurophysiol.* **26:**365–378.

Thuma, B. D., 1928, Studies on the diencephalon of the cat. I. The cyto-architecture of the corpus geniculatum laterale, *J. Comp. Neurol.* **46:173–199.**

Tigges, J., 1970, Retinal projections to subcortical optic nuclei in diurnal and nocturnal squirrels, *Brain Behav. Evol.* **3:**121–134.

Tigges, J., and O'Steen, W. K., 1974, Termination of retinofugal fibers in squirrel monkey: A reinvestigation using autoradiographic methods, *Brain Res.* **79:**489–495.

Tigges, J., Tigges, M., and Kalaga, C. S., 1973, Efferent connections of area 17 in *Galago*, *Am. J. Phys. Anthropol.* **38:**393–398.

Tigges, J., Bos, J., and Tigges, M., 1977a, An autoradiographic investigation of the subcortical visual system in chimpanzee, *J. Comp. Neurol.* **172:**367–380.

Tigges, J., Tigges, M., and Perachio, A. A., 1977b, Complementary laminar terminations of afferents to area 17 originating in area 18 and in the lateral geniculate nucleus in squirrel monkey, *J. Comp. Neurol.* **176:**87–100.

Tigges, M., and Tigges, J., 1970, The retinofugal fibers and their terminal nuclei in *Galago crassicaudatus* (primates), *J. Comp. Neurol.* **138:**87–101.

Tiwari, R. K., and King, R. B., 1974, Fiber projections from trigeminal nucleus caudalis in primate (squirrel monkey and baboon), *J. Comp. Neurol.* **158:**191–206.

Tobias, T. J., 1975, Afferents to prefrontal cortex from the thalamic mediodorsal nucleus in the rhesus monkey, *Brain Res.* **83:**191–212.

Tobias, T. J., and Ebner, F. F., 1973, Thalamocortical projections from the mediodorsal nucleus in the Virginia opossum, *Brain Res.* **52:**79–96.

Tohyama, M., Toshihiro, M., Hashimoto, J., Shrestha, G. R., Tamura, O., and Shimizu, N., 1974, Comparative anatomy of the locus coeruleus. I. Organizations and ascending projections of the catecholamine-containing neurons in the pontine region of the bird, *Melopsittacus undulatus*, *J. Hirnforsch.* **15:**319–330.

Tokunaga, A., and Otani, K., 1978, Fine structure of the medial habenular nucleus in the rat, *Brain Res.* **150:**600–606.

Tömböl, T., 1967, Short neurons and their synaptic relations in the specific thalamic nuclei, *Brain Res.* **3:**307–326.

Tömböl, T., 1969, Two types of short axon (Golgi 2nd) interneurons in the specific thalamic nuclei, *Acta Morphol. Acad. Sci. Hung.* **17:**285–297.

Tömböl, T., Ungvary G., Hajdu, F., Madarász, M., and Somogyi, G., 1969, Quantitative aspects of neuron arrangement in the specific thalamic nuclei, *Acta Morphol. Acad. Sci. Hung.* **17:**299–313.

Toncray, J. E., and Krieg, W. J. S., 1946, The nuclei of the human thalamus: A comparative approach, *J. Comp. Neurol.* **85:**421–459.

Tong, L., Kalil, R. E., and Spear, P. D., 1982, Thalamic projections to visual areas of the middle suprasylvian sulcus in the cat, *J. Comp. Neurol.* **212:**103–117.

Tootell, R. B. H., Silverman, M. S., Swirkes, E., and DeValois, R. L., 1982, Deoxyglucose analysis of retinotopic organization in primate striate cortex, *Science* **218:**902–903.

Tootell, R. B. H., Silverman, M. S., De Valois, R. L., and Jacobs, G. H., 1983, Functional organization of the second cortical visual area in primates, *Science* **220:**737–739.

Torrealba, F., Partlow, G. D., and Guillery, R. W., 1981, Organization of the projection from the superior colliculus to the dorsal lateral geniculate nucleus of the cat, *Neuroscience* **6:**1341–1360.

Torvik, A., 1957, The ascending fibers from the main trigeminal sensory nucleus: An experimental study in the cat, *Am. J. Anat.* **100:**1–16.

Towns, L. C., Burton, S. L., Kimberly, C. J., and Fetterman, M. R., 1982, Projections of the dorsal lateral geniculate and lateral posterior nuclei to visual cortex in the rabbit, *J. Comp. Neurol.* **210:**87–98.

Tracey, D. J., 1980, The projection of joint receptors to the cuneate nucleus in the cat, *J. Physiol. (London)* **305:**433–449.

Tracey, D. J., Asanuma, C., Jones, E. G., and Porter, R., 1980, Thalamic relay to motor cortex: Afferent pathways from brain stem, cerebellum and spinal cord in monkeys, *J. Neurophysiol.* **44:**532–554.

Trachtenberg, M. D., and Ingle, D., 1974, Thalamo-tectal projections in the frog, *Brain Res.* **79:**419–430.

Trevarthen, C. B., and Sperry, R. W., 1973, Perceptual unity of the ambient visual field in human commissurotomy patients, *Brain* **96:**547–570.

Trevino, D. L., 1976, The origin and projections of a spinal nociceptive and thermoreceptive pathway, in: *Sensory Functions of the Skin in Primates, with Special Reference to Man* (Y. Zotterman, ed.), Pergamon Press, Elmsford, N. Y., pp. 367–379.

Trevino, D. L., and Carstens, E., 1975, Confirmation of the location of spinothalamic neurons in the cat and monkey by the retrograde transport of horseradish peroxidase, *Brain Res.* **98:**177–182.

Trevino, D. L., Maunz, R. A., Bryan, R. N., and Willis, W. D., 1972, Location of cells of origin of the spinothalamic tract in the lumbar enlargement of cat, *Exp. Neurol.* **34:**64–77.

Trevino, D. L., Coulter, J. D., and Willis, W. D., 1973, Location of cells of origin of spinothalamic tract in lumbar enlargement of the monkey, *J. Neurophysiol.* **36:**750–761.

Trojanowski, J. Q., and Jacobson, S., 1974, Medial pulvinar afferents to frontal eye fields in rhesus monkey demonstrated by horseradish peroxidase, *Brain Res.* **80:**395–411.

Trojanowski, J. Q., and Jacobson, S., 1975, A combined horseradish peroxidase–autoradiographic investigation of reciprocal connections between superior temporal gyrus and pulvinar in squirrel monkey, *Brain Res.* **85:**347–353.

Trojanowski, J. Q., and Jacobson, S., 1976, Areal and laminar distribution of some pulvinar cortical efferents in rhesus monkey, *J. Comp. Neurol.* **169:**371–391.

Trojanowski, J. Q., and Jacobson, S., 1977, The morphology and laminar distribution of cortico-pulvinar neurons in the rhesus monkey, *Exp. Brain Res.* **28:**51–62.

Truex, R. C., Taylor, J. H., Smythe, M. Q., and Gildenberg, P. L., 1970, The lateral cervical nuclei of cat, dog and man, *J. Comp. Neurol.* **139:**93–104.

Tsai, C., 1925, The optic tracts and centers of the opossum, *Didelphis virginiana, J. Comp. Neurol.* **39:**173–216.

Tsang, Y., 1937, Visual centers in blinded rats, *J. Comp. Neurol.* **66:**211–261.

Tsumoto, T., 1974, Characteristics of the thalamic ventrobasal relay neurons as a function of conduction velocities of medial lemniscal fibers, *Exp. Brain Res.* **21:**211–224.

Tsumoto, T., and Nakamura, S., 1974, Inhibitory organization of the thalamic ventrobasal neurons with different peripheral representations, *Exp. Brain Res.* **21:**195–210.

Tsumoto, T., and Suda, K., 1980, Three groups of cortico-geniculate neurons and their distribution in binocular and monocular segments of cat striate cortex, *J. Comp. Neurol.* **193:**223–236.

Türck, L., 1859a, Über die Beziehung gewisser Krankheitsherde des grossen Gehirnes zur Anästhesie, *Sitzungsber. Akad. Wiss. Wien Math.-Naturwiss. Kl.* **36:**191–199.

Türck, L., 1859b, Über die Beziehungen gewisser Krankheitsherde des grossen Gehirns zur Anästhesie, *Jahrb. Psychiatr. Neurol.* **31:**186–194.

Tusa, R. J., and Palmer, L. A., 1980, Retinotopic organization of areas 20 and 21 in the cat, *J. Comp. Neurol.* **193:**147–164.

Tusa, R. J., Rosenquist, A. C., and Palmer, L. A., 1979, Retinotopic organization of areas 18 and 19 in the cat, *J. Comp. Neurol.* **185:**657–678.

Ueki, A., Uno, M., Anderson, M., and Yoshida, M., 1977, Monosynaptic inhibition of thalamic neurons produced by stimulation of substantia nigra, *Experientia* **33:**1480–1481.

Uhl, G. R., Kuhar, M. J., and Snyder, S. H., 1977, Neurotensin: Immunohistochemical localization in rat central nervous system, *Proc. Natl. Acad. Sci. USA* **74:**4059–4063.

Ulinski, P. S., 1977, Tectal efferents in the banded water snake, *Natrix, sipedon, J. Comp. Neurol.* **173:**251–274.

Ulinski, P. S., 1979, Intrinsic organization of snake dorsomedial cortex: An electron microscopic and Golgi study, *J. Morphol.* **161:**185–210.

Ulinski, P. S., 1981, Thick caliber projections from brainstem to cerebral cortex in the snakes, *Thamnophis sirtalis* and *Natrix sipedon, Neuroscience* **6:**1725–1743.

Ulinski, P. S., 1983, *Dorsal Ventricular Ridge,* Wiley, New York.

Ulinski, P. S., 1984, Thalamic projections to the somatosensory cortex of the echidna, *Tachyglossus aculeatus, J. Comp. Neurol.* **229:**153–170.

Ulinski, P. S., and Rainey, W. T., 1980, Intrinsic organization of snake lateral cortex, *J. Morphol.* **165:**85–116.

Ungerleider, L. G., and Christensen, C. A., 1977, Pulvinar lesions in monkeys produce abnormal eye movements during visual discrimination training, *Brain Res.* **136:**189–196.

Ungerleider, L. G., and Christensen, C. A., 1979, Pulvinar lesions in monkey produce abnormal scanning of a complex visual array, *Neuropsychologia* **17:**493–501.

Ungerleider, L. G., Galkin, T. W., and Mishkin, M., 1983, Visuotopic organization of projections from striate cortex to inferior and lateral pulvinar in rhesus monkey, *J. Comp. Neurol.* **217:**137–157.

Ungerstedt, U., 1971, Stereotaxic mapping of the monoamine pathways in the rat brain, *Acta Physiol. Scand. Suppl.* **367:**1–48.

Uno, M., Yoshida, M., and Hirota, I., 1970, The mode of cerebello-thalamic relay transmission investigated with intracellular recordings from cells of the ventrolateral nucleus of cat's thalamus, *Exp. Brain Res.* **10:**121–139.

Updyke, B. V., 1975, The patterns of projection of cortical areas 17, 18 and 19 onto the laminae of the dorsal lateral geniculate nucleus in the cat, *J. Comp. Neurol.* **163:**377–395.

Updyke, B. V., 1977, Topographic organization of the projections from cortical areas 17, 18 and 19 onto the thalamus, pretectum and superior colliculus in the cat, *J. Comp. Neurol.* **173:**81–122.

Updyke, B. V., 1979, A Golgi study of the class V cell in the visual thalamus of the cat, *J. Comp. Neurol.* **186:**603–620.

Updyke, B. V., 1981a, Projections from visual areas of the middle suprasylvian sulcus onto the lateral posterior complex and adjacent thalamic nuclei in cat, *J. Comp. Neurol.* **201:**477–506.

Updyke, B. V., 1981b, Multiple representations of the visual field: Corticothalamic and thalamic organization in the cat, in: *Cortical Sensory Organization*, Volume 2 (C. N. Woolsey, ed.), Humana Press, Clifton, N. J., pp. 83–101.

Updyke, B. V., 1983, A reevaluation of the functional organization and cytoarchitecture of the feline lateral posterior complex, with observations on adjoining cell groups, *J. Comp. Neurol.* **219:**143–181.

Vahle-Hinz, C., and Gottschaldt, K.-M., 1983, Principal differences in the organization of the thalamic face representation in rodents and felids, in: *Somatosensory Integration in the Thalamus* (G. Macchi, A. Rustioni, and R. Spreafico, eds.), Elsevier, Amsterdam, pp. 125–145.

Valenstein, E. S., and Nauta, W. J. H., 1959, A comparison of the distribution of the fornix system in the rat, guinea pig, cat and monkey, *J. Comp. Neurol.* **113:**337–364.

Van der Kooy, D., and Carter, D. A., 1981, The organization of the efferent projections and striatal afferents of the entopenduncular nucleus and adjacent areas in the rat, *Brain Res.* **211:**15–36.

Van der Kooy, D., Kuypers, H. G. J. M., and Catsman-Berrevoets, C. E., 1978, Single mammillary body cells with divergent axon collaterals: Demonstration by a simple, fluorescent retrograde double labeling technique in the rat, *Brain Res.* **158:**189–196.

Van der Loos, H., 1976, Barreloids in mouse somatosensory thalamus, *Neurosci. Lett.* **2:**1–6.

Van der Loos, H., and Woolsey, T. A., 1973, Somato-sensory cortex: Structural alterations following early injury to sense organs, *Science* **179:**395–398.

Vanegas, H., and Ebbesson, S. O. E., 1976, Telencephalic projections in two teleost species, *J. Comp. Neurol.* **165:**181–196.

Vanegas, H., and Ito, H., 1983, Morphological aspects of the teleostean visual system: A review, *Brain Res. Rev.* **6:**117–138.

Van Essen, D. C., and Zeki, S. M., 1978, The topographic organization of rhesus monkey prestriate cortex, *J. Physiol. (London)* **277:**193–226.

Van Essen, D. C., Newsome, W. T., and Bixby, J. L., 1982, The pattern of interhemispheric connections and its relation to extrastriate visual areas in the macaque monkey, *J. Neurosci.* **2:**265–283.

Vaney, D. I., Levick, W. R., and Thibos, L. N., 1981, Rabbit retinal ganglion cells—receptive field classification and axonal conduction properties, *Exp. Brain Res.* **44:**27–33.

Van Gehuchten, A., 1894a, Contribution a l'étude du faisceau de Meynert ou faisceau rétro-réflexe, *Bull. Acad. R. Med. Belg.* 1–5.

Van Gehuchten, A., 1894b, Contribution a l'étude du système nerveux des téléostéens: Communication préliminaire, *Cellule* **10:**253–295.

Van Gehuchten, A., 1897, *Anatomie du Système Nerveux de L'Homme, Leçons Professées à l'Université de Louvain*, 2nd ed., Uystpruyst-Dieudonne, Louvain.

Van Gehuchten, A., 1901, Recherches sur les voies sensitives centrales: La voie centrale du trijumeau, *Névraxe* **3**:235–261.

Vastola, E. F., 1961, A direct pathway from lateral geniculate body to association cortex, *J. Neurophysiol.* **24**:469–487.

Veazey, R. B., Amaral, D. G., and Cowan, W. M., 1982, The morphology and connections of the posterior hypothalamus in the cynomolgus monkey *(Macaca fascicularis)*. II. Efferent connections, *J. Comp. Neurol.* **207**:135–156.

Velayos, J. L., and Reinoso-Suarez, F., 1982, Topographic organization of the brainstem afferents to the mediodorsal thalamic nucleus, *J. Comp. Neurol.* **206**:17–27.

Veraart, C., Meulders, M., and Godfraind, J. M., 1972, Visual properties of neurons in pulvinar, nucleus lateralis posterior and nucleus suprageniculatus thalami in the cat. II. Quantitative investigation, *Brain Res.* **44**:527–546.

Verley, R., 1977, The post-natal development of the functional relationships between the thalamus and the cerebral cortex in rats and rabbits, *Electroencephalogr. Clin. Neurophysiol.* **43**:679–690.

Verley, R., and Axelrad, H., 1975, Postnatal ontogenesis of potentials elicited in the cerebral cortex by afferent stimulation, *Neurosci. Lett.* **1**:99–104.

Verley, R., and Gaillard, P., 1978, Postnatal ontogenesis of potentials elicited in the barrelfield of mice by stimulation of the maxillary nerve, *Neurosci. Lett.* **10**:121–127.

Verley, R., and Onnen, I., 1981, Somatotopic organization of the tactile thalamus in normal adult and developing mice and in adult mice dewhiskered since birth, *Exp. Neurol.* **72**:462–474.

Verley, R., and Pidoux, B., 1981, Electrophysiological study of the topography of the vibrissae projections to the tactile thalamus and cerebral cortex in mutant mice with hair defects, *Exp. Neurol.* **72**:475–485.

Verzeano, M., and Calma, I., 1954, Unit activity in spindle bursts, *J. Neurophysiol.* **17**:417–428.

Vesalius, A., 1543, *De Humani Corporis Fabrica Libri Septem*, Oporinus, Basel.

Veyssiére, R., 1874, *Recherches Cliniques et Expérimentales sur l'Hémi-anesthésie de Cause Cérébrale*, Delahaye, Paris.

Vicq d'Azyr, F., 1786, *Traite d'Anatomie et de Physiologie*, 3 vols., Didot, Paris.

Victor, M., Adams, R., and Collins, G., 1971, *The Wernicke–Korsakoff Syndrome*, Davis, Philadelphia.

Vierck, C. J., Jr., 1973, Alterations of spatio-tactile discrimination after lesions of primate spinal cord, *Brain Res.* **58**:69–79.

Vierck, C. J., Jr., 1974, Tactile movement detection and discrimination following dorsal column lesions in monkeys, *Exp. Brain Res.* **20**:331–346.

Vierck, C. J., Jr., and Luck, M. M., 1979, Loss and recovery of reactivity to noxious stimuli in monkeys with primary spinothalamic cordotomies, followed by secondary and tertiary lesions of other cord sectors, *Brain* **102**:233–248.

Vieussens, R., 1684, *Neurographia Universalis*, Certe, Lyons.

Vincent, C. R., Staines, W. A., McGeer, E. G., and Fibiger, H. C., 1980, Transmitters contained in the efferents of the habenula, *Brain Res.* **195**:479–484.

Vital-Durand, F., Garey, L. J., and Blakemore, C., 1978, Monocular binocular deprivation in the monkey: Morphological effects and reversibility, *Brain Res.* **158**:45–64.

Vogt, B. A., Rosene, D. L., and Pandya, D. N., 1979, Thalamic and cortical afferents differentiate anterior from posterior cingulate cortex in the monkey, *Science* **204**:205–207.

Vogt, C., 1909, La myéloarchitecture du thalamus du cercopithèque, *J. Psychol. Neurol.* **12**:285–324.

Vogt, C., and Vogt, O., 1919, Allgemeinere Ergebnisse unserer Hirnforschung, *J. Psychol. Neurol.* **25**:277–462.

Vogt, C., and Vogt, O., 1920, Zur Lehre der Erkrankungen des striären Systems, *J. Psychol. Neurol.* **25**(Suppl. 3.):627–846.

Vogt, C., and Vogt, O., 1941, Thalamusstudien I–III: I. Zur Einführung. II. Homogenität und Grenzgestaltung der Grisea des Thalamus. III. Das Griseum centrale (Centrum medianum Luys), *J. Psychol. Neurol.* **50**:32–154.

Vogt, O., 1895, Ueber Fasersysteme in den mitteren und caudalen Balkenabschnitten, *Neurol. Zentralbl.* **14**:208–218, 253–260.

Vogt, O., 1898, Sur un faisceau septo-thalamique, *C. R. Soc. Biol. Ser. 19* **5**:206–207.

Von Bonin, G., and Bailey, P., 1947, *The Neocortex of Macaca Mulatta*, University of Illinois Press, Urbana.

von Economo, C., 1911, Über dissoziierte Empfindungslähmung bei Ponstumoren und über die zentralen Bahnen des sensiblen Trigeminus, *Jahrb. Psychiatr. Neurol.* **32**:107–138.

von Economo, C., and Koskinas, G. N., 1925, *Die Cytoarchitektonik der Hirnrinde des Erwachsenen Menschen,* Springer, Berlin.

von Gudden, B., 1870, Experimentaluntersuchungen über das peripherische und central Nervensystem, *Arch. Psychiatr. Nervenkr.* **2**:693–723.

von Gudden, B., 1875, Über ein neues Microtom, *Arch. Psychiatr. Nervenkr.* **5**:229–234.

von Gudden, B., 1881a, Mittheilung über das Ganglion interpendunculare, *Arch. Psychiatr. Nervenkr.* **11**:424–427.

von Gudden, B., 1881b, Beitrag zur Kenntniss des Corpus mammillare und der sogenannten Schenkel des Fornix, *Arch. Psychiatr. Nervenkr.* **11**:428–452.

von Gudden, B., 1896, Klinische und anatomische Beiträge zur Kenntnis der multiphen Alkoholneuritis nebst Bemerkungen über die Regenerationsvorgange im peripheren Nervensystem, *Arch. Psychiatr. Nervenkr.* **28**:643–741.

von Kölliker, A., 1896, *Handbuch der Gewebelehre des Menschen,* 6th ed., Volume 2, Engelmann, Leipzig.

von Monakow, C., 1882, Ueber einige durch Exstirpation circumscripter Hirnrindenregionen bedingte Entwickelungshemmungen des Kaninchengehirns, *Arch. Psychiatr. Nervenkr.* **12**:141–156, 535–549.

von Monakow, C., 1885, Experimentelle und pathologisch-anatomische Untersuchungen über die Beziehungen der sogennannten Sehsphäre zu den infracorticalen Opticuscentren und zum N. opticus, *Arch. Psychiatr. Nervenkr.* **16**:151–199, 317–352.

von Monakow, C., 1889, Experimentelle und pathologisch-anatomische Untersuchungen über die optischen Centren und Bahnen, *Arch. Psychiatr. Nervenkr.* **20**:714–787.

von Monakow, C., 1895, Experimentelle und pathologisch-anatomische Untersuchungen über die Haubenregion, den Sehhügel und die Regio subthalamica, nebst Beiträgen zur Kenntniss früh erworbener Gross- und Kleinhirndefecte, *Arch. Psychiatr. Nervenkr.* **27**:1–128, 386–478.

von Monakow, C., 1909–1910, Der rote Kern, die Haube und die Regio subthalamica bei einigen Säugetieren und beim Menschen: Vgl. anatomische, normal-anatomische experimentelle und pathologisch-anatomische Untersuchungen, *Arb. Hirnanat. Inst. Zürich* **3**:49–267; **4**:103–225.

von Monakow, C., 1914, *Die Lokalisation im Grosshirn und der Abbau der Funktion durch Kortikale Herde,* Bergmann, Wiesbaden.

Von Noorden, G. K., 1973, Histological studies of the visual system in monkey with experimental amblyopia, *Invest. Ophthalmol.* **12**:727–738.

Von Noorden, G. K., and Middleditch, P. R., 1975, Histology of the monkey's lateral geniculate nucleus after unilateral lid closure and experimental strabismus: Further observations, *Invest. Ophthalmol.* **14**:674–683.

von Sölder, F., 1897, Degenerirte Bahnen im Hirnstamme bei Läsion des unteren Cervicalmarks, *Neurol. Centralbl.* **16**:308–312.

von Volkmann, R., 1928, Vergleichende Cytoarchitektonik der Regio occipitalis Kleiner Nager und ihre Beziehung zur Sehleistung, *Z. Anat. Entwicklungsgesch.* **85**:561–657.

Voshart, K., and Van der Kooy, P., 1981, The organization of the efferent projections of the parabrachial nucleus to the forebrain in the rat: A retrograde fluorescent double-labeling study, *Brain Res.* **212**:271–286.

Vulpian, A., 1866, *Leçons sur la Physiologie Génerale et Comparée du Système Nerveux Faites à Museum d'Histoire Naturelle,* Germer-Baillière, Paris.

Wahren, W., 1957, Das Zwischenhirn des Kaninchens, *J. Hirnforsch.* **3**:143–242.

Waite, P. M. E., 1973a, Somatotopic organization of vibrissal responses in the ventro-basal complex of the rat thalamus, *J. Physiol. (London)* **228**:527–540.

Waite, P. M. E., 1973b, The responses of cells in the rat thalamus to mechanical movements of the whiskers, *J. Physiol. (London)* **228**:541–561.

Walker, A. E., 1934, The thalamic projection to the central gyri in *Macacus rhesus, J. Comp. Neurol.* **60**:161–184.

Walker, A. E., 1935, The retrograde cell degeneration in the thalamus of *Macacus rhesus* following hemidecortication, *J. Comp. Neurol.* **62**:407–419.

Walker, A. E., 1936, An experimental study of the thalamocortical projection of the macaque monkey, *J. Comp. Neurol.* **64**:1–39.

Walker, A. E., 1937, A note on the thalamic nuclei of *Macaca mulatta, J. Comp. Neurol.* **66**:145–155.

Walker, A. E., 1938a, *The Primate Thalamus*, University of Chicago Press, Chicago.

Walker, A. E., 1938b, The thalamus of the chimpanzee. I. Terminations of the somatic afferent systems, *Confin. Neurol.* **1**:99–127.

Walker, A. E., 1938c, The thalamus of the chimpanzee. II. Its nuclear structure, normal and following hemidecortication, *J. Comp. Neurol.* **69**:487–507.

Walker, A. E., 1938d, The thalamus of the chimpanzee. IV. Thalamic projections to the cerebral cortex, *J. Anat.* **73**:59–86.

Walker, A. E., 1940a, A cytoarchitectural study of the prefrontal area of the macaque monkey, *J. Comp. Neurol.* **73**:59–86.

Walker, A. E., 1940b, The medial thalamic nucleus: A comparative anatomical, physiological and clinical study of the nucleus medialis dorsalis thalami, *J. Comp. Neurol.* **73**:87–115.

Walker, A. E., 1949, Afferent connections, in: *The Precentral Motor Cortex* (P. C. Bucy, ed.), 2nd ed., University of Illinois Press, Urbana, pp. 111–132.

Walker, A. E., 1966, Internal structure and afferent–efferent relations of the thalamus, in: *The Thalamus* (D. P. Purpura and M. D. Yahr, eds.), Columbia University Press, New York, pp. 1–12.

Walker, A. E., and Fulton, J. F., 1938, The thalamus of the chimpanzee. III. Metathalamus, normal structure and cortical connexions, *Brain* **61**:250–268.

Wall, J. T., Symonds, L. L., and Kaas, J. H., 1982, Cortical and subcortical projections of the middle temporal area (MT) and adjacent cortex in galagos, *J. Comp. Neurol.* **211**:193–214.

Wall, P. D., 1967, The laminar organization of dorsal horn and effects of descending impulses, *J. Physiol. (London)* **188**:403–423.

Wallenberg, A., 1895, Acute Bulbaraffection (Embolie der Art. cerebellar. post. inf. sinistr.), *Arch. Psychiatr. Nervenkr.* **27**:504–540.

Wallenberg, A., 1896, Die sekundäre Bahn des sensiblen Trigeminus, *Anat. Anz.* **12**:95–110.

Wallenberg, A., 1900, Secundäre sensible Bahnen im Gehirnstamme des Kaninchens, ihre gegenseitige Lage und ihre Bedeutung für den Aufbau des Thalamus, *Anat. Anz.* **18**:81–105.

Wallenberg, A., 1906, Die basalen Aste des Scheidewandbundels der Vögel, *Anat. Anz.* **28**:394–400.

Waller, W. H., 1934, Topographical relations of cortical lesions to thalamic nuclei in the albino rat, *J. Comp. Neurol.* **60**:237–269.

Waller, W. H., 1938, Thalamus of the cat after hemidecortication, *J. Anat.* **72**:475–487.

Waller, W. H., 1940a, Thalamic degeneration induced by temporal lesions in the cat, *J. Anat.* **74**:528–536.

Waller, W. H., 1940b, Thalamic connections of the frontal cortex of the cat, *J. Comp. Neurol.* **73**:117–138.

Waller, W. H., and Barris, R. W., 1937, Relationships of thalamic nuclei to the cerebral cortex in the cat, *J. Comp. Neurol.* **67**:317–341.

Walls, G. L., 1953, The lateral geniculate nucleus and visual histophysiology, *Univ. Calif. Berkeley Publ. Physiol.* **9**:1–100.

Walsh, C., Polley, E. H., Hickey, T. L. and Guillery, R. W., 1983, Generation of cat retinal ganglion cells in relation to central pathways, *Nature (London)* **302**:611–614.

Walsh, T. M., and Ebner, F. F., 1973, Distribution of cerebellar and somatic lemniscal projections in the ventral nuclear complex of the Virginia opossum, *J. Comp. Neurol.* **147**:427–446.

Walzl, E. M., and Mountcastle, V. B., 1949, Projection of vestibular nerve to cerebral cortex of cat, *Am. J. Physiol.* **159**:595.

Wamsley, J. K., Zarbin, M., Birdsall, N., and Kuhar, M. J., 1980, Muscarinic cholinergic receptors: Autoradiographic localization of high and low affinity agonist binding sites, *Brain Res.* **200**:1–12.

Wamsley, J. K., Zarbin, M. A., Young, W. S., III, and Kuhar, M. J., 1982, Distribution of opiate receptors in the monkey brain: An autoradiographic study, *Neuroscience* **7**:595–613.

Wang, R. T., and Halpern, M., 1977, Afferent and efferent connections of thalamic nuclei of the visual system of garter snakes, *Anat. Rec.* **187**:741–742.

Wang, R. Y., and Aghajanian, G. K., 1977, Physiological evidence for habenula as a major link between forebrain and midbrain raphe, *Science* **197**:89–91.

Ware, C. B., Diamond, I. T., and Casagrande, V. A., 1974, Effects of ablating the striate cortex on a successive pattern discrimination: Further study of the visual system in the tree shrew (*Tupaia glis*), *Brain Behav. Evol.* **9**:264–279.

Warren, J. M., Warren, H. B., and Akert, K., 1962, Orbitofrontal cortical lesions and learning in cats, *J. Comp. Neurol.* **118:**17–41.

Wässle, H., Boycott, B. B., and Illing, R. B., 1981a, Morphology and mosaic of on- and off-beta cells in the cat retina and some functional considerations, *Proc. R. Soc. London Ser. B* **212:**177–195.

Wässle, H., Peichel, L., and Boycott, B. B., 1981b, Morphology and topography of on- and off-alpha cells in the cat retina, *Proc. R. Soc. London Ser. B* **212:**157–175.

Waszak, M., 1974, Firing pattern of neurons in the rostral and ventral part of nucleus reticularis thalami during EEG-spindles, *Exp. Neurol.* **43:**38–59.

Watanabe, T., Yanagisawa, K., Kanaki, J., and Katsuki, Y., 1966, Cortical efferent flow influencing unit responses of medial geniculate body to sound stimulation, *Exp. Brain Res.* **2:**302–317.

Watson, S., Akil, H., Sullivan, S., and Barchas, J., 1977, Immunocytochemical localization of methionine enkephalin: Preliminary observations, *Life Sci.* **21:**733–738.

Weber, A. J., and Kalil, R. E., 1983, The percentage of interneurons in the dorsal lateral geniculate nucleus of the cat and observations on several variables that affect the sensitivity of horseradish peroxidase as a retrograde marker, *J. Comp. Neurol.* **220:**336–346.

Weber, J. T., and Harting, J. K., 1980, The efferent projections of the pretectal complex: An autoradiographic and horseradish peroxidase analysis, *Brain Res.* **194:**1–28.

Weber, J. T., Casagrande, V. A., and Harting, J. K., 1977, Transneuronal transport of [³H]proline within the visual system of the grey squirrel, *Brain Res.* **129:**346–352.

Weber, J. T., Huerta, M. F., Kaas, J. H., and Harting, J. K., 1983, The projections of the lateral geniculate nucleus of the squirrel monkey: Studies of the interlaminar zones and the S layers, *J. Comp. Neurol.* **213:**135–145.

Weigert, C., 1882, Über eine neue Untersuchungsmethode des Zentralnervensystems, in: *Gesammelte Abhandlungen* (1906), Volume 2, Springer, Berlin, pp. 533–538.

Weinrich, M., and Wise, S. P., 1982, The premotor cortex of the monkey, *J. Neurosci.* **2:**1329–1342.

Weisbach, W., and Schwartzkopf, J., 1967, Nervöse Antworten auf Schallreiz im Grosshirn von Krokodilen, *Naturwissenschaften* **54:**650.

Weiskrantz, L., 1972, Behavioural analysis of the monkey's visual nervous system, *Proc. R. Soc. London Ser. B* **182:**427–455.

Weiskrantz, L., 1974, The interaction between occipital and temporal cortex in vision: An overview, in: *The Neurosciences: Third Study Program* (F. O. Schmitt and F. G. Worden, eds.), MIT Press, Cambridge, Mass., pp. 189–204.

Weiskrantz, L., Warrington, E. K., Sanders, M. D., and Marshall, J., 1974, Visual capacity in the hemianopic field following a restricted cortical ablation, *Brain* **97:**709–728.

Welker, C., 1971, Microelectrode delineation of fine grain somatotopic organization of SmI cerebral neocortex in albino rat, *Brain Res.* **26:**259–275.

Welker, C., 1976, Receptive fields of barrels in the somatosensory neocortex of the rat, *J. Comp. Neurol.* **166:**173–190.

Welker, C., and Sinha, M. M., 1972, Somatotopic organization of SmII cerebral neocortex in albino rat, *Brain Res.* **37:**132–136.

Welker, W. I., 1974, Principles of organization of the ventrobasal complex in mammals, *Brain Behav. Evol.* **7:**253–336.

Welker, W. I., and Johnson, J., 1965, Correlation between nuclear morphology and somatotopic organization in ventro-basal complex of the raccoon's thalamus, *J. Anat.* **99:**761–790.

Welker, W. I., and Lende, R. A., 1980, Thalamocortical relationships in echidna (*Tachyglossus aculeatus*), in: *Comparative Neurology of the Telencephalon* (S. O. E. Ebbesson, ed.), Plenum Press, New York, pp. 449–481.

Welker, W. I., Adrian, H. O., Lifschitz, W., Kaulen, R., Caviedes, E., and Gutman, W., 1976, Somatic sensory cortex of llama (*Lama glama*), *Brain Behav. Evol.* **13:**284–293.

Werner, G., and Mountcastle, V. B., 1963, The variability of central neural activity in a sensory system, and its implications for the central reflection of sensory events, *J. Neurophysiol.* **26:**958–977.

White, E. L., 1979, Thalamocortical synaptic relations: A review with emphasis on the projections of specific thalamic nuclei to the primary sensory areas of the neocortex, *Brain Res. Rev.* **1:**275–312.

White, E. L., and Hersch, S. M., 1982, A quantitative study of thalamocortical and other synapses involving apical dendrites of corticothalamic projection cells in mouse SmI cortex, *J. Neurocytol.* **11:**137–157.

White, J. C., and Sweet, W. H., 1969, *Pain and the Neurosurgeon: A Forty-Year Experience,* Thomas, Springfield, Ill.

Whitehouse, P. J., Price, D. L., Stuble, R. G., Clark, A. W., Coyle, J. T., and DeLong, M. R., 1982, Alzheimer's disease and senile dementia: Loss of neurons in the basal forebrain, *Science* **215:**1237–1239.

Whitfield, I. C., Cranford, J., Ravizza, R., and Diamond, I. T., 1972, Effects of unilateral ablation of auditory cortex in cat on complex sound localization, *J. Neurophysiol.* **35:**718–731.

Whitlock, D. G., and Perl, E. R., 1959, Afferent projections through ventrolateral funiculi to thalamus of cat, *J. Neurophysiol.* **22:**133–148.

Whitlock, D. G., and Perl, E. R., 1961, Thalamic projections of spinothalamic pathways in monkey, *Exp. Neurol.* **3:**240–255.

Whitsel, B. L., Petrucelli, L. M., and Sapiro, G., 1969a, Modality representation in the lumbar and cervical fasciculus gracilis of squirrel monkeys, *Brain Res.* **15:**67–78.

Whitsel, B. L., Petrucelli, L. M., and Werner, G., 1969b, Symmetry and connectivity in the map of the body surface in somatosensory area II of primates, *J. Neurophysiol.* **32:**170–183.

Whittier, J. R., and Mettler, F. A., 1949, Studies on the subthalamus of the rhesus monkey. I. Anatomy and fiber connections of the subthalamic nucleus of Luys, *J. Comp. Neurol.* **90:**281–318.

Widén, K., and Ajmone-Marsan, C., 1960, Effects of corticopetal and corticofugal impulses upon single elements of the dorsolateral geniculate nucleus, *Exp. Neurol.* **2:**468–502.

Wiesel, T. N., 1960, Receptive fields of ganglion cells in the cat's retina, *J. Physiol. (London)* **153:**583–594.

Wiesel, T. N., and Hubel, D. H., 1963a, Single cell responses in striate cortex of kittens deprived of vision in one eye, *J. Neurophysiol.* **26:**1003–1017.

Wiesel, T. N., and Hubel, D. H., 1963b, Effects of visual deprivation on morphology and physiology of cells in the cat's lateral geniculate body, *J. Neurophysiol.* **26:**978–993.

Wiesel, T. N., and Hubel, D. H., 1965a, Comparison of the effects of unilateral and bilateral eye closure on cortical unit responses in kittens, *J. Neurophysiol.* **28:**1029–1040.

Wiesel, T. N., and Hubel, D. H., 1965b, Extent of recovery from the effects of visual deprivation in kittens, *J. Neurophysiol.* **28:**1060–1072.

Wiesel, T. N., and Hubel, D. H., 1966, Spatial and chromatic interactions in the lateral geniculate body of the rhesus monkey, *J. Neurophysiol.* **29:**1115–1156.

Wiesel, T. N., Hubel, D. H., and Lam, D., 1974, Autoradiographic demonstration of ocular-dominance columns in the monkey striate cortex by means of transneuronal transport, *Brain Res.* **79:**273–279.

Wiesendanger, M., Rüegg, D. C., and Lucier, G. E., 1976, The influence from stretch receptors on cortical cells of area 3a and 4 in monkey, *Exp. Brain Res. Suppl.* **1:**437–439.

Williams, M., and Pennybacker, J., 1954, Memory disturbances in third ventricle tumors, *J. Neurol. Neurosurg. Psychiatry* **17:**115–123.

Willis, T., 1664, *Cerbri Anatome: Cui Accessit Nervorum Descriptio et Usus,* Martyn & Allestry, London.

Willis, T., 1681, *The Remaining Medical Works of that Famous and Renowned Physician, Dr. Thomas Willis* (translated by S. Pordage), Dring, Harper & Leigh, London.

Willis, W. D., and Coggeshall, R. E., 1978, *Sensory Mechanisms of the Spinal Cord,* Plenum Press, New York.

Willis, W. D., Trevino, D. L., Coulter, J. D., and Maunz, R. A., 1974, Responses of primate spinothalamic tract neurons to natural stimulation of hindlimb, *J. Neurophysiol.* **37:**358–372.

Willis, W. D., Maunz, R. A., Foreman, R. D., and Coulter, J. D., 1975, Static and dynamic responses of spinothalamic tract neurons to mechanical stimuli, *J. Neurophysiol.* **38:**587–600.

Willis, W. D., Leonard, R. B., and Kenshalo, D. R., Jr., 1978, Spinothalamic tract neurons in the substantia gelatinosa, *Science* **202:**986–988.

Willis, W. D., Kenshalo, D. R., Jr., and Leonard, R. B., 1979, The cells of origin of the primate spinothalamic tract, *J. Comp. Neurol.* **188:**543–574.

Wilson, C. J., Chang, H. T., and Kitai, S. T., 1983, Disfacilitation and long-lasting inhibition of neostriatal neurons in the rat, *Exp. Brain Res.* **51:**227–235.

Wilson, J. R., and Sherman, S. M., 1977, Differential effects of early monocular deprivation on binocular and monocular segments of cat striate cortex, *J. Neurophysiol.* **40:**891–903.

Wilson, J. R., Tessin, D. E., and Sherman, S. M., 1982, Development of the electrophysiological properties of Y-cells in the kitten's medial interlaminar nucleus, *J. Neurosci.* **2:**562–571.

Wilson, M. E., and Cragg, B. G., 1967, Projections from the lateral geniculate nucleus in the cat and monkey, *J. Anat.* **101**:677–692.

Wilson, M. E., and Cragg, B. G., 1969, Projections from the medial geniculate body to the cerebral cortex in the cat, *Brain Res.* **13**:462–475.

Wilson, P. D., and Stone, J., 1975, Evidence of W-cell input to the cat's visual cortex via the C laminae of the lateral geniculate nucleus, *Brain Res.* **92**:472–478.

Wilson, P. D., Pecci-Saavedra, J., and Doty, R. W., 1973, Mesencephalic control of lateral geniculate nucleus in primates. II. Effective loci, *Exp. Brain Res.* **18**:204–213.

Wilson, P. D., Rowe, M. H., and Stone, J., 1976, Properties of relay cells in cat's lateral geniculate nucleus: A comparison of W-cells with X- and Y-cells, *J. Neurophysiol.* **39**:1193–1209.

Wilson, S. A. K., 1914, An experimental research into the anatomy and physiology of the corpus striatum, *Brain* **36**:427–492.

Winer, J. A., and Morest, D. K., 1983, The neuronal architecture of the dorsal division of the medial geniculate body of the cat: A study with the rapid Golgi method, *J. Comp. Neurol.* **221**:1–30.

Winer, J. A., Diamond, I. T., and Raczkowski, D., 1977, Subdivisions of the auditory cortex of the cat: The retrograde transport of horseradish peroxidase to the medial geniculate body and posterior thalamic nuclei, *J. Comp. Neurol.* **176**:387–418.

Winfield, D. A., and Powell, T. P. S., 1980, An electron microscopic study of the postnatal development of the lateral geniculate nucleus in the normal kitten and after eyelid suture, *Proc. R. Soc. London Ser. B* **210**:197–210.

Winfield, D. A., Gatter, K. C., and Powell, T. P. S., 1975, Certain connections of the visual cortex of the monkey shown by the use of horseradish peroxidase, *Brain Res.* **92**:456–461.

Winfield, D. A., Headon, M. P., and Powell, T. P. S., 1976, Postnatal development of the synaptic organization of the lateral geniculate nucleus in the kitten with unilateral eye closure, *Nature (London)* **263**:591–594.

Winfield, D. A., Hiorns, R. W., and Powell, T. P. S., 1980, A quantitative electron microscopic study of the postnatal development of the lateral geniculate nucleus in normal kittens and in kittens with eyelid suture, *Proc. R. Soc. London Ser. B* **210**:211–234.

Winkler, C., and Potter, A., 1911, *An Anatomical Guide to Experimental Researches on the Rabbit's Brain*, Versluys, Amsterdam.

Winkler, C., and Potter, A., 1914, *An Anatomical Guide to Experimental Researches on the Cat's Brain*, Versluys, Amsterdam.

Winter, D. L., 1965, N. gracilis of cat: Functional organization and corticofugal effects, *J. Neurophysiol.* **28**:48–70.

Winterkorn, J. M. S., Shapley, R., and Kaplan, E., 1981, The effect of monocular paralysis on the lateral geniculate nucleus of the cat, *Exp. Brain Res.* **42**:117–121.

Wise, R. P., and Lund, R. D., 1976, The retina and central projections of heterochromic rats, *Exp. Neurol.* **51**:68–77.

Wise, S. P., 1975, The laminar organization of certain afferent and efferent fiber systems in the rat somatosensory cortex, *Brain Res.* **90**:139–142.

Wise, S. P., and Jones, E. G., 1976, the organization and postnatal development of the commissural projection of the rat somatic sensory cortex, *J. Comp. Neurol.* **168**:313–343.

Wise, S. P., and Jones, E. G., 1977a, Cells of origin and terminal distribution of descending projections of the rat somatic sensory cortex, *J. Comp. Neurol.* **175**:129–157.

Wise, S. P., and Jones, E. G., 1977b, Topographic and columnar organization of the corticotectal projection of the rat somatic sensory cortex, *Brain Res.* **133**:223–235.

Wise, S. P., and Jones, E. G., 1978, Developmental studies of thalamocortical and commissural connections in the rat somatic sensory cortex, *J. Comp. Neurol.* **178**:187–208.

Wise, S. P., Hendry, S. H. C., and Jones, E. G., 1977, Prenatal development of sensorimotor cortical projections in cats, *Brain Res.* **138**:538–544.

Wise, S. P., Fleshman, J. W., Jr., and Jones, E. G., 1979, Maturation of pyramidal cell form in relation to developing afferent and efferent connections of rat somatic sensory cortex, *Neuroscience* **4**:1275–1297.

Wong-Riley, M. T. T., 1972, Neuronal and synaptic organization of the normal dorsal lateral geniculate nucleus of the squirrel monkey, *Saimiri sciureus*, *J. Comp. Neurol.* **144**:25–59.

Wong-Riley, M. T. T., 1974, Demonstration of geniculocortical and callosal projection neurons in the squirrel monkey by means of retrograde axonal transport of horseradish peroxidase, *Brain Res.* **79:**267–272.

Wong-Riley, M. T. T., 1976, Projections from the dorsal lateral geniculate nucleus to prestriate cortex in the squirrel monkey as demonstrated by retrograde transport of horseradish peroxidase, *Brain Res.* **109:**595–600.

Wong-Riley, M. T. T., 1977, Connections between the pulvinar nucleus and the prestriate cortex in the squirrel monkey as revealed by peroxidase histochemistry and autoradiographic, *Brain Res.* **134:**249–268.

Wong-Riley, M., 1979, Changes in the visual system of monocularly sutured or enucleated cats demonstrable with cytochrome oxidase histochemistry, *Brain Res.* **171:**11–28.

Woodburne, R. T., Crosby, E. C., and McCotter, R. E., 1946, The mammalian midbrain and isthmus regions. Part II. The fiber connections. A. The relations of the tegmentum of the midbrain with the basal ganglia in *Macaca mulatta, J. Comp. Neurol.* **85:**67–92.

Woollard, H. H., and Beattie, J., 1927, The comparative anatomy of the lateral geniculate body, *J. Anat.* **61:**414–423.

Woollard, H. H., and Harpman, J. A., 1940, The connexions of the inferior colliculus and of the dorsal nucleus of the lateral lemniscus, *J. Anat.* **74:**441–458.

Woolsey, C. N., 1943, "Second" somatic receiving areas in the cerebral cortex of cat, dog and monkey, *Fed. Proc.* **2:**55.

Woolsey, C. N., 1958, Organization of somatic sensory and motor areas of the cerebral cortex, in: *Biological and Biochemical Bases of Behavior* (H. F. Harlow and C. N. Woolsey, eds.), University of Wisconsin Press, Madison, pp. 63–81.

Woolsey, C. N., 1964, Electrophysiological studies on thalamocortical relations in the auditory system, in: *Unfinished Tasks in the Behavioral Sciences* (A. Abrams, H. H. Garner, and J. E. P. Toman, eds.), William & Wilkins, Baltimore, pp. 45–57.

Woolsey, C. N., and Fairman, D., 1946, Contralateral, ipsilateral and bilateral representation of cutaneous receptors in somatic areas I and II of the cerebral cortex of pig, sheep and other mammals, *Surgery* **19:**684–702.

Woolsey, C. N., and Walzl, E. M., 1942, Topical projection of nerve fibers from local regions of the cochlea to the cerebral cortex of the cat, *Bull. Johns Hopkins Hosp.* **71:**315–344.

Woolsey, C. N., and Wang, G. H., 1945, Somatic areas I and II of the cerebral cortex of the rabbit, *Fed. Proc.* **4:**79.

Woolsey, C. N., Marshall, W. H., and Bard, P., 1942, Representation of cutaneous tactile sensibility in the cerebral cortex of the monkey as indicated by evoked potentials, *Bull. Johns Hopkins Hosp.* **70:**399–441.

Woolsey, T. A., and Van der Loos, H., 1970, The structural organization of layer IV in the somatosensory region (SI) of mouse cerebral cortex, *Brain Res.* **17:**205–242.

Woolsey, T. A., and Wann, J. R., 1976, Areal changes in mouse cortical barrels following vibrissal damage at different postnatal ages, *J. Comp. Neurol.* **170:**53–66.

Woolsey, T. A., Anderson, J. R., Wann, J. R., and Stanfield, B. B., 1979, Effects of early vibrissae damage on neurons in the ventrobasal (VB) thalamus of the mouse, *J. Comp. Neurol.* **184:**363–380.

Wyss, J. M., Swanson, L. W., and Cowan, W. M., 1979, A study of subcortical afferents to the hippocampal formation in the rat, *Neuroscience* **4:**463–476.

Yamamura, H. I., Kuhar, M. J., Greenberg, D., and Snyder, S. H., 1974, Muscarinic cholinergic receptor binding: Regional distribution in monkey brain, *Brain Res.* **66:**541–546.

Yarita, H., Iino, M., Tanabe, T., Kogure, S., and Takagi, S. F., 1980, A transthalamic olfactory pathway to orbitofrontal cortex in the monkey, *J. Neurophysiol.* **43:**69–85.

Yazulla, S., and Granda, A. M., 1973, Opponent-color units in the thalamus of the pigeon *(Columba livia)*, *Vision Res.* **13:**1555–1563.

Yen, C.-T., and Jones, E. G., 1983, Intracellular staining of physiologically identified neurons and axons in the somatosensory thalamus of the cat, *Brain Res.* **280:**148–154.

Yen, C.-T., Conley, M., and Jones, E. G., 1984, Morphological and functional types of neurons in cat ventral posterior thalamic nucleus, *J. Neurosci.* (in press).

Yen, C.-T., Hendry, S.H.C., and Jones, E. G., 1985, The morphology of physiologically identified GABAergic neurons in the somatic sensory part of the thalamic reticular nucleus in the cat, *J. Neurosci.* (in press).

Yin, T. C. T., and Mountcastle, V. B., 1977, Visual input to the visuomotor mechanisms of the monkey's parietal lobe, *Science* **197:**1381–1383.

Yin, T. C. T., and Williams, W. J., 1976, Dynamic response and transfer characteristics of joint neurons in somatosensory thalamus of the cat, *J. Neurophysiol.* **39:**582–600.

Yingling, C. D., and Skinner, J. E., 1976, Selective regulation of thalamic sensory relay nuclei by nucleus reticularis thalami, *Electroencephalogr. Clin. Neurophysiol.* **41:**476–482.

Yokota, T., and Matsumoto, N., 1983a, Location and functional organization of trigeminal wide dynamic range neurons within the nucleus ventralis posteromedialis of the cat, *Neurosci. Lett.* **39:**231–236.

Yokota, T., and Matsumoto, N., 1983b, Somatotopic distribution of trigeminal nociceptive specific neurons within the caudal somatosensory thalamus of cat, *Neurosci. Lett.* **39:**125–130.

Yukie, M., and Iwai, E., 1981, Direct projection from the dorsal lateral geniculate nucleus to the prestriate cortex in macaque monkeys, *J. Comp. Neurol.* **201:**81–97.

Yurkewicz, L., Valentino, K. L., Floeter, M. K., Fleshman, J. W., Jr., and Jones, E. G., 1984, Effects of cytotoxic deletions of somatic sensory cortex in fetal rats, *Somatosensory Res.* **1:**303–327.

Zarbin, M. A., Innis, R. B., Wamsley, J. K., Snyder, S. H., and Kuhar, M. J., 1983, Autoradiographic localization of cholecystokinin receptors in rodent brain, *J. Neurosci.* **3:**877–906.

Zaretsky, M. D., 1978, A new auditory area of the songbird forebrain, *Exp. Brain Res.* **32:**267–273.

Zaretsky, M. D., and Konishi, M., 1976, Tonotopic organization in the avian telencephalon, *Brain Res.* **111:**167–171.

Zeigler, H. P., and Karten, H. J., 1973, Brain mechanisms and feeding behavior in the pigeon *(Columba livia):* Quinto-frontal structures, *J. Comp. Neurol.* **152:**59–81.

Zeki, S. M., 1973, Functional specialisation in the visual cortex of the rhesus monkey, *Nature (London)* **274:**423–428.

Zeki, S. M., 1978, Uniformity and diversity of structure and function in rhesus monkey prestriate visual cortex, *J. Physiol. (London)* **277:**273–290.

Ziehen, T., 1908, *Das Centralnervensystem der Monotremen und Marsupialier, Semon's Zool. Forschungs,* Volume 3, Part 2, Fischer, Jena, pp. 789–921.

Index